P9-DIE-216

PRINCIPLES OF

Environmental Science

Inquiry & Applications

PRINCIPLES OF

Environmental Science

Inquiry & Applications

Ninth Edition

William P. Cunningham
University of Minnesota

Mary Ann Cunningham
Vassar College

PRINCIPLES OF ENVIRONMENTAL SCIENCE: INQUIRY AND APPLICATIONS, NINTH EDITION

This book is printed on acid-free paper.

1 2 3 4 5 6 7 8 9 LWI 21 20 19

ISBN 978-1-260-21971-5 (bound edition)
MHID 1-260-21971-2 (bound edition)
ISBN 978-1-260-49283-5 (loose-leaf edition)
MHID 1-260-49283-4 (loose-leaf edition)

Senior Portfolio Manager: *Michael Ivanov, PhD.*
Senior Product Developer: *Jodi Rhomberg*
Senior Marketing Manager: *Britney Ross*
Senior Content Project Manager: *Vicki Krug*
Lead Content Project Manager: *Tammy Juran*
Senior Buyer: *Laura Fuller*
Design: *Egzon Shaqiri*
Content Licensing Specialist: *Lorraine Buczek*
Cover Image: ©*naturalv/123RF*
Compositor: *Aptara®, Inc.*

All credits appearing on page or at the end of the book are considered to be an extension of the copyright page.

Library of Congress Cataloging-in-Publication Data

Names: Cunningham, William P., author. | Cunningham, Mary Ann, author.
Title: Principles of environmental science : inquiry & applications / William P. Cunningham, University of Minnesota, Mary Ann Cunningham, Vassar College.
Description: Ninth Edition. | New York, NY : McGraw-Hill Education, [2020]
Identifiers: LCCN 2018034427| ISBN 9781260219715 (alk. paper) | ISBN 1260219712 (alk. paper)
Subjects: LCSH: Environmental sciences–Textbooks.
Classification: LCC GE105 .C865 2020 | DDC 363.7–dc23 LC record available at https://lccn.loc.gov/2018034427

The Internet addresses listed in the text were accurate at the time of publication. The inclusion of a website does not indicate an endorsement by the authors or McGraw-Hill Education, and McGraw-Hill Education does not guarantee the accuracy of the information presented at these sites.

Logo applies to the text stock only

mheducation.com/highered

About the Authors

WILLIAM P. CUNNINGHAM

William P. Cunningham is an emeritus professor at the University of Minnesota. In his 38-year career at the university, he taught a variety of biology courses, including Environmental Science, Conservation Biology, Environmental Health, Environmental Ethics, Plant Physiology, General Biology, and Cell Biology. He is a member of the Academy of Distinguished Teachers, the highest teaching award granted at the University of Minnesota. He was a member of a number of interdisciplinary programs for international students, teachers, and nontraditional students. He also carried out research or taught in Sweden, Norway, Brazil, New Zealand, China, and Indonesia.

Courtesy Tom Finkle

Professor Cunningham has participated in a number of governmental and nongovernmental organizations over the past 40 years. He was chair of the Minnesota chapter of the Sierra Club, a member of the Sierra Club national committee on energy policy, vice president of the Friends of the Boundary Waters Canoe Area, chair of the Minnesota governor's task force on energy policy, and a citizen member of the Minnesota Legislative Commission on Energy.

In addition to environmental science textbooks, Professor Cunningham edited three editions of *Environmental Encyclopedia* published by Thompson-Gale Press. He has also authored or co-authored about 50 scientific articles, mostly in the fields of cell biology and conservation biology as well as several invited chapters or reports in the areas of energy policy and environmental health. His Ph.D. from the University of Texas was in botany.

His hobbies include birding, hiking, gardening, traveling, and video production. He lives in St. Paul, Minnesota, with his wife, Mary. He has three children (one of whom is co-author of this book) and seven grandchildren.

Courtesy Tom Finkle

MARY ANN CUNNINGHAM

Mary Ann Cunningham is a professor of geography at Vassar College, in New York's Hudson Valley. A biogeographer with interests in landscape ecology, geographic information systems (GIS), and land use change, she teaches environmental science, natural resource conservation, and land use planning, as well as GIS and spatial data analysis. Field research methods, statistical methods, and scientific methods in data analysis are regular components of her teaching. As a scientist and educator, she enjoys teaching and conducting research with both science students and non-science liberal arts students. As a geographer, she likes to engage students with the ways their physical surroundings and social context shape their world experience. In addition to teaching at a liberal arts college, she has taught at community colleges and research universities. She has participated in Environmental Studies and Environmental Science programs and has led community and college field research projects at Vassar.

Mary Ann has been writing in environmental science for nearly two decades, and she is also co-author of *Environmental Science: A Global Concern*, now in its fourteenth edition. She has published work on habitat and landcover change, on water quality and urbanization, and other topics in environmental science. She has also done research with students and colleagues on climate change, its impacts, and carbon mitigation strategies.

Research and teaching activities have included work in the Great Plains, the Adirondack Mountains, and northern Europe, as well as in New York's Hudson Valley, where she lives and teaches. In her spare time she loves to travel, hike, and watch birds. She holds a bachelor's degree from Carleton College, a master's degree from the University of Oregon, and a Ph.D. from the University of Minnesota.

Brief Contents

©Stocktrek/Getty Images

©Martin Kubat/Shutterstock

Contents

©Jesse Kraft/123RF

3

©Fotos593/Shutterstock

4

5

6

©Kostic Dusan/123RF

Food and Agriculture 152

©Joe Raedle/Getty Images

8

Environmental Health and Toxicology 180

©Pat Bonish/Alamy Stock Photo

Climate 205

©Saurav022/Shutterstock

10

©Justin Sullivan/Getty Images

11

©Felt Soul Media

12

©William P. Cunningham

13

Source: NOAA Photo Library/NOAA's Fisheries Collection/National Oceanic and Atmospheric Administration (NOAA)

14

©Pierre Leclerc Photography/Getty Images

15

©Wang Chengyun/Newscom

16

Environmental Policy and Sustainability 380

List of Case Studies

Over 200 additional Case Studies can be found online on the instructor's resource page at www.mcgrawhill-connect.com.

©Martin Kubat/Shutterstock

Preface

UNDERSTANDING CRISIS AND OPPORTUNITY

Environmental science often emphasizes that while we are surrounded by challenges, we also have tremendous opportunities. We face critical challenges in biodiversity loss, clean water protection, climate change, population growth, sustainable food systems, and many other areas. But we also have tremendous opportunities to take action to protect and improve our environment. By studying environmental science, you have the opportunity to gain the tools and the knowledge to make intelligent choices on these and countless other questions.

Because of its emphasis on problem solving, environmental science is often a hopeful field. Even while we face burgeoning cities, warming climates, looming water crises, we can observe solutions in global expansion in access to education, health care, information, even political participation and human rights. Birth rates are falling almost everywhere, as women's rights gradually improve. Creative individuals are inventing new ideas for alternative energy and transportation systems that were undreamed of a generation ago. We are rethinking our assumptions about how to improve cities, food production, water use, and air quality. Local action is rewriting our expectations, and even economic and political powers feel increasingly compelled to show cooperation in improving environmental quality

Climate change is a central theme in this book and in environmental science generally. As in other topics, we face dire risks but also surprising new developments and new paths toward sustainability. China, the world's largest emitter of carbon dioxide, expects to begin reducing its emissions within a decade, much sooner than predicted. Many countries are starting to show declining emissions, and there is clear evidence that economic growth no longer depends on carbon fossil fuels. Greenhouse gas emissions continue to rise, but nations are showing unexpected willingness to cooperate in striving to reduce emissions. Much of this cooperation is driven by growing acknowledgment of the widespread economic and humanitarian costs of climate change. Additional driving forces, though, are the growing list of alternatives that make carbon reductions far easier to envision, or even to achieve, than a few years ago.

Sustainability, also a central idea in this book, has grown from a fringe notion to a widely shared framework for daily actions (recycling, reducing consumption) and civic planning (building energy-efficient buildings, investing in public transit and bicycle routes). Sustainability isn't just about the environment anymore. Increasingly we know that sustainability is also smart economics and that it is essential for social equity. Energy efficiency saves money. Alternative energy can reduce our reliance on fuel sources in politically unstable regions. Healthier food options reduce medical costs. Accounting for the public costs and burdens of pollution and waste disposal helps us rethink the ways we dispose of our garbage and protect public health. Growing awareness of these co-benefits helps us understand the broad importance of sustainability.

Students are providing leadership

Students are leading the way in reimagining our possible futures. Student movements have led innovation in technology and science, in sustainability planning, in environmental governance, and in environmental justice around the world. They have energized local communities to join the public debate on how to seek a sustainable future. Students have the vision and the motivation to create better paths toward sustainability and social justice, at home and globally.

You may be like many students who find environmental science an empowering field. It provides the knowledge needed to use your efforts more effectively. Environmental science applies to our everyday lives and the places where we live, and we can apply ideas learned in this discipline to any place or occupation in which we find ourselves. And environmental science can connect to any set of interests or skills you might bring to it: Progress in the field involves biology, chemistry, geography, and geology. Communicating and translating ideas to the public, who are impacted by changes in environmental quality, requires writing, arts, media, and other communication skills. Devising policies to protect resources and enhance cooperation involves policy, anthropology, culture, and history. What this means is that while there is much to learn, this field can also connect with whatever passions you bring to the course.

WHAT SETS THIS BOOK APART?

Solid science and an emphasis on sustainability: This book reflects the authors' decades of experience in the field and in the classroom, which make it up-to-date in approach, in data, and in applications of critical thinking. The authors have been deeply involved in sustainability, environmental science, and conservation programs at the University of Minnesota and at Vassar College. Their experience and courses on these topics have strongly influenced the way ideas in this book are presented and explained.

Demystifying science: We make science accessible by showing how and why data collection is done and by giving examples, practice, and exercises that demonstrate central principles. *Exploring Science* readings empower students by helping them understand how scientists do their work. These readings give examples of technology and methods in environmental science.

Quantitative reasoning: Students need to become comfortable with graphs, data, and comparing numbers. We provide focused discussions on why scientists answer questions with numbers, the nature of statistics, of probability, and how to interpret the message in a graph. We give accessible details on population models, GIS (mapping and spatial analysis), remote sensing, and other quantitative techniques. In-text applications and online, testable *Data Analysis* questions give students opportunities to practice with ideas, rather than just reading about them.

Critical thinking: We provide a focus on critical thinking, one of the most essential skills for citizens, as well as for students. Starting with a focused discussion of critical thinking in chapter 1, we offer abundant opportunities for students to weigh contrasting evidence and evaluate assumptions and arguments, including *What Do You Think?* readings.

Up-to-date concepts and data: Throughout the text we introduce emerging ideas and issues such as ecosystem services, cooperative ecological relationships, epigenetics, and the economics of air pollution control, in addition to basic principles such as population biology, the nature of systems, and climate processes. Current approaches to climate change mitigation, campus sustainability, sustainable food production, and other issues give students current insights into major issues in environmental science and its applications. We introduce students to current developments such as ecosystem services, coevolution, strategic targeting of Marine Protected Areas, impacts of urbanization, challenges of REDD (reducing emissions through deforestation and degradation), renewable energy development in China and Europe, fertility declines in the developing world, and the impact of global food trade on world hunger.

Active learning: Learning how scientists approach problems can help students develop habits of independent, orderly, and objective thought. But it takes active involvement to master these skills. This book integrates a range of learning aids–*Active Learning* exercises, *Critical Thinking and Discussion* questions, and *Data Analysis* exercises–that push students to think for themselves. Data and interpretations are presented not as immutable truths but rather as evidence to be examined and tested, as they should be in the real world. Taking time to look closely at figures, compare information in multiple figures, or apply ideas in text is an important way to solidify and deepen understanding of key ideas.

Synthesis: Students come to environmental science from a multitude of fields and interests. We emphasize that most of our pressing problems, from global hunger or climate change to conservation of biodiversity, draw on sciences and economics and policy. This synthesis shows students that they can be engaged in environmental science, no matter what their interests or career path.

A global perspective: Environmental science is a globally interconnected discipline. Case studies, data, and examples from around the world give opportunities to examine international questions. Nearly half of the opening case studies, and many of the boxed readings, examine international issues of global importance, such as forest conservation in Indonesia, air quality in India, or family planning in Thailand. In addition, Google Earth place marks take students virtually to locations where they can see and learn the context of the issues they read.

Key concepts: In each chapter this section draws together compelling illustrations and succinct text to create a summary "take-home" message. These key concepts draw together the major ideas, questions, and debates in the chapter but give students a central idea on which to focus. These can also serve as starting points for lectures, student projects, or discussions.

Positive perspective: All the ideas noted here can empower students to do more effective work for the issues they believe in. While we don't shy away from the bad news, we highlight positive ways in which groups and individuals are working to improve their environment. *What Can You Do?* features in every chapter offer practical examples of things everyone can do to make progress toward sustainability.

Thorough coverage: No other book in the field addresses the multifaceted nature of environmental questions such as climate policy, sustainability, or population change with the thoroughness this book has. We cover not just climate change but also the nature of climate and weather systems that influence our day-to-day experience of climate conditions. We explore both food shortages and the emerging causes of hunger–such as political conflict, biofuels, and global commodity trading–as well as the relationship between food insecurity and the growing pandemic of obesity-related illness. In these and other examples, this book is a leader in in-depth coverage of key topics.

Student empowerment: Our aim is to help students understand that they can make a difference. From campus sustainability assessments (chapter 16) to public activism (chapter 13) we show ways that student actions have led to policy changes on all scales. In all chapters we emphasize ways that students can take action to practice the ideas they learn and to play a role in the policy issues they care about. *What Can You Do?* boxed features give steps students can take to make a difference.

Exceptional online support: Online resources integrated with readings encourage students to pause, review, practice, and explore ideas, as well as to practice quizzing themselves on information presented. McGraw-Hill's ConnectPlus (www.mcgrawhillconnect.com) is a web-based assignment and assessment platform that gives students the means to better connect with their coursework, with their instructors, and with the important concepts that they will need to know for success now and in the future. Valuable assets such as LearnSmart (an adaptive learning system), an interactive ebook, *Data Analysis* exercises, the extensive case study library, and Google Earth exercises are all available in Connect.

WHAT'S NEW IN THIS EDITION?

This edition continues our focus on two major themes, **climate protection and sustainability.** These topics are evolving rapidly, often with student leadership, and they greatly impact the future and the career paths of students. We explore **emerging ideas and examples** to help students consider these dominant issues of our time. The climate chapter (chapter 9), for example, provides up-to-date data from the Paris Accord to the latest Intergovernmental Panel on

Climate Change (IPCC) as well as in-depth explanations of climate dynamics, including positive feedbacks and how greenhouse gases capture energy. The energy chapter (chapter 13) explores the rapidly changing landscape of energy production, in which fossil fuels still dominate, but explosive growth of renewables in China, India, and Europe have altered what we think is possible for renewable energy systems.

We also provide a new emphasis on **science and citizenship.** In a world overflowing with conflicting views and arguments, students today need to understand the importance of being able to evaluate evidence, to think about data, to understand environmental systems, and to see linkages among systems we exploit and depend on. And they need to understand their responsibility, as voters and members of civil society, to apply these abilities to decision making and participation in their communities.

Many topics in environmental science are shifting rapidly, and so much of the material in this edition is updated. Nearly two-thirds of the chapters have new opening case studies, and data and figures have been updated throughout the book. Brief **learning objectives** have been added to every A head to help students focus on the most important topics in each major section.

We also recognize that students have a lot to remember from each chapter. As teachers, we have found it is helpful to provide a few key reference ideas, which students can focus on and even compare to other data they encounter. So in this edition, we have provided short lists of **benchmark data,** selected to help students anchor key ideas and to understand the big picture. Specific chapter changes include the following.

In **Chapter 1,** a new opening case study describes an important development in renewable energy on the Navajo Reservation in Arizona. In a dramatic shift, the tribe has decided to move away from a reliance on dirty fossil fuels and to turn instead to clean, renewable solar energy. This shift will protect precious water resources, improve air quality for the whole region, reduce health risks from mining and burning coal, and help fight climate change for all of us. The chapter also has a new *Exploring Science* box on recent United Nations Sustainable Development Goals and the most current Human Development Index. We also have added text and a figure explaining planetary boundaries for critical resources and ecosystem services as well as how we may transgress crucial systems on which we all depend. We introduce a new feature in this chapter on science and citizenship with a focus on evidence and critical thinking.

Chapter 2 opens with a case study on the Gulf of Mexico's "dead zone," which continues to grow in size despite the good intentions of many stake-holders. This example shows the importance of understanding principles of chemistry and biogeochemical cycles in ecology. We expand on the discussion of trophic levels in biological communities with an essay on how overexploitation of Antarctic krill is disrupting the entire Antarctic Ocean food chain.

Chapter 3 provides new insights into the importance of the microbiome in chronic diseases and the possible effects of chronic exposure to antimicrobial compounds on our microbiological symbionts.

Chapter 4 features a new opening case study on the success of family planning in Thailand, where total fertility rates have fallen from 7 children per woman on average in 1974 to 1.5 in 2017. This dramatic change is linked to a new section later in the chapter describing how about half the world's countries are now at or below the replacement rate. The *What Do You Think?* essay on China's one-child policy has been updated to reflect emerging worries about a birth dearth in China. Population data have been updated throughout the chapter, reflecting ongoing demographic changes in many regions of the world.

Chapter 5 has a new opening case study on the growing threat of bark beetles in forest destruction and the frequency and cost of wild fires. This is a major case of ecosystem disturbance, state shift, and resource management policy, as well as a dramatic illustration of how climate shapes biomes. The *Exploring Science* essay in this chapter describes efforts to restore coral reefs, including breeding experiments that seek to create coral strains that can grow in warmer, more acidic sea water. Successful recovery of protected species under the Endangered Species Act is highlighted, along with the benefits of habitat protection.

Chapter 6 provides new data on the effects of palm oil plantations on biodiversity, including endangered orangutans, in the opening case study. Although many major food companies and oil traders have pledged to stop using or selling oil from recently deforested areas, compliance is difficult to monitor. In the meantime, orangs and people who try to protect them continue to be killed. Adding to this discussion, we have added a new *Exploring Science* essay on how we can use remote sensing to assess forest loss. We also have an updated *What Can You Do?* box with suggestions for individual actions to reduce forest impacts. Habitat loss isn't just a problem in other countries; the U.S. also has continued threats to natural areas. We address threats to the Alaska National Wildlife Refuge and to recently created national monuments in two new boxes for this edition.

Chapter 7 opens with a new case study about introduction of crop varieties engineered to tolerate multiple herbicides, and herbicide "cocktails" containing mixtures of different herbicides. This innovation is meant to combat pesticide resistance, but will it simply accelerate evolution of super weeds? And what are the potential human health effects and the ecological consequences of ever greater exposure to these compounds? Fuel consumption in crop production is addressed in light of concern about global climate change, along with questions about how we'll feed a growing human population in a changing world. Low-input, sustainable farming is discussed as an alternative to modern industrial-scale farming methods.

Chapter 8 introduces environmental health with a new case study about the toxic floods that inundated Houston after Hurricane Harvey in 2017. The long-term effects of flooding thousands of chemical plants and Superfund sites remain to be seen, but this is an excellent example of a growing threat from pollutants and synthetic chemicals, especially in vulnerable coastal cities. Our discussion of global health burdens is updated to reflect the threats of chronic conditions. Many new outbreaks of emergent diseases are noted. And we provide a new profile of important persistent organic pollutants (POPs).

Chapter 9's focus on the causes and consequences of climate change remains among the most important topics in the book. An extensive new section on the potential effects of a 2-degree average global temperature updates this discussion. Because no one can take action without hope, we emphasize the many, readily available strategies we can take to avoid these changes. A thorough examination of possible solutions, including goals and accomplishments of the Paris Accord, shows the many options that we have right now to solve our climate challenges. This chapter also contains updated discussions of basic climate processes and feedbacks.

Chapter 10 begins with a new case study about air quality in Delhi, India, which is now worse than that in Beijing, China. We amplify this case study with a new discussion in the text about health effects of air pollution, using Asia as an example. We also note that more than half of the 3 billion air pollution-related deaths worldwide are thought to be caused by indoor air. This is elaborated on in a new *Exploring Science* box about black carbon from combustion and its effects on health and climate.

Chapter 11 is a rare example in which the opening case study hasn't changed because water emergencies in California remain a critical long-term problem. Other topics, such as inexpensive water purification techniques and water recycling, also remain relevant and current.

Chapter 12 introduces a new case study on the Pebble mine, a proposed giant strip mine at the headwaters of rivers flowing into Bristol Bay, Alaska. This mine, which had been blocked during the Obama administration, is now in play again with a new regime in Washington. It threatens the largest remaining sockeye salmon fishery on the planet along with thousands of fish-related jobs and traditional native ways of life. It's an example of the many controversies about mining and mineral production. We update the discussion of induced seismicity with a new *Exploring Science* box about saltwater injection wells associated with oil and gas production in Oklahoma. Surface mining and coal sludge storage remain a serious problem in many places, so we've incorporated a new section into the text about these topics. And discussion of 2017 floods in South Asia, which displaced more than 40 million people and killed at least 1,200, illustrates the dangers of global climate change for geological hazards.

Chapter 13, which focuses on energy, is a focal chapter for climate solutions and sustainability. The opening case study on New York City's commitment to 80 percent reduction of greenhouse gas reductions becomes even more important with the 2017 announcement that both the city and state of New York would divest $5 billion in fossil fuel investments from their retirement funds (discussed in chapter 16). The chapter also reviews dramatic shifts in the price and efficiency of solar and wind power, which have made renewable energy cheaper than fossil fuels or nuclear even for existing facilities. An extensive new section on an energy transition explores future options for generating, storing, and transmitting energy. Drawing on the work of Jacobson and Delucchi, and Pawl Hawken's recent *Drawdown* study, we show how sustainable energy could supply all our power needs.

Chapter 14 starts with a new opening case study about the huge problem of plastic trash accumulating in the oceans. In particular, the estimated 100 million tons of plastic circulating in a massive gyre the size of California just northwest of Hawaii is a threat both to fish and to oceanic birds. A new *What Do You Think?* essay examines new Chinese policies that outlaw shipment of two dozen kinds of low-quality or dangerous solid waste and threaten to upend waste disposal practices throughout the world.

Chapter 15 opens with an important new case study on British Columbia's groundbreaking carbon tax. This revenue-neutral use tax has been a tremendous environmental and economic success and has provided millions to decrease corporate and personal taxes as well as to accomplish broader social goals while fostering an economic boom. This is an excellent and positive application of environmental economics. The section on cities and city planning in this chapter builds on the discussion in chapter 10 on New Delhi air pollution. We also return to the Human Development Index and the problems of massive urban agglomerations in developing countries, some of which, like Lagos, Nigeria, could reach 100 million inhabitants by the end of this century. Valuation of nature is discussed in a new *Exploring Science* essay, which examines a new estimate that raises the value of all global ecological services from $33 trillion to as much as $173 trillion, or more than twice the current global GDP.

Chapter 16 commences with a new case study on fossil fuel divestment pledges by New York City and New York State. Decarbonization of these huge economies is inspired by the damage done by Hurricane Sandy, which resulted in more than $70 billion in damages. Even more notable than its divestment pledge, New York City is suing the world's five largest publicly traded oil companies for their role in climate change. The divestment movement in colleges, universities, and other entities represents more than $6 trillion in assets. We support this discussion with a new section on policy making at both the individual and collective levels. We discuss the creation and implementation of some of our most important environmental laws, but we also examine how those rules and laws are now under attack by the current administration. We also have added an extensive new section on how colleges and universities can be powerful catalysts for change. Finally, we end with a review of the 2016 UN Sustainable Development Goals.

ACKNOWLEDGMENTS

We are sincerely grateful to Jodi Rhomberg and Michael Ivanov who oversaw the development of this edition, and to Vicki Krug who shepherded the project through production.

We would like to thank the following individuals who wrote and/or reviewed learning goal-oriented content for **LearnSmart.**
Brookdale Community College, Juliette Goulet
Broward College, Nilo Marin
Broward College, David Serrano
College of the Desert, Kurt Leuschner
Des Moines Area Community College, Curtis Eckerman
Georgia Southern University, Ed Mondor
Harrisburg Area Community College, Geremea Fioravanti
Kennesaw State University, Karyn A. Alme
Miami Dade College, Kendall College, David Moore
Northern Arizona University, Sylvester Allred
Oakland Community College, Shannon J. Flynn
Ozarks Technical Community College, Rebecca Gehringer
Ozarks Technical Community College, Michael S. Martin
Palm Beach State College, Jessica Miles
Roane State Community College, Arthur C. Lee
Rutgers University, Craig Phelps
St. Petersburg College, Amanda H. Gilleland
The University of Texas at San Antonio, Terri Matiella
University of North Carolina-Asheville, David Gillette
University of North Carolina at Chapel Hill, Trent McDowell
University of Wisconsin–Milwaukee, Gina S. Szablewski
University of Wisconsin–River Falls, Eric Sanden
Wilmington University, Milton Muldrow Jr.
Wilmington University, Scott V. Lynch

Input from instructors teaching this course is invaluable to the development of each new edition. Our thanks and gratitude go out to the following individuals who either completed detailed chapter reviews or provided market feedback for this course.
American University, Priti P. Brahma
Antelope Valley College, Zia Nisani
Arizona Western College, Alyssa Haygood
Aurora University, Carrie Milne-Zelman
Baker College, Sandi B. Gardner
Boston University, Kari L. Lavalli
Bowling Green State University, Daniel M. Pavuk
Bradley University, Sherri J. Morris
Broward College, Elena Cainas
Broward College, Nilo Marin
California Energy Commission, James W. Reede
California State University, Natalie Zayas
California State University–East Bay, Gary Li
Carthage College, Tracy B. Gartner
Central Carolina Community College, Scott Byington
Central State University, Omokere E. Odje
Clark College, Kathleen Perillo
Clemson University, Scott Brame
College of DuPage, Shamili Ajgaonkar Sandiford
College of Lake County, Kelly S. Cartwright
College of Southern Nevada, Barry Perlmutter
College of the Desert, Tracy Albrecht
Community College of Baltimore County, Katherine M. Van de Wal
Connecticut College, Jane I. Dawson
Connecticut College, Chad Jones
Connors State College, Stuart H. Woods
Cuesta College, Nancy Jean Mann
Dalton State College, David DesRochers
Dalton State College, Gina M. Kertulis-Tartar
East Tennessee State University, Alan Redmond
Eastern Oklahoma State College, Patricia C. Bolin Ratliff
Edison State College, Cheryl Black
Elgin Community College, Mary O'Sullivan
Erie Community College, Gary Poon
Estrella Mountain Community College, Rachel Smith
Farmingdale State College, Paul R. Kramer
Fashion Institute of Technology, Arthur H. Kopelman
Flagler College, Barbara Blonder
Florida State College at Jacksonville, Catherine Hurlbut
Franklin Pierce University, Susan Rolke
Galveston College, James J. Salazar
Gannon University, Amy L. Buechel
Gardner-Webb University, Emma Sandol Johnson
Gateway Community College, Ramon Esponda
Geneva College, Marjory Tobias
Georgia Perimeter College, M. Carmen Hall
Georgia Perimeter College, Michael L. Denniston
Gila Community College, Joseph Shannon
Golden West College, Tom Hersh
Gulf Coast State College, Kelley Hodges
Gulf Coast State College, Linda Mueller Fitzhugh
Heidelberg University, Susan Carty
Holy Family University, Robert E. Cordero
Houston Community College, Yiyan Bai
Hudson Valley Community College, Janet Wolkenstein
Illinois Mathematics and Science Academy, C. Robyn Fischer
Illinois State University, Christy N. Bazan
Indiana University of Pennsylvania, Holly J. Travis
Indiana Wesleyan University, Stephen D. Conrad
James Madison University, Mary Handley
James Madison University, Wayne S. Teel
John A. Logan College, Julia Schroeder
Kentucky Community & Technical College System–Big Sandy District, John G. Shiber
Lake Land College, Jeff White
Lane College, Satish Mahajan
Lansing Community College, Lu Anne Clark
Lewis University, Jerry H. Kavouras
Lindenwood University, David M. Knotts
Longwood University, Kelsey N. Scheitlin
Louisiana State University, Jill C. Trepanier
Lynchburg College, David Perault
Marshall University, Terry R. Shank
Menlo College, Neil Marshall
Millersville University of Pennsylvania, Angela Cuthbert

Minneapolis Community and Technical College, Robert R. Ruliffson
Minnesota State College-Southeast Technical, Roger Skugrud
Minnesota West Community and Technical College, Ann M. Mills
Mt. San Jacinto College, Shauni Calhoun
Mt. San Jacinto College, Jason Hlebakos
New Jersey City University, Deborah Freile
New Jersey Institute of Technology, Michael P. Bonchonsky
Niagara University, William J. Edwards
North Carolina State University, Robert I. Bruck
North Georgia College & State University, Kelly West
North Greenville University, Jeffrey O. French
Northeast Lakeview College, Diane B. Beechinor
Northeastern University, Jennifer Rivers Cole
Northern Virginia Community College, Jill Caporale
Northwestern College, Dale Gentry
Northwestern Connecticut Community College, Tara Jo Holmberg
Northwood University Midland, Stelian Grigoras
Notre Dame College, Judy Santmire
Oakton Community College, David Arieti
Parkland College, Heidi K. Leuszler
Penn State Beaver, Matthew Grunstra
Philadelphia University, Anne Bower
Pierce College, Thomas Broxson
Purdue University Calumet, Diane Trgovcich-Zacok
Queens University of Charlotte, Greg D. Pillar
Raritan Valley Community College, Jay F. Kelly
Reading Area Community College, Kathy McCann Evans
Rutgers University, Craig Phelps
Saddleback College, Morgan Barrows
Santa Monica College, Dorna S. Sakurai
Shasta College, Morgan Akin
Shasta College, Allison Lee Breedveld
Southeast Kentucky Community and Technical College, Sheila Miracle
Southern Connecticut State University, Scott M. Graves
Southern New Hampshire University, Sue Cooke
Southern New Hampshire University, Michele L. Goldsmith
Southwest Minnesota State University, Emily Deaver
Spartanburg Community College, Jeffrey N. Crisp
Spelman College, Victor Ibeanusi
St. Johns River State College, Christopher J. Farrell
Stonehill College, Susan M. Mooney
Tabor College, Andrew T. Sensenig
Temple College, John McClain
Terra State Community College, Andrew J. Shella
Texas A&M University-Corpus Christi, Alberto M. Mestas-Nuñez
Tusculum College, Kimberly Carter
Univeristy of Nebraska, James R. Brandle
University of Akron, Nicholas D. Frankovits
University of Denver, Shamim Ahsan
University of Kansas, Kathleen R. Nuckolls
University of Miami, *Kathleen* Sullivan Sealey
University of Missouri at Columbia, Douglas C. Gayou
University of Missouri-Kansas City, James B. Murowchick
University of North Carolina-Wilmington, Jack C. Hall
University of North Texas, Samuel Atkinson
University of Tampa, Yasoma Hulathduwa
University of Tennessee, Michael McKinney
University of Utah, Lindsey Christensen Nesbitt
University of Wisconsin-Stevens Point, Holly A Petrillo
University of Wisconsin-Stout, Charles R. Bomar
Valencia College, Patricia Smith
Vance Granville Community College, Joshua Eckenrode
Villanova University, Lisa J. Rodrigues
Virginia Tech, Matthew Eick
Viterbo University, Christopher Iremonger
Waubonsee Community College, Dani DuCharme
Wayne County Community College District, Nina Abubakari
West Chester University of Pennsylvania, Robin C. Leonard
Westminster College, Christine Stracey
Worcester Polytechnic Institute, Theodore C. Crusberg
Wright State University, Sarah Harris

Students—study more efficiently, retain more and achieve better outcomes. Instructors—focus on what you love—teaching.

SUCCESSFUL SEMESTERS INCLUDE CONNECT

For Instructors

You're in the driver's seat.

Want to build your own course? No problem. Prefer to use our turnkey, prebuilt course? Easy. Want to make changes throughout the semester? Sure. And you'll save time with Connect's auto-grading too.

65%

Less Time Grading

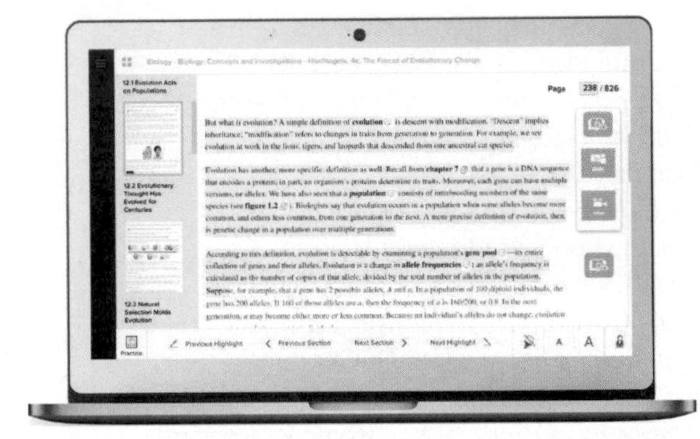

They'll thank you for it.

Adaptive study resources like SmartBook® help your students be better prepared in less time. You can transform your class time from dull definitions to dynamic debates. Hear from your peers about the benefits of Connect at **www.mheducation.com/highered/connect**

Make it simple, make it affordable.

Connect makes it easy with seamless integration using any of the major Learning Management Systems—Blackboard®, Canvas, and D2L, among others—to let you organize your course in one convenient location. Give your students access to digital materials at a discount with our inclusive access program. Ask your McGraw-Hill representative for more information.

©Hill Street Studios/Tobin Rogers/Blend Images LLC

Solutions for your challenges.

A product isn't a solution. Real solutions are affordable, reliable, and come with training and ongoing support when you need it and how you want it. Our Customer Experience Group can also help you troubleshoot tech problems—although Connect's 99% uptime means you might not need to call them. See for yourself at **status.mheducation.com**

For Students

©Shutterstock/wavebreakmedia

Effective, efficient studying.

Connect helps you be more productive with your study time and get better grades using tools like SmartBook, which highlights key concepts and creates a personalized study plan. Connect sets you up for success, so you walk into class with confidence and walk out with better grades.

"I really liked this app—it made it easy to study when you don't have your textbook in front of you."

- Jordan Cunningham,
Eastern Washington University

Study anytime, anywhere.

Download the free ReadAnywhere app and access your online eBook when it's convenient, even if you're offline. And since the app automatically syncs with your eBook in Connect, all of your notes are available every time you open it. Find out more at **www.mheducation.com/readanywhere**

No surprises.

The Connect Calendar and Reports tools keep you on track with the work you need to get done and your assignment scores. Life gets busy; Connect tools help you keep learning through it all.

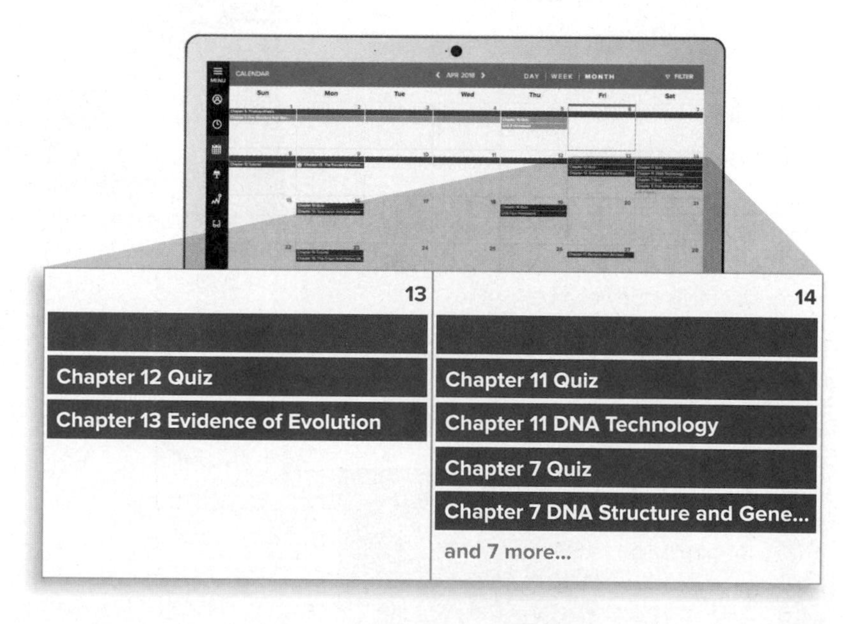

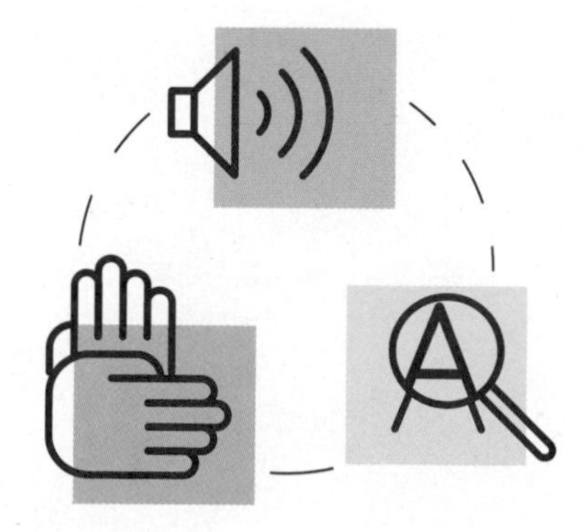

Learning for everyone.

McGraw-Hill works directly with Accessibility Services Departments and faculty to meet the learning needs of all students. Please contact your Accessibility Services office and ask them to email accessibility@mheducation.com, or visit **www.mheducation.com/accessibility** for more information.

Guided Tour

Application-based learning contributes to engaged scientific investigation.

Key Concepts

Key concepts from each chapter are presented in a beautifully arranged layout to guide the student through the often complex network issues.

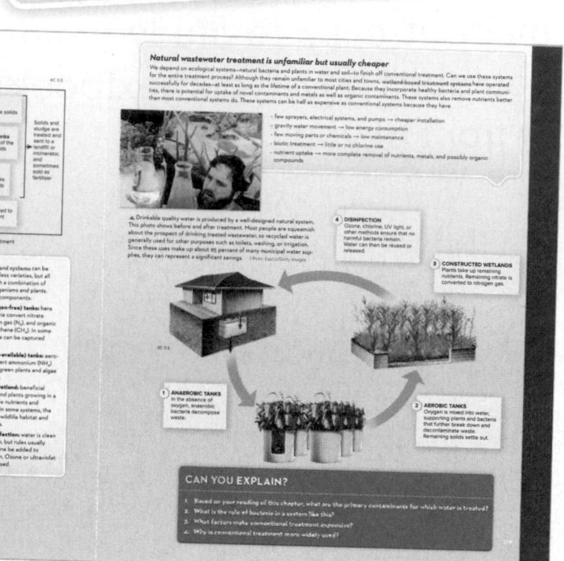

Case Studies

All chapters open with a real-world case study to help students appreciate and understand how environmental science impacts lives and how scientists study complex issues.

CASE STUDY

Palm Oil and Endangered Species

Are your donuts, toothpaste, and shampoo killing critically endangered orangutans in Sumatra and Borneo? It seems remote, but they might be. Palm oil, a key ingredient in at least half of the packaged foods, cosmetics, and soaps in the supermarket, is almost entirely sourced from plantations that just 20 years ago were moist tropical forest. In Indonesia and Malaysia these forests were the habitat of orangutans, Sumatran tigers and rhinos, and other endangered species. As palm oil has become the world's most widely used vegetable oil, expanding palm oil plantations have become one of the greatest causes of tropical deforestation.

▲ **FIGURE 6.1** Over the past 15 years, palm plantation area in Southeast Asia has grown to more than 14 million hectares (34 million acres), replacing some of the world's richest primary forest. This rapid growth has destroyed habitat and displaced many critically endangered species. ©KhunJompol/Getty Images

A 2017 study of orangutan populations in Borneo, an island owned partly by Malaysia and partly by Indonesia, estimated that at least 100,000 of these rare and reclusive forest primates were killed in just 15 years, between 1999 and 2015. This represents over half of the region's orangutans. By 2050 the population is expected to be only around 50,000, many of them in tiny, dispersed, and nonviable populations. The main reasons for this decline are the rapid conversion of primary forest to palm plantations, deforestation for wood products, and the increasing density of human populations as settlements expand to serve these industries. Habitat loss is a driving factor, but actual mortality in this study was attributed mainly to hunting in the forests around plantations, made possible by the expansion of the plantations and logging roads deep into the primary forest.

In Indonesian, *orang utan* means person of the forest. Orangutans are among our closest primate relatives, sharing at least 97 percent of our genes. Traditional cultures in Borneo may recognize this relationship, because taboos have discouraged hunting

composed mainly of ancient, undecomposed plant material, so draining and burning of a hectare of peatland can release 15,000 tons of CO_2. More than 70 percent of the carbon released from Sumatran forests is from burning peat. Indonesia, which has the third largest area of rainforest in the world as well as the highest rate of deforestation, is now the world's third highest emitter of greenhouse gases. Smoke from burning peat often blankets Singapore, Malaysia, and surrounding regions.

At the 2014 UN Climate Summit in New York, 150 companies, including McDonald's, Nestlé, General Mills, Kraft, and Procter & Gamble, promised to stop using palm oil from recently cleared rainforest and to protect human rights in forest regions. Several logging companies, including the giant Asia Pulp and Paper, pledged to stop draining peat lands and to reduce deforestation by 50 percent by 2020.

Will these be effective promises or empty ones? It is difficult to trace oil origins or to monitor remote areas, but at least this movement sets a baseline for acceptable practices. In 2017 two of the world's largest palm oil traders, Wilmar International and Cargill, announced they would no longer do business with a Guatemalan company, Reforestadora de Palmas del Petén S.A. (REPSA), because of environmental and human rights abuses. REPSA was implicated in the murder of Rigoberto Lilma Choc, a 28-year-old Guatemalan schoolteacher who had protested when effluent from a REPSA palm oil operation poisoned the Pasión River, killing millions of fish. When a Guatemalan judge ordered REPSA to stop operations for 6 months, the ruling was quickly followed by the kidnappings of three human rights activists and by Choc's murder. Cargill then cut ties with REPSA, citing its failure to

©Martin Kubat/Shutterstock

Active Learning

Students will be encouraged to practice critical thinking skills and apply their understanding of newly learned concepts and to propose possible solutions.

Active LEARNING

Comparing Biome Climates

Look back at the climate graphs for San Diego, California, an arid region, and Belém, Brazil, in the Amazon rainforest (see fig. 5.6). How much colder is San Diego than Belém in January? in July? Which location has the greater range of temperature through the year? How much do the two locations differ in precipitation during their wettest months?

Compare the temperature and precipitation in these two places with those in the other biomes shown in the pages that follow. How wet are the wettest biomes? Which biomes have distinct dry seasons? How do rainfall and length of warm seasons explain vegetation conditions in these biomes?

ANSWERS: San Diego is about 13°C colder in January, about 6°C colder in July; San Diego has the greater range of temperature; there is about 250 mm difference in precipitation in December–February.

What Can YOU DO?

Working Locally for Ecological Diversity

You might think that the diversity and complexity of ecological systems are too large or too abstract for you to have any influence. But you can contribute to a complex, resilient, and interesting ecosystem, whether you live in the inner city, a suburb, or a rural area.

- Take walks. The best way to learn about ecological systems in your area is to take walks and practice observing your environment. Go with friends, and try to identify some of the species and trophic relationships in your area.
- Keep your cat indoors. Our lovable domestic cats are also very successful predators. Migratory birds, especially those nesting on the ground, have not evolved defenses against these predators.
- Plant a butterfly garden. Use native plants that support a diverse insect population. Native trees with berries or fruit also support birds. (Be sure to avoid non-native invasive species.) Allow structural diversity (open areas, shrubs, and trees) to support a range of species.
- Join a local environmental organization. Often the best way to be effective is to concentrate your efforts close to home. City parks and neighborhoods support ecological communities, as do farming and rural areas. Join an organization work... aintain ecosystem health; start by looking for environmental clubs

What Can You Do?

Students can employ these practical ideas to make a positive difference in our environment.

EXPLORING *Science* — *Inexpensive Water Purification*

When Ashok Gadgil was a child in Bombay, India, five of his cousins died in infancy from diarrhea spread by contaminated water. Although he didn't understand the implications of those deaths at the time, as an adult he realized how heartbreaking and preventable those deaths were. After earning a degree in physics from the University of Bombay, Gadgil moved to the University of California at Berkeley, where he was awarded a PhD in 1979. He's now senior staff scientist in the Environmental Energy Technology Division, where he works on solar energy and

▲ A woman fills her jug with clean water from the village WaterHealth kiosk. More than 6 million people's lives have been improved by this innovative system of water purification. ©Waterhealth International

mount the UV source above the water where it couldn't develop mineral deposits. He designed a system in which water flows through a shallow, stainless steel trough. The apparatus can be gravity fed and requires only a car battery as an energy source.

The system can disinfect 15 liters (4 gallons) of water per minute, killing more than 99.9 percent of all bacteria and viruses. This produces enough clean water for a village of 1,000 people. This simple system costs only about 5 cents per ton (950 liter). Of course, removing pathogens doesn't do anything about minerals, such as arsenic,

Exploring Science

Current environmental issues exemplify the principles of scientific observation and data-gathering techniques to promote scientific literacy.

What Do You Think?

Students are presented with challenging environmental studies that offer an opportunity to consider contradictory data, special interest topics, and conflicting interpretations within a real scenario.

What Do YOU THINK?

Shade-Grown Coffee and Cocoa

Do your purchases of coffee and chocolate help to protect or destroy tropical forests? Coffee and cocoa are two of the many products grown exclusively in developing countries but consumed almost entirely in the wealthier, developed nations. Coffee grows in cool, mountain areas of the tropics, while cocoa is native to the warm, moist lowlands. What sets these two apart is that both come from small trees adapted to grow in low light, in the shady understory of a mature forest. **Shade-grown** coffee and cocoa (grown beneath an understory of taller trees) allow farmers to produce a crop at the same time as forest habitat remains for birds, butterflies, and other wild species.

Until a few decades ago, most of the world's coffee and cocoa were shade-grown. But new varieties of both crops have been developed that can be grown in full sun. Growing in full sun, trees can be crowded together more closely. With more sunshine, photosyn-

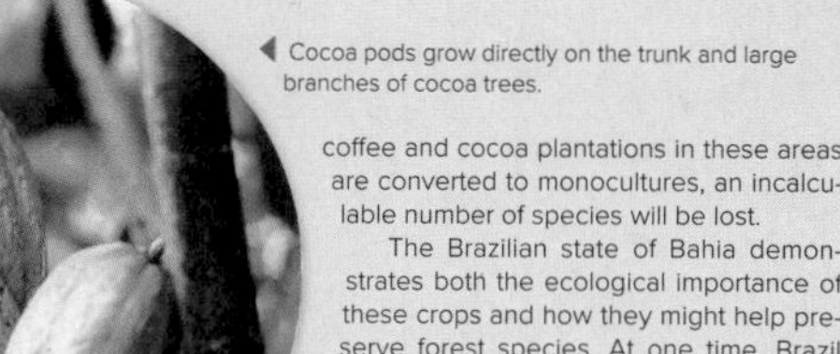

◀ Cocoa pods grow directly on the trunk and large branches of cocoa trees. ©William P. Cunningham

coffee and cocoa plantations in these areas are converted to monocultures, an incalculable number of species will be lost.

The Brazilian state of Bahia demonstrates both the ecological importance of these crops and how they might help preserve forest species. At one time, Brazil produced much of the world's cocoa, but in the early 1900s, the crop was introduced into West Africa. Now Côte d'Ivoire alone grows more than 40 percent of the world total. Rapid increases in global supplies have made prices plummet, and the value of Brazil's harvest has dropped by 90 percent. Côte d'Ivoire is aided in this competition by a labor system that reportedly includes widespread child slavery. Even adult workers in Côte d'Ivoire get only about $165 (U.S.) per year (if they get paid at all), compared with a minimum wage of $850 (U.S.) per year in Brazil. As African cocoa

Pedagogical Features Facilitate Student Understanding of Environmental Science

Learning Outcomes

Questions at the beginning of each chapter challenge students to find their own answers.

CHAPTER

6 **Environmental Conservation: Forests, Grasslands, Parks, and Nature Preserves**

LEARNING OUTCOMES

After studying this chapter, you should be able to answer the following questions:

- What portion of the world's original forests remains?
- What activities threaten global forests? What steps can be taken to preserve them?
- Why is road construction a challenge to forest conservation?
- Where are the world's most extensive grasslands?
- How are the world's grasslands distributed, and what activities degrade grasslands?
- What are the original purposes of parks and nature preserves in North America?
- What are some steps to help restore natural areas?

Orangutans are among the most critically endangered of all the great apes. Over the past 20 years, about 90 percent of their rainforest habitat in Borneo and Sumatra has been destroyed by logging and conversion to palm oil plantations.

Practice Quiz

Short-answer questions allow students to check their knowledge of chapter concepts.

PRACTICE QUIZ

1. What are the two most important nutrients causing eutrophication in the Gulf of Mexico?
2. What are systems, and how do feedback loops regulate them?
3. Your body contains vast numbers of carbon atoms. How is it possible that some of these carbons may have been part of the body of a prehistoric creature?
4. List six unique properties of water. Describe, briefly, how each of these properties makes water essential to life as we know it.
5. What is DNA, and why is it important?
6. The oceans store a vast amount of heat, but this huge reservoir of energy is of little use to humans. Explain the difference between high-quality and low-quality energy.
7. In the biosphere, matter follows circular pathways, while energy flows in a linear fashion. Explain.
8. To which wavelengths do our eyes respond, and why? (Refer to fig. 2.13.) About how long are short ultraviolet wavelengths compared to microwave lengths?
9. Where do extremophiles live? How do they get the energy they need for survival?
10. Ecosystems require energy to function. From where does most of this energy come? Where does it go?
11. How do green plants capture energy, and what do they do with it?
12. Define the terms *species, population,* and *biological community.*
13. Why are big, fierce animals rare?
14. Most ecosystems can be visualized as a pyramid with many organisms in the lowest trophic levels and only a few individuals at the top. Give an example of an inverted numbers pyramid.
15. What is the ratio of human-caused carbon releases into the atmosphere shown in figure 2.18 compared to the amount released by terrestrial respiration?

CRITICAL THINKING AND DISCUSSION

Apply the principles you have learned in this chapter to discuss these questions with other students.

1. Ecosystems are often defined as a matter of convenience because we can't study everything at once. How would you describe the characteristics and boundaries of the ecosystem in which you live? In what respects is your ecosystem an open one?
2. Think of some practical examples of increasing entropy in everyday life. Is a messy room really evidence of thermodynamics at work or merely personal preference?
3. Some chemical bonds are weak and have a very short half-life (fractions of a second, in some cases); others are strong and stable, lasting for years or even centuries. What would our world be like if all chemical bonds were either very weak or extremely strong?
4. If you had to design a research project to evaluate the relative biomass of producers and consumers in an ecosystem, what would you measure? (*Note:* This could be a natural system or a human-made one.)
5. Understanding storage compartments is essential to understanding material cycles, such as the carbon cycle. If you look around your backyard, how many carbon storage compartments are there? Which ones are the biggest? Which ones are the longest lasting?

Critical Thinking and Discussion Questions

Brief scenarios of everyday occurrences or ideas challenge students to apply what they have learned to their lives.

Data Analysis

At the end of each chapter, these exercises give students further opportunities to apply critical thinking skills and analyze data. These are assigned through Connect in an interactive online environment. Students are asked to analyze data in the form of documents, videos, and animations.

DATA ANALYSIS: A Closer Look at Nitrogen Cycling

1. Which forms of N do plants take up? Can they capture N_2 from the air?
2. Refer to section 2.5. How is N_2 captured, or fixed, from the air into the food web?
3. Most of the processes are hard to quantify, but the figure shown here gives approximate amounts for fossil fuel burning and commercial N fixation, and for N fixing by bacteria. What do these terms mean? What is the magnitude of each? What is the difference?
4. If anthropogenic processes introduce increasing amounts of atmospheric N to the biosphere and hydrosphere, where does that N go? (*Hint:* Refer to the opening case study.)
5. Why is N so important for living organisms?
6. In marine systems, N is often a limiting factor. What is a "limiting factor"? What is a consequence of increasing the supply of N in a marine system?

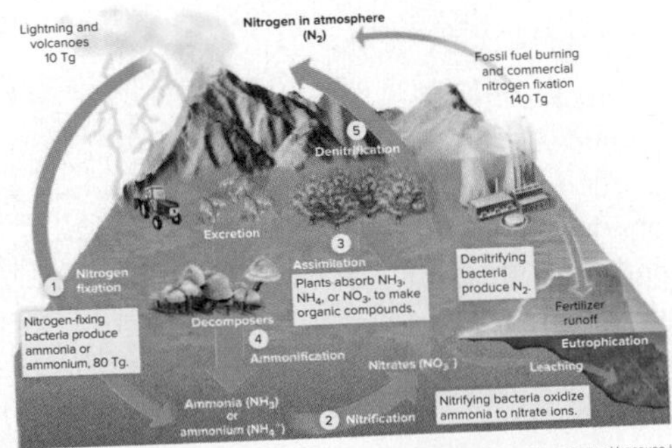

▲ Nitrogen cycles through living and nonliving systems. This biogeochemical cycle is important to understand because it strongly influences how ecosystems function.

CHAPTER

1 Understanding Our Environment

LEARNING OUTCOMES

After studying this chapter, you should be able to answer the following questions:

- List several major environmental challenges and some ways to address them.
- Explain the idea of sustainability and some of its aims.
- Why are scientists cautious about claiming absolute proof of particular theories?
- What is critical thinking, and why is it important in environmental science?
- Why do we use graphs and data to answer questions in science?
- Identify several people who helped shape our ideas of resource conservation and preservation—why did they promote these ideas when they did?

The Kayenta solar plant in Monument Valley, AZ is the first step for the Navajo Tribe towards renewable energy, water conservation, clean air, green-collar jobs, and climate protection.

CASE STUDY

Sustainability and Power on the Reservation

Sustainable development is a challenge faced by all developing nations and regions. How can they ensure a healthy, safe environment and also provide jobs for young people? Can they reduce air, water, and soil pollution and simultaneously reduce poverty?

These are questions members of the Navajo, or Diné, Nation have been asking. The largest tribe in the United States, they are a nation within another nation, but they share challenges of most developing areas. They have half the per-capita income and twice the unemployment of the rest of Arizona. Rural poverty, lack of water and sanitation, and inadequate electricity connection are chronic conditions that hinder education and health care.

Also like other developing nations, the Navajo are debating their energy future. Since 1973 one of the most important employers on the reservation has been the Navajo Generating Station, a coal-powered plant that produces 16 percent of Arizona's electricity and employs about 500 people, 90 percent of them Navajo. The power plant is also an environmental liability. It produces 30 percent of Arizona's carbon dioxide and 25 percent of the state's sulfur dioxide, a source of smog and acid rain, as well as airborne mercury and cadmium. For over 45 years, the plant has been one of Arizona's worst polluters, often obscuring visibility in the nearby Grand Canyon. The Kayenta coal mine, which supplies the plant, produces dust and other airborne pollutants and threatens local waterways with acidic runoff. The multinational Peabody Energy, one of the world's largest coal companies, owns the mine. The plant and mine also consume water extravagantly: about 33 million m^3 every year for steam, cooling, and dust control, with most of it from the declining Colorado River. Filters and other equipment capture much of the pollution at the Navajo station, but ongoing upgrades and maintenance are costly. At the same time, other sources of power are becoming cheaper to produce. Despite opposition from Peabody and other interests, owners of the plant and Navajo leaders agreed that it was time to transition away from coal. They agreed to shutter the facility by 2020.

The decision has been controversial, as closing the plant eliminates hundreds of steady jobs. But many members of the Navajo Nation want independence from coal and they want to diversify energy and the economy, with more local ownership. They want to develop in green jobs that don't undermine their children's health. They are motivated to provide energy while protecting the land they live on and their scarce water resources. And they want to address climate change, to which coal is the worst single contributor. Financial cost doomed the plant, but these social and environmental costs also weighed heavily in the decision.

An important step in the energy transition was the Kayenta photovoltaic solar plant, owned by the Navajo Tribal Utility Authority and the first utility-scale solar power plant on the reservation. Kayenta began delivering clean electricity in June 2017. Rated for 27.5 megawatts (MW) of electricity, the solar plant produces far less than the 1,700 MW delivered by the Navajo Generating Station. (A megawatt is a million watts, enough to power 100,000 10-watt lightbulbs simultaneously or about 500 U.S. households.) But it was just the beginning. Six months after Kayenta opened, tribal authorities signed an agreement to build Kayenta II, doubling production to over 50 MW. Tribal officials have planned another 500 MW of solar in the next few years.

Constructing the Kayenta site took only about 6 months, which is good for energy production but employed its 275 workers for only a short time. As installations scale up, however, employment is expected to increase and stabilize. Increasing investment in solar could also aid remote rural access to electricity. Hooking up a household on the reservation to the electric grid can cost $50,000, far more than solar panels and battery storage.

A solar plant is cleaner than coal, but what about space and financial costs? These are similar: The 120-hectare (300-acre) Kayenta plant uses about 4.5 hectares/MW (11 acres/MW), while the Navajo Generating Station, including its active coal mines (but excluding closed, spent mining areas), comes to about 4–5 hectares/MW (10–12 acres/MW). The $64 million cost of the Kayenta plant's first phase amounted to about $2.3 million/MW. Adjusted for inflation, the coal plant cost about $2.5 million/MW, plus the cost of continuously supplying coal, at a rate of 240 100-ton train car loads every day.

Access to clean energy is often central to sustainable development. Electric lights help you study and learn. Water pumps can improve sanitation. "Green-collar" jobs can transform lives and livelihoods. These aspects of sustainable development are goals for communities around the world. In this chapter we explore some of the ways environmental science contributes to understanding and addressing the widespread need for more equitable economies, societies, and environmental quality. ■

▲ **FIGURE 1.1** The Navajo Generating Station has been a major source of revenue and of pollution for almost 50 years.

Today we are faced with a challenge that calls for a shift in our thinking, so that humanity stops threatening its life-support system.

—WANGARI MAATHAI,
WINNER OF 2004 NOBEL PEACE PRIZE

1.1 WHAT IS ENVIRONMENTAL SCIENCE?

- *Environmental science draws on diverse disciplines, skills, and interests.*
- *A global perspective helps us understand environmental systems.*
- *The scientific method makes inquiry orderly.*

Environmental science uses scientific approaches to understand the complex systems in which we live. Often environmental science focuses on finding basic explanations for how systems function: How does biodiversity affect the ways an ecosystem functions, or how does land use affect a river system? But because human decisions about resources, land use, or waste management affect environmental systems, decisions and policies about resources are also a part of environmental science.

In this chapter we examine some central ideas and approaches in environmental science. You will explore these themes in greater depth in later chapters. We focus on issues of sustainability, environmental justice, and the scientific method that underlies our understanding of these ideas. We also examine some key ideas that have influenced our understanding of environmental science.

Environmental science integrates many fields

We inhabit both a natural world of biological diversity and physical processes and a human environment of ideas and practices. Environmental science involves both these natural and human worlds. Because environmental systems are complex and interconnected, the field also draws on a wide range of disciplines and skills, and multiple ways of knowing are often helpful for finding answers (fig. 1.2). Biology, chemistry, earth science, and geography contribute ideas and evidence of basic science. Political science, economics, communications, and arts help us understand how people share resources, compete for them, and evaluate their impacts on society. One of your tasks in this course may be to understand where your own knowledge and interests contribute (Active Learning, p. 4). Identifying your particular interest will help you do better in this class, because you'll have more reason to explore the ideas you encounter.

Environmental science often informs policy, because it provides information for decision making about resources and the living systems we occupy. This doesn't imply particular policy positions, but it does provide an analytical approach to using observable evidence, rather than assumptions or hearsay, in making decisions.

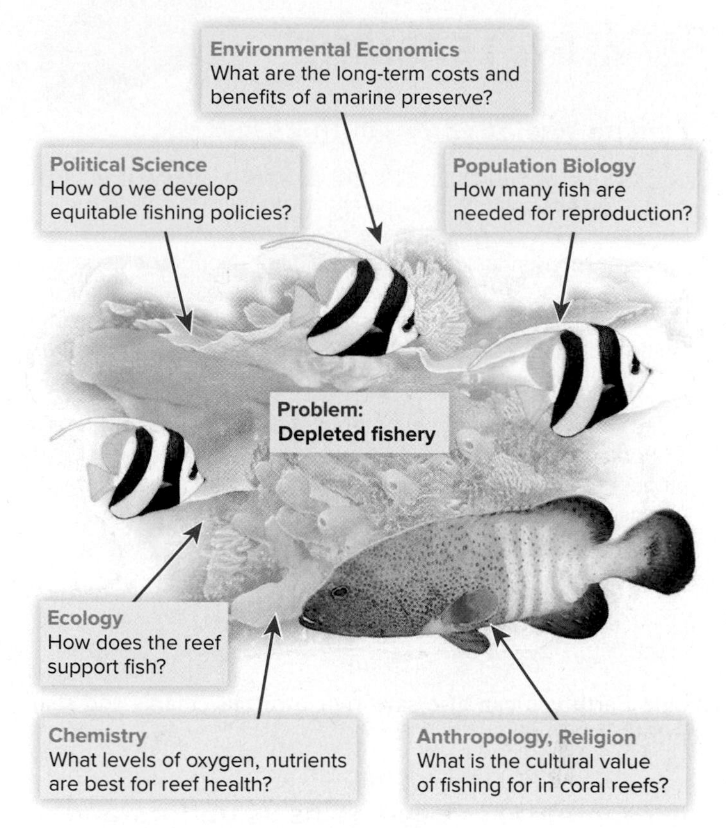

▲ **FIGURE 1.2** Many types of knowledge are needed in environmental science. A few examples are shown here.

Environmental science is global

You are already aware of our global dependence on resources and people in faraway places, from computers built in China to oil extracted in Iraq or Venezuela. These interdependencies become clearer as we learn more about global and regional environmental systems. Often the best way to learn environmental science is to see how principles play out in real places. Familiarity with the world around us will help you understand the problems and their context. Throughout this book we've provided links to places you can see in Google Earth, a free online mapping program that you can download from googleearth.com. When you see a blue globe in the margin of this text, like the one at left, you can go to Connect and find placemarks that let you virtually visit places discussed. In

Benchmark Data	
Among the ideas and values in this chapter, these are a few worth remembering.	
280 ppm	Pre-industrial concentration of CO_2 in the atmosphere, in parts per million
410 ppm	Approximate concentration of CO_2 now
6 billion	Global population 2000
9 billion	Global population in 2050 (projected)
5	Average number of children per woman in 1950
2	Average number by 2050 (projected)

Active LEARNING

Finding Your Strengths in This Class

A key strategy for doing well in this class is to figure out where your strengths and interests intersect with the subjects you will be reading about. As you have read, environmental science draws on many kinds of knowledge (fig. 1.2). Nobody is good at all of these, but everyone is good at some of them. Form a small group of students; then select one of the questions in section 1.2. Explain how each of the following might contribute to understanding or solving that problem:

artist, writer, politician, negotiator, chemist, mathematician, hunter, angler, truck driver, cook, parent, builder, planner, economist, speaker of multiple languages, musician, businessperson

ANSWERS: All of these provide multiple insights; answers will vary.

Google Earth you can also save your own placemarks and share them with your class.

Environmental science helps us understand our remarkable planet

Imagine that you were an astronaut returning home after a trip to the moon or Mars. What a relief it would be, after the silent void of outer space, to return to this beautiful, bountiful planet (fig. 1.3). We live in an incredibly prolific and colorful world that is, as far as we know, unique in the universe. Compared with other planets in our solar system, temperatures on the earth are mild and relatively constant. Plentiful supplies of clean air, fresh water, and fertile soil are regenerated endlessly and spontaneously by biogeochemical cycles and biological communities (discussed in chapters 2 and 3). The value of these ecological services is almost incalculable, although economists estimate that they account for a substantial proportion of global economic activity (see chapter 15).

Perhaps the most amazing feature of our planet is its rich diversity of life. Millions of beautiful and intriguing species populate the earth and help sustain a habitable environment (fig. 1.4). This vast multitude of life creates complex, interrelated communities where towering trees and huge animals live together with, and depend upon, such tiny life-forms as viruses, bacteria, and fungi. Together, all these organisms make up delightfully diverse, self-sustaining ecosystems, including dense, moist forests; vast, sunny savannas; and richly colorful coral reefs.

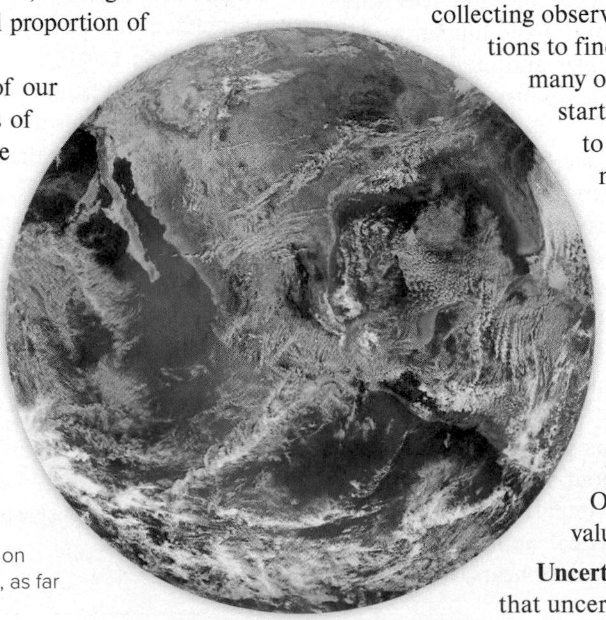

▶ **FIGURE 1.3** The life-sustaining ecosystems on which we all depend are unique in the universe, as far as we know. Source: Norman Kuring/NASA

▲ **FIGURE 1.4** Perhaps the most amazing feature of our planet is its rich diversity of life. ©Fuse/Getty Images

From time to time we should pause to remember that, in spite of the challenges of life on earth, we are incredibly lucky to be here. Because environmental scientists observe this beauty around us, we often ask what we can do, and what we *ought* to do, to ensure that future generations have the same opportunities to enjoy this bounty.

Methods in environmental science

Keep an eye open for the ideas that follow as you read this book. These are a few of the methods that you will find in science generally. They reflect the fact that environmental science is based on careful, considered observation of the world around us.

Observation: A first step in understanding our environment is careful, detailed observation and evaluation of factors involved in pollution, environmental health, conservation, population, resources, and other issues. Knowing about the world we inhabit helps us understand where our resources originate, and why.

The scientific method: Discussed later in this chapter, the scientific method is an orderly approach to asking questions, collecting observations, and interpreting those observations to find an answer to a question. In daily life, many of us have prior expectations when we start an investigation, and it takes discipline to avoid selecting evidence that conveniently supports our prior assumptions. In contrast, the scientific method aims to be rigorous, using statistics, blind tests, and careful replication to avoid simply confirming the investigator's biases and expectations.

Quantitative reasoning: This means understanding how to compare numbers and interpret graphs, to perceive what they show about problems that matter. Often this means interpreting changes in values, such as population size over time.

Uncertainty: A repeating theme in this book is that uncertainty is an essential part of science.

Science is based on observation and testable hypotheses, but we know that we cannot make all observations in the universe, and we have not asked all possible questions. We know there are limits to our knowledge. Understanding how much we *don't* know, ironically, can improve our confidence in what we *do* know.

Critical and analytical thinking: The practice of stepping back to examine what you think and why you think it, or why someone says or believes a particular idea, is known generally as critical thinking. Acknowledging uncertainty is one part of critical thinking. This is a skill you can practice in all your academic pursuits as you make sense of the complexity of the world we inhabit.

1.2 MAJOR THEMES IN ENVIRONMENTAL SCIENCE

- *Water, air quality, and climate change are key concerns.*
- *Population growth has slowed, as food resources and education have improved.*
- *Natural resource depletion is a major concern.*

In this section we review some of the main themes in this book. All of these are serious problems, but they are also subjects of dramatic innovation. Often solutions lie in policy and economics, but environmental scientists provide the evidence on which policy decisions can be made.

We often say that crisis and opportunity go hand in hand. Serious problems can drive us to seek better solutions. As you read, ask yourself what factors influence these conditions and what steps might be taken to resolve them.

Environmental quality

Climate Change The atmosphere retains heat near the earth's surface, which is why it is warmer here than in space. But concentrations of heat-trapping "greenhouse gases," especially CO_2, increased dramatically, from 280 parts per million (ppm) 200 years ago to about 410 ppm in 2019. Burning fossil fuels, clearing forests and farmlands, and raising billions of methane-producing cattle are some of the main causes. Climate models indicate that by 2100, if current trends continue, global mean temperatures will probably increase by 2° to 6°C compared to 1990 temperatures (3.6° to 12.8°F; fig. 1.5), far warmer than the earth has been since the beginning of human civilization. For comparison, the last ice age was about 4°C cooler than now. Increasingly severe droughts and heat waves are expected in many areas. Greater storm intensity and flooding are expected in many regions. Disappearing glaciers and snowfields threaten the water supplies on which cities such as Los Angeles and Delhi depend.

Military experts argue that climate change is among our greatest threats, contributing to refugee crises and terrorism. Already, climate change has forced hundreds of millions of people from farmlands that have become too dry or hot to produce crops. Storms, floods, and rising sea levels, threaten villages in many regions. Climate refugees in Syria, Nigeria, Pakistan, and other regions are vulnerable to terrorist activity and sometimes carry it abroad.

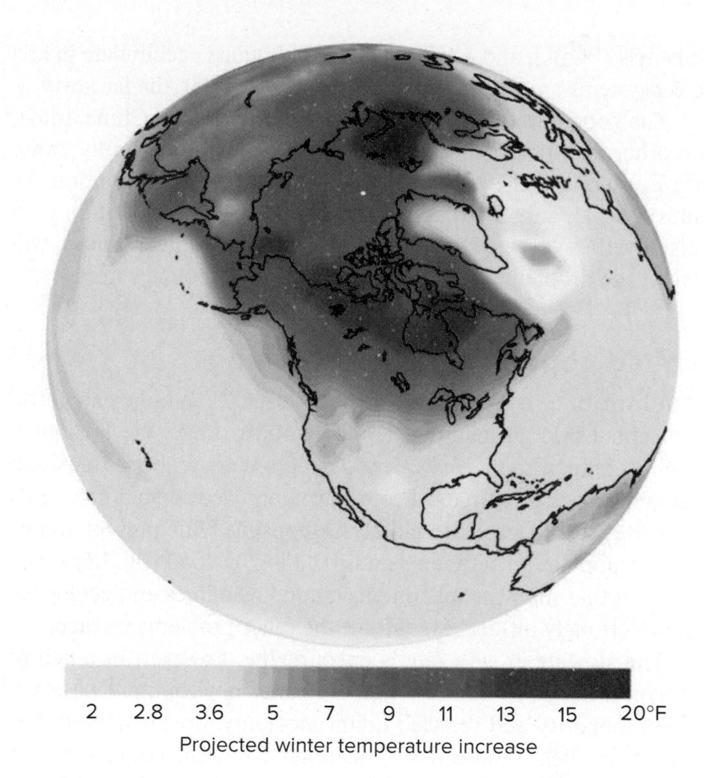

▲ **FIGURE 1.5** Climate change is projected to raise temperatures, especially in northern winter months. Source: NOAA, 2010.

On the other hand, efforts to find solutions to climate change may force new kinds of international cooperation. New strategies for energy production could reduce conflicts over oil and promote economic progress for the world's poorest populations.

Clean Water Water may be the most critical resource in the twenty-first century. At least 1.1 billion people lack access to safe drinking water, and twice that many don't have adequate sanitation. Polluted water contributes to the death of more than 15 million people every year, most of them children under age 5. About 40 percent of the world population lives in countries where water demands now exceed supplies, and the United Nations projects that by 2025 as many as three-fourths of us could live under similar conditions. Despite ongoing challenges, more than 800 million people have gained access to treated water supplies and modern sanitation since 1990.

Air Quality Air quality has worsened dramatically in newly industrializing areas, especially in much of China and India. In Beijing and Delhi, wealthy residents keep their children indoors on bad days and install air filters in their apartments. Poor residents become ill, and cancer rates are rising in many areas. Millions of early deaths and many more illnesses are triggered by air pollution each year. Worldwide, the United Nations estimates, more than 2 billion metric tons of air pollutants (not including carbon dioxide or windblown soil) are released each year. These air pollutants travel easily around the globe. On some days 75 percent of the smog and airborne particulates in California originate in Asia; mercury, polychlorinated biphenyls (PCBs), and other industrial pollutants accumulate in arctic ecosystems and in the tissues of native peoples in the far north.

The good news is that environmental scientists in China, India, and other countries suffering from poor air quality are fully aware that Europe and the United States faced deadly air pollution decades ago. They know that enforceable policies on pollution controls, together with newer, safer, and more efficient technology, will correct the problem, if they can just get needed policies in place.

Human population and well-being

Population Growth There are now over 7.7 billion people on earth, about twice as many as there were 40 years ago. We are adding about 80 million more each year. Demographers report a transition to slower growth rates in most countries: Improved education for girls and better health care are chiefly responsible. But present trends project a population between 8 and 10 billion by 2050 (fig. 1.6a). The impact of that many people on our natural resources and ecological systems strongly influences many of the other problems we face.

The slowing growth rate is encouraging, however. In much of the world, better health care and a cleaner environment have improved longevity and reduced infant mortality. Social stability has allowed families to have fewer, healthier children. Population has stabilized in most industrialized countries and even in some very poor countries where social security, education, and democracy have been established. Since 1960 the average number of children born per woman worldwide has decreased from 5 to 2.45 (fig. 1.6b). By 2050, the UN Population Division predicts, most countries will have fertility rates below the replacement rate of 2.1 children per woman. If this happens, the world population will stabilize at about 8.9 billion rather than the 9.3 billion previously expected.

Infant mortality in particular has declined in most countries, as vaccines and safe water supplies have become more widely available. Smallpox has been completely eradicated, and polio has been vanquished except in a few countries, where violent conflict has contributed to a resurgence of the disease. Life expectancies have nearly doubled, on average (fig. 1.7a).

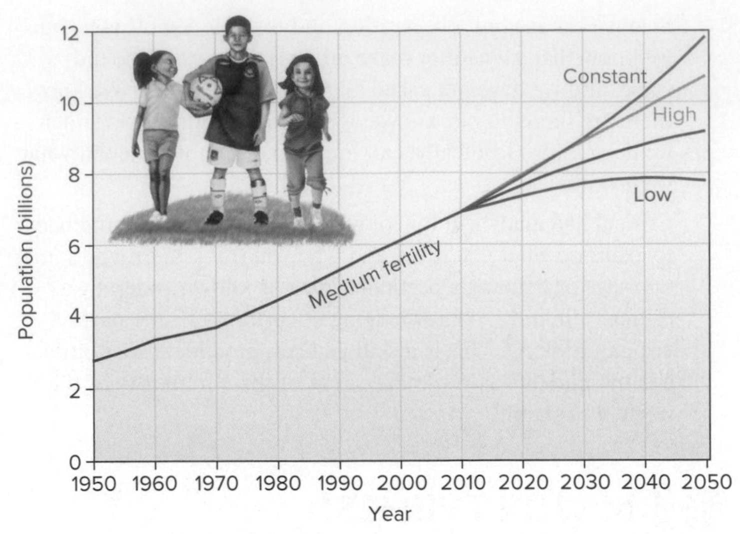

(a) Possible population trends

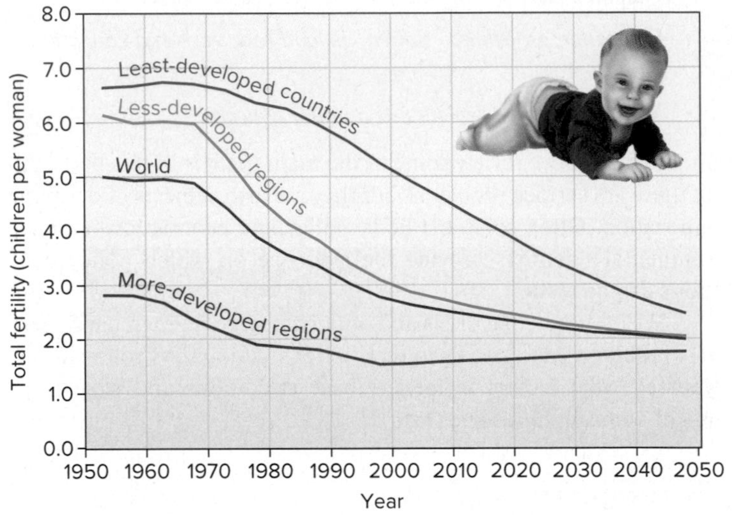

(b) Fertility rates

▲ **FIGURE 1.6** Bad news and good news: Globally, populations continue to rise (a), but our rate of growth has plummeted (b). Some countries are below the replacement rate of about two children per woman. Source: United Nations Population Program, 2011.

Hunger and Food Over the past century, global food production has increased faster than human population growth. We now produce about half again as much food as we need to survive, and consumption of protein has increased worldwide. In most countries weight-related diseases are far more prevalent than hunger-related illnesses. In spite of population growth that added nearly a billion people to the world during the 1990s, the number of people facing food insecurity and chronic hunger during this period actually declined by about 40 million.

Despite this abundance, hunger remains a chronic problem worldwide because food resources are unevenly distributed. In a world of food surpluses, currently more than 850 million people are chronically undernourished, and at least 60 million people face acute food shortages due to weather, politics, or war (fig. 1.7b). At the same time, soil scientists report that about two-thirds of all agricultural lands show signs of degradation. The biotechnology and intensive farming techniques responsible for much of our recent production gains are too expensive for many poor farmers. How can we produce food sustainably and distribute it fairly? These are key questions in environmental science.

Information and Education Because so many environmental issues can be fixed by new ideas, technologies, and strategies, expanding access to knowledge is essential to progress. The increased speed at which information now moves around the world offers unprecedented opportunities for sharing ideas. At the same time, literacy and access to education are expanding in most regions of the world (fig. 1.7c). Rapid exchange of information on the Internet also makes it easier to quickly raise global awareness of environmental problems, such as deforestation or pollution, that historically would have proceeded unobserved and unhindered. Improved access to education is helping to release many of the world's population from cycles of poverty and vulnerability.

Expanding education for girls is a primary driver for declining birth rates worldwide.

Natural resources

Biodiversity Loss Biologists report that habitat destruction, overexploitation, pollution, and the introduction of exotic organisms are eliminating species as quickly as the great extinction that marked the end of the age of dinosaurs. The United Nations Environment Programme reports that over the past century more than 800 species have disappeared and at least 10,000 species are now considered threatened. This includes about half of all primates and freshwater fish, together with around 10 percent of all plant species. Top predators, including nearly all the big cats in the world, are particularly rare and endangered. A 2017 study in Germany found that populations of insects, key pollinators and components of the food web, had declined 75 percent since 1990, and bird populations were 15 percent lower. At least half of the forests existing before the introduction of agriculture have been cleared, and many of the ancient forests, which harbor some of the greatest biodiversity, are rapidly being cut for timber, for oil extraction, or for agricultural production of globally traded commodities such as palm oil or soybeans.

Conservation of Forests and Nature Preserves Although exploitation continues, the rate of deforestation has slowed in many regions. Brazil, which led global deforestation rates for decades, has

(a) Health care

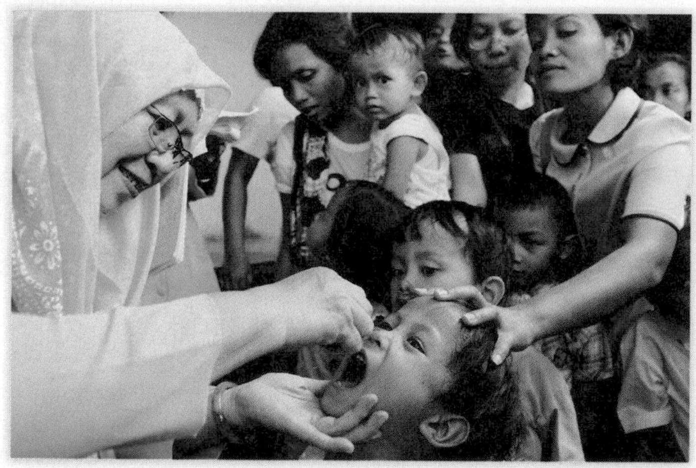

(b) Hunger

(c) Education

(d) Sustainable resource use

▲ **FIGURE 1.7** Human welfare is improving in some ways and stubbornly difficult in others. Health care is improving in many areas (a). Some 800 million people lack adequate nutrition. Hunger persists, especially in areas of violent conflict (b). Access to education is improving, including for girls (c), and local control of fishery resources is improving food security in some places (d). (a): ©Dimas Ardian/Getty Images; (b): ©Jonas Gratzer/Getty Images; (C): ©Anjo Kan/Shutterstock; (d): ©William P. Cunningham

dramatically reduced deforestation rates. Nature preserves and protected areas have increased sharply over the past few decades. Ecoregion and habitat protection remains uneven, and some areas are protected only on paper. Still, this is dramatic progress in biodiversity protection.

Marine Resources The ocean provides irreplaceable and imperiled food resources. More than a billion people in developing countries depend on seafood for their main source of animal protein, but most commercial fisheries around the world are in steep decline. According to the World Resources Institute, more than three-quarters of the 441 fish stocks for which information is available are severely depleted or in urgent need of better management. Some marine biologists estimate that 90 percent of all the large predators, including bluefin tuna, marlin, swordfish, sharks, cod, and halibut, have been removed from the ocean.

Despite this ongoing overexploitation, many countries are beginning to acknowledge the problem and find solutions. Marine protected areas and improved monitoring of fisheries provide opportunities for sustainable management (fig. 1.7d). The strategy of protecting fish nurseries is an altogether new approach to sustaining ocean systems and the people who depend on them. Marine reserves have been established in California, Hawaii, New Zealand, Great Britain, and many other areas.

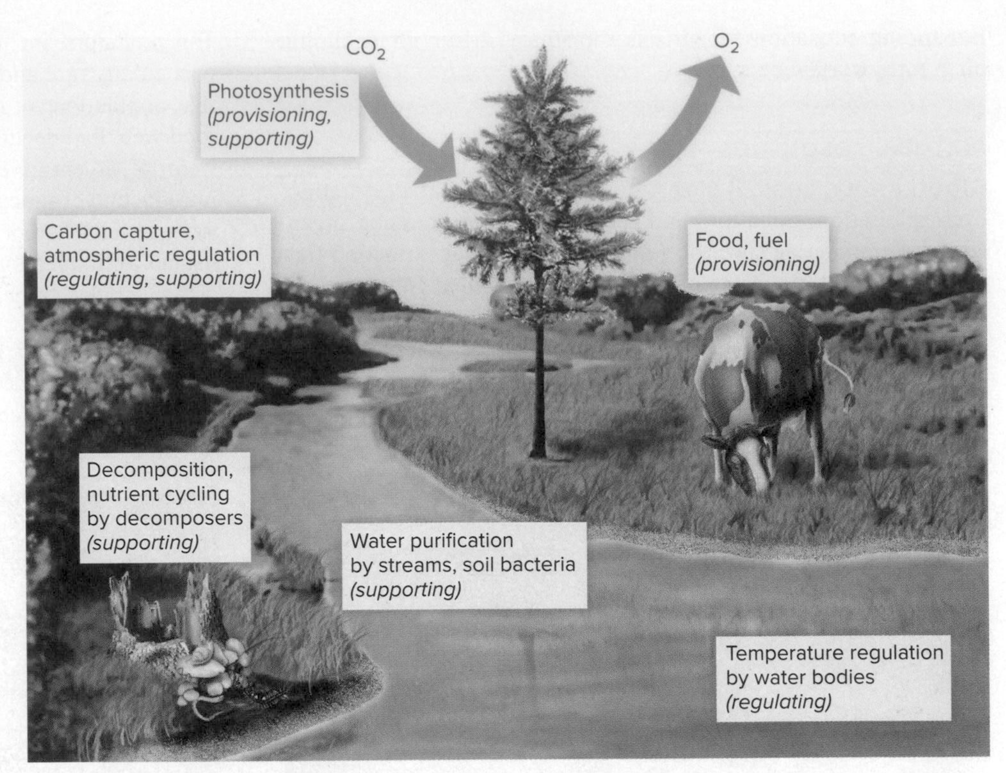

▲ **FIGURE 1.8** Ecosystem services we depend on are countless and often invisible.

Energy Resources How we obtain and use energy will greatly affect our environmental future. Fossil fuels (oil, coal, and natural gas) presently provide around 80 percent of the energy used in industrialized countries. The costs of extracting and burning these fuels are among our most serious environmental challenges. Costs include air and water pollution, mining damage, and violent conflicts, in addition to climate change.

At the same time, improving alternatives and greater efficiency are beginning to reduce reliance on fossil fuels. As noted in the opening case study, renewable energy is an increasingly available and attractive option. The cost of solar power has plummeted, and in many areas solar costs the same as conventional electricity over time. Solar and wind power are now far cheaper, easier, and faster to install than nuclear power or new coal plants.

1.3 HUMAN DIMENSIONS OF ENVIRONMENTAL SCIENCE

- *Ecosystem services are important in evaluating system values.*
- *Sustainable development goals identify key needs.*
- *Both poverty and wealth produce environmental challenges.*

Aldo Leopold, one of the greatest thinkers on conservation, observed that the great challenges in conservation have less to do with managing resources than with managing people and our demands on resources. Foresters have learned much about how to grow trees, but still we struggle to establish conditions under which villagers in developing countries can manage plantations for themselves. Engineers know how to control pollution but not how to persuade factories to install the necessary equipment. City planners know how to design urban areas, but not how to make them affordable for everyone. In this section we'll review some key ideas that guide our understanding of human dimensions of environmental science and resource use. These ideas will be useful throughout the rest of this book.

How do we describe resource use and conservation?

The natural world supplies the water, food, metals, energy, and other resources we use. Some of these resources are finite; some are constantly renewed (see chapter 14). Often, renewable resources can be destroyed by excessive exploitation, as in the case of fisheries or forest resources (see section 1.2). When we consider resource consumption, an important idea is **throughput,** the amount of resources we use and dispose of. A household that consumes abundant consumer goods, foods, and energy brings in a great deal of natural resource-based materials; that household also disposes of a great deal of materials. Conversely a household that consumes very little also produces little waste (see chapter 2).

Ecosystem services, another key idea, refers to services or resources provided by environmental systems (fig. 1.8). *Provisioning* of resources, such as the fuels we burn, may be the most obvious service we require. *Supporting* services are less obvious until you start listing them: These include water purification, production of food and atmospheric oxygen by plants, and decomposition of waste by fungi and bacteria. *Regulating* services include maintenance of temperatures suitable for life by the earth's atmosphere and carbon capture by green plants, which maintains a stable atmospheric composition. Cultural services include a diverse range of recreation, aesthetic, and other nonmaterial benefits.

Global ecosystem services amounted to a value of about $124 trillion to $145 trillion per year in 2011, according to ecological economist Robert Costanza, far more than the $65 trillion global economy in that year. These services support most other economic activity, but we tend to forget our reliance on them, and conventional economics has little ability to value them.

Planetary boundaries

Another way to think about environmental services is planetary boundaries, or thresholds of abrupt or irreversible environmental change. Studies by Johan Rockström and colleagues at the Stockholm Resilience Centre have identified nine major systems with these critical thresholds: climate change, biodiversity, land system change, freshwater use, biogeochemical flows (nitrogen and phosphorus), ocean acidification, atmospheric aerosols, stratospheric ozone loss, and "novel entities," including chemical pollution and other factors (fig. 1.9). Calculations are that we have already passed the planetary boundaries for three of these–climate change, biodiversity loss, and nitrogen cycling. We are approaching the limits for freshwater supplies, land use, ocean acidification, and phosphorus loading.

These ecosystem services are tightly coupled. Destruction of tropical forests in Southeast Asia, for example, can influence heat and drought in North America. Drought and fires in North America enhance climate warming and sea ice loss in the Arctic. A planetary perspective helps us see interconnections in global systems and their effects on human well-being. What it means to pass these boundaries remains uncertain.

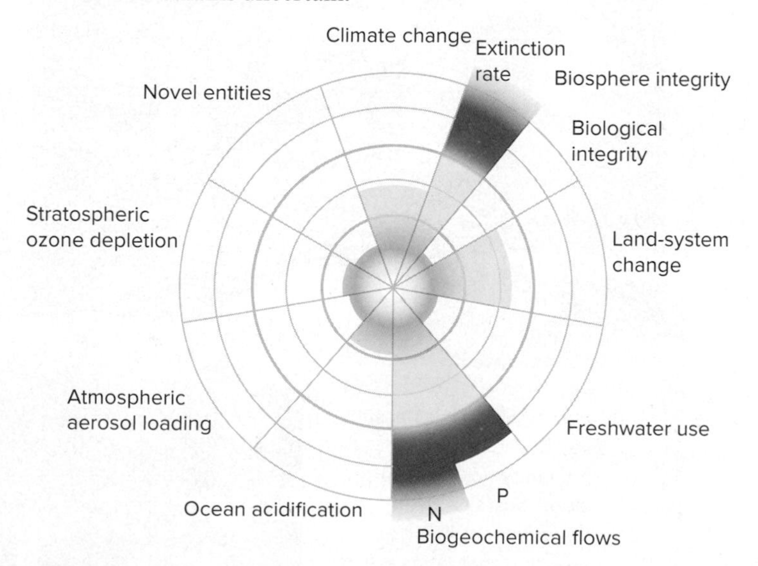

▲ **FIGURE 1.9** Calculated planetary boundaries, or thresholds beyond which irreversible change is likely. Green shading represents safe ranges; yellow represents a zone of increasing risk; red wedges represent factors exceeding boundaries. Source: Will Steffen, Katherine Richardson, Johan Rockström, et al. 2015. Planetary boundaries: Guiding human development on a changing planet. *Science* 15 Jan 2015: 1259855 DOI: 10.1126/science.1259855.

Sustainability requires environmental and social progress

Sustainability is a search for ecological stability and human progress that can last over the long term. Of course, neither ecological systems nor human institutions can continue forever. We can work, however, to protect the best aspects of both realms and to encourage resiliency and adaptability in both of them. World Health Organization director Gro Harlem Brundtland has defined **sustainable development** as "meeting the needs of the present without compromising the ability of future generations to meet their own needs." In these terms, development means bettering people's lives. Sustainable development, then, means progress in human well-being that we can extend or prolong over many generations, rather than just a few years.

In 2016 the United Nations initiated a 15-year program to promote 17 **Sustainable Development Goals** (SDGs). Ambitious and global, the goals include eliminating the most severe poverty and hunger; promoting health, education, and gender equality; providing safe water and clean energy; and preserving biodiversity. This global effort seeks to coordinate data gathering and reporting, so that countries can monitor their progress, and to promote sustainable investment in developing areas.

For each of the 17 goals, organizers identified targets: some quantifiable, some more general. For example, Goal 1, "End poverty," includes targets to eradicate extreme poverty, defined as less than $1.90 per day, and to ensure that all people have rights to basic services, ownership and inheritance of property, and other necessities for economic stability. Goal 7, "Ensure access to affordable, sustainable energy," includes targets of doubling energy efficiency and enhancing international investment in clean energy. Goal 12, "Ensure sustainable consumption and production," calls for cutting food waste in half and phasing out fossil fuel subsidies that encourage wasteful consumption. These goals may not be accomplished by 2030, but having a target to aim for improves the odds of success. And targets allow us to measure how far we have fallen short.

The SDGs also include targets for economic and social equity and for better governance. To most economists and policymakers it seems clear that economic growth is the only way to improve the lot of all people: As former U.S. president John F. Kennedy put it, "a rising tide lifts all boats." But history shows that equity is also essential. Extreme inequality undermines democracy, opportunity, and political stability. Economic and social equality, on the other hand, can promote economic growth by ensuring that extreme poverty and political unrest don't impede progress.

These ambitious goals might appear unrealistic, but they build on the remarkable (though not complete) successes of the **Millennium Development Goals** program, from 2000 to 2015. Targets included an end to poverty and hunger, universal education, gender equity, child health, maternal health, combating of HIV/AIDS, environmental sustainability, and global cooperation in development efforts. While only modest progress was achieved on some goals, UN Secretary General Ban Ki-Moon called that effort "the most successful anti-poverty movement in history." Extreme poverty dropped from nearly half the population of developing countries to just

Sustainable development

What does it mean? What does it have to do with environmental science?

Sustainable development is a goal. The aim is to meet the needs of people today without compromising resources and environmental systems for future generations. In this context, the term *development* refers to improving access to health care, education, and other conditions necessary for a healthy and productive life, especially in regions of extreme poverty. Meeting the needs of people now, while also guarding those resources for their great-great grandchildren, is both a steep challenge and a good idea.

What parts of it are achievable, and how? In general, development means equitable economic growth, which supports better education, housing, and health care. Often development involves accelerated extraction of natural resources, such as more mining, forestry, or conversion of forests and wetlands to farmlands. Sometimes development involves more efficient use of resources or growth in parts of the economy that don't depend on resource extraction, such as education, health care, or knowledge-based economic activities.

Some resources can be enhanced, for example, through reforestation, maintaining fish nurseries, or careful management of soil resources, to use them without depletion for future generations.

Here are ten key factors necessary for sustainable development, according to the United Nations agreement on development, Agenda 21.

©Digital Vision/PunchStock

KC 1.1

1. **Combating poverty** is a central goal because poverty reduces access to health care, education, and other essential components of development.

2. **Reducing resource consumption** is a global consideration, but wealthy regions are responsible for most of the world's consumption. For example, the United States and Europe have less than 15 percent of the world's population, but these regions consume about half of the world's metals, food, energy, and other resources.
3. **Population growth** leads to ever-greater resource demands, because all people need some resources. Better family planning, ensuring that all children are wanted, is a matter of justice, resource supply, and economic and social stability for states as well as for families.

KC 1.2 ©Cynthia Shaw

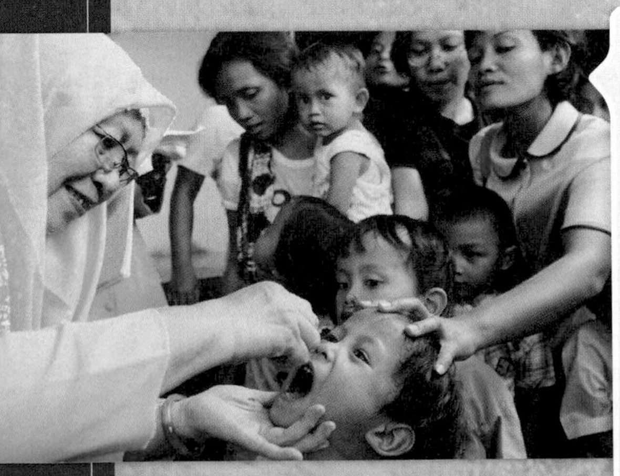

KC 1.3 ©Dimas Ardian/Getty Images

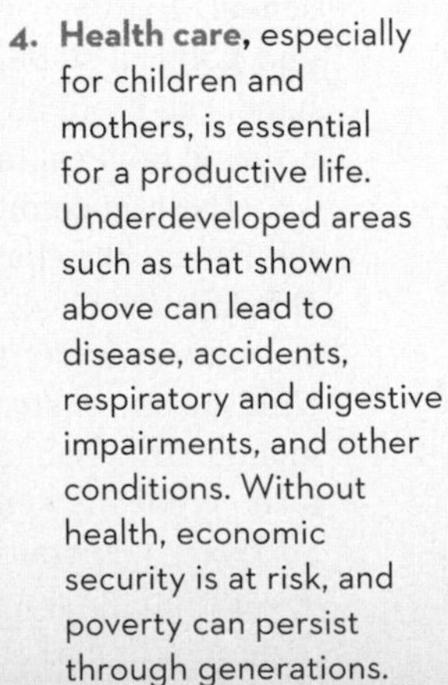

4. **Health care,** especially for children and mothers, is essential for a productive life. Underdeveloped areas such as that shown above can lead to disease, accidents, respiratory and digestive impairments, and other conditions. Without health, economic security is at risk, and poverty can persist through generations.

5. **Sustainable cities** are key because over half of humanity now lives in cities. Sustainable development involves ensuring that cities are healthy places to live and that they cause minimal environmental impact.

KC 1.4

©William P. Cunningham

Environmental science is essential to sustainable development because it helps us understand how environmental systems work, how they are degraded, and what factors can help restore them. Studying environmental science can prepare you to aid human development and environmental quality, both at home and abroad, through better policies, resource protection, and planning.

KC 1.7

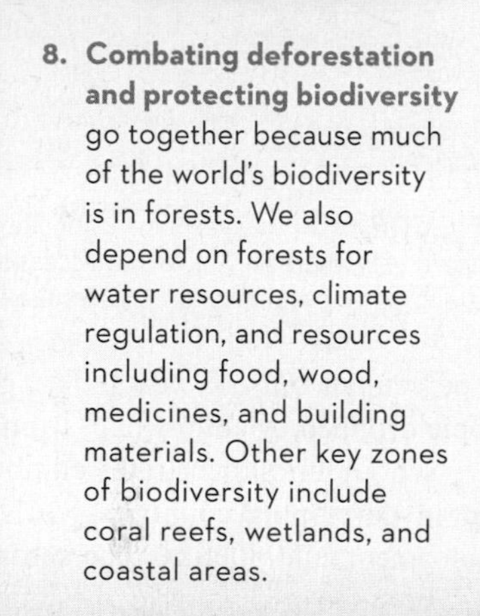

6. **Environmental policy** needs to guide decision making in local and national governments, to ensure that environmental quality is protected before it gets damaged and to set agreed-upon rules for resource use.
7. **Protection of the atmosphere** is essential for minimizing the rate of climate change and for reducing impacts of air pollution on people, plants, and infrastructure.

©Design Pics/Getty Images KC 1.8

8. **Combating deforestation and protecting biodiversity** go together because much of the world's biodiversity is in forests. We also depend on forests for water resources, climate regulation, and resources including food, wood, medicines, and building materials. Other key zones of biodiversity include coral reefs, wetlands, and coastal areas.

9. **Combating desertification and drought** through better management of water resources can save farms, ecosystems, and lives. Often removal of vegetation and soil loss make drought worse, and a few bad rainfall years can convert a landscape to desertlike conditions.

©Eye Ubiquitous/Sudan/Kordofan Province/Farming/Newscom

KC 1.6

10. **Agriculture and rural development** affect the lives of the nearly half of humanity who don't live in cities. Improving conditions for billions of rural people, including more sustainable farming systems, soil stewardship to help stabilize yields, and access to land, can help reduce populations in urban slums.

KC 1.5

©Santokh Kochar/Photodisc/Getty Images

These ten ideas and others were described in Agenda 21 of the United Nations Conference on Environment and Development (the "Earth Summit") in Rio de Janeiro, Brazil, in 1992. Laying out priorities for stewardship of resources and equity in development, the document known as Agenda 21 was a statement of principles for guiding development policies. This document has no legal power, but it does represent an agreement in principle by the more than 200 countries participating in that 1992 conference.

CAN YOU EXPLAIN?

1. **What is the relationship between environmental quality and health?**
2. **Why is sustainable development an issue for people in wealthy countries to consider?**
3. **Examine the top left photo carefully. What health risks might affect the people you see? What do you suppose the rate of material consumption is here, compared to your neighborhood? Why?**

©William P. Cunningham

▲ **FIGURE 1.10** The very poor often are forced to live in degraded or unproductive areas, where they have little access to sufficient clean water, diet, medical care, and other essentials for a humane existence. Courtesy Tom Finkle

14 percent in only 15 years. The proportion of undernourished people dropped by almost half, from 23 percent to 13 percent. Primary school enrollment rates climbed from 83 percent to 91 percent in developing countries. Girls gained access to education, employment, and political representation in national parliaments.

The value of having clearly stated goals, especially with quantifiable targets, is that they help people agree on what to work for. With so many simultaneous problems in developing areas, it can be hard for leaders to know where to focus first. Agreed-upon targets, especially when they are shared and monitored by many countries, can strongly motivate action. International agreement on goals can also help motivate financial and planning assistance, both often badly needed in developing areas.

What is the state of poverty and wealth today?

Policymakers are becoming aware that eliminating poverty and protecting our common environment are inextricably interlinked. The poorest people are often forced to meet short-term survival needs at the cost of long-term sustainability. The good news is that between 1990 and 2015 more than 1 billion people moved out of extreme poverty, mostly in China and India. But the World Bank estimates that at least 760 million people (10 percent of the world population) live below an international poverty line of (U.S.) $1.90 per day. Seventy percent of those poorest people are women and children.

The human suffering engendered by poverty is tragic. The very poor often lack access to an adequate diet, decent housing, basic sanitation, clean water, education, medical care, and other essentials for a humane existence (fig. 1.10). Poverty is both a cause and a consequence of poor health. Every year tens of millions of poor people die from malnutrition, infectious diseases, accidents, and developmental defects that could be avoided fairly easily. People too ill to work become trapped in a cycle of poverty.

TABLE 1.1 Quality-of-Life Indicators

	LEAST-DEVELOPED COUNTRIES	MOST-DEVELOPED COUNTRIES
GDP/person[1]	$615	$40,677
Poverty index[2]	71.8%	0
Life expectancy	59.2 years	82.8 years
Adult literacy	34.8%	99%
Female primary education	10%	95%
Total fertility[3]	6.3	1.3
Infant mortality[4]	74.7	4.3
Improved sanitation	19.8%	100%
Improved water	50.8%	100%
CO_2/capita[5]	0.3 tons	11 tons

[1]ANNUAL gross domestic product (U.S. $).
[2]PERCENT living on less than (U.S.)$2/day.
[3]AVERAGE births/woman.
[4]PER 1,000 live births.
[5]METRIC tons/yr/person.
Source: UNDP Human Development Index, 2017.

The status of well-being in different countries is reflected in **quality-of-life indicators** monitored by the United Nations (table 1.1). More than 1 billion people have insufficient access to clean water, and 2.6 billion lack basic sanitation. These measures are summarized in the Human Development Index (HDI), calculated each year by the United Nations Development Fund (fig. 1.11). The HDI represents a wide variety of factors, such as life expectancy, years of school, gross national income, and income equity. The bottom 20 HDI rankings are generally in Sub-Saharan Africa, former colonies

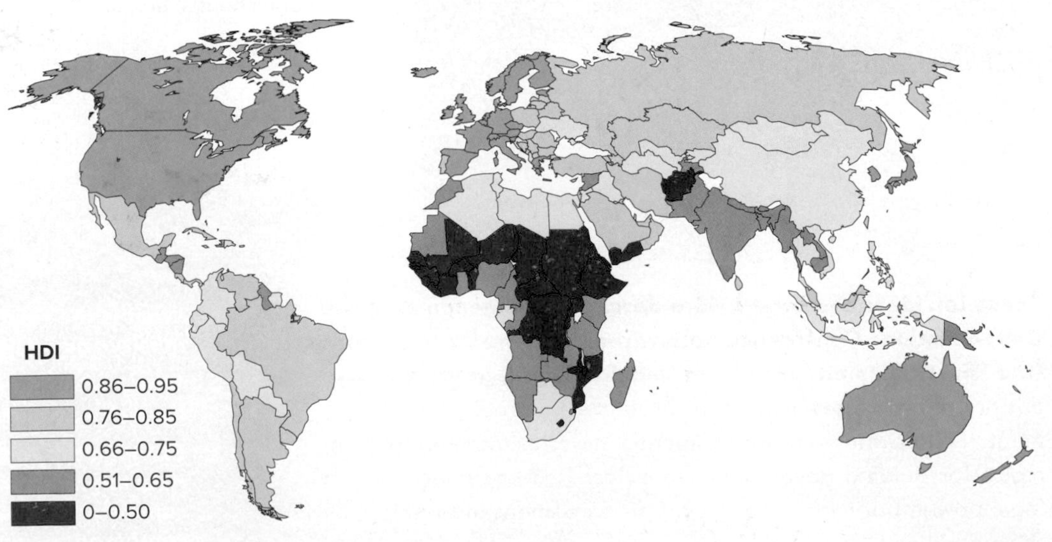

▲ **FIGURE 1.11** Human Development Index. Values near 1 represent strong health, education, and quality of life indicators. Data Source: UNEP 2016.

of European powers. The highest HDI scores aren't usually in the richest countries–these often have repressive monarchies, a few very wealthy citizens, and large populations with few rights. The happiest and healthiest countries have high levels of economic equality, education, and human rights.

Inequality is increasingly recognized as a key concern in economic development. We used to think of the world as divided between a few rich nations, the "First World," and the vast majority of desperately poor countries, the "Third World." (The "Second World" was a group of socialist countries.) Globalization and the Internet have dramatically changed that view. Incomes have risen, but so have wealth disparities. China, for example, has more billionaires and a larger middle class than any other country, but it also has millions of extremely impoverished people. On a global scale, inequality is even more extreme: The most affluent 1 percent of the world now owns more wealth than the other 99 percent. Even more startling, the richest 62 individuals in the world own more wealth than the poorest half (3.8 billion) of the world's population.

Indigenous peoples safeguard biodiversity

In both rich and poor countries, native, or **indigenous, peoples** are generally the least powerful, most neglected groups. Typically descendants of the original inhabitants of an area taken over by more powerful outsiders, native people often are distinct from their country's dominant language, culture, religion, and racial communities. Of the world's nearly 6,000 recognized cultures, 5,000 are indigenous, and these account for only about 10 percent of the total world population. In many countries, traditional caste systems, discriminatory laws, economics, and prejudice repress indigenous people. At least half of the world's 6,000 distinct languages are dying because they are no longer taught to children. When the last elders who still speak the language die, so will much of the culture that was its origin. Lost with those cultures will be a rich repertoire of knowledge about nature and a keen understanding of a particular environment and way of life.

Nonetheless, the 500 million indigenous people who remain in traditional homelands still possess valuable ecological wisdom and remain the guardians of little-disturbed habitats that are refuges for rare and endangered species and undamaged ecosystems. The eminent ecologist E. O. Wilson argues that the cheapest and most effective way to preserve species is to protect the natural ecosystems in which they now live. As the Kuna Indians of Panama say, "Where there are forests, there are native people, and where there are native people, there are forests."

Native people also are playing a valuable role in protecting their homelands. From the Amazon jungles, where members of the Suri tribe are using smartphones and computers to track information about illegal logging, to far-northern Alaska, where the Gwich'in tribe is resisting oil drilling in the Arctic National Wildlife Refuge, indigenous people have been effective in environmental protection.

Canada's Idle No More movement, one of the largest of these, has mobilized thousands of First Nations, Métis, and Inuit people across the country to protest environmentally destructive projects and land use issues. A particular focus has been the water pollution and destruction of boreal forest and wetlands caused by tar sands mining in Alberta, as well as the dangers of pipeline spills in transporting this dirty fuel to markets (fig. 1.12).

▲ **FIGURE 1.12** Native American tribes and representatives from Canada's Idle No More movement march to protest tar sands pipelines. ©William P. Cunningham

Canada's First Nations have linked with native groups in the United States who share their concerns about the dangers of oil pipelines crossing their territories and threatening natural resources. In 2016, construction of the Dakota Access Pipeline in North Dakota prompted thousands of people representing hundreds of native tribes to gather where the pipeline route crossed treaty lands and beneath the Missouri River just upstream from the Standing Rock Reservation. The stand-off lasted for months and attracted the attention of millions of people on social media.

And as the opening case study for this chapter shows, native people are taking important steps to fight global climate change and unemployment by developing clean energy resources. Often they lead the way and help the rest of us envision alternatives to business as usual.

1.4 SCIENCE HELPS US UNDERSTAND OUR WORLD

- *The scientific method is an orderly way to ask questions.*
- *Understanding probability reduces uncertainty.*
- *Science is a cumulative process.*

Because environmental questions are complex, we need orderly methods of examining and understanding them. Environmental science provides such an approach. In this section we'll investigate what science is, what the scientific method is, and why that method is important.

What is science? **Science** (from *scire,* Latin, to know) is a process for producing knowledge based on observations (fig. 1.13). We develop or test theories (proposed explanations of how a process

▲ **FIGURE 1.13** Scientific studies rely on repeated, careful observations to establish confidence in their findings. Source: Dave Partee/Alaska Sea Grant/NOAA

works) using these observations. *Science* also refers to the cumulative body of knowledge produced by many scientists. Science is valuable because it helps us understand the world and meet practical needs, such as finding new medicines, new energy sources, or new foods. In this section we'll investigate how and why science follows standard methods.

Science rests on the assumption that the world is knowable and that we can learn about it by careful observation and logical reasoning (table 1.2). For early philosophers of science, this assumption was a radical departure from religious and philosophical approaches. In the Middle Ages the ultimate sources of knowledge about matters such as how crops grow, how diseases spread, or how the stars move were religious authorities or cultural traditions. Although these sources provided many useful insights, there was no way to test their explanations independently and objectively. The benefit of scientific thinking is that it searches for testable evidence. As evidence improves, we can seek better answers to important questions.

TABLE 1.2 Basic Principles of Science

1. *Empiricism:* We can learn about the world by careful observation of empirical (real, observable) phenomena; we can expect to understand fundamental processes and natural laws by observation.
2. *Uniformitarianism:* Basic patterns and processes are uniform across time and space; the forces at work today are the same as those that shaped the world in the past, and they will continue to do so in the future.
3. *Parsimony:* When two plausible explanations are reasonable, the simpler (more parsimonious) one is preferable. This rule is also known as Ockham's razor, after the English philosopher who proposed it.
4. *Uncertainty:* Knowledge changes as new evidence appears, and explanations (theories) change with new evidence. Theories based on current evidence should be tested on additional evidence, with the understanding that new data may disprove the best theories.
5. *Repeatability:* Tests and experiments should be repeatable; if the same results cannot be reproduced, then the conclusions are probably incorrect.
6. *Proof is elusive:* We rarely expect science to provide absolute proof that a theory is correct, because new evidence may always improve on our current explanations. Even evolution, the cornerstone of modern biology, ecology, and other sciences, is referred to as a "theory" because of this principle.
7. *Testable questions:* To find out whether a theory is correct, it must be tested; we formulate testable statements (hypotheses) to test theories.

Science depends on skepticism and reproducibility

Ideally scientists are skeptical. They are cautious about accepting a proposed explanation until there is substantial evidence to support it. Even then, every explanation is considered only provisionally true, because there is always a possibility that some additional evidence will appear to disprove it. Scientists also aim to be methodical and unbiased. Because bias and methodical errors are hard to avoid, scientific tests are subject to review by informed peers, who can evaluate results and conclusions (fig. 1.14). The peer review process is an essential part of ensuring that scientists maintain good standards in study design, data collection, and interpretation of results.

Scientists demand **reproducibility** because they are cautious about accepting conclusions. Making an observation or obtaining a result just once doesn't count for much. You have to produce the same result consistently to be sure that your first outcome wasn't a fluke. Even more important, you must be able to describe the conditions of your study, so that someone else can reproduce your findings. Repeating studies or tests is known as **replication.**

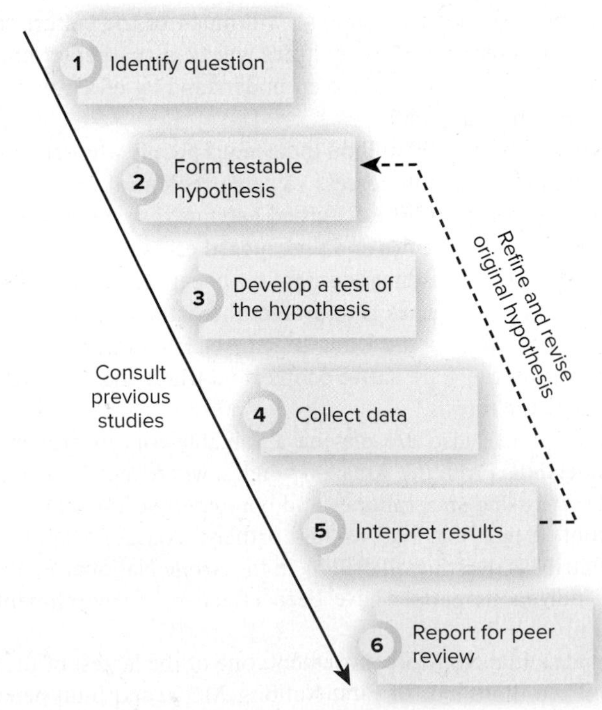

▲ **FIGURE 1.14** Ideally, scientific investigation follows a series of logical, orderly steps to formulate and test hypotheses.

We use both deductive and inductive reasoning

Ideally, scientists deduce conclusions from general laws that they know to be true. For example, if we know that massive objects attract each other (because of gravity), then it follows that an apple will fall to the ground when it releases from the tree. This logical reasoning from general to specific is known as **deductive reasoning.** Often, however, we do not know general laws that guide natural systems. Then we must rely on observations to find general rules. We observe, for example, that birds appear and disappear as a year goes by. Through many repeated observations in different places, we can infer that the birds move from place to place in the spring and fall. We can develop a general rule that birds migrate seasonally. Reasoning from many observations to produce a general rule is **inductive reasoning.** Although deductive reasoning is more logically sound than inductive reasoning, it only works when our general laws are correct. We often rely on inductive reasoning to understand the world because we have few absolute laws.

Insight, creativity, and experience can also be essential in science. Often discoveries are made by investigators who are passionately interested in their subjects and who pursue hunches that appear unreasonable to other scientists. For example, some of our most basic understanding of plant genetics comes from the intuitive guesses of Barbara McClintock, a geneticist who discovered that genes in corn can move and recombine spontaneously. Where other corn geneticists saw random patterns of color and kernel size, McClintock's years of experience in corn breeding and her uncanny ability to recognize patterns led her to guess that genes can recombine in ways that no one had previously imagined. This intuition helped to transform our understanding of genetics.

The scientific method is an orderly way to examine problems

You may use the scientific method even if you don't think about it. Suppose you have a flashlight that doesn't work. The flashlight has several components (switch, bulb, batteries) that could be faulty. If you change all the components at once, your flashlight might work, but a more methodical series of tests will tell you more about what was wrong with the system–knowledge that may be useful next time you have a faulty flashlight. So you decide to follow the standard scientific steps:

1. *Observe* that your flashlight doesn't light and that there are three main components of the lighting system (batteries, bulb, and switch).
2. Propose a **hypothesis,** a testable explanation: "The flashlight doesn't work because the batteries are dead."
3. Develop a *test* of the hypothesis and *predict* the result that would indicate your hypothesis was correct: "I will replace the batteries; the light should then turn on."
4. Gather *data* from your test: After you replaced the batteries, did the light turn on?
5. *Interpret* your results: If the light works now, then your hypothesis was right; if not, then you should formulate a new hypothesis–perhaps that the bulb is faulty–and develop a new test for that hypothesis.

In systems more complex than a flashlight, it is almost always easier to prove a hypothesis wrong than to prove it unquestionably true. This is because we usually test our hypotheses with observations but there is no way to make every possible observation. The philosopher Ludwig Wittgenstein illustrated this problem as follows: Suppose you saw hundreds of swans, and all were white. These observations might lead you to hypothesize that all swans were white. You could test your hypothesis by viewing thousands of swans, and each observation might support your hypothesis, but you could never be entirely sure that it was correct. On the other hand, if you saw just one black swan, you would know with certainty that your hypothesis was wrong.

As you'll read in later chapters, the elusiveness of absolute proof is a persistent problem in environmental policy and law. Rarely can you absolutely prove that the toxic waste dump up the street is making you sick. You could collect evidence to show that it is very probable that the waste has made you and your neighbors sick (fig. 1.15). But scientific uncertainty is often used as an excuse to avoid environmental protection.

When an explanation has been supported by a large number of tests, and when a majority of experts have reached a general consensus that it is a reliable description or explanation, we call it a **scientific theory.** Note that scientists' use of this term is very different from the way the public uses it. To many people, a theory is speculative and unsupported by facts. To a scientist, it means just the opposite: While all explanations are tentative and open to revision and correction, an explanation that counts as a scientific theory is supported by an overwhelming body of data and experience, and it is generally accepted by the scientific community, at least for the present.

Understanding probability reduces uncertainty

One strategy to improve confidence in the face of uncertainty is to focus on probability. **Probability** is a measure of how likely something is to occur. Usually probability estimates are based on a set of previous observations or on standard statistical measures. Probability

▲ **FIGURE 1.15** Careful, repeated measurements, and well-formed hypotheses are essential for good science.

Active LEARNING

Calculating Probability

An understanding of probability (the likelihood of an event) is fundamental in most areas of modern science. Working with these concepts is critical to your ability to comprehend scientific information.

Every time you flip a coin, the chance that heads will end up on top is 1 in 2 (50 percent, assuming you have a normal coin). The odds of getting heads two times in a row is 1/2 × 1/2, or 1/4.

1. What are the odds of getting heads five times in a row?
2. As you start the fifth flip, what are the odds of getting heads?
3. If there are 100 students in your class and everybody flips a coin five times, how many people are likely to get five heads in a row?

ANSWERS: 1. 1/2 × 1/2 × 1/2 × 1/2 × 1/2 = 1/32; 2. 1 in 2; 3. 100 students × 1/32 = about 3.

does not tell you what *will* happen, but it tells you what *is likely* to happen. If you hear on the news that you have a 20 percent chance of catching a cold this winter, that means that 20 of every 100 people are likely to catch a cold. This doesn't mean that *you* will catch one. In fact, it's more likely, an 80 percent chance, that you *won't* catch a cold. If you hear that 80 out of every 100 people will catch a cold, you still don't know whether you'll get sick, but there's a much higher chance that you will.

Science often involves probability, so it is important to be familiar with the idea. Sometimes probability has to do with random chance: If you flip a coin, you have a random chance of getting heads or tails. Every time you flip, you have the same 50 percent probability of getting heads. The chance of getting ten heads in a row is small (in fact, the chance is 1 in 2^{10}, or 1 in 1,024), but on any individual flip, you have exactly the same 50 percent chance, since this is a random test. Sometimes probability is weighted by circumstances: Suppose that about 10 percent of the students in your class earn an A each semester. Your likelihood of being in that 10 percent depends a great deal on how much time you spend studying, how many questions you ask in class, and other factors. Sometimes there is a combination of chance and circumstances: The probability that you will catch a cold this winter depends partly on whether you encounter someone who is sick (largely random chance) and on whether you take steps to stay healthy (get enough rest, wash your hands frequently, eat a healthy diet, and so on).

Probability is often a more useful idea than proof. This is because absolute proof is hard to achieve, but we can frequently demonstrate a strong trend or relationship, one that is unlikely to be achieved by chance. For example, suppose you flipped a coin and got heads 20 times in a row. That could happen by chance, but it would be pretty unlikely. You might consider it very likely that there was a causal explanation, such as that the coin was weighted toward heads. Often we consider a causal explanation reliable (or "significant") if there is less than 5 percent probability that it happened by random chance.

Experimental design can reduce bias

Many research problems in environmental science involve observational experiments, in which you observe natural events and interpret a causal relationship between the variables. This kind of study is also called a **natural experiment,** one that involves observation of events that have already happened. Many scientists depend on natural experiments: A geologist, for instance, might want to study mountain building, or an ecologist might want to learn about how species evolve, but neither scientist can spend millions of years watching the process happen. Similarly, a toxicologist cannot give people a disease just to see how lethal it is.

Other scientists can use **manipulative experiments,** in which conditions are deliberately altered and all other variables are held constant. Most manipulative experiments are done in the laboratory, where conditions can be carefully controlled. Suppose you are interested in studying whether lawn chemicals contribute to deformities in tadpoles. You might keep two groups of tadpoles in fish tanks and expose one to chemicals. In the lab you can ensure that both tanks have identical temperatures, light, food, and oxygen. By comparing a treatment (exposed) group and a control (unexposed) group, you also make this a **controlled study.**

Often there is a risk of experimenter bias. Suppose the researcher sees a tadpole with a small nub that looks like it might become an extra leg. Whether she calls this nub a deformity might depend on whether she knows that the tadpole is in the treatment group or the control group. To avoid this bias, **blind experiments** are often used, in which the researcher doesn't know which group is treated until after the data have been analyzed. In health studies, such as tests of new drugs, **double-blind experiments** are used, in which neither the subject (who receives a drug or a placebo) nor the researcher knows who is in the treatment group and who is in the control group.

In each of these studies there is one **dependent variable** and one, or perhaps more, **independent variables.** The dependent variable, also known as a response variable, is affected by the independent variables. In a graph, the dependent variable is on the vertical (Y) axis, by convention. Independent variables are rarely really independent (they may be affected by the same environmental conditions as the dependent variable, for example). Often we call them **explanatory variables** because we hope they will explain differences in a dependent variable (Exploring Science, p. 17).

Science is a cumulative process

The scientific method outlined in figure 1.14 is the process used to carry out individual studies. Larger-scale accumulation of scientific knowledge involves cooperation and contributions from countless people. Good science is rarely carried out by a single individual working in isolation. Instead, a community of scientists collaborates in a cumulative, self-correcting process. You often hear about big breakthroughs and dramatic discoveries that change our understanding

EXPLORING Science — Understanding sustainable development with statistics

In environmental science, we know sustainable development is important, but how do we evaluate it? Mainly with statistics. Distilling complex problems to a few numbers can allow you to see the state of a group, compare groups, and see change over time. One key statistic for understanding poverty is the Human Development Index (HDI), a measure that combines national scores for income, education, health care, and other measures (Key Concepts, p. 10).

Suppose you want to know how India is doing on human development and environmental conditions. You might recall that India has a growing population—soon to be the world's largest—and that poverty remains a persistent problem there. If you look up India's HDI score on the website of the United Nations Development Programme (UNDP), you can find that India's HDI is 0.59, on a scale from 0 to 1.0. Does this mean India is doing well? Or not?

Finding the center and distribution of a data set Many statistics mean little without context. To understand an HDI of 0.59, you can compare it to those of other countries. To start, you can compare India's HDI to the mid-point of the group. One common measure of the mid-point is the mean (or average): Add up all the HDI values for the 182 countries with reported scores; then divide the sum by the number of countries (182). It turns out that the mean HDI among these countries is 0.69. Evidently India is slightly below average in development.

Many of us understand visual patterns more readily than numbers. A **histogram,** for example, is a graph that shows the distribution of a data set at a glance. To make a histogram, we first specify ranges of HDI values—say, 0.3 to 0.4, 0.4 to 0.5, and so on. Then we count up the number of countries that fall in each value range. The resulting distribution appears in figure 1.

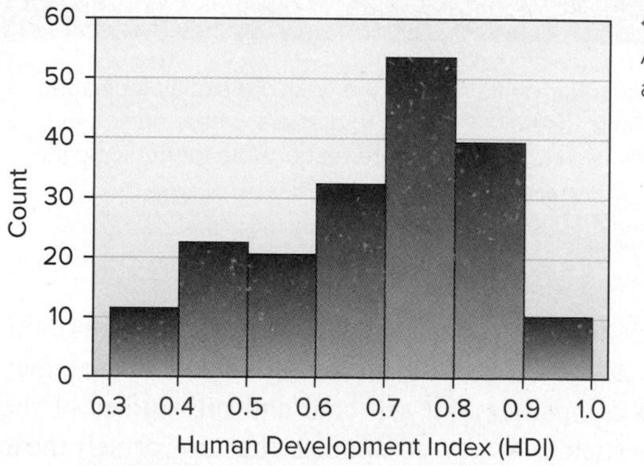

FIGURE 1 A histogram shows a distribution.

Plotting relationships among variables You may recall from earlier in this chapter that many developing areas lack access to safe drinking water and that young children, especially, are vulnerable to waterborne illness. How strong is the relationship between pollution-related deaths and HDI? The UNDP keeps data on estimated numbers of children under 5 years old who die each year from unsafe water. You can use a scatterplot to show the relationship between this variable and HDI (fig. 2). Each point represents a country.

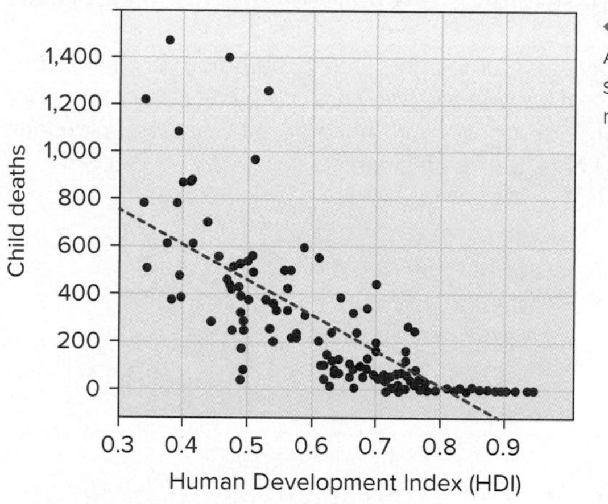

FIGURE 2 A scatter plot shows the relationship.

The scatterplot shows a pattern that generally declines from left to right. Look carefully at the axis labels: Number of deaths generally *decreases* (vertical axis) as HDI *increases* (horizontal axis). This is a **negative relationship.** A straight line shows the approximate trend in the data.

The points don't fit the straight line very tightly, though. Countries with low HDI, around 0.3 to 0.5, have a very wide range of infant deaths, from about 400 to 1,400. Some countries clearly have better success than others in controlling this risk factor. Almost every country with an HDI above about 0.75 has near zero infant deaths from unsafe water. It appears that while there is a negative relationship, countries don't need perfect HDI scores to see much improved infant health.

Error bars improve confidence When you calculate a mean of a *sample* (a portion of all possible observations), your calculated mean is properly considered an approximation of the universal *population* mean. In our case, we have HDI numbers for most countries but not all, and the data set excludes dozens of regions that were once independent states. So we basically have a sample of a larger population. To be confident that we have a reasonable approximation of the population mean from our sample, it is best to estimate the range of *likely* values for the actual (universal) population mean. One approach is *standard error bars*, which use sample size (were there many observations or few?) and variation in the data (all similar? widely different?) to calculate the likely range of the population mean.

This becomes important if you are comparing groups. Suppose you are concerned that affluence is associated with environmental harm, such as climate-changing greenhouse gas emissions. You could compare the average emissions and see if the high-HDI countries also tend to have high CO_2 emissions.

Continued ▶

◀ Continued

Figure 3 shows that your hunch was correct. Not only is the mean higher for high-HDI countries, the standard error bars (the likely range of means) don't overlap. You can say with confidence that these groups are really different in their climate impacts.

Statistics give useful insights into problems we care about. Like any source of knowledge, they are often just part of the story, but they can provide good confidence about what we know, and what we don't know, about an issue.

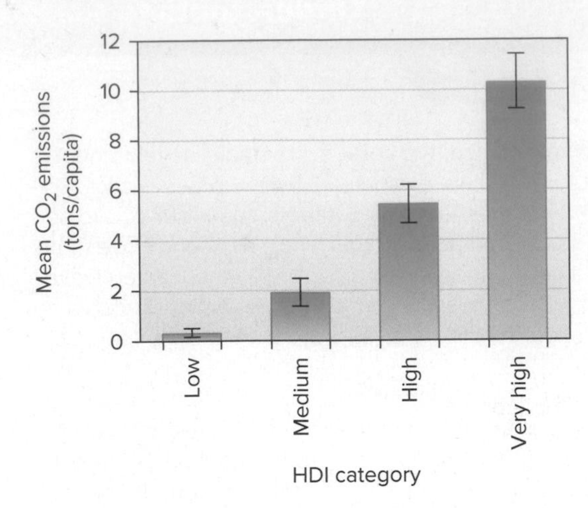

◀ **FIGURE 3** Standard error bars show whether groups differ meaningfully.

overnight, but in reality these changes are usually the culmination of the labor of many people, each working on different aspects of a common problem, each adding small insights to solve a problem. Ideas and information are exchanged, debated, tested, and retested to arrive at **scientific consensus,** or general agreement among informed scholars.

The idea of consensus is important. For those not deeply involved in a subject, the multitude of contradictory results can be bewildering: Are coral reefs declining, and does it matter? Is climate changing, and how much? Among those who have done many studies and read many reports, there tends to emerge a general agreement about the state of a problem. Scientific consensus now holds that many coral reefs are in danger, as a result of pollution, physical damage, and warming seas. Consensus is that global climate conditions are changing, though models differ somewhat on how rapidly they will change in different regions.

Sometimes new ideas emerge that cause major shifts in scientific consensus. These great changes in explanatory frameworks were termed **paradigm shifts** by Thomas Kuhn (1967), who studied revolutions in scientific thought. According to Kuhn, paradigm shifts occur when a majority of scientists accept that the old explanation no longer describes new observations very well. For example, two centuries ago geologists explained many of the earth's features in terms of Noah's flood. The best scientists held that the flood created beaches well above modern sea level, scattered boulders erratically across the landscape, and gouged enormous valleys where there is no water now (fig. 1.16). Then the Swiss glaciologist Louis Agassiz and others suggested that the earth had once been much colder and that glaciers had covered large areas. Periodic ice ages proved to be a more durable explanation for geologic features than did a flood, and this new idea completely altered the way geologists explained their subject. Similarly, the idea of tectonic plate movement, in which continents shift slowly around the earth's surface (see chapter 11), revolutionized the ways geologists, biogeographers, ecologists, and others explained the development of the earth and its life-forms.

▲ **FIGURE 1.16** Paradigm shifts change the ways we explain our world. Geologists now attribute Yosemite's valleys to glaciers, where once they believed catastrophes like Noah's flood were responsible for geological features like these. ©John A. Karachewski

What is sound science?

Environmental science often deals with questions that are emotionally or politically charged. Scientific studies of climate change may be threatening to companies that sell coal and oil; studies of the health costs of pesticides worry companies that use or sell these chemicals. When controversy surrounds science, claims about **sound science** and accusations of "junk science" often arise. What do these terms mean, and how can you evaluate who is right?

When you hear arguments about whose science is valid, you need to remember the basic principles of science: Are the disputed studies reproducible? Are conclusions drawn with caution and skepticism? Are samples large and random? Are conclusions supported by a majority of scholars who have studied the problem? Do any of the experts have an economic interest in the outcome?

What Do YOU THINK?

Science and Citizenship: Evidence-Based Policy vs. Policy-Based Evidence?

[This] was not evidence-based policymaking—this was policy-based evidence-making.

—Michael Greenstone, University of Chicago economist, describing Energy Secretary Scott Pruitt's effort to recalculate the cost of carbon emissions

Suppose you observe that your local lake has become brown and smelly, and kids swimming in it are starting to get sick. You want to find out why and what to do about it. Of course, you want the best available information and a reliable solution to the problem. Who wouldn't want to use the best available evidence to keep people safe?

It turns out that it is surprisingly common for people to ignore evidence in decision making. Sometimes it's easier to pretend the lake is still clean enough. Sometimes we assume there is no alternative, and we just put up with getting sick. Sometimes we like the person responsible for polluting the lake, so we don't want to complain. Sometimes the polluter is skilled in suppressing information. He or she might even present creative "alternative facts" to explain or distract from the problem.

What to do? As an educated member of the community, you know to look for reliable, impartial information about water contaminants and to evaluate that information in the context of other studies. This is what much of environmental science is about. It uses the methods of science (orderly collection and analysis of evidence) to understand how environmental systems function, to evaluate environmental conditions, and to address issues of environmental quality.

We collect and evaluate information in decision making all the time. You might evaluate weather data every day, as you decide how to dress or if it's likely to rain. You might closely evaluate your budget when deciding if you should get a new cell phone or a new car. If you don't evaluate the evidence in your budget, there might be painful consequences.

Of course, it's easy to disregard consequences if they only affect someone else, especially someone far away or in the future. If your neighbor is polluting the lake and *you* suffer, should he really care? Why? If the pollution has a delayed effect—say, a pesticide that gradually degrades the lake ecosystem—but the damage doesn't become evident for years, whose problem is that (fig. 1)?

In a larger society, if we want to minimize conflict, then we try to ensure that one group doesn't systematically harm another group. So policymakers, who influence policies about water quality, health, or environmental releases of pesticides, need the best possible information (that is, data) and analysis for decision making. We set up agencies to collect, store, and analyze information. The Environmental Protection Agency has responsibility for monitoring pollution in order to protect public health. The Centers for Disease Control and Prevention monitors illnesses and environmental health, to catch and control the spread of diseases. Dozens of agencies collect data, and they share it with the public because data are so critical to public health and well-being.

Science and citizenship often go hand in hand. A functioning society depends on informed, thoughtful members who look out for the community's interests. Being educated in environmental science helps you develop a number of useful analytical skills, such as these:

▲ **FIGURE 1** Should pollutants be regulated? How? These and other policy questions involve your knowledge of science and of civil society. ©Leks_Laputin/Getty Images

- Critically analyzing data, ideas, and arguments
- Evaluating complex systems—understanding that interconnections are complex and diverse and indirect consequences can be important
- Knowing how to gather and weigh evidence, including data visualization (graphs and maps)
- Understanding the environmental context of policies or events
- Understanding the logic and scientific evidence invoked in policymaking

Being aware of complex relationships and systems helps you understand your interrelationships with your community. Part of citizenship, of course, is to consider consequences beyond ourselves and to minimize harm to others in the community. As a well-informed citizen, you can decide whether or not you want to vote for a candidate who has promised to dismantle water monitoring systems or air quality protections, or whether you should support a policymaker who has eliminated funding for chemical safety.

Most of us are interested in staying healthy and living in a healthy environment. Whether we act or vote accordingly depends on whether we are thinking about evidence when decision time comes around, or how near and personal the consequences of our decisions might be.

Many policymakers today criticize science, scientific agencies like the EPA, and even the colleges and universities that educate the next generation of scientists and citizens. As you listen to them, consider why. Are data standing in the way of their intentions? Does evidence contradict their statements? Are "alternative facts" more convenient than impartial and reproducible ones? Are they practicing "policy-based evidence-making"? As you read this book and as you listen to the news, these are good questions to keep in mind.

TABLE 1.3 Questions for Baloney Detection

1. How reliable are the sources of this claim? Is there reason to believe that they might have an agenda to pursue in this case?
2. Have the claims been verified by other sources? What data are presented in support of this opinion?
3. What position does the majority of the scientific community hold in this issue?
4. How does this claim fit with what we know about how the world works? Is this a reasonable assertion, or does it contradict established theories?
5. Are the arguments balanced and logical? Have proponents of a particular position considered alternate points of view or only selected supportive evidence for their particular beliefs?
6. What do you know about the sources of funding for a particular position? Are studies financed by groups with partisan goals?
7. Where was evidence for competing theories published? Has it undergone impartial peer review, or is it only in proprietary publication?

Source: Carl Sagan, *The Demon Haunted World: Science as a Candle in the Dark*, 1997.

Often media figures on television or radio will take a position contrary to the scientific majority. A contrarian position gains them publicity and political allies (and sometimes money). This strategy has been especially popular around large issues such as climate change. For decades now, almost all climate scientists have agreed that human activities, such as fossil fuel burning and land clearing, are causing climate change. But it is always possible to find a contrarian scientist who is happy to contradict the majority of evidence. Especially when political favors, publicity, or money is involved, there are always "expert" witnesses who will testify on opposite sides of any case.

If you see claims of fake news and junk science, how can you evaluate them? How can you identify bogus analysis that is dressed up in quasi-scientific jargon but that has no objectivity? This is such an important question that astronomer Carl Sagan proposed a "Baloney Detection Kit" (table 1.3) to help you out.

Uncertainty, proof, and group identity

Scientific uncertainty is frequently invoked as a reason to postpone actions that a vast majority of informed scientists consider to be prudent. In questions of chemical safety, energy conservation, climate change, or air pollution control, opponents of change may charge that the evidence doesn't constitute absolute proof, so that no action needs to be taken. You will see examples of this in later chapters on environmental health, climate, air and water pollution, and other topics.

Similarly, disputes over evolution often hinge on the concept of uncertainty and proof in science. Opponents of teaching evolution in public schools often charge that because scientists call evolution a "theory," evolution is just a matter of conjecture. This is a confused use of terminology. The theory of evolution is supported by overwhelming evidence, but we still call it a theory because scientists prefer to be precise about the idea of proof.

In recent years sociologists have pointed out that our decisions to accept or dispute scientific evidence often depend on group identity. We like to associate with like-minded people, so we tend to adhere to a group viewpoint. Subconsciously we may ask, "Does the community I belong to agree with evolution? Does it accept the evidence for climate change?" Our urge to be agreeable to our group can be surprisingly strong, compared to our interest in critically analyzing evidence. Expectations of group behavior can shift over time, though. In decades past, you might have asked, "Am I the kind of person who recycles?" Today recycling is normal for most people, and few people probably decline to recycle just because their friends don't. Resolving differences on environmental policy sometimes requires recognition of group identity in our attitudes toward science, as well as our attitudes toward policies and issues beyond science. In these ways, you are often integrating your education in environmental science with your actions as a member of society (What Do You Think?, p. 19).

1.5 CRITICAL THINKING

- *Critical thinking helps us analyze information.*
- *There are many aspects of critical thinking.*

In science we frequently ask, "How do I know that what you just said is true?" Part of the way we evaluate arguments in science has to do with observable evidence, or data. Logical reasoning from evidence is also essential. And part of the answer lies in critical evaluation of evidence.

An ability to think critically, clearly, and analytically about a problem may be the most valuable skill you can learn in any of your classes. As you know by now, many issues in environmental science are hotly disputed, with firm opinions and plenty of evidence on both sides. How do you evaluate contradictory evidence and viewpoints? **Critical thinking** is a term we use to describe logical, orderly, analytical assessment of ideas, evidence, and arguments. Developing this skill is essential for the course you are taking now. Critical thinking is also an extremely important skill for your life in general. You can use it when you evaluate the claims of a car salesman, a credit card offer, or the campaign rhetoric of a political candidate.

Critical thinking helps us understand why prominent authorities can vehemently disagree about a topic. Disagreements may be based on contradictory data, on different interpretations of the same data, or on different priorities. One expert might consider economic health the overriding priority; another might prioritize environmental quality. A third might worry only about company stock prices, which might depend on the outcome of an environmental policy debate. You can examine the validity of contradictory claims by practicing critical thinking.

Critical thinking is part of science and of citizenship

We evaluate many claims every day, in class, in TV advertising, in understanding public affairs and polices, in reading or watching the news. It is worth pausing to think about what critical thinking means. In general, it means examining sources and considering

TABLE 1.4 Steps in Critical Thinking

1. What is the purpose of my thinking?
2. What precise question am I trying to answer?
3. Within what point of view am I thinking?
4. What information am I using?
5. How am I interpreting that information?
6. What concepts or ideas are central to my thinking?
7. What conclusions am I aiming toward?
8. What am I taking for granted; what assumptions am I making?
9. If I accept the conclusions, what are the implications?
10. What would the consequences be if I put my thoughts into action?

Source: Paul, R. (1993). Critical Thinking. Foundation for Critical Thinking.

how a source influences statements or ideas. But you can also distinguish among different kinds of critical thinking: **Analytical thinking** involves breaking down a problem into its constituent parts. **Creative thinking** means envisioning new, different approaches to a problem. **Logical thinking** examines the structure of an argument, from premises to conclusions. **Reflective thinking** means asking, "What does it all mean?"

These processes are often self-reflective and self-correcting. They encourage you to ask, "How do I know that what I just said is true?" Developing habits of critical thinking can help you identify unspoken assumptions, biases, beliefs, priorities, or motives (table 1.4 and fig. 1.17). These habits will also help you do well in class, and they can help you be an informed, thoughtful reader of the world around you.

▲ **FIGURE 1.17** Critical thinking evaluates premises, contradictions, and assumptions. Is this sign, in the middle of a popular beach near Chicago, the only way to reduce human exposure to bacteria? What other strategies might there be? Why was this one chosen? Who might be affected?
©Mary Ann Cunningham

Here are some steps to practice in critical thinking:

1. *Identify and evaluate premises and conclusions in an argument.* What is the basis for the claims made? What evidence is presented to support these claims, and what conclusions are drawn from this evidence? If premises and evidence are reasonable, do the conclusions truly follow from them?
2. *Acknowledge and clarify uncertainties, vagueness, equivocation, and contradictions.* Are terms used in vague or ambiguous ways? Are all participants in an argument using the same meanings? Is ambiguity or equivocation deliberate?
3. *Distinguish between facts and values.* Can claims be tested, or are statements based on untestable assumptions and beliefs? Are claims made about the worth or lack of worth of something? (If so, these are value statements or opinions and probably cannot be verified objectively.)
4. *Recognize and assess assumptions.* Consider the backgrounds and views behind an argument: What underlying reasons might there be for the premises, evidence, or conclusions presented? Does anyone have a personal agenda in this issue? What does he or she think you know, need, want, or believe? Do hidden biases based on race, gender, ethnicity, economics, or belief systems distort arguments?
5. *Distinguish source reliability from unreliability.* What qualifies the experts on this issue? Is that qualification sufficient for you to believe them? Why or why not?
6. *Recognize and understand conceptual frameworks.* What basic beliefs, attitudes, and values underlie an argument or action? How do these beliefs and values affect the way people view themselves and the world around them?

In this book you will have many opportunities to practice critical thinking skills. Every chapter includes facts, figures, opinions, and theories. Are all of them true? Probably not. They were the best information available when this text was written, but new evidence is always emerging. Data change constantly, as does our interpretation of data.

You'll find more on critical thinking, as well as some useful tips on how to study effectively, on our website at www.connect.mheducation.com.

1.6 WHERE DO OUR IDEAS ABOUT THE ENVIRONMENT COME FROM?

- *Utilitarian conservation focuses on usable resources.*
- *Preservation of nature recognizes the rights of other species.*
- *Modern environmentalism focuses on health and social justice.*

Historically, many societies have degraded the resources on which they depended, while others have lived in relative harmony with their surroundings. Today our burgeoning population and our technologies that accelerate resource exploitation have given the problems of environmental degradation increased urgency.

Many of our current responses to these changes are rooted in the writings of relatively recent environmental thinkers. For simplicity, their work can be grouped into four distinct stages: (1) resource conservation for optimal use, (2) nature preservation for moral and aesthetic reasons, (3) concern over health and ecological consequences of pollution, and (4) global environmental citizenship. These stages are not mutually exclusive. You might embrace them all simultaneously. As you read this section, consider why you agree with those you find most appealing.

Environmental protection has historic roots

Recognizing human misuse of nature is not unique to modern times. Plato complained in the fourth century B.C. that Greece once was blessed with fertile soil and clothed with abundant forests of fine trees. After the trees were cut to build houses and ships, however, heavy rains washed the soil into the sea, leaving only a rocky "skeleton of a body wasted by disease." Springs and rivers dried up, and farming became all but impossible. Despite these early observations, most modern environmental ideas developed in response to resource depletion associated with more recent agricultural and industrial revolutions.

Some of the earliest recorded scientific studies of environmental damage were carried out in the eighteenth century by French or British colonial administrators, many of whom were trained scientists and who observed rapid soil loss and drying wells that resulted from intensive colonial production of sugar and other commodities. Some of these colonial administrators considered responsible environmental stewardship as an aesthetic and moral priority, as well as an economic necessity. These early conservationists observed and understood the connections among deforestation, soil erosion, and local climate change. The pioneering British plant physiologist Stephen Hales, for instance, suggested that conserving green plants preserves rainfall. His ideas were put into practice in 1764 on the Caribbean island of Tobago, where about 20 percent of the land was marked as "reserved in wood for rains."

Pierre Poivre, an early French governor of Mauritius, an island in the Indian Ocean, was appalled at the environmental and social devastation caused by the destruction of wildlife (such as the flightless dodo) and the felling of ebony forests on the island by early European settlers. In 1769 Poivre ordered that one-quarter of the island be preserved in forests, particularly on steep mountain slopes and along waterways. Mauritius remains a model for balancing nature and human needs. Its forest reserves shelter more original species than are found on most other populated islands.

Resource waste triggered pragmatic resource conservation (stage 1)

Many historians consider the publication of *Man and Nature* in 1864 by geographer George Perkins Marsh as the wellspring of environmental protection in North America. Marsh, who also was a lawyer, politician, and diplomat, traveled widely around the Mediterranean as part of his diplomatic duties in Turkey and Italy. He read widely in the classics (including Plato) and personally observed the damage caused by excessive grazing by goats and sheep and by the deforestation of steep hillsides. Alarmed by the wanton destruction and profligate waste of resources still occurring on the American frontier in his lifetime, he warned of its ecological consequences. Largely because of his book, national forest reserves were established in the United States in 1873 to protect dwindling timber supplies and endangered watersheds.

Among those influenced by Marsh's warnings were U.S. President Theodore Roosevelt and his chief conservation adviser, Gifford Pinchot (fig. 1.18a,b). In 1905 Roosevelt, who was the leader of the populist Progressive movement, moved forest management out of the corruption-filled Interior Department into the Department of Agriculture. Pinchot, who was the first American-born professional forester, became the first chief of the new Forest Service. He put resource management on an honest, rational, and scientific basis for the first time in American history. Together with naturalists and activists such as John Muir, Roosevelt and Pinchot established the framework of the national forest, park, and wildlife refuge system. They passed game protection laws and tried to stop some of the most flagrant abuses of the public domain. In 1908 Pinchot organized and chaired the White House Conference on Natural Resources, perhaps the most prestigious and influential environmental meeting ever held in the United States. Pinchot also was governor of Pennsylvania and founding head of the Tennessee Valley Authority, which provided inexpensive power to the southeastern United States.

The basis of Roosevelt's and Pinchot's policies was pragmatic **utilitarian conservation.** They argued that the forests should be

(a) President Teddy Roosevelt

(b) Gifford Pinchot

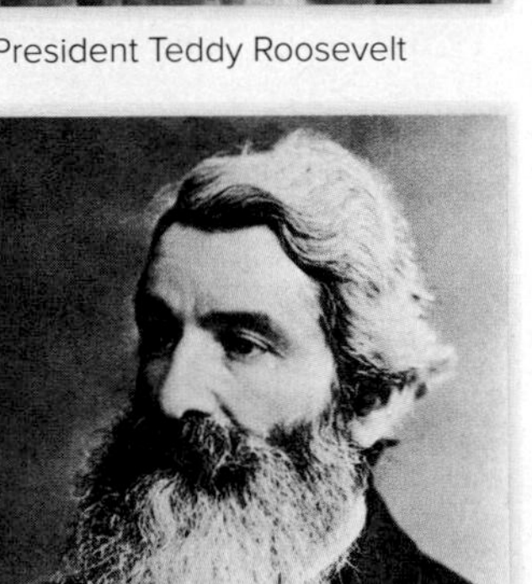

(c) John Muir

(d) Aldo Leopold

▲ **FIGURE 1.18** Some early pioneers of the American conservation movement. (a) President Teddy Roosevelt and his main adviser, (b) Gifford Pinchot, emphasized pragmatic resource conservation, whereas (c) John Muir and (d) Aldo Leopold focused on ethical and aesthetic relationships.
(a): Source: Underwood & Underwood, Library of Congress, LC-USZC4-4698; (b): ©Grey Towers National Historic Landmark; (c): ©Bettmann/Getty Images; (d): ©AP Images

saved "not because they are beautiful or because they shelter wild creatures of the wilderness, but only to provide homes and jobs for people." Resources should be used "for the greatest good, for the greatest number, for the longest time." "There has been a fundamental misconception," Pinchot wrote, "that conservation means nothing but husbanding of resources for future generations. Nothing could be further from the truth. The first principle of conservation is development and use of the natural resources now existing on this continent for the benefit of the people who live here now. There may be just as much waste in neglecting the development and use of certain natural resources as there is in their destruction." This pragmatic approach still can be seen in the multiple-use policies of the U.S. Forest Service.

Ethical and aesthetic concerns inspired the preservation movement (stage 2)

John Muir (fig. 1.18c), amateur geologist, popular author, and first president of the Sierra Club, strenuously opposed Pinchot's utilitarian policies. Muir argued that nature deserves to exist for its own sake, regardless of its usefulness to us. Aesthetic and spiritual values formed the core of his philosophy of nature protection. This outlook prioritizes **preservation** because it emphasizes the fundamental right of other organisms—and nature as a whole—to exist and to pursue their own interests (fig. 1.19). Muir wrote, "The world, we are told, was made for man. A presumption that is totally unsupported by the facts. . . . Nature's object in making animals and plants might possibly be first of all the happiness of each one of them. . . . Why ought man to value himself as more than an infinitely small unit of the one great unit of creation?"

Muir, who was an early explorer and interpreter of California's Sierra Nevada range, fought long and hard for establishment of Yosemite and Kings Canyon national parks. The National Park Service, established in 1916, was first headed by Muir's disciple, Stephen Mather, and has always been oriented toward preservation of nature rather than consumptive uses. Muir's preservationist ideas have often been at odds with Pinchot's utilitarian approach. One of Muir and Pinchot's biggest battles was over the damming of Hetch Hetchy Valley in Yosemite. Muir regarded flooding the valley a sacrilege against nature. Pinchot, who championed publicly owned utilities, viewed the dam as a way to free San Francisco residents from the clutches of greedy water and power monopolies.

In 1935, pioneering wildlife ecologist Aldo Leopold (fig. 1.18d) bought a small, worn-out farm in central Wisconsin. A dilapidated chicken shack, the only remaining building, was remodeled into a rustic cabin. Working together with his children, Leopold planted thousands of trees in a practical experiment in restoring the health and beauty of the land. "Conservation," he wrote, "is the positive exercise of skill and insight, not merely a negative exercise of abstinence or caution." The shack became a writing refuge and the main focus of *A Sand County Almanac,* a much beloved collection of essays about our relation with nature. In it, Leopold wrote, "We abuse land because we regard it as a commodity belonging to us. When we see land as a community to which we belong, we may begin to use it with love and respect." Together with Bob Marshall and two others, Leopold was a founder of the Wilderness Society.

▲ **FIGURE 1.19** A conservationist might say this forest is valuable as a supplier of useful resources, including timber and fresh water. A preservationist might argue that this ecosystem is important for its own sake. Many people are sympathetic with both outlooks. ©Altrendo nature/Getty Images

Rising pollution levels led to the modern environmental movement (stage 3)

The undesirable effects of pollution probably have been recognized as long as people have been building smoky fires. In 1723 the acrid coal smoke in London was so severe that King Edward I threatened to hang anyone who burned coal in the city. In 1661 the English diarist John Evelyn complained about the noxious air pollution caused by coal fires and factories and suggested that sweet-smelling trees be planted to purify city air. Increasingly dangerous smog attacks in Britain led, in 1880, to formation of a national Fog and Smoke Committee to combat this problem. But nearly a century later, London's air (like that of many cities) was still bad. In 1952 an especially bad episode turned midday skies dark and may have caused 12,000 deaths (see chapter 10). This event was extreme, but noxious air was common in many large cities.

The tremendous expansion of chemical industries during and after World War II added a new set of concerns to the environmental agenda. *Silent Spring,* written by Rachel Carson (fig. 1.20a) and published in 1962, awakened the public to the threats of pollution and toxic chemicals to humans as well as other species. The movement she engendered might be called **modern environmentalism** because its concerns extended to include both natural resources and environmental pollution.

Under the leadership of a number of other brilliant and dedicated activists and scientists, the environmental agenda was expanded in the 1970s to most of the issues addressed in this textbook, such as human population growth, atomic weapons testing and atomic power, fossil fuel extraction and use, recycling, air and water pollution, and wilderness protection. Environmentalism has become well established in the public agenda since the first national Earth Day in 1970.

As environmental concerns have expanded to climate action, one of the new leaders has been Bill McKibben (fig. 1.20b), an author, educator, and environmentalist who has written extensively

(a) Rachel Carson

(b) Bill McKibben

(c) Van Jones

(d) Wangari Maathai

▲ **FIGURE 1.20** Among many distinguished environmental leaders in modern times, (a) Rachel Carson, (b) Bill McKibben, (c) Van Jones, and (d) Wangari Maathai stand out for their dedication, innovation, and bravery. (a): ©RHS/AP Images; (b): ©Cindy Ord/Getty Images; (c): ©Ryan Rodrick Beiler/Shutterstock; (d): ©s_bukley/Shutterstock

about climate change and has led campaigns to demand political action on this existential threat. As scholar in residence at Middlebury College, he worked with a group of undergraduate students to create 350.org, an organization that has sponsored thousands of demonstrations in 181 countries to raise public awareness about climate change and has sparked actions for fossil fuel divestment on many campuses. The group has been widely praised for its creative use of social media and public organization. McKibben and 350.org led the opposition to the Keystone XL pipeline project, which was designed to transport crude oil from Alberta's tar sands to export terminals in Texas.

Environmental quality is tied to social progress (stage 4)

In recent years some people have argued that the roots of the environmental movement are elitist—promoting the interests of a wealthy minority who can afford to vacation in wilderness. In fact, most environmental leaders have seen social justice and environmental equity as closely intertwined. Gifford Pinchot, Teddy Roosevelt, and John Muir all strove to keep nature accessible to everyone, at a time when public lands, forests, and waterways were increasingly controlled by a few wealthy individuals and private corporations. The idea of national parks, one of our principal strategies for nature conservation, is to provide public access to natural beauty and outdoor recreation. Aldo Leopold, a founder of the Wilderness Society, promoted ideas of land stewardship among farmers, fishers, and hunters. Robert Marshall, also a founder of the Wilderness Society, campaigned all his life for social and economic justice for low-income groups.

Increasingly, environmental activists are making explicit the links between environmental quality and social progress on a global scale (fig. 1.21). But issues of sustainable development are also being recognized across economic divides in wealthy countries. Anthony Kapel "Van" Jones (fig. 1.20c) is one of those who has been a powerful voice for social and environmental progress, and he has helped bring visibility to the role of people of color in environmental action. As both a social justice and environmental activist, Jones has fought poverty and racial injustice by creating hundreds of thousands of "green-collar" jobs installing solar systems and upgrading the energy efficiency of millions of American homes. He served as President Barack Obama's Special Advisor for Green Jobs and has worked to build a "green economy for everyone." He has also brought artists, athletes, and local leaders into national dialogues and engagement around social and environmental issues.

Some of today's leading environmental thinkers come from developing nations, where poverty and environmental degradation together plague hundreds of millions of people. Dr. Wangari Maathai of Kenya (1940–2011) was a notable example. In 1977 Dr. Maathai (see fig. 1.20d) founded the Green Belt Movement in her native Kenya as a way to both organize poor rural women and restore their environment. Beginning at a small, local scale, this organization has grown to more than 600 grassroots networks across Kenya. They have planted more than 30 million trees while mobilizing communities for self-determination, justice, equity, poverty reduction, and environmental conservation. Dr. Maathai was elected to the Kenyan Parliament and served as Assistant Minister for Environment and Natural Resources. Her leadership helped bring

▲ **FIGURE 1.21** Environmental scientists increasingly try to address both public health and environmental quality. The poorest populations often suffer most from environmental degradation. ©Kaetana/Shutterstock

democracy and good government to her country. In 2004 she received the Nobel Peace Prize for her work, the first time a Nobel has been awarded for environmental action. In her acceptance speech she said, "Working together, we have proven that sustainable development is possible; that reforestation of degraded land is possible; and that exemplary governance is possible when ordinary citizens are informed, sensitized, mobilized and involved in direct action for their environment."

Photographs of the earth from space (see fig. 1.3) provide powerful icons for the fourth wave of ecological concern, which might be called **global environmentalism.** Such photos remind us how small, fragile, beautiful, and rare our home planet is. We all share an environment at this global scale. As Ambassador Adlai Stevenson noted in his 1965 farewell address to the United Nations, we now need to worry about the life-support systems of the planet as a whole: "We cannot maintain it half fortunate, half miserable, half confident, half despairing, half slave to the ancient enemies of mankind and half free in a liberation of resources undreamed of until this day. No craft, no crew, can travel with such vast contradictions. On their resolution depends the security of us all."

CONCLUSION

Environmental science gives us useful tools and ideas for understanding environmental problems and for finding new solutions to those problems. Environmental science draws on many disciplines, and on people with diverse interests, to understand the persistent problems we face, including human population growth, contaminated water and air, climate change, and biodiversity losses. There are also encouraging examples of progress. Population growth has slowed, the extent of habitat preserves has expanded greatly in recent years, we have promising new energy options, and in many regions we have made improvements in air and water quality.

The scientific method provides an orderly way to examine these issues. Ideally, scientists are skeptical about evidence and cautious about conclusions. These practices are much like critical thinking, which is also emphasized in environmental science.

Environmental science also is concerned with sustainable development because both poverty and affluence contribute to environmental degradation. Impoverished populations often overexploit land and water supplies, while wealthy populations consume or degrade extraordinary amounts of energy, water, forest products, food, and other resources. Differences in wealth lead to contrasts in life expectancy, infant mortality, and other measures of well-being. Resolving these multiple problems together is the challenge for sustainability.

Our ideas about conservation and environment have evolved in response to environmental conditions, from a focus on conservation of usable resources to preservation of nature for its own sake. Throughout these ideas has been a concern for social equity, for the rights of low-income people to have access to resources and to a healthy environment. In recent years these twin concerns have expanded to recognize the possibilities of change in developing countries and the global interconnections of environmental and social concerns.

PRACTICE QUIZ

1. Describe how global fertility rates and populations are changing (see fig. 1.6).
2. What is the idea of "ecological services"? Give an example.
3. Distinguish between a hypothesis and a theory.
4. Describe the steps in the scientific method.
5. What is probability? Give an example.
6. Why are scientists generally skeptical? Why do tests require replication?
7. What is the first step in critical thinking, according to table 1.4?
8. Distinguish between utilitarian conservation and preservation. Name two environmental leaders associated with each of these philosophies.
9. Why do some experts regard water as the most critical natural resource for the twenty-first century?
10. Where in figure 1.5 does the most dramatic warming occur?
11. What are the HDI ranges for the United States, India, and China (see fig. 1.11)?
12. What is the link between poverty and environmental quality?
13. Define *sustainability* and *sustainable development*.

CRITICAL THINKING AND DISCUSSION

Apply the principles you have learned in this chapter to discuss these questions with other students.

1. Changing fertility rates are often explained in terms of better education for girls and women. What might be some reasons for this association?
2. The analytical approaches of science are suitable for answering many questions. Are there some questions that science cannot answer? Why or why not?
3. Often opinions diverge sharply in controversial topics, such as the allowable size of fish catches or the balance of environmental and economic priorities in land management. Think of a controversial topic with which you are familiar. What steps can you take to maintain objectivity and impartiality in evaluating the issue?
4. Environmental activists often focus on questions of social justice and environmental justice. Consider an issue such as air or water quality. Why does it affect different groups unequally?
5. Suppose you wanted to study the environmental impacts of a rich versus a poor country. What factors would you examine, and how would you compare them?

DATA ANALYSIS Working with Graphs

To understand trends and compare values in environmental science, we need to examine a great many numbers. Most people find it hard to quickly assess large amounts of data in a table. Graphing a set of data makes it easier to see patterns, trends, and relationships. For example, scatter plots show relationships between two variables, while bar graphs show the range of values in a set (figs. 1 and 2). Reading graphs takes practice, but it is an essential skill that will serve you well in this course and others.

You will encounter several common types of graphs in this book. Go to the Data Analysis exercise on Connect to practice these skills and demonstrate your knowledge of how to read and use graphs.

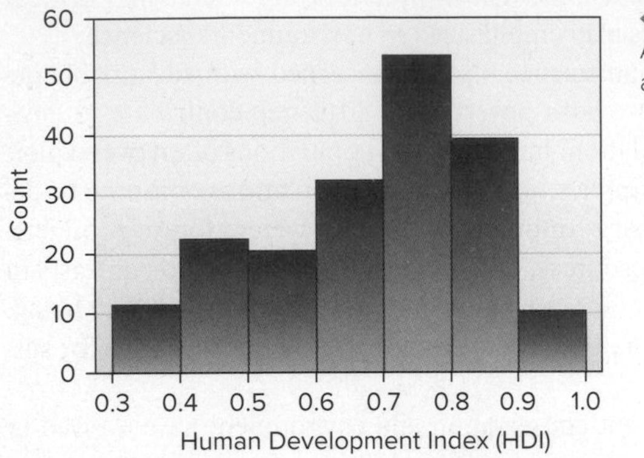

FIGURE 1 A histogram shows a distribution.

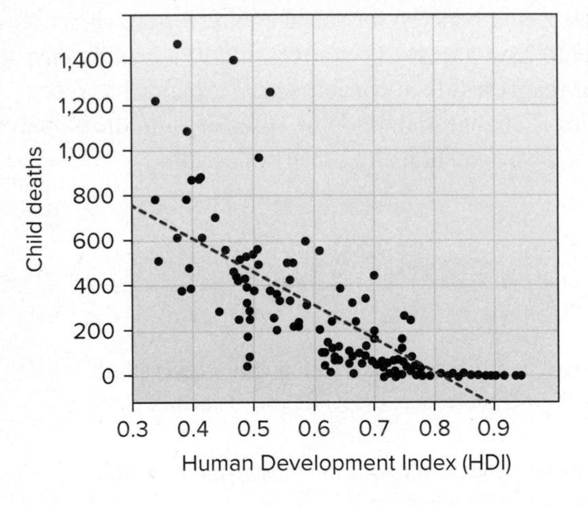

FIGURE 2 A scatter plot shows the relationship.

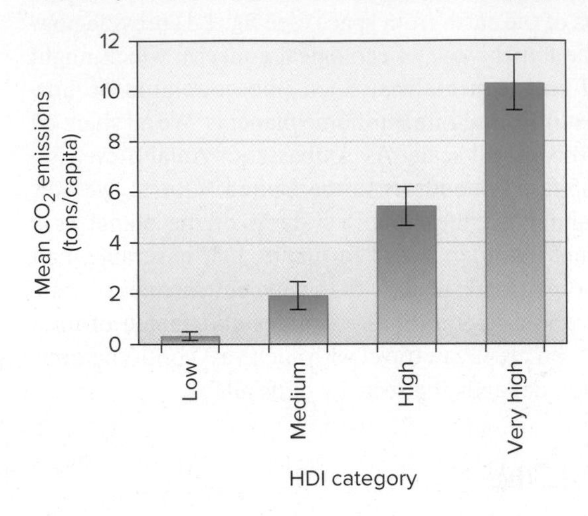

FIGURE 3 Standard error bars show whether groups differ meaningfully.

Design Elements: Active Learning (Toad): ©Gaertner/Alamy Stock Photo; Case Study (Globe): ©McGraw-Hill Education; Google Earth: ©McGraw-Hill Education; Abstract Background: ©Martin Kubat/Shutterstock; What do you think (Students using tablets): ©McGraw-Hill Education/Richard Hutchings, photographer; What can you do (Hand holding Globe): ©Christoph Weihs/Shutterstock

CHAPTER

2 Environmental Systems: Matter, Energy, and Life

LEARNING OUTCOMES

After studying this chapter, you should be able to answer the following questions:

- What are systems, and how do feedback loops affect them?
- Explain the first and second laws of thermodynamics.
- Ecologists say there is no "away" to throw things to, and that everything in the universe tends to slow down and fall apart. What do they mean?
- Explain the processes of photosynthesis and respiration.
- What qualities make water so unique and essential for life as we know it?
- Why are big, fierce animals rare?
- How and why do elements such as carbon, nitrogen, phosphate and sulfur cycle through ecosystems?

Fishing crews come home with empty nets along much of the Louisiana coast in recent years, as low oxygen levels are killing or driving away marine life.

CASE STUDY

Death by Fertilizer: Hypoxia in the Gulf of Mexico

In the 1980s, fishing crews began observing large areas in the Gulf of Mexico, near the Mississippi River mouth, that were nearly devoid of aquatic life in early summer (fig. 2.1). This region supports shrimp, fish, and oyster fisheries worth $250 to $450 million per year, so this "dead zone" was an economic disaster, as well as an ecological one. Marine biologists suspected that the Gulf ecosystem was collapsing because of oxygen deprivation.

To evaluate the problem, marine scientist Nancy Rabelais began mapping areas of low oxygen concentrations along the Louisiana coast in 1985. Every summer since then, she has found vast areas with oxygen concentration below 2 parts per million (ppm). At 2 ppm, nearly all aquatic life, other than microorganisms and primitive worms, is eliminated. In 2017 the Gulf's hypoxic (oxygen-starved) area was the largest ever, at 22,730 km^2 (8,776 mi^2), an area the size of New Jersey.

What causes this huge dead zone? The familiar process of eutrophication, visible in golf course ponds and city parks, is responsible. Eutrophication is the explosive growth of algae and phytoplankton (tiny, floating plants) that occurs when scarce nutrients become available. Normally, scarcity of key nutrients limits plants and algae, but a flush of nutrients allows explosive growth. The overabundance of plants then dies and decays, and decomposers use up nearly all the available oxygen, especially near the sea bed where dead matter falls and collects.

Rabelais and her team observed that each year, 7 to 10 days after large spring rains in the farmlands of the upper Mississippi watershed, oxygen concentrations in the Gulf would drop from 5 ppm to less than 2 ppm. Spring rains are known to wash nutrient-rich soil, organic debris, and fertilizers from farm fields. Pulses of agricultural runoff into the Gulf are followed by a profuse growth of algae and phytoplankton, which drifts to the sea bed. Normally, shrimp, clams, oysters, and other filter feeders consume this debris, but they can't keep up with the sudden flood of material. Instead, decomposing bacteria in the sediment multiply and consume the dead material, using up most of the dissolved oxygen in the process. Putrefying sediments also produce hydrogen sulfide, which further poisons the water near the seafloor.

In well-mixed water bodies, such as the open ocean, oxygen from the surface mixes down into lower layers. But warm, protected water bodies like the Gulf are often stratified: Abundant sunlight keeps the upper layers warmer and less dense than lower layers; cold, dense layers lie stable at depth, and fresh oxygen from the surface can't mix downward. Fish may be able to swim away from the hypoxic zone, but bottom dwellers often simply die. Widespread fish kills are also associated with hypoxia in enclosed waters.

First observed in the 1970s, dead zones now occur along the coast of nearly every major populated region. The number increases almost every year, and more than 400 are now known. They occur mainly in enclosed coastal waters, which tend to be stratified and are vulnerable to nutrient influxes, such as Chesapeake Bay, Long Island Sound, the Mediterranean Sea, the Black Sea, and China's Bohai Bay. But they have also been observed on open coastlines.

Can dead zones recover? Yes. If the influx of nitrogen stops, the system can return to normal. In 1996 in the Black Sea region, farmers in collapsing communist economies were forced to cut nitrogen fertilizer use in half, as fertilizer subsidies collapsed. The Black Sea dead zone disappeared, while farmers saw little decline in their crop yields. But in the Mississippi River watershed, farmers upstream are far from the Gulf and its fisheries. Midwestern policymakers have shown little interest in what happens to fisheries in Louisiana.

The flow of nitrogen reaching U.S. coastal waters has grown by eightfold since the 1950s. Phosphorus, another key nutrient, has tripled. Despite decades of efforts to control nutrients upstream, the dead zone has continued to grow, as intensification of agriculture upstream continues.

The movement of nutrients and energy determines how ecosystems function and how organisms and biological communities flourish or collapse. These topics set the stage for much of the rest of our study of environmental science. In this chapter we examine terms of matter and energy, key elements in living systems, and how they contribute to ecosystems and communities. ■

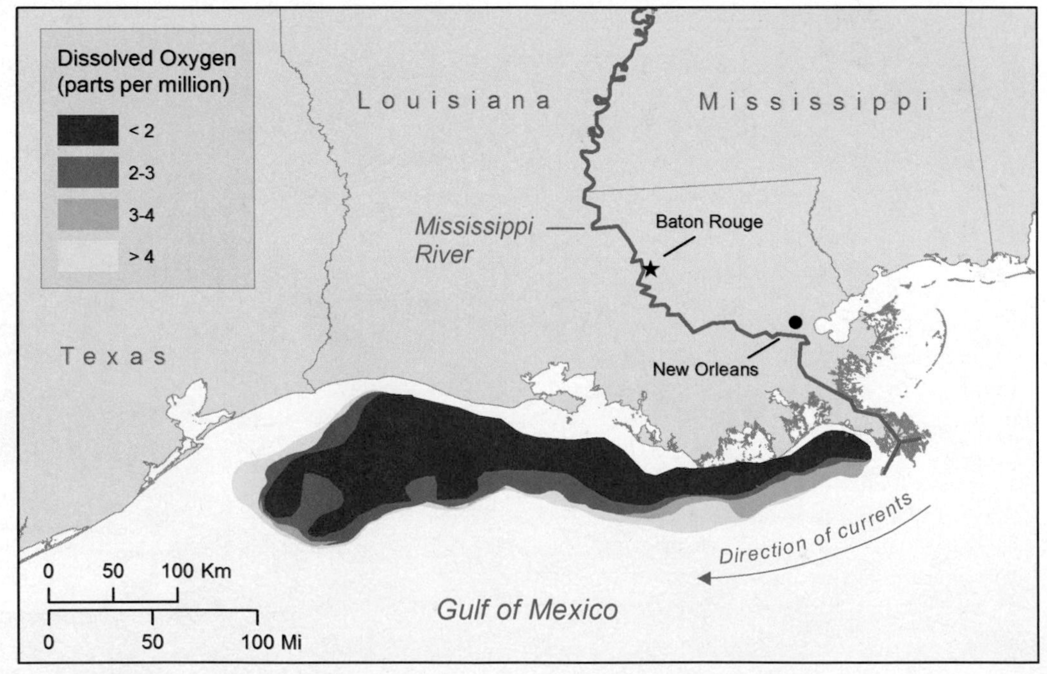

▲ **FIGURE 2.1** A hypoxic "dead zone" about the size of New Jersey forms in the Gulf of Mexico each summer, the result of nutrients from the Mississippi River. Source: N. Rabalais, LSU/LUMCON, http://www.noaa.gov/media-release/gulf-of-mexico-dead-zone-is-largest-ever-measured

> *Most institutions demand unqualified faith; but the institution of science makes skepticism a virtue.*
>
> —ROBERT KING MERTON

2.1 SYSTEMS DESCRIBE INTERACTIONS

- *Matter and energy move through ecosystems.*
- *Throughput is the amount of matter or energy entering and leaving a system.*
- *Positive feedbacks enhance a process; negative feedbacks diminish it.*

Managing nutrients in the Gulf of Mexico is an effort to restore a stable system, one with equal amounts of inputs and outputs and with balanced populations of plants, animals, and microorganisms. This balance maintains overall stability and prevents dramatic change or collapse. In general, a **system** is a network of interdependent components and processes, with materials and energy flowing from one component of the system to another. The term *ecosystem* is probably familiar to you. This simple word represents complex assemblages of animals, plants, and their environment, through which materials and energy move. In a sense, *you* are a system consisting of trillions of cells and thousands of species that live in or on your body, as well as the energy and matter that move through you.

The idea of systems is useful because it helps us organize our thoughts about the inconceivably complex phenomena around us. For example, an ecosystem might consist of countless animals, plants, and their physical surroundings. Keeping track of all the elements and relationships in an ecosystem would probably be an impossible task. But if we step back and think about components in terms of their roles—plants, herbivores, carnivores, and decomposers—and the relationships among them, then we can start to comprehend how the system works (fig. 2.2).

We can use some general terms to describe the components of a system. A simple system consists of compartments (also called state variables), which store resources such as energy or matter, and the flows, or pathways, by which those resources move from one compartment to another. In figure 2.2, the plants and animals represent elements in this cycle of life. We can describe the flows in terms of photosynthesis, herbivory, predation, and decomposition. The plants use photosynthesis to create carbohydrates from carbon, water, and sunlight. The rabbit is an herbivore that consumes plants and uses the energy and chemicals stored there for its own life functions. The fox, in turn, is a carnivore that eats other animals, while the decomposers recycle wastes from all previous compartments.

It may seem cold and analytical to describe a rabbit or a flower as a state variable, but it is also helpful to do so. The energy and matter in the flower or rabbit are really the same; they just change their physical location (or state) when moving from one organism to another.

Understanding the characteristics of ecological systems can help us diagnose disturbances or changes in those systems: For example, if rabbits become too numerous, herbivory can become too rapid for plants to sustain. This could lead to collapse of the system. Similarly, in the Gulf of Mexico, excess nitrogen and phosphorus have led to excess algal growth—this results in reduced oxygen levels in the water, which kills fish and other aquatic organisms. Let's examine some of the common characteristics we can find in systems.

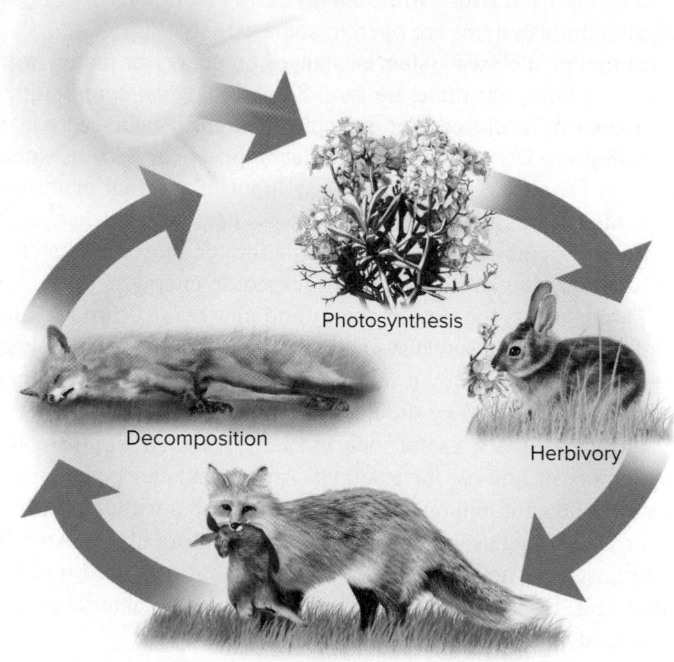

▲ **FIGURE 2.2** A system can be described in very simple terms.

Systems can be described in terms of their characteristics

Open systems are those that receive inputs from their surroundings and produce outputs that leave the system. The Gulf of Mexico is an open system. Fresh water, nutrients, and some organisms flow

Benchmark Data

Among the ideas and values in this chapter, the following are a few worth remembering.

2 ppm	Dissolved oxygen minimum for most aquatic life
1:10	Ratio of H^+ ions at successive pH values
1:10	Rule of thumb: energy retained at successive trophic levels
1,372 W/m^2	Solar energy at the top of the atmosphere, the energy for nearly all ecosystems and biogeochemical cycles
80 Tg	Amount of N fixed by natural processes per year
140 Tg	Amount of N released by agriculture and industry per year
7 GT	Amount of C released by land clearing and fossil fuel combustion

into the Gulf from rivers, while energy comes from the sun. In general, all natural systems are open to some extent.

In theory, a **closed system** exchanges no energy or matter with its surroundings, but these are rare. Some biological communities may appear to be closed. For example, a carefully balanced aquarium with just a few plants, fish, and decomposers or detritus-eaters may exist for a long time without any input of external materials, but plants need light to survive, and excess heat must be dissipated into the surrounding environment. So although closed to material flow, the aquarium is still open with respect to energy.

Throughput is the flow of energy and matter into, through, and out of a system. In a stable system, the amount of matter or energy entering equals the amount leaving: If 100 kg of nitrogen enters a system, ultimately 100 kg should be exported.

Throughput is a useful idea in many systems. Compare the throughputs in houses, for example. An uninsulated house might consume and emit millions of calories of heat in a winter month as the furnace struggles to keep a steady temperature inside. An efficiently insulated house might consume (and lose) a small fraction of that heat. Households also vary in throughput of material goods, water, food, and other resources.

Ideally, systems tend to be stable over time. If the Gulf's nitrogen inputs equal outputs, then the amount of phytoplankton, algae, and other organisms should remain relatively constant. When a system is in a stable balance, we say it is in **equilibrium.** Systems can change, though, sometimes suddenly. Often there are **thresholds,** or "tipping points," where rapid change suddenly occurs if you pass certain limits. The Gulf's oxygen levels may decrease gradually without much visible effect until they reach a level at which organisms can't survive. Then suddenly, decaying organic material releases toxins and uses up even more oxygen, which throws the whole system into a downward spiral from which it's difficult to recover.

Feedback loops help stabilize systems

Systems function in cycles, with each component eventually feeding back to influence the size or rate of itself. A **positive feedback loop** tends to increase a process or component, whereas a **negative feedback loop** diminishes it. Feedbacks occur in countless familiar systems. Think of a sound system, when a microphone picks up the sound coming from speakers. The microphone amplifies the sound and sends it back to the speakers, which get louder and louder (a positive feedback) until the speakers blow out or someone cuts the power. Your body has feedback loops that regulate everything from growth and development to internal temperature. When you exercise and get hot, you sweat, and evaporation cools your skin (a negative feedback) to maintain a stable temperature. When you are cold, you shiver, and that activity helps return your temperature to normal.

Positive and negative feedbacks are a fundamental part of ecosystems. Consider the simple system in figure 2.2. A pair of rabbits produces several baby rabbits, and reproduction can lead to greater and greater numbers of rabbits (a positive feedback). If there were an unlimited food supply, the rabbit population would increase indefinitely (fig. 2.3). But if the growing rabbit population depletes its food supply, starvation will slow or reverse population growth

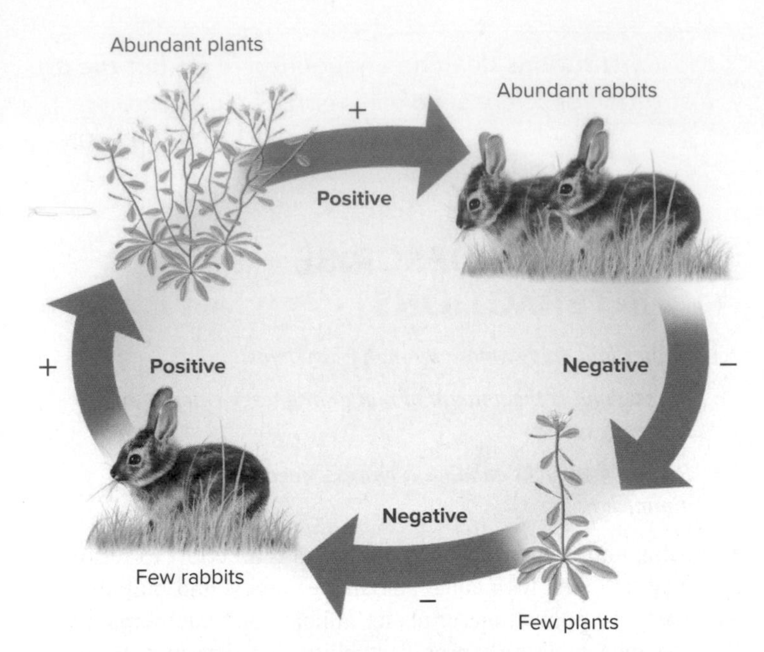

▲ **FIGURE 2.3** Feedback loops (both positive and negative) help regulate and stabilize processes in ecosystems.

(a negative feedback). When there are fewer rabbits, the plants recover and the cycle starts again.

Multiple loops interact in ecosystems. As the number of rabbits rises, the fox in figure 2.2 has more to eat and will raise more healthy fox kits. A growing fox population reduces the number of rabbits, which in turn allows plants to increase. In general, the number of plants, rabbits, and foxes should oscillate, or vary, around an average that remains stable over time. Often we call this variation around a stable average a "dynamic equilibrium." It's not quite equilibrium because numbers vary, but the overall trend is relatively stable.

One of the more important global feedback systems now being disrupted is Arctic sea ice. As you can read in chapter 9, over the past 200 years, we have greatly increased the amount of heat-trapping greenhouse gases in the atmosphere, and the resulting warming is melting much of the ice in the Arctic Ocean. Normally Arctic sea ice reflects most sunlight back to space, but open water is warmed by absorbed sunlight. Melting thus leads to more warming, which leads to more melting, in a positive feedback loop that is affecting temperatures globally.

Disturbances, events such as fire, flooding, climate change, invasion by new species, or destructive human activities, also interrupt normal feedback loops. Sometimes systems are resilient, so that they return to something like their previous states after a disturbance. Sometimes disturbances cause a system to shift to a new state, so that conditions become permanently altered.

Emergent properties are another interesting aspect of systems, in which the characteristics of a whole system are greater than the sum of its parts. Consider mangrove forests, which grow along coastlines throughout the tropical world (fig. 2.4). Because their direct economic value can be modest, thousands of kilometers of mangrove-lined coasts have been cleared to make room for shrimp ponds or beach resorts. But the mangrove forest provides countless indirect economic values. Prop-roots trap sediment and stabilize mudflats, improving water clarity and protecting offshore coral reefs from silt.

▲ FIGURE 2.4 Mangrove forests, like many complex ecosystems, have emergent properties beyond being mere collections of trees. They provide many valuable ecosystem services, such as stabilizing shorelines, providing habitat, and protecting the land from tidal waves and storms.
©William P. Cunningham

Mangroves also protect small fish and other marine species. Many commercial ocean fish and shrimp species spend juvenile stages hiding among the mangrove roots. And the forest protects the shoreline from damaging storms and tsunamis. Collectively these services can be worth hundreds of times the value of a shrimp farm, but they exist only when the many parts of the system function together.

2.2 ELEMENTS OF LIFE

- *Matter is recycled but doesn't disappear.*
- *The elements O, C, H, and N account for most mass of living organisms.*
- *Water has unique properties that make life possible.*

What exactly are the materials that flow through a system like the Gulf of Mexico? If a principal concern is the control of nutrients, such as NO_3 and PO_4, as well as the maintenance of O_2 levels, just what exactly are these elements, and why are they important? What does "O_2" or "NO_3" mean? In this section we will examine matter and the elements and compounds on which all life depends. In the sections that follow, we'll consider how organisms use those elements and compounds to capture and store solar energy, and how materials cycle through global systems as well as ecosystems.

To understand how these compounds form and move, we need to begin with some of the fundamental properties of matter and energy.

Matter is recycled but not destroyed

Everything that takes up space and has mass is **matter.** Matter exists in three distinct states—solid, liquid, and gas—due to differences in the arrangement of its constitutive particles. Water, for example, can exist as ice (solid), as liquid water, or as water vapor (gas).

Matter also behaves according to the principle of **conservation of matter:** Under ordinary circumstances, matter is neither created nor destroyed but rather is recycled over and over again. It can be transformed or recombined, but it doesn't disappear; everything goes somewhere. Some of the molecules that make up your body probably contain atoms that once were part of the body of a dinosaur and most certainly were part of many smaller prehistoric organisms, as chemical elements have been used and reused by living organisms.

How does this principle apply to human relationships with the biosphere? In affluent societies, we use natural resources to produce an incredible amount of "disposable" consumer goods. If everything goes somewhere, where do the things we dispose of go after the garbage truck leaves? As the sheer amount of "disposed-of stuff" increases, we are having greater problems finding places to put it. Ultimately, there is no "away" where we can throw things we don't want anymore.

Elements have predictable characteristics

Matter consists of **elements** such as P (phosphorus) or N (nitrogen), which are substances that cannot be broken down into simpler forms by ordinary chemical reactions. Each of the 118 accepted elements (92 natural, plus 26 created under special conditions) has distinct chemical characteristics.

Just four elements—oxygen, carbon, hydrogen, and nitrogen (symbolized as O, C, H, and N)—make up more than 96 percent of the mass of most living organisms. Water is composed of two H atoms and one O atom (written H_2O). All the elements are listed in the Periodic Table of the Elements, which you can find on the foldout map in the back of this book. Often, though, it's enough to pay attention to just a few (table 2.1).

Atoms are the smallest particles that exhibit the characteristics of an element. As difficult as it may be to imagine when you look at a solid object, all matter is composed of tiny, moving particles, separated by space and held together by energy. Atoms are composed of a nucleus, made of positively charged protons and electrically neutral neutrons, circled constantly by negatively charged

TABLE 2.1 Functions of Some Common Elements

FUNCTION	ELEMENTS	COMMENTS
Fertilizers	N nitrogen P phosphorus K potassium	Essential components of proteins, cells, other biological compounds; crucial fertilizers for plants
Organic compounds	C carbon O oxygen H hydrogen	Form the basic structure of cells and other components of living things, in combination with many other elements
Metals	Fe iron Al aluminum Au gold	Generally malleable; most (not all) react readily with other elements
Toxic elements	Pb lead Hg mercury As arsenic	Many are metals that can interfere with processes in nervous systems

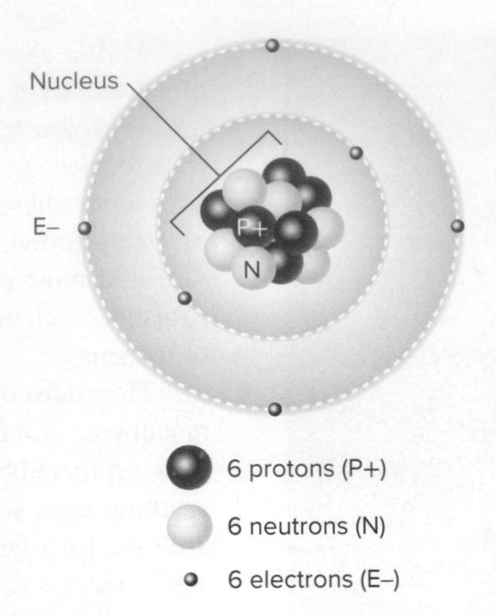

▶ **FIGURE 2.5** This model represents a carbon-12 atom, containing six protons and six neutrons in its nucleus, as well as six electrons orbiting in outer shells. Many physicists say electrons should be shown as a fuzzy cloud of wave energy rather than individual particles.

electrons (fig. 2.5). Electrons, which are tiny in comparison to the other particles, orbit the nucleus at the speed of light.

Each element is listed in the periodic table according to the number of protons per atom, called its **atomic number.** The number of neutrons in the atoms of an element can vary slightly. Thus, the atomic mass, which is the sum of the protons and neutrons in each nucleus, also can vary. We call forms of a single element that differ in atomic mass **isotopes.** For example, hydrogen, the lightest element, normally has only one proton (and no neutrons) in its nucleus. A small percentage of hydrogen atoms have one proton and one neutron. We call this isotope deuterium (^2H). An even smaller percentage of natural hydrogen occurs as the isotope tritium (^3H), with one proton plus two neutrons. The heavy form of nitrogen (^{15}N) has one more neutron in its nucleus than does the more common ^{14}N. Both of these nitrogen isotopes are stable, but some isotopes are unstable—that is, they may spontaneously emit electromagnetic energy, subatomic particles, or both. Radioactive waste and nuclear energy result from unstable isotopes of elements such as uranium and plutonium.

Why should you know about isotopes? Although you might not often think of it this way, both stable and unstable types are in the news every day. Unstable, radioactive isotopes have been distributed in the environment by both radioactive waste from nuclear power plants and nuclear bombs. Every time you hear debates about nuclear power plants—which might produce the energy that powers the lights you are using right now—or about nuclear weapons in international politics, the core issue is radioactive isotopes. Understanding why they are dangerous—because emitted particles damage living cells—is important as you form your opinions about these policy issues.

Stable isotopes—those that do not change mass by losing neutrons—are important because they help us understand climate history and many other environmental processes. This is because lightweight isotopes move differently than heavier ones do. For example, oxygen occurs as both ^{16}O (a lighter isotope) and ^{18}O (a heavier isotope). Some water molecules (H_2O) contain the lightweight oxygen isotopes. These lightweight molecules evaporate, and turn into rain or snowfall, *slightly* more easily than heavier water molecules, especially in cool weather. Consequently, ice stored in glaciers during cool periods in the earth's history will contain slightly higher proportions of light-isotope water, with ^{16}O. Ice from warm periods contains relatively higher proportions of ^{18}O. By examining the proportions of isotopes in ancient ice layers, climate scientists can deduce the earth's temperature from hundreds of thousands of years ago (see chapter 9).

So although atoms, elements, and isotopes may seem a little arcane, they actually are fundamental to understanding climate change, nuclear weapons, and energy policy—all issues you can read about in the news almost any day of the week.

Electrical charges keep atoms together

Atoms frequently gain or lose electrons, acquiring a negative or positive electrical charge. Charged atoms (or combinations of atoms) are called **ions.** Negatively charged ions (with one or more extra electrons) are *anions.* Positively charged ions are *cations.* A sodium (Na) atom, for example, can give up an electron to become a sodium ion (Na^+). Chlorine (Cl) readily gains electrons, forming chlorine ions (Cl^-).

Atoms often join to form **compounds,** or substances composed of different kinds of atoms (fig. 2.6). A pair or group of atoms that can exist as a single unit is known as a **molecule.** Some elements commonly occur as molecules, such as molecular oxygen (O_2) or molecular nitrogen (N_2), and some compounds can exist as molecules, such as glucose ($C_6H_{12}O_6$). In contrast to these molecules, sodium chloride (NaCl, table salt) is a compound that cannot exist as a single pair of atoms. Instead, it occurs in a large mass of Na and Cl atoms or as two ions, Na^+ and Cl^-, in solution. Most molecules consist of only a few atoms. Others, such as proteins, can include millions or even billions of atoms.

When ions with opposite charges form a compound, the electrical attraction holding them together is an *ionic bond.* Sometimes atoms form bonds by *sharing* electrons. For example, two hydrogen

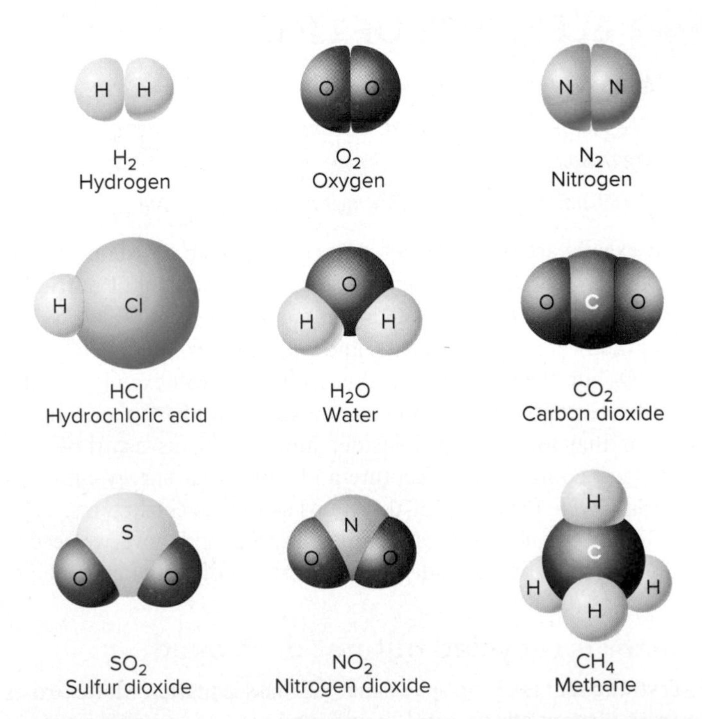

▲ **FIGURE 2.6** These common molecules, with atoms held together by covalent bonds, are important components of the atmosphere or are important pollutants.

atoms can bond by sharing a single electron—it orbits the two hydrogen nuclei equally and holds the atoms together. Such electron-sharing bonds are known as *covalent bonds*. Carbon (C) can form covalent bonds simultaneously with four other atoms, so carbon can create complex structures such as sugars and proteins. Atoms in covalent bonds do not always share electrons evenly. An important example in environmental science is the covalent bonds in water (H_2O). The oxygen atom attracts the shared electrons more strongly than do the two hydrogen atoms. Consequently, the hydrogen portion of the molecule has a slight positive charge, while the oxygen has a slight negative charge. These charges create a mild attraction between water molecules, so that water tends to be somewhat cohesive. This fact helps explain some of the remarkable properties of water.

When an atom gives up one or more electrons, we say it is *oxidized* (because it is very often oxygen that takes the electron, as bonds are formed with this very common and highly reactive element). When an atom gains electrons, we say it is *reduced*. Chemical reactions necessary for life involve oxidation and reduction: Oxidation of sugar and starch molecules, for example, is an important part of how you gain energy from food.

Breaking bonds requires energy, whereas forming bonds generally releases energy. Burning wood in a fireplace breaks up large molecules, such as cellulose, and forms many smaller ones, such as carbon dioxide and water. The net result is a release of energy (heat). Generally, some energy input (activation energy) is needed to initiate these reactions. In your fireplace, a match might provide the needed activation energy. In your car, a spark from the spark-plug provides activation energy to initiate the oxidation (burning) of gasoline.

Water has unique properties

Water is almost miraculously distinctive. It has properties shared by few other compounds, and these properties are key to supporting life as we know it. This is why astronomers examining other planets or solar systems explore for evidence of water. The following are some of the notable aspects that make water so important and unique:

1. Water molecules are polar—that is, they have a slight positive charge on one side and a slight negative charge on the other side. Therefore, water readily dissolves polar or ionic substances, including sugars and nutrients, and carries materials to and from cells.
2. Water is the only inorganic liquid that occurs in nature under normal conditions at temperatures suitable for life. Organisms synthesize organic compounds, such as oils and alcohols, that remain liquid at ambient temperatures and that are therefore extremely valuable to life, but the predominant liquid in nature is water.
3. Water molecules are cohesive, tending to stick together. Water has the highest surface tension of any common, natural liquid. Water also adheres to surfaces. As a result, water is subject to capillary action: It can be drawn into small channels. Without this property, the movement of water and nutrients through living organisms might not be possible.
4. Water is unique in that it expands when it crystallizes. Most substances shrink as they change from liquid to solid. Ice floats because it is less dense than liquid water. When temperatures fall below freezing, the surface layers of lakes, rivers, and oceans cool faster than and freeze before deeper water. Floating ice then insulates underlying layers, keeping most water bodies liquid (and aquatic organisms alive) throughout the winter in most places. Without this feature, many aquatic systems would freeze solid in winter.
5. Water has a high heat of vaporization, using a great deal of heat to convert from liquid to vapor. Consequently, evaporating water is an effective way for organisms to shed excess heat. Many animals pant or sweat to moisten evaporative cooling surfaces.
6. Water also has a high specific heat; that is, water absorbs a great deal of heat before it changes temperature. Water's slow response to temperature change helps moderate global temperatures, keeping the environment warm in winter and cool in summer.

All these properties make water the most critical component of living things, of ecosystems, and of our climate system. Water is key to cycles that move matter and energy through earth systems, which make life possible on earth.

Acids and bases release reactive H^+ and OH^-

Substances that readily give up hydrogen ions in water are known as **acids.** Hydrochloric acid, for example, dissociates in water to form H^+ and Cl^- ions. In later chapters you may read about acid rain (which has an abundance of H^+ ions), acid mine drainage, and many other environmental problems involving acids. In general, acids cause environmental damage because the H^+ ions react readily with living tissues (such as your skin or tissues of fish larvae) and with nonliving substances (such as the limestone on buildings, which erodes under acid rain).

Substances that readily bond with H^+ ions are called **bases** or alkaline substances. Sodium hydroxide (NaOH), for example, releases hydroxide ions (OH^-) that bond with H^+ ions in water. Bases can be highly reactive, so they also cause significant environmental problems. Acids and bases can also be essential to living things: The acids in your stomach help you digest food, for example, and acids in soil help make nutrients available to growing plants.

We describe acids and bases in terms of **pH,** the negative logarithm of its concentration of H^+ ions (fig. 2.7). Acids have a pH below 7; bases have a pH greater than 7. A solution of exactly pH 7 is "neutral." Because the pH scale is logarithmic, pH 6 represents *ten times* more hydrogen ions in solution than pH 7.

A solution can be neutralized by adding buffers, or substances that accept or release hydrogen ions. In the environment, for example, alkaline rock can buffer acidic precipitation, decreasing its acidity. Lakes with acidic bedrock, such as granite, are especially vulnerable to acid rain because they have little buffering capacity.

Organic compounds have a carbon backbone

Organisms use some elements in abundance, others in trace amounts, and others not at all. Certain vital substances are concentrated within

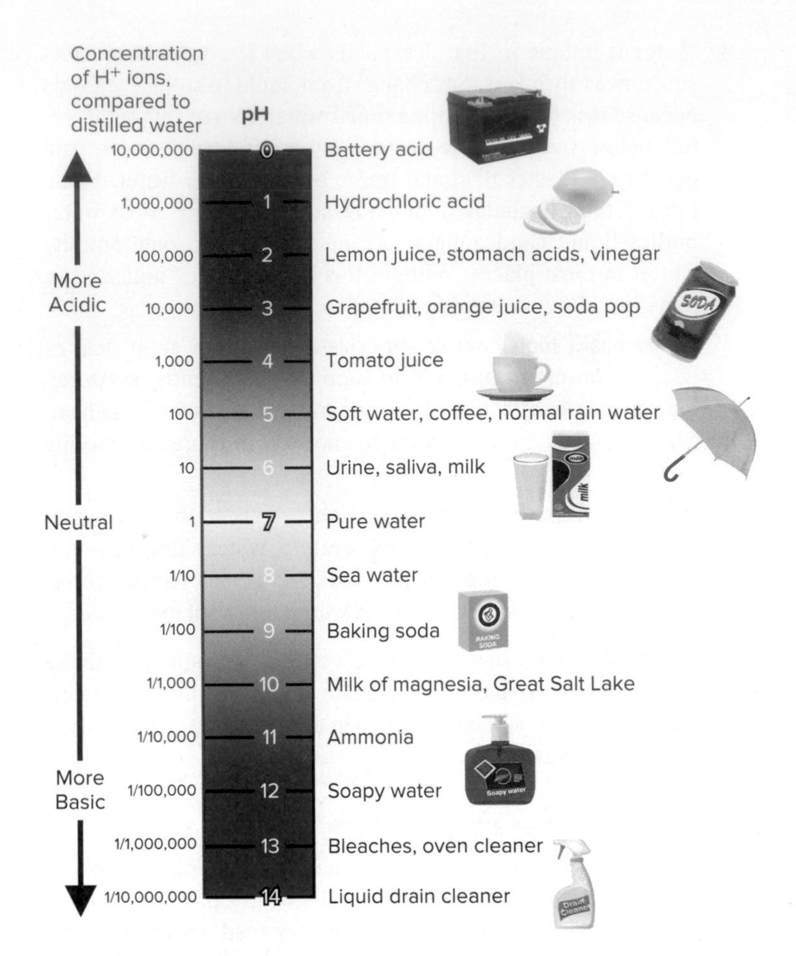

▲ **FIGURE 2.7** The pH scale. The numbers represent the negative logarithm of the hydrogen ion concentration in water. Alkaline (basic) solutions have a pH greater than 7. Acids (pH less than 7) have high concentrations of reactive H^+ ions.

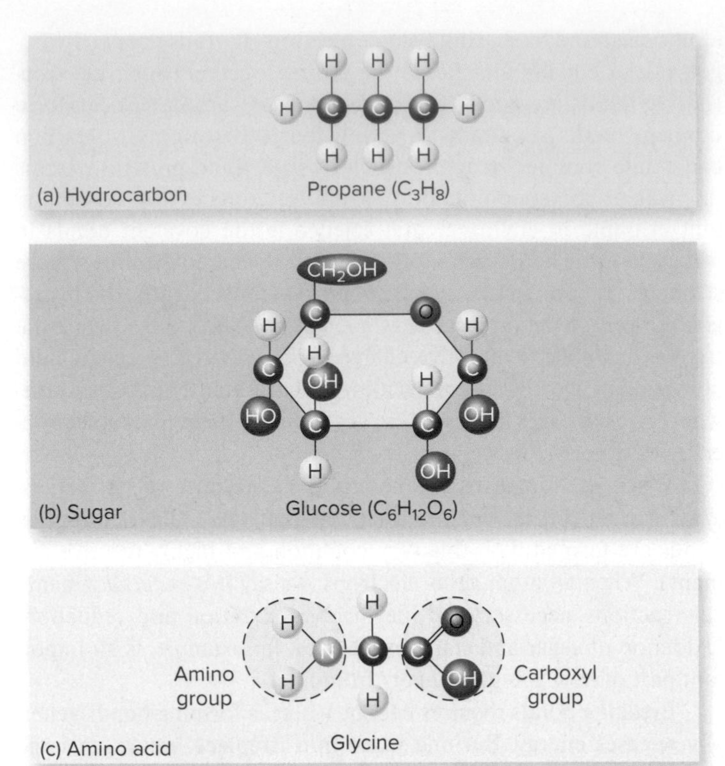

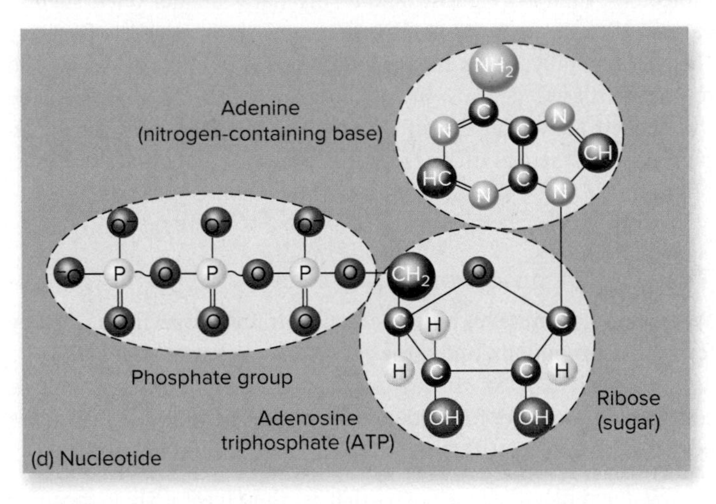

▲ **FIGURE 2.8** The four major groups of biologically important organic molecules are based on repeating subunits of these carbon-based structures. Basic structures are shown for (a) butyric acid (a building block of lipids) and a hydrocarbon, (b) a simple carbohydrate, (c) a protein, and (d) a nucleotide (a component of nucleic acids).

cells, while others are actively excluded. Carbon is a particularly important element because chains and rings of carbon atoms form the skeletons of **organic compounds,** the material of which biomolecules, and therefore living organisms, are made.

The four major categories of organic compounds in living things ("bio-organic compounds") are lipids, carbohydrates, proteins, and nucleic acids. Lipids (including fats and oils) store energy for cells, and they provide the core of cell membranes and other structures. Many hormones are also lipids. Lipids do not readily dissolve in water, and their structure is a chain of carbon atoms with attached hydrogen atoms. This structure makes them part of the family of hydrocarbons (fig. 2.8a). Carbohydrates (including sugars, starches, and cellulose) also store energy and provide structure to cells. Like lipids, carbohydrates have a basic structure of carbon atoms, but hydroxyl (OH) groups replace half the hydrogen atoms in their basic structure, and they usually consist of long chains of simple sugars. Glucose (fig. 2.8b) is an example of a very simple sugar.

Proteins are composed of chains of subunits called amino acids (fig. 2.8c). Folded into complex, three-dimensional shapes, proteins provide structure to cells and are used for countless cell functions. Enzymes, such as those that release energy from lipids and carbohydrates, are proteins. Proteins also help identify disease-causing microbes, make muscles move, transport oxygen to cells, and regulate cell activity.

Nucleotides are complex molecules made of a five-carbon sugar (ribose or deoxyribose), one or more phosphate groups, and an organic nitrogen-containing base (fig. 2.8d). Nucleotides are extremely important as signaling molecules (they carry information between cells, tissues, and organs) and as sources of energy within cells. They also form long chains called *r*ibo*n*ucleic *a*cid (RNA) or ***d*eoxyribo*n*ucleic *a*cid (DNA),** which are essential for storing and expressing genetic information. Only four kinds of nucleotides (adenine, guanine, cytosine, and thymine) occur in DNA, but DNA contains millions of these molecules arranged in very specific sequences. These sequences of nucleotides provide genetic information, or instructions, for cells. These instructions direct the growth and development of an organism. They also direct the formation of proteins or other compounds, such as those in melanin, a pigment that protects your skin from sunlight.

Long chains of DNA bind together to form a double helix (a two-stranded spiral; fig. 2.9). These chains replicate themselves when cells divide, so that as you grow, your DNA is reproduced in all your cells, from blood cells to hair cells. Everyone (even identical twins) has a distinctive DNA pattern, making DNA useful in identifying individuals in forensics. Because DNA is passed down to us from our ancestors, it has allowed scientists to establish relationships, such as evolutionary pathways and taxonomic links, as well as the presence of specific species in a landscape. Furthermore, DNA editing is allowing breakthroughs in many biological fields (What Do You Think?, p. 36).

Cells are the fundamental units of life

All living organisms are composed of **cells,** minute compartments within which the processes of life are carried out (fig. 2.10). Microscopic organisms such as bacteria, some algae, and protozoa are composed of single cells. Most higher organisms are multicellular, usually with many different cell varieties. Your body, for instance, is composed of several trillion cells of about 200 distinct types. Every cell is surrounded by a thin but dynamic membrane of lipid and protein that receives information about the exterior world and regulates the flow of materials between the cell and its environment. Inside, cells are subdivided into tiny organelles and subcellular particles that provide the machinery for life. Some of these organelles store and release energy. Others manage and distribute information. Still others create the internal structure that gives the cell its shape and allows it to fulfill its role.

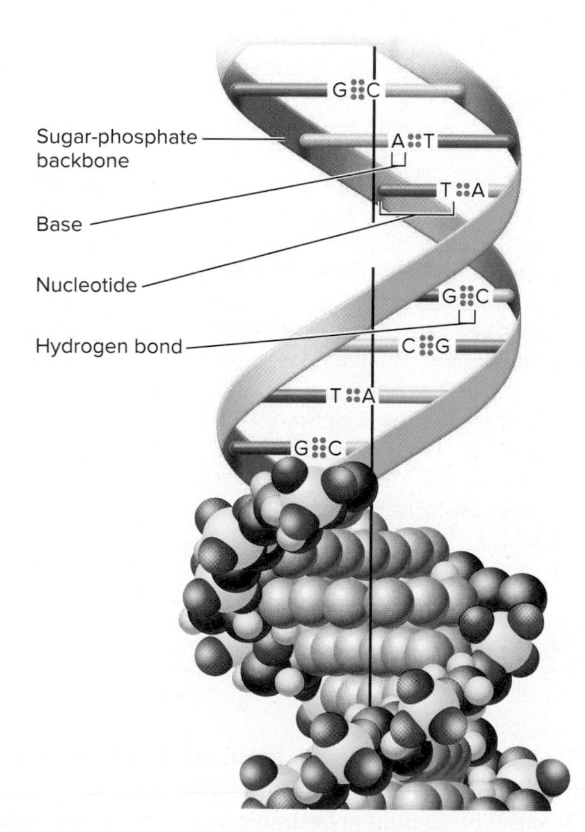

▲ **FIGURE 2.9** A composite molecular model of DNA. The lower part shows individual atoms, while the upper part has been simplified to show the strands of the double helix held together by hydrogen bonds (small dots) between matching nucleotides (A, T, G, and C). A complete DNA molecule contains millions of nucleotides and carries genetic information for many specific, inheritable traits.

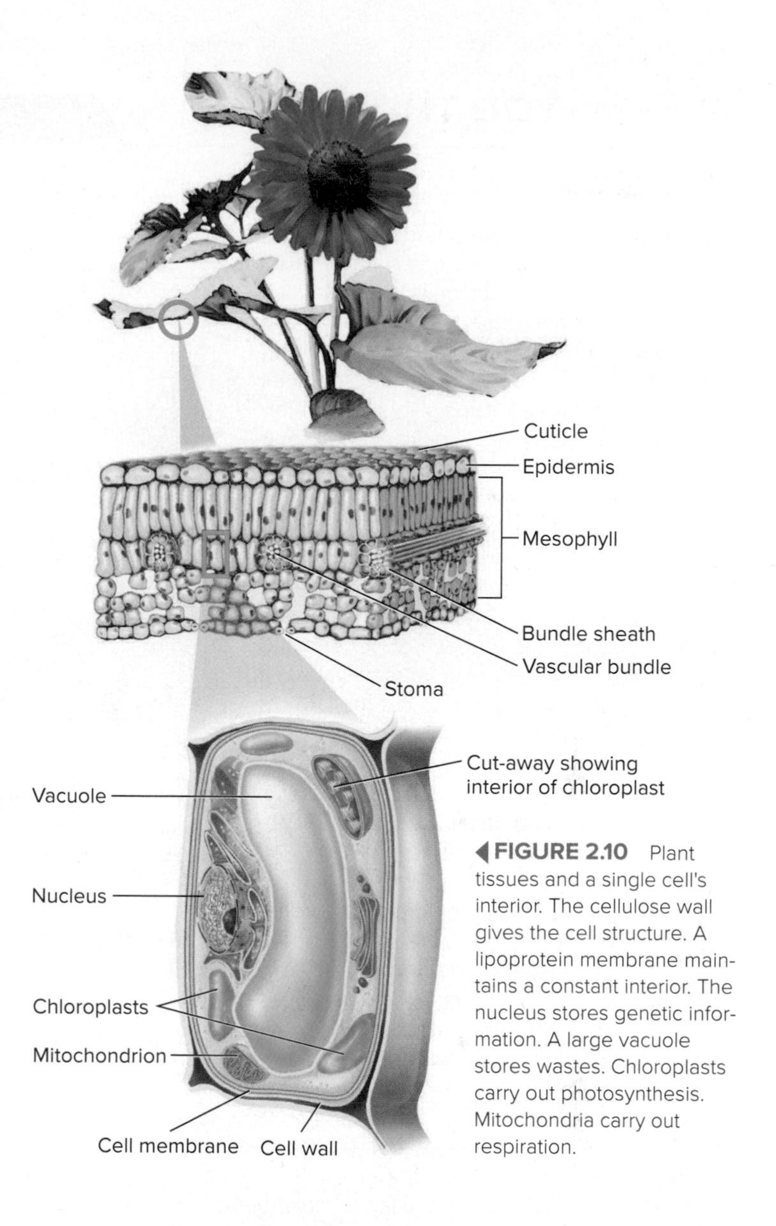

◀ **FIGURE 2.10** Plant tissues and a single cell's interior. The cellulose wall gives the cell structure. A lipoprotein membrane maintains a constant interior. The nucleus stores genetic information. A large vacuole stores wastes. Chloroplasts carry out photosynthesis. Mitochondria carry out respiration.

Nitrogen and phosphorus are key nutrients

As you know from the opening case study, nitrogen and phosphorus are key components of ecosystems. These are limiting elements because they are essential for plant and animal growth, but normally they are not abundant in ecosystems.

What Do YOU THINK?

Gene Editing

Humans have known for centuries that selective breeding can improve the characteristics of domestic plants and animals. But selective breeding is slow and rather unpredictable. The development of molecular genetics, inserting pieces of DNA from one species into another, dramatically improved our ability to tailor organisms, but this process is difficult and prone to errors. The discovery of a bacterial system for editing genes, however, may unleash a gold rush in genetic engineering.

This gene editing system is called CRISPR, short for "clustered regularly interspaced short palindromic repeats." CRISPR uses short sequences (palindromic repeats) of genetic material to attach to specific sections of DNA in a cell; it then uses an enzyme to cut the DNA in specific places. Bacteria use this process to cut and disable the DNA of invading viruses. But geneticists have realized that this bacterial process could be used to recognize and modify, or "edit," genes in higher plants and animals. Just as bacteria target the DNA of an invading virus, CRISPR uses molecules synthesized to bind to and cut any gene we want to edit. When a target DNA sequence is identified and cut, it can inactivate the gene. This gives us important information about the gene's functions or expressions. Alternatively, as the cell tries to repair broken DNA, CRISPR can supply a template for new versions of the target gene, to replace the original sequence. Think of this as a molecular version of the search and replace function in your word processor.

▲ **FIGURE 1** Experiments with CRISPR have modified genes and edited inherited traits in lab mice and other organisms. ©Nikolay Suslov/123RF

The tool is being used in the lab to make human cells impervious to HIV, to correct a mutation that leads to blindness, and to cure mice of muscular dystrophy, cataracts, and a hereditary liver disease (fig. 1). It has been used to improve wheat, rice, soybeans, tomatoes, and oranges. Libraries of tens of thousands of DNA sequences are now available to target and activate, or silence, specific genes. One of the most exciting features of CRISPR is that it can modify multiple genes at the same time in a single cell. This may make it possible to study and eventually treat complex diseases, such as Alzheimer's or Parkinson's, that are regulated by many genes.

In 2017, scientists tried editing a gene inside a living human, in an attempt to permanently cure an inherited metabolic disorder called Hunter syndrome. In this syndrome, cells cannot produce an enzyme needed to break down complex sugars, so these sugar molecules accumulate in cells, blood, and tissues. Consequences can include nerve degeneration and mental impairment. The patient in this experiment received an intravenous transmission of billions of copies of a corrective gene, along with complexes designed to repair the DNA and restore his ability to produce the missing enzyme.

CRISPR has also become the latest question in debates about genetically modified organisms. On one hand, CRISPR makes it easier, quicker, and cheaper to produce dramatically more modified organisms, with unknown consequences. On the other hand, CRISPR edits genes much more precisely than other tools, making gene editing more predictable and potentially safer. The ability to precisely edit a single gene, or even a single nucleotide in a gene, makes it unnecessary to move DNA between species, a process that has worried many observers.

Why is this important in environmental science? A persistent worry with genetic engineering is that a superbug might emerge, causing new epidemic diseases or disrupting ecosystems. This outcome seems less likely with CRISPR's precision. CRISPR has been used in efforts to control invasive *Aedes aegypti* mosquitoes, which carry Zika virus and Dengue fever, by producing and releasing modified mosquitoes that mate with wild relatives and produce nonviable larvae. In Brazilian tests the number of observed *Aedes aegypti* larvae dropped by 82 percent in only 8 months. This approach could aid disease control without blasting ecosystems with powerful pesticides. We also might be able to engineer some plant and animal species to help them tolerate changing climate and water availability.

What do you think? If you were appointed to a regulatory panel commissioned to oversee gene editing, what limits—if any—would you impose on this new technology?

Notice that in figure 2.8 there are only a few different elements that are especially common. Carbon (C) is captured from air by green plants, and oxygen (O) and hydrogen (H) derive from air or water. The most important additional elements are nitrogen (N) and phosphorus (P), which are essential parts of the complex proteins, lipids, sugars, and nucleic acids that keep you alive. Of course, your cells use many other elements, but these are the most abundant. You derive all these elements by consuming molecules produced by green plants. Plants, however, must extract these elements from their environment. Low levels of N and P often limit growth in ecosystems where they are scarce. Abundance of N and P can cause runaway growth. In fertilizers, these elements often occur in the form of nitrate (NO_3), ammonium (NH_4), and phosphate (PO_4). Later in this chapter you will read more about how C, H_2O, N, and P circulate in our environment.

2.3 ENERGY AND LIVING SYSTEMS

- *Energy is transformed, but not created or destroyed.*
- *In every energy exchange, some energy is degraded to less useful forms.*
- *Primary producers capture energy.*

If matter is the material of which things are made, energy provides the force to hold structures together, tear them apart, and move them from one place to another. Fundamental principles of energy determine how all systems function. These principles also explain the movement of energy that allows all organisms and ecosystems to exist.

Energy occurs in different types and qualities

Energy is the ability to do work, such as moving matter over a distance or causing a heat transfer between two objects at different temperatures. Energy can take many different forms. Heat, light, electricity, and chemical energy are examples that we all experience. The energy contained in moving objects is called **kinetic energy.** A rock rolling down a hill, the wind blowing through the trees, water flowing over a dam (fig. 2.11), and electrons speeding around the nucleus of an atom are all examples of kinetic energy. **Potential energy** is stored energy that is available for use. A rock poised at the top of a hill and water stored behind a dam are examples of potential energy. **Chemical energy** stored in the food you eat and the gasoline you put into your car are also examples of potential energy that can be released to do useful work. Energy is often measured in units of heat (calories) or work (joules). One joule (J) is the work done when 1 kilogram is accelerated at 1 meter per second per second. One calorie is the amount of energy needed to heat 1 gram of pure water 1 degree Celsius. A calorie can also be measured as 4.184 J.

Heat describes the energy that can be transferred between objects of different temperature. When a substance absorbs heat, the kinetic energy of its molecules increases, or it may change state: A solid may become a liquid, or a liquid may become a gas. We sense change in heat content as change in temperature (unless the substance changes state).

▲ **FIGURE 2.11** Water stored behind this dam represents potential energy. Water flowing over the dam has kinetic energy, some of which is converted to heat. ©William P. Cunningham

An object can have a high heat content but a low temperature, such as a lake that freezes slowly in the fall. Other objects, like a burning match, have a high temperature but little heat content. Heat storage in lakes and oceans is essential to moderating climates and maintaining biological communities. Heat absorbed in changing states is also critical. As you will read in chapter 9, the evaporation and condensation of water in the atmosphere help distribute heat around the globe.

Energy that is diffused, dispersed, and low in temperature is considered low-quality energy because it is difficult to gather and use for productive purposes. The heat stored in the oceans, for instance, is immense but hard to capture and use, so it is low–quality. Conversely, energy that is intense, concentrated, and high in temperature is high-quality energy because of its usefulness in carrying out work. The intense flames of a very hot fire and high-voltage electrical energy are examples of high-quality forms that are easy to use. Many of our alternative energy sources (such as wind) are diffuse compared to the higher-quality, more concentrated chemical energy in oil, coal, or gas.

Thermodynamics describes the conservation and degradation of energy

Atoms and molecules cycle endlessly through organisms and their environment, but energy flows in a one-way path. A constant supply of energy—nearly all of it from the sun—is needed to keep biological processes running. Energy can be used repeatedly as it flows through the system, and it can be stored temporarily in the chemical bonds of organic molecules, but eventually it is released and dissipated.

The study of thermodynamics deals with how energy is transferred in natural processes. More specifically, it deals with the rates of flow and the transformation of energy from one form or quality to another. Thermodynamics is a complex, quantitative discipline, but you don't need a great deal of math to understand some of the broad principles that shape our world and our lives.

The **first law of thermodynamics** states that energy is *conserved;* that is, it is neither created nor destroyed under normal conditions. Energy may be transformed—for example, from the energy in a chemical bond to heat energy—but the total amount does not change.

The **second law of thermodynamics** states that with each successive energy transfer or transformation in a system, less energy is available to do work. That is, as energy is used, it is degraded to lower-quality forms, or it dissipates and is lost. When you drive a car, for example, the chemical energy of the gas is degraded to kinetic energy and heat, which dissipates, eventually, to space. The second law recognizes that disorder, or **entropy,** tends to increase in all natural systems. Consequently, there is always less *useful* energy available when you finish a process than there was before you started. Because of this loss, everything in the universe tends to fall apart, slow down, and get more disorganized.

How does the second law of thermodynamics apply to organisms and biological systems? Organisms are highly organized, both structurally and metabolically. Constant care and maintenance are required to keep up this organization, and a continual supply of energy is required to maintain these processes. Every time a cell uses some energy to do work, some of that energy is dissipated or lost as heat. If cellular energy supplies are interrupted or depleted, the result—sooner or later—is death.

Organisms live by capturing energy

Where does the energy needed by living organisms come from? How is it captured and transferred among organisms? For nearly all life on earth, the sun is the ultimate energy source, and the sun's energy is captured by green plants. Green plants are often called **primary producers** because they create carbohydrates and other compounds using just sunlight, air, and water.

There are organisms that get energy in other ways, and these are interesting because they are exceptions to the normal rule. Deep in the earth's crust, deep on the ocean floor, and in hot springs, such as those in Yellowstone National Park, we can find extremophiles, organisms that gain their energy from **chemosynthesis,** or the extraction of energy from inorganic chemical compounds, such as hydrogen sulfide (H_2S). Until 30 years ago we knew almost nothing about these organisms and their ecosystems. Recent deep-sea exploration has shown that an abundance of astonishingly varied life occurs hundreds of meters deep on the ocean floor. These ecosystems cluster around thermal vents. Thermal vents are cracks where boiling-hot water, heated by magma in the earth's crust, escapes from the ocean floor. Here microorganisms grow by oxidizing hydrogen sulfide; bacteria support an ecosystem that includes blind shrimp, giant tube worms, hairy crabs, strange clams, and other unusual organisms (fig. 2.12).

These fascinating systems are exciting and mysterious because we have discovered them so recently. They are also interesting because of their contrast to the incredible profusion of photosynthesis-based life we enjoy here at the earth's surface.

▲ **FIGURE 2.12** A colony of tube worms and mussels clusters over a cool, deep-sea methane seep in the Gulf of Mexico. Source: NOAA

Green plants get energy from the sun

Our sun is a star, a fiery ball of exploding hydrogen gas. Its thermonuclear reactions emit powerful forms of radiation, including potentially deadly ultraviolet and nuclear radiation (fig. 2.13), yet life here is nurtured by, and dependent upon, this searing energy source.

Solar energy is essential to life for two main reasons. First, the sun provides warmth. Most organisms survive within a relatively narrow temperature range. In fact, each species has its own range of temperatures within which it can function normally. At high temperatures (above 40°C), most biomolecules begin to break down or become distorted and nonfunctional. At low temperatures (near 0°C), some chemical reactions of metabolism occur too slowly to enable organisms to grow and reproduce. Other planets in our solar system are either too hot or too cold to support life as we know it. The earth's water and atmosphere help to moderate, maintain, and distribute the sun's heat.

Second, nearly all organisms on the earth's surface depend on solar radiation for life-sustaining energy, which is

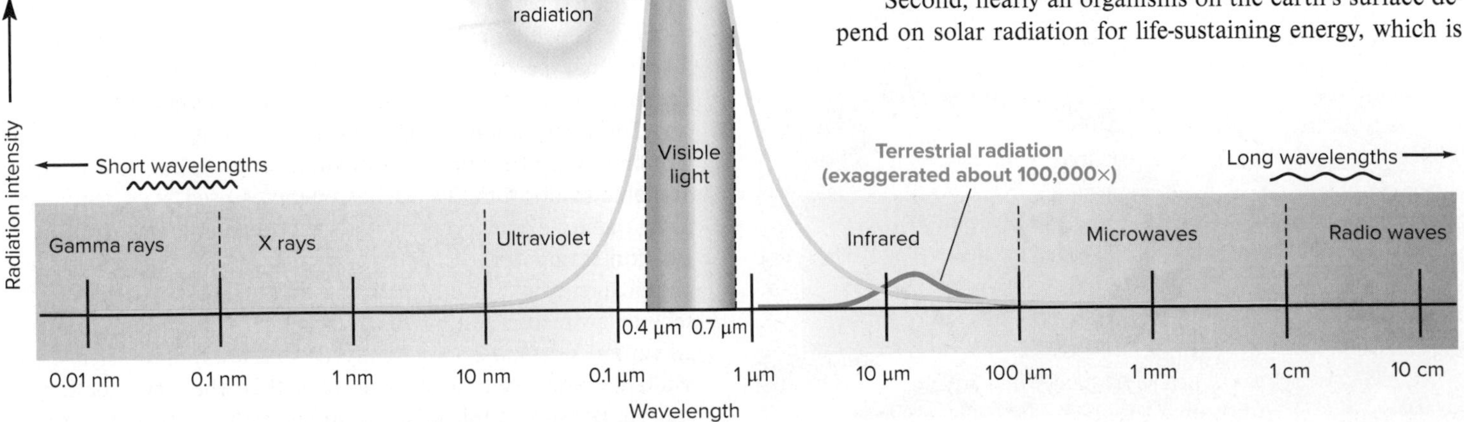

▲ **FIGURE 2.13** The electromagnetic spectrum. Our eyes are sensitive to visible-light wavelengths, which make up nearly half the energy that reaches the earth's surface (represented by the area under the "solar radiation" curve). Photosynthesizing plants use the most abundant solar wavelengths (light and infrared). The earth reemits lower-energy, longer wavelengths (shown by the "terrestrial radiation" curve), mainly the infrared part of the spectrum.

captured by green plants, algae, and some bacteria in a process called **photosynthesis.** Photosynthesis converts radiant energy into useful, high-quality chemical energy in the bonds that hold together organic molecules.

How much of the available solar energy is actually used by organisms? The amount of incoming solar radiation is enormous, about 1,372 watts/m^2 at the top of the atmosphere (imagine thirteen 100-watt lightbulbs on every square meter of your ceiling). However, more than half of the incoming sunlight is reflected or absorbed by atmospheric clouds, dust, and gases. In particular, harmful, short wavelengths are filtered out by gases (such as ozone) in the upper atmosphere; thus, the atmosphere is a valuable shield, protecting life-forms from harmful doses of ultraviolet and other forms of radiation. Even with these energy reductions, however, the sun provides much more energy than biological systems can harness, and more than enough for all our energy needs if technology could enable us to tap it efficiently.

Of the solar radiation that does reach the earth's surface, about 10 percent is ultraviolet, 45 percent is visible, and 45 percent is infrared. Most of that energy is absorbed by land or water or is reflected into space by water, snow, and land surfaces. (Seen from outer space, the earth shines about as brightly as Venus.)

Of the energy that reaches the earth's surface, photosynthesis uses only certain wavelengths, mainly red and blue light. Most plants reflect green wavelengths, so that is the color they appear to us. Half of the energy that plants absorb is used in evaporating water. In the end, only 1 to 2 percent of the sunlight falling on plants is available for photosynthesis. This small percentage represents the energy base for virtually all life in the biosphere!

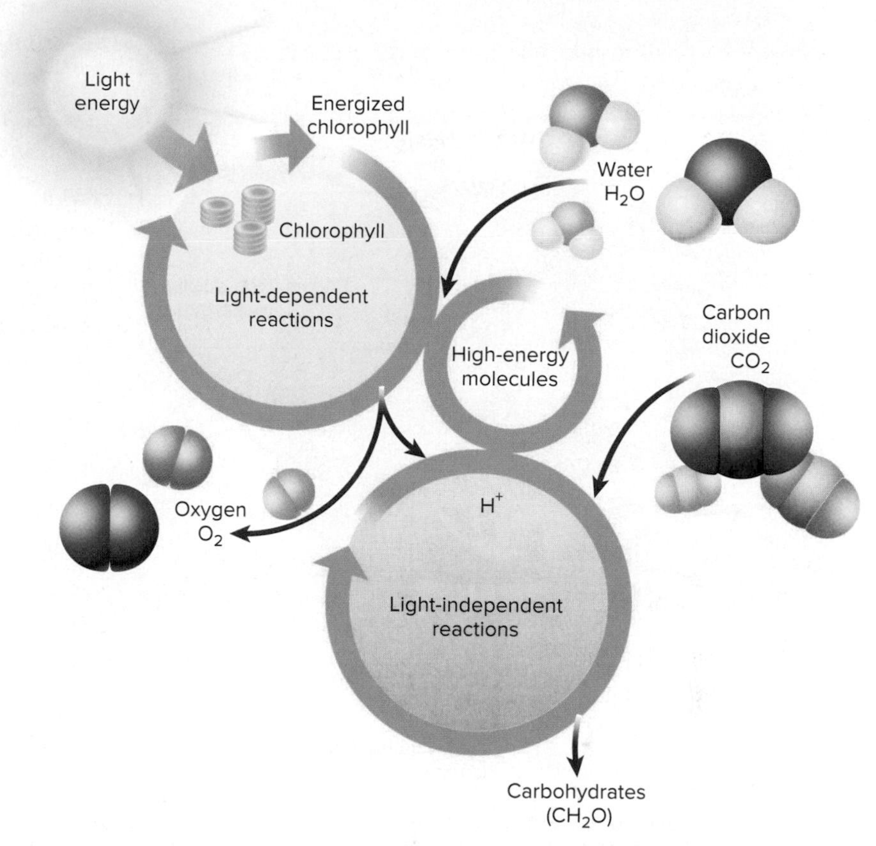

▲ **FIGURE 2.14** Photosynthesis involves a series of reactions in which chlorophyll captures light energy and forms high-energy molecules, ATP and NADPH. Light-independent reactions then use energy from ATP and NADPH to fix carbon (from air) in organic molecules.

How does photosynthesis capture energy?

Photosynthesis occurs in tiny organelles called chloroplasts that reside within plant cells (see fig. 2.10). The most important key to this process is chlorophyll, a unique green molecule that can absorb light energy and use it to create high-energy chemical bonds in compounds that serve as the fuel for all subsequent cellular metabolism. Chlorophyll doesn't do this important job all alone, however. It is assisted by a large group of other lipid, sugar, protein, and nucleotide molecules. Together these components carry out two interconnected, cyclic sets of reactions (fig. 2.14).

Photosynthesis begins with a series of steps called light-dependent reactions: These occur only while the chloroplast is receiving light. Enzymes split water molecules and release molecular oxygen (O_2). This is the source of nearly all the oxygen in the atmosphere on which all animals, including you, depend for life. The light-dependent reactions also create mobile, high-energy molecules (adenosine triphosphate, or ATP, and nicotinamide adenine dinucleotide phosphate, or NADPH), which provide energy for the next set of processes, the light-independent reactions. As their name implies, these reactions do not use light directly. Here enzymes extract energy from ATP and NADPH to add carbon atoms (from carbon dioxide) to simple sugar molecules, such as glucose. These molecules provide the building blocks for larger, more complex organic molecules.

In most temperate-zone plants, photosynthesis can be summarized in the following equation:

$$6H_2O + 6CO_2 + \text{solar energy} \xrightarrow[\text{chlorophyll}]{} C_6H_{12}O_6\ \text{(sugar)} + 6O_2$$

We read this equation as "water plus carbon dioxide plus energy produces sugar plus oxygen." The reason the equation uses six water and six carbon dioxide molecules is that it takes six carbon atoms to make the sugar product. If you look closely, you will see that all the atoms in the reactants balance with those in the products. This is an example of conservation of matter.

You might wonder how making a simple sugar benefits the plant. The answer is that glucose is an energy-rich compound that serves as the central, primary fuel for all metabolic processes of cells. The energy in its chemical bonds—the ones created by photosynthesis—can be released by other enzymes and used to make other molecules (lipids, proteins, nucleic acids, or other carbohydrates), or it can drive kinetic processes such as movement of ions across membranes, transmission of messages, changes in cellular shape or structure, or movement of the cell itself in some

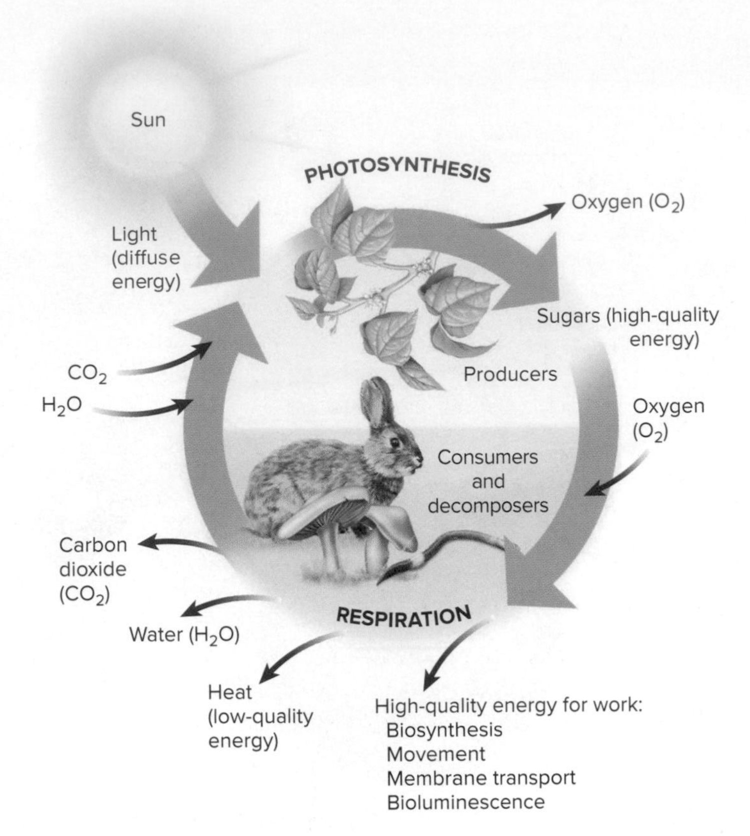

▲ **FIGURE 2.15** Energy exchange in ecosystems. Plants use sunlight, water, and carbon dioxide to produce sugars and other organic molecules. Consumers use oxygen and break down sugars during cellular respiration. Plants also carry out respiration, but during the day, if light, water, and CO_2 are available, they have a net production of O_2 and carbohydrates.

cases. This process of releasing chemical energy, called **cellular respiration,** involves splitting carbon and hydrogen atoms from the sugar molecule and recombining them with oxygen to re-create carbon dioxide and water. The net chemical reaction, then, is the reverse of photosynthesis:

$$C_6H_{12}O_6 + 6O_2 \longrightarrow 6H_2O + 6CO_2 + \text{released energy}$$

Note that in photosynthesis, energy is *captured,* whereas in respiration, energy is *released*. Similarly, photosynthesis *uses* water and carbon dioxide to *produce* sugar and oxygen, whereas respiration does just the opposite. In both sets of reactions, energy is stored temporarily in chemical bonds, which constitute a kind of energy currency for the cell. Plants carry out both photosynthesis and respiration, but during the day, if light, water, and CO_2 are available, they have a net production of O_2 and carbohydrates.

We animals don't have chlorophyll and can't carry out photosynthetic food production. We do have the components for cellular respiration, however. In fact, this is how we get all our energy for life. We eat plants—or other animals that have eaten plants—and break down the organic molecules in our food to obtain energy (fig. 2.15). Later in this chapter we'll see how these feeding relationships work.

2.4 FROM SPECIES TO ECOSYSTEMS

- *Different trophic levels—producers and consumers—make up a food web.*
- *Ecological pyramids reflect the many producers and few consumers in an ecosystem.*

While many biologists study life at the cellular and molecular levels, ecologists study interactions at the species, population, biotic community, or ecosystem level. In Latin, *species* literally means kind. In biology, **species** refers to all organisms of the same kind that are genetically similar enough to breed in nature and produce live, fertile offspring. There are several qualifications and some important exceptions to this definition of species (especially among bacteria and plants), but for our purposes this is a useful working definition.

Organisms occur in populations, communities, and ecosystems

A **population** consists of all the members of a species living in a given area at the same time. Chapter 4 deals further with population growth and dynamics. All of the populations of organisms living and interacting in a particular area make up a **biological community.** What populations make up the biological community of which you are a part? The population sign marking your city limits announces only the number of humans who live there, disregarding the other populations of animals, plants, fungi, and microorganisms that are part of the biological community within the city's boundaries. Characteristics of biological communities are discussed in more detail in chapter 3.

An ecological system, or **ecosystem,** is composed of a biological community and its physical environment. The Gulf ecosystem, for example, is a complex community of different species that rely on water, sunlight, and nutrients from the surrounding environment. It is useful to think about the biological community and its environment together, because energy and matter flow through both. Understanding how those flows work is a major theme in ecology.

Food chains, food webs, and trophic levels define species relationships

Photosynthesis (and rarely chemosynthesis) is the base of all ecosystems. Organisms that produce organic material by photosynthesis, mainly green plants and algae, are therefore known as **producers.** One of the most important properties of an ecosystem is its **productivity,** the amount of **biomass** (biological material) produced in a given area during a given period of time. Photosynthesis is described as *primary productivity* because it is the basis for almost all other growth in an ecosystem. A given ecosystem may have very high total productivity, but if decomposers consume organic material as rapidly as it is formed, the *net primary productivity* will be low.

In ecosystems, some consumers feed on a single species, but most consumers have multiple food sources. Similarly, some species are prey to a single kind of predator, but many species in an ecosystem are beset by several types of predators and parasites. In this way, individual food chains become interconnected to form a **food web**. Figure 2.16 shows feeding relationships among some of the larger organisms in an African savanna. If we were to add all the insects, worms, and microscopic organisms that belong in this picture, however, we would have overwhelming complexity. Perhaps you can imagine the challenge ecologists face in trying to quantify and interpret the precise matter and energy transfers that occur in a natural ecosystem!

An organism's feeding status in an ecosystem can be expressed as its **trophic level** (from the Greek *trophe*, food). In a savanna, grasses and trees are the primary producers (see bottom level of fig. 2.16). We call them autotrophs because they feed themselves using only sunlight, water, carbon dioxide, and minerals. Other organisms in the ecosystem are **consumers** of the chemical energy harnessed by the producers. An organism that eats primary producers is a primary consumer. Organisms that eat primary consumers are secondary consumers, which may in turn be eaten by a tertiary consumer, and so on. The highest trophic level is called the top predator. The complexity of a food chain depends on both the number of species available and the physical characteristics of a particular ecosystem. A harsh arctic landscape generally has a much simpler food chain than a temperate or tropical one.

Active LEARNING

Food Webs

To what food webs do you belong? Make a list of what you have eaten today, and trace the energy it contained back to its photosynthetic source. Are you at the same trophic level in all the food webs in which you participate? Are there ways that you could change your ecological role? Might that make more food available for other people? Why or why not?

Organisms can be identified both by the trophic level at which they feed and by the *kinds* of food they eat. **Herbivores** are plant eaters, **carnivores** are flesh eaters, and **omnivores** eat both plant and animal matter.

One of the most important feeding categories is made up of the parasites, scavengers, and decomposers that remove and recycle the dead bodies and waste products of others. Like omnivores, these recyclers feed on all the trophic levels. **Scavengers,** such as jackals and vultures, clean up dead carcasses of larger animals. **Detritivores,** such as ants and beetles, consume litter, debris, and dung,

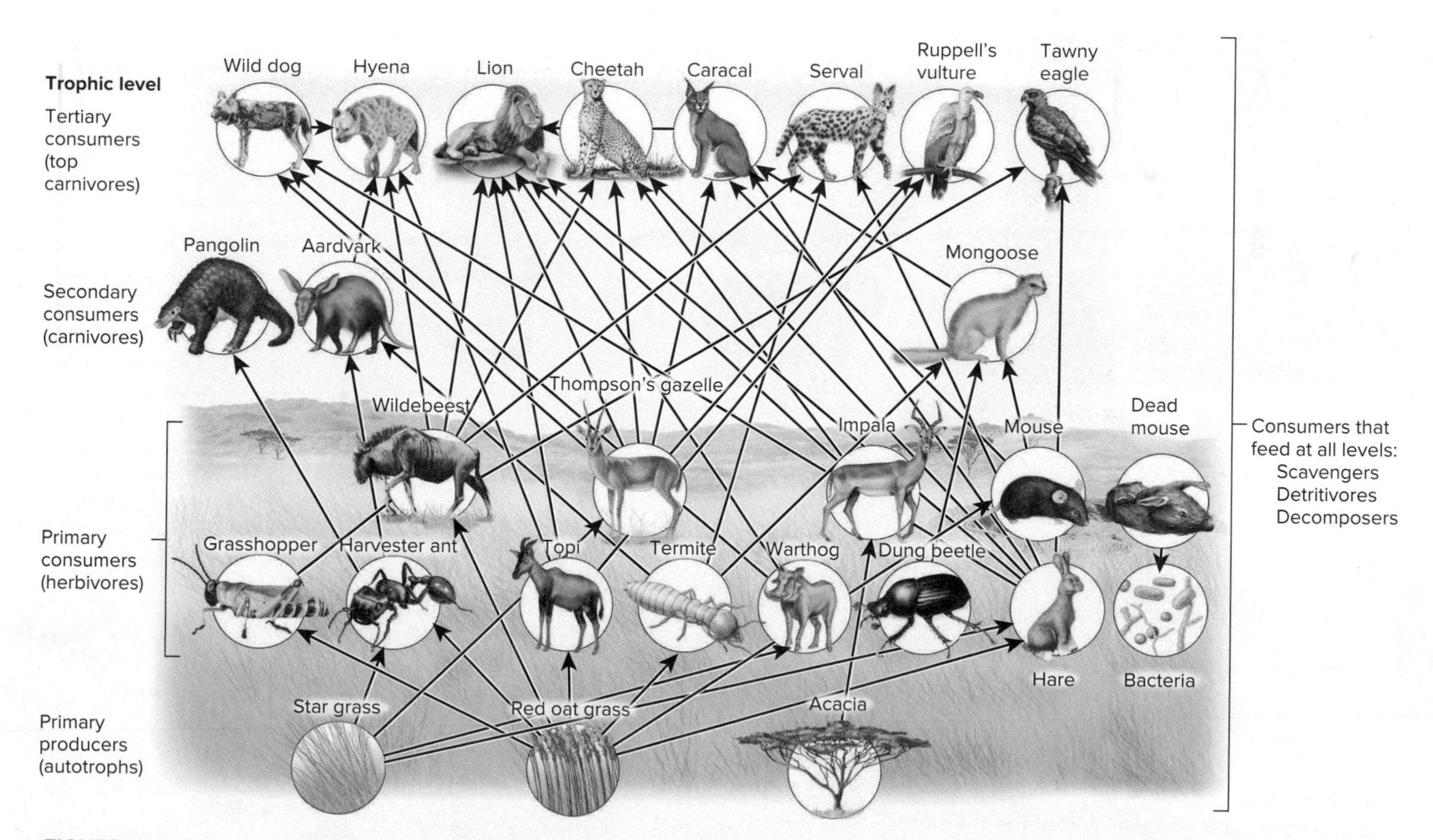

▲ **FIGURE 2.16** Each time an organism feeds, it becomes a link in a food chain. In an ecosystem, food chains become interconnected when predators feed on more than one kind of prey, thus forming a food web. The arrows in this diagram indicate the directions in which matter and energy are transferred through feeding relationships.

How do energy and matter move through systems?

Movement of energy and matter unites the parts of a system. In the Gulf of Mexico (opening case study), movement of water and nutrients supports photosynthesis, which supports the ecosystem. But excess nutrients can trigger so much plant and bacterial growth that the system becomes unstable.

For ecosystems in general, it is helpful to group organisms by **trophic levels** (feeding levels). In general, *primary producers* (organisms that produce organic matter, mainly green plants) are consumed by *herbivores* (plant eaters), which are consumed by *primary carnivores* (meat eaters), which are consumed by *secondary carnivores. Decomposers* consume at all levels and provide energy and matter to producers.

KC 2.1

Why do we find a pyramid of biomass?

Each trophic level requires a great deal of biomass at lower levels because energy is lost through growth, heat, respiration, and movement. This inefficiency is consistent with the second principle of thermodynamics, that energy dissipates and degrades to lower levels as it moves through a system.

A **general rule of thumb** is that only about 10 percent of the energy in one trophic level is represented in the next higher level. For example, it takes roughly 100 kg of clover to make 10 kg of rabbit, and 10 kg of rabbit to make 1 kg of fox.

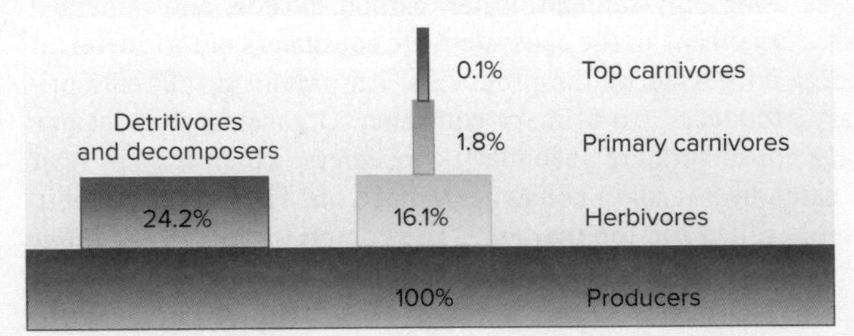

▲ In this example, numbers show the percentage of energy that is incorporated into biomass at the next level. Here decomposers are grouped with producers.

©TTphoto/Shutterstock

Why is there so much less energy in each successive trophic level?

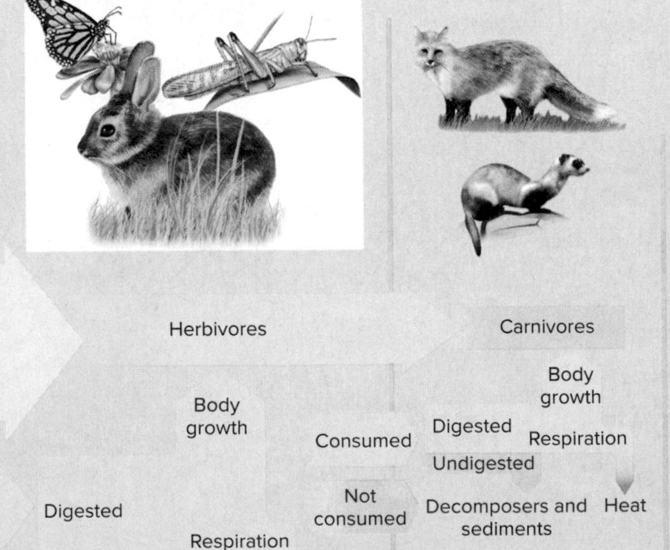

KC 2.2

2. Some chemical energy (food) is converted to movement (kinetic energy) or to heat energy, which dissipates to the environment. Energy used in growth—for example, in accumulation of muscle tissue—remains available for consumption at the next level.

1. Some of the food that organisms eat is undigested and doesn't provide usable energy.

©TTphoto/Shutterstock

What happens if the pyramid is disrupted?

Ecosystems undergo many types of disturbances and disruptions. Often ecosystems recover in time; sometimes they shift to a new type of system structure. Forest fire is a disturbance that eliminates primary production for a short time. Fire also accelerates movement of nutrients through the system, so that nutrients once locked up in standing trees become available to support a burst of new growth.

Removal of other trophic levels also disturbs an ecosystem. If there are too many predators, prey species will decline or disappear. An overabundance of foxes, for example, may eliminate the rabbit population. With too few rabbits, the foxes may die off, or they may find alternate prey, which can further destabilize the system.

On the other hand, removal of a higher trophic level can also destabilize a system: If foxes were removed, rabbits could become overabundant and overgraze the primary producers (plants).

Sometimes a pyramid can be temporarily inverted. The biomass pyramid, for instance, can be inverted by periodic fluctuations in producer populations. For example, low plant and algal biomass is present during winter in temperate aquatic ecosystems.

©Larry Mayer/Getty Images KC 2.3

KC 2.4

Grassland in summer ©Weldon Schloneger/Shutterstock

By the numbers

▲ We often think of a pyramid in terms of the number of organisms, rather than amount of biomass in each level. The pyramid is a general model. In this pyramid, many smaller organisms support one organism at the next trophic level. So 1,000 m^2 of grassland might contain 1,500,000 producers (plants), which support 200,000 herbivores, which support 90,000 primary carnivores, which support 1 top carnivore.

Don't forget the little things.

A single gram of soil can contain hundreds of millions of bacteria, algae, fungi, and insects.

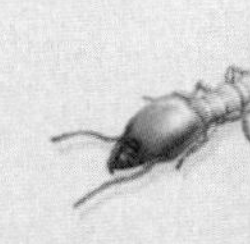

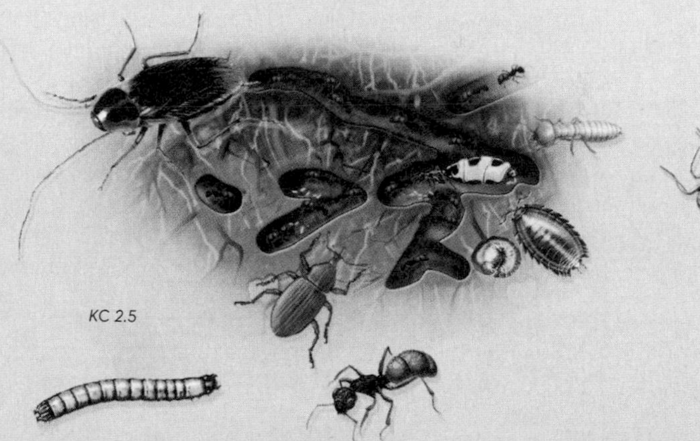

KC 2.5

CAN YOU EXPLAIN?

1. **How many trophic levels do you eat? Is your food pyramid large or small?**
2. **Does your trophic level matter in terms of the structure and stability of the ecosystems you occupy?**
3. **Explain the food pyramid in terms of the two principles of thermodynamics.**

EXPLORING Science — *Who Cares About Krill?*

Krill are small, shrimplike animals of immense importance in marine ecology. They occur in all the oceans of the world, but they are especially abundant in the Southern Ocean around Antarctica. Krill are considered one of the most abundant species in the world. Although each individual is tiny, their combined biomass is estimated to exceed that of any other species on the planet. The supply has seemed inexhaustible, but now marine biologists are concerned that a combination of overfishing and climate change could decimate the krill population—and the entire Antarctic ecosystem.

▲ Small, shrimplike animals called krill may be the most abundant organisms in the world. They are the most important prey species in the Antarctic marine ecosystem. How should we regulate harvest of this resource? ©pilipenkoD/Getty Images

Krill flourish in the cold, nutrient-rich waters of Antarctica, where they subsist on photosynthetic algae under and around the ice. Directly or indirectly, they support all higher trophic levels, including penguins, seals, fish, squid, seabirds, and whales. A single blue whale can eat as much as 4 tons of krill per day. But Antarctic krill are also increasingly sought as a source of protein and oil for farmed seafood, poultry, pets, and livestock. Krill are rich in protein (40 percent or more) and lipids (up to 20 percent). You probably won't find krill on your menu, but their high levels of omega-3 fatty acids make them a popular source of "fish oil" supplements and health products. And if you eat farmed salmon, there is a good chance it was fed with krill or krill-consuming fish. That would make you a top consumer in a krill-based ecosystem.

Between 1970 and 1990 the world catch of Antarctic krill grew from almost nothing to as much as half a million tons per year. The Soviet Union was the first nation to establish large-scale commercial krill harvests. After the demise of the USSR in 1991 the Soviet fishery collapsed, but other nations, notably Norway, South Korea, and China, have developed extensive krill fisheries. The Commission for the Conservation of Antarctic Marine Living Resources (CCAMLR) was created in 1982 in an effort to manage this obscure fishery and to prevent overfishing.

The proposed CCAMLR catch limit for the Scotia Sea (around Antarctica) is 5.6 million tons per year, but biologists fear that even the current annual harvest of nearly 300,000 tons is unsustainable. They estimate that krill populations have dropped 80 percent since the 1970s.

Krill are also threatened by the loss of floating ice shelves, the habitat for algae on which they feed. As the climate warms, ice has been declining, even while heavy fishing is depleting krill populations. At the same time, the ocean is becoming more acidic, as it absorbs CO_2 from the atmosphere, and acid conditions are likely to diminish krill reproduction. As krill have disappeared, other species far up the food web, such as Emperor and Adelie penguins, have shown corresponding declines.

Remarkably, this species is finally gaining legal protection with the establishment of the world's largest marine protected area. Following years of impasses, a 2016 agreement protected 1.5 million km^2 of the Ross Sea, including 1.1 million km^2 where fishing is banned altogether. The agreement allows the same levels of fishing as previously, but it bans fishing in critical habitat near the Antarctic continent, where krill reproduce, along with many of the fish and marine mammal species they support. The preserve is temporary, though, only 35 years. Russia, China, and other fishing interests would not budge on longer protections.

All food webs rest on a foundation of small, seemingly insignificant species. Especially near the poles, there are often few species in vast, teeming numbers. But these small organisms are vitally important. As the eminent naturalist E. O. Wilson said, "the little things rule the world."

---

while **decomposer** organisms, such as fungi and bacteria, complete the final breakdown and recycling of organic materials. It could be argued that these cleanup organisms are second in importance only to producers, because without their activity, nutrients would remain locked up in the organic compounds of dead organisms and discarded body wastes, rather than being made available to successive generations of organisms.

Ecological pyramids describe trophic levels

If we consider organisms according to trophic levels, they often form a pyramid, with a broad base of primary producers and only a few individuals in the highest trophic levels. Top predators are generally large, fierce animals, such as wolves, bears, sharks, and big cats. It usually takes a huge number of organisms at lower trophic levels (and thus a very large territory) to support a few top carnivores. While there is endless variation in the organization of ecosystems, the pyramid idea helps us describe generally how energy and matter move through ecosystems (Key Concepts, pp. 42–43).

We tend to focus our attention on species at the top of the food web (especially those things that might eat us), but sometimes the most important species are at or near the bottom of the pyramid (Exploring Science, above).

2.5 BIOGEOCHEMICAL CYCLES AND LIFE PROCESSES

- *Key elements cycle continuously through living and nonliving systems.*
- *Carbon cycles through all living organisms, the atmosphere, and the oceans.*
- *Nitrogen, key to life, is abundant but not easily captured.*

The elements and compounds that sustain us are cycled endlessly through living things and through the environment. As the great naturalist John Muir said, "When one tugs at a single thing in nature, he finds it attached to the rest of the world." On a global scale, this movement is referred to as biogeochemical cycling. Substances can move quickly or slowly: Carbon might reside in a plant for days or weeks, in the atmosphere for days or months, or in your body for hours, days, or years. The earth stores carbon (in coal or oil, for example) for millions of years. When human activities increase flow rates or reduce storage time, these materials can become pollutants. Here we will explore some of the pathways involved in cycling several important substances: water, carbon, nitrogen, sulfur, and phosphorus.

The hydrologic cycle

The path of water through our environment is probably the most familiar material cycle, and it is discussed in greater detail in chapter 10. Most of the earth's water is stored in the oceans, but solar energy continually evaporates this water, and winds distribute water vapor around the globe (fig. 2.17). Water that condenses over land surfaces, in the form of rain, snow, or fog, supports all terrestrial (land-based) ecosystems. Living organisms emit the moisture they have consumed through respiration and perspiration. Eventually, this moisture reenters the atmosphere or enters lakes and streams, from which it ultimately returns to the ocean again.

As it moves through living things and through the atmosphere, water is responsible for metabolic processes within cells, for maintaining the flows of key nutrients through ecosystems, and for global-scale distribution of heat and energy (see chapter 9). Water performs countless services because of its unusual properties.

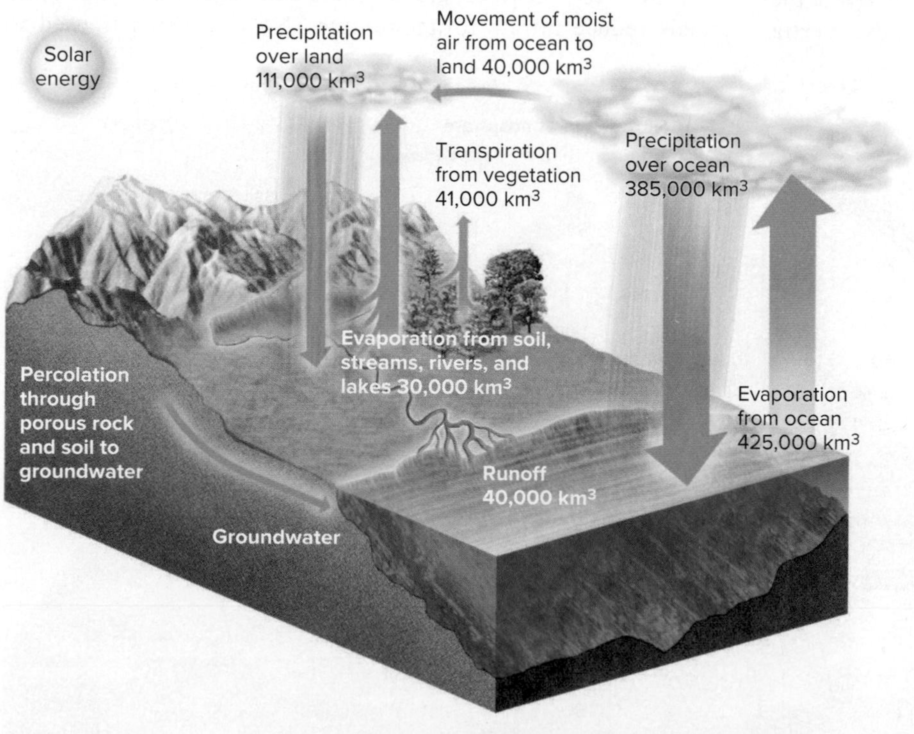

▲ **FIGURE 2.17** The hydrologic cycle. Most exchange occurs with evaporation from oceans and precipitation back to oceans. About one-tenth of water evaporated from oceans falls over land, is recycled through terrestrial systems, and eventually drains back to oceans via rivers.

The carbon cycle

Carbon serves a dual purpose for organisms: (1) It is a structural component of organic molecules, and (2) chemical bonds in carbon compounds provide metabolic energy. The **carbon cycle** begins with photosynthetic organisms taking up carbon dioxide (CO_2) (fig. 2.18). This is called carbon fixation, because carbon is changed from gaseous CO_2 to less mobile organic molecules. Once a carbon atom is incorporated into organic compounds, its path to recycling may be very quick or extremely slow. Imagine what happens to a simple sugar molecule you swallow in a glass of fruit juice. The sugar molecule is absorbed into your bloodstream, where it is made available to your cells for cellular respiration or the production of more complex biomolecules. If it is used in respiration, you may exhale the same carbon atom as CO_2 in an hour or less, and a plant could take up that exhaled CO_2 the same afternoon.

Alternatively, your body may use that sugar molecule to make larger organic molecules that become part of your cellular structure. The carbon atoms in the sugar molecule could remain a part of your body until it decays after death. Similarly, carbon in the wood of a thousand-year-old tree will be released only when fungi and bacteria digest the wood and release carbon dioxide as a by-product of their respiration.

Sometimes recycling takes a very long time. Coal and oil are the compressed, chemically altered remains of plants and microorganisms that lived millions of years ago. Their carbon atoms (and hydrogen, oxygen, nitrogen, sulfur, etc.) are not released until the coal and oil are burned. Enormous amounts of carbon also are locked up as calcium carbonate ($CaCO_3$) in the shells and skeletons of marine organisms, from tiny protozoans to corals. The world's extensive surface limestone deposits are biologically formed calcium carbonate from ancient oceans, exposed by geological events. The carbon in limestone has been locked away for millennia, which is probably the fate of carbon currently being deposited in ocean sediments. Eventually, even the deep-ocean deposits are recycled as they are drawn into deep molten layers and released via volcanic activity. Geologists estimate that every carbon atom on the earth has made about 30 such round-trips over the past 4 billion years.

atmospheric CO_2 could support faster plant growth, speeding some of the recycling processes.

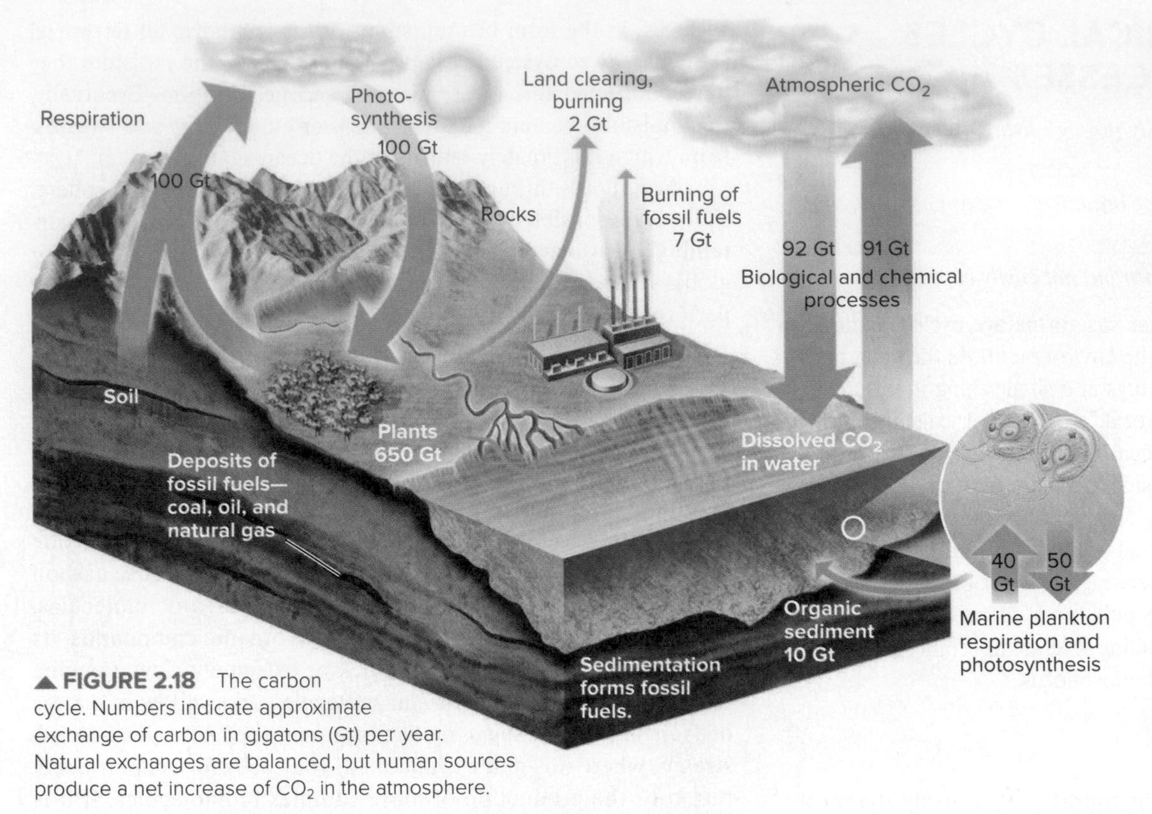

▲ **FIGURE 2.18** The carbon cycle. Numbers indicate approximate exchange of carbon in gigatons (Gt) per year. Natural exchanges are balanced, but human sources produce a net increase of CO_2 in the atmosphere.

Materials that store carbon, including geological formations and standing forests, are known as carbon sinks. When carbon is released from these sinks, as when we burn fossil fuels and inject CO_2 into the atmosphere, or when we clear extensive forests, natural recycling systems may not be able to keep up. This is the root of the global warming problem, discussed in chapter 9. Alternatively, extra atmospheric CO_2 could support faster plant growth, speeding some of the recycling processes.

The nitrogen cycle

Organisms cannot exist without amino acids, peptides, and proteins, all of which are organic molecules that contain nitrogen. Nitrogen is therefore an extremely important nutrient for living things. This is why nitrogen is a primary component of household and agricultural fertilizers. Nitrogen makes up about 78 percent of the air around us.

Plants cannot use N_2, the stable two-atom form most common in air. But bacteria can. So plants acquire nitrogen from nitrogen-fixing bacteria (including some blue-green algae or cyanobacteria) that live in and around their roots. These bacteria can "fix" nitrogen, or combine gaseous N_2 with hydrogen to make ammonia (NH_3) and ammonium (NH_4^+). Nitrogen fixing by bacteria is a key part of the **nitrogen cycle** (fig. 2.19).

Other bacteria then combine ammonia with oxygen to form nitrite (NO_2^-). Another group of bacteria converts nitrites to nitrate (NO_3^-), which green plants can absorb and use. Plant cells reduce nitrate to ammonium (NH_4^+), which is used to

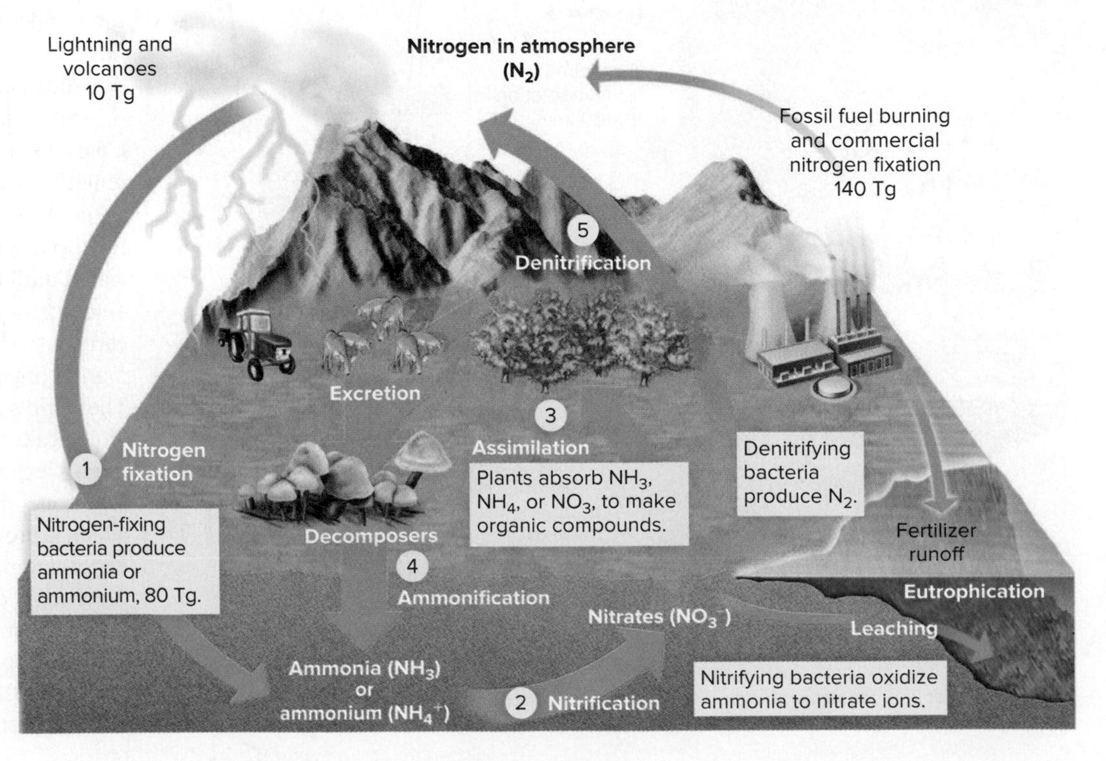

▶ **FIGURE 2.19** The nitrogen cycle. Human sources of nitrogen fixation (conversion of molecular nitrogen to ammonia or ammonium) are now about 50 percent greater than natural sources. Bacteria convert ammonia to nitrates, which plants use to create organic nitrogen. Eventually, nitrogen is stored in sediments or converted back to molecular nitrogen (1 Tg = 10^{12}g, or 0.001 Gt).

▲ **FIGURE 2.20** Nitrogen molecules (N_2) are converted to usable forms in the bumps (nodules) on the roots of this bean plant. Each nodule is a mass of root tissue containing many bacteria that help convert nitrogen in the soil to a form that the bean plant can assimilate and use to manufacture amino acids. ©Nigel Cattlin/Alamy Stock Photo

build amino acids that become the building blocks for peptides and proteins.

Members of the bean family (legumes) and a few other kinds of plants are especially useful in agriculture because nitrogen-fixing bacteria actually live in their root tissues (fig. 2.20). Legumes and their associated bacteria add nitrogen to the soil, so interplanting and rotating legumes with crops, such as corn, that use but cannot replace soil nitrates are beneficial farming practices that take practical advantage of this relationship.

Nitrogen reenters the environment in several ways. The most obvious path is through the death of organisms. Fungi and bacteria decompose dead organisms, releasing ammonia and ammonium ions, which then are available for nitrate formation. Organisms don't have to die to donate proteins to the environment, however. Plants shed their leaves, needles, flowers, fruits, and cones; animals shed hair, feathers, skin, exoskeletons, pupal cases, and silk. Animals also produce excrement and urinary wastes that contain nitrogenous compounds. Urine is especially high in nitrogen because it contains the detoxified wastes of protein metabolism. All of these by-products of living organisms decompose, replenishing soil fertility.

How does nitrogen reenter the atmosphere, completing the cycle? Denitrifying bacteria break down nitrates (NO_3^-) into N_2 and nitrous oxide (N_2O), gases that return to the atmosphere. Thus, denitrifying bacteria compete with plant roots for available nitrates. Denitrification occurs mainly in waterlogged soils that have low oxygen availability and a large amount of decomposable organic matter. These are suitable growing conditions for many wild plant species in swamps and marshes, but not for most cultivated crop species, except for rice, a domesticated wetland grass.

In recent years humans have profoundly altered the nitrogen cycle. By using synthetic fertilizers, cultivating nitrogen-fixing crops, and burning fossil fuels, we now convert more nitrogen to ammonia and nitrates through industrial reactions than all natural land processes combined. This excess nitrogen input causes algal blooms and excess plant growth in water bodies, called eutrophication, which we will discuss in more detail in chapter 10. Excess nitrogen also causes serious loss of soil nutrients such as calcium and potassium; acidification of rivers and lakes; and rising atmospheric concentrations of nitrous oxide, a greenhouse gas. It also encourages the spread of weeds into areas such as prairies, where native plants are adapted to nitrogen-poor environments.

Phosphorus eventually washes to the sea

Minerals become available to organisms after they are released from rocks or salts (which are ancient sea deposits). Two mineral cycles of particular significance to organisms are phosphorus and sulfur. At the cellular level, energy-rich, phosphorus-containing compounds are primary participants in energy-transfer reactions.

Phosphorus is usually transported in water. Producer organisms take in inorganic phosphorus, incorporate it into organic molecules, and then pass it on to consumers. In this way, phosphorus cycles through ecosystems (fig. 2.21).

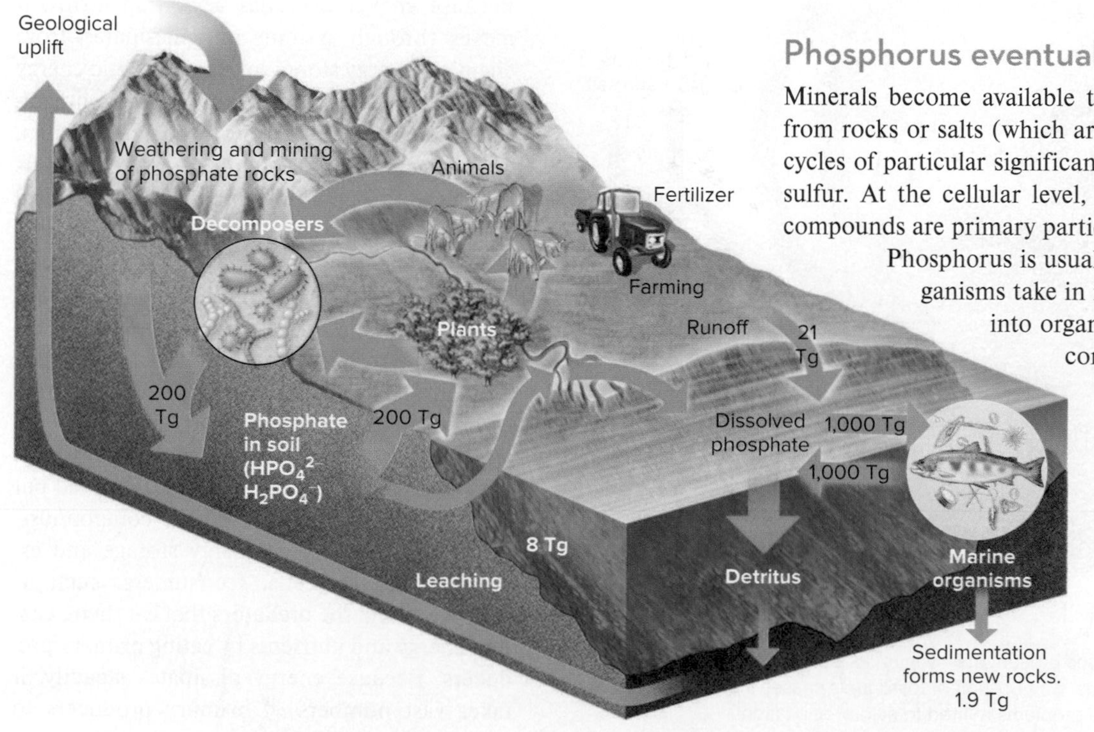

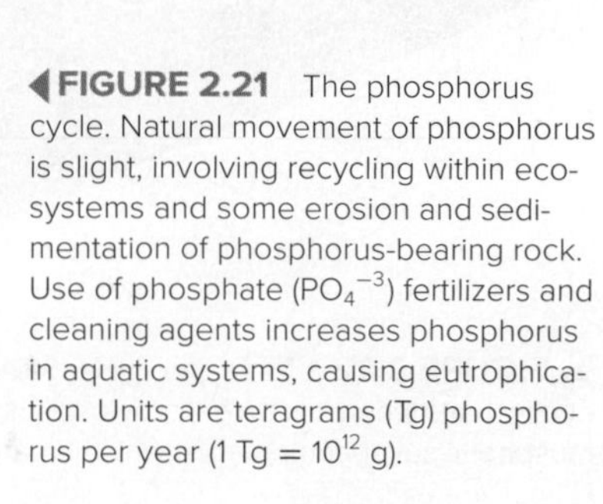

◀ **FIGURE 2.21** The phosphorus cycle. Natural movement of phosphorus is slight, involving recycling within ecosystems and some erosion and sedimentation of phosphorus-bearing rock. Use of phosphate (PO_4^{-3}) fertilizers and cleaning agents increases phosphorus in aquatic systems, causing eutrophication. Units are teragrams (Tg) phosphorus per year (1 Tg = 10^{12} g).

The release of phosphorus from rocks and mineral compounds is normally very slow, but mining of fertilizers has greatly speeded the use and movement of phosphorus in the environment. Most phosphate ores used for detergents and inorganic fertilizers come from salt deposits from ancient, shallow sea beds. Most of the phosphorus used in agriculture winds up in the ocean again, from field runoff or through human and animal waste that is released to rivers. Over millions of years this phosphorus will become part of mineral deposits, but on shorter timescales many earth scientists worry that we could use up our available sources of phosphorus, putting our agricultural systems at risk.

Like nitrogen, phosphorus is a critical limiting nutrient. This makes phosphorus an important water pollutant, when excessive amounts stimulate eutrophication. While nitrogen is a limiting factor in marine systems, like the Gulf of Mexico, phosphorus is more commonly a problem in freshwater bodies, such as Lake Erie.

The sulfur cycle

Sulfur plays a vital role in organisms, especially as a minor but essential component of proteins. Sulfur compounds are important determinants of the acidity of rainfall, surface water, and soil. In addition, sulfur in particles and tiny, airborne droplets may act as critical regulators of global climate. Most of the earth's sulfur is tied up underground in rocks and minerals, such as iron disulfide (pyrite) and calcium sulfate (gypsum). Weathering, emissions from deep seafloor vents, and volcanic eruptions release this inorganic sulfur into the air and water (fig. 2.22).

The sulfur cycle is complicated by the large number of oxidation states the element can assume, producing hydrogen sulfide (H_2S), sulfur dioxide (SO_2), sulfate ion (SO_4^{2-}), and others. Inorganic processes are responsible for many of these transformations, but living organisms, especially bacteria, also sequester sulfur in biogenic deposits or release it into the environment. Which of the several kinds of sulfur bacteria prevails in any given situation depends on oxygen concentrations, pH level, and light level.

Human activities also release large quantities of sulfur, primarily through burning fossil fuels. Total yearly anthropogenic sulfur emissions rival those of natural processes, and acid rain (caused by sulfuric acid produced as a result of fossil fuel use) is a serious problem in many areas (see chapter 9). Sulfur dioxide and sulfate aerosols cause human health problems, damage buildings and vegetation, and reduce visibility. They also absorb ultraviolet (UV) radiation and create cloud cover that cools cities and may be offsetting greenhouse effects of rising CO_2 concentrations.

Interestingly, the biogenic sulfur emissions of oceanic phytoplankton may play a role in global climate regulation. When ocean water is warm, tiny, single-celled organisms release dimethylsulfide (DMS), which is oxidized to SO_2 and then SO_4^{2-} in the atmosphere. Acting as cloud droplet condensation nuclei, these sulfate aerosols increase the earth's albedo (reflectivity) and cool the earth. As ocean temperatures drop because less sunlight gets through, phytoplankton activity decreases, DMS production falls, and clouds disappear. Thus, DMS, which may account for half of all biogenic sulfur emissions, is one of the feedback mechanisms that keep temperature within a suitable range for all life.

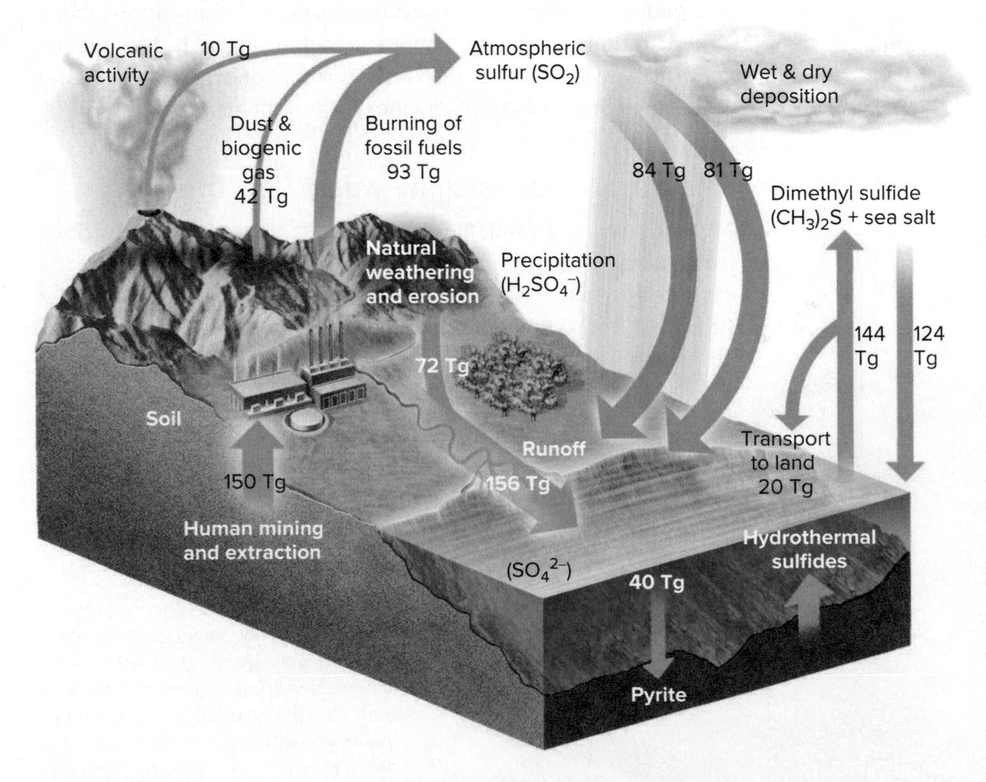

▲ **FIGURE 2.22** The sulfur cycle. Sulfur is present mainly in rocks, soil, and water. It cycles through ecosystems when it is taken in by organisms. Combustion of fossil fuels causes increased levels of atmospheric sulfur compounds, which create problems related to acid precipitation.

CONCLUSION

The movement and capture of matter and energy maintain the world's living environments. Because energy degrades as it transforms, it moves through systems and dissipates—from chemical energy stored in cells to kinetic energy and heat energy released to the environment. But matter recycles constantly in an ecosystem. The elements that make up cells and organisms also cycle through atmospheric and geological systems, in what we call biogeochemical (living-earth-chemical) cycles.

Primary producers capture the energy and matter that support an ecosystem. Nearly all ecosystems rely on green plants, which capture radiant solar energy and bind it into organic compounds, using carbon from the air and water and nutrients mainly from soil. Nutrients, such as nitrogen and phosphorus, are small but essential components of organic compounds—for example, aiding in energy storage and exchange between cells. Consumers, such as herbivores and the predators that eat them, capture energy and nutrients by eating primary producers. Because energy dissipates steadily, it takes vast numbers of primary producers to support a consumer. The food pyramid describes the rapidly diminishing number of organisms that can exist at each successive trophic level.

Principles of how energy and matter cycle through earth systems and ecosystems are the foundation of much of environmental science. These principles help us understand why the Gulf of Mexico is destabilized by influxes of nitrogen from farmlands in the Mississippi River basin; they also help us understand why teeming billions of tiny organisms, like krill, are so essential to all other organisms in an ecosystem. These principles also help us understand issues of population dynamics, water quality, and biodiversity, all topics we will explore in the chapters ahead.

PRACTICE QUIZ

1. What are the two most important nutrients causing eutrophication in the Gulf of Mexico?
2. What are systems, and how do feedback loops regulate them?
3. Your body contains vast numbers of carbon atoms. How is it possible that some of these carbons may have been part of the body of a prehistoric creature?
4. List six unique properties of water. Describe, briefly, how each of these properties makes water essential to life as we know it.
5. What is DNA, and why is it important?
6. The oceans store a vast amount of heat, but this huge reservoir of energy is of little use to humans. Explain the difference between high-quality and low-quality energy.
7. In the biosphere, matter follows circular pathways, while energy flows in a linear fashion. Explain.
8. To which wavelengths do our eyes respond, and why? (Refer to fig. 2.13.) About how long are short ultraviolet wavelengths compared to microwave lengths?
9. Where do extremophiles live? How do they get the energy they need for survival?
10. Ecosystems require energy to function. From where does most of this energy come? Where does it go?
11. How do green plants capture energy, and what do they do with it?
12. Define the terms *species*, *population*, and *biological community*.
13. Why are big, fierce animals rare?
14. Most ecosystems can be visualized as a pyramid with many organisms in the lowest trophic levels and only a few individuals at the top. Give an example of an inverted numbers pyramid.
15. What is the ratio of human-caused carbon releases into the atmosphere shown in figure 2.18 compared to the amount released by terrestrial respiration?

CRITICAL THINKING AND DISCUSSION

Apply the principles you have learned in this chapter to discuss these questions with other students.

1. Ecosystems are often defined as a matter of convenience because we can't study everything at once. How would you describe the characteristics and boundaries of the ecosystem in which you live? In what respects is your ecosystem an open one?
2. Think of some practical examples of increasing entropy in everyday life. Is a messy room really evidence of thermodynamics at work or merely personal preference?
3. Some chemical bonds are weak and have a very short half-life (fractions of a second, in some cases); others are strong and stable, lasting for years or even centuries. What would our world be like if all chemical bonds were either very weak or extremely strong?
4. If you had to design a research project to evaluate the relative biomass of producers and consumers in an ecosystem, what would you measure? (*Note:* This could be a natural system or a human-made one.)
5. Understanding storage compartments is essential to understanding material cycles, such as the carbon cycle. If you look around your backyard, how many carbon storage compartments are there? Which ones are the biggest? Which ones are the longest lasting?

DATA ANALYSIS: A Closer Look at Nitrogen Cycling

1. Which forms of N do plants take up? Can they capture N_2 from the air?
2. Refer to section 2.5. How is N_2 captured, or fixed, from the air into the food web?
3. Most of the processes are hard to quantify, but the figure shown here gives approximate amounts for fossil fuel burning and commercial N fixation, and for N fixing by bacteria. What do these terms mean? What is the magnitude of each? What is the difference?
4. If anthropogenic processes introduce increasing amounts of atmospheric N to the biosphere and hydrosphere, where does that N go? (*Hint:* Refer to the opening case study.)
5. Why is N so important for living organisms?
6. In marine systems, N is often a limiting factor. What is a "limiting factor"? What is a consequence of increasing the supply of N in a marine system?

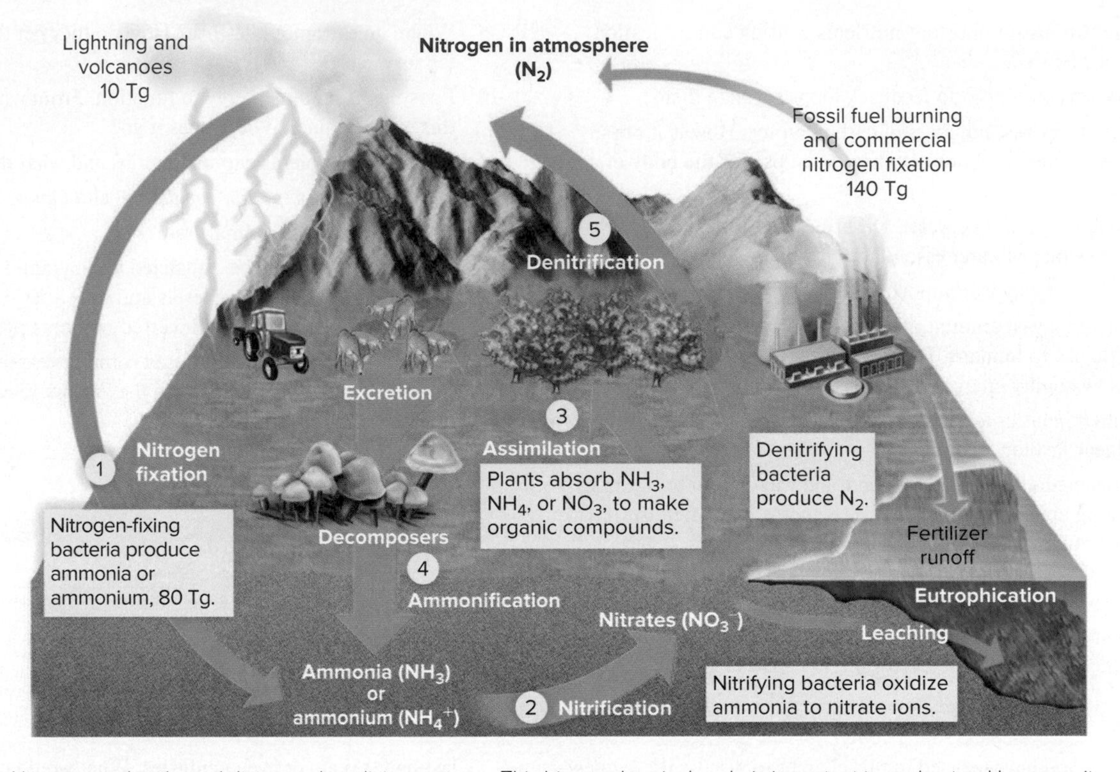

▲ Nitrogen cycles through living and nonliving systems. This biogeochemical cycle is important to understand because it strongly influences how ecosystems function.

Design Elements: Active Learning (Toad): ©Gaertner/Alamy Stock Photo; Case Study (Globe): ©McGraw-Hill Education; Google Earth: ©McGraw-Hill Education; Abstract Background: ©Martin Kubat/Shutterstock; What do you think (Students using tablets): ©McGraw-Hill Education/Richard Hutchings, photographer; What can you do (Hand holding Globe): ©Christoph Weihs/Shutterstock

CHAPTER

3 Evolution, Species Interactions, and Biological Communities

LEARNING OUTCOMES

After studying this chapter, you should be able to answer the following questions:

- How does species diversity arise?
- What do we mean by tolerance limits? Give examples.
- How do interactions both aid and hinder species?
- Why don't species always reproduce up to their biotic potential?
- What is the relationship between species diversity and community stability?
- What is disturbance, and how does it affect communities?
- Explain ecological succession, and give examples of its stages.

The Galápagos Islands have provided an accidental laboratory for examining biological diversity and species interactions.

CASE STUDY

Natural Selection and the Galápagos Finches

The Galápagos Islands are a small, archipelago of arid volcanic islands, isolated and remote—nearly 1,000 km from mainland South America. These small, rocky islands lack the profusion of life seen on many tropical islands, yet they are renowned as the place where Charles Darwin revolutionized our understanding of evolution, biodiversity, and biology in general. Why is this?

Charles Darwin (1809–1882) visited the islands in 1835 while serving as ship's naturalist on the *Beagle*. His job was to collect specimens and record observations for general interest. He found there a variety of unusual creatures, most occurring only in the Galápagos and some on just one or two islands. Giant land tortoises fed on tree-size cacti. Unique marine iguanas lived by grazing on algae scraped from rocky shoals. Sea birds were so unafraid of humans that Darwin could pick them off their nests. The islands also had a variety of small, brown finches that differed markedly in appearance, food preferences, and habitat. Most finches forage for small seeds, but in the Galápagos there were fruit eaters with thick, parrot-like beaks; seed eaters with heavy, crushing beaks; and insect eaters with thin, probing beaks to catch their prey. The woodpecker finch even pecked at tree bark for hidden insects. Lacking the woodpecker's long tongue, the finch used a cactus spine as a tool to extract insects.

Like other naturalists, Darwin was intrigued by the question of how such variety came to be. Most Europeans at the time believed that all living things had existed, unchanged, since a moment of divine creation, just a few thousand years ago. But Darwin had observed fossils of vanished creatures in South America. And he had read the new theories of geologist Charles Lyell (1797–1875), who argued that fossils showed that the world was much older than previously thought, and that species could undergo gradual but profound change over time. After his return to England, Darwin continued to ponder his Galápagos specimens. They seemed to be adapted to particular food resources on the different islands. It seemed likely that these birds were related, but somehow they had been modified to survive under different conditions.

Observing that dog breeders created new varieties of dogs, from Dachshunds to Great Danes, by selecting for certain traits, Darwin proposed that "natural selection" could explain the origins of species. Just as dog breeders favored individuals with particular characteristics, such as long legs or short noses, environmental conditions in an area could favor certain characteristics. On an island where only large seeds were available, finches with larger-than-average beaks could have more success in feeding—and reproducing—than smaller-beaked individuals. Thus, competition for limited food resources could explain the prevalence of particular traits in a population. Darwin derived this idea from Thomas Malthus, a minister whose *Essay on the Principle of Population* (1798) argued that growth of human populations is always held in check by food scarcity (together with war and disease). Only the best competitors are likely to survive, according to Malthus. Darwin proposed that individuals with traits suitable for their environment are the best competitors for scarce resources. As those individuals survive and reproduce, their traits eventually become common in the population.

The explanation of evolution by natural selection has been supported by 150 years of observations and experiments, and Darwin's theory has provided explanation for countless examples of species variations. In a now-classic study of competition, for example, ecologist David Lack carefully measured the sizes and shapes of finches' beaks on several islands, to see how beaks varied with resources. Lack showed not only that beaks vary with resources but that they specialize further in cases where multiple species compete for those resources. When two finch species, *Geospiza fuliginosa* and *Geospiza fortis*, occurred separately on the islands of Daphne Major and Los Hermanos, both had beaks of moderate depth (thickness). But on islands where the two species coexisted, beak sizes shifted to the extremes, a shift that minimized competition for food resources (fig. 3.1). Where three species coexisted, traits again shifted to minimize competition. Thus, competition among species had led to shifts in beak traits. Among Darwin's Galápagos finches, 13 modern species are now recognized, probably descended from a few seed-eating ancestors that blew to the islands from South America, where a similar species still exists.

Evolution of species through natural selection is now a cornerstone of biology and its many subfields, from ecology to medicine and health care. Subsequent discoveries have filled in many details. The discovery of DNA in the 1950s, in particular, allowed us to understand how random mutations (changes) in genes can account

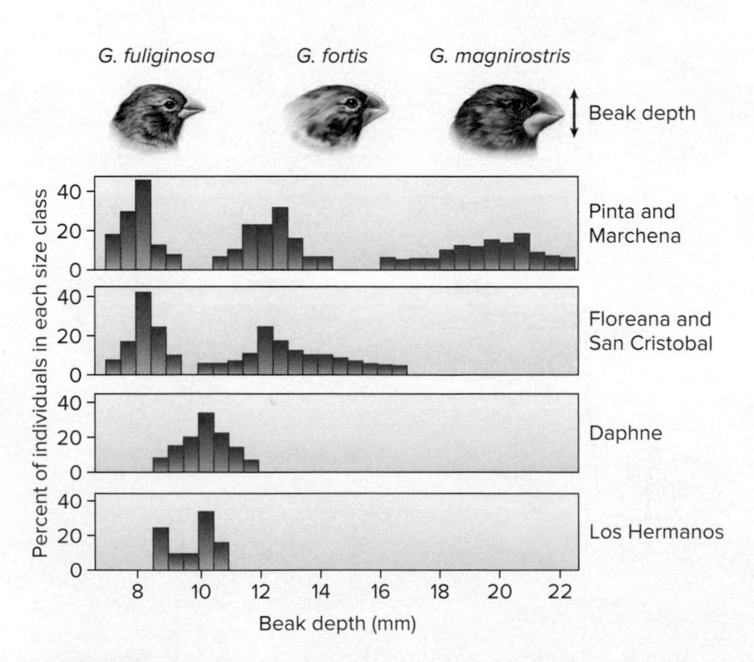

▲ **FIGURE 3.1** Two finch species have similar beaks when they occur separately (on the islands of Daphne and Los Hermanos), but beak sizes shift when the species occur together, as they specialize in different feeding strategies. When three finches coexist, feeding strategies and beak sizes are further differentiated. Source: Lack, D., *Darwin's Finches*. Cambridge University Press, 1947.

(continued)

CASE STUDY continued

for the development of the variation in a population on which natural selection acts. In this chapter we look at some of the ways species and communities adapt to their environments. We also consider the ways populations interact, and the adaptations that make some species abundant and those that make others rare.

FURTHER READING

Lack, D. 1947. *Darwin's Finches*. Cambridge University Press.

Stix, Gary. 2009. Darwin's living legacy. *Scientific American* 300(1): 138–43. ■

> *When I view all beings not as special creations, but as lineal descendents of some few beings which have lived long before the first bed of the Cambrian system was deposited, they seem to me to become ennobled.*
>
> —CHARLES DARWIN

3.1 EVOLUTION LEADS TO DIVERSITY

- *Evolution results from random mutations and natural selection.*
- *Limiting factors often cause natural selection.*
- *A species' niche is its role in an ecosystem, as well as the factors that determine distribution.*

Why does the earth support the astonishing biological diversity that we see around us? What determines which species will survive in one environment or another, and why are some species abundant, while others are rare? These are fundamental questions in ecology and biology, and in subfields such as population biology, the study of why populations grow and decline. In this chapter we examine the mechanisms that have produced the extraordinary diversity of life that surrounds us. We also consider how certain traits can lead to some species' being rare and specialized, while others are overabundant "weeds," as well as the processes that lead to changes in population range or abundance. First, we'll start with the basics: How do species arise?

Natural selection and adaptation modify species

How does a polar bear endure the long, sunless, super-cold arctic winter? How does the saguaro cactus survive blistering temperatures and extreme drought of the desert? Each species has inherited characteristics, or traits, that help it survive. The polar bear has heat-capturing fur, insulating fat layers, wide feet for swimming, and white hair that make it nearly invisible to the seals on which it preys. The saguaro has specially adapted leaves (spines), water-retaining cells, water-saving mechanisms in photosynthesis, and other traits that help it survive conditions that would kill most plants. We refer to the acquisition of these advantageous traits in a species as **adaptation.**

Adaptation involves changes in a population, with characteristics that are passed from one generation to the next. This is different from acclimation–an individual organism's changes in response to an altered environment. For example, if you spend the summer outside, you may acclimate to the sunlight: Your skin will increase its concentration of dark pigments that protect you from the sun. This is a temporary change, and you won't pass the temporary change on to future generations. However, the *capacity* to produce skin pigments is inherited. For populations living in intensely sunny environments, individuals with a good ability to produce skin pigments are more likely to thrive, or to survive, than people with a poor ability to produce pigments, and that trait becomes increasingly common in subsequent generations. If you look around, you can find countless examples of adaptation. The distinctive long neck of a giraffe, for example, developed as individuals that happened to have longer necks had an advantage in browsing on the leaves of tall trees (fig. 3.2).

This process was explored in detail in Charles Darwin's 1859 book *On the Origin of Species by Means of Natural Selection.* Darwin was one of many who observed and pondered the origins of natural variation, one of the great scientific questions of his time. He concluded

▼ **FIGURE 3.2** Giraffes have long necks because in previous generations long-necked individuals happened to have an advantage in finding food, and in reproducing. This trait has become fixed in the population. ©Westend61/getty images

that species change over generations because individuals compete for limited resources. Better competitors in a population are more likely to survive–giving them greater potential to produce offspring. We use the term **natural selection** to refer to the process in which individuals with useful traits pass on those traits to the next generation, while others reproduce less successfully.

We now know that these traits are encoded in genes, which are portions of an individual's DNA. Every organism has thousands of genes. Occasionally, random changes occur as DNA is replicated. If those random changes ("mutations") occur in reproductive cells, then they are passed on to offspring. (Mutations in nonreproductive cells, such as cancers, are not inherited.) Most mutations have little effect on fitness, and some can have a negative effect. But some happen to be useful in helping individuals exploit new resources or survive more successfully in new environmental conditions. We now understand the development of new species to result from many small mutations accumulating over time.

▲ **FIGURE 3.3** The northern limit of Saguaro cactus is partly controlled by its low tolerance of freezing temperatures. In some cases, frost damage may not be visible in adult plants, but it can limit distribution by reducing reproduction. ©William P. Cunningham

Limiting factors influence species distributions

An organism's physiology and behavior allow it to survive only in certain environments. Temperature, moisture level, nutrient supply, soil and water chemistry, living space, and other environmental factors must be at appropriate levels for organisms to persist. Generally, a critical limiting factor keeps an organism from expanding everywhere. Limitations can include (1) physiological stress due to inappropriate levels of a critical environmental factor, such as moisture, light, temperature, pH, or specific nutrients; (2) competition with other species; and (3) predation, including parasitism and disease. Any one of these could constrain a species in different circumstances. For example, nutrients are often a limiting factor in aquatic environments such as the Gulf of Mexico (see chapter 2). An infusion of nitrogen or phosphorus allows an explosion of algae, which continues until clouds of algae block sunlight in the water column. Then solar energy becomes a limiting factor, and the algae, unable to photosynthesize, die off rapidly.

In 1840 the chemist Justus von Liebig proposed that the single factor in shortest supply relative to demand is the **critical factor** determining where a species lives. The giant saguaro cactus (*Carnegiea gigantea*), which grows in the dry, hot Sonoran Desert of southern Arizona and northern Mexico, for example, is tolerant of extreme heat and drought, intense sunlight, and nutrient-poor soils, but it is extremely sensitive to freezing temperatures (fig. 3.3). A single winter night with temperatures below freezing for 12 or more hours kills growing tips on the branches, preventing further development. Because of this sensitivity, the northern edge of the saguaro's range corresponds to a zone where freezing temperatures last less than half a day at any time.

Ecologist Victor Shelford (1877–1968) later expanded Liebig's principle by stating that each environmental factor has both minimum and maximum levels, called **tolerance limits,** beyond which a particular species cannot survive or is unable to reproduce (fig. 3.4). The single factor closest to these survival limits, Shelford postulated, is the critical factor that limits where a particular organism can live. At one time ecologists tried to identify unique factors limiting the growth of every plant and animal population. We now know that several factors working together usually determine a species' distribution. If you have ever explored the rocky coasts of New England or the Pacific Northwest, you have probably noticed that mussels and barnacles grow thickly in the intertidal zone, the place between high and low tides. Multiple factors, including temperature extremes, drying time between tides, salt concentrations, competitors, and food availability, determine the distribution of these animals.

In some species, tolerance limits affect the distribution of young differently than adults. The desert pupfish, for instance, lives in small, isolated populations in warm springs in the northern Sonoran Desert. Adult pupfish can survive temperatures between 0° and 42°C (a remarkable temperature range for a fish) and tolerate an equally wide range of salt concentrations. Eggs and juvenile fish, however, can survive only between 20° and 36°C and are killed by high salt levels. Reproduction, therefore, is limited to a small part of the range of the adult fish. Many species have greater sensitivity to salinity and other factors in young (or larvae or seedlings) than in adults.

Sometimes the requirements and tolerances of species are useful indicators of general environmental characteristics. Lichens and eastern white pine, for example, are highly sensitive to sulfur dioxide and ozone, respectively. The presence or absence of lichens or white pines, then, can indicate whether these pollutants are abundant in an area. **Indicator species** is a general term for organisms whose sensitivities can tell about environmental conditions in an area. Similarly, anglers know that trout require cool, clean, well-oxygenated water. The presence or absence of trout is used as an indicator of water quality. Trout streams are often protected with special care, because the presence of trout indicates that the stream ecosystem as a whole is healthy.

Benchmark Data

Among the ideas and values in this chapter, these are a few worth remembering.

r; K	Rate of growth; carrying capacity
$dN/dt = rN$	Exponential rate of growth
$dN/dt = rN(K - N)/K$	Logistic rate of growth
20 kcal/m^2/year	Primary productivity in a tropical rainforest
2 kcal/m^2/year	In tundra or temperate grassland

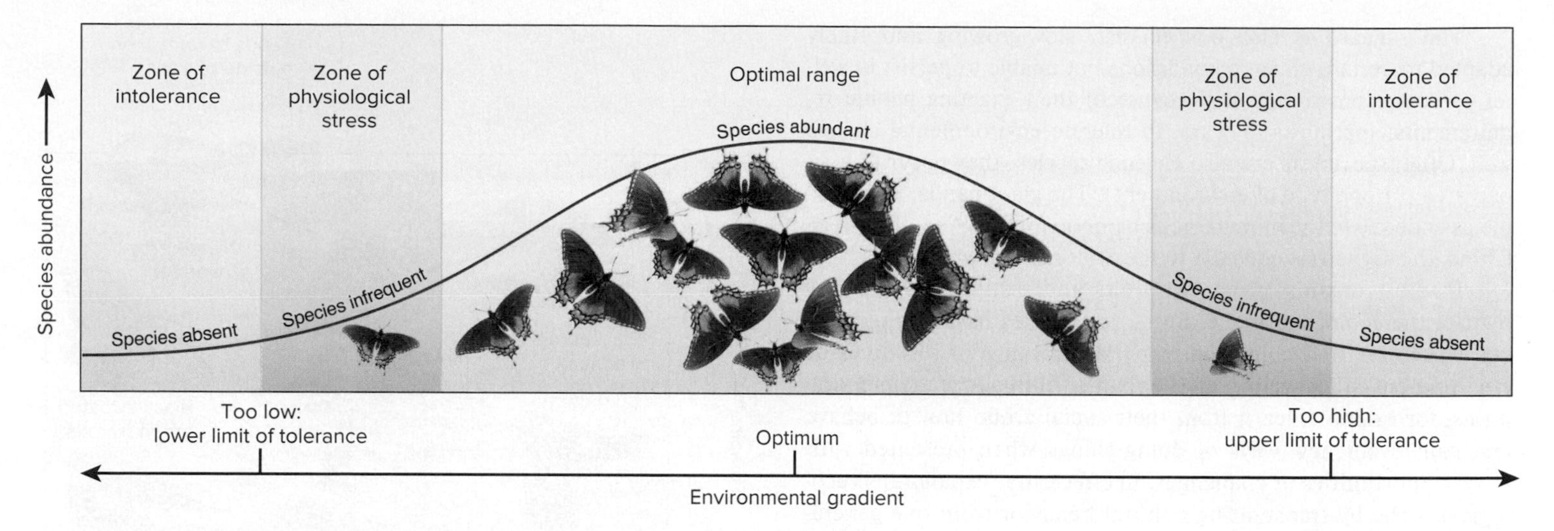

▲ **FIGURE 3.4** Tolerance limits affect species distributions. For every environmental factor, there is an optimal range within which a species lives or reproduces most easily, so abundance is high. The horizontal axis here could represent a factor such as temperature, rainfall, vegetation height, or availability of some critical resource.

A niche is a species' role and environment

Habitat describes the place or set of environmental conditions in which a particular organism lives. The term **ecological niche** is more functional, describing both the role played by a species in a biological community and the set of environmental factors that determine its distribution. The concept of niche was first defined in 1927 by the British ecologist Charles Elton (1900–1991). To Elton, each species had a role in a community of species, and the niche defined its way of obtaining food, the relationships it had with other species, and the services it provided to its community.

Thirty years later, the American limnologist G. E. Hutchinson (1903–1991) proposed a more physical as well as biological definition of the niche. Every species, he pointed out, exists within a range of physical and chemical conditions, such as temperature, light levels, acidity, humidity, or salinity. It also exists within a set of biological interactions, such as the presence of predators or prey or the availability of nutritional resources.

Some species tolerate a wide range of conditions or exploit a wide range of resources. These species are known as **generalists** (fig. 3.5). Generalists often have large geographic ranges, as in the case of the black bear (*Ursus americana*), which is omnivorous and abundant, ranging across most of North America's forested regions. Sometimes generalists are also "weedy" species or pests, such as rats, cockroaches, or dandelions, because they thrive in a broad variety of environments.

Other species, such as the giant panda (*Ailuropoda melanoleuca*), are **specialists** and have a narrow ecological niche (fig. 3.6). Pandas have evolved from carnivorous ancestors to subsist almost entirely on bamboo, a large but low-nutrient grass. To acquire enough nutrients, pandas must spend as much as 16 hours a day eating. The panda's slow metabolism, slow movements, and low reproductive rate help it survive on this highly specialized diet. Its narrow niche now endangers the giant panda, though, as recent destruction of most of its native bamboo forest has reduced its range and its population to the margin of survival.

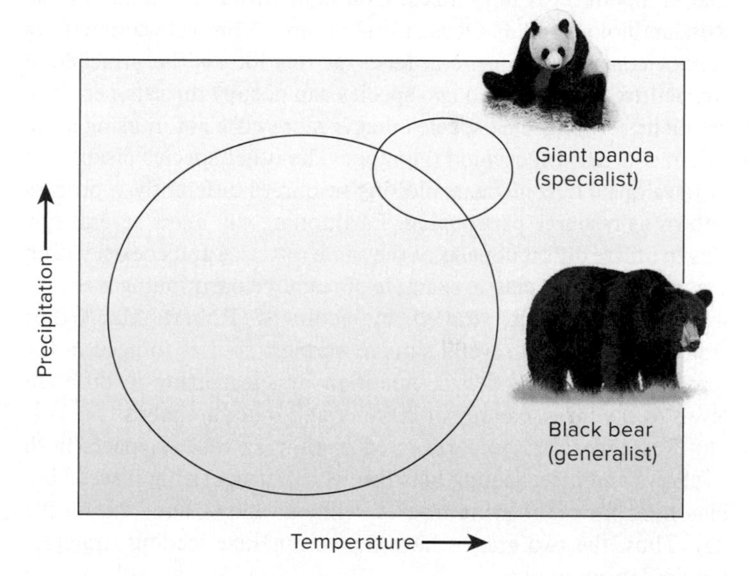

▲ **FIGURE 3.5** Generalists, such as the American black bear, tolerate a wide range of environmental conditions. Specialists, in contrast, have narrower tolerance of environmental conditions.

▶ **FIGURE 3.6** The giant panda has evolved from carnivorous ancestors to live on a diet composed almost exclusively of bamboo. Adaptations include "thumbs" that help it grasp bamboo leaves and teeth that help it chew the grass.

The saguaro is also a specialist, slow-growing and finely adapted to certain climatic conditions but unable to persist in wetter or cooler environments. Because of their exacting habitat requirements, specialists tend not to tolerate environmental change well. Often specialists are also **endemic species**—they occur only in one area (or one type of environment). The giant panda, for example, is endemic to the mountainous bamboo forests of southwestern China; the saguaro is endemic to the Sonoran Desert.

In most organisms, genetic traits and instinctive behaviors restrict the ecological niche. But some species have complex social structures that help them expand the range of resources or environments they can use. Elephants, chimpanzees, and dolphins, for example, learn from their social group how to behave and can invent new ways of doing things when presented with novel opportunities or challenges. In effect, they expand their ecological niche by transmitting cultural behavior from one generation to the next.

When two species compete for limited resources, one eventually gains the larger share, while the other finds different habitat, dies out, or experiences a change in its behavior or physiology, so that competition is minimized. Consequently, as explained by the Russian biologist G. F. Gause (1910–1986), "complete competitors cannot coexist." The general term for this idea is the **principle of competitive exclusion:** No two species can occupy the same ecological niche for long. The species that is more efficient in using available resources will exclude the other. The other species disappears or develops a new niche, exploiting resources differently, a process known as **resource partitioning.** Partitioning can allow several species to utilize different parts of the same resource and coexist within a single habitat. A classic example of resource partitioning is that of woodland warblers, studied by ecologist Robert MacArthur (fig. 3.7). Although several similar warblers species foraged in the same trees, they avoided competition by specializing in different levels of the forest canopy or in inner and outer branches.

Resources can be partitioned in time as well as space. Both swallows and insect-eating bats live by capturing flying insects, but bats hunt for night-flying insects, while swallows hunt during the day. Thus, the two groups have noncompetitive feeding strategies for similar insect prey.

▲ **FIGURE 3.7** Several species of insect-eating wood warblers occupy the same forests in eastern North America. The competitive exclusion principle predicts that the warblers should partition the resource—insect food—in order to reduce competition. And in fact, the warblers feed in different parts of the forest. Source: R. H. MacArthur (1958) *Ecology* 39:599–619.

Speciation leads to species diversity

As a population becomes more adapted to its ecological niche, it may develop specialized or distinctive traits that eventually differentiate it entirely from its biological cousins. The development of a new species is called **speciation** (Key Concepts, p. 58). In the case of Galápagos finches, evidence from body shape, behavior, and genetic similarity suggests that the 13 current species of finch derive from an original seed-eating finch species that probably blew to the islands from the mainland, perhaps in a storm, since finches are land birds. Accidental invasions, such as those by storms, winds, or ocean currents are probably rare. In the Galápagos, though, all land plants and animals (except those introduced by humans) derive from a few accidental colonizers. We know this because, as volcanic seamounts, the islands were never connected to a continental source of species.

In the Galápagos finches, speciation occurred largely because of **geographic isolation.** The islands are far enough apart that, in many cases, populations were genetically isolated: They couldn't interbreed with populations on other islands. These isolated populations gradually changed in response to their individual environments, some of which were extremely dry, with sparse vegetation, while others were relatively moist, with greater abundance and diversity of food resources. Speciation that occurs when populations are geographically separated is known as **allopatric speciation.**

The barriers that divide subpopulations aren't always physical. For example, two virtually identical tree frogs (*Hyla versicolor, H. chrysoscelis*) live in similar habitats of eastern North America but have different mating calls. This difference, which is enough to prevent interbreeding, is known as behavioral isolation. Speciation that occurs within one geographic area is known as **sympatric speciation.** Fern species and other plants sometimes undergo sympatric speciation by doubling or quadrupling the chromosome number of their ancestors, which makes them reproductively incompatible.

A general term for factors that make certain mutations advantageous is **selection pressure.** Normally, we assume this selection happens slowly, but some degree of selection can happen in just a few years. A study of the finch species *Geospiza fortis* on the Galápagos island of Daphne Major, for example, showed selection after a 2-year drought. As the drought reduced the availability of small seeds on the island, large-billed individuals in the finch population had better success in opening the remaining larger seeds. Within 2 years, the large-billed trait came to dominate the population. This shift toward one extreme of a trait is known as

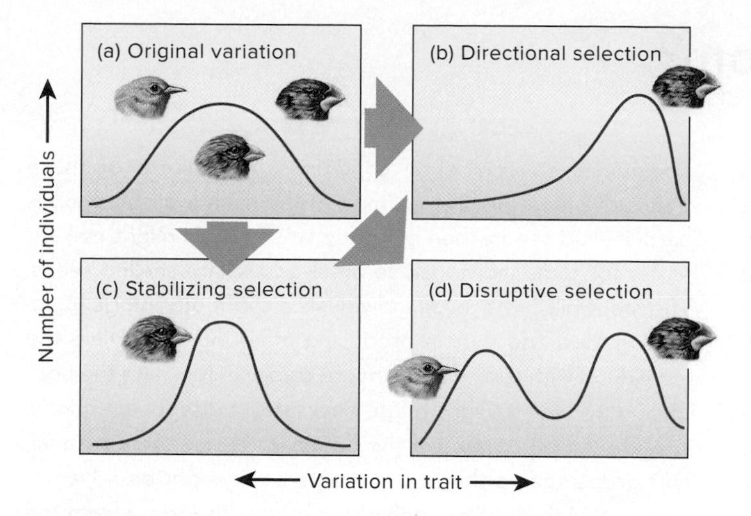

▲ **FIGURE 3.8** A trait such as beak shape can change as changing environmental conditions make one type more advantageous for survival. From an original population with wide variation in the trait (a), environmental conditions might favor one extreme (b) or neither extreme (c). Where a population occupies contrasting conditions, traits in the different areas may diverge (d), producing two distinct populations.

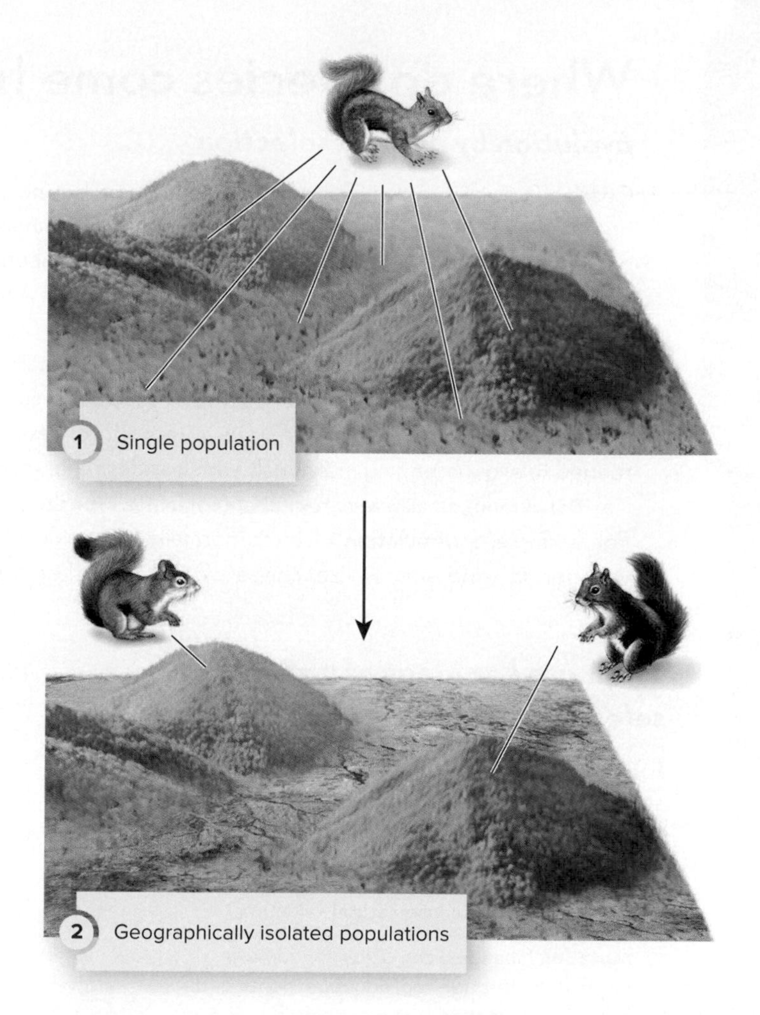

▲ **FIGURE 3.9** Geographic isolation can lead to allopatric (geographically separated) speciation. For example, in cool, moist glacial periods, Arizona was forested and red squirrels interbred freely (1). As the climate warmed and dried, desert replaced forests on the plains. Isolated in remnant forest on mountains, populations became reproductively isolated and began to develop different traits (2).

directional selection (fig. 3.8). In this case the changes were not dramatic enough to result in speciation. When the drought ended, the population shifted back toward moderate-size beaks, which aided exploitation of a wider range of seeds.

Sometimes environmental conditions can reduce variation in a trait (stabilizing selection), or they can cause traits to diverge to the extremes (disruptive selection). Competition can cause disruptive selection, which allows for better partitioning of a resource (see fig. 3.1). Directional selection can be observed in the emergence of antibiotic-resistant bacteria (see chapter 8) and of pesticide-tolerant insects (see chapter 7). In both cases, individuals that happen to have better-than-average tolerance of these compounds tend to survive in a population, while other individuals die off. Resistant survivors produce new generations, leading to a population with the resistant traits.

New environmental conditions often lead to speciation, as new opportunities become available–as in the case of finch diversification on the previously unoccupied Galápagos Islands. Time is marked by periods of tremendous diversification that have followed the sudden extinctions of species (see chapter 5). The end of the age of dinosaurs, for example, was followed by dramatic diversification of mammals, which expanded to fill newly available niches. The fossil record is one of ever-increasing species diversity, despite several events that wiped out large numbers of species.

We generally believe that species arise slowly, and many have existed unchanged for tens of millions of years. (For example, the American alligator has existed unchanged for over 150 million years.) But some organisms evolve swiftly. New flu viruses, for instance, evolve every season. Fruit flies in Hawaii mutate frequently and can give rise to new species in just a few years. Reproductive isolation has led to the development of new traits in populations of red squirrels in Arizona in just the past 10,000 years or so, since the last glacial period (fig. 3.9).

Taxonomy describes relationships among species

Taxonomy is the study of types of organisms and their evolutionary relationships. Taxonomists are biologists who use various lines of evidence, such as cell structures or genetic similarities, to trace how organisms have descended from common ancestors. Taxonomists are continually refining their understanding, but most now classify all life into three general domains according to cell structure. These domains are Bacteria (whose cells have no membrane around the nucleus), Archaea (whose DNA differs from bacteria, and whose cell functions allow them to survive in extreme environments, such as hot springs), and Eukarya (whose cells do have a membrane around the nucleus). The organisms most familiar to us are in the four kingdoms of the Eukarya (fig. 3.10). These kingdoms include animals, plants, fungi (molds and mushrooms), and protists (algae, protozoans, slime molds). Within these kingdoms

Where do species come from?

Evolution by natural selection

Charles Darwin is known for explaining evolution, but he was one of many in the nineteenth century who pondered the great question **how do new species appear?** Darwin's contribution, which he developed simultaneously with biogeographer Alfred Russell Wallace, was the idea of **natural selection.** Darwin noted that horse breeders and dog breeders selected for certain advantageous traits in their animals—speed, strength, and so on—by allowing individuals with those traits to reproduce. Breeders tended to avoid breeding individuals with less desirable traits.

Darwin suggested that natural chance might act the same way. For example, a population of birds normally has some slight variation in wing shape, size, shape of bill, and other traits. Sometimes environmental conditions make some of those traits advantageous: For example, if the main available food is hard-shelled seeds, then a slightly stronger bill might make it easier for some individuals to break open hard-shelled seeds. The stout bill gives those individuals a slight advantage in acquiring food and thus in producing offspring. Over time the individuals with strong beaks might come to dominate the population. Those with lightweight beaks might disappear entirely.

On the other hand, if the dominant food source is small, soft grubs, then a thin and movable beak might be advantageous, and thick-billed individuals might disappear from the population.

What are key ideas of natural selection?

1. **Natural selection** occurs when circumstances make one type of trait especially advantageous, so that individuals with that trait reproduce more frequently or successfully than do others. Food resources, climate, the presence of predators, or other factors can cause natural selection.
2. **Mutation** (changes) can occur randomly in a population, through reproduction. Some birds have slightly longer or shorter bills, for example.
3. **Genetic drift,** or shifting traits in a population, is more likely in a small or isolated population. If an unusual trait (for example, red hair in humans) is relatively common in a small population, that trait has a relatively high probability of passing to later generations.
4. **Isolation** separates populations, making genetic drift more likely. Isolation of the Galápagos meant that the island ecosystems had few birds species and few food sources. Species have specialized for the types of foods on different islands. Geographic isolation also reduces interbreeding with a larger population—which might mix other traits into island populations.

The Galápagos Islands, nearly 1,000 km from South America, are renowned as the place whose unusual species helped Darwin form his ideas of natural selection. Finches, mockingbirds, and other species on the isolated islands showed distinctive adaptations to different food sources and conditions. Yet they had similarities that pointed to common ancestors.

Craters show these islands are volcanic. They have been isolated since they emerged above the ocean's surface. Relatively few species have reached the Galápagos from the distant mainland. So it's relatively easy to detect divergence from a common ancestor.

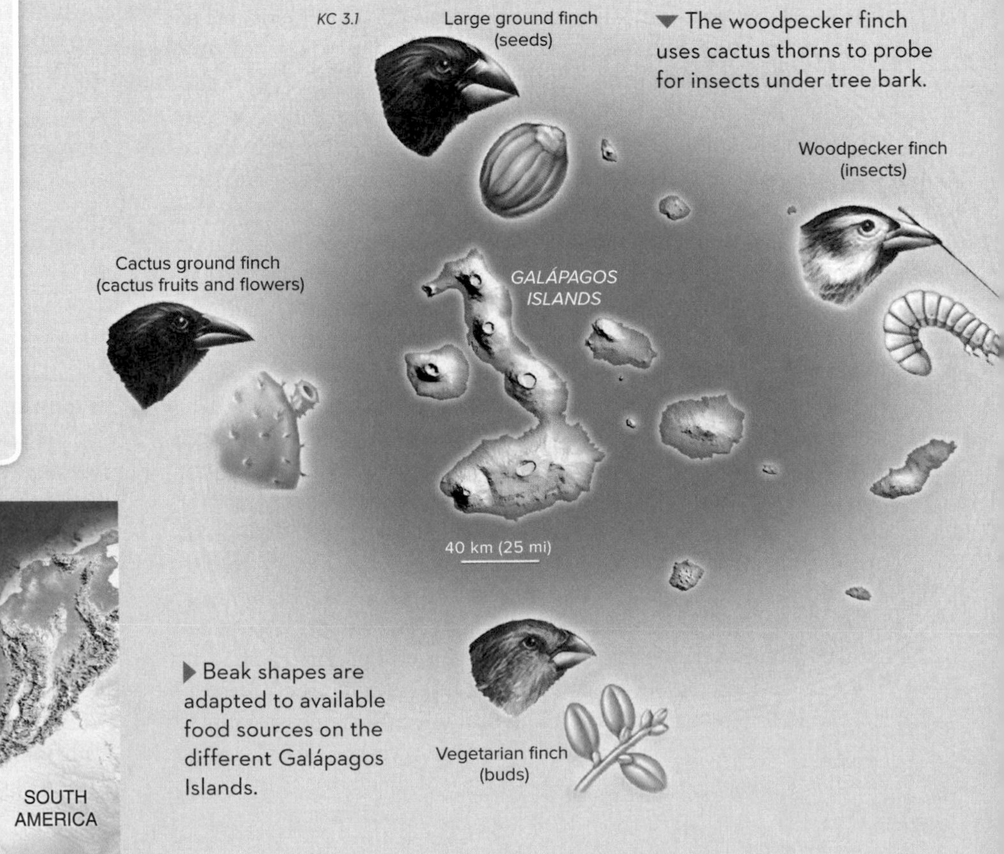

▼ The woodpecker finch uses cactus thorns to probe for insects under tree bark.

▶ Beak shapes are adapted to available food sources on the different Galápagos Islands.

KC 3.2

CENTRAL AMERICA

The Galápagos Islands are nearly 1,000 km (600 mi) from the mainland.

SOUTH AMERICA

Selective pressure is a general term for factors that modify species' traits. **Competition** over resources can exert selective pressure by causing species to partition, or separate, their use of the resource. Where there is overlap in resource use, individuals that share the resource (orange shading, **a**) should be at a disadvantage, and individuals that specialize should be more abundant. Over time the traits of the populations diverge, leading to specialization, narrower niche breadth, and less competition between species. **(b) Speciation,** or separation into entirely separate species, can result from competition or from isolation. ▶

KC 3.3

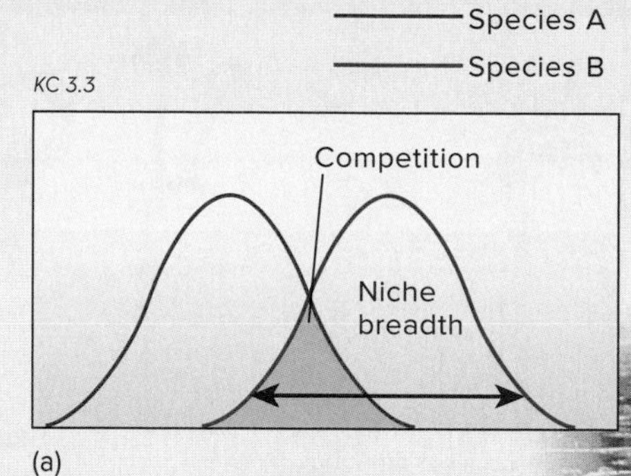

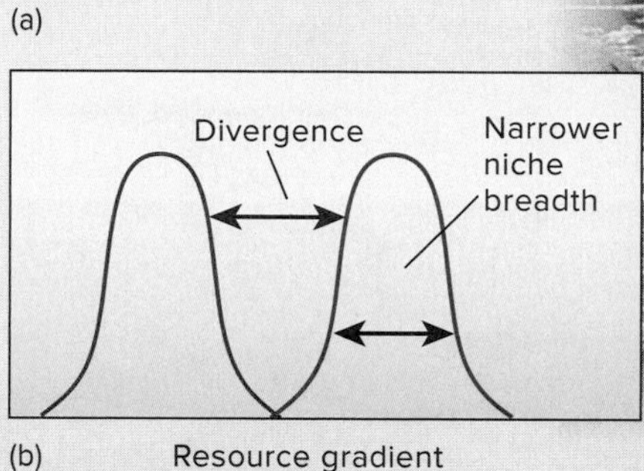

▼ Two wading birds **partition** a mudflat: The northern jacana captures insects at the surface with its short bill. The long-legged black-necked stilt probes deeper with its long bill.

©Mary Ann Cunningham KC 3.4

▶ **Intraspecific competition** (within a species) can lead to colorful and surprising traits.

©Kevin Schafer/Alamy Stock Photo

KC 3.5

KC 3.7

©DC_Colombia/Getty Images

▼ Galápagos tortoises have been shaped by the arid conditions of these isolated islands.

KC 3.6

©fominayaphoto/123RF

Why does this matter?

Since Darwin published *On the Origin of Species* in 1859, countless studies have supported his **theory of evolution.** The idea of evolution by natural selection, or "descent with modification," as Darwin called it, has allowed us to describe evolutionary relationships among the millions of organisms around us, and to understand why they feed, breathe, reproduce, and live as they do.

We now understand most biological processes in terms of the mechanism of natural selection. Every time you get a flu shot, that vaccine is created by careful observation of adaptations in the rapidly evolving flu virus, in efforts to create a vaccine suited to the latest variety of the virus. ▼

KC 3.9

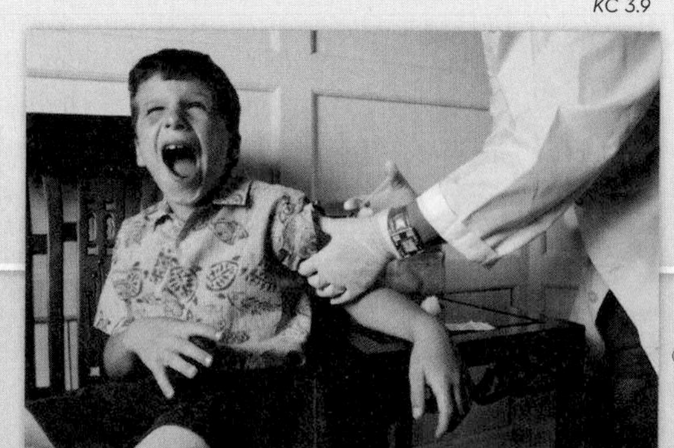

©McGraw-Hill Education/Jill Braaten, photographer

KC 3.8

©Tom Cooper

CAN YOU EXPLAIN?

1. **What factors made natural selection relatively likely to occur in the Galápagos?**
2. **Think of several birds where you live. What kinds of food resources or feeding strategies might be reflected in their bill shapes?**
3. **Think of an example of resource partitioning between two organisms with which you are familiar.**

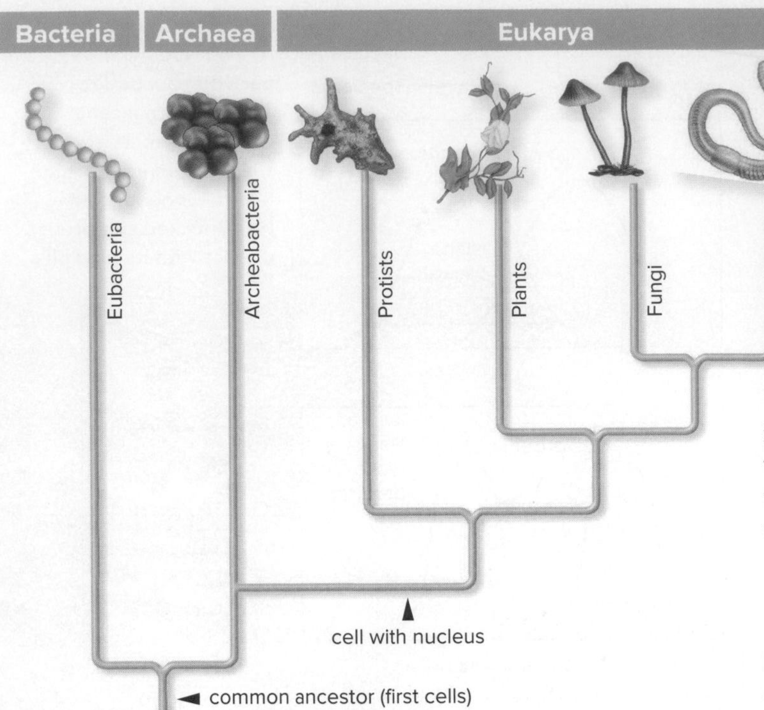

▲ **FIGURE 3.10** Taxonomists divide living organisms into three domains (Bacteria, Archaea, and Eukarya) based on fundamental cell structures. These arose from a common ancestor over the past 4 billion years. The Animal, Fungi, Plant, and Protist kingdoms are all part of the domain Eukarya.

are millions of different species, which we discuss further in chapter 5. The taxonomic tree groups organisms at multiple levels of specificity. Two familiar examples–one you know and one you eat–are shown in table 3.1.

Botanists, ecologists, and other scientists often use the two most specific levels of taxonomy, genus and species, to refer to species. These two-part binomial names, also called scientific or Latin names, are more precise than common names. Many different types of yellow flowers, growing in different places, are called "buttercups," for example. This can lead to serious confusion when we try to describe these flowers. At the same time, a single species might have multiple common names. The binomial name *Pinus resinosa,* however, always refers to the same tree, whether you call it red pine, Norway pine, or just pine.

TABLE 3.1 | Taxonomy of Two Common Species

TAXONOMIC LEVEL	HUMANS	CORN
Kingdom	Animalia	Plantae
Phylum	Chordata	Anthophyta
Class	Mammalia	Monocotyledons
Order	Primates	Commenales
Family	Hominidae	Poaceae
Genus	*Homo*	*Zea*
Species	*Homo sapiens*	*Zea mays*
Subspecies	*H. sapiens sapiens*	*Z. mays mays*

3.2 SPECIES INTERACTIONS

- *Competition occurs between individuals and between species.*
- *Selective pressures, such as competition and predation, lead to evolution and adaptation.*
- *Keystone species play critical roles in ecosystems.*

Resource limitations influence a species' adaptation to its environment and its ecological niche, but interactions with other species also help shape species' traits and behaviors. Competition and predation cause species to evolve in response to each other's attributes; cooperative interactions and even interdependent relationships also confer advantages, which can lead to evolution of traits. In this section we will look at the interactions within and between species that affect their success and shape biological communities.

Competition leads to resource allocation

Competition is a type of antagonistic relationship within a biological community. Organisms compete for resources that are in limited supply: energy and matter in usable forms, living space, and specific sites to carry out life's activities. Plants compete for growing space to develop roots and leaves, so that they can absorb and process sunlight, water, and nutrients (fig. 3.11). Animals compete for living, nesting, and feeding sites, as well as for mates. Competition among members of the same species is called **intraspecific competition,** whereas competition between members of different species is called **interspecific competition.** Recall the competitive exclusion principle, that no two species can occupy the same ecological niche for long. Competition causes individuals and species to shift their use of a common resource; thus, warblers that compete with each other for insect food in deciduous forests tend to specialize on different areas of the trees, reducing or avoiding competition. There have been hundreds of interspecific competition studies in natural populations showing a wide variety of evolutionary adaptations.

When different species use a common resource, competition often leads to resource partitioning, as in the case of some of the Galápagos finches (see fig. 3.1). Alternatively, one species can exclude another. When two species compete, the one living in the center of its tolerance limits has an advantage and, more often than not, prevails in competition with another species living at the margins of its optimal environmental conditions.

In intraspecific competition, members of the same species compete directly with each other for resources. Competition can hone a species' attributes–for example, the fastest cheetahs prosper by catching more prey than slower ones, or warblers with the best insect-foraging bills prosper more than poorer foragers. Direct intraspecific fighting for resources can occur, as when two bull elk battle for territory and for females. But most animals avoid fighting if possible. They posture and challenge each other, but usually the weaker individual eventually backs off. Physical injury is too high a cost to risk in most cases of intraspecific competition.

Several avenues exist to reduce competition within a species. First, the young of the year can disperse. Young animals move in

▲ **FIGURE 3.11** In this tangled Indonesian rainforest, plants compete for light and space. Special adaptations allow ferns, mosses, and bromeliads to find light by perching high on tree trunks and limbs; vines climb toward the canopy. These and other adaptations lead to profuse speciation.
©William P. Cunningham

search of new territories; in plants, seeds travel with wind, water, or passing animals and at least some land in available habitat away from the parent plants. Second, by exhibiting strong territoriality, many animals force their offspring or other trespassers out of their vicinity. In this way territorial species, such as bears, songbirds, ungulates, and fish, minimize competition between individuals and generations. A third way to reduce intraspecific competition is resource partitioning between generations. The adults and juveniles of these species occupy different ecological niches. For instance, monarch caterpillars munch on milkweed leaves, while metamorphosed butterflies sip nectar. Crabs begin as swimming larvae and do not compete with bottom-dwelling adult crabs.

Predation affects species relationships

All organisms need food to live. Producers make their own food, whereas consumers eat organic matter created by other organisms. As we saw in chapter 2, photosynthetic plants and algae are the producers in nearly all communities. Consumers include herbivores, carnivores, omnivores, scavengers, detritivores, and decomposers. Often we think only of carnivores as predators, but ecologically a predator is any organism that feeds directly on another living organism, whether or not this kills the prey (fig. 3.12). Herbivores, carnivores, and omnivores, which feed on live prey, are predators, but scavengers, detritivores, and decomposers, which feed on dead things, are not. In this sense, parasites (organisms that feed on a host organism or steal resources from it without necessarily killing it) and even pathogens (disease-causing organisms) can be considered predator organisms. Herbivory is the type of predation practiced by grazing and browsing animals on plants.

Predation is a powerful but complex influence on species populations in communities. It affects (1) all stages in the life cycles of predator and prey species; (2) many specialized food-obtaining mechanisms; and (3) the evolutionary adjustments in behavior and body characteristics that help prey escape being eaten and help predators more efficiently catch their prey. Predation also interacts with competition. In **predator-mediated competition,** a superior competitor in a habitat builds up a larger population than its competing species; predators take note and increase their hunting pressure on the superior species, reducing its abundance and allowing the weaker competitor to increase its numbers. To test this idea, scientists remove predators from communities of competing species. Often the superior competitors eliminate other species from the habitat. In a classic example, the ochre starfish (*Pisaster ochraceus*) was removed from Pacific tidal zones, and its main prey, the common mussel (*Mytilus californicus*), exploded in numbers and crowded out other species.

Knowing how predators affect prey populations has direct application to human needs, such as pest control in cropland. The cyclamen mite (*Phytonemus pallidus*), for example, is a pest of California strawberry crops. Predatory mites (*Typhlodromus* and *Neoseiulus*), which arrive naturally or are introduced into fields, reduce the damage caused by the cyclamen mite. Spraying pesticides to control the cyclamen mite can actually increase the infestation because it also kills the beneficial predatory mites.

Predatory relationships may change as the life stage of an organism changes. In marine ecosystems, crustaceans, mollusks, and worms release eggs directly into the water, where they and hatchling larvae join the floating plankton community (fig. 3.13). Planktonic animals eat each other and are food for larger carnivores, including fish. As prey species mature, their predators change. Barnacle larvae are planktonic

▲ **FIGURE 3.12** Insect herbivores are predators as much as lions and tigers. Insects consume the vast majority of biomass in the world. Complex predation and defense mechanisms have evolved between insects and their plant prey.
©Amble Design/Shutterstock

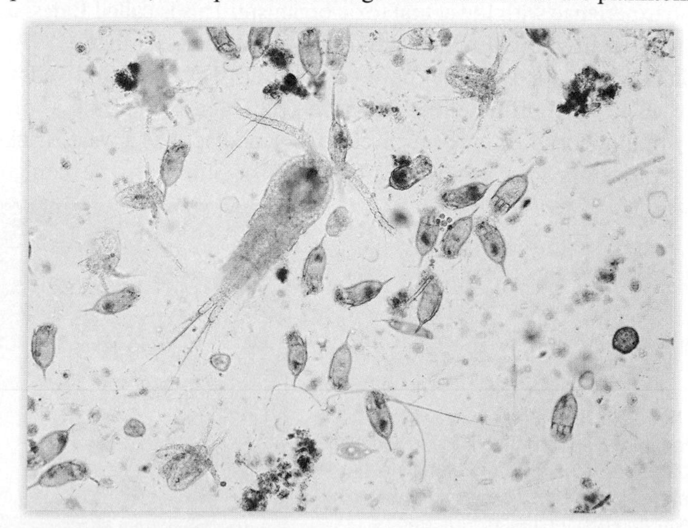

▲ **FIGURE 3.13** Microscopic plants and animals form the basic levels of many aquatic food chains and account for a large percentage of total world biomass.
©Choksawatdikorn/Shutterstock

and are eaten by small fish, but as adults their hard shells protect them from fish, but not from starfish and predatory snails. Predators often switch prey in the course of their lives. Carnivorous adult frogs usually begin their lives as plant-eating tadpoles. Some predators also switch prey easily: House cats, for example, prey readily on a wide variety of small birds, mammals, lizards, and even insects. Other predators, such as the polar bear, are highly specialized in their prey preferences—a loss of access to seals can lead to polar bear starvation.

Predation leads to adaptation

Predator-prey relationships exert selection pressures that favor evolutionary adaptation. Predators become more efficient at searching and feeding, and prey become more effective at escape and avoidance. Prey organisms have developed countless strategies to avoid predation, including toxic or bad-tasting compounds, body armor, extraordinary speed, and the ability to hide. Plants have evolved thick bark, spines, thorns, or distasteful and even harmful chemicals in tissues—poison ivy and stinging nettle are examples. In response, animals have found strategies for avoiding spines, eating through thick bark, or tolerating chemicals. Arthropods, amphibians, snakes, and some mammals produce noxious odors or poisonous or venemous secretions that cause other species to leave them alone. Speed is a common defense against predation. On the Serengeti Plain of East Africa, the swift Thomson's gazelle and even swifter cheetah are engaged in an arms race of speed and endurance. The cheetah has an edge in a surprise attack, because it can accelerate from 0 to 72 kph in 2 seconds. But the gazelle often escapes because the cheetah lacks stamina. A general term for this close adaptation of two species is **coevolution.**

Species with chemical defenses often display distinct coloration and patterns to warn away enemies (fig. 3.14). Species also display forms, colors, and patterns that help them hide. Insects that look exactly like dead leaves or twigs are among the most remarkable examples (fig. 3.15). Predators also use camouflage to conceal themselves as they lie in wait for their next meal. In a neat evolutionary twist, certain species that are harmless resemble poisonous or distasteful ones, gaining protection against predators that remember a bad experience with the actual toxic organism. This is called **Batesian mimicry,** after the English naturalist H. W. Bates (1825–1892). Many wasps, for example, have bold patterns of black and yellow stripes to warn off potential predators (fig. 3.16a). A harmless variety of longhorn beetle has evolved to look and act like a wasp, tricking predators into avoiding it (fig. 3.16b). Similarly, the benign viceroy butterfly has evolved to closely resemble the distasteful monarch butterfly. When two unpalatable or dangerous species look alike, we call it **Müllerian mimicry** (after the biologist Fritz Müller). When predators learn to avoid either species, both benefit.

▶ **FIGURE 3.14** Poison arrow frogs of the family Dendrobatidae display striking patterns and brilliant colors that alert potential predators to the extremely toxic secretions on their skin. Indigenous people in Latin America use the toxin to arm blowgun darts. ©Allison Isztok/EyeEm/Getty Images

▲ **FIGURE 3.15** This walking stick is highly camouflaged to blend in with the forest floor, a remarkable case of selection and adaptation. Courtesy Tom Finkle

(a) Wasp ©blickwinkel/Alamy Stock Photo

(b) Beetle ©Ger Bosma/Getty Images

▲ **FIGURE 3.16** In Batesian mimicry, a stinging wasp (a) has bold yellow and black bands, which a harmless long-horned beetle mimics (b) to avoid predators.

TABLE 3.2 Types of Species Interactions

INTERACTION BETWEEN TWO SPECIES	EFFECT ON FIRST SPECIES*	EFFECT ON SECOND SPECIES*
Mutualism	+	+
Commensalism	+	0
Parasitism	+	–
Predation	+	–
Competition	±	±

*Effects on individuals of each species: + beneficial; – harmful; ± varies.

▲ **FIGURE 3.17** Coevolution has led to close evolutionary relationships between many species, as in this star orchid and the specially adapted hawk moth that pollinates it.

Symbiosis involves cooperation

In contrast to predation and competition, some interactions between organisms can be nonantagonistic, even beneficial (table 3.2). In such relationships, called **symbiosis,** two or more species live intimately together, with their fates linked. Symbiotic relationships often involve coevolution. Many plants and pollinators have forms and behaviors that benefit each other. Many moths, for example, are adapted to pollinate particular flowering plants (fig. 3.17).

Symbiotic relationships often enhance the survival of one or both partners. In lichens, a fungus and a photosynthetic partner (either an alga or a cyanobacterium) combine tissues to mutual benefit. A symbiotic relationship such as this, in which both species clearly benefit, is also called **mutualism** (fig. 3.18). Competition and predation were long thought to drive most adaptation and speciation, but ecologists increasingly recognize the frequency and importance of cooperative and mutualistic relationships. You have trillions of symbiotic microorganisms living in and on your body (Exploring Science, p. 64).

The interdependence of coral polyps and algae in coral reefs is a globally important form of mutualism, in which the polyp provides structure and safety for algae, while the photosynthetic algae provide nutrients to the coral polyp as it builds a coral reef system. Another widespread mutualistic relationship is that between ants and acacia trees in Central and South America. Colonies of ants live inside protective cover of hollow thorns on the acacia tree branches. Ants feed on nectar that is produced in glands at the leaf bases and eat protein-rich structures that are produced on leaflet tips. The acacias thus provide shelter and food for the ants. What do the acacias get in return? Ants aggressively defend their territories, driving away herbivorous insects that might feed on the acacias. Ants also trim away vegetation that grows around the tree, reducing competition by other plants for water and nutrients. This mutualistic relationship thus affects the biological community around acacias, just as competition or predation shapes communities.

Commensalism is a type of symbiosis in which one member clearly benefits and the other apparently is neither benefited nor harmed. Many mosses, bromeliads, and other plants growing on trees in the moist tropics are considered commensals (fig. 3.18c). These epiphytes are watered by rain and obtain nutrients from leaf litter and falling dust, and often they neither help nor hurt the trees on which they grow. **Parasitism,** a form of predation, may also be considered symbiosis because of the dependency of the parasite on its host.

(a) Symbiosis

(b) Mutualism

(c) Commensalism

▲ **FIGURE 3.18** *Symbiosis* refers to species living together: For example, lichens (a) consist of a fungus, which gives structure, and an alga or a cyanobacterium, which photosynthesizes. Mutualism is a symbiotic relationship that benefits both species, such as a lichen or a parasite-eating red-billed oxpicker and a parasite-infested impala (b). Commensalism benefits one species but has little evident effect on the other, as with a tropical tree and a free-loading bromeliad (c).
(a): ©William P. Cunningham; (b): ©P. de Graaf/Getty Images; (c): ©William P. Cunningham

EXPLORING Science

Say Hello to Your 90 Trillion Little Friends

Have you ever thought of yourself as a biological community or an ecosystem? You should. Researchers estimate that each of us has about 90 trillion bacteria, fungi, protozoans, and other organisms living in or on our bodies. The largest group—around 2 kg worth—inhabit your gut, but there are thousands of species living in every orifice, gland, pore, and crevice of your anatomy. Although the 10 trillion or so mammalian cells make up more than 95 percent of the volume of your body, they represent less than 10 percent of all the cell types that occupy that space.

We call the collection of cells that inhabit us our *microbiome*. The species composition of your own microbial community will be very similar to that of other people and pets with whom you live, but each of us has a unique collection of species that may be as distinctive as our fingerprints.

As you'll learn elsewhere in this chapter, symbiotic relationships can be mutualistic (both benefit), commensal (one benefits while the other is unaffected), or parasitic (one harms the other). We used to think of *all* microorganisms as germs to be eliminated as quickly and thoroughly as possible. Current research suggests, however, that many of our fellow travelers are beneficial, perhaps even indispensable, to our good health and survival.

Your microbiome is essential, for example, in the digestion and absorption of nutrients. Symbiotic bacteria in your gut supply essential nutrients (important amino acids and short-chain fatty acids), vitamins (such as K and some B varieties), hormones and neurotransmitters (such as serotonin), and a host of other signaling molecules that communicate with, and modulate, your immune and metabolic systems. They help exclude pathogens by competing with them for living space, or by creating an environment in which the bad species can't grow or prosper.

In contrast, the inhabitants of different organs can have important roles in specific diseases. Oral bacteria, for example, have been implicated in cardiovascular disease, pancreatic cancer, rheumatoid arthritis, and preterm birth, among other things. Symbionts in the lung have been linked to cystic fibrosis and chronic obstructive pulmonary disease (COPD). And the gut community seems to play a role in obesity, diabetes, colitis, susceptibility to infections, and allergies, as well as posssibly playing a role in neurological diseases, such as multiple sclerosis, Parkinson's, Alzheimer's, Huntington's, and amyotrophic lateral sclerosis (or Lou Gehrig's disease). A healthy biome seems to be critical in controlling chronic inflammation that triggers many important long-term health problems.

As in many ecosystems, the diversity of your microbiome may play an important role in its stability and resilience. Having a community rich in good microbes will not only help you resist infection by pathogens but will allow faster recovery after a catastrophic event. People who eat a wide variety of whole grains, raw fruits and vegetables, and unprocessed meat and dairy products tend to have a much greater species blend than those of us who have a diet full of simple sugars and highly processed foods. Widespread use of antibiotics to treat illnesses can limit diversity in our symbiotic community. Constant exposure to antimicrobial soaps, lotions, and other consumer products can also impoverish our microbiome and possibly our resilience.

A growing problem in many places is antibiotic-resistant, hospital-acquired infections. One of the most intractable of these is *Clostridium difficile*, or *C. diff*, which infects 250,000 and kills 14,000 Americans every year. An effective treatment for this superpathogen is fecal transplants. Either a sample of the microbiome from a healthy person is implanted directly through a feeding tube into the patient's stomach or frozen, encapsulated pellets of feces are delivered orally. In one trial, 18 of 20 patients who received fecal transplants recovered from *C. diff.*

Similarly, obese mice given fecal transplants from lean mice lose weight, while lean mice that receive samples of gut bacteria from obese mice gain weight. The microbiome may even regulate mood and behavior. When microbes from easygoing, adventurous mice are transplanted into the gut of anxious, timid mice, they become bolder and more adventurous.

So it may pay to take care of your garden of microbes. If you keep them happy, they may help keep you healthy and happy as well.

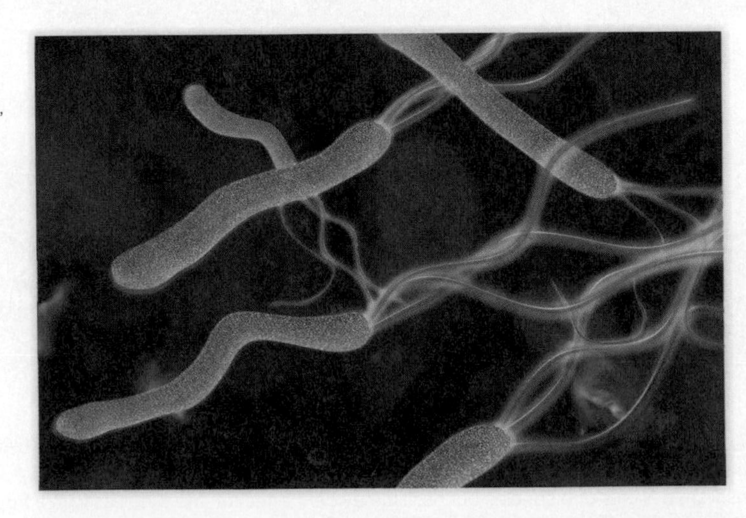

▶ Intestinal bacteria, such as these, help crowd out pathogens, aid in digestion, supply your body with essential nutrients, and may play a role in obesity, diabetes, colitis, allergies, and chronic inflammation, along with a host of other critical diseases. ©imageBROKER/Alamy Stock Photo

Keystone species play critical roles

A **keystone species** plays a critical role in a biological community that is out of proportion to its abundance. Originally, keystone species were thought to be only top predators—lions, wolves, tigers—which limited herbivore abundance and reduced the herbivory of plants. Scientists now recognize that less conspicuous species also play keystone roles. Tropical fig trees, for example, bear fruit year-round at a low but steady rate. If figs were removed from a forest, many fruit-eating animals (frugivores) would starve in the dry season when fruit of other species is scarce. In turn, the disappearance of frugivores would affect plants that depend on them for pollination and seed dispersal. The effect of a keystone species on communities ripples across multiple trophic levels.

Off the northern Pacific coast, a giant brown alga (*Macrocystis pyrifera*) forms dense "kelp forests," which shelter fish and shellfish species from predators, allowing them to become established in the community. Within this kelp forest are also sea urchins, which graze on the kelp on the seafloor, and sea otters, which eat the sea urchins. When sea otters have been eliminated—by trapping or by predation, for example—the urchins overgraze and diminish the kelp forests, potentially causing collapse of this complex system (fig. 3.19). Because of their critical role in supporting the entire kelp forest, otters are seen as a classic example of a keystone species.

Keystone functions have been documented for vegetation-clearing elephants, predatory ochre sea stars, and frog-eating salamanders in coastal North Carolina. Even microorganisms can play keystone roles. In many temperate forest ecosystems, groups of fungi that are associated with tree roots (mycorrhizae) facilitate the uptake of essential minerals. When fungi are absent, trees grow poorly or not at all.

(a) Kelp shelter fish, seals, and other species.

(b) Sea urchins graze on kelp.

(c) Sea otters protect kelp ecosystem by preying on urchins.

▲ **FIGURE 3.19** Sea otters protect kelp ecosystems on the Pacific coast by eating sea urchins, which could otherwise destroy the kelp. (a): ©Gregory Ochocki/Science Source; (b): ©Medioimages/Photodisc/Getty Images; (c): ©Kirsten Wahlquist/Shutterstock

3.3 POPULATION GROWTH

- *Exponential (J-curve) growth leads to overshoot and dieback cycles.*
- *Logistic (S-curve) growth slows as it approaches carrying capacity (K).*
- *Species with rapid reproduction (r-selected) tend toward exponential growth; those regulated by internal factors (K-selected) tend toward logistic growth.*

Apart from their interactions with other species, organisms have an inherent rate of reproduction that influences population size. Many species have the potential to produce almost unbelievable numbers of offspring. Consider a single female housefly (*Musca domestica*), which can lay 120 eggs. In 56 days those eggs become mature adults, and each female—suppose half are female—can lay another 120 eggs. At this rate, there can be seven generations of flies in a year, and that original fly would be the proud grandparent of 5.6 trillion offspring. If this rate of reproduction continued for 10 years, the entire earth would be covered in several meters of housefly bodies. Luckily, housefly reproduction, as for most organisms, is constrained in a variety of ways—scarcity of resources, competition, predation, disease, accident. The housefly merely demonstrates the remarkable amplification—the **biotic potential**—of unrestrained biological reproduction.

Growth without limits is exponential

Understanding population dynamics, or the rise and fall of populations in an area, is essential for understanding how species interact and use resources. As discussed in chapter 2, a population consists of all the members of a single species living in a specific area at the same time. The growth of the housefly population just described is **exponential,** having no limit and possessing a distinctive shape when graphed over time. An exponential growth rate (increase in numbers per unit of time) is expressed as a constant fraction, or exponent, which is used as a multiplier of the existing population. The mathematical equation for exponential growth is

$$\frac{dN}{dt} = rN$$

Here d means "change," so the change in number of individuals (dN) per change in time (dt) equals the rate of growth (r) times the number of individuals in the population (N). The r term (intrinsic capacity for increase) is a fraction representing the average individual contribution to population growth. If r is positive, the population is increasing. If r is negative, the population is shrinking. If r is zero, there is no change, and $dN/dt = 0$.

A graph of exponential population growth is described as a **J curve** (fig. 3.20) because of its shape. As you can see, the number of individuals added to a population at the

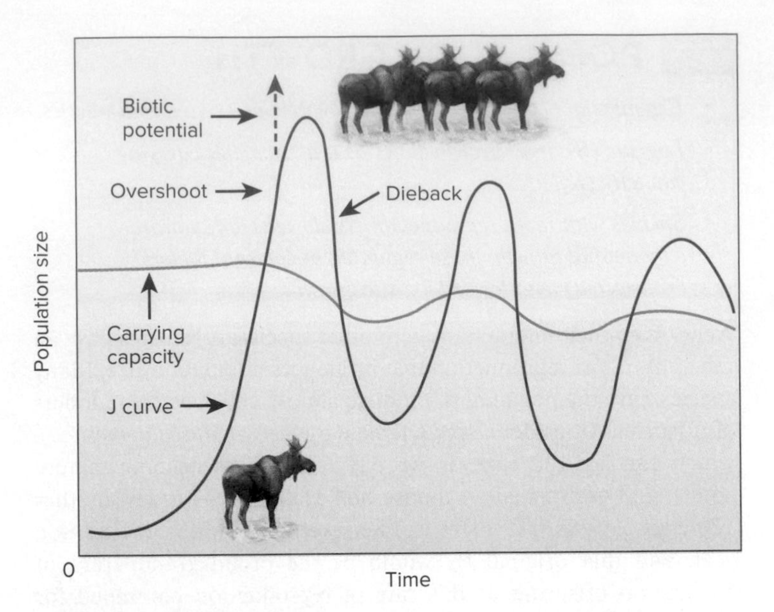

▲ **FIGURE 3.20** A J curve, or exponential growth curve, leads to repeated overshoot and dieback cycles. The environment's ability to support the species (carrying capacity) may diminish as overuse degrades habitat. Moose on Isle Royales in Lake Superior seem to have exhibited this pattern.

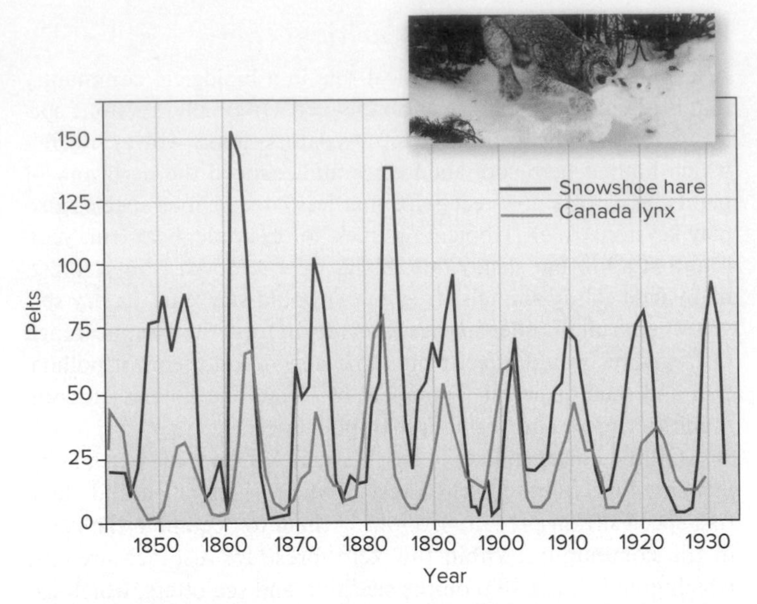

▲ **FIGURE 3.21** Ten-year oscillations in the populations of snowshoe hares and lynx in Canada suggest a close linkage of predator and prey. These data are based on the number of pelts received by the Hudson Bay Company from fur traders. Source: Data from D. A. MacLulich. *Fluctuations in the numbers of the Varying hare (Lepus americanus)*. University of Toronto Press, 1937, reprinted 1974. ©Ed Cesar/Science Source

beginning of an exponential growth curve can be rather small. But the numbers begin to increase quickly with a fixed growth rate. For example, when a population has just 100 individuals, a 2 percent growth rate adds just 2 individuals. For a population of 10,000, that 2 percent growth adds 200 individuals.

The exponential growth equation is a very simple model; it is an idealized description of a real process. The same equation is used to calculate growth in your bank account due to compounded interest rates; achieving the maximum growth potential requires that you never withdraw any money. But, in fact, some money probably will be withdrawn. Similarly, not all individuals in a population survive, so actual growth rates are something less than the full biotic potential.

Carrying capacity limits growth

In the real world there are limits to growth. Around 1970, ecologists developed the concept of **carrying capacity** to mean the number or biomass of animals that can be supported (without harvest) in a certain area of habitat. The concept is now used more generally to suggest a limit of sustainability that an environment has in relation to the size of a species' population. Carrying capacity is helpful in understanding the population dynamics of some species, perhaps even humans.

When a population overshoots, or exceeds, the carrying capacity of its environment, resources become limited and death rates rise. If deaths exceed births, the growth rate becomes negative and the population may suddenly decrease, a change called a population crash or dieback (fig. 3.20). Populations may oscillate from high to low levels around the habitat's carrying capacity, which may be lowered if the habitat is damaged. Moose and other browsers or grazers sometimes overgraze their food plants, so future populations in the same habitat find less preferred food to sustain them, at least until the habitat recovers. Some species go through predictable cycles if simple factors are involved, such as the seasonal light- and temperature-dependent bloom of algae in a lake. Cycles can be irregular if complex environmental and biotic relationships exist. Irregular cycles include outbreaks of migratory locusts in the Sahara and tent caterpillars in temperate forests—these represent irruptive population growth. Often immigration of a species into an area, or emigration from an area, also affects population growth and declines.

Sometimes predator and prey populations oscillate in synchrony with each other. One classic study employed the 200-year record of furs sold at Hudson Bay Company trading posts in Canada (fig. 3.21 shows a portion of that record). The ecologist Charles Elton showed that numbers of Canada lynx (*Lynx canadensis*) fluctuate on about a 10-year cycle that mirrors, slightly out of phase, the population peaks of snowshoe hares (*Lepus americanus*). When the hare population is high, the lynx prosper on abundant prey; they reproduce well, and their population grows. Eventually, the abundant hares overgraze the vegetation, decreasing their food supplies, and the hare populations shrink. For a while the lynx benefits because starving hares are easier to catch than healthy ones. As hares become scarce, however, so do lynx. When hares are at their lowest levels, their food supply recovers and the whole cycle starts over again. This predator-prey oscillation is described mathematically in the Lotka-Volterra model, named for the scientists who developed it.

Environmental limits lead to logistic growth

Not all biological populations cycle through exponential overshoot and catastrophic dieback. Many species are regulated by both internal and external factors and come into equilibrium with their environmental resources while maintaining relatively stable population sizes. When resources are unlimited, they may even grow exponentially, but this growth slows as the carrying capacity of the environment is approached. This population dynamic is called **logistic growth** because of its changes in growth rate over time.

Mathematically, this growth pattern is described by the following equation, which adds a feedback term for carrying capacity (*K*) to the exponential growth equation:

$$\frac{dN}{dt} = rN\frac{(K - N)}{K}$$

The logistic growth equation says that the change in numbers over time (*dN*/*dt*) equals the exponential growth rate (*rN*) times the portion of the carrying capacity (*K*) not already taken by the current population size (*N*). The term (*K* – *N*)/*K* establishes the relationship between population size at any given time and the carrying capacity (*K*). If *N* is less than *K*, the rate of population change will be positive. If *N* is greater than *K*, then change will be negative (Active Learning, below).

The logistic growth curve has a different shape than the exponential growth curve. It is a sigmoidal-shaped, or **S, curve** (fig. 3.22). It describes a population whose growth rate decreases if its numbers approach or exceed the carrying capacity of the environment.

Population growth rates are affected by external and internal factors. External factors include habitat quality, food availability, and interactions with other organisms. As populations grow, food becomes scarcer and competition for resources more intense. With a larger population, there is an increased risk that disease or parasites will spread, or that predators will be attracted to the area. Internal factors, such as slow growth and maturity, body size, metabolism, or hormonal status, can reduce reproductive output. Often crowding increases these factors. Overcrowded house mice ($>1{,}600/m^3$), for instance, average 5.1 babies per litter, while uncrowded house mice ($<34/m^3$) produce 6.2 babies per litter. All these factors are **density-dependent:** As population size increases, the effect intensifies. With **density-independent** factors, a population is affected no matter what its size. Drought, an early killing frost, flooding, landslide, habitat destruction by people—all increase mortality rates regardless of the population size. Density-independent limits to population are often nonbiological, capricious acts of nature.

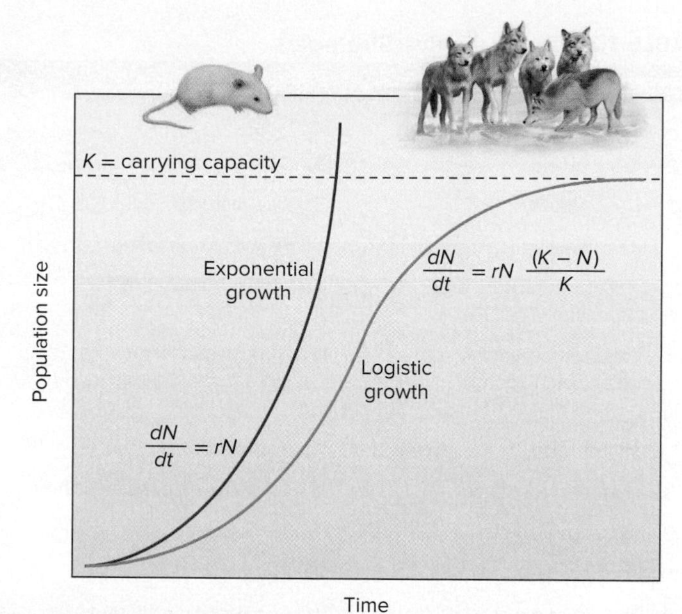

▲ **FIGURE 3.22** Exponential growth rises in a J-shaped curve. In contrast, logistic growth rates form an S-shaped curve as carrying capacity slows or stops population growth.

Species respond to limits differently: *r*- and *K*-selected species

Some organisms, such as dandelions and barnacles, depend on a high rate of reproduction and growth (*r*) to secure a place in the environment. These organisms are called ***r*-selected species** because they are adapted to employ a high reproductive rate to overcome the high mortality of virtually ignored offspring. These species may even overshoot carrying capacity and experience population crashes, but as long as vast quantities of young are produced, a few will survive. Other organisms that reproduce more conservatively—longer generation times, late sexual maturity, fewer young—are referred to as ***K*-selected species** because they are adapted for slower growth conditions near the carrying capacity (*K*) of their environment.

Many species blend exponential (*r*-selected) and logistic (*K*-selected) growth characteristics. Still, it's useful to contrast the advantages and disadvantages of organisms at the extremes of the continuum. It also helps if we view differences in terms of "strategies" of adaptation and the "logic" of different reproductive modes (table 3.3).

So-called *K*-selected organisms are usually large live long lives, mature slowly, produce few offspring in each generation, and have few natural predators (fig. 3.23, Type I). Elephants, for example, are not reproductively mature until they are 18 to 20 years old. In youth and adolescence, a young elephant belongs to an extended family that cares for it, protects it, and teaches it how to behave. A female elephant normally conceives only once every 4 or 5 years. The gestation period is about 18 months; thus, an elephant herd doesn't produce many babies in any year. Because elephants have few enemies and live a long life (60 to 70 years), this low reproductive rate produces enough elephants to keep the population stable, given good environmental conditions and no poachers.

Organisms with *r*-selected, or exponential, growth patterns tend to occupy low trophic levels in their ecosystems (see chapter 2) or they are successional pioneers. Niche generalists occupy disturbed or new environments, grow rapidly, mature early, and produce many offspring with excellent dispersal abilities. As individual parents, they do little to care for their offspring or protect them from predation. They invest their energy in producing huge numbers of young and count on some surviving to adulthood (fig. 3.23, Type III).

Active LEARNING

Effect of K on Population Growth Rate (rN)

In logistic growth, the term (*K* – *N*)/*K* creates a fraction that is multiplied by the growth rate (*rN*). Suppose carrying capacity (*K*) is 100. If *N* is 150, then is the term (*K* – *N*)/*K* positive or negative? Is population change positive or negative? What if *N* is 50? if *N* is 100?

ANSWERS: negative, positive, no growth

TABLE 3.3 | Reproductive Strategies

r-SELECTED SPECIES	*K*-SELECTED SPECIES
1. Short life	1. Long life
2. Rapid growth	2. Slower growth
3. Early maturity	3. Late maturity
4. Many, small offspring	4. Few, large offspring
5. Little parental care and protection	5. High parental care or protection
6. Little investment in individual offspring	6. High investment in individual offspring
7. Adapted to unstable environment	7. Adapted to stable environment
8. Pioneers, colonizers	8. Later stages of succession
9. Niche generalists	9. Niche specialists
10. Prey	10. Predators
11. Regulated mainly by intrinsic factors	11. Regulated mainly by extrinsic factors
12. Low trophic level	12. High trophic level

A female clam, for example, can release up to 1 million eggs in her short lifetime. The vast majority of young clams die before reaching maturity, but if even a few survive, the species will continue. Many marine invertebrates, parasites, insects, rodents, and annual plants follow this reproductive strategy. Also included in this group are most invasive and pioneer organisms, weeds, and pests.

Many species exhibit some characteristics of both *r* and *K* selection. These might approximately follow a Type II survivorship pattern (fig. 3.23), with similar likelihood of mortality at multiple life stages.

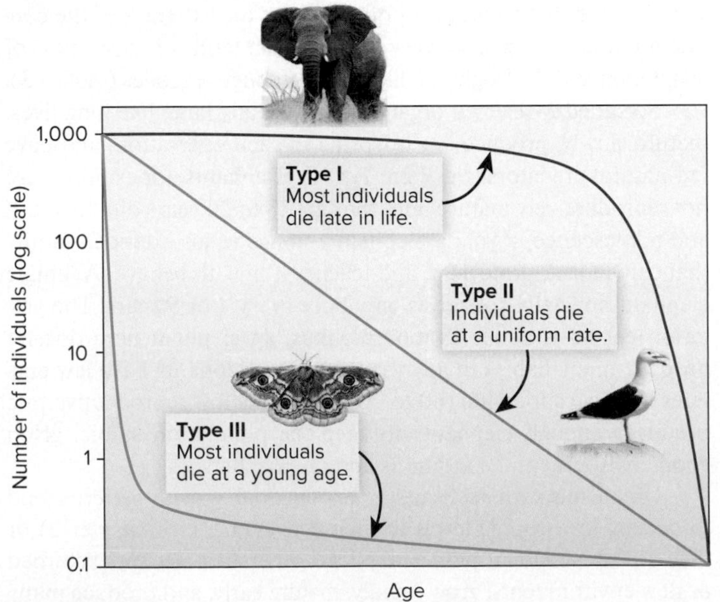

▲ **FIGURE 3.23** Idealized survivorship curves include high survival rates for juveniles (Type I), uniform mortality risk throughout life (Type II), and high mortality for juveniles, with long survival for a few individuals that reach maturity (Type III). (a): ©Digital Vision./Getty Images; (b): ©Stockbyte/Getty Images; (c): ©Simon Kovacic/123RF

When you consider the species you recognize from around the world, can you classify them as *r*- or *K*-selected species? What strategies seem to be operating for ants, bald eagles, cheetahs, clams, giraffes, pandas, or sharks? An important question in the back of your mind might be, where do people fit? Are we like wolves and elephants or more similar to clams and rabbits in our population growth strategy?

3.4 COMMUNITY DIVERSITY

- *Diversity is the number of species; abundance is the number of individuals in an area.*
- *Diversity is often associated with resilience and stability.*
- *Primary productivity is highest in warm, wet environments.*

No species is an island. It always lives with other species in a biological community in a particular environment. You've seen how interactions among species affect biological communities. In this section we'll consider how the many species in an area together produce the basic properties of biological communities and ecosystems. The main properties of interest to ecologists are (1) diversity and abundance; (2) community structure and patchiness; and (3) complexity, resilience, productivity, and stability.

Diversity and abundance

Diversity is the number of different species in an area, or the number per unit area. The number of herbaceous plant species per square meter of African savanna, the number of bird species in Costa Rica, all the insect species on earth (tens of millions of them!)–all describe diversity. Diversity is important because it indicates the variety of ecological niches and genetic variation in a community. **Abundance** refers to the number of individuals of a particular species (or of a group) in an area. Diversity and abundance are often related. Communities with high diversity often have few individuals of any one species, because there are many different species sharing available resources. Most communities contain a few common species and many rarer ones.

As a general rule, species diversity is greatest at the equator and drops toward the poles. Though the number of species is lower near the poles, the abundance of particular species can be very high. In the Arctic there are few insect species overall, but the abundance of one or two, especially mosquitoes, is almost incalculable. In the tropics, on the other hand, hundreds of thousands of different insect species coexist. Highly specialized forms and behaviors allow them to exploit narrow niches, but at any one location only a few individuals of a species might occur. The same pattern is seen in trees of high-latitude boreal forests versus tropical rainforests, and in bird populations. In Greenland there are 56 species of breeding birds, whereas in Colombia, with one-fifth the area, there are 1,395.

Climate explains much of this difference in diversity. Greenland has a harsh climate and a short, cool growing season that restricts the biological activity that can take place. During half the year, solar energy is abundant, but during winter incoming energy is almost absent. Only species mobile enough to take advantage of

(a) Random

(b) Unifrom

(c) Clustered

▲ **FIGURE 3.24** Distribution of a population can be random (a), uniform (b), or clustered (c). (a): ©Jim Zuckerman/Getty Images; (b): ©Eric and David Hosking/Getty Images; (c): ©anopdesignstock/Getty Images

fleeting resources can survive. History also matters: Greenland's coast has been free of glaciers for only about 10,000 years, so new species have had little time to develop.

Many areas in the tropics, by contrast, were never covered by glacial ice and have abundant rainfall and warm temperatures year-round, so ecosystems there are highly productive. The year-round availability of food, moisture, and warmth supports an exuberance of life and allows a high degree of specialization in physical shape and behavior. Many niches exist in small areas, with associated high species diversity. Coral reefs are similarly stable, productive, and conducive to proliferation of diverse and exotic life-forms. An enormous abundance of brightly colored and fantastically shaped fishes, corals, sponges, and arthropods live in the reef community. Increasingly, human activities also influence biological diversity today. The cumulative effects of our local actions can dramatically alter biodiversity (What Can You Do?, at right). We discuss this issue in chapter 5.

Patterns produce community structure

The spatial distribution of individuals, species, and populations can influence diversity, productivity, and stability in a community. Niche diversity and species diversity can increase as the complexity increases at the landscape scale, for example. **Community structure** is a general term we use for spatial patterns. Ecologists focus on several aspects of community structure, which we discuss here.

Distribution can be random, ordered, or patchy Even in a relatively uniform environment, individuals of a species' population can be distributed randomly, arranged in uniform patterns, or clustered together. In randomly distributed populations, individuals live wherever resources are available and chance events allow them to settle (fig. 3.24a). Uniform patterns arise from the physical environment also but more often are caused by competition and territoriality. For example, penguins or seabirds compete fiercely for nesting sites in their colonies. Each nest tends to be just out of reach of neighbors sitting on their own nests. Constant squabbling produces a highly regular pattern (fig. 3.24b). Plants also compete, producing a uniform pattern. Sagebrush releases toxins from roots and fallen leaves, which inhibit the growth of competitors and create a circle of bare ground around each bush. Neighbors grow up to the limit of this chemical barrier, and regular spacing results.

What Can YOU DO?

Working Locally for Ecological Diversity

You might think that the diversity and complexity of ecological systems are too large or too abstract for you to have any influence. But you can contribute to a complex, resilient, and interesting ecosystem, whether you live in the inner city, a suburb, or a rural area.

- Take walks. The best way to learn about ecological systems in your area is to take walks and practice observing your environment. Go with friends, and try to identify some of the species and trophic relationships in your area.
- Keep your cat indoors. Our lovable domestic cats are also very successful predators. Migratory birds, especially those nesting on the ground, have not evolved defenses against these predators.
- Plant a butterfly garden. Use native plants that support a diverse insect population. Native trees with berries or fruit also support birds. (Be sure to avoid non-native invasive species.) Allow structural diversity (open areas, shrubs, and trees) to support a range of species.
- Join a local environmental organization. Often the best way to be effective is to concentrate your efforts close to home. City parks and neighborhoods support ecological communities, as do farming and rural areas. Join an organization working to maintain ecosystem health; start by looking for environmental clubs at your school, park organizations, a local Audubon chapter, or a local Nature Conservancy branch.
- Live in town. Suburban sprawl consumes wildlife habitat and reduces ecosystem complexity by removing many specialized plants and animals. Replacing forests and grasslands with lawns and streets is the surest way to simplify, or eliminate, ecosystems.

Other species cluster together for protection, mutual assistance, reproduction, or access to an environmental resource. Ocean and freshwater fish form dense schools, increasing their chances of detecting and escaping predators (fig. 3.24c). Meanwhile, many predators—whether wolves or humans—hunt in packs. When blackbirds flock in a cornfield, or baboons troop across the African savanna, their group size helps them evade predators and find food efficiently. Plants also cluster for protection in harsh environments. You often see groves of wind-sheared evergreens at mountain treelines or behind foredunes at seashores. These treelines protect the plants from wind damage and incidentally shelter other animals and plants, creating a cluster of communities.

Individuals can also be distributed vertically in a community. Forests, for instance, have many layers, each with different environmental conditions and combinations of species. Distinct communities of plants, animals, and microbes live in the treetops, at mid-canopy level, and near the ground. This layering, known as vertical stratification, is best developed in tropical rainforests (fig. 3.25).

▲ **FIGURE 3.25** Vertical layering of plants and animals is an important type of community structure and is especially evident in tropical rainforests.

Aquatic communities also are often stratified into layers formed by species responding to varying levels of light, temperature, salinity, nutrients, and pressure.

Communities form patterns in landscapes If you fly in an airplane, you can see that the landscape consists of patches of different colors and shapes (fig. 3.26). Some appear long and narrow (hedgerows or rivers), while others are rectangular (pastures and cropfields) or green and lumpy in summer (forests). Each patch represents a biological community with its own set of species and environmental conditions. Most landscapes exhibit patchiness in some way. The largest patches might contain **core habitat**, a relatively uniform environment that is free of the influence of edges. Often we consider generalist species to occur on edges, while many specialists may require the more consistent conditions of core habitat. The northern spotted owl, for example, nests in the interior of large patches of mature coniferous forest in the Pacific Northwest (see chapter 6). In smaller patches the owl fares poorly, possibly due to competition with the closely related barred owl. A single pair of northern spotted owls may require over 1,000 hectares (ha) of core habitat to survive.

Where communities meet, the environmental conditions blend and the species and microclimate of one community can penetrate the other. Called **edge effects,** the penetrating influences may extend hundreds of meters into an adjacent community. A forest edge adjacent to open grassland is sunnier, drier, hotter, and more susceptible to storm damage than the center of the forest. Generalist grassland species, including weeds and predators, may move into the forest and negatively affect forest species. The shape of a patch also can affect the amount of core habitat. In a narrow, irregularly shaped patch, far-reaching edge effects would leave no core habitat.

▲ **FIGURE 3.26** Complex landscapes include contrasting environments, edges where they meet, and corridors connecting larger patches. Edges are biologically rich, but core areas are critical for many species. ©Fuse/Getty Images

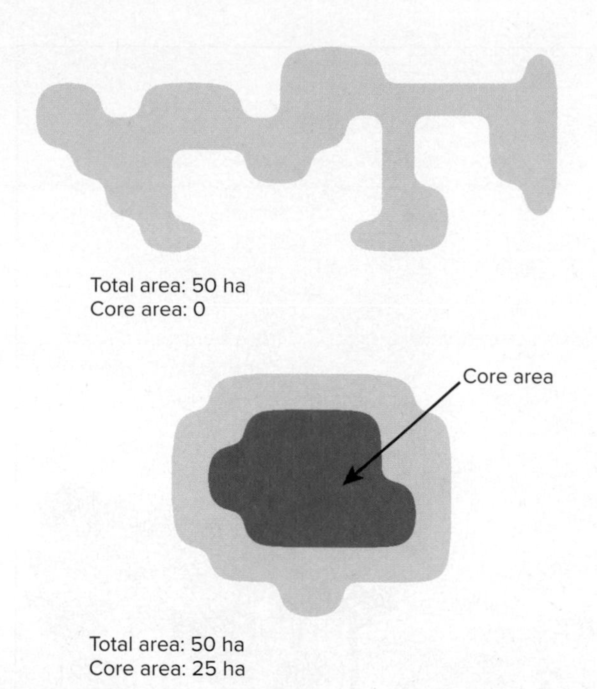

▲ **FIGURE 3.27** Shape can influence the availability of core area in a habitat area or a preserve.

In a similar-sized square patch, however, interior species would still find core habitat (fig. 3.27). Human activities often produce fragmented habitat with increased edge effects and decreased or loss of core habitat. Suburban expansion and forest clearing, for example, tend to occur in patchy patterns with a high density of edges in the landscape.

Edges are often rich in species because individuals from both environments occupy the boundary area. Many species prefer edges and use the resources of both environments. Many game animals, such as deer and pheasants, are most plentiful on edges. To boost these populations, North American game managers in the 1930s worked to create openings in forests and to plant trees and shrubs in grasslands. More recently, wildlife biologists have recognized that these edges reduce habitat for species that need interior conditions. Habitat managers have therefore worked to preserve larger patches and to connect smaller patches where possible (fig. 3.27).

Resilience seems related to complexity

The relationship between complexity and resilience has long been an important question in ecology. The issue was framed most famously by ecologist Robert MacArthur (1930–1972), who proposed that the more complexity a community possesses, the more resilient it is when disturbance strikes. He reasoned that if many different species occupy each trophic level, some can fill in if others are stressed or eliminated by external forces. The whole community has **resilience** and either resists or recovers quickly from disturbance. For example, the diversity of your intestinal microbial community appears to play a vital role in its stability and resilience (Exploring Science, p. 64).

Often we think of diversity in terms of species counts, but another aspect is community **complexity.** *Complexity* refers to the number of trophic levels in a community and the number of species at each of those trophic levels. A complex community might have many trophic levels and groups of species performing the same functions. In tropical rainforests and many other communities, herbivores form guilds based on the specialized ways they feed on plants. There may be fruit-eaters, leaf-nibblers, root-borers, seed-gnawers, and sap-suckers—each guild is composed of species of different sizes, shapes, and even biological kingdoms, but they feed in similar ways.

Similarly, we can see community complexity in an Antarctic ecosystem. The sun powers the system, with floating algae performing photosynthesis. These feed tiny crustaceans called krill, which in turn support subsequent trophic levels (table 3.4 and see What Do You Think? in chapter 2). Most ecosystems exhibit even greater diversity of species in the multiple trophic levels.

Another factor that can contribute to resilience is productivity. **Primary productivity** is the production of biomass by photosynthesis. Plants, algae, and some bacteria produce biomass by converting solar energy into chemical energy, which they use or pass on to other organisms. We can measure primary productivity in terms of units of biomass per unit area per year—for example, in grams per m^2 per year. Because cellular respiration in producing organisms uses much of that energy, a more useful term is **net primary productivity,** or the amount of biomass stored after respiration. Productivity depends on light levels, temperature, moisture, and nutrient availability, so it varies dramatically among different ecosystem types (fig. 3.28). Tropical forests, coral reefs, and the bays and estuaries (where rivers meet the ocean) have high productivity because abundant energy and moisture are available. In deserts, a lack of water limits photosynthesis, and productivity is low. On the arctic tundra or on high mountains, low temperatures and low solar energy inputs inhibit plant growth and productivity. In the open ocean, a lack of nutrients reduces the ability of algae to make use of plentiful sunshine and water.

TABLE 3.4 Community Complexity in the Antarctic Ocean

TYPE OF FUNCTION	MEMBERS OF FUNCTIONAL GROUP
Top ocean predator	Sperm and killer whales, leopard and elephant seals
Aerial predator	Albatross, skuas
Other ocean predator	Weddell and Ross seals, king penguin, pelagic fish
Krill/plankton-feeder	Minke, humpback, fin, blue, and sei whales
Ocean herbivore	Krill, zooplankton (many species)
Ocean-bottom predator	Many species of octopods and bottom-feeding fish
Ocean-bottom herbivore	Many species of echinoderms, crustaceans, mollusks
Photosynthesizer	Many species of phytoplankton and algae

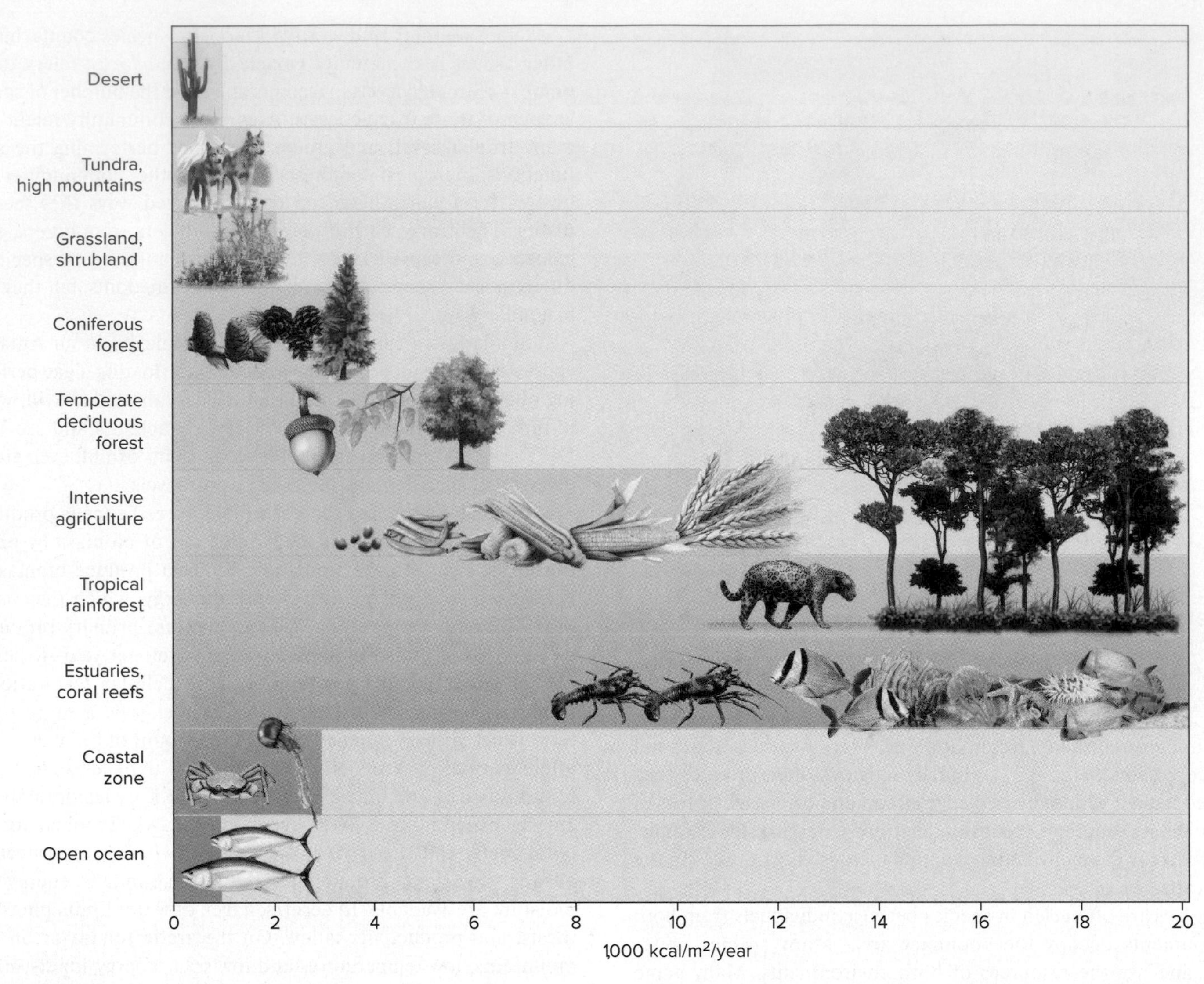

▲ **FIGURE 3.28** Biomass accumulates at different rates in the world's major ecosystem types. Differences in net primary production is chiefly due to temperature, rainfall, and nutrients. Interactions among species also boost productivity.

Even the most photosynthetically active ecosystems capture only a small percentage of the available sunlight and use it to make energy-rich compounds. In a temperate-climate oak forest, for example, leaves absorb only about half the available light on a midsummer day. Of this absorbed energy, 99 percent is used in respiration and in evaporating water to cool leaves. A large oak tree can transpire (evaporate) several thousand liters of water on a warm, dry, sunny day while making only a few kilograms of sugars and other energy-rich organic compounds.

Stability is another property that varies among ecosystems. When we say a community or an ecosystem is stable, we mean it resists changes despite disturbance, springs back resiliently after disturbance, and supports the same species in about the same numbers as before the disturbance. We often think of tropical rainforests as stable over long periods of time. They may have annual rainy and dry seasons, but we don't expect dramatic changes in species composition. In contrast, many coniferous forests are periodically "reset" by fires. There is a turnover in dominant species that might last for decades.

Sometimes disturbance dramatically changes an ecosystem. When the Great Plains experienced devastating droughts in the 1930s and early 1940s, overall productivity declined, some species' populations practically vanished, and others held their own but at lower abundance. When the rains returned, the plant communities were different, not just because of drought but also because millions of cattle had overgrazed the range—an additional, simultaneous disturbance. Today's remnant Great Plains grasslands are very different from the grasslands of two centuries ago. Yet they remains grasslands, produce forage, and feed cattle. Are the Great Plains grasslands stable because they are still relatively productive, or are they unstable because disturbances have dramatically altered species diversity and abundance? If the range were grazed according to ecological principles, would that restore the original species diversity and abundance, and raise productivity? Asking questions about diversity, productivity, resilience, and stability helps us consider what aspects of an ecosystem we value most, and what properties help maintain those characteristics of the community.

3.5 COMMUNITIES ARE DYNAMIC AND CHANGE OVER TIME

- *Succession is a shift in community composition over time.*
- *Disturbance, such as fire or forest clearing, can alter community composition.*
- *Many communities and species are adapted to tolerate disturbance.*

If fire sweeps through a forest, we often say that the forest was destroyed. But usually that is an inaccurate description of what happened. Often fire is good for a community—releasing nutrients in a burned grassland or allowing for regeneration of aging trees in a coniferous forest. The idea that dramatic, periodic change can be a part of normal ecosystems is relatively new in ecology. We used to consider stability the optimal state of ecosystems. But with more observations and more studies, we have learned that communities can be dynamic, with dramatic changes over time.

Are communities organismal or individualistic?

For several decades starting in the early 1900s, ecologists in North America and Europe argued about the basic nature of communities. This debate doesn't make great party conversation (unless you're an ecologist), but it has long influenced the ways we study and understand communities, the ways we view changes in a community, and ultimately the ways we use them. On one side of the debate was an idea proposed by J. E. B. Warming (1841–1924) in Denmark and Henry Chandler Cowles (1869–1939) in the United States. These two proposed that communities develop in a sequence of stages, starting either from new land or after a severe disturbance. Working in sand dunes, they examined the changes as plants first took root in bare sand and, with further development, ultimately created a forest. The community that developed last and lasted the longest was called the **climax community.**

The importance of climax communities was championed by the biogeographer F. E. Clements (1874–1945). He viewed the process as a relay—species replace each other in predictable groups and in a fixed, regular order. He argued that every landscape has a characteristic climax community, determined mainly by climate. If left undisturbed, this community would mature to a characteristic set of species, each performing its optimal functions. A climax community represented to Clements the maximum possible complexity and stability in a given situation. He and others made the analogy that the development of a climax community resembled the maturation of an organism. Like organisms, they argued, communities began simply and primitively, maturing until a highly integrated, complex, and stable condition developed.

On the other side of the debate was an individualistic view of community change, which was championed by Clements's contemporary, H. A. Gleason (1882–1975). Gleason saw community history as an unpredictable process. He argued that species are individualistic, each establishing in an environment according to its own ability to colonize, tolerate the environmental conditions, and reproduce there. This idea allows for myriad temporary associations of plants and animals to form, fall apart, and reconstitute in slightly different forms, depending on environmental conditions and the species in the neighborhood. Imagine a time-lapse movie of a busy airport terminal. Passengers come and go; groups form and dissipate. Patterns and assemblages that seem significant may not mean much a year later.

In nature, general growth forms might be predictable—grasses grow in some conditions, while trees grow in others—but the exact composition of a community is not necessarily predictable. Gleason suggested that we think ecosystems are uniform and stable only because our lifetimes are too short and our geographic scope too limited to understand their actual dynamic nature. Most ecologists now find that Gleason's explanation fits observed communities better than Clements's does.

Succession describes community change

Succession is a process in which organisms occupy a site and change its environmental conditions, gradually making way for another type of community. In **primary succession,** land that is bare of soil—a sandbar, rock surface, volcanic flow—is colonized by living organisms where none lived before (fig. 3.29). **Secondary succession,** meanwhile, occurs after a disturbance, when a new community develops from the biological legacy of the previous one. In both kinds of succession, organisms change the environment by

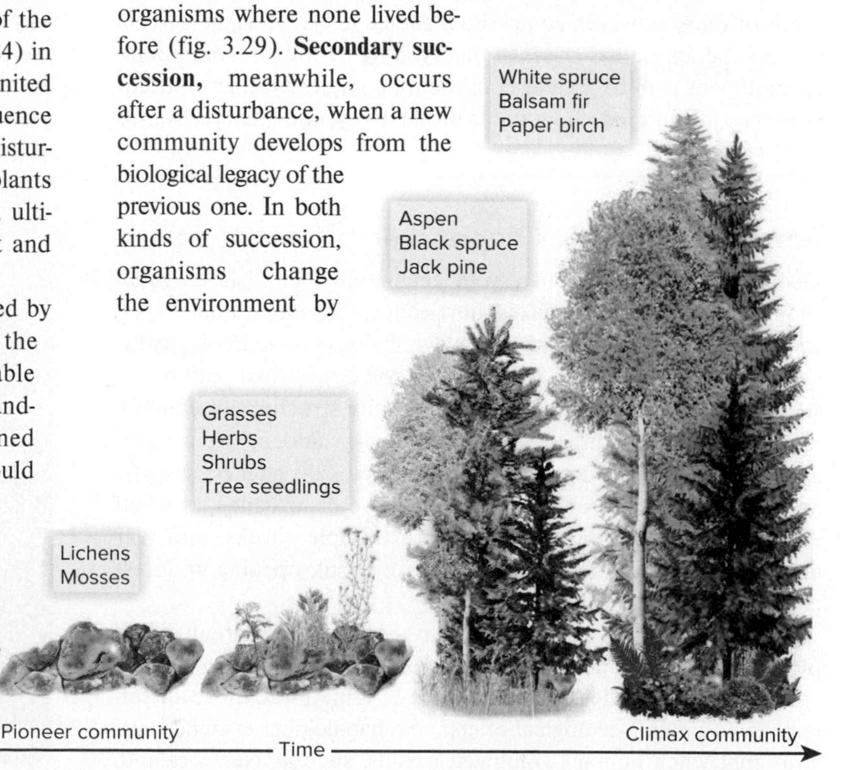

▲ **FIGURE 3.29** Primary succession in boreal forest involves five main stages. Exposed rocks are colonized by lichens and mosses, which trap moisture and build soil for grasses, shrubs, and eventually trees. Periodic fires can set back trees, initiating secondary succession. Aspens spring back from roots, pines germinate in openings, and a new forest begins.

modifying soil, light levels, food and water supplies, and microclimate. This change permits new species to colonize and eventually replace the previous species, a process known as ecological development or facilitation.

In primary succession on land, the first colonists are hardy **pioneer species,** often microbes, mosses, and lichens that can withstand a harsh environment with few resources. Pioneer species create patches of organic matter and debris that accumulate in pockets and crevices, retaining water and creating soil where seeds of more plants lodge and grow. As succession proceeds, the community becomes more diverse and interspecies competition arises. Larger plants grow, creating vertical structure. Pioneers disappear as the environment favors new colonizers that have competitive abilities more suited to the new environment.

You can see secondary succession all around you, in abandoned farm fields, in clear-cut forests, and in disturbed suburbs and lots. Soil, seeds, and residual plant roots may be present. Because disturbed soil lacks vegetation, plants that live 1 or 2 years (annuals and biennials) do well. Their lightweight seeds travel far on the wind, and their seedlings tolerate full sun and extreme heat. When they die, they lay down organic material that improves the soil's fertility and shelters other seedlings. Soon long-lived and deep-rooted perennial grasses, herbs, shrubs, and trees take hold, building up the soil's organic matter and increasing its ability to store moisture. Forest species that cannot survive bare, dry, sunny ground eventually find ample food, a diverse community structure, and shelter from drying winds and low humidity.

Generalists figure prominently in early succession. Over thousands of years, however, competition should decrease as niches proliferate and specialists arise. In theory, long periods of community development lead to greater community complexity, high nutrient conservation and recycling, stable productivity, and great resistance to disturbance.

Some communities depend on disturbance

Disturbances occur frequently in ecosystems. Landslides, mudslides, hailstorms, earthquakes, hurricanes, tornadoes, tidal waves, wildfires, and volcanoes are just a few obvious types. Ecologically, a **disturbance** is any force that disrupts the established patterns of species diversity and abundance, community structure, or community properties. Disturbances are not always sudden, like a hurricane. Slow events like drought can also dramatically alter ecosystems and their functions. Animals can also cause disturbance, as when African elephants rip out small trees, trample shrubs, and tear down tree limbs as they forage and move about, opening up forest communities and creating savannas.

People also cause disturbances in many ways. Aboriginal people set fires, introduced new species, harvested resources, or changed communities in many places. Sometimes those disturbances had major ecological effects. Archaeological evidence suggests that when humans colonized islands, such as New Zealand, Hawaii, and Madagascar, large-scale extinction of many animal species followed. Although there is debate about whether something similar happened in the Americas at the end of the last ice age, about the same time that humans arrived, many of the large mammal species disappeared. In some cases the landscapes created by aboriginal people may have been maintained for so long that we assume that those conditions always existed there. In eastern North America, for example, fires set by native people maintained grassy, open savannas that early explorers assumed were the natural state of the landscape.

The disturbances caused by modern technological societies are often much more obvious and irreversible. You may have seen scars from road building, mining, clear-cut logging, or other disruptive activities. Obviously, once a landscape has been turned into a highway, a parking lot, or a huge hole in the ground, it will be a long time before it returns to its former condition. But sometimes even if the disturbance seems relatively slight, it can have profound effects.

Consider the Kingston Plains in Michigan's Upper Peninsula, for example. Clear-cut logging at the end of the nineteenth century removed the white pine forest that once grew there. Repeated burning by pioneers, who hoped to establish farms, removed nutrients from the sandy soil and changed ecological conditions so that more than a century later, the forest still hasn't regenerated (fig. 3.30). Given extensive changes by either humans or nature, it may take centuries for a site to return to its predisturbance state, and if climate or other conditions change in the meantime, it may never recover.

Clements's organismal perspective would suggest that most disturbance is similarly harmful. In the early 1900s this view guided efforts to protect timber supplies in the American West from ubiquitous wildfires. Establishing stability and reducing disturbance was also an argument for building dams on rivers all across the West—to prevent disturbance from flooding and to store water for dry seasons. Fire suppression and flood control became the central policies in American natural resource management for most of the twentieth century.

▲ **FIGURE 3.30** Sometimes recovery from disturbance is extremely slow, or the system shifts to a new type of ecosystem. At Kingston Plains in Michigan's eastern Upper Peninsula, stumps remain a century after the trees were cleared by logging and fire. ©William P. Cunningham

Recently, new ideas about natural disturbances have entered land management discussions and brought change to some land management policies. Grasslands and some forests are now considered "fire-adapted," and fires are allowed to burn in them if weather conditions are appropriate. Floods also are seen as crucial for maintaining floodplain and river health. Policymakers and managers increasingly consider ecological information when deciding on new dams and levee construction projects.

Ecologists have found that disturbance benefits many species, much as predation does, because it sets back the most competitors and allows less competitive species to persist. In northern deciduous forests, maples (especially sugar maple and the more recently arrived Norway maple) are more prolific seeders and more shade tolerant than most other tree species. After decades without disturbance, maples outcompete other trees for a place in the forest canopy. The dense shade of maples basically starves other species for light. Most species of oak, hickory, and other light-requiring trees diminish in abundance, as do forest understory plants. When windstorms, tornadoes, wildfires, or ice storms hit a maple-dominated forest, trees are toppled, branches are broken, and light once again reaches the forest floor. This stimulates seedlings of oaks and hickories, as well as understory plants. Breaking the grip of the strongest competitor is the helpful role disturbances often play. The 1988 Yellowstone fires set back the lodgepole pine (*Pinus contorta*), which had expanded its acreage in the park. After a few years, successional processes created dense forests in some areas and open savannas in others (fig. 3.31). This resulted in a greater variety of plant species, and wildlife responded vigorously to the greater habitat diversity.

▲ **FIGURE 3.31** The Yellowstone National Park fires of 1988 burned 1.4 million acres. Recovery of the region, which was 80 percent lodgepole pine before the fire, has produced greater diversity of vegetation, including savannas and more mixed forest types. ©William P. Cunningham

In some landscapes, periodic disturbance is a basic characteristic, and communities are made up of **disturbance-adapted species.** Some survive fires by hiding underground; others reseed quickly after fires. Grasslands, the chaparral scrubland of California and the Mediterranean region, savannas, and some kinds of coniferous forests are shaped and maintained by periodic fires that have long been a part of their history. In fact, many of the dominant plant species in these communities need fire to suppress competitors, to prepare the ground for seeds to germinate, or to pop open cones or split thick seed coats and release seeds. Without fire, community structure would be quite different.

From another view, disturbance resets the successional clock that always operates in every community. Even though all seems chaotic after a disturbance, it may be that preserving species diversity by allowing in natural disturbances (or judiciously applied human disturbances) actually ensures stability over the long run, just as diverse prairies managed with fire recover after drought.

CONCLUSION

Evolution is one of the key organizing principles of biology. It explains how species diversity originates and how organisms are able to live in highly specialized ecological niches. Natural selection, in which beneficial traits are passed from survivors in one generation to their progeny, is the mechanism by which evolution occurs. Species interactions—competition, predation, symbiosis, and coevolution—are important factors in natural selection. The unique set of organisms and environmental conditions in an ecological community gives rise to important properties, such as productivity, abundance, diversity, structure, complexity, connectedness, resilience, and succession. Human-caused introduction of new species as well as removal of existing ones can cause profound changes in biological communities and can compromise the life-supporting ecological services on which we all depend. These community ecology principles are useful in understanding many aspects of population change and ecosystem function.

PRACTICE QUIZ

1. Explain how tolerance limits to environmental factors determine distribution of a highly specialized species such as the saguaro cactus.
2. What is allopatric speciation? sympatric speciation?
3. Define *selective pressure*, and explain how it can alter traits in a species.
4. Explain the three types of survivorship curves shown in figure 3.23.
5. Describe several types of symbiotic relationships. Give an example of a symbiotic relationship that benefits both species.
6. What is coevolution? Give an example, either in a predatory relationship or in a symbiotic relationship.
7. Competition for a limited quantity of resources occurs in all ecosystems. This competition can be interspecific or intraspecific. Explain some of the ways an organism might deal with these different types of competition.

8. Explain the idea of *K*-selected and *r*-selected species. Give an example of each.
9. Describe the process of succession that occurs after a forest fire destroys an existing biological community. Why may periodic fire be beneficial to a community?
10. Which world ecosystems are most productive in terms of biomass (see fig. 3.28)? Which are least productive? What units are used in this figure to quantify biomass accumulation?
11. What do ecologists mean by the term *resilience*? In what ways might diversity contribute to resilience in an ecosystem?

CRITICAL THINKING AND DISCUSSION

Apply the principles you have learned in this chapter to discuss these questions with other students.

1. The concepts of natural selection and evolution are central to how most biologists understand and interpret the world. Why is Darwin's explanation so useful in biology? Does this explanation necessarily challenge traditional religious views? In what ways?
2. What is the difference between saying that a duck has webbed feet because it needs them to swim and saying that a duck is able to swim because it has webbed feet?
3. Given the information you've learned from this chapter, how would you explain the idea of speciation through natural selection and adaptation to a nonscientist?
4. Productivity, diversity, complexity, resilience, and structure are exhibited to some extent by all communities and ecosystems. Describe how these characteristics apply to an ecosystem with which you are familiar.
5. In what ways is disturbance good or bad in an ecosystem? Can it be both? Give an example from an ecosystem you know, and consider some of the disturbances that affect it. What are some negative and positive effects?
6. Ecologists debate whether biological communities have self-sustaining, self-regulating characteristics or are highly variable, accidental assemblages of individually acting species. What outlook or worldview might lead scientists to favor one or the other of these theories?
7. Many rare and endangered species are specialists. Explain what this means. As environments change, should we worry about losing specialists? Why or why not?

DATA ANALYSIS Competitive Exclusion

The principle of competitive exclusion is one of the observations that can be explained by the process of evolution by natural selection. David Lack's classic study of finches on several Galápagos Islands is one example of this. These species are all closely related, in a genus of ground finches, *Gespiza*. (Often biologists abbreviate the genus name when it is used repeatedly: For example, *Gespiza fuliginosa* is shortened, for convenience, to *G. fuliginosa*). Go to Connect to demonstrate your understanding of these graphs and what they tell us about how Lack demonstrated competitive exclusion in these species.

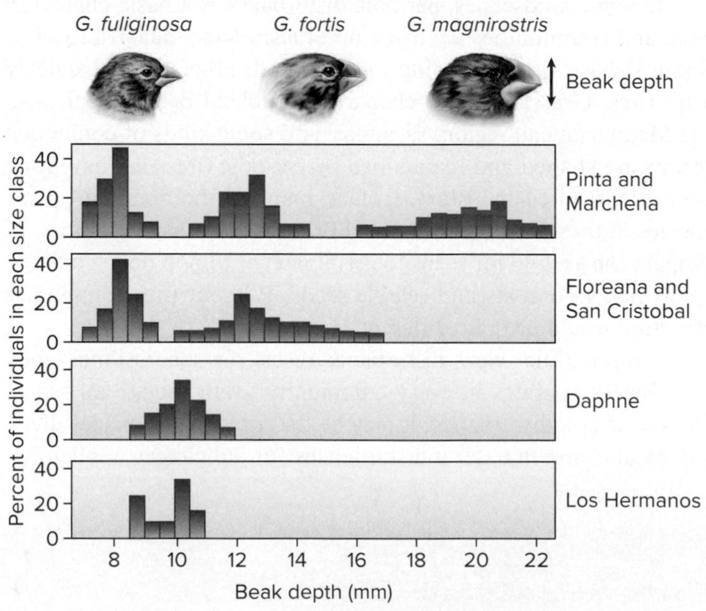

Plots of beak size on different islands. After D. Lack, 1947.

Design Elements: Active Learning (Toad): ©Gaertner/Alamy Stock Photo; Case Study (Globe): ©McGraw-Hill Education; Google Earth: ©McGraw-Hill Education; Abstract Background: ©Martin Kubat/Shutterstock; What do you think (Students using tablets): ©McGraw-Hill Education/Richard Hutchings, photographer; What can you do (Hand holding Globe): ©Christoph Weihs/Shutterstock

CHAPTER

4 Human Populations

LEARNING OUTCOMES

After studying this chapter, you should be able to answer the following questions:

- Why are we concerned about human population growth?
- Will the world's population triple in the twenty-first century, as it did in the twentieth?
- What is the relationship between population growth and environmental impact?
- Why did human populations grow so rapidly in the last century?
- How is human population growth changing in different parts of the world?
- How does population growth change as a society develops?
- What factors slow down or speed up human population growth?

Thailand's highly successful family planning program combines humor and education with economic development.

CASE STUDY

Family Planning in Thailand: A Success Story

Down a narrow lane off Bangkok's busy Sukhumvit Road is a most unusual café, Cabbages and Condoms. Not only is it highly rated for its spicy Thai food but it's also part of the only restaurant chain in the world dedicated to birth control. In an adjoining gift shop, baskets of condoms stand next to decorative handicrafts of the northern hill tribes. Piles of T-shirts carry messages such as "A condom a day keeps the doctor away" and "Our food is guaranteed not to cause pregnancy." The businesses are run by the Population and Community Development Association (PDA), Thailand's largest and most influential nongovernmental organization.

The PDA was founded in 1974 by Mechai Viravaidya, a genial and fun-loving former Thai Minister of Health, who is a genius at public relations and human motivation (fig. 4.1). While traveling around Thailand in the early 1970s, Mechai recognized that rapid population growth—particularly in poor rural areas—was an obstacle to community development. Although Thailand's economy was growing, poor families struggled to feed and educate their children. Rural communities remained depressed. And cities couldn't keep up with growing demand for housing, transportation, sanitation, and other services.

Mechai decided the most appealing way to teach family planning, and address the pains of population growth, was with humor. PDA workers handed out condoms at theaters, in traffic jams, anywhere a crowd gathered. They challenged government officials to condom balloon–blowing contests and taught youngsters Mechai's family planning song: "Too Many Children Make You Poor." The PDA even pays farmers to paint birth control ads on the sides of their water buffalo.

▲ **FIGURE 4.1** Mechai Viravaidya, right, is joined by Peter Piot, executive director of UNAIDS, left, in passing out free condoms on family planning and AIDS awareness day in Bangkok. ©APICHART WEERAWONG/AP Images

This campaign has been extremely successful in slowing population growth. Birth control and family planning, once taboo topics in polite society, became normal and unembarrassing. Although condoms—now commonly called "mechais" in Thailand—are the trademark of PDA, other contraceptives, such as pills, spermicidal foam, and IUDs, are promoted as well. Thailand was one of the first countries to allow the use of the injectable contraceptive DMPA and remains a major user. Free nonscalpel vasectomies are available on the king's birthday. Sterilization has become the most widely used form of contraception in the country. The campaign to encourage condom use and to discuss safe sex has also been helpful in combating AIDS.

In 1974, when PDA started, Thailand's growth rate was 3.3 percent per year. In just 15 years, the growth rate dropped to 1.6 percent, one of the most dramatic birth rate declines ever recorded. Contraceptive use among married couples increased from 15 to 70 percent, and all this was done without coercive programs. Now Thailand's growth rate is 0.5 percent, lower than that of the United States. The fertility rate (average number of children per woman) decreased from 7 in 1974 to 1.5 in 2017. The PDA is credited with the fact that Thailand's population is 20 million less than it would have been if it had followed its pre-1970 trajectory.

In addition to Mechai's creative genius and flair for showmanship, there are several reasons for this success story. Thai people love humor and are more egalitarian than most developing countries. Thai spouses share in decisions regarding children, family life, and contraception. The government recognizes the need for family planning and is willing to work with volunteer organizations, such as the PDA. And Buddhism, the religion of 95 percent of Thais, doesn't object to family planning.

The PDA isn't limited to family planning and condom distribution. It has expanded into a variety of economic development projects. Microlending provides money for a couple of pigs, or a bicycle, or a small supply of goods to sell at the market. Thousands of water-storage jars and cement rainwater-catchment basins have been distributed. Larger-scale community development grants include road-building, rural electrification, and irrigation projects. Like many population planning organizations, the PDA recognizes that human development and economic security are keys to successful population programs.

Resource limits aren't simply a matter of total number of people on the planet; they also depend on consumption levels and the types of technology we use. But population numbers have a critical impact on the environment as well as on human well-being. In this chapter we explore a number of key population questions: What have population trends been, and what might be the effects of exponential growth? What are the links among poverty, birth rates, and our common environment? What are the most effective, and most fair, ways to manage fertility and population growth? ■

> *For every complex problem there is an answer that is clear, simple, and wrong.*
>
> —H. L. MENCKEN

4.1 PAST AND CURRENT POPULATION GROWTH ARE VERY DIFFERENT

- *A fundamental debate is whether overpopulation will inevitably destroy resources or innovation will reduce our impacts.*
- *Doubling times describe changing rates of growth.*
- *Both environment and culture influence growth rates.*
- *Impacts are a function of population, affluence, and technology (I = PAT).*

There are a lot more people in the world today than when your grandparents were born. Over the twentieth century the global population more than tripled from 1.9 billion in 1900 to 6 billion in 2000. By the time you read this, there will be more than 7.7 billion humans on the globe. Two of the most important questions in environmental science are (1) how much more will the population grow? and (2) how many of us can our planet support sustainably?

According to the Population Reference Bureau, we're currently growing at an annual rate of 1.12 percent. That means we're adding about 83 million more people to the planet every year. Humans are now one of the most numerous vertebrate species on the earth. We also are more widely distributed and manifestly have a greater global environmental impact than any other species. For the families to whom these children are born, however, each birth may be a joyous and long-awaited event (fig. 4.2).

Birth rates are falling nearly everywhere in the world except sub-Saharan Africa. As the case study of Thailand shows, population control can be successful in developing countries even with relatively low standards of living. But many people worry that population will cause—or perhaps already is causing—resource depletion and environmental degradation that threaten the ecological life-support systems on which we all depend. These fears often lead to demands for immediate, worldwide birth control programs to reduce fertility rates and to eventually stabilize or even shrink the total number of humans.

Others believe that human ingenuity, technology, and enterprise can expand the world's carrying capacity and allow us to overcome any problems we encounter. From this perspective, more people may be beneficial, rather than disastrous. A larger population means a larger workforce, more geniuses, and more ideas about what to do. Along with every new mouth comes a pair of hands. Proponents of this worldview argue that continued economic and technological growth can both feed the world's billions and enrich everyone enough to end the population explosion voluntarily.

Still another opinion on this subject derives from social justice concerns. According to this worldview, resources are sufficient for everyone. Current shortages are only signs of greed, waste, and oppression. The root cause of environmental degradation, in this perspective, is inequitable distribution of wealth and power rather than merely population size. Fostering democracy, increasing rights for women and minorities, and improving the lives of the poor are essential for sustainability (fig. 4.2). A narrow focus on population growth fosters racism and blames the poor for their problems, while ignoring the high resource consumption of richer individuals and nations, according to this view.

What these different paradigms or worldviews imply for environmental quality and human life are among the most central and pressing questions in environmental science. Will the example of Thailand apply to other developing countries, or is it a unique situation? In this chapter we look at some causes of population growth, as well as at how populations are measured and described. Family planning and birth control are essential for stabilizing populations. The number of children a couple decides to have and the methods they use to regulate fertility, however, are strongly influenced by culture, religion, politics, and economics, as well as basic biological and medical considerations.

▲ **FIGURE 4.2** Population growth rates depend on many factors, including human rights and educational opportunities for girls. ©CandyraiN/Shutterstock

Benchmark Data

Among the ideas and values in this chapter, these are a few worth remembering.

7.7 billion	Global population, mid-2018
70/*n*	Doubling time (*n* = percentage growth rate)
33 years	Population doubling time in 1965
215 years	Doubling time in 2050
2.45	Global average fertility rate
2.1	Replacement fertility rate (children/woman)
1.9	Fertility rate in the United States
1.6	Fertility rate in China
9.5 ha	Ecological footprint in the United States, per person
2.2 ha	Footprint in Thailand

Human populations grew slowly until recently

For most of our history, humans were not very numerous, compared with many other species. Studies of hunting-and-gathering societies suggest that the total world population was probably only a few million people before the invention of agriculture and the domestication of animals around 10,000 years ago. The agricultural revolution produced a larger and more secure food supply and allowed the human population to grow, reaching perhaps 50 million people by 5000 B.C. For thousands of years, the number of humans increased very slowly. Archaeological evidence and historical descriptions suggest that only about 300 million people were living in the first century A.D. (table 4.1).

Active LEARNING

Population Doubling Time

If the world population is growing at 1.1 percent per year and continues at that rate, how long before it doubles? The "rule of 70" is a useful way to calculate the approximate doubling time in years for anything growing exponentially. For example, a savings account (or biological population) growing at a compound interest rate of 1 percent per year will double in about 70 years. Using the formula below, calculate doubling time for the world (growth rate = 1.1 percent/year), Uganda (3.2 percent), Nicaragua (2.7 percent), India (1.7 percent), United States (0.6 percent), Japan (0.1 percent), and Russia (−0.6 percent).

$$\text{Example: } \frac{\text{70 years}}{\text{(Growth percent)}} = \text{doubling time in years}$$

ANSWERS: World = 64 years; Uganda = 22 years; Nicaragua = 26 years; India = 41 years; United States = 117 years; Japan = 700 years; Russia = never

TABLE 4.1 World Population Growth and Doubling Times

DATE	POPULATION	DOUBLING TIME
5000 B.C.	50 million	?
800 B.C.	100 million	4,200 years
200 B.C.	200 million	600 years
A.D. 1200	400 million	1,400 years
A.D. 1700	800 million	500 years
A.D. 1900	1,600 million	200 years
A.D. 1965	3,200 million	33 years
A.D. 2000	6,100 million	51 years
A.D. 2050 (estimate)	8,920 million	215 years

Source: Data from United Nations Population Division.

As you can see in figure 4.3, human populations began to increase rapidly after about A.D. 1600. Many factors contributed to this rapid growth. Increased sailing and navigating skills stimulated commerce and communication among nations. Agricultural developments, better sources of power, and improved health care and hygiene also played a role. We are now in an exponential, or J-curve, pattern of growth, described in chapter 3.

It took all of human history to reach 1 billion people in 1800 but only 156 more years to get to 3 billion in 1960. It took us about 12 years to add the seventh billion. Another way to look at population growth is that the number of living humans tripled during the twentieth century. Will it do so again in the twenty-first century? If it does, will we overshoot our environment's carrying capacity and experience a catastrophic dieback similar to those described in chapter 3? As you will see later in this chapter, there is evidence that population growth already is slowing, but whether we will reach equilibrium soon enough and at a size that can be sustained over the long term remains a difficult but vital question.

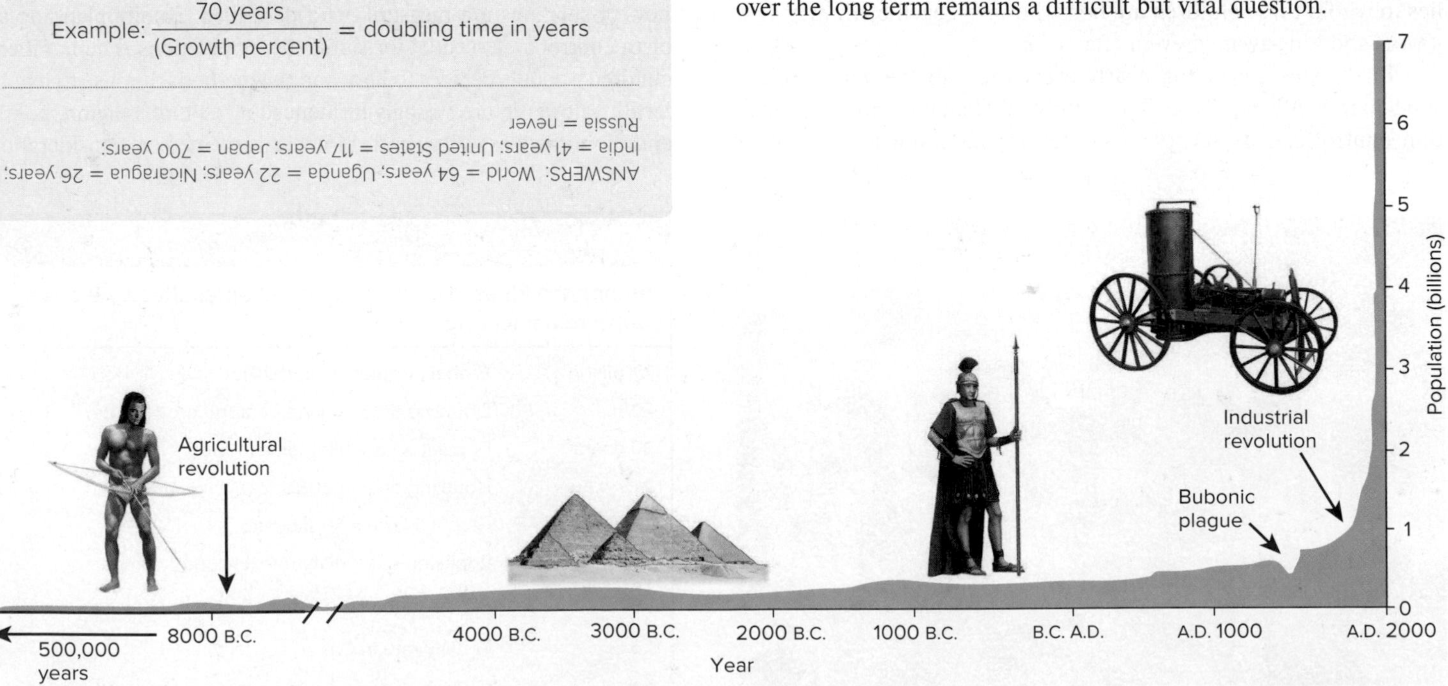

▲ **FIGURE 4.3** Human population levels throughout history. It is clear from the J-shaped growth curve that the human population is growing exponentially. When will the growth curve assume an S shape and population growth level off? Many factors influence ideal family sizes.

Does environment or culture control human population growth?

As with many topics in environmental science, people have widely differing opinions about population and resources. Some believe that population growth is the ultimate cause of poverty and environmental degradation. Others argue that poverty, environmental degradation, and overpopulation are all merely symptoms of deeper social and political factors. The worldview we choose to believe will profoundly affect our approach to population issues.

Since the time of the Industrial Revolution, when the world population began growing rapidly, individuals have argued about the causes and consequences of population growth. In 1798 Thomas Malthus (1766–1834) wrote *An Essay on the Principle of Population*, changing the way European leaders thought about population growth. Malthus collected data to show that populations tended to increase at an exponential, or compound, rate, whereas food production either remained stable or increased only slowly. Eventually, he argued, human populations would outstrip their food supply and collapse into starvation, crime, and misery. He converted most economists of the day from believing that high fertility increased industrial output and national wealth to believing that per-capita output actually fell with rapidly rising population.

In Malthusian terms, growing human populations are limited only by disease, famine, or social constraints that compel people to reduce birth rates—late marriage, insufficient resources, celibacy, and "moral restraint." However, the economist Karl Marx (1818–1883) presented an opposing view that population growth results from poverty, resource depletion, pollution, and other social ills. Slowing population growth, claimed Marx, requires that people be treated justly, and that exploitation and oppression be eliminated from social arrangements.

Both Marx and Malthus developed their theories about human population growth when the world, technology, and society were understood much differently than they are today. Some believe that we are approaching, or may have surpassed, the earth's carrying capacity. Joel Cohen, a mathematical biologist at Rockefeller University, reviewed published estimates of the maximum human population size the planet can sustain. The estimates, spanning 300 years of thinking, converged on a median value of 10–12 billion. We are about 7.7 billion strong today and still growing, an alarming prospect for some (fig. 4.4). Cornell University entomologist David Pimental, for example, has said, "By 2100, if current trends continue, twelve billion miserable humans will suffer a difficult life on Earth." In this view birth control should be our top priority.

Technology increases carrying capacity for humans

Optimists argue that Malthus was wrong in his predictions of famine and disaster 200 years ago because he failed to account for scientific and technical progress. In fact, food supplies have increased faster than population growth since Malthus's time. For example, according to the UN FAO Statistics Division, in 1970 world food supplies provided 2,435 calories of food per person per day, while

▲ **FIGURE 4.4** Is the world overcrowded already, or are people a resource? In large part the answer depends on the kinds of resources we use and how we use them. It also depends on democracy, equity, and justice in our social systems. ©William P. Cunningham

in 2015 there was enough for 3,150 calories per person. Even poorer, developing countries saw a rise, from an average of 2,135 calories per day in 1970 to 2,850 in 2015. In that same period the world population grew from 3.7 to more than 7 billion people. Certainly, terrible famines have stricken various locations in the past 200 years, but many observers argue they were caused more by politics and economics than by lack of resources or population size. Whether the world can continue to feed its growing population remains to be seen, but technological advances have vastly increased human carrying capacity—so far (see chapter 7).

The burst of world population growth that began 200 years ago was stimulated by scientific and industrial revolutions. Progress in agricultural productivity, engineering, information technology, commerce, medicine, sanitation, and other achievements of modern life have made it possible to support approximately 1,000 times as many people per unit area as was possible 10,000 years ago. Economist Stephen Moore of the Cato Institute in Washington, D.C., regards this achievement as "a real tribute to human ingenuity and our ability to innovate." There is no reason, he argues, to think that our ability to find technological solutions to our problems will diminish in the future.

Much of our rising standard of living in the past two centuries, however, has been based on easily acquired natural resources, especially cheap, abundant fossil fuels. Many people are concerned about whether limited supplies of these fuels or adverse consequences of their use will result in a crisis in food production, transportation, or some other critical factor in human society.

Moreover, technology can be a double-edged sword. Our environmental effects aren't just a matter of sheer population size; they also depend on what kinds of resources we use and how we use them. This concept is summarized as the **I = PAT** formula. It says that our environmental impacts (I) are the product of our population size (P) times affluence (A) and the technology (T) used to produce the goods and services we consume (fig. 4.5).

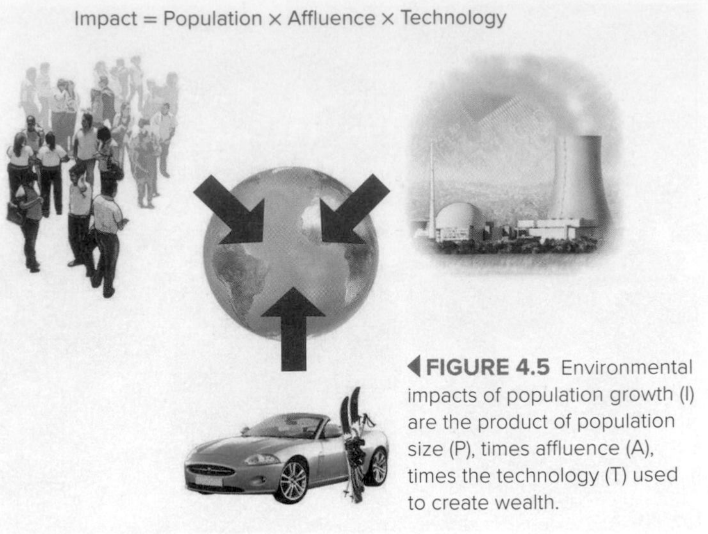

Impact = Population × Affluence × Technology

◀ **FIGURE 4.5** Environmental impacts of population growth (I) are the product of population size (P), times affluence (A), times the technology (T) used to create wealth.

While increased standards of living in the United States, for example, have helped stabilize population, they also bring about higher technological impacts. A family living an affluent lifestyle that depends on high levels of energy and material consumption, and that produces excessive amounts of pollution, could cause greater environmental damage than a whole village of hunters and gatherers or subsistence farmers.

Put another way, if the billions of people in Asia, Africa, and Latin America were to reach the levels of consumption now enjoyed by rich people in North America or Europe, using the same technology that provides that lifestyle today, the environmental effects would undoubtedly be disastrous. Growing wealth in China–its middle class is now estimated to be above 300 million, or nearly the entire population of the United States–is already stressing world resources, for example, and has made China the largest emitter of CO_2. There are now more millionaires in China than in all of Europe, and China has passed the United States in annual automobile production.

But China has also become the global leader in renewable energy and is now promoting electric vehicles. Ideally, all of us will begin to use nonpolluting, renewable energy and material sources. Better yet, we'll extend the benefits of environmentally friendly technology to the poorer people of the world, so that everyone can enjoy the benefits of a better standard of living without degrading our shared environment.

One way to estimate our environmental impacts is to express our consumption choices in the equivalent amount of land required to produce goods and services. This gives us a single number, called our **ecological footprint,** which estimates the relative amount of productive land required to support each of us. Services provided by nature make up a large proportion of our ecological footprint. For example, forests and grasslands store carbon, protect watersheds, purify air and water, and provide wildlife habitat.

Footprint calculations are imperfect, but they give us a way to compare different lifestyle effects (Key Concepts, pp. 84–85). The average resident of the United States, for instance, lives at a level of consumption that requires 9.7 ha of bioproductive land, whereas the average Malawian has an ecological footprint of less than 0.5 ha. Worldwide, we're currently using about one-third more resources than the planet can provide on a sustainable basis. That means we're running up an ecological debt that future generations will have to pay. Another way to look at it is that it would take 3.5 more earths to support the world at a current American lifestyle. If everyone lived like Malawians, on the other hand, the planet could sustainably house more than 20 billion people.

Population can push economic growth

Think of the gigantic economic engines that large countries, such as the United States and China, represent. More people mean larger markets, more workers, and efficiencies of scale in mass production of goods. Moreover, adding people boosts human ingenuity and intelligence that can create new resources by finding new materials and discovering new ways of doing things. Economist Julian Simon (1932–1998), a champion of this rosy view of human history, believed that people are the "ultimate resource" and that no evidence shows that pollution, crime, unemployment, crowding, the loss of species, or any other resource limitations will worsen with population growth.

In a famous wager in 1980, Simon challenged Paul Ehrlich, author of *The Population Bomb,* to pick five commodities that would become more expensive by the end of the decade. Ehrlich chose a group of metals that actually became cheaper, and he lost the bet. Leaders of many developing countries insist that, instead of being obsessed with population growth, we should focus on the inordinate consumption of the world's resources by people in richer countries. For his part, Ehrlich did not really want to bet on metals but on renewable resources or certain critical measures of environmental health. He and Simon were negotiating a second bet just before Simon's death.

4.2 MANY FACTORS DETERMINE POPULATION GROWTH

- *We live in two demographic worlds: one poor, young, and growing and one old, rich, and shrinking.*
- *Fertility is decreasing in most countries.*
- *Life expectancy is increasing in most places.*

Demography (derived from the Greek words *demos,* people, and *graphein,* to write or to measure) encompasses vital statistics about people, such as births, deaths, and where they live, as well as total population size. In this section we will investigate ways to measure and describe human populations and discuss demographic factors that contribute to population growth.

How many of us are there?

The United Nations estimate of 7.7 billion people in 2018 is only an estimate. Even in this age of information technology and advanced communication, counting the number of people in the world is an inexact science. Some countries have never even taken a census, and some that have been done may not be accurate. Governments

▲ **FIGURE 4.6** We live in two demographic worlds. One is rich, is technologically advanced, and has an elderly population that is growing slowly, if at all. The other is poor, crowded, and underdeveloped and is growing rapidly. ©Frans Lemmens/Getty Images

TABLE 4.2 | The World's Largest Countries

2010		2050	
COUNTRY	**POPULATION (MILLIONS)**	**COUNTRY**	**POPULATION (MILLIONS)**
China	1,388	India	1,628
India	1,325	China	1,437
United States	326	United States	420
Indonesia	262	Nigeria	299
Brazil	210	Pakistan	295
Pakistan	210	Indonesia	285
Nigeria	189	Bangladesh	231
Bangladesh	164	Brazil	220
Russia	147	Dem. Rep. of Congo	183

Source: Data from the U.S. Census Bureau, 2012.

overstate or understate their populations to make their countries appear larger and more important or smaller and more stable than they really are. Some individuals, especially if they are homeless persons, refugees, or illegal migrants, may not want to be counted or identified.

We live in two very different demographic worlds. One of these worlds is poor, young, and growing rapidly, while the other is rich, old, and shrinking in population size. The poorer world is occupied by the vast majority of people who live in the less-developed countries of Africa, Asia, and Latin America (fig. 4.6). These countries represent 80 percent of the world population but will contribute more than 90 percent of all projected future growth. The richer world is made up of North America, Western Europe, Japan, Australia, and New Zealand. The average age in richer countries is 40, and life expectancy of their residents may exceed 90 by 2050. With many couples choosing to have either one or no children, the populations of these countries are expected to decline over the next century.

The highest population growth rates occur in a few "hot spots" in the developing world, such as sub-Saharan Africa and the Middle East, where economics, politics, religion, and civil unrest keep birth rates high and contraceptive use low. In Niger, for example, annual population growth is currently 3.2 percent. Less than 15 percent of all couples use any form of birth control, women average 5.5 children each, and nearly half the population is less than 15 years old. Even faster growth is occurring in Qatar, where the population has doubled in the past 10 years.

Some countries in the developing world are growing so fast that they will reach immense population sizes by the middle of the twenty-first century (table 4.2). China was the most populous country throughout the twentieth century; India is expected to pass China in the twenty-first century. Nigeria, which had only 33 million residents in 1950, is forecast to have 299 million in 2050. Ethiopia, with about 18 million people 50 years ago, is likely to grow nearly tenfold over a century. In many of these countries, rapid population growth poses huge challenges to food supplies and stability. Bangladesh, about the size of Iowa, is already crowded with 164 million people. If rising sea levels flood one-third of the country by 2050, as some climatologists predict, adding another 70 million people will make its resource needs dire.

On the other hand, some richer countries have shrinking populations. Japan, which has 128 million residents now, is expected to shrink to about 90 million by 2050. Europe, which now makes up about 12 percent of the world population, will constitute less than 7 percent in 50 years, if current trends continue. Even the United States and Canada would have stable populations if immigration were stopped. For the moment, the U.S. population continues to grow. In 2018 the U.S. population was about 327 million and growing at 0.86 percent per year.

Birth rates and life expectancies can decline as a result of political or economic instability. For instance, between about 1990 and 2006, Russia declined steadily as death rates soared and birth rates fell to 1.2 children per woman. These changes were generally attributed to a collapsing economy, hyperinflation, crime, and corruption. Since then, the birth rate has recovered to 1.6, life expectancy has increased, and the population is approximately stable.

Epidemics can also diminish populations, at least temporarily. The most serious recent case is that of HIV/AIDS, which has affected many African countries especially severely. According to UNICEF, AIDS is the number one cause of death for adolescents in Africa. In South Africa, Zimbabwe, Botswana, and Zambia, for example, over 10 percent of the adult population has AIDS or is HIV positive. Botswana, where 23 percent of adults are living with HIV/AIDS, saw life expectancy decline from 64 to a low of 49 years in 2002; access to antiretroviral treatment has helped survival rates increase since then. The world's highest rate is in Swaziland, where 26.5 percent of adults are living with HIV/AIDS.

The world population density map in appendix 2 on p. A-3 shows human population distribution around the world. Notice the high densities supported by fertile river valleys of the Nile, Ganges, Yellow, Yangtze, and Rhine Rivers and the well-watered coastal plains of India, China, and Europe. Historic factors, such as technology diffusion and geopolitical power, also play a role in population distribution.

How big is your footprint?

Human populations are rising, and our resource use per person is growing. How can we assess the ways our resource consumption is changing the world? One approach is **ecological footprint analysis**—estimating the amount of territory needed to support all our consumption of food, paper, computers, energy, water, and other resources. This analysis obviously simplifies and approximates our real use, but the aggregate measure allows us to compare resource use among places or over time.

Perhaps the most comprehensive analysis has been done by the Worldwide Fund for Nature (WWF). The summary you see here gives key points.

Where is population growing?

Poorer less-developed countries are expected to account for 90 percent of all population growth in this century.

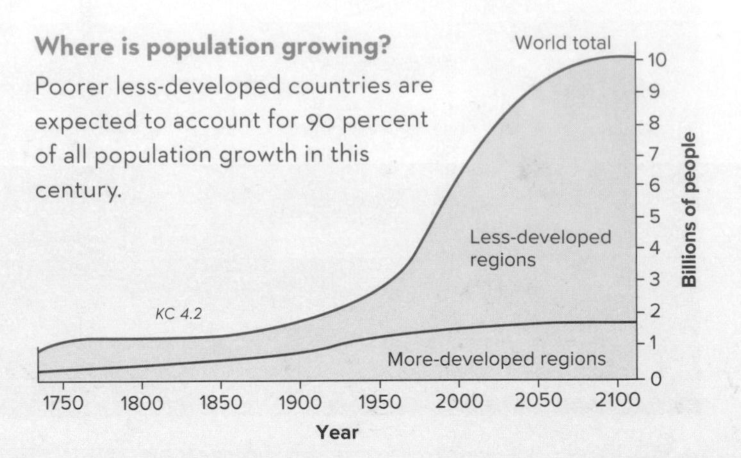

Terms to note:

Biocapacity is the capacity of living systems to provide for our needs. Both biocapacity and global footprints can be measured in gigahectares (gha). **One ha = 2.59 acres. One gha = 1 billion ha**. The WWF calculated that an average hectare of land could store carbon equivalent to 1,450 liters of gasoline.

We consume more than the earth's biocapacity by mining ancient energy, soil, and other resources at a rate faster than these resources can be reproduced.

Which component of the global footprint is changing most rapidly?

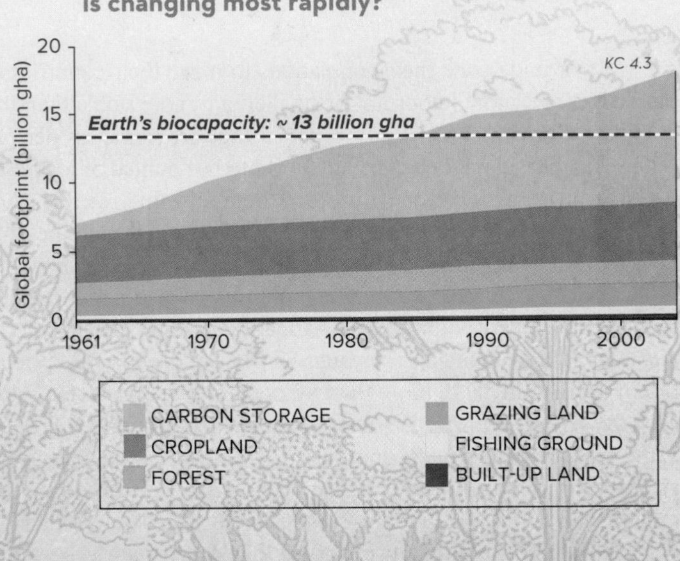

What's our average footprint?

Each person in the top-consuming countries has a footprint of nearly 10 ha. Half of the 171 countries evaluated by the WWF have a footprint of less than 2 ha/person. If everyone had a typical American lifestyle, it would take 3.5 more earths to support us all.

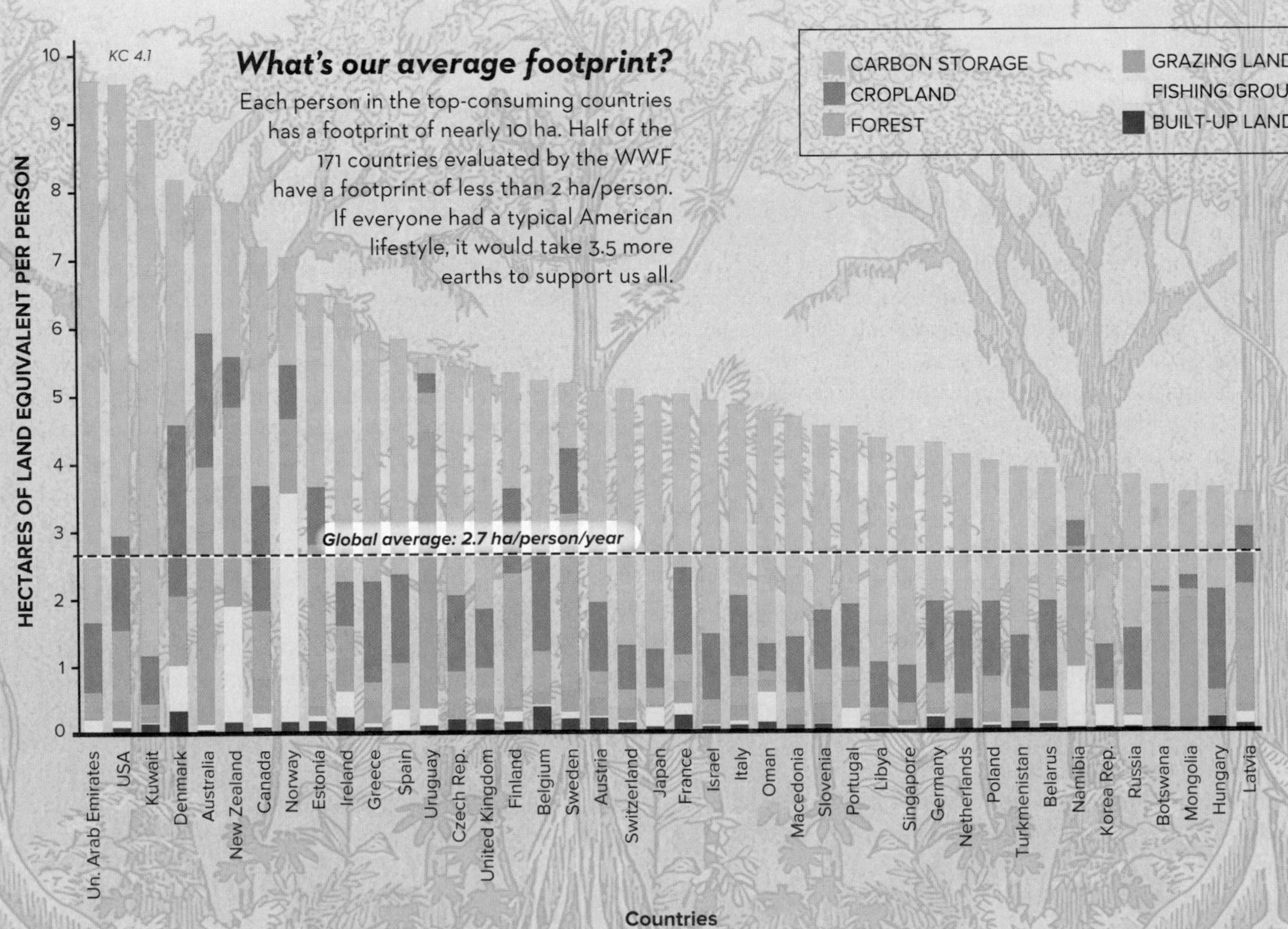

©Nigel Hicks/Alamy Stock Photo

1. CARBON STORAGE **Why are carbon emissions the biggest part of our footprint?** Burning of fossil fuels, clearing of forests, and oxidation of agricultural soils emit climate-changing gases. These gases account for nearly all of our rising carbon footprint in the past 50 years. Wealthy countries vary greatly in their carbon emissions: Compare, for example, Sweden and the United States, which have similar wealth and standards of living. At nearly 10 gha, our carbon footprint alone preempts a majority of the earth's biocapacity.

KC 4.4

©David Frazier/AGE Fotostock *KC 4.5*

2. CROPLAND **What are some of the resources used in farming?** Costs vary greatly. Some farming systems deplete soil and depend on fossil fuels; others build soil and require few inputs. Grain-fed beef is perhaps our most costly agricultural product.

6. BUILT-UP LAND **How much land is occupied by roads and buildings?** They take up less space than other uses but preempt important ecological services.

©William P. Cunningham *KC 4.9*

3. FORESTLAND **What benefits do we get from forests?** Forestry provides our wood and paper products, and many other useful products. Forests also protect watersheds, provide wildlife habitat, and purify and store water.

KC 4.6

©Amy Johansson/Shutterstock

4. GRAZING LAND **Can grazing deplete land?** Expansive area is needed for grazing. Overgrazing can badly degrade biodiversity and cause soil erosion. Less intense grazing can be an efficient way to convert grass to protein.

©branislavpudar/Shutterstock *KC 4.7*

5. FISHING GROUND **How much sea do we depend on?** Fishery impact is large for some countries. Globally, 90 percent of all large marine predators are gone, and 13 of 17 major fisheries are exhausted (see chapter 9).

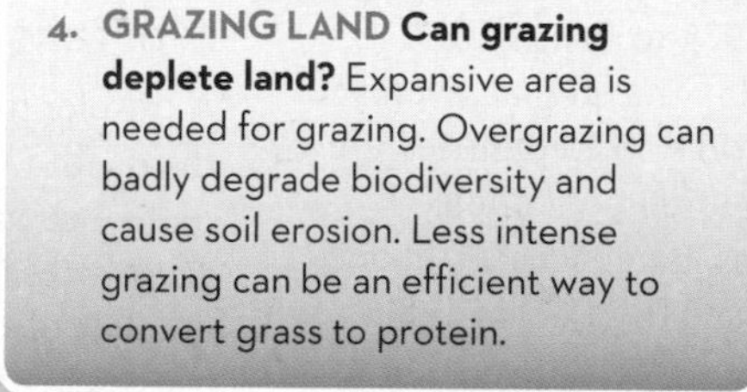

KC 4.8

Nassau grouper

©Cynthia Shaw

CAN YOU EXPLAIN?

1. **Which factors are largest for the United States? Why?**
2. **Which countries have the greatest footprint per person in forestry? fishing? carbon? grazing? Why?**

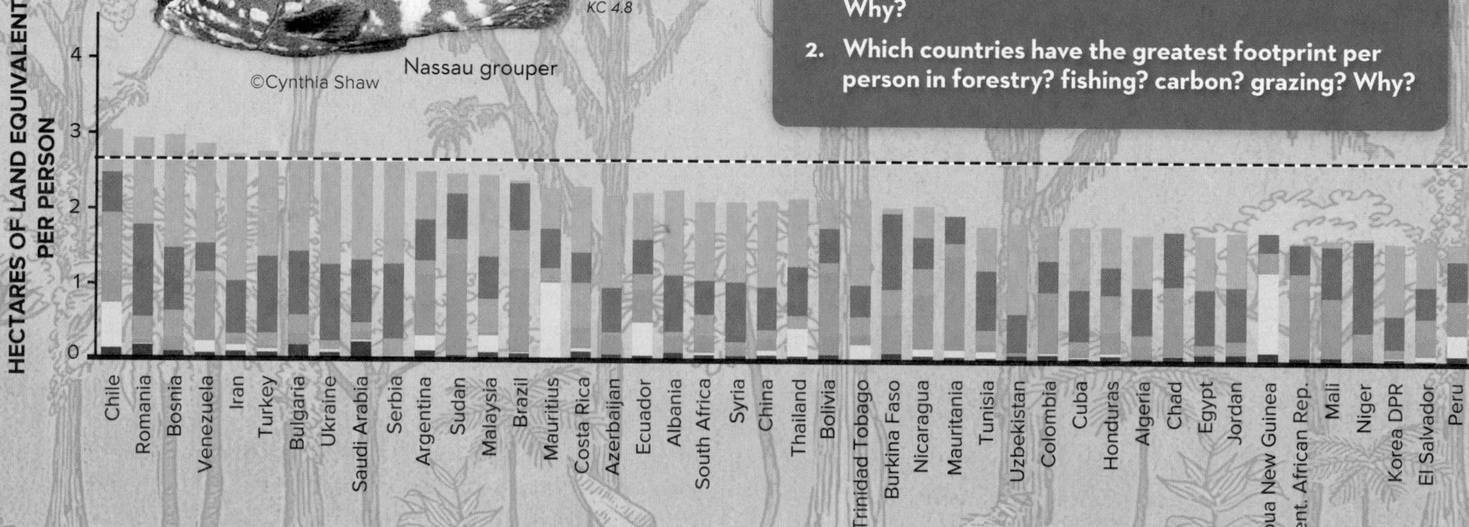

Low Impact Countries

Fertility has declined in recent decades

Fecundity is the physical ability to reproduce, whereas **fertility** is the actual production of offspring. A common statistic used to describe fertility in a population is the **crude birth rate,** the number of births in a year per thousand persons. It is statistically "crude" in the sense that it is not adjusted for population characteristics, such as the number of women of reproductive age.

The **total fertility rate,** the average number of children per woman, is sometimes easier to remember. Historically, there have been some extraordinary cases of fertility. Some upper-class women in seventeenth- and eighteenth-century Europe, whose babies were given to wet nurses immediately after birth, had 25 or more pregnancies. Among working-class people, the highest recorded fertility rates were in North American Anabaptist agricultural communities, which averaged up to 12 children per woman. In most tribal or traditional societies, though, food resources, health, long nursing periods, and cultural practices have limited fertility to about 6 or 7 children or fewer per woman, and high rates of child mortality have often kept populations fairly stable.

In the last 50 years, fertility rates have declined dramatically almost everywhere except sub-Saharan Africa, where poverty and other factors remain persistently entrenched (fig. 4.7). In 1975 the average family in Mexico, for instance, had 7 children. By 2017, however, the average Mexican woman had only 2.17 children. Similarly, in Iran total fertility fell from 6.5 in 1975 to 1.8 in 2017. According to the World Health Organization, the global average fertility rate is 2.45, and about half the world's 192 countries are now at or below a **replacement rate** of 2.1 children per couple.

This decline cuts across economic regions. Bangladesh, still one of the poorest countries, reduced its fertility rate from 6.9 in 1980 to only 2.1 children per woman in 2017. China's one-child-per-family policy decreased the fertility rate from 6 in 1970 to 1.6 in 2017. This program was remarkably successful in reducing population growth, but China decided to end it in 2015 (What Do You Think?, p. 87).

Despite these dramatic declines, population growth will continue because much of the world's population is very young. Brazil, for example, now has a fertility rate of only 1.8 children per woman. But 26 percent of its population is under 14, so the population will continue to grow for some decades. Demographers call this **population momentum.**

Since the 1960s, which saw the fastest growth and shortest population doubling times ever (see table 4.1), many people concerned about resource availability have been eager to see **zero population growth (ZPG).** Zero growth occurs when number of deaths exactly equals number of births plus immigration.

Ironically, now that many populations are falling below replacement levels, it is emerging that economists are unable to accommodate zero population growth. States and businesses need constantly growing numbers of workers, and especially consumers, to maintain economic growth. One of the strong forces promoting China's abandonment of the one-child policy was this economic growth imperative. It remains unclear how these contrasting priorities of environmental conservation, economic growth, and social stability will be resolved.

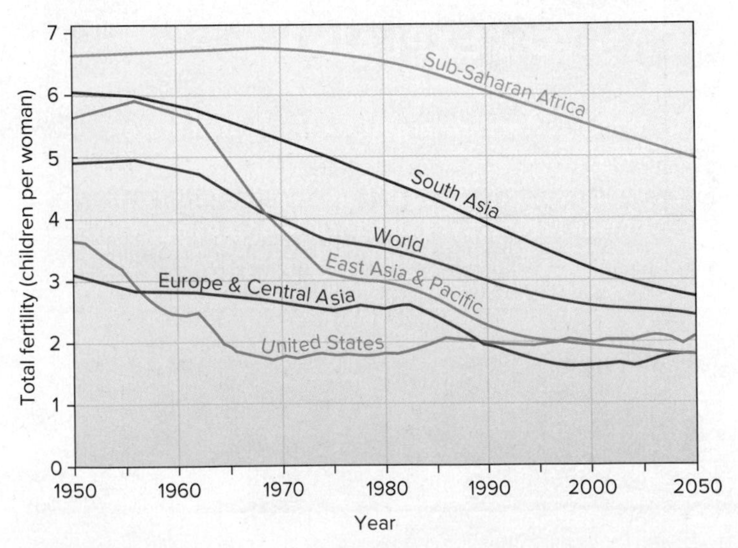

▲ **FIGURE 4.7** Total fertility rates for the whole world have fallen by more than half over the past 50 years. Much of this dramatic change has occurred in China and India. Progress has lagged in sub-Saharan Africa, but by 2050 the world average should be approaching the replacement rate of 2.1 children per woman on reproductive age. Source: World Bank, 2012.

Mortality offsets births

A traveler to a foreign country once asked a local resident, "What's the death rate around here?" "Oh, the same as anywhere," was the reply, "about one per person." In demographics, however, **crude death rates** (or crude mortality rates) are expressed in terms of the number of deaths per thousand persons in any given year. Countries in Africa where health care and sanitation are limited may have mortality rates of 20 or more per 1,000 people. Wealthier countries generally have mortality rates around 10 per 1,000. The number of deaths in a population is sensitive to the population's age structure.

Rapidly growing, developing countries, such as Qatar, often have lower crude death rates (1.5 per 1,000 currently) than do the more-developed, slowly growing countries, such as Germany (11.7 per 1,000), even though their life expectancies are considerably lower. This is because a rapidly growing country has proportionately more youths and fewer elderly than a more slowly growing country.

Life expectancy is rising worldwide

Life span is the oldest age to which a species is known to survive. Although there are many claims in ancient literature of kings living a thousand years or more, the oldest age that can be certified by written records was that of Jeanne Louise Calment of Arles, France, who was 122 years old at her death in 1997. Though modern medicine has made it possible for many of us to survive much longer than our ancestors, it doesn't appear that the maximum life span has increased much at all. Apparently, cells in our bodies have a limited ability to repair damage and produce new components.

What Do YOU THINK?

China's One-Child Policy

When the People's Republic of China was founded in 1949, it had about 540 million residents, and official government policy encouraged large families. The republic's first chairman, Mao Zedong, proclaimed, "Of all things in the world, people are the most precious." He reasoned that more workers would mean greater output, increasing national wealth, and higher prestige for the country. This optimistic outlook was challenged, however, in the 1960s, when a series of disastrous government policies triggered massive famines and resulted in at least 30 million deaths.

When Deng Xiaoping became chairman in 1978, he reversed many of Mao's policies, privatizing farms, encouraging private enterprise, and discouraging large families. Deng recognized that with an annual growth rate of 2.5 percent, China's population, which had already reached 975 million, would double in only 28 years. China might have 2 billion residents today if that growth had continued. Feeding, housing, educating, and employing all those people would put a severe strain on China's (and the world's) limited resources.

China's one-child-per-family policy was remarkably successful in reducing birth rates, but social costs were also high. ©David Pollack/Getty Images

Deng introduced a highly successful, but also controversial, one-child-per-family policy. Rural families and ethnic minorities were exempt from the rule, but local authorities often were capricious and tyrannical in applying sanctions. Ordinary families were punished harshly for having unauthorized children, but government officials and other powerful individuals often had multiple children. There were many reports of bribery, forced abortions, coerced sterilizations, and even infanticide as a result of this policy.

Another result of China's one-child policy is called the 4:2:1 problem: There are now often four grandparents and two parents doting on a single child. Social scientists often refer to this highly spoiled generation as "little emperors." And it is not clear how those single children can care for so many parents and grandparents.

The Chinese government is also worried now about a "birth dearth." If the population declines, will there be enough workers, and enough consumers, to keep the economy growing? Will there be enough farmers, soldiers, scientists, inventors, and other productive individuals to keep society functioning in the future?

The one-child policy was officially abandoned in 2015. All couples are now allowed to have two children if they want and can afford them. But many demographers doubt that birth rates will change dramatically. They argue that industrialization, urbanization, and economic growth might have brought about a demographic transition without any Draconian policies. Look at neighboring countries, they say. The fertility rate in Singapore is 0.8, Macau is 0.9, Taiwan is 1.11, South Korea is 1.25, and Japan is 1.4—all without policies to restrict childbearing.

Nevertheless, it's important that the Chinese birth rate has declined. If China had stayed on its 1978 trajectory, the population, instead of 1.367 billion, would now be at least 660 million higher. Its annual growth rate is now 0.4 percent, considerably less than the 0.7 percent annual growth in the United States. Thailand, by comparison, has achieved a growth rate of 0.5 percent per year, and Hong Kong and Taiwan have growth rates of 0.3 percent per year, lower than that of mainland China.

China has also been much more successful in controlling population growth than India. At about the same time that Deng introduced his one-child plan, India, under Indira Gandhi, started a program of compulsory sterilization in an effort to reduce population growth. This coercive policy was often deceptive and coercive, and it caused so much public outrage that the Indian government decided to delegate family planning to individual states. Some states have since been highly successful in their family planning efforts. Others have not. The net effect, however, is that India is expected to grow to nearly 1.7 billion by 2050, while China is expected to reach zero population growth by 2030.

What do you think? Was the rapid reduction in Chinese population growth worth the social disruption and abuses it caused? If you were in charge of family planning, what policies would you pursue?

Sooner or later they simply wear out, and we fall victim to disease, degeneration, accidents, or senility.

Life expectancy is the average age that a newborn infant can be expected to attain in any given society. It is another way of expressing the average age at death. For most of human history, life expectancy in most societies probably was 35 to 40 years. This does not mean that no one lived past age 40 but instead that many people died at earlier ages (mostly early childhood), which balanced out those who managed to live longer.

The twentieth century saw a global transformation in human health unmatched in history. This revolution can be seen in the dramatic increases in life expectancy in most places (table 4.3). Worldwide, the average life expectancy rose from about 40 to 67.2 years over the past 100 years. The greatest progress was in developing

TABLE 4.3 Life Expectancy at Birth for Selected Countries in 1900 and 2017

	1900		2017	
COUNTRY	MALES	FEMALES	MALES	FEMALES
India	23	23	64	71
Russia	31	33	68	78
United States	46	48	79	81
Sweden	57	60	81	84
Japan	42	44	81	87

Source: World Health Organization, 2014.

countries. For example, in 1900 the average Indian man or woman could expect to live about 23 years. A century later, although India had an annual per capita income of only $3,500 (U.S.), the average life expectancy for both men and women had nearly tripled and was very close to that of countries with ten times its income level. Longer lives were due primarily to better nutrition, improved sanitation, clean water, and education, rather than to miracle drugs or high-tech medicine.

Although the gains were not as great for the already industrialized countries, residents of the United States and Japan, for example, now live nearly twice as long as they did at the beginning of the twentieth century, and they can expect to enjoy much of that life in relatively good health. The Disability Adjusted Life Years (DALYs, a measure of disease burden that combines premature death with loss of healthy life resulting from illness or disability) that someone living in Japan can expect is now 74.5 years, compared with only 64.5 DALYs two decades ago.

As figure 4.8 shows, annual income and life expectancy are strongly correlated up to about US$5,000 (U.S.) per person. Beyond

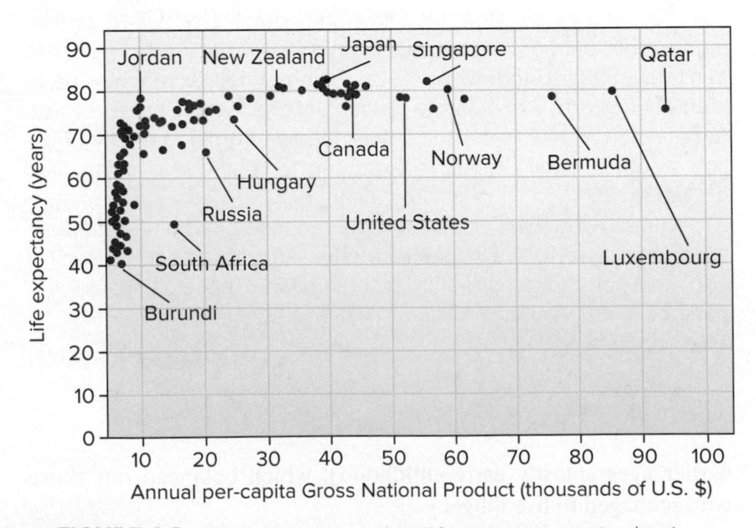

▲ **FIGURE 4.8** As incomes rise, so does life expectancy up to about $5,000 (U.S.). Above that amount the curve levels off. Some countries, such as South Africa and Russia, have far lower life expectancies than their GDP would suggest. Jordan, on the other hand, which has only one-tenth the per-capita GDP of the United States, actually has a higher life expectancy. Source: World Bank, 2015.

that level—which is generally enough for adequate food, shelter, and sanitation for most people—life expectancies level out at about 75 years for men and 85 for women.

Large discrepancies in how the benefits of modernization and social investment are distributed within countries are revealed in differential longevities of various groups. The greatest life expectancy reported anywhere in the United States is for Asian American women in New Jersey, who live to an average age of 91. By contrast, Native American men on the Pine Ridge Indian Reservation in South Dakota live, on average, only to age 48. Two-thirds of the countries in Africa have a higher life expectancy. The Pine Ridge Reservation is the poorest area in America, with an unemployment rate near 75 percent and high rates of poverty, alcoholism, drug use, and alienation. Similarly, African-American men in Washington, D.C., live, on average, only 57.9 years, which is less than the life expectancy in Lesotho or Swaziland.

Living longer has profound social implications

A population growing rapidly by natural increase has more young people than does a stable population. One way to show these differences is to graph age classes in histograms (fig. 4.9). In Niger, which was growing at a rate of 3.9 percent per year when these data were collected, about half the population was in the prereproductive category (below age 15). Even if total fertility rates fell abruptly, the total number of births, and the population size, would continue to grow for some years as these young people entered reproductive age (an example of population momentum).

By contrast, a country, such as Sweden, with a relatively stable population, will have nearly the same number in most cohorts. Notice that females outnumber males in Sweden's oldest group because of differences in longevity between sexes. A country that has only recently reached zero population growth, such as Singapore, can have a pronounced bulge in middle-age cohorts as fewer children are born than in their parents' generation.

Both rapidly growing countries and slowly growing countries can have a problem with their **dependency ratio,** or the number of nonworking compared with working individuals in a population. In Niger, for example, each working person supports a high number of children. In the United States, by contrast, a declining working population is now supporting an ever larger number of retired persons.

These changing age structures and shifting dependency ratios are occurring worldwide (fig. 4.10). In 1950 there were only 130 million people in the world over 65 years old. In 2017 more than 650 million had reached this age. By 2050, the UN predicts, there could be three older persons for every child in the world. Countries such as Japan, France, and Germany already are concerned that they don't have enough young people to fill jobs and support their retirement system. They are encouraging couples to have more children. Immigrants can reduce the average age of the population and create a fresh supply of workers. But nativist groups in many countries resist the integration of

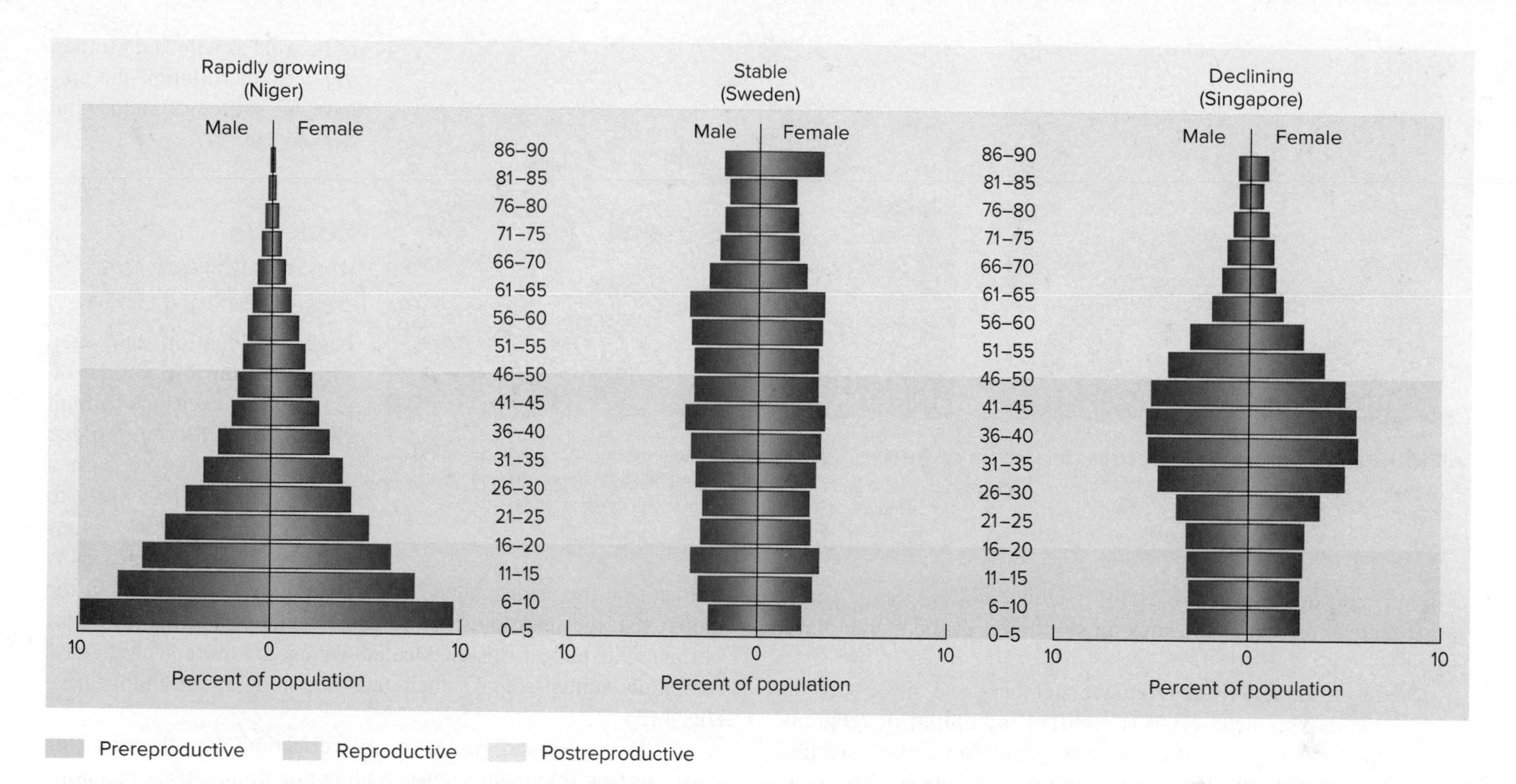

▲ **FIGURE 4.9** The shape of each age-class histogram is distinctive for a population that is rapidly growing (Niger), stable (Sweden), or declining (Singapore). Horizontal bars represent the percentage of the country's population in consecutive age classes (0–5 yrs., 6–10 yrs., etc.). Source: U.S. Census Bureau, 2003.

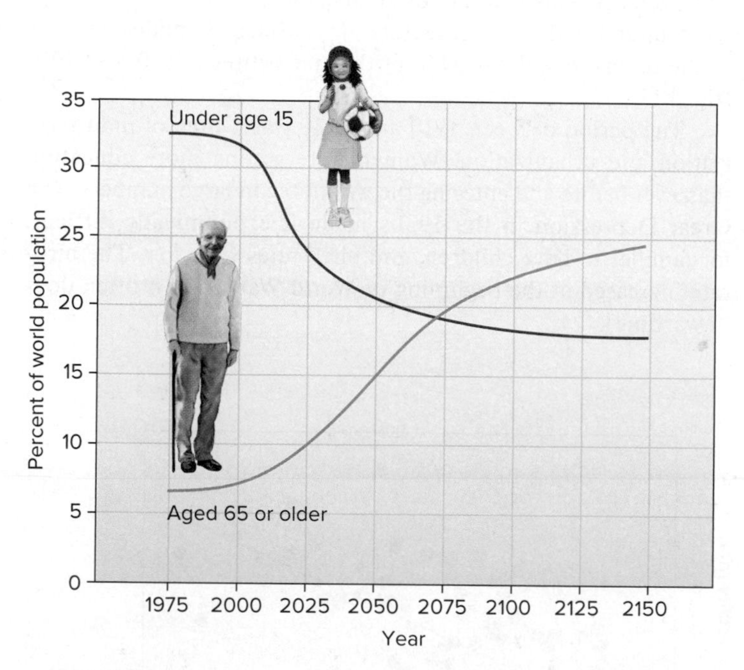

▲ **FIGURE 4.10** By the mid-twenty-first century, children under age 15 will make up a smaller percentage of world population, whereas people over age 65 will contribute an increasing share of the population.

foreigners who may not share their culture, religion, or language. Others, however, argue that immigrants can rejuvenate and revitalize an aging society. Where do you stand on this contentious issue?

4.3 FERTILITY IS INFLUENCED BY CULTURE

- *Children are desirable for happiness, pride, and old-age support.*
- *Education and opportunity influence childbearing decisions.*

A number of social and economic pressures affect decisions about family size, which in turn affect the population at large. In this section we will examine both positive and negative pressures on reproduction.

People want children for many reasons

Factors that increase people's desires to have babies are called **pronatalist pressures.** Raising a family may be the most enjoyable and rewarding part of many people's lives. Children can be a source of pleasure, pride, and comfort. They may be the only source of support for elderly parents in countries without a social security system. Where infant mortality rates are high, couples may need to have many children to ensure that at least a few will survive to take care of them when they are old. Where there is little opportunity for upward mobility, children give status in society, express parental creativity, and provide a sense of continuity and accomplishment otherwise missing from life.

Often children are valuable to the family not only for future income but even more as a source of current income and help with household chores. In much of the developing world, small

(a)

(b)

▲ **FIGURE 4.11** In rural areas with little mechanized agriculture (a), children are needed to tend livestock, care for younger children, and help parents with household chores. Where agriculture is mechanized (b), rural families view children just as urban families do—helpful, but not critical to survival. This affects the decision about how many children to have.
(a): ©William P. Cunningham; (b): ©Reed Kaestner/Getty Images

children tend domestic animals and younger siblings, fetch water, gather firewood, help grow crops, or sell things in the marketplace (fig. 4.11).

Society also has a need to replace members who die or become incapacitated. This need often is codified in cultural or religious values that encourage bearing and raising children. Some societies look upon families with few or no children with pity or contempt, and for them the idea of deliberately controlling fertility may be shocking, even taboo. Women who are pregnant or have small children have special status and protection. Boys frequently are more valued than girls because they carry on the family name and are expected to support their parents in old age. Couples may have more children than they desire in an attempt to produce a son who lives to maturity.

Male pride often is linked to having as many children as possible. In Niger and Cameroon, for example, men on average want 12.6 and 11.2 children, respectively. Women in these countries want only 5 or 6 on average. Even though a woman might desire fewer children, however, she may have few choices and little control over her own fertility. In many societies a woman has no status outside of her role as wife and mother. Yet without children, she may have no source of support in her old age.

Education and income affect the desire for children

Higher education and personal freedom for women often result in decisions to limit childbearing. When women have opportunities to earn a salary, they are less likely to stay home and have many children. Not only do many women find the challenge and variety of a career attractive, but the money that they earn outside the home becomes an important part of the family budget. Also, educated women are more likely to have the family status to make their own decisions about childbearing (fig. 4.12).

In less-developed countries, where feeding and clothing children can be a minimal expense, adding one more child to a family usually doesn't cost much. By contrast, raising a child in a developed country can cost hundreds of thousands of dollars, and it can be hard for parents to afford more than one or two.

Cultural and political factors also influence childbearing, as in the dramatic shifts in U.S. birth rates between 1910 and 2010 (fig. 4.13).

The period between 1910 and 1930 was a time of industrialization and urbanization. Women were getting more education than ever before and entering the workforce in large numbers. The Great Depression in the 1930s made it economically difficult for families to have children, and birth rates were low. The birth rate increased at the beginning of World War II (as it often does in wartime).

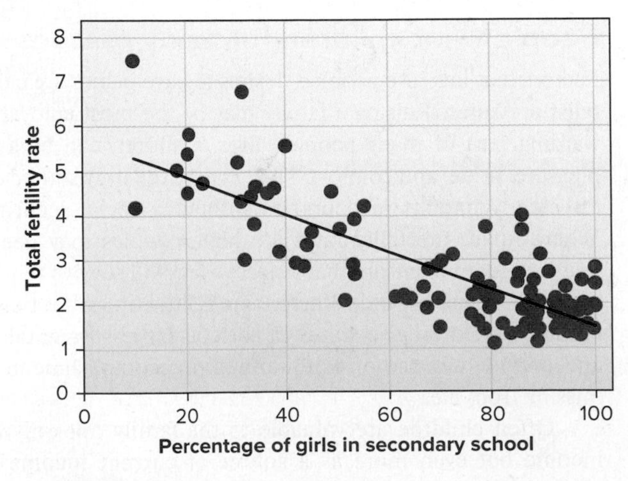

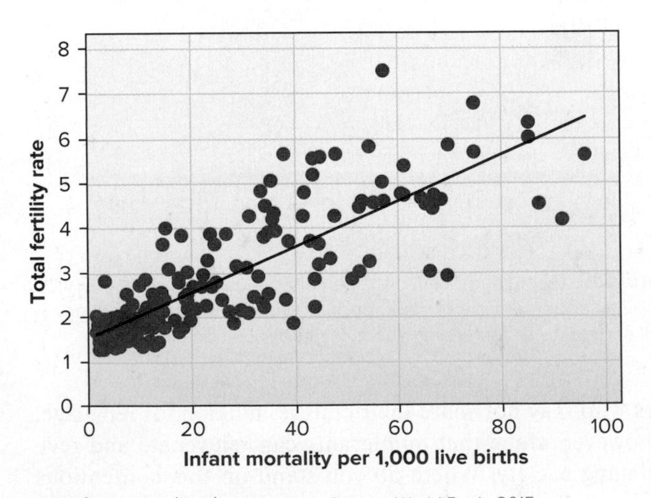

▲ **FIGURE 4.12** Total fertility declines as girls' education increases and infant mortality decreases. Source: World Bank, 2015.

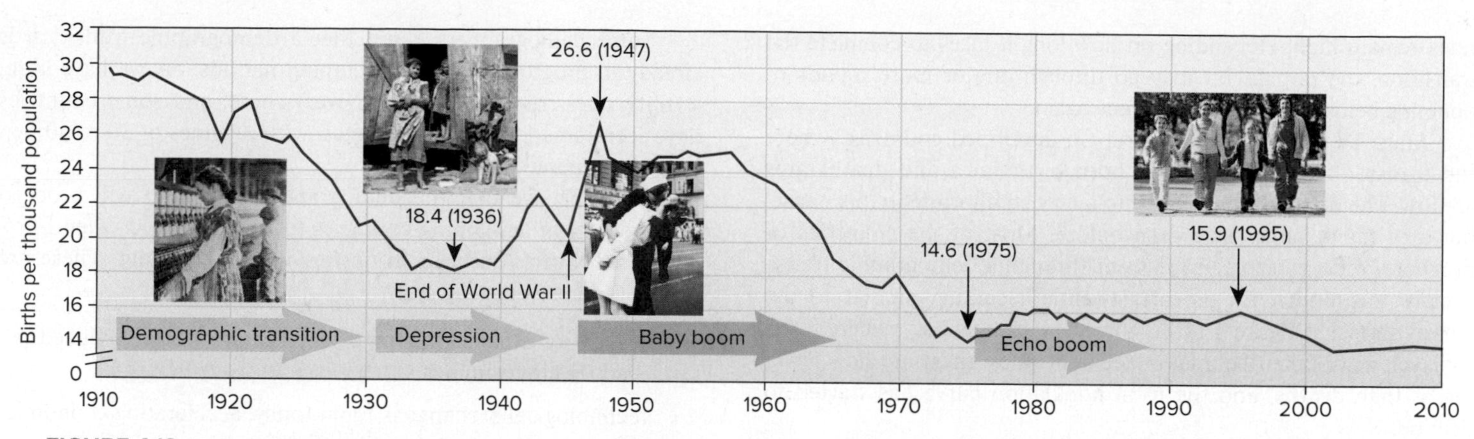

▲ **FIGURE 4.13** Birth rates in the United States, 1910–2010. The falling birth rate from 1910 to 1929 represents a demographic transition from an agricultural to an industrial society. The baby boom following World War II lasted from 1945 to 1965. A much smaller "echo boom" occurred around 1980 when the baby boomers started to reproduce. Sources: Data from Population Reference Bureau and U.S. Bureau of the Census. (a): Hine/Lewis Wickes/Library of Congress, Prints and Photographs Division [LC-DIG-nclc-01366]; (b): U.S. Department of Agriculture (USDA); (c): Department of the Navy/Naval Photographic Center/Victor Jorgensen (US Navy photo journalist)/National Archives and Records Administration (NWDNS-80-G-377094); (d): ©Stockbyte/Getty Images

A "baby boom" followed World War II, as couples were reunited and new families started. During this time the government encouraged women to leave their wartime jobs and stay home. A high birth rate persisted through the times of prosperity and optimism of the 1950s but began to fall in the 1960s. Part of this decline was caused by the small number of babies born in the 1930s, which resulted in fewer young adults to give birth in the 1960s. Part was due to changed perceptions of the ideal family size. Whereas in the 1950s women typically wanted four children or more, the norm dropped to one or two (or no) children in the 1970s. A small "echo boom" occurred in the 1980s, as baby boomers began to have children, but changing economics and attitudes seem to have altered our view of ideal family size in the United States.

4.4 THE DEMOGRAPHIC TRANSITION

- *The demographic transition model predicts declining birth and death rates with economic development.*
- *Population growth occurs as death rates fall faster than births.*
- *Incentives and equity often promote demographic transition.*

In 1945 demographer Frank Notestein pointed out that a typical pattern of falling death rates and birth rates due to improved living conditions usually accompanies economic development. He called this pattern the **demographic transition** from high birth and death rates to lower birth and death rates. Figure 4.14 shows an idealized model of a demographic transition. This model is often used to explain connections between population growth and economic development.

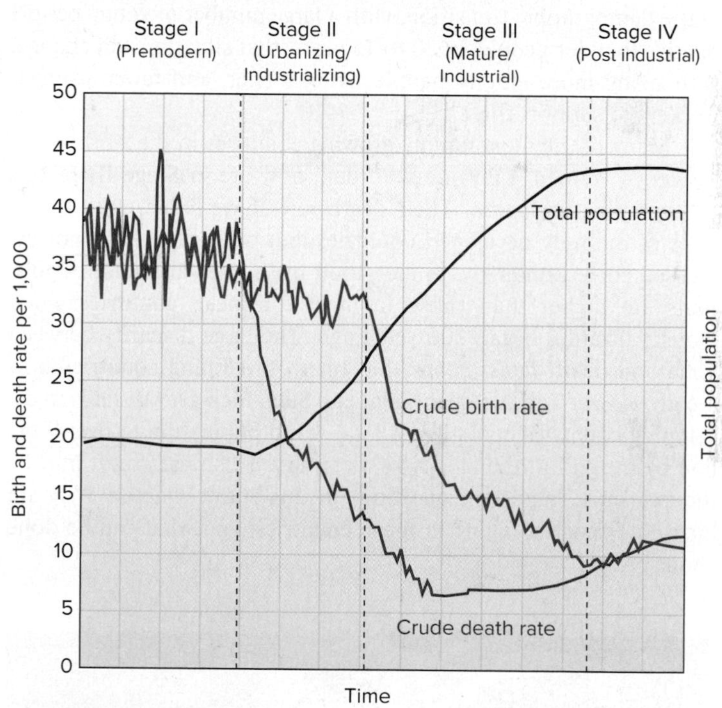

▲ **FIGURE 4.14** Theoretical birth, death, and population growth rates in a demographic transition accompanying economic and social development. In a predevelopment society, birth and death rates are both high, and total population remains relatively stable. During development, death rates tend to fall first, followed in a generation or two by falling birth rates. Total population grows rapidly until both birth and death rates stabilize in a fully developed society.

Economic and social conditions change mortality and births

Stage I in figure 4.14 represents the conditions in a premodern society. Malnutrition, illness, accidents, and other hazards keep death rates high, around 30 per 1,000 people. But high birth rates keep the population relatively constant. Economic development in Stage II brings better jobs, medical care, sanitation, and a generally improved standard of living, and death rates often fall very rapidly. Birth rates may rise at first as more money and better nutrition allow people to have the children they always wanted. In a couple of generations, however, birth rates fall as people see that all their children are more likely to survive and that the whole family benefits from concentrating more resources on fewer children. Note that populations grow rapidly during Stages II and III, when death rates have already fallen but birth rates remain high. Depending on how long it takes to complete the transition, the population may go through one or more rounds of doubling before coming into balance again.

Stage IV represents conditions in developed countries, where the transition is complete and both birth rates and death rates are low. The population comes into a new equilibrium in this phase, but at a much larger size than before. Most of the countries of northern and western Europe went through a demographic transition in the nineteenth or early twentieth century similar to the curves shown in figure 4.14. In countries such as Italy, where fertility levels have fallen below replacement rates, there are now fewer births than deaths, and the total population curve has started to decline.

A huge challenge facing countries in the final stage of the demographic transition is the imbalance between people in their most productive years and people who are retired or in their declining years. The continuing debate in the U.S. Congress about how to fund the Social Security system is due to the fact that when this program was established, the United States was in the middle of the demographic transition, with a large number of young people relative to older people. In 10 to 15 years, that situation will change, with many more elderly people living longer, and fewer younger workers to support them.

Many of the most rapidly growing countries in the world, such as Kenya, Yemen, Libya, and Jordan, now are in Stage III of this demographic transition. Their death rates have fallen close to the rates of the fully developed countries, but birth rates have not decreased correspondingly. In fact, their birth rates and total populations are higher than those in most European countries when industrialization began 300 years ago. The large disparity between birth and death rates means that many developing countries now are growing at 2 to 3 percent per year. Such high growth rates in developing countries could boost total world population to over 9 billion by the end of the twenty-first century. This raises what may be the two most important questions in this entire chapter: Why are birth rates not yet falling in these countries, and what can be done about it?

Some demographers claim that a demographic transition is already in progress in most developing nations. As we have seen, fertility rates have fallen almost everywhere, and some countries have seen remarkably rapid changes, with declines of 30 to 60 percent in a generation.

Many believe that the demographic transition will proceed more rapidly in developing countries today than it did when European and North American countries were developing. These are some of the reasons they see:

- Prosperity, urbanization, and social reforms often reduce the need for large families.
- Technological exchange is rapid today, accelerating economic advances in developing areas.
- Developing areas can take advantage of trade and information networks that are stronger now than in the past.
- Communication technology enhances cultural exchange, economic development, and social change more rapidly than in the past.

Two ways to complete the demographic transition

The Indian states of Kerala and Andra Pradesh exemplify two very different approaches to regulating population growth. In Kerala, providing a fair share of social benefits to everyone is seen as the key to family planning. This social justice strategy assumes that the world has enough resources for everyone but inequitable social and economic systems cause maldistributions of those resources. Hunger, poverty, violence, environmental degradation, and overpopulation are symptoms of a lack of justice, rather than a lack of resources. Although overpopulation exacerbates other problems, a focus on growth rates alone encourages racism and hostility toward the poor. Proponents of this perspective argue that richer people should recognize the impacts their exorbitant consumption has on others (fig. 4.15).

▲ **FIGURE 4.15** Population growth, resulting from high birth rates, has important impacts on resource consumption. But small, wealthy families consume far more resources per person, usually sourced from around the globe. Many people debate which impact is more important, but ultimately there are good reasons for reducing both family size and consumption rates. (above): ©Andre Babiak/Alamy Stock Photo; (left): ©Duncan Vere Green/Alamy Stock Photo

The leaders of Andra Pradesh, on the other hand, have adopted a strategy of aggressively emphasizing birth control, rather than promoting social justice. This strategy depends on policies, similar to those in China, that assume providing carrots (economic rewards for reducing births) along with sticks (mandates for limiting reproduction together with punishment for exceeding limits) are the only effective ways to regulate population size.

Both states have slowed population growth significantly. And though they employ very different strategies, both aim to avoid a "demographic trap" in which rapidly growing populations exceed the sustainable yield of local forests, grasslands, croplands, and water resources. The environmental deterioration, economic decline, and political instability caused by resource shortages may prevent countries caught in this trap from ever completing modernization. Their populations may continue to grow until catastrophe intervenes.

What do you think? If you were advising developing countries on their population policies, which approach would you adopt?

Improving women's lives helps reduce birth rates

As it is increasingly understood that empowering women is key to achieving population stability. The 1994 International Conference on Population and Development in Cairo, Egypt, supported this approach to population issues. A broad consensus reached by the 180 participating countries agreed that responsible economic development, education, and women's rights, along with high-quality health care (including family planning services), must be accessible to everyone if population growth is to be slowed. Child survival is one of the most critical factors in stabilizing population. When infant and child mortality rates are high, as they are in much of the developing world, parents tend to have high numbers of children to ensure that some will survive to adulthood. There has never been a sustained drop in birth rates that was not first preceded by a sustained drop in infant and child mortality.

However, increasing family income doesn't always translate into better welfare for children, since men in many cultures control most financial assets. As the UN Conference in Cairo noted, often the best way to improve child survival is to ensure the rights of mothers. Opportunities for women's education, for instance, as well as land reform, political rights, opportunities to earn an independent income, and improved health status of women often are better indicators of family welfare than is rising gross national product (fig. 4.16).

Family planning gives us choices

Family planning allows couples to determine the number and spacing of their children. It doesn't necessarily mean fewer children—people could use family planning to have the maximum number of children possible—but it does imply that the parents will control their reproductive lives and make conscious decisions about how many children they will have and when those children will be born, rather than leaving it to chance. As the desire for smaller families becomes more common, birth control often becomes an essential part of family planning. In this context, **birth control** usually means any method used to reduce births, including celibacy, delayed marriage, contraception, methods that prevent embryo implantation, and induced abortions.

Humans have always regulated their fertility. The high birth rate of the last two centuries is not the norm, compared to previous millennia of human existence. Evidence suggests that people in every culture and every historic period used a variety of techniques to control population size. Studies of hunting-and-gathering people, such as the !Kung, or San, of the Kalahari Desert in southwest Africa, indicate that our early ancestors had stable population densities, not because they killed each other or starved to death regularly but because they controlled fertility.

For instance, San women breast-feed children for 3 or 4 years. When calories are limited, lactation depletes body fat stores and suppresses ovulation. Coupled with taboos against intercourse while breast-feeding, this is an effective way of spacing children. (However, breast-feeding among well-nourished women in modern societies doesn't necessarily suppress ovulation or prevent conception.) Other ancient techniques to control population size include celibacy, folk medicines, abortion, and infanticide. We may find some or all of these techniques unpleasant or morally unacceptable, but we shouldn't assume that other people are too ignorant or too primitive to make decisions about fertility.

Modern medicine gives us many more options for controlling fertility than were available to our ancestors. More than 100 new

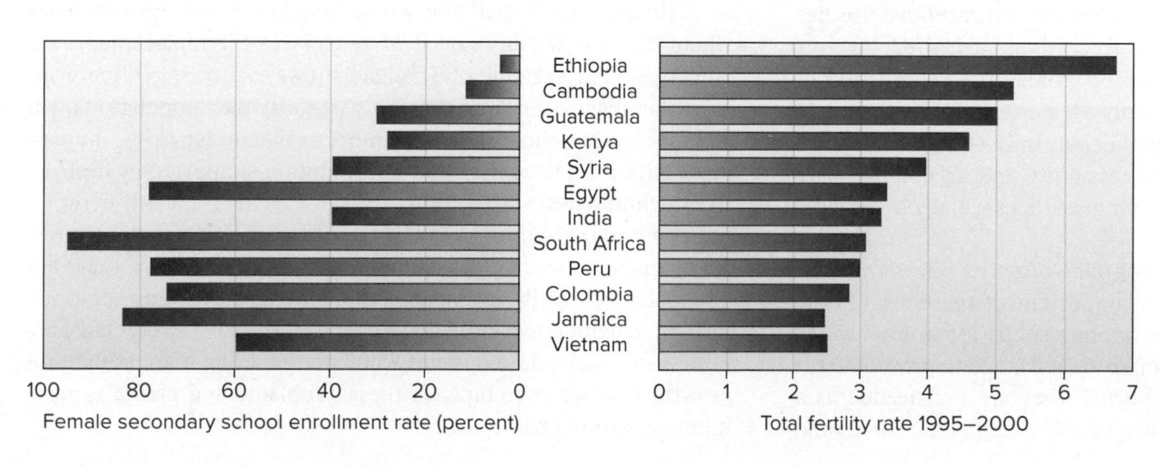

FIGURE 4.16 Total fertility declines as women's education increases. Source: Data from Worldwatch Institute, 2003.

contraceptive methods are now being studied, and some appear to have great promise. Nearly all are hormonal rather than mechanical (e.g., condom, IUD).

Other methods are years away from use but take a new direction entirely. Vaccines for women are being developed that either will prepare the immune system to reject the hormone chorionic gonadotropin, which maintains the uterine lining and allows egg implantation, or will cause an immune reaction against sperm. Injections for men that reduce sperm production without reducing testosterone levels are now in clinical trials. Without a doubt, the contemporary couple has access to many more birth control options than their grandparents had.

4.5 WHAT KIND OF FUTURE ARE WE CREATING NOW?

- *Population projections vary from 8 billion to over 12 billion.*
- *Family planning still lags developing areas, especially in Africa.*
- *Half of our population lives in countries where fertility is below 2.1.*

Because there's often a lag between the time when a society reaches replacement birth rate and the end of population growth, we are deciding now what the world will look like in a hundred years. How many people will be in the world a century from now? Most demographers believe that world population will stabilize sometime during the twenty-first century. When we reach that equilibrium, the total number of humans is likely to be around 8 to 10 billion, depending on the multitude of factors affecting human populations. The United Nations Population Division projects four population scenarios (fig. 4.17). The optimistic (low) projection suggests that world population might stabilize just below 8 billion by the end of the century. This doesn't seem likely. The medium projection shows a population of about 9.4 billion in 35 years, while the high projection will reach nearly 12 billion by midcentury.

Which of these scenarios will we follow? As you have seen in this chapter, population growth is a complex subject. Stabilizing or reducing human populations will require substantial changes from business as usual.

An encouraging sign is that worldwide contraceptive use has increased sharply in recent years. According to United Nations data, 64 percent of married or in-union women used some form of contraception in 2015, up from only 10 percent in 1970. But in least-developed countries, that number was only 40 percent, and in Africa only 33 percent. Clearly, accessibility and acceptability are improving, but there is great room for growth, especially in developing areas.

Successful family planning programs often require significant societal changes. Among the most important of these are (1) improved social, educational, and economic status for women (birth control and women's rights are often linked); (2) improved status for children (fewer children are born if they are not needed as a cheap labor source); (3) acceptance of calculated choice as a valid

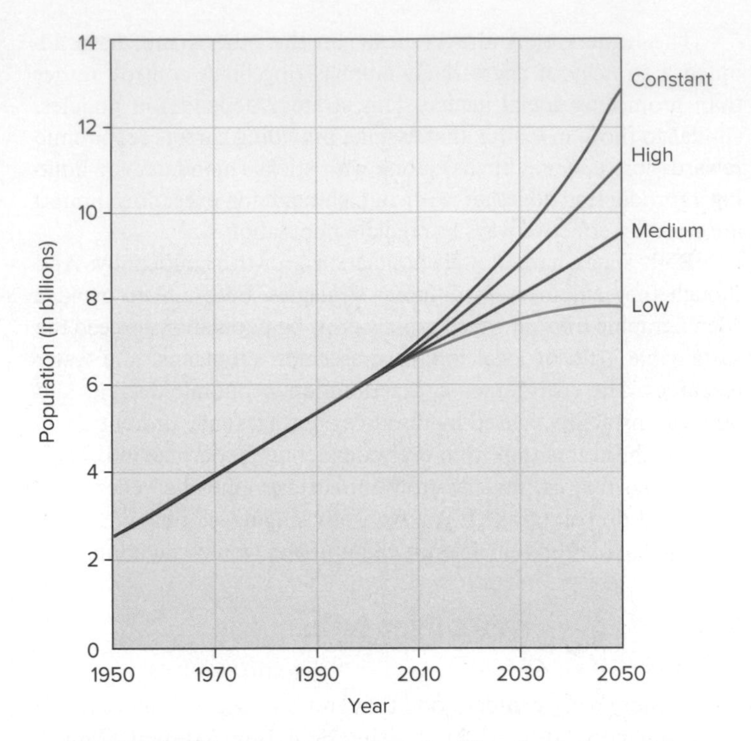

▲ **FIGURE 4.17** Population projections for different growth scenarios. Recent progress in family planning and economic development have led to significantly reduced estimates compared to a few years ago. The medium projection is 9.4 billion in 2050, compared to previous estimates of over 10 billion for that date. Source: UN Population Division, 2015.

element in life in general and in fertility in particular (the belief that we have no control over our lives discourages a sense of responsibility); (4) social security and political stability that give people the means and the confidence to plan for the future; and (5) the knowledge, availability, and use of effective and acceptable means of birth control.

The current world average fertility rate of 2.45 births per woman is less than half what it was 50 years ago. If similar progress could be sustained for the next half century, fertility rates could fall to the replacement rate of 2.1 children per woman. Whether this scenario comes true or not depends on choices that all of us make.

Already, nearly half the world population lives in countries where the total fertility rate is at or close to the replacement rate (fig. 4.18). The example of Thailand shows that people often want small families, given a choice. This suggests that populations will stabilize as that choice becomes more available. However, increasing wealth creates worries that consumption supported by destructive technologies will be unsustainable. The trade-off between population size and affluence may still create unacceptable environmental conditions. Furthermore, as figure 4.18 shows, there are countries, especially in Africa, where wars, corruption, colonial history, religious tensions, and other factors have prevented economic and social development while perpetuating high population growth. Can we overcome all these problems and create a more humane, sustainable world?

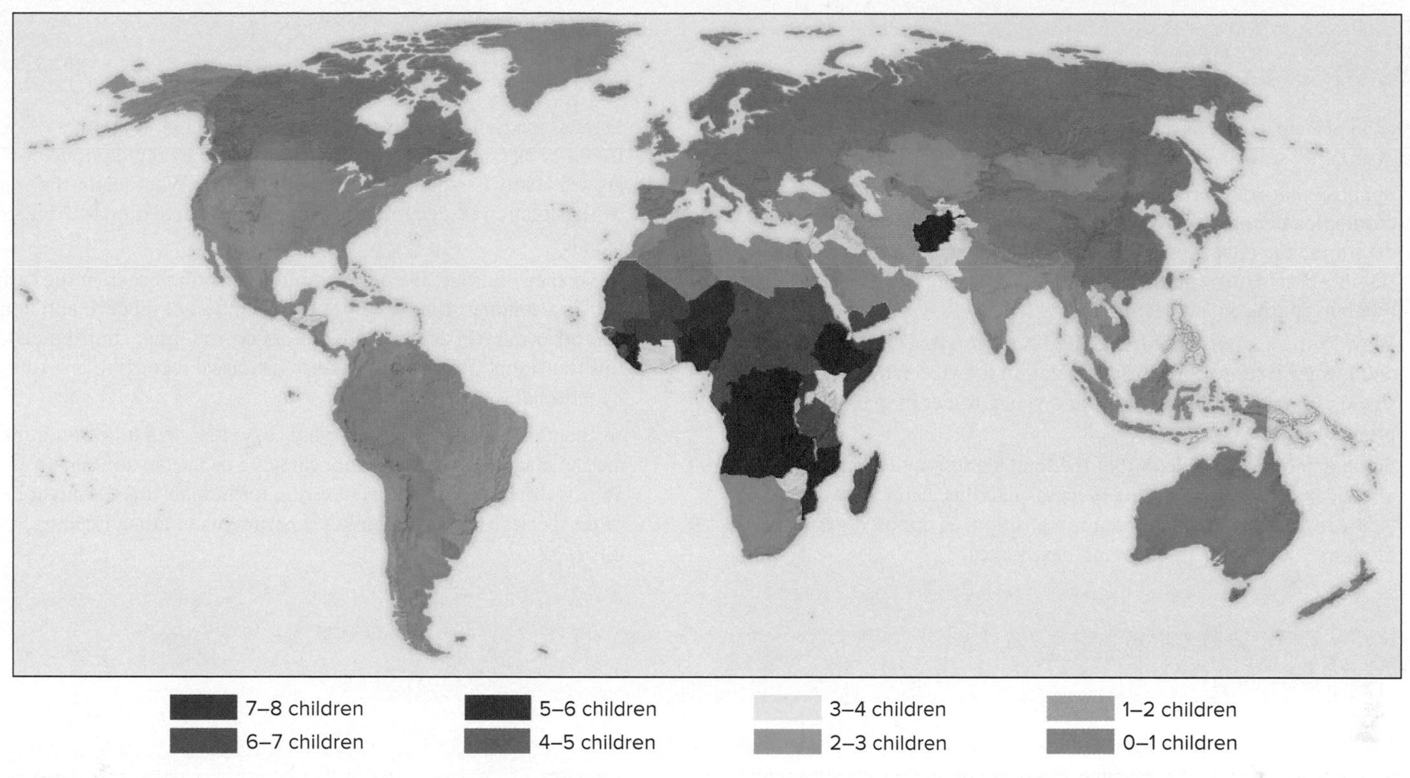

▲ **FIGURE 4.18** Fertility rates by country. Although average fertility in the United States is currently 2.06, it's below the replacement rate of 2.1 children per woman.

CONCLUSION

A few decades ago, we were warned that a human population explosion would soon cause global starvation and ecological collapse. Coercive policies in China and India were some of the responses to this threat. We still don't know how populations will grow, but birth rates have fallen dramatically and widely in recent decades. Most demographers now believe our population will stabilize at around 8–10 billion before the end of this century.

The trajectory we follow depends a great deal on political and social reforms, on access to family planning and health care, and on factors like political stability and peace, which help ensure stability of families.

The demographic transition model projects that populations should stabilize as development proceeds. But what stable population can our planet support in the long run? If we all try to live at the level of material comfort and affluence currently enjoyed by residents of the wealthiest nations, using the old, polluting, inefficient technologies, the answer is almost certain that even 7 billion people is too many in the long run. If we find more sustainable ways to live, however, it may be that many more of us could live happy, productive lives. But what about the other species with which we share the planet? Will we leave room for wild species and natural ecosystems in our efforts to achieve comfort and security? We'll discuss pollution problems, energy sources, and sustainability in subsequent chapters of this book.

PRACTICE QUIZ

1. About how many years of human existence passed before the world population reached its first billion? What factors restricted population before that time, and what factors contributed to growth after that point?
2. Describe the pattern of human population growth over the past 200 years. What is the shape of the growth curve (recall chapter 3)?
3. Define *ecological footprint*. How many more earths would it take if all of us tried to live at the same level of affluence as the average North American?
4. Why do some economists consider human resources more important than natural resources in determining a country's future?
5. In which regions of the world will most population growth occur during the twenty-first century? What conditions contribute to rapid population growth in these locations?
6. Define *crude birth rate, total fertility rate, crude death rate,* and *zero population growth*.
7. What is the difference between life expectancy and life span? Why are they different?
8. What is the dependency ratio, and how might it affect the United States in the future?
9. What factors increase or decrease people's desire to have babies?
10. Describe the conditions that lead to a demographic transition.

CRITICAL THINKING AND DISCUSSION

Apply the principles you have learned in this chapter to discuss these questions with other students.

1. Suppose that you were head of a family planning agency in a developing country. How would you design a scientific study to determine the effectiveness of different approaches to population stabilization? How would you account for factors such as culture, religion, education, and economics?
2. Why do you suppose that the United Nations gives high, medium, and low projections for future population growth? Why not give a single estimate? What factors would you consider in making these projections?
3. Some demographers claim that the total world population has already begun to slow, while others dispute this claim. How would you recognize a true demographic transition, as opposed to mere random fluctuations in birth and death rates?
4. Discuss the ramifications of China's one-child policy with a friend or classmate. Do the problems caused by rapid population growth justify harsh measures to limit births? What might the world situation be like today if China had a population of 2 billion people?
5. In northern Europe, the demographic transition began in the early 1800s, a century or more before the invention of modern antibiotics and other miracle drugs. What factors do you think contributed to this transition? How would you use historical records to test your hypothesis?
6. In chapter 3, we discussed carrying capacities. What do you think are the maximum and optimum carrying capacities for humans? Why is this a more complex question for humans than it might be for other species? Why is designing experiments in human demography difficult?

DATA ANALYSIS Population Change over Time

Thailand's population trends have shifted dramatically in recent years. Is Thailand unusual in this change? Take a look at population size and trends in different regions and at some of the factors that influence growth rates. **Gapminder.org** is a rich source of data on global population, health, and development, including animated graphs showing change over time. Go to Connect to find a link to Gapminder graphs, and answer questions about what they tell you.

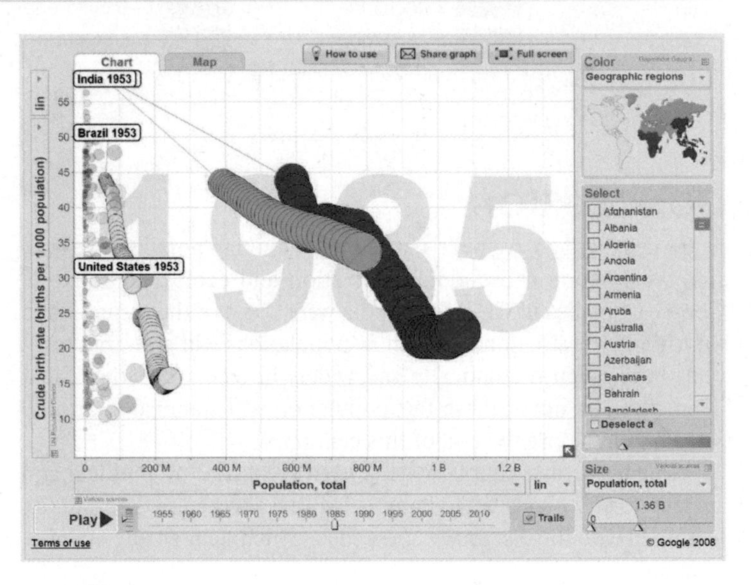

▲ **FIGURE 1** Go to Connect to examine changes in population trends over time.

Design Elements: Active Learning (Toad): ©Gaertner/Alamy Stock Photo; Case Study (Globe): ©McGraw-Hill Education; Google Earth: ©McGraw-Hill Education; Abstract Background: ©Martin Kubat/Shutterstock; What do you think (Students using tablets): ©McGraw-Hill Education/Richard Hutchings, photographer; What can you do (Hand holding Globe): ©Christoph Weihs/Shutterstock

CHAPTER

5 Biomes and Biodiversity

LEARNING OUTCOMES

After studying this chapter, you should be able to answer the following questions:

- What are nine major terrestrial biomes, and what environmental conditions control their distribution?
- How does vertical stratification differentiate life zones in oceans?
- Why are coral reefs, mangroves, estuaries, and wetlands biologically important?
- What do we mean by *biodiversity*? List several regions of high biodiversity.
- What are the major benefits of biodiversity?
- What are the major human-caused threats to biodiversity?
- How can we reduce these threats to biodiversity?

Drought, longer fire seasons, and bark beetle infestations have combined to make forest fires bigger and more destructive across the American West in recent years.

©Kari Greer

CASE STUDY

Ecosystems in Transition

Recent years have seen the most extensive—and most expensive—forest fire seasons ever in the American West. The fire season has lengthened by nearly 3 months, and fires have become larger and more intense. They are also more expensive, rising from around $250 million in 1985 to nearly $2.5 billion in 2017. What has caused the changes? A combination of factors, including longer, hotter summers and drought. On top of these is the explosive growth of billions of tiny beetles, the size of a grain of rice, that have been blamed for damaging or destroying nearly 17 million ha (42 million acres) of western forests. All of these factors are a natural part of western ecosystems. The question is, as all of them accelerate with climate change, what will become of the region's vast forests?

Fires are common in the region, periodically burning patches of spruce, fir, and pine forests in hot, dry summers and allowing young trees to emerge in aging stands. After a fire the forest in a particular area may grow for decades or centuries before burning again. Fire doesn't necessarily kill trees—prolonged or repeated drought is the more important cause of tree mortality. Many trees survive and grow, with bark gradually covering over scars from multiple fires.

Also common are bark beetles, mainly species of the genus *Dendroctonus* and the genus *Ips,* which burrow below the bark of conifers to consume the living inner bark, or phloem, and deposit their eggs. Bark beetles are especially attracted to damaged or drought-stressed trees: These cannot defend themselves, as a healthy tree can, by overwhelming the insects with oozing sap. Normally, long and intensely cold winters kill off many of the beetle larvae, so widespread destruction is infrequent.

In recent decades longer and hotter summers have allowed beetles to thrive. Larvae develop more rapidly in hot weather, and several generations may now occur in a summer. More beetles survive the shorter, milder winters. Outbreaks are now far more extensive and more devastating than western states have seen in the past, and the pests are moving north. Ever-larger areas of forest have become standing dry tinder, waiting for the next lightning strike for ignition (fig 5.1).

Forests can recover from periodic disturbance, but increasingly ecologists are warning that a growing frequency and severity of beetle outbreaks and of fires, both driven by climate warming, are likely to lead to a state transition. What we have known as a forested western landscape is likely to transition to something new in many areas (fig. 5.2). Perhaps former forests will become patchy savannahs, shrublands, or even grasslands. It depends on how much a given area warms and how patterns of rain, snow, and below-freezing temperatures change.

To an ecologist, a transition from forest to some other type of ecosystem makes sense. Temperature and precipitation determine the general type of biome that occurs in an area, often aided by related factors such as insect pests. We have usually considered biome regions stable, but shifting patterns of temperature and

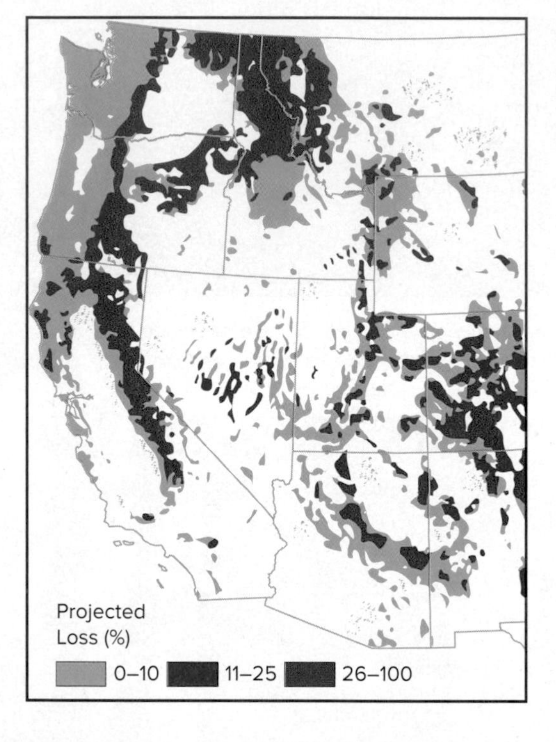

FIGURE 5.1 Projected forest losses from bark beetles and other insect pests by 2027. Source: Data from US Forest Service

FIGURE 5.2 Dead trees killed by bark beetles are interspersed with green, still-living trees. ©WAVE/ Science Source

(continued)

CASE STUDY continued

precipitation should cause a different type of ecosystem to develop in a place. So as climate regions change in coming decades, we can expect that the map of biomes and ecological regions will shift. Where will shifts occur, and how dramatic will they be? That remains to be seen.

To residents of the West, a state transition may mean serious disruption. Much of the region is defined by its forested mountain landscapes. Economies depend on historically reliable forest resources, from timber to the scenery that supports tourism. Ecosystem services, especially provision of drinking water and irrigation from forested watersheds, support cities and farms throughout the region. A shift from abundant forests to another biome type could have dramatic impacts on the lives and livelihoods of people in the region.

In this chapter we examine some of the factors that help us understand why ecosystems and biomes look and function as they do. Understanding these factors can help us understand why the environments we live in look as they do, and what changes we are likely to see in these systems where we live. ■

In the end, we conserve only what we love. We will love only what we understand. We will understand only what we are taught.

—BABA DIOUM

5.1 TERRESTRIAL BIOMES

- *Biomes are broad categories of living systems defined mainly by climate.*
- *Biomes vary in biodiversity, productivity, and structure.*

Biological communities are strongly shaped by temperature ranges and the availability of moisture. As we evaluate potential effects of environmental change, it's useful to start by examining how these factors influence general patterns of plant and animal distribution. **Biome** is a term we use to describe these broad classes of biological communities. If we know the range of temperature and precipitation in a particular place, we can generally predict what kind of biome is likely to occur there, in the absence of human disturbance (fig. 5.3).

An important characteristic of each biome is its **biodiversity,** or the number and variety of biological species that live there. Species not only create much of the structure and functions of an ecosystem but, as we discussed in chapter 2, also generate emergent properties, such as productivity, homeostasis, and resilience. Productivity, the rate at which plants produce biomass, varies a great

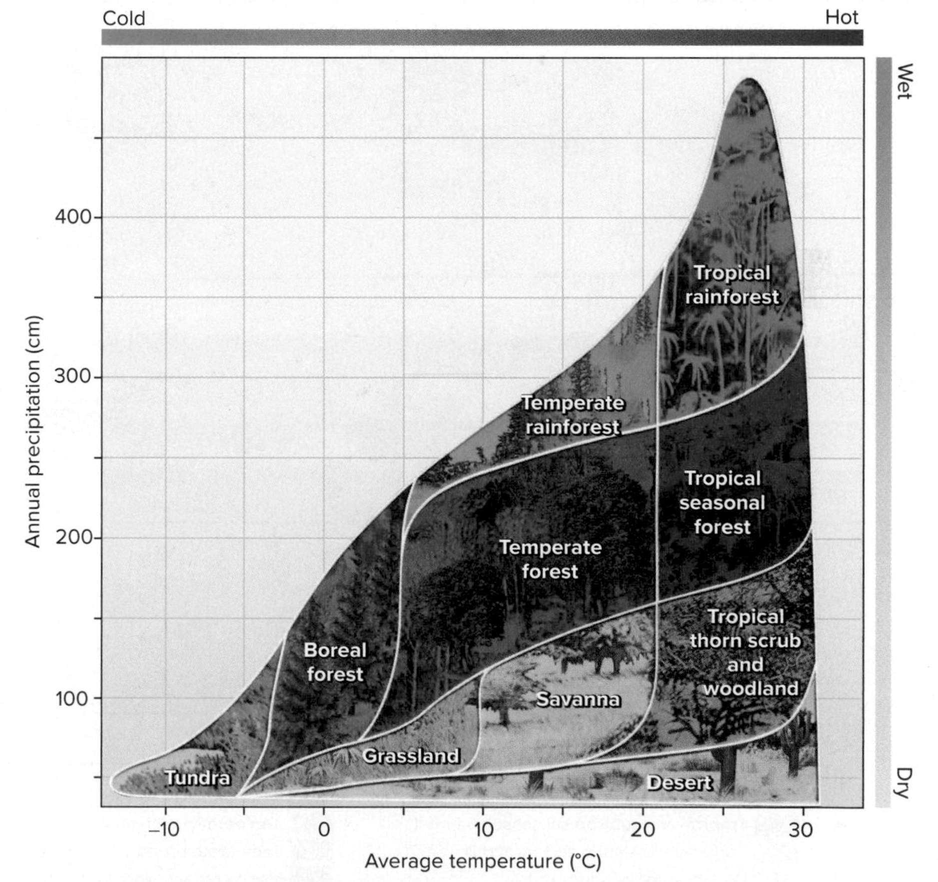

▲ **FIGURE 5.3** Biomes most likely to occur in the absence of human disturbance or other disruptions, according to average annual temperature and precipitation. Changes in temperature or precipitation could shift a region into a different place on this chart. Note that this diagram does not consider soil type, topography, wind speed, or other important environmental factors.

Benchmark Data	
Among the ideas and values in this chapter, these are a few worth remembering.	
200 mm	Annual precipitation in tropical rainforests
< 50 mm	Annual precipitation in deserts
> 5,000	Number of mammal species
21%	Mammal species endangered
> 280,000	Number of flowering plant species
73%	Flowering plant species threatened
2 million	Number of reptiles imported illegally into the United States per year

deal from warm to cold climates, and from wet to dry environments. The amount of resources we can extract, such as timber, fish, or crops, depends largely on a biome's biological productivity. Similarly, homeostasis (stability) and resilience (the ability to recover from disturbance) also depend on biodiversity and productivity. Clear-cut forests, for example, regrow very quickly in the warm, moist Amazon but slowly, if at all, in northern Canada's harsh cold, dry climate.

In the sections that follow, you will learn about nine major biome types. These nine can be further divided into smaller classes: For example, some temperate forests have mainly cone-bearing trees (conifers), such as pines or fir, while others have mainly broad-leaf trees, such as maples or oaks (fig. 5.4).

Many temperature-controlled biomes occur in latitudinal bands. A band of boreal forest crosses Canada, Europe, and Siberia; tropical forests occur near the equator; and expansive grasslands lie near—or just beyond—the tropics. Some biomes are named for their latitudes: Tropical rainforests occur between the Tropic of Cancer (23° north) and the Tropic of Capricorn (23° south); arctic tundra lies near or above the Arctic Circle (66.6° north).

Temperature and precipitation change with elevation as well as with latitude. In mountainous regions, temperatures are cooler and precipitation is usually greater at high elevations. Mountains are cooler, and often wetter, than low elevations. **Vertical zonation** is a term applied to vegetation zones defined by altitude. A 100-km transect from California's Central Valley up to Mt. Whitney, for example, crosses as many vegetation zones as you would find on a journey from southern California to northern Canada (fig. 5.5).

As you consider the terrestrial biomes, compare the climatic conditions that help shape them. To begin, examine the three climate graphs in figure 5.6. These graphs show annual trends in temperature and precipitation (rainfall and snowfall). They also indicate the relationship between potential evaporation, which depends on temperature, and precipitation. When evaporation exceeds precipitation, dry conditions result (marked yellow). Moist climates may vary in precipitation rates, but evaporation rarely

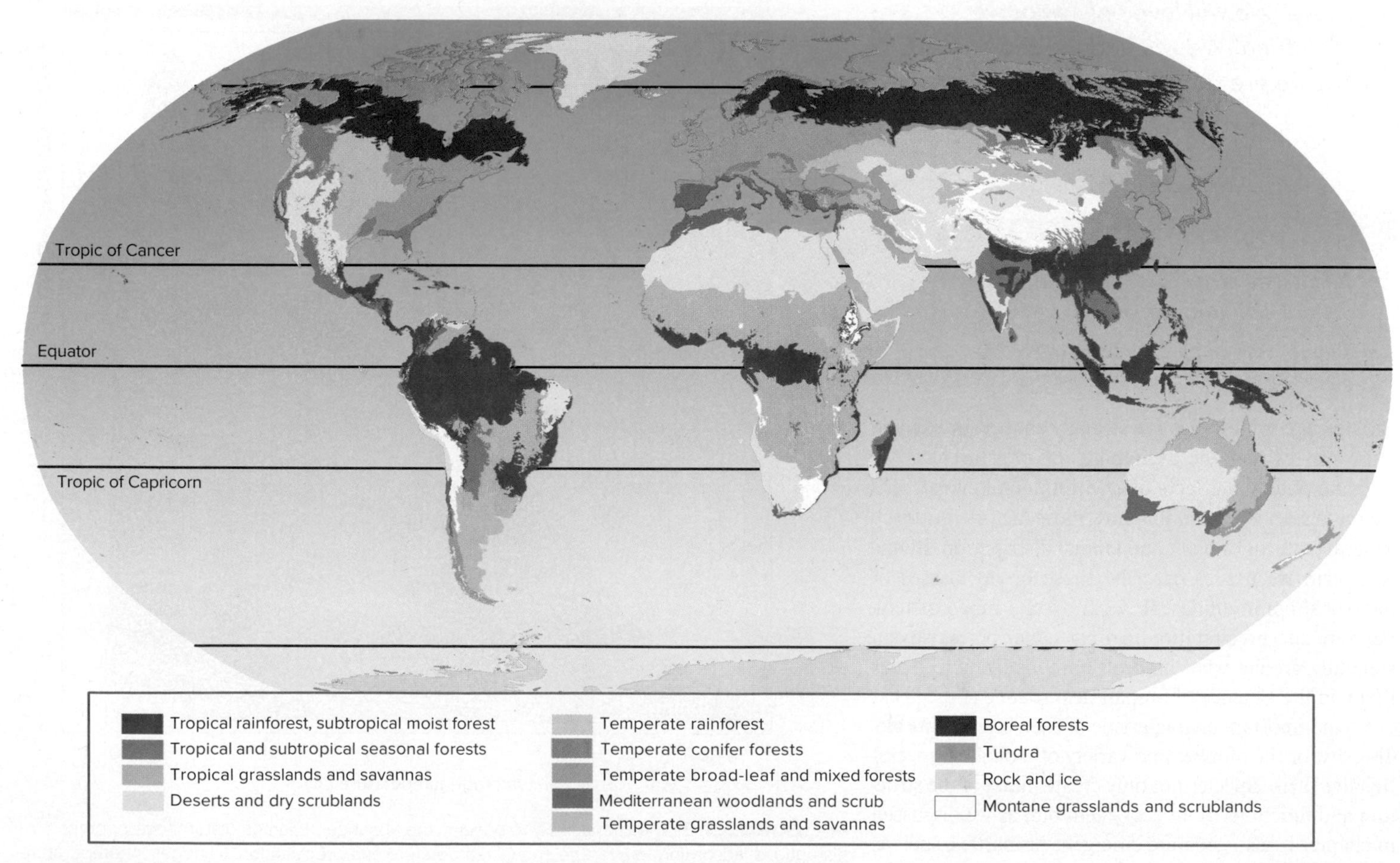

▲ **FIGURE 5.4** Major world biomes. Compare this map with figure 5.3 for generalized temperature and moisture conditions that control biome distribution. Also compare it with the satellite image of biological productivity (see fig. 5.15). Source: WWF Ecoregions.

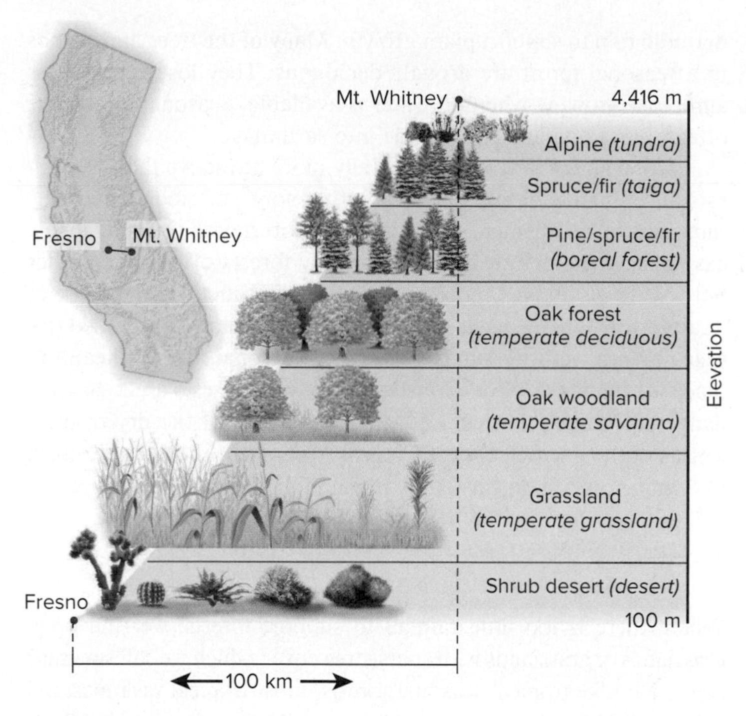

▲ **FIGURE 5.5** Vegetation changes with elevation because temperatures are lower and there is more precipitation high on a mountainside. A 100-km transect from Fresno, California, to Mt. Whitney (California's highest point) crosses vegetation zones similar to about seven different biome types.

exceeds precipitation. Months above freezing temperature (marked brown) have most evaporation. Comparing these climate graphs helps us understand the different seasonal conditions that control plant and animal growth in the different biomes.

Tropical moist forests are warm and wet year-round

The humid tropical regions support one of the most complex and biologically rich biome types in the world (fig. 5.7). Although there are several kinds of moist tropical forests, all have ample rainfall and uniform temperatures. Cool **cloud forests** are found high in the mountains where fog and mist keep vegetation wet all the time. **Tropical rainforests** occur where rainfall is abundant—more than 200 cm (80 in.) per year—and temperatures are warm to hot year-round.

The soil of both these tropical moist forest types tends to be thin, acidic, and nutrient-poor, yet the number of species present can be mind-boggling. For example, the number of insect species in the canopy of tropical rainforests has been estimated to be in the millions! It is thought that one-half to two-thirds of all species of terrestrial plants and insects live in tropical forests.

The nutrient cycles of these forests also are distinctive. About 90 percent of all the nutrients in the rainforest are contained in the bodies of the living organisms. This is a striking contrast to temperate forests, where nutrients are held within the soil and made available for new plant growth. The luxuriant growth in tropical rainforests depends on rapid decomposition and recycling of dead organic material. Leaves and branches that fall to the forest floor decay and are incorporated almost immediately back into living biomass.

When the forest is removed for logging, agriculture, and mineral extraction, the thin soil cannot support continued cropping and cannot resist erosion from the abundant rains. Recovery of the forest could take centuries.

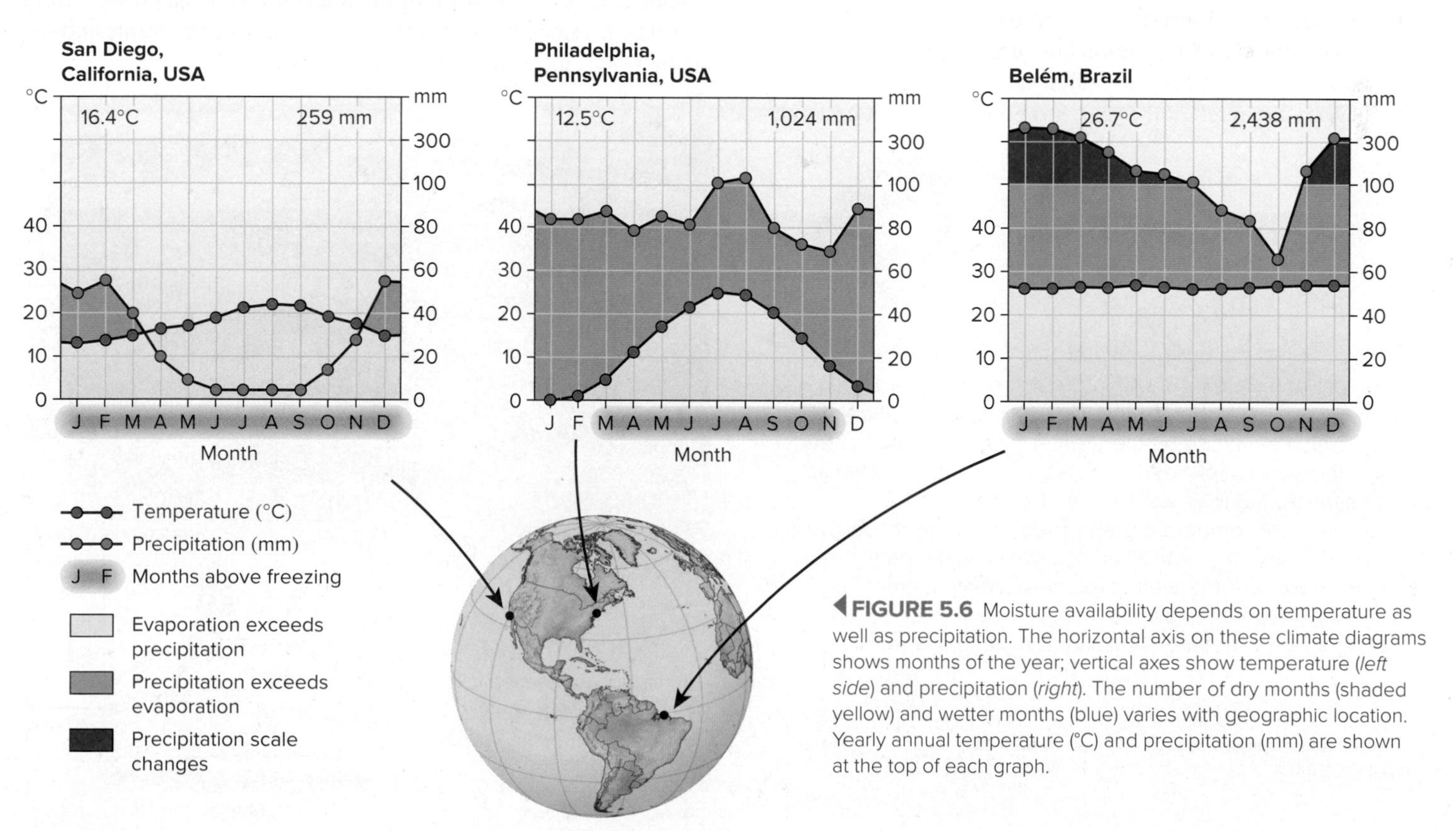

◀ **FIGURE 5.6** Moisture availability depends on temperature as well as precipitation. The horizontal axis on these climate diagrams shows months of the year; vertical axes show temperature (*left side*) and precipitation (*right*). The number of dry months (shaded yellow) and wetter months (blue) varies with geographic location. Yearly annual temperature (°C) and precipitation (mm) are shown at the top of each graph.

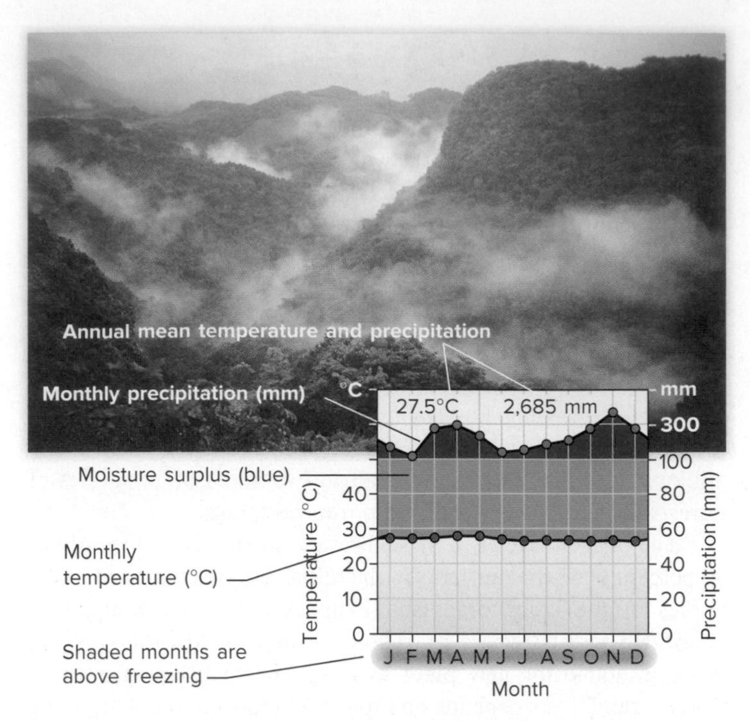

▲ **FIGURE 5.7** Tropical rainforests have luxuriant and diverse plant growth. Heavy rainfall in most months, shown in the climate graph, supports this growth.
©Adalberto Rios Szalay/Sexto Sol/Getty Images

Tropical seasonal forests have annual dry seasons

Many tropical regions are characterized by distinct wet and dry seasons, although temperatures remain hot year-round. These areas support **tropical seasonal forests:** drought-tolerant forests that look brown and dormant in the dry season but burst into vivid green during rainy months. These forests are often called dry tropical forests because they are dry much of the year; however, there must be some periodic rain to support plant growth. Many of the trees and shrubs in a seasonal forest are drought-deciduous: They lose their leaves and cease growing when no water is available. Seasonal forests are often open woodlands that grade into savannas.

Tropical dry forests are generally more attractive than wet forests for human habitation and have, therefore, suffered greater degradation from settlement. Clearing a dry forest with fire is relatively easy during the dry season. Soils of dry forests often have higher nutrient levels and are more agriculturally productive than those of a rainforest. Finally, having fewer insects, parasites, and fungal diseases than a wet forest makes a dry or seasonal forest a healthier place for humans to live. Consequently, these forests are highly endangered in many places. Less than 1 percent of the dry tropical forests of the Pacific coast of Central America or the Atlantic coast of South America, for instance, remain in an undisturbed state.

Active LEARNING

Comparing Biome Climates

Look back at the climate graphs for San Diego, California, an arid region, and Belém, Brazil, in the Amazon rainforest (see fig. 5.6). How much colder is San Diego than Belém in January? in July? Which location has the greater range of temperature through the year? How much do the two locations differ in precipitation during their wettest months?

Compare the temperature and precipitation in these two places with those in the other biomes shown in the pages that follow. How wet are the wettest biomes? Which biomes have distinct dry seasons? How do rainfall and length of warm seasons explain vegetation conditions in these biomes?

ANSWERS: San Diego is about 13°C colder in January, about 6°C colder in July; San Diego has the greater range of temperature; there is about 250 mm difference in precipitation in December–February.

Tropical savannas and grasslands are dry most of the year

Where there is too little rainfall to support forests, we find open **grasslands** or grasslands with sparse tree cover, which we call **savannas** (fig. 5.8). Like tropical seasonal forests, most tropical savannas and grasslands have a rainy season, but generally the rains are less abundant or less dependable than in a forest. During dry seasons, fires can sweep across a grassland, killing off young trees and keeping the landscape open. Savanna and grassland plants have many adaptations to survive drought, heat, and fires. Many have deep, long-lived roots that seek groundwater and that persist when leaves and stems above the ground die back. After a fire or drought, fresh, green shoots grow quickly from the roots. Migratory grazers, such as wildebeest, antelope, or bison, thrive on this new growth. Grazing pressure from domestic livestock is an important threat to both the plants and the animals of tropical grasslands and savannas.

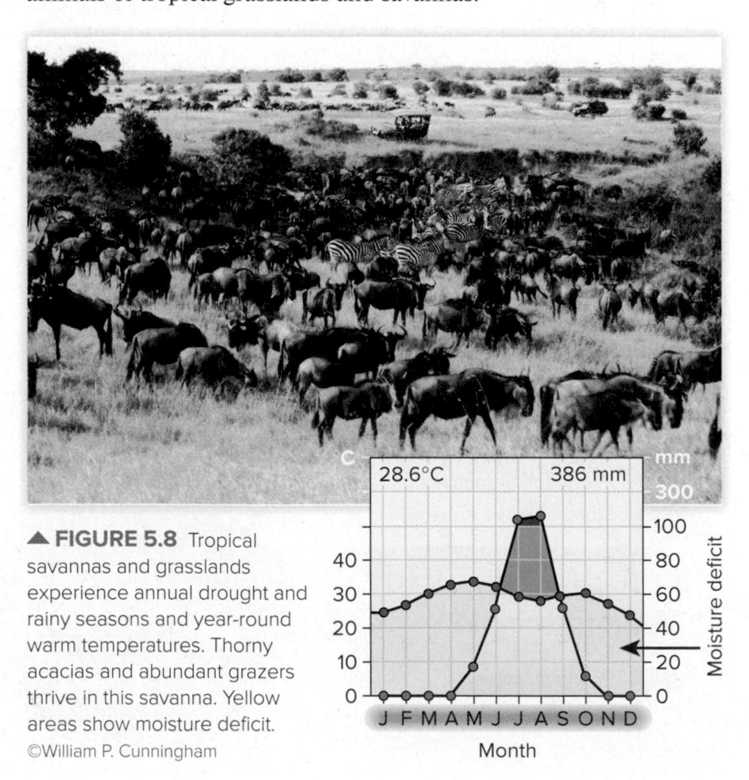

▲ **FIGURE 5.8** Tropical savannas and grasslands experience annual drought and rainy seasons and year-round warm temperatures. Thorny acacias and abundant grazers thrive in this savanna. Yellow areas show moisture deficit.
©William P. Cunningham

Deserts are hot or cold, but always dry

You may think of deserts as barren and biologically impoverished. Their vegetation is sparse, but it can be surprisingly diverse, and most desert plants and animals are highly adapted to survive long droughts, extreme heat, and often extreme cold. **Deserts** occur where precipitation is sporadic and low, usually with less than 30 cm of rain per year. Adaptations to these conditions include water-storing leaves and stems, thick epidermal layers to reduce water loss, and salt tolerance. As in other dry environments, many plants are drought-deciduous. Most desert plants also bloom and set seed quickly when rain does fall.

Warm, dry, high-pressure climate conditions (see chapter 9) create desert regions at about 30° north and south. Extensive deserts occur in continental interiors (where rain is rare and evaporation rates are high) of North America, Central Asia, Africa, and Australia (fig. 5.9). The rain shadow of the Andes produces the world's driest desert in coastal Chile. Deserts can also be cold. Most of Antarctica is a desert; some inland valleys apparently get almost no precipitation at all.

Like plants, animals in deserts are specially adapted. Many are nocturnal, spending their days in burrows to avoid the sun's heat and desiccation. Pocket mice, kangaroo rats, and gerbils can get most of their moisture from seeds and plants. Desert rodents also have highly concentrated urine and nearly dry feces, which allow them to eliminate body waste without losing precious moisture.

Deserts are more vulnerable than you might imagine. Sparse, slow-growing vegetation is quickly damaged by off-road vehicles. Desert soils recover slowly. Tracks left by army tanks practicing in California deserts during World War II can still be seen today.

Deserts are also vulnerable to overgrazing. In Africa's vast Sahel (the southern edge of the Sahara Desert), livestock are destroying much of the plant cover. Bare, dry soil becomes drifting sand, and restabilization is extremely difficult. Without plant roots and organic matter, the soil loses its ability to retain what rain does fall, and the land becomes progressively drier and more bare. Similar degradation of dryland vegetation is happening in many desert areas, including Central Asia, India, and the American Southwest and Plains states.

Temperate grasslands have rich soils

As in tropical latitudes, temperate (midlatitude) grasslands occur where there is enough rain to support abundant grass but not enough for forests (fig. 5.10). Usually, grasslands are a complex, diverse mix of grasses and flowering herbaceous plants, generally known as forbs. Myriad flowering forbs make a grassland colorful and lovely in summer. In dry grasslands, vegetation may be less than a meter tall. In more humid areas, grasses can exceed 2 m. Where scattered trees occur in a grassland, we call it a savanna.

Deep roots help plants in temperate grasslands and savannas survive drought, fire, and extreme heat and cold. These roots, together with an annual winter accumulation of dead leaves on the surface, produce thick, organic-rich soils in temperate grasslands. Because of this rich soil, many grasslands have been converted to farmland. The legendary tallgrass prairies of the central United States and Canada are almost completely replaced by corn, soybeans, wheat, and other crops. Most remaining grasslands in this region are too dry to support agriculture, and their greatest threat is overgrazing. Excessive grazing eventually kills even deep-rooted plants. As ground cover dies off, soil erosion results, and unpalatable weeds, such as cheatgrass or leafy spurge, spread.

Temperate scrublands have summer drought

Often, dry environments support drought-adapted shrubs and trees, as well as grass. These mixed environments can be highly variable. They can also be very rich biologically. Such conditions

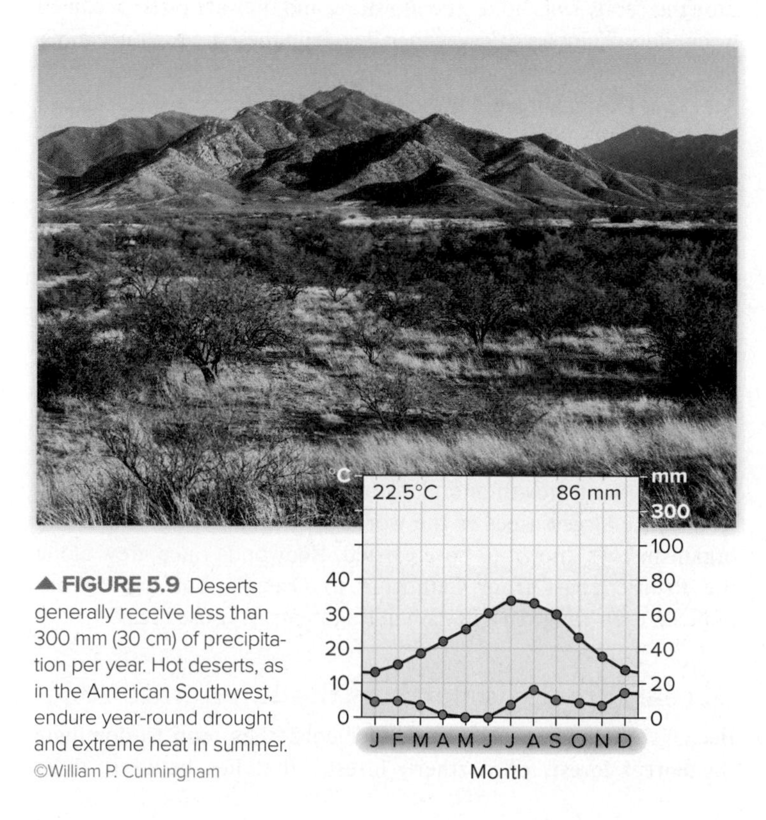

▲ **FIGURE 5.9** Deserts generally receive less than 300 mm (30 cm) of precipitation per year. Hot deserts, as in the American Southwest, endure year-round drought and extreme heat in summer.
©William P. Cunningham

▲ **FIGURE 5.10** Grasslands occur at midlatitudes on all continents. Kept open by extreme temperatures, dry conditions, and periodic fires, grasslands can have surprisingly high plant and animal diversity.
©Mary Ann Cunningham

are often described as Mediterranean (where the hot season coincides with the dry season, producing hot, dry summers and cool, moist winters). Evergreen shrubs with small, leathery, sclerophyllous (hard, waxy) leaves form dense thickets. Scrub oaks, drought-resistant pines, or other small trees often cluster in sheltered valleys. Periodic fires burn fiercely in this fuel-rich plant assemblage and are a major factor in plant succession. Annual spring flowers often bloom profusely, especially after fires. In California this landscape is called **chaparral,** Spanish for thicket. Resident animals are drought-tolerant species, such as jackrabbits, kangaroo rats, mule deer, chipmunks, lizards, and many bird species. Very similar landscapes are found along the Mediterranean coast as well as southwestern Australia, central Chile, and South Africa. Although this biome doesn't cover a very large total area, it contains a high number of unique species and is often considered a "hot-spot" for biodiversity. It also is highly desired for human habitation, often leading to conflicts with rare and endangered plant and animal species.

▲ **FIGURE 5.11** Temperate deciduous forests have year-round precipitation and winters near or below freezing.
©William P. Cunningham

Temperate forests can be evergreen or deciduous

Temperate, or midlatitude, forests occupy a wide range of precipitation conditions but occur mainly between about 30° and 55° latitude (see fig. 5.4). In general we can group these forests by tree type, which can be broad-leaf **deciduous** (losing leaves seasonally) or evergreen **coniferous** (cone-bearing).

Deciduous Forests Broad-leaf forests occur throughout the world where rainfall is plentiful. In midlatitudes, these forests are deciduous and lose their leaves in winter. The loss of green chlorophyll pigments can produce brilliant colors in these forests in autumn (fig. 5.11). At lower latitudes broad-leaf forests may be evergreen or drought-deciduous. Southern live oaks, for example, are broad-leaf evergreen trees.

Although these forests have a dense canopy in summer, they have a diverse understory that blooms in spring, before the trees leaf out. Spring ephemeral (short-lived) plants produce lovely flowers, and vernal (springtime) pools support amphibians and insects. These forests also shelter a great diversity of songbirds.

North American deciduous forests once covered most of what is now the eastern half of the United States and southern Canada. Most of western Europe was once deciduous forest but was cleared a thousand years ago. When European settlers first came to North America, they quickly settled and cut most of the eastern deciduous forests for firewood, lumber, and industrial uses, as well as to clear farmland. Many of those regions have now returned to deciduous forest, though the dominant species may have changed.

Deciduous forests can regrow quickly because they occupy moist, moderate climates. But most of these forests have been occupied so long that human impacts are extensive, and most native species are at least somewhat threatened. The greatest current threat to temperate deciduous forests is in eastern Siberia, where deforestation is proceeding rapidly. Siberia may have the highest deforestation rate in the world. As forests disappear, so do Siberian tigers, bears, cranes, and a host of other endangered species.

Coniferous Forests Evergreen forests of pines, spruce, fir, and other species grow in a wide range of environmental conditions. Often they occur where moisture is limited: In cold climates, moisture is unavailable (frozen) in winter; hot climates may have seasonal drought; sandy soils hold little moisture, and they are often occupied by conifers. Thin, waxy leaves (needles) help these trees reduce moisture loss. Coniferous forests provide most wood products in North America. Dominant wood production regions include the southern Atlantic and Gulf coast states, the mountain West, and the Pacific Northwest (northern California to Alaska), but coniferous forests support forestry in many regions. However, climate change, insects, and other stressors, as described in the opening case study for this chapter, can degrade or dramatically change these forests.

The coniferous forests of the Pacific coast grow in extremely wet conditions. The wettest coastal forests are known as **temperate rainforests,** a cool, rainy forest often enshrouded in fog (fig. 5.12). Condensation in the canopy (leaf drip) is a major form of precipitation in the understory. Mild year-round temperatures and abundant rainfall, up to 250 cm (100 in.) per year, result in luxuriant plant growth and giant trees such as the California redwoods, the largest trees in the world and the largest aboveground organism ever known to have existed. Redwoods once grew along the Pacific coast from California to Oregon, but logging has reduced their range to a few small fragments of those areas.

Boreal forests lie north of the temperate zone

Because conifers can survive winter cold, they tend to dominate the **boreal forest,** or northern forests, that lies between about

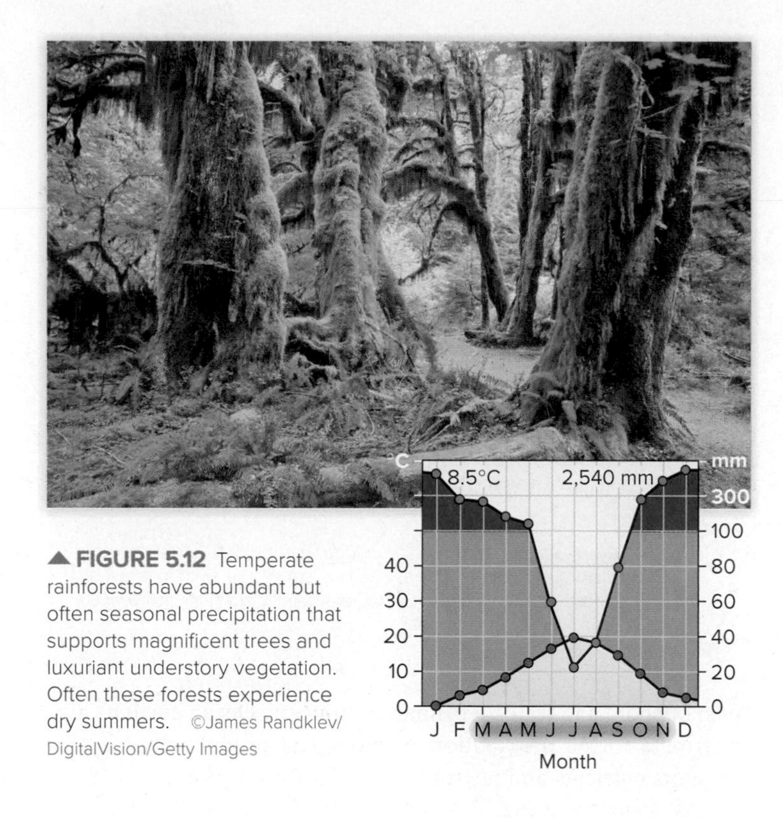

▲ **FIGURE 5.12** Temperate rainforests have abundant but often seasonal precipitation that supports magnificent trees and luxuriant understory vegetation. Often these forests experience dry summers. ©James Randklev/DigitalVision/Getty Images

50° and 60° north (fig. 5.13). Mountainous areas at lower latitudes may also have many characteristics and species of the boreal forest. Dominant trees are pines, hemlocks, spruce, cedar, and fir. Some deciduous trees are also present, such as maples, birch, aspen, and alder. These forests are slow-growing because of the cold temperatures and short frost-free growing season, but they are still an expansive resource. In Siberia, Canada, and the western United States, large regional economies depend on boreal forests. They are favorite places for hunting, fishing, and recreation as well as extractive resource use. If those forests disappear as a result of global warming, both human uses and the unique species, such as moose, lynx, and otter, that depend on that biome will be sorely missed.

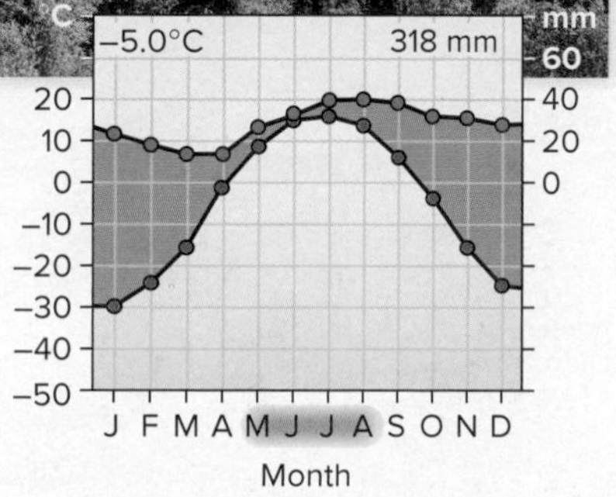

▲ **FIGURE 5.13** Boreal forests have moderate precipitation but are often moist because temperatures are cold most of the year. Cold-tolerant and drought-tolerant conifers dominate boreal forests and taiga, at the forest fringe. ©William P. Cunningham

The extreme, ragged edge of the boreal forest, where forest gradually gives way to open tundra, is known by its Russian name, **taiga.** Here extreme cold and short summer limit the growth rate of trees. A 10-cm-diameter tree may be over 200 years old in the far north.

Tundra can freeze in any month

Where temperatures are below freezing most of the year, only small, hardy vegetation can survive. **Tundra,** a treeless landscape that occurs at high latitudes or on mountaintops, has a growing season of only 2 to 3 months, and it may have frost any month of the year. Some people consider tundra a variant of grasslands because it has no trees; others consider it a very cold desert because water is unavailable (frozen) most of the year.

Arctic tundra is an expansive biome that has low productivity because it has a short growing season. During midsummer, however, 24-hour sunshine supports a burst of plant growth and an explosion of insect life. Tens of millions of waterfowl, shorebirds, terns, and songbirds migrate to the Arctic every year to feast on the abundant invertebrate and plant life and to raise their young on the brief bounty. These birds then migrate to wintering grounds, where they may be eaten by local predators—effectively, they carry energy and protein from high latitudes to low latitudes. Arctic tundra is essential for global biodiversity, especially for birds.

Alpine tundra, occurring on or near mountaintops, has environmental conditions and vegetation similar to arctic tundra (fig. 5.14). These areas have a short, intense growing season. Often one sees a splendid profusion of flowers in alpine tundra; everything must flower at once in order to produce seeds in a few weeks before the snow comes again. Many alpine tundra plants also have deep pigmentation and leathery leaves to protect against the strong ultraviolet light in the thin mountain atmosphere.

Compared to other biomes, tundra has relatively low diversity. Dwarf shrubs, such as willows, sedges, grasses, mosses, and lichens, tend to dominate the vegetation. Migratory musk ox, caribou, or alpine mountain sheep and mountain goats can live on the vegetation because they move frequently to new pastures.

Because these environments are too cold for most human activities, they are not as badly threatened as other biomes. There are important problems, however. Global climate change may be altering the balance of some tundra ecosystems, and air pollution from distant cities tends to accumulate at high latitudes (see chapter 9). In eastern Canada, coastal tundra is being badly depleted by overabundant populations of snow geese, whose numbers have exploded due to winter grazing on the rice fields of Arkansas and Louisiana. Oil and gas drilling—and associated truck traffic—threatens tundra in Alaska and Siberia. Clearly, this remote biome is not independent of human activities at lower latitudes.

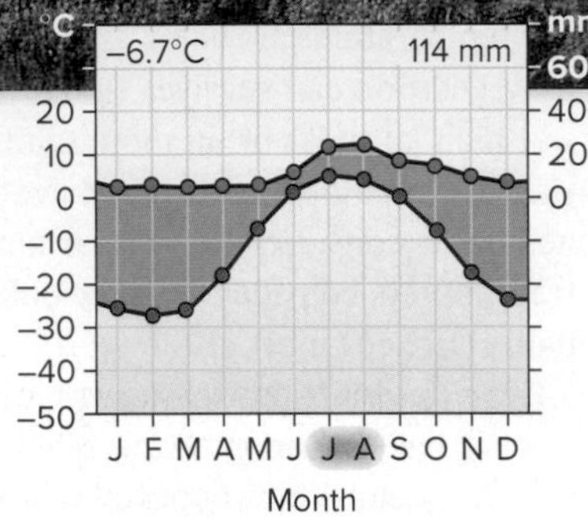

▲ **FIGURE 5.14** This landscape in Canada's Northwest Territories has both alpine and arctic tundra. Plant diversity is relatively low, and frost can occur even in summer. ©Mary Ann Cunningham

−6.7°C 114 mm

Month

Active LEARNING

Examining Climate Graphs

Among the nine types of terrestrial biomes you've just read about, one of the important factors is the number of months when the average temperature is below freezing (0°C). This is because most plants photosynthesize most actively when daytime temperatures are well above freezing—and when water is fluid, not frozen (see chapter 2). Among the biome examples shown, how many sites have fewer than 3 months when the average temperature is above 0°? How many sites have all months above freezing? Look at figure 5.3. Do all deserts have average yearly temperatures above freezing? Now look at figure 5.4. Which biome do you live in? Which biome do most Americans live in?

ANSWERS: Only the tundra site has less than 3 months above freezing. Three sites have all months above freezing. No. Answers will vary. Most Americans live in temperate coniferous or broad-leaf forest biomes.

5.2 MARINE ENVIRONMENTS

- *Nutrients and living organisms are sparse in open oceans.*
- *Shallow, near-shore communities can have very high productivity and biodiversity.*
- *Near-shore communities are vulnerable to human disturbances.*

The biological communities in oceans and seas are poorly understood, but many are as diverse and complex as terrestrial biomes. In this section we will explore a few facets of these fascinating environments. Oceans cover nearly three-fourths of the earth's surface, and they contribute in important, although often unrecognized, ways to terrestrial ecosystems. Like land-based systems, most marine communities depend on photosynthetic organisms. Often it is algae or tiny, free-floating photosynthetic plants (**phytoplankton**) that support a marine food web, rather than the trees and grasses we see on land. In oceans, photosynthetic activity tends to be greatest near coastlines, where nitrogen, phosphorus, and other nutrients wash offshore and fertilize primary producers. Ocean currents also contribute to the distribution of biological productivity, as they transport nutrients and phytoplankton far from shore (fig. 5.15).

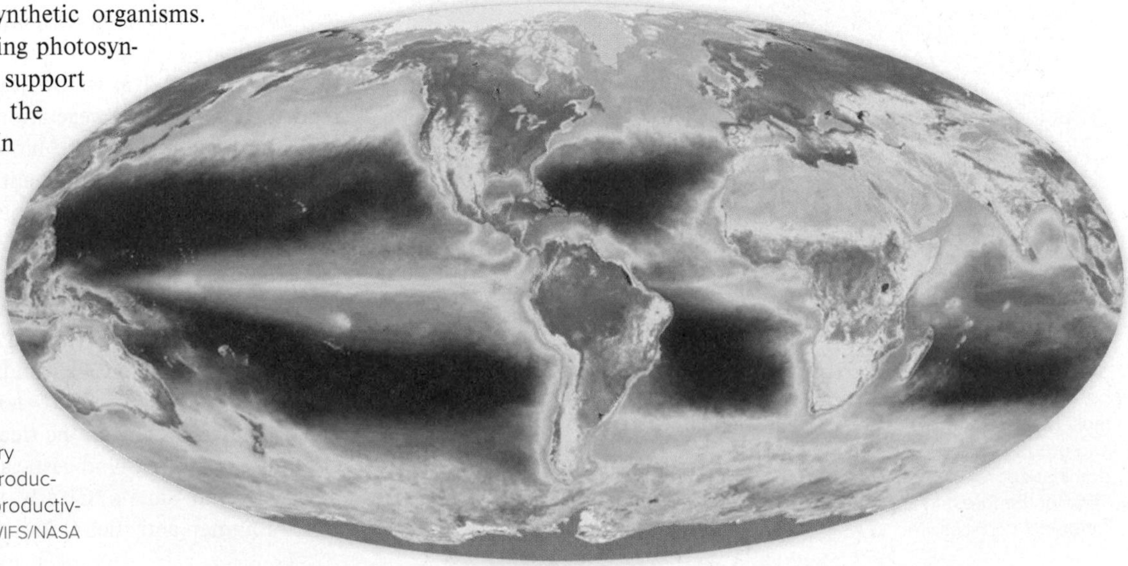

As plankton, algae, fish, and other organisms die, they sink toward the ocean floor. Deep-ocean ecosystems, consisting of crabs, filter-feeding organisms, strange phosphorescent fish, and many other life-forms, often rely on this "marine snow" as a primary nutrient source. Surface communities also depend on this material. Upwelling currents circulate nutrients from the ocean floor back to the surface. Along the coasts of South America, Africa, and Europe, these currents support rich fisheries.

Vertical stratification is a key feature of aquatic ecosystems. Light decreases rapidly with depth, and communities below the photic zone (light zone, often reaching about 20 m deep) must rely on energy sources other than photosynthesis to persist. Temperature also decreases with depth. Deep-ocean species often grow slowly, in part because metabolism is reduced in cold conditions. In contrast, warm,

▶ **FIGURE 5.15** Biological productivity, from satellite measurements of chlorophyll concentrations. In oceans, productivity is highest (red to green) near coastlines and where currents carry nutrients. Mid-ocean basins have low productivity. On land, dark greens show high productivity; yellow-orange is low. Source: Sea WIFS/NASA

bright, near-surface communities, such as coral reefs and estuaries, are among the world's most biologically productive environments. Temperature also affects the amount of oxygen and other elements that can be absorbed in water. Cold water holds abundant oxygen, so productivity is often high in cold oceans, as in the North Atlantic, North Pacific, and Antarctic.

Open ocean communities vary from surface to hadal zone

Ocean systems can be described by depth and proximity to shore (fig. 5.16). In general, **benthic** communities occur on the bottom, and **pelagic** (sea in Greek) zones are the water column. The epipelagic zone (*epi* = on top) has photosynthetic organisms. Below this are the mesopelagic (*meso* = medium) and bathypelagic (*bathos* = deep) zones. The deepest layers are the abyssal zone (to 4,000 m) and hadal zone (deeper than 6,000 m). Shorelines are known as littoral zones, and the area exposed by low tides is known as the intertidal zone. Often there is a broad, relatively shallow region along a continent's coast, which may reach a few kilometers or hundreds of kilometers from shore. This undersea area is the continental shelf.

We know relatively little about marine ecosystems and habitats, and much of what we know we have learned only recently. The open ocean has long been known as a biological desert, because it has relatively low productivity, or biomass production. Fish and plankton abound in many areas, however. Sea mounts, or undersea mountain chains and islands, support many commercial fisheries and much newly discovered biodiversity. In the equatorial Pacific and Antarctic oceans, currents carry nutrients far from shore, supporting biological productivity. The Sargasso Sea, a large region of the Atlantic near Bermuda, is known for its free-floating mats of brown algae. These algae mats support a phenomenal diversity of animals, including sea turtles, fish, and other species. Eels that hatch amid the algae eventually migrate up rivers along the Atlantic coasts of North America and Europe.

Deep-sea thermal vent communities are another remarkable type of marine system that was completely unknown until 1977, when the deep-sea submarine *Alvin* descended to the deep-ocean floor. These communities are based on microbes that capture chemical energy, mainly from sulfur compounds released from thermal vents—jets of hot water and minerals on the ocean floor (fig. 5.17). Magma below the ocean crust heats these vents. Tube worms, mussels, and microbes on the vents are adapted to survive both extreme temperatures, often above 350°C (700°F), and intense water pressure at depths of 7,000 m (20,000 ft) or more. Oceanographers have discovered thousands of different types of organisms, most of them microscopic, in these communities. Some estimate that the total mass of microbes on the seafloor represents one-third of all biomass on the planet.

Tidal shores support rich, diverse communities

As in the open ocean, shoreline communities vary with depth, light, and temperature. Some shoreline communities, such as estuaries, have high biological productivity and diversity because they are enriched by nutrients washing from the land. Others, such as coral

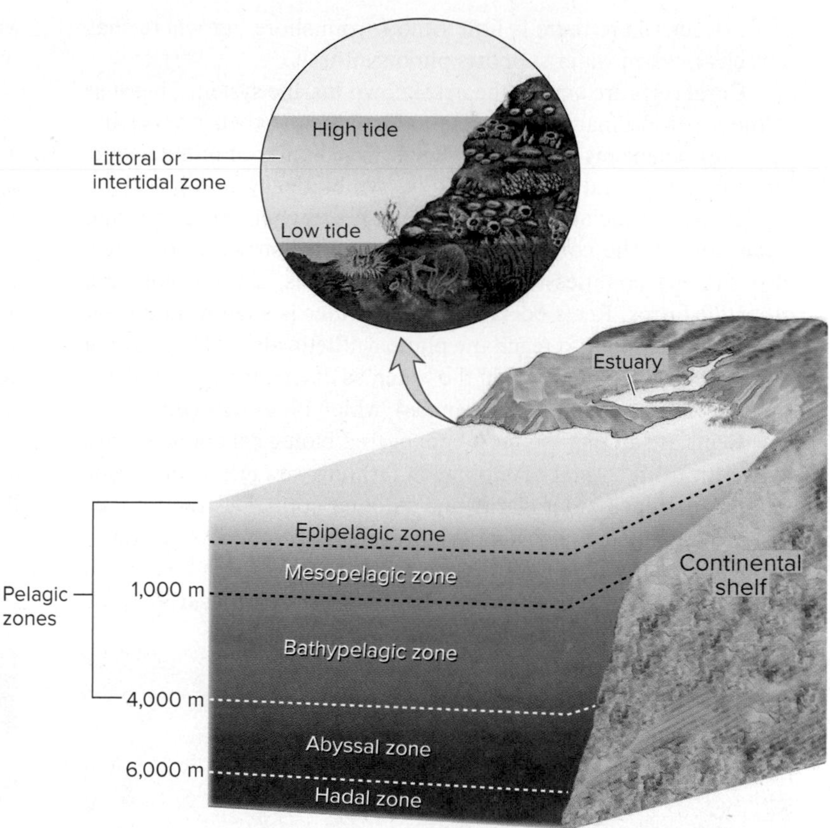

▲ **FIGURE 5.16** Light penetrates only the top 10–20 m of the ocean. Below this level, temperatures drop and pressure increases. Nearshore environments include the intertidal zone and estuaries.

▼ **FIGURE 5.17** Deep-ocean thermal vent communities have great diversity and are unusual because they rely on chemosynthesis, not photosynthesis, for energy. Source: National Oceanic and Atmospheric Administration (NOAA)

reefs, occur where there is little runoff from shore but where shallow, clear, warm water supports photosynthesis.

Coral reefs are among the best-known marine systems, because of their extraordinary biological productivity and their diverse and beautiful organisms (fig. 5.18a). Reefs are colonies of minute, colonial animals ("coral polyps") that live symbiotically with photosynthetic algae. Calcium-rich coral skeletons shelter the algae, and algae nourish the coral animals. The complex structure of a reef also shelters countless species of fish, worms, crustaceans, and other life-forms. Reefs occur where the water is shallow and clear enough for sunlight to reach the photosynthetic algae. They cannot tolerate abundant nutrients in the water, as nutrients support plankton (tiny floating plants and animals), which block sunlight.

Reefs are among the most endangered biological communities. Sediment from coastal development, farming, sewage, or other pollution can reduce water clarity and smother coral. Destructive fishing practices, including dynamite and cyanide poison, have destroyed many Asian reefs. Reefs can also be damaged or killed by changes in temperature, by invasive fish, and by diseases. **Coral bleaching,** the whitening of reefs due to stress, often followed by coral death, is a growing and spreading problem that worries marine biologists.

Sea-grass beds, occupy shallow, warm, sandy coastlines. Like reefs, these support rich communities of grazers, from snails to turtles to Florida's manatees. Also like coral reefs, sea-grass systems are very vulnerable to ocean warming.

Mangroves are a diverse group of salt-tolerant trees that grow along warm, calm marine coasts around the world (fig. 5.18b). Growing in shallow, tidal mudflats, mangroves help stabilize shorelines, blunt the force of storms, and build land by trapping sediment and organic material. After the devastating Indonesian tsunami in 2004, studies showed that mangroves, where they still stood, helped reduce the speed, height, and turbulence of the tsunami waves. Detritus, including fallen leaves, collects below mangroves and provides nutrients for a diverse community of animals and plants. Both marine species (such as crabs and fish) and terrestrial species (such as birds and bats) rely on mangroves for shelter and food.

Like coral reefs and sea-grass beds, mangrove forests provide sheltered nurseries for juvenile fish, crabs, shrimp, and other

(a) Coral reefs and islands

(b) Mangroves

(c) Estuary and salt marsh

(d) Tide pool

▲ **FIGURE 5.18** Coastal environments support incredible diversity and help stabilize shorelines. Coral reefs (a), mangroves (b), and estuaries (c) also provide critical nurseries for marine ecosystems. Tide pools (d) shelter highly specialized organisms. (a) ©Glen Allison/Getty Images; (b) ©Mary Ann Cunningham; (c) ©Andrew F. Kazmierski/123RF; (d) ©Bill Coster

marine species on which human economies depend. However, like reefs and sea-grass beds, mangroves have been devastated by human activities. More than half of the world's mangroves that stood a century ago, perhaps 22 million ha, have been destroyed or degraded. The leading cause of mangrove destruction has been clearing for coastal shrimp or fish farms. Timber production, urban development, municipal sewage, and industrial waste also destroy mangroves. Parts of Southeast Asia and South America have lost 90 percent of their mangrove forests.

Estuaries are bays where rivers empty into the sea, mixing fresh water with salt water. **Salt marshes,** shallow wetlands flooded regularly or occasionally with seawater, occur on shallow coastlines, including estuaries (fig. 5.18c). Usually calm, warm, and nutrient-rich, estuaries and salt marshes are biologically diverse and productive. Rivers provide nutrients and sediments, and a muddy bottom supports emergent plants (whose leaves emerge above the water surface), as well as the young forms of crustaceans, such as crabs and shrimp, and mollusks, such as clams and oysters. Nearly two-thirds of all marine fish and shellfish rely on estuaries and saline wetlands for spawning and juvenile development.

Estuaries near major American cities once supported an enormous wealth of seafood. Oyster beds and clam banks in the waters adjacent to New York, Boston, and Baltimore provided free and easy food to early residents. Sewage and other contaminants long ago eliminated most of these resources, however. Recently, major efforts have been made to revive Chesapeake Bay, America's largest and most productive estuary. These efforts have shown some success, but many challenges remain.

In contrast to the shallow, calm conditions of estuaries, coral reefs, and mangroves, there are violent, wave-blasted shorelines that support fascinating life-forms in **tide pools.** Tide pools are depressions in a rocky shoreline that are flooded at high tide but retain some water at low tide. These areas remain rocky where wave action prevents most plant growth or sediment (mud) accumulation. Extreme conditions, with frigid flooding at high tide and hot, desiccating sunshine at low tide, make life impossible for most species. But the specialized animals and plants that do occur in this rocky intertidal zone are astonishingly diverse and beautiful (fig. 5.18d).

5.3 FRESHWATER ECOSYSTEMS

- *Freshwater systems vary in depth, nutrients, and circulation.*
- *Wetlands are disproportionately important for biodiversity.*

Freshwater environments are far less extensive than marine environments, but they are centers of biodiversity. Most terrestrial communities rely, to some extent, on freshwater environments. Isolated pools, streams, and even underground water systems support astonishing biodiversity in deserts, as well as provide water to land animals. In Arizona, for example, many birds are found in trees and bushes surrounding the few available rivers and streams.

Lakes have extensive open water

Freshwater lakes, like marine environments, have distinct vertical zones (fig. 5.19). Near the surface a subcommunity of plankton, mainly microscopic plants, animals, and protists (single-celled organisms, such as amoebae), floats freely in the water column. Insects such as water striders and mosquitoes also live at the air-water interface. Fish move through the water column, sometimes near the surface and sometimes at depth.

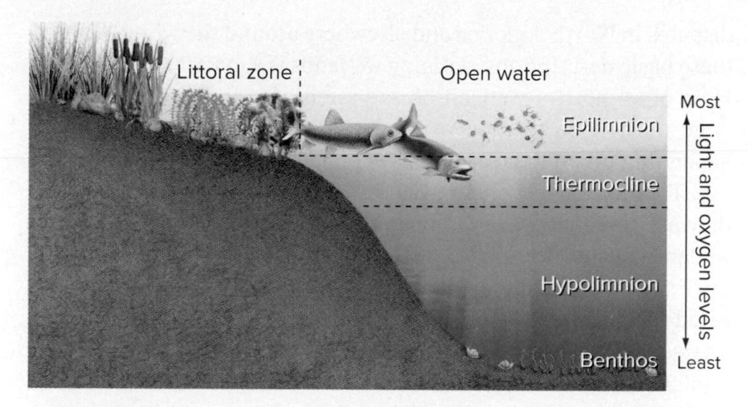

▲ **FIGURE 5.19** The layers of a deep lake are determined mainly by gradients of light, oxygen, and temperature. The epilimnion is affected by surface mixing from wind and thermal convections, while mixing between the hypolimnion and epilimnion is inhibited by a sharp temperature and density difference at the thermocline.

Finally, the bottom, or *benthos,* is occupied by a variety of snails, burrowing worms, fish, and other organisms. These make up the benthic community. Oxygen levels are lowest in the benthic environment, mainly because there is little mixing to introduce oxygen to this zone. Anaerobic (not using oxygen) bacteria may live in low-oxygen sediments. In the littoral zone, emergent plants, such as cattails and rushes, grow in the bottom sediment. These plants create important functional links between layers of an aquatic ecosystem, and they may provide the greatest primary productivity to the system.

Lakes, unless they are shallow, have a warmer upper layer that is mixed by wind and warmed by the sun. This layer is the *epilimnion.* Below the epilimnion is the hypolimnion (*hypo* = below), a colder, deeper layer that is not mixed. If you have gone swimming in a moderately deep lake, you may have discovered the sharp temperature boundary, known as the **thermocline,** between these layers. Below this boundary, the water is much colder. This boundary is also called the mesolimnion.

Local conditions that affect the characteristics of an aquatic community include (1) nutrient availability (or excess), such as nitrates and phosphates; (2) suspended matter, such as silt, that affects light penetration; (3) depth; (4) temperature; (5) currents; (6) bottom characteristics, such as muddy, sandy, or rocky floor; (7) internal currents; and (8) connections to, or isolation from, other aquatic and terrestrial systems.

Wetlands are shallow and productive

Wetlands are shallow ecosystems in which the land surface is saturated or submerged at least part of the year. Wetlands have vegetation that is adapted to grow under saturated conditions. These legal definitions are important because, although wetlands make up only a small part of most countries, they are disproportionately important in conservation debates and are the focus of continual legal disputes in North America and elsewhere around the world. Beyond these basic descriptions, defining wetlands is a matter of hot debate. How often must a wetland be saturated, and for how long? How large must it be to deserve legal protection? Answers can vary, depending on political as well as ecological concerns.

These relatively small systems support rich biodiversity, and they are essential for both breeding and migrating birds. Although wetlands occupy less than 5 percent of the land in the United States, the Fish and Wildlife Service estimates that one-third of all endangered species spend at least part of their lives in wetlands. Wetlands retain storm water and reduce flooding by slowing the rate at which rainfall reaches river systems. Floodwater storage is worth $3 billion to $4 billion per year in the United States. As water stands in wetlands, it also seeps into the ground, replenishing groundwater supplies. Wetlands filter, and even purify, urban and farm runoff, as bacteria and plants take up nutrients and contaminants in water. They are also in great demand for filling and development. They are often near cities or farms, where land is valuable, and, once drained, wetlands are easily converted to more lucrative uses. At least half of all the wetlands that existed in the United States when Europeans first arrived have been drained, filled, or degraded. In some major farming states, losses have been even greater. Iowa, for example, has lost 99 percent of its original wetlands.

Wetlands are described by their vegetation (fig. 5.20). **Swamps,** also called wooded wetlands, are wetlands with trees. **Marshes** are wetlands without trees. **Bogs** are areas of water-saturated ground, and usually the ground is composed of deep layers of accumulated, undecayed vegetation known as peat. **Fens** are similar to bogs except that they are mainly fed by groundwater, so that they have mineral-rich water and specially adapted plant species. Many bogs are fed mainly by precipitation. Swamps and marshes have high biological productivity. Bogs and fens, which are often nutrient-poor, have low biological productivity. They may have unusual and interesting species, though, such as sundews and pitcher plants, which are adapted to capture nutrients from insects rather than from soil.

The water in marshes and swamps usually is shallow enough to allow full penetration of sunlight and seasonal warming. These mild conditions favor great photosynthetic activity, resulting in high productivity at all trophic levels. In short, life is abundant and varied. Wetlands are major breeding, nesting, and migration staging areas for waterfowl and shorebirds.

Streams and rivers are open systems

Streams form wherever precipitation exceeds evaporation and surplus water drains from the land. Within small streams, ecologists distinguish areas of riffles, where water runs rapidly over a rocky substrate, and pools, which are deeper stretches of slowly moving current. Water tends to be well mixed and oxygenated in riffles; pools tend to collect silt and organic matter. If deep enough, pools can have vertical zones similar to those of lakes. As streams collect water and merge, they form rivers, although there isn't a universal definition of when one turns into the other. Ecologists consider a river system to be a continuum of constantly changing environmental conditions and community inhabitants, from the headwaters to the mouth of a drainage or watershed. The biggest distinction between stream and lake ecosystems is that, in a stream, materials, including plants, animals, and water, are continually moved downstream by flowing currents. This downstream drift is offset by active movement of animals upstream, productivity in the stream itself, and input of materials from adjacent wetlands or uplands.

(a) Swamp, or wooded wetland

(b) Marsh

(c) Bog

▲ **FIGURE 5.20** Wetlands provide irreplaceable ecological services, including water filtration, water storage and flood reduction, and habitat. Forested wetlands (a) are often called swamps; marshes (b) have no trees; bogs (c) are acidic and accumulate peat. (a) ©William P. Cunningham; (b) ©William P. Cunningham; (c) ©Mary Ann Cunningham

5.4 BIODIVERSITY

- *Genetic, species, and ecological diversity influence ecosystem function.*
- *Amphibians and gymnosperms are widely threatened, and many groups are poorly known.*
- *Biodiversity is important for ecosystem stability and for resources we use.*

The biomes you've just learned about shelter an astounding variety of living organisms. From the driest desert to the dripping rainforests, from the highest mountain peaks to the deepest ocean trenches, life occurs in a marvelous spectrum of sizes, colors, shapes, life cycles, and interrelationships. The varieties of organisms and complex ecological relationships give the biosphere its unique, productive characteristics. **Biodiversity,** the variety of living things, also makes the world a more beautiful and exciting place to live. Three kinds of biodiversity are essential to preserve ecological systems and functions: (1) *Genetic diversity* is a measure of the variety of versions of the same genes within individual species; (2) *species diversity* describes the number of different kinds of organisms within individual communities or ecosystems; and (3) *ecological diversity* specifies the number of niches, trophic levels, and ecological processes that capture energy, sustain food webs, and recycle materials within this system. Redundancy in each of these categories enhances resiliency in a biome.

Diversity can have many benefits. It may help biological communities withstand environmental stress and recover from disturbances, such as fire, drought, storms, or pest invasions. In a diverse community some species are likely to survive disturbance, so that ecological functions persist, even if some resident species disappear. It is estimated that 95 percent of the potential pests and disease-carrying organisms in the world are controlled by natural predators and competitors. Maintaining biodiversity can be essential for pest control and other ecological functions.

Biodiversity also has cultural and aesthetic value. The U.S. Fish and Wildlife Service estimates that Americans spend $104 billion every year on wildlife-related recreation. This is 25 percent more than the $81 billion spent each year on new automobiles. Often recreation is worth far more than the resources that can be extracted from an area. Fishing, hunting, camping, hiking, and other nature-based activities also have cultural value. These activities provide exercise, and contact with nature can be emotionally restorative. Often the idea of biodiversity is important: Even if you will never see a tiger or a blue whale, it may be important or gratifying to know these animals exist. This idea is termed *existence value.* In many cultures nature carries spiritual connotations, and observing and protecting nature have religious or moral significance.

TABLE 5.1 Estimated Number of Species

GROUP	KNOWN	ENDANGERED	PERCENTAGE ENDANGERED
Mammals	5,560	1,194	22
Birds	11,121	1,460	13
Reptiles	10,450	1,090	10
Amphibians	7,635	2,067	27
Fish	33,500	2,359	7
Insects	10,00,000	1,298	0.1
Molluscs	85,000	1,984	2
Crustaceans	47,000	732	1.6
Other animals	173,250	539	0.003
Mosses	16,236	76	0.5
Ferns and allies	12,000	217	2
Gymnosperms	1,052	400	38
Flowering plants	268,000	10,972	4
Fungi, lichens, protists	65,434	43	0.001

Source: Data from IUCN RedList, 2017.

FIGURE 5.21 Insects and other invertebrates make up more than half of all known species. Many, like this blue morpho butterfly, are beautiful as well as ecologically important. ©McGraw-Hill Education/ Barry Barker, photographer

Increasingly we identify species by genetic similarity

The concept of a species is fundamental in understanding biodiversity, but what is a species? In general, species are distinct organisms that persist because they can produce fertile offspring. But many organisms reproduce asexually; others don't reproduce in nature just because they don't normally encounter one another. Because of such ambiguities, evolutionary biologists favor the **phylogenetic species concept,** which identifies genetic similarity. Alternatively, the **evolutionary species concept** defines species according to evolutionary history and common ancestors. Both of these approaches rely on DNA analysis to define similarity among organisms.

How many species are there? Biologists have identified about 1.7 million species, but these probably represent only a small fraction of the actual number (table 5.1). Based on the rate of new discoveries by research expeditions—especially in the tropics—taxonomists estimate that between 3 million and 50 million different species may be alive today. About 70 percent of all known species are invertebrates (animals without backbones, such as insects, sponges, clams, and worms) (fig. 5.21). This group probably makes up the vast majority of organisms yet to be discovered and may constitute 90 percent of all species.

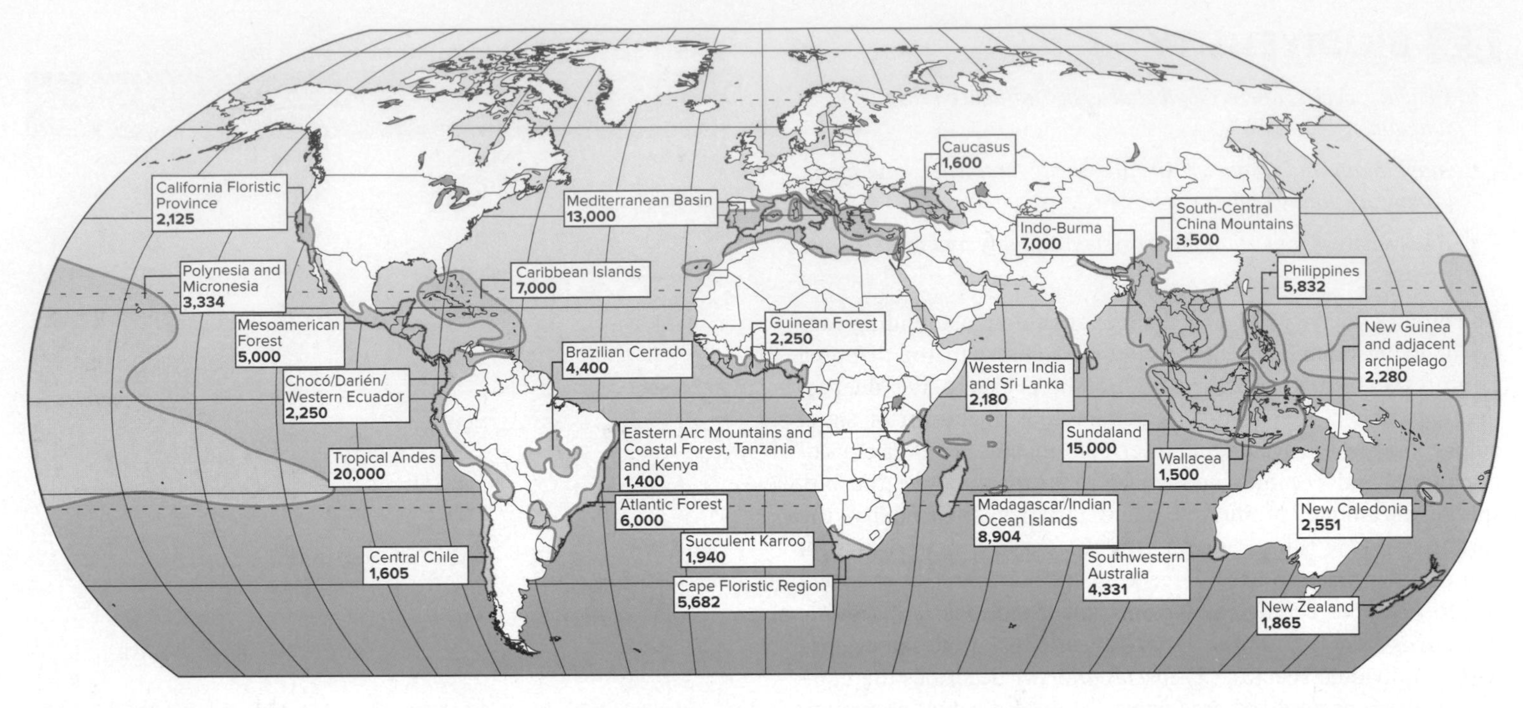

▲ **FIGURE 5.22** Biodiversity "hot spots" identified by Conservation International tend to be in tropical or Mediterranean climates and on islands, coastlines, or mountains where many habitats exist and physical barriers encourage speciation. Numbers represent estimated endemic (locally unique) species in each area. Source: Data from Conservation International.

Biodiversity hot spots are rich and threatened

Most of the world's biodiversity concentrations are near the equator, especially tropical rainforests and coral reefs (fig. 5.22). Of all the world's species, only 10 to 15 percent live in North America and Europe. Many of the organisms in megadiversity countries have never been studied by scientists. The Malaysian Peninsula, for instance, has at least 8,000 species of flowering plants, while Britain, with an area twice as large, has only 1,400 species. There may be more botanists in Britain than there are species of higher plants. South America, on the other hand, has fewer than 100 botanists to study perhaps 200,000 species of plants.

Areas isolated by water, deserts, or mountains can also have high concentrations of unique species and biodiversity. Madagascar, New Zealand, South Africa, and California are all midlatitude areas isolated by barriers that prevent mixing with biological communities from other regions and produce rich, unusual collections of species.

Biodiversity provides food and medicines

Wild plant species make important contributions to human food supplies. Genetic material from wild plants has been used to improve domestic crops. Noted tropical ecologist Norman Myers estimates that as many as 80,000 edible wild plant species could be utilized by humans. Villagers in Indonesia, for instance, are thought to use some 4,000 native plant and animal species for food, medicine, and other products. Few of these species have been explored for possible domestication or more widespread cultivation. Wild bees, moths, bats, and other organisms provide pollination for most of the world's crops. Without these we would have little agriculture in much of the world.

Pharmaceutical products derived from developing world plants, animals, and microbes have a value of more than $30 billion per year, according to the United Nations Development Programme (Key Concepts, p. 114). Consider the success story of vinblastine and vincristine. These anticancer alkaloids are derived from the Madagascar periwinkle (*Catharanthus roseus,* fig. 5.23). They inhibit the growth of cancer cells and are very effective in treating certain kinds of cancer. Twenty years ago, before these drugs were

▲ **FIGURE 5.23** The rosy periwinkle (*Catharanthus roseus*) from Madagascar provides anticancer drugs that now make childhood leukemias and Hodgkin's disease highly remissible. ©William P. Cunningham

introduced, childhood leukemias were invariably fatal. Now the remission rate for some childhood leukemias is 99 percent. Hodgkin's disease was 98 percent fatal a few years ago but is now only 40 percent fatal, thanks to these compounds. The total value of the periwinkle crop is roughly $17 million per year, although Madagascar gets little of those profits.

5.5 WHAT THREATENS BIODIVERSITY?

- *Extinction rates are far higher now than in the past.*
- *Habitat loss, climate change, invasive species, pollution, and overharvesting, all enhanced by population growth, are the main threats to biodiversity.*
- *Invasive species seriously threaten islands and specialized habitats.*

Extinction, the elimination of a species, is a normal process of the natural world. Species die out and are replaced by others, often their own descendants, as part of evolutionary change. In undisturbed ecosystems the rate of extinction appears to be about one species lost every decade. Over the past century, however, human impacts on populations and ecosystems have accelerated that rate, possibly causing untold thousands of species, subspecies, and varieties to become extinct every year. Many of these are probably invertebrates, fungi, and microbes, which are unstudied but may perform critical functions for ecosystems. The very high proportion of higher plants and animals listed as endangered in table 5.1 probably reveals more about what we like to study than about what species are really most endangered. Large or charismatic species frequently receive more attention and research funding.

In geologic history, extinctions are common. Studies of the fossil record suggest that more than 99 percent of all species that ever existed are now extinct. Most of those species were gone long before humans came on the scene. Periodically, mass extinctions have wiped out vast numbers of species and even whole families (fig. 5.24). The best-studied of these events occurred at the end of the Cretaceous period, when dinosaurs disappeared, along with at least 50 percent of existing species. An even greater disaster occurred at the end of the Permian period, about 250 million years ago, when 95 percent of species and perhaps half of all families died out over a period of about 10,000 years—a mere moment in geologic time. Current theories suggest that these catastrophes were caused by climate changes, perhaps triggered by volcanic eruptions or large asteroids striking the earth. Many ecologists worry that global climate change caused by our release of greenhouse gases into the atmosphere could have similarly catastrophic effects (see chapter 9). Notice that we name these geologic ages (and the extinctions that occurred in them) after their dominant life-forms. Many scientists believe that the present age should be named the Anthropocene and that the current extinction crisis ranks with those of the past.

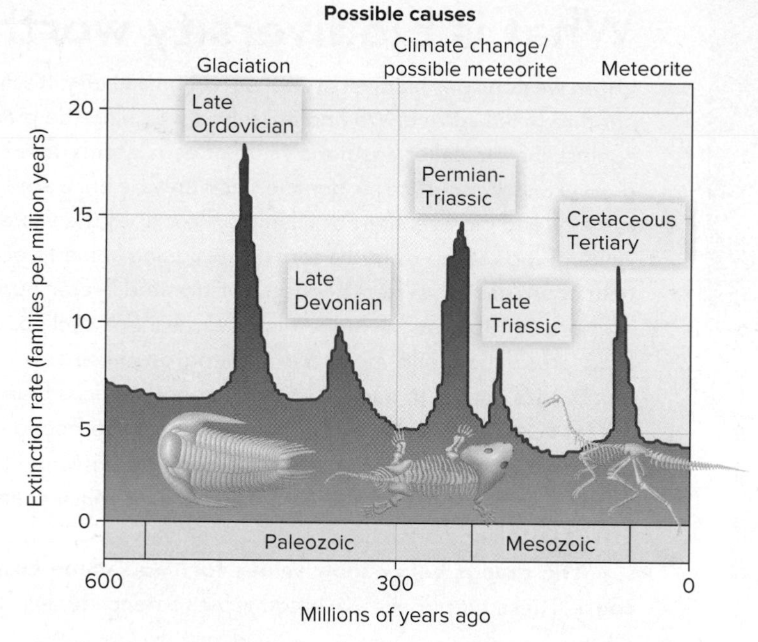

▲ **FIGURE 5.24** Major mass extinctions through history. We may be in a sixth mass extinction now, caused by human activities.

HIPPO summarizes human impacts

The rate at which species are disappearing has increased dramatically over the past 150 years. Between A.D. 1600 and 1850, human activities appear to have eliminated two or three species per decade, about double the natural extinction rate. In the past 150 years the extinction rate has increased to thousands per decade. Conservation biologists call this the sixth mass extinction but note that this time it's not asteroids or volcanoes but human impacts that are responsible. E. O. Wilson summarizes human threats to biodiversity with the acronym **HIPPO**, which stands for ***H***abitat destruction, ***I***nvasive species, ***P***ollution, ***P***opulation of humans, and ***O***verharvesting. Let's look in more detail at each of these issues.

Habitat destruction is usually the main threat

The most important extinction threat for most species—especially terrestrial ones—is habitat loss. Perhaps the most obvious example of habitat destruction is conversion of forests and grasslands to farmland (fig. 5.25). Over the past 10,000 years, humans have transformed billions of hectares of former forests and grasslands to croplands, cities, roads, and other uses. These human-dominated spaces aren't devoid of wild organisms, but they generally favor weedy species adapted to coexist with us.

Today forests cover less than half the area they once did, and only around one-fifth of the original forest retains its old-growth characteristics. Species that depend on the varied structure and resources of old-growth forest, such as the northern spotted owl (*Strix occidentalis caurina*), vanish as their habitat disappears. Grasslands currently occupy about 4 billion ha (roughly equal to the area of closed-canopy forests). Much of the most highly productive and species-rich grasslands—for example, the tallgrass prairie that once covered the U.S. Corn Belt—has been converted to cropland. Much more may need to be used as farmland or pasture if human populations continue to expand.

(a) ©William P. Cunningham; (b) ©pniesen/Getty Images

What is biodiversity worth?

Often we consider biodiversity conservation a luxury: It's nice if you can afford it, but most of us need to make a living. We find ourselves weighing the pragmatic economic value of resources against the ethical or aesthetic value of ecosystems. **Is conservation necessarily contradictory to good economic sense?** This question can only be answered if we can calculate the value of ecosystems and biodiversity. For example, how does the value of a standing forest compare to the value of logs taken from the forest? Assigning value to ecosystems has always been hard. We take countless ecosystem services for granted: water purification, prevention of flooding and erosion, soil formation, waste disposal, nutrient cycling, climate regulation, crop pollination, food production, and more. We depend on these services, but because nobody sells them directly, it's harder to name a price for these services than for a truckload of timber.

In 2009–2010, a series of studies called The Economics of Ecosystems and Biodiversity (TEEB) compiled available research findings on valuing ecosystem services. TEEB reports found that **the value of ecological services is more than double the total world GNP,** or at least $33 trillion per year.

The graphs below show values for two sample ecosystems: tropical forests and coral reefs. These graphs show average values among studies, because values vary widely by region.

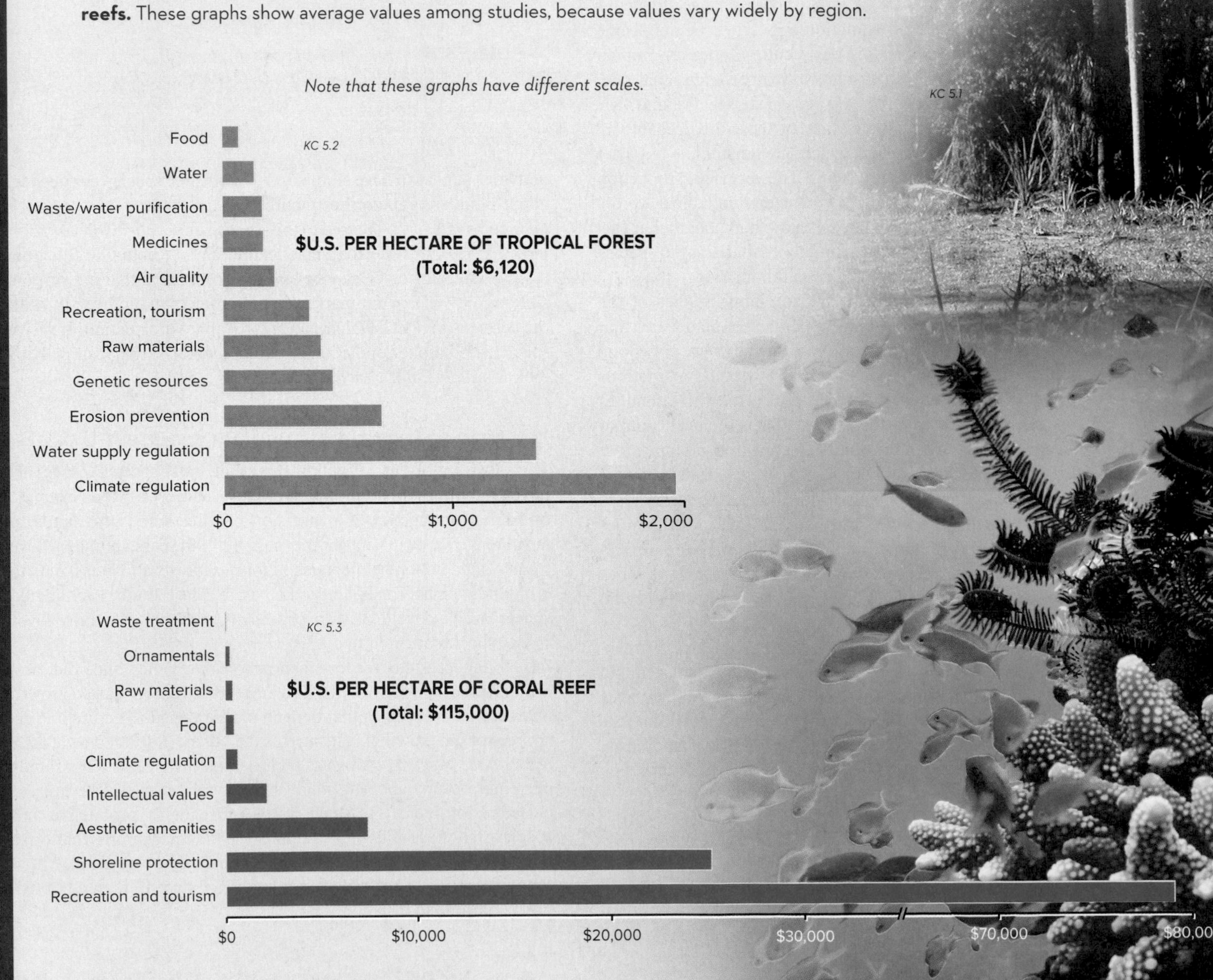

Can we afford to restore biodiversity?

It's harder to find money to restore ecosystems than to destroy them. But the benefits derived over time greatly exceed average restoration costs, according to TEEB calculations.

Tropical forests
Lakes/rivers
Inland wetlands
Mangroves
Coastal wetlands
Coral reefs

$0 | $200,000 | $400,000 | $600,000 | $800,000 | $1,000,000 | $1,200,000

Restoration cost
Benefits over 40 years ($U.S. per hectare)

KC 5.4

©Stockbyte/Getty Images

Foods and wood products These are easy to imagine but much lower in value than erosion prevention, climate controls, and water supplies provided by forested ecosystems. Still, we depend on biodiversity for foods. By one estimate, Indonesia produces 250 different edible fruits. All but 43, including this mangosteen, are little known outside the region.

KC 5.5

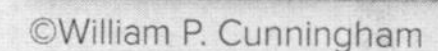

©William P. Cunningham

©IT Stock/Age Fotostock

Pollination Most of the world is completely dependent on wild insects to pollinate crops. Natural ecosystems support populations year-round, so they are available when we need them.

KC 5.6

Medicines More than half of all prescriptions contain some natural products. The United Nations Development Programme estimates the value of pharmaceutical products derived from developing world plants, animals, and microbes to be more than $30 billion per year.

Some natural medicine products

KC 5.8

PRODUCT	SOURCE	USE
Penicillin	Fungus	Antibiotic
Bacitracin	Bacterium	Antibiotic
Tetracycline	Bacterium	Antibiotic
Erythromycin	Bacterium	Antibiotic
Digitalis	Foxglove	Heart stimulant
Quinine	Chincona bank	Malaria treatment
Diosgenin	Mexican yam	Birth control drug
Cortisone	Mexican yam	Anti-inflammation treatment
Cytarabine	Sponge	Leukemia cure
Vinblastine, vincristine	Periwinkle plant	Anticancer drugs
Reserpine	Rauwolfia	Hypertension drugs
Bee venom	Bee	Arthritis relief
Allantoin	Blowfly larva	Wound healer
Morphine	Poppy	Analgesic

Climate and water supplies These may be the most valuable aspects of forests. Effects of these services impact areas far beyond forests themselves.

©Cynthia Shaw

Fish nurseries As discussed in chapter 1, the biodiversity of reefs and mangroves is necessary for reproduction of the fisheries on which hundreds of millions of people depend. Marine fisheries, including most farmed fish, depend entirely on wild food sources. These fish are worth a great deal as food, but they are worth far more for their recreation and tourism value.

KC 5.7

CAN YOU EXPLAIN?

1. **Do the relative costs and benefits justify restoring a coral reef? a tropical forest?**
2. **Identify the primary economic benefits of tropical forest and reef systems. Can you explain how each works?**

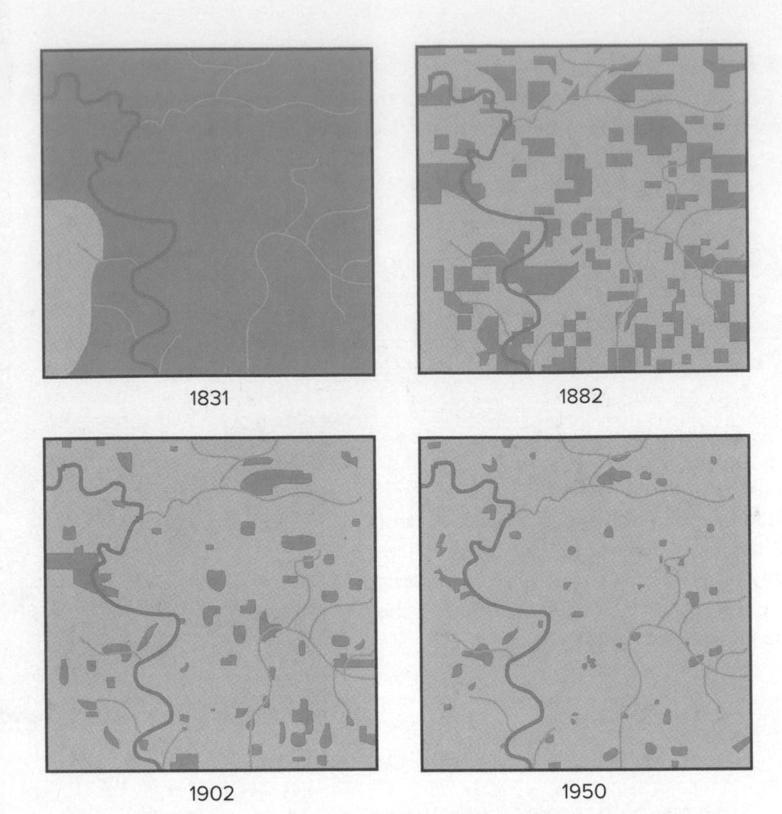

▲ **FIGURE 5.25** Decrease in wooded area of Cadiz Township in southern Wisconsin during European settlement. Green areas represent the amount of land in forest each year. Republished with permission of the University of Chicago Press, from Curtis, J. in William L. Thomas (ed.), *Man's Role in Changing the Face of the Earth*, 1956; permission conveyed through Copyright Clearance Center, Inc.

Sometimes we destroy habitat as a side effect of resource extraction, such as mining, dam-building, and indiscriminate fishing methods. Surface mining, for example, strips off the land covering along with everything growing on it. Waste from mining operations can bury valleys and poison streams with toxic material. Dam-building floods vital stream habitat under deep reservoirs and eliminates food sources and breeding habitat for some aquatic species. Our current fishing methods are highly unsustainable. One of the most destructive fishing techniques is bottom trawling, in which heavy nets are dragged across the ocean floor, scooping up every living thing and crushing the bottom structure to lifeless rubble. Marine biologist Jan Lubechenco says that trawling is "like collecting forest mushrooms with a bulldozer."

Fragmentation reduces habitat to small, isolated areas

In addition to the loss of total habitat area, the loss of large, contiguous areas is a serious problem. A general term for this is habitat **fragmentation**—the reduction of habitat into small, isolated patches. Breaking up habitat reduces biodiversity because many species, such as bears and large cats, require large territories to subsist. Other species, such as forest interior birds, reproduce successfully only in deep forest far from edges and human settlement. Predators and invasive species often spread quickly into new regions following fragment edges.

Fragmentation also divides populations into isolated groups, making them much more vulnerable to catastrophic events, such as storms or diseases. A very small population may not have enough breeding adults to be viable even under normal circumstances. An important question in conservation biology is, what is the **minimum viable population** size for a species, and when have dwindling populations grown too small to survive?

Much of our understanding of fragmentation was outlined in the theory of **island biogeography,** developed by R. H. MacArthur and E. O. Wilson in the 1960s. Noticing that small islands far from a mainland have fewer terrestrial species than larger islands, or those nearer a continent, MacArthur and Wilson proposed that species diversity is a balance between colonization and extinction rates. A remote island is hard for terrestrial organisms to reach, so new species rarely arrive and establish new populations. An island near other land may have new arrivals frequently. At the same time, the population of any single species on a small island is likely to be small and therefore vulnerable to extinction. By contrast, a large island can support more individuals of a given species and is, therefore, less vulnerable to natural disasters or genetic problems.

Large islands also tend to have more variation in habitat types than small islands do, and this contributes to higher species counts in large islands.

The effect of island size has been observed in many places. Cuba, for instance, is 100 times as large and has about 10 times as many amphibian species as its Caribbean neighbor Montserrat. Similarly, in a study of bird species on the California Channel Islands, Jared Diamond observed that on islands with fewer than 10 breeding pairs, 39 percent of the populations went extinct over an 80-year period, whereas only 10 percent of populations with 10 to 100 pairs went extinct in the same time (fig. 5.26). Only one species numbering between 100 and 1,000 pairs went extinct, and no species with over 1,000 pairs disappeared over this time.

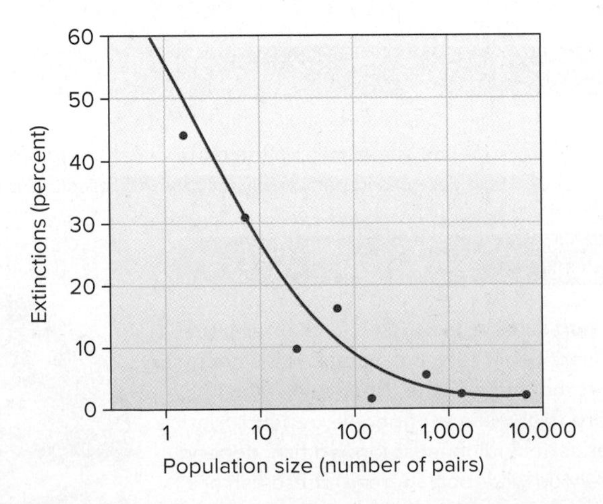

▲ **FIGURE 5.26** Extinction rates of bird species on the California Channel Islands as a function of population size over 80 years. Source: Data from Jones, H. L., and Diamond, J., "Short-term-base Studies of Turnover in Breeding Bird Populations on the California Coast Island," in *Condor*, vol. 78, 1976, 526–549.

The island idea has informed our understanding of parks and wildlife refuges, which are effectively islands of habitat surrounded by oceans of inhospitable territory. Like small, remote islands, they may be too isolated to be reached by new migrants, and they can't support large enough populations to survive catastrophic events or genetic problems. Often they are also too small for species that require large territories. Tigers and wolves, for example, need large expanses of contiguous range relatively free of human incursion to survive. Glacier National Park in Montana, for example, is excellent habitat for grizzly bears. It can support only about 100 bears, however, and if there isn't migration at least occasionally from other areas, this probably isn't a large enough population to survive in the long run.

Invasive species are a growing threat

A major threat to native biodiversity in many places is from accidentally or deliberately introduced species. Called a variety of names—alien, exotic, non-native, nonindigenous, pests—**invasive species** are organisms that thrive in new territory where they are free of predators, diseases, or resource limitations that may have controlled their population in their native habitat. Note that not all exotic species expand after introduction: For example, peacocks have been introduced to many city zoos, but they tend not to escape and survive in the wild in most places. At the same time, not all invasive species are foreign. But most of the uncontrollable invasive species are introduced from elsewhere.

Humans have always transported organisms into new habitats, but the rate of movement has risen sharply in recent years with the huge increase in speed and volume of travel by air, water, and land. Some species are deliberately released because people believe they will be aesthetically pleasing or economically beneficial. Some of the worst are pets released to the wild when owners tire of them. Many hitch a ride in ship ballast water, in the wood of packing crates, inside suitcases or shipping containers, or in the soil of potted plants (fig. 5.27). Sometimes we introduce invasive species into new habitats thinking that we're being kind and compassionate without being aware of the ecological consequences.

Over the past 300 years, approximately 50,000 non-native species have become established in the United States. Many of these introductions, such as corn, wheat, rice, soybeans, cattle, poultry, and honeybees, were intentional and mostly beneficial. At least 4,500 of these species have established wild populations, of which 15 percent cause environmental or economic damage (fig. 5.28). Invasive species are estimated to cost the United States $138 billion annually and are forever changing a variety of ecosystems (What Can You Do?, p. 119).

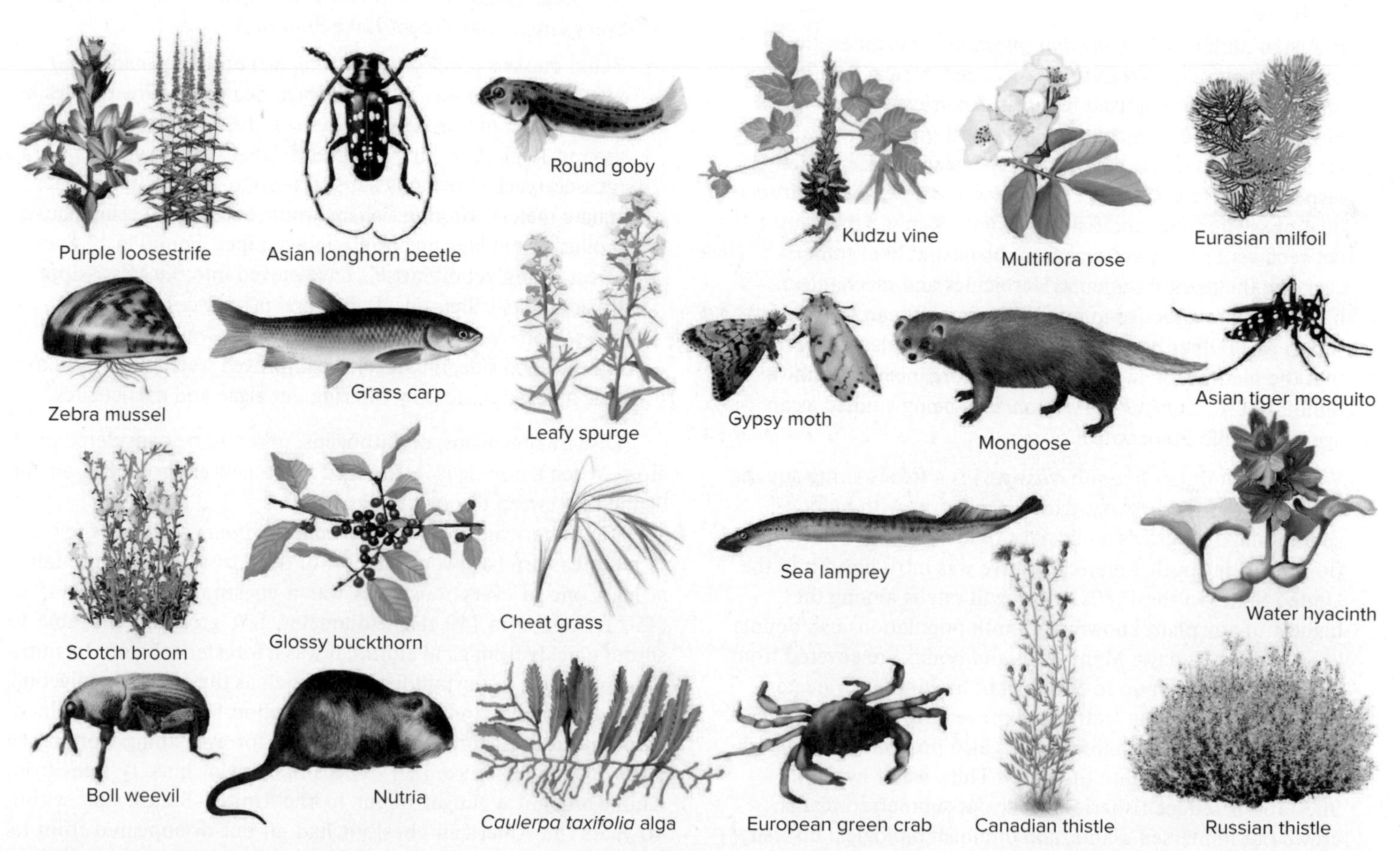

▲ **FIGURE 5.27** A few of the approximately 50,000 invasive species in North America. Do you recognize any that occur where you live? What others can you think of?

▲ **FIGURE 5.28** Invasive leafy spurge (*Euphorbia esula*) blankets a formerly diverse pasture. Introduced accidentally, and inedible for most herbivores, this plant costs hundreds of millions of dollars each year in lost grazing value and in weed control. ©Mary Ann Cunningham

A few important examples of invasive species include the following:

- Eurasian milfoil (*Myriophyllum spicatum* L.) is an exotic aquatic plant native to Europe, Asia, and Africa. Scientists believe that milfoil arrived in North America during the late nineteenth century in shipping ballast. It grows rapidly and tends to form a dense canopy on the water surface, which displaces native vegetation, inhibits water flow, and obstructs boating, swimming, and fishing. Humans spread the plant between water body systems from boats and boat trailers carrying the plant fragments. Herbicides and mechanical harvesting are effective in milfoil control but can be expensive (up to $5,000 per hectare per year). There is also concern that the methods may harm nontarget organisms. A native milfoil weevil, *Euhrychiopsis lecontei*, is being studied as an agent for milfoil biocontrol.
- Water hyacinth (*Eichhornia crassipes*) is a free-floating aquatic plant that has thick, waxy, dark green leaves with bulbous, spongy stalks. It grows a tall spike of lovely blue or purple flowers. This South American native was introduced into the United States in the 1880s. Its growth rate is among the highest of any plant known: Hyacinth populations can double in as little as 12 days. Many lakes and ponds are covered from shore to shore with up to 500 tons of hyacinths per hectare. Besides blocking boat traffic and preventing swimming and fishing, water hyacinth infestations also prevent sunlight and oxygen from getting into the water. Thus, water hyacinth infestations reduce fisheries, shade-out submersed plants, crowd-out immersed plants, and diminish biological diversity. Water hyacinth is controlled with herbicides, machines, and biocontrol insects.
- The emerald ash borer (*Agrilus planipennis*) is an invasive wood-boring beetle from Siberia and northern China. It was first identified in North America in the summer of 2002 in southeast Michigan and in Windsor, Ontario. It's believed to have been introduced into North America in shipping pallets and wooden containers from Asia. In just 8 years the beetle spread into 13 states from West Virginia to Minnesota. Adult emerald ash borers have golden or reddish-green bodies with dark metallic emerald green wing covers. More than 40 million ash trees have died or are dying from emerald ash borer attack in the United States, and more than 7.5 billion trees are at risk.
- In the 1970s several carp species, including bighead carp (*Hypophthalmichthys nobilis*), grass carp (*Ctenopharyngo donidella*), and silver carp (*Hypophthalmichthys molitrix*), were imported from China to control algae in aquaculture ponds. Unfortunately, they escaped from captivity and have become established–often in very dense populations–throughout the Mississippi River Basin. Silver carp can grow to 100 pounds (45 kg). They are notorious for being easily frightened by boats and personal watercraft, which causes them to leap as much as 8–10 feet (2.5–3 m) into the air. Getting hit in the face by a large carp when you're traveling at high speed in a boat can be life threatening. Large amounts of money have been spent trying to prevent Asian carp from spreading into the Great Lakes, but carp DNA has already been detected in every Great Lake except Lake Superior.
- Zebra mussels (*Dreissena polymorpha*) probably made their way from their home in the Caspian Sea to the Great Lakes in ballast water of transatlantic cargo ships, arriving sometime around 1985. Attaching themselves to any solid surface, zebra mussels reach enormous densities–up to 70,000 animals per square meter–covering fish spawning beds, smothering native mollusks, and clogging utility intake pipes. Found in all the Great Lakes, zebra mussels have moved into the Mississippi River and its tributaries. Public and private costs for zebra mussel removal now amount to some $400 million per year. On the good side, mussels have improved water clarity in Lake Erie at least fourfold by filtering out algae and particulates.

Disease organisms, or pathogens, may also be considered predators. When a disease is introduced into a new environment, an epidemic may sweep through the area.

The American chestnut (*Castanea dentata*) was once the heart of many eastern hardwood forests. In the Appalachian Mountains, at least one of every four trees was a chestnut. Often over 45 m (150 ft) tall, 3 m (10 ft) in diameter, fast growing, and able to sprout quickly from a cut stump, it was a forester's dream. Its nutritious nuts were important for birds (such as the passenger pigeon), forest mammals, and humans. The wood was straight-grained, light, and rot-resistant, and it was used for everything from fence posts to fine furniture. In 1904 a shipment of nursery trees from China brought a fungal blight to the United States, and within 40 years the American chestnut had all but disappeared from its native range. Efforts are now under way to transfer blight-resistant genes into the few remaining American chestnuts that weren't

What Can YOU DO?

You Can Help Preserve Biodiversity

Our individual actions are some of the most important obstacles—and most important opportunities—in conserving biodiversity.

Pets and Plants

- Help control invasive species. Never release fish or vegetation from fish tanks into waterways or sewers. Pet birds, cats, dogs, snakes, lizards, and other animals, released by well-meaning owners, are widespread invasive predators.
- Keep your cat indoors. House cats are major predators of woodland birds and other animals.
- Plant native species in your garden. Exotic nursery plants often spread from gardens, compete with native species, and introduce parasites, insects, or diseases that threaten ecosystems. Local-origin species are an excellent, and educational, alternative.
- Don't buy exotic birds, fish, turtles, reptiles, or other pets. These animals are often captured, unsustainably, in the wild. The exotic pet trade harms ecosystems and animals.
- Don't buy rare or exotic houseplants. Rare orchids, cacti, and other plants are often collected and sold illegally and unsustainably.

Food and Products

- When buying seafood, inquire about the source. Try to buy species from stable populations. Farm-raised catfish, tilapia, trout, Pacific pollack, Pacific salmon, mahimahi, squid, crabs, and crayfish are some of the stable or managed species that are good to buy. Avoid slow-growing top predators, such as swordfish, marlin, bluefin tuna, and albacore tuna.
- Buy shade-grown coffee and chocolate. These are also organic and often "fair trade" varieties that support workers' families as well as biodiversity in growing regions.
- Buy sustainably harvested wood products. Your local stores will start carrying sustainable wood products if you and your friends ask for them. Persistent consumers are amazingly effective forces of change!

reached by the fungus or to find biological controls for the fungus that causes the disease.

The flow of organisms happens everywhere. The Leidy's comb jelly (*Mnemiopsis leidyi*), native to North American coastal areas, has devastated the Black Sea, where it now makes up more than 90 percent of all biomass at certain times of the year. Similarly, the bristle worm from North America has invaded the coast of Poland and now is almost the only thing living on the bottom of some of Poland's bays and lagoons. A tropical seaweed named *Caulerpa taxifolia*, originally grown for the aquarium trade, has escaped into the northern Mediterranean, where it covers the shallow seafloor with a dense, meter-deep shag carpet from Spain to Croatia. Producing more than 5,000 leafy fronds per square meter, this aggressive weed crowds out everything in its path. This type of algae grows low and sparse in its native habitat, but aquarium growers transformed it into a robust competitor that has transformed much of the Mediterranean.

Pollution poses many types of risk

We have long known that toxic pollutants can have disastrous effects on local populations of organisms. The links between pesticides and the declines of fish-eating birds were well documented in the 1970s (fig. 5.29). Population declines are especially likely in species high in the food chain, such as marine mammals, alligators, fish, and fish-eating birds. Mysterious, widespread deaths of thousands of Arctic seals are thought to be linked to an accumulation of persistent chlorinated hydrocarbons, such as DDT, PCBs, and dioxins, in the food chain. These chemicals accumulate in fat and cause weakened immune systems. Mortality of Pacific sea lions, beluga whales in the St. Lawrence estuary, and striped dolphins in the Mediterranean is similarly thought to be caused by accumulation of toxic pollutants.

▲ **FIGURE 5.29** Bald eagles, and other bird species at the top of the food chain, were decimated by DDT in the 1960s. Many such species have recovered since DDT was banned in the United States and because of protection under the Endangered Species Act. Source: Dave Menke/U.S. Fish and Wildlife Service

Lead poisoning is another major cause of mortality for many species of wildlife. Bottom-feeding waterfowl, such as ducks, swans, and cranes, ingest spent shotgun pellets that fall into lakes and marshes. They store the pellets, instead of stones, in their gizzards and the lead slowly accumulates in their blood and other tissues. The U.S. Fish and Wildlife Service (USFWS) estimates that 3,000 metric tons of lead shot are deposited annually in wetlands and that between 2 and 3 million waterfowl die each year from lead poisoning (fig. 5.30). Copper bullets and shot cost about the same, have just as good ballistics, and fragment much less than lead.

Population growth consumes space, resources

Even if per-capita consumption patterns remain constant, more people will require more timber harvesting, fishing, farmland, and extraction of fossil fuels and minerals. In the past 40 years, the global population has doubled from about 3.5 billion to about 7 billion. In that time, according to calculations of the Worldwide Fund for Nature (WWF), our consumption of global resources has grown from 60 percent of what the earth can support over the long term to 150 percent. At the same time, global wildlife populations have declined by more than a third because of expanding agriculture, urbanization, and other human activities.

The human population growth curve is leveling off (see chapter 4), but it remains unclear whether we can reduce poverty and provide a tolerable life for all humans while preserving healthy natural ecosystems and a high level of biodiversity.

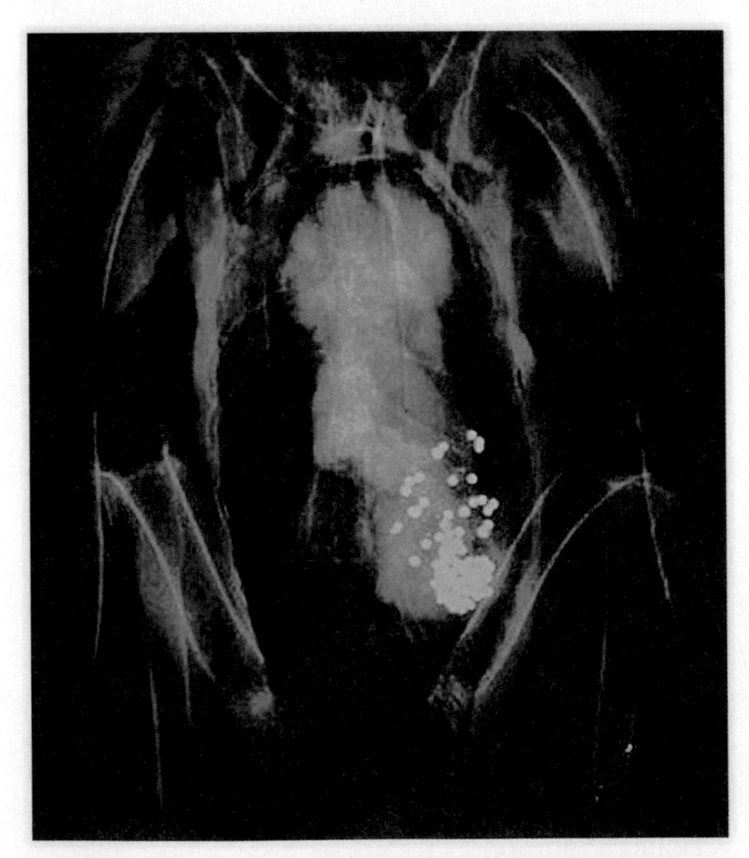

▲ **FIGURE 5.30** Lead shot shown in the stomach of a bald eagle, consumed along with its prey. Fishing weights and shot remain a major cause of lead poisoning in aquatic and fish-eating birds. Source: Dave Menke/U.S. Fish and Wildlife Service

▲ **FIGURE 5.31** A pair of stuffed passenger pigeons (*Ectopistes migratorius*). The last member of this species died in the Cincinnati Zoo in 1914. ©Mary Ann Cunningham

Overharvesting depletes or eliminates species

Overharvesting involves taking more individuals than reproduction can replace. A classic example is the extermination of the American passenger pigeon (*Ectopistes migratorius*). Even though it inhabited only eastern North America, 200 years ago this was the world's most abundant bird, with a population of 3 to 5 billion animals (fig. 5.31). It once accounted for about one-quarter of all birds in North America. In 1830 John James Audubon saw a single flock of birds estimated to be 10 miles wide and hundreds of miles long and thought to contain perhaps a billion birds. In spite of this vast abundance, market hunting and habitat destruction caused the entire population to crash in only about 20 years between 1870 and 1890. The last known wild bird was shot in 1900 and the last existing passenger pigeon, a female named Martha, died in 1914 in the Cincinnati Zoo.

At about the same time that passenger pigeons were being extirpated, the American bison, or buffalo (*Bison bison*), was being hunted to near extinction on the Great Plains. In 1850 some 60 million bison roamed the western plains. Many were killed only for their hides or tongues, leaving millions of carcasses to rot. Much of the bison's destruction was carried out by the U.S. Army to eliminate them as a source of food, clothing, and shelter for Indians, thereby forcing Indians onto reservations. After 40 years, there were only about 150 wild bison left and another 250 in captivity. Today the herd has regrown to about half a million animals, but many are cross-breeds with domestic cattle.

Fish stocks have been seriously depleted by overharvesting in many parts of the world. A huge increase in fishing fleet size and efficiency in recent years has led to a crash of many oceanic populations. Worldwide, 13 of 17 principal fishing zones are now reported to be commercially exhausted or in steep decline. At least three-quarters of all commercial oceanic species are overharvested. Canadian fisheries biologists estimate that only 10 percent of the top predators, such as swordfish, marlin, tuna, and shark, remain in the Atlantic Ocean. Groundfish, such as cod, flounder, halibut, and hake, also are severely depleted. You can avoid adding to this overharvest by eating only abundant, sustainably harvested varieties.

Perhaps the most destructive example of harvesting terrestrial wild animal species today is the African bushmeat trade. Wildlife biologists estimate that 1 million tons of bushmeat, including antelope, elephants, primates, and other animals, are sold in African markets every year. Thousands of tons more are probably sold illegally in New York, Paris, and other global cities each year. For many poor Africans, this is the only source of animal protein in their diet. If we hope to protect the animals targeted by bushmeat hunters, we will need to help them find alternative livelihoods and replacement sources of high-quality protein. Bushmeat can endanger those who eat it, too. Outbreaks of ebola and other diseases are thought to result from capturing and eating wild monkeys and other primates. The emergence of the respiratory disease known as SARS in 2003 (see chapter 8) resulted from the wild food trade in China and Southeast Asia, where millions of civets, monkeys, snakes, turtles, and other animals are consumed each year as luxury foods.

Collectors Serve Medicinal and Pet Trades In addition to harvesting wild species for food, we also obtain a variety of valuable commercial products from nature. Much of this represents sustainable harvest, but some forms of commercial exploitation are highly destructive and a serious threat to certain rare species (fig. 5.32). Despite international bans on trade in products from endangered species, smuggling of furs, hides, horns, live specimens, and folk medicines amounts to millions of dollars each year.

Developing countries in Asia, Africa, and Latin America with the richest biodiversity in the world are the main sources of wild animals and animal products, while Europe, North America, and some of the wealthy Asian countries are the principal importers. Japan and China (including Taiwan and Hong Kong) buy three-quarters of all cat and snake skins, for instance, while European countries buy most live wild birds, such as South American parrots. The United States imports 99 percent of all live cacti and 75 percent of all orchids sold each year.

The profits to be made in wildlife smuggling are enormous. Tiger or leopard fur coats can bring $700,000 in China. The population of African black rhinos dropped from approximately 100,000 in the 1960s to about 5,000 today because of a demand for their horns. In Asia, where it is prized for its supposed medicinal properties, powdered rhino horn fetches as much as $100,000 per kilogram. The entire species is classified as critically endangered, and the western population is officially extinct.

▲ **FIGURE 5.32** Parts from rare and endangered species for sale on the street in China. Use of animal products in traditional medicine and prestige diets is a major threat to many species. ©William P. Cunningham

Plants also are threatened by overharvesting. Wild ginseng has been nearly eliminated in many areas because of the Asian demand for the roots, which are used as an aphrodisiac and folk medicine. Cactus "rustlers" steal cacti by the ton from the American Southwest and Mexico. With prices as high as $1,000 for rare specimens, it's not surprising that many are now endangered.

The trade in wild species for pets is a vast global business. Worldwide, some 5 million live birds are sold each year for pets, mostly in Europe and North America. Pet traders import (often illegally) into the United States some 2 million reptiles, 1 million amphibians and mammals, 500,000 birds, and 128 million tropical fish each year. About 75 percent of all saltwater tropical aquarium fish sold come from coral reefs of the Philippines and Indonesia. Some of these wild animal harvests are sustainable; many are not.

Many of these fish are caught by divers using plastic squeeze bottles of cyanide to stun their prey (fig. 5.33). Far more fish die with this technique than are caught. Worst of all, it kills the coral animals that create the reef. A single diver can destroy all of the

▲ **FIGURE 5.33** A diver uses cyanide to stun tropical fish being caught for the aquarium trade. Many fish are killed by the method itself, while others die later during shipment. Even worse is the fact that cyanide kills the coral reef itself. ©WILDLIFE GmbH/Alamy Stock Photo

EXPLORING Science — *Restoring Coral Reefs*

Coral reefs are among the richest and most endangered biological communities on earth, the marine equivalent of tropical rainforests in diversity, productivity, and complexity. One-quarter of all marine species are thought to spend some or all of their life cycle in the shelter of coral reefs. Globally, at least 17 percent of the protein we eat depends on coral reef systems. In some coastal areas that number is 70 percent. Reefs protect shorelines from storms, and they support tourism economies worldwide.

▲ **FIGURE 1** Fragments of staghorn and elkhorn coral can be cultivated in nurseries and used to replenish damaged reef systems. ©ZUMA Press Inc/Alamy Stock Photo

But reefs are in serious trouble. We've already lost about 30 percent of coral worldwide to pollution, destructive fishing methods, and especially climate warming and ocean acidification. We are likely to lose another 60 percent in coming decades. Some researchers warn that if current trends continue there won't be any viable coral reefs anywhere in the world by the end of this century.

Can knowledge of corals' reproduction and genetic diversity help us find ways to regrow corals, or even help them adapt to warming seas? Some branched corals, such as staghorn and elkhorn, which are among the most threatened of all species, can grow and reproduce through fragmentation. If a branch breaks off and conditions are favorable, it can reattach to the rock substrate and begin to grow a new colony. Taking advantage of this trait, conservationists are harvesting coral fragments and growing them in underwater nurseries until they're large enough to be relocated to suitable areas (fig. 1). Dozens of these nurseries now operate worldwide, and tens of thousands of small corals have been transplanted to damaged or depleted reefs.

Focusing on genetic variants might increase the success of restoration efforts. In a lagoon in American Samoa, corals have been found that can survive much warmer water than most corals can tolerate. Natural selection has produced this unusual population, as heat-tolerant individuals survived but intolerant ones did not. Might other coral populations also contain tolerant traits? Similarly, in the Miami ship channel, a colony of corals was discovered growing in highly contaminated, turbid water. Here again, natural selection produced a population capable of withstanding badly polluted conditions. If geneticists can pinpoint the traits that make these specimens so tough, that knowledge might help improve other colonies.

Some biologists aren't waiting for nature to produce resistant coral strains. Labs in Australia and Hawaii are working on "assisted evolution." By growing corals in warm and acidic water, conditions similar to those we expect to see in the future, they are attempting to select for advantageous genes. These breeding experiments are assisted by using species that reproduce continuously rather than only once a year, as most corals do. Transplanting new, improved varieties could help restore or re-create these critically important systems.

Saving coral reefs, if it is possible at all, will involve insights from a variety of sciences, from evolution and genetics to ecology and conservation biology. Applying this knowledge just might make it possible to save these wonderful, endangered communities.

---

life on 200 m^2 of reef in a day. Altogether, thousands of divers destroy about 50 km^2 of reefs each year. Net fishing would prevent this destruction, and it could be enforced if pet owners would insist on net-caught fish. More than half the world's coral reefs are potentially threatened by human activities, with up to 80 percent at risk in the most populated areas. But efforts are currently under way to breed corals in captivity and to restore reefs (Exploring Science, p. 122).

Predator and Pest Control Is Expensive but Widely Practiced Some animal populations have been greatly reduced, or even deliberately exterminated, because they are regarded as dangerous to humans or livestock or because they compete with our use of resources. Every year U.S. government animal-control agents trap, poison, or shoot thousands of coyotes, bobcats, prairie dogs, and other species considered threats to people, domestic livestock, or crops.

This animal-control effort costs about $20 million in federal and state funds each year and kills some 700,000 birds and mammals, about 100,000 of which are coyotes. Defenders of wildlife regard this program as cruel, callous, and mostly ineffective in reducing livestock losses. Protecting flocks and herds with guard dogs or herders or keeping livestock out of areas that are the home range of wild species would be a better solution, they believe. Ranchers, on the other hand, argue that without predator control western livestock ranching would be impossible.

5.6 BIODIVERSITY PROTECTION

- *The Endangered Species Act is one of our most important environmental laws.*
- *Protecting a keystone or flagship species protects its ecosystem.*
- *Species protection can be controversial.*

We have gradually become aware of our damage to biological resources and of reasons for conserving them. Slowly, we are adopting national legislation and international treaties to protect these irreplaceable assets. Parks, wildlife refuges, nature preserves, zoos, and restoration programs have been established to protect nature and rebuild depleted populations. There has been encouraging progress in this area, but much remains to be done.

In this section we examine legal protections for species in the United States, but keep in mind that this is only a small part of species protection measures worldwide. Most countries now have laws protecting endangered species (though many laws remain unenforced), and dozens of international treaties aim to reduce the decline of biodiversity worldwide.

Hunting and fishing laws protect useful species

In 1874 a bill was introduced in the U.S. Congress to protect the American bison, whose numbers were falling dramatically. This initiative failed, partly because most legislators could not imagine that wildlife that was so abundant and prolific could ever be depleted by human activity. By the end of the nineteenth century, bison numbers had plunged from some 60 million to only a few hundred animals.

By the 1890s, though, most states had enacted some hunting and fishing restrictions. The general idea behind these laws was to conserve the resource for future human use rather than to preserve wildlife for its own sake. The wildlife regulations and refuges established since that time have been remarkably successful for many species. A hundred years ago, there were an estimated half a million white-tailed deer in the United States; now there are some 14 million—more in some places than the environment can support. Wild turkeys and wood ducks were nearly gone 50 years ago. By restoring habitat, planting food crops, transplanting breeding stock, building shelters or houses, protecting these birds during breeding season, and using other conservation measures, we have restored populations of these beautiful and iconic birds to several million each. Snowy egrets, which were almost wiped out by plume hunters 80 years ago, are now common again.

The Endangered Species Act protects habitat and species

The Endangered Species Act (ESA) is one of our most powerful tools for protecting biodiversity and environmental quality. It not only defends rare and endangered organisms but also helps protect habitat that benefits a whole biological community and safeguards valuable ecological services.

What does the ESA do? It provides (1) criteria for identifying species at risk, (2) directions for planning for their recovery, (3) assistance to landowners to help them find ways to meet both economic needs and the needs of a rare species, and (4) enforcement of measures for protecting species and their habitat.

The act identifies three degrees of risk: **Endangered species** are those considered in imminent danger of extinction; **threatened species** are likely to become endangered, at least locally, within the foreseeable future. **Vulnerable species** are naturally rare or have been locally depleted by human activities to a level that puts them at risk. Vulnerable species are often candidates for future listing as endangered species. For vertebrates, a protected subspecies or a local race or ecotype can be listed, as well as an entire species.

Currently, the United States has 1,372 species on the endangered and threatened species lists and about 386 candidate species waiting to be considered. The number of listed species in different taxonomic groups reflects much more about the kinds of organisms that humans consider interesting and desirable than the actual number in each group. In the United States, invertebrates make up about three-quarters of all known species but only 9 percent of those considered worthy of protection.

Worldwide, the International Union for Conservation of Nature and Natural Resources (IUCN) lists 24,431 endangered and threatened species, including nearly one-fifth of mammals; nearly one-third of amphibians, reptiles, and fish; and most of the few mosses and flowering plants that have been evaluated (see table 5.1). IUCN has no direct jurisdiction for slowing the loss of those species. Within the United States, the ESA provides mechanisms for reducing species losses.

Recovery plans aim to rebuild populations

Once a species is listed, the Fish and Wildlife Service (FWS) is given the task of preparing a recovery plan. This plan details how populations will be stabilized or rebuilt to sustainable levels. A recovery plan could include many different kinds of strategies, such as buying habitat areas, restoring habitat, reintroducing a species to its historic ranges (as with Yellowstone's gray wolves), instituting captive breeding programs, and negotiating the needs of a species and the people who live in an area.

The FWS can then help landowners prepare habitat conservation plans. These plans are specific management approaches that identify steps to conserve particular pieces of critical habitat. For example, the red-cockaded woodpecker is an endangered species that preys on insects in damaged pine forests from North Carolina to Texas. Few suitable forests remain on public lands, so much of the remaining population occurs on privately owned lands that are actively managed for timber production.

International Paper and other corporations have collaborated with the FWS to devise management strategies that conserve specified amounts of damaged tree stands while harvesting other areas. These plans restrict cutting of some trees, but they also ensure that the FWS will not interfere further with management of the timber, as long as the provisions of the plan continue to protect the woodpecker. This approach has helped to stabilize populations of the red-cockaded woodpecker, and timber companies have gained goodwill and sustainable forestry certification for their products. Habitat conservation plans are not always perfect, but often they can produce mutually satisfactory solutions.

Restoration can be slow and expensive, because it tries to undo decades or centuries of damage to species and ecosystems. About half of all funding is spent on a dozen charismatic species, such as the California condor, the Florida panther, and the grizzly bear, which receive around $13 million per year. By contrast, the 137 endangered invertebrates and 532 endangered plants get less than $5 million per year altogether. This disproportionate funding results from political and emotional preferences for large, charismatic species (fig. 5.34). There are also scientifically established designations that make some species merit special attention:

- *Keystone species* are those with major effects on ecological functions and whose elimination would affect many other members of the biological community; examples are prairie dogs (*Cynomys ludovicianus*) and bison (*Bison bison*).
- *Indicator species* are those tied to specific biotic communities or successional stages or environmental conditions. They can be reliably found under certain conditions but not others; an example is brook trout (*Salvelinus fontinalis*).
- *Umbrella species* require large blocks of relatively undisturbed habitat to maintain viable populations. Saving this habitat also benefits other species. Examples of umbrella species are the northern spotted owl (*Strix occidentalis caurina*), tiger (*Panthera tigris*), and gray wolf (*Canis lupus*).
- *Flagship species* are especially interesting or attractive organisms to which people react emotionally. These species can motivate the public to preserve biodiversity and contribute to conservation; an example is the giant panda (*Ailuropoda melanoleuca*).

Landowner collaboration is key

Two-thirds of listed species occur on privately owned lands, so cooperation between federal, state, and local agencies and private and tribal landowners is critical for progress. Often the ESA is controversial because protecting a species is legally enforceable, and that protection can require that landowners change their plans for their property. On the other hand, many landowners and communities appreciate the value of biodiversity on their land and like the idea of preserving species for their grandchildren to see. Others, like International Paper, have decided they can afford to allow some dying trees for woodpeckers, and they benefit from goodwill generated by preserving biodiversity.

A number of provisions protect landowners, and these serve as incentives for them to participate in developing habitat conservation plans. For example, permits can be issued to protect landowners from liability if a listed species is accidentally harmed during normal land use activities. In a Candidate Conservation Agreement, the FWS helps landowners reduce threats to a species in an effort to avoid listing it at all. A Safe Harbor Agreement is a promise that, if landowners voluntarily implement conservation measures, the FWS will not require additional actions that could limit future management options. For example, suppose a landowner's efforts to improve red-cockaded woodpecker habitat lead to population increases. A Safe Harbor Agreement ensures that the landowners would not be required to do further management for additional woodpeckers attracted to the improved habitat.

▲ **FIGURE 5.34** The Endangered Species Act seeks to restore populations of species, such as the bighorn sheep, which has been listed as endangered in much of its range. Charismatic species are easier to get listed than obscure ones. ©2009 J.T. Oris/Miami University

The U.S. Fish and Wildlife Service is currently using innovative agreements with private landholders to create "pop-up wildlife refuges." In California's Central Valley, for example, extensive wetlands used to provide food and resting areas for vast flocks of migratory birds. However, much of those wetlands have been drained for agriculture, and there isn't money to buy land for permanent refuges. But for relatively modest payments, many farmers are willing to flood rice fields in the spring and fall to provide temporary habitat. Millions of birds benefit, along with other wildlife and the people who like to watch them.

The ESA has seen successes and controversies

The ESA has held off the extinction of hundreds of species. Some have recovered and been delisted, including the brown pelican, the peregrine falcon, and the bald eagle, which was delisted in 2007. In 1967, before the ESA was passed, only about 800 bald eagles remained in the contiguous United States. DDT poisoning, which prevented the hatching of young eagles, was the main cause. By 1994, after the banning of DDT, the population had rebounded to 8,000 birds; there are now an estimated 70,000 bald eagles in the United States and Canada. The habitat appears to be fully occupied in most places. Similarly, peregrine falcons, which had been down to 39 breeding pairs in the 1970s, had rebounded to 1,650 pairs by 1999 and were taken off the list. The American alligator was listed as endangered in 1967 because hunting and habitat destruction had reduced populations to precarious levels. Protection has been so effective that the species is now plentiful throughout its entire southern range. Florida alone may have a population of 1 million or more.

Many people are dissatisfied with the slow pace of listing new species, however. Hundreds of species are classified as "warranted but precluded," or deserving of protection but lacking funding or local support. At least 18 species have gone extinct since being nominated for protection.

Part of the reason listing is slow is that political and legal debates can drag on for years. Political opposition is especially fierce when large profits are at stake. An important test of the ESA occurred in 1978 in Tennessee, when construction of the Tellico Dam threatened a tiny fish called the snail darter. The powerful Tennessee Valley Authority, which was building the dam, argued to the Supreme Court that the dam was more important than the fish. After this case a new federal committee was given power to override the ESA for economic reasons. (This committee subsequently became known as the God Squad because of its power over the life and death of a species.)

Another important debate over the economics of endangered species protection has been that of the northern spotted owl (see chapter 6). Preserving this owl requires the conservation of expansive, undisturbed areas of old-growth temperate rainforest in the Pacific Northwest, where old-growth timber is extremely valuable and increasingly scarce (fig. 5.35). Timber industry economists calculated the cost of conserving a population of 1,600 to 2,400 owls at $33 billion. Ecologists countered that this number was highly inflated; moreover, forest conservation would preserve countless other species and ecosystem services, whose values are almost impossible to calculate.

Sometimes the value of conserving a species is easier to calculate. Salmon and steelhead in the Columbia River are endangered by hydropower dams and water storage reservoirs that block their migration to the sea. Opening the floodgates could allow young fish to run downriver and adults to return to spawning grounds, but at high costs to electricity consumers, barge traffic, and farmers who depend on cheap water and electricity. On the other hand, commercial and sport fishing for salmon is worth over $1 billion per year and employs about 60,000 people directly or indirectly.

▲ **FIGURE 5.35** Endangered species often serve as a barometer for the health of an entire ecosystem and as surrogate protector for a myriad of less well-known creatures. "DAMN SPOTTED OWL" A 1990 Herblock Cartoon, copyright by The Herb Block Foundation.

Many countries have species protection laws

In the past 25 years or so, many countries have recognized the importance of legal protection for endangered species. Rules for listing and protecting endangered species are established by Canada's Committee on the Status of Endangered Wildlife in Canada (COSEWIC) of 1977, the European Union's Birds Directive (1979) and Habitat Directive (1991), and Australia's Endangered Species Protection Act (1992). International agreements have also been developed, including the Convention on Biological Diversity (1992).

The Convention on International Trade in Endangered Species (CITES) of 1975 provides a critical conservation strategy by blocking the international sale of wildlife and their parts. The Convention makes it illegal to export or import elephant ivory, rhino horns, tiger skins, or live endangered birds, lizards, fish, and orchids. CITES enforcement has been far from perfect: Smugglers hide live animals in their clothing and luggage; the volume of international shipping makes it impossible to inspect transport containers and ships; documents may be falsified. The high price of these products in North America and Europe, and increasingly in wealthy cities of China, makes the risk of smuggling worthwhile: A single rare parrot can be worth tens of thousands of dollars, even though its sale is illegal. Even so, CITES provides a legal structure for restricting this trade, and it raises public awareness of the real costs of the trade in endangered species.

A striking example of success in endangered species under CITES is the recovery of the southern white rhino in Africa. A century ago these huge herbivores were near extinction, but now there are at least 17,500 white rhinos in parks and game preserves.

Habitat protection may be better than individual species protection

Growing numbers of scientists, land managers, policymakers, and developers are arguing that we need a rational, continent-wide preservation of ecosystems that supports maximum biological diversity. They argue that this would be more effective than species-by-species battles for desperate cases. By concentrating on individual species,

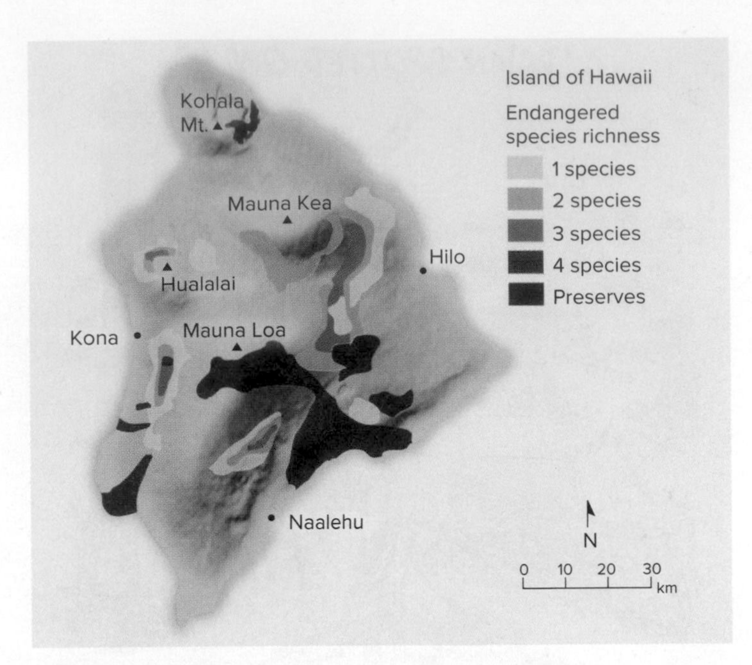

▲ **FIGURE 5.36** Protected lands (green) are often different from biologically diverse areas (red shades), as shown here on the island of Hawaii.

we spend millions of dollars to breed plants or animals in captivity that have no natural habitat where they can be released. While flagship species, such as mountain gorillas and Indian tigers, are reproducing well in zoos and wild animal parks, the ecosystems they formerly inhabited have largely disappeared.

A leader of this new form of conservation has been J. Michael Scott, who was project leader of the California condor recovery program in the mid-1980s and had previously spent 10 years working on endangered species in Hawaii. In making maps of endangered species, Scott discovered that even Hawaii, where more than 50 percent of the land is federally owned, has many vegetation types completely outside of natural preserves (fig. 5.36). The gaps between protected areas may contain more endangered species than are preserved within them.

This observation has led to an approach called **gap analysis,** in which conservationists and wildlife managers look for unprotected landscapes, or gaps in the network of protected lands, that are rich in species. Gap analysis involves mapping protected conservation areas and high-biodiversity areas. Overlaying the two makes it easy to identify priority spots for conservation efforts. Maps also help biologists and land use planners communicate about threats to biodiversity. This broad-scale, holistic approach seems likely to save more species than a piecemeal approach.

Conservation biologist R.-E. Grumbine suggests four remanagement principles for protecting biodiversity in a large-scale, long-range approach:

1. Protect enough habitat for viable populations of all native species in a given region.
2. Manage at regional scales large enough to accommodate natural disturbances (fire, wind, climate change, etc.).
3. Plan over a period of centuries, so that species and ecosystems can continue to evolve.
4. Allow for human use and occupancy at levels that do not result in significant ecological degradation.

CONCLUSION

Biodiversity can be understood in terms of the environmental conditions where different organisms live, in terms of biomes, and in terms of habitat types. Knowing the influence of climate conditions is an important place to start describing biodiversity and to conserving biological communities. Although we know a great deal about many species (especially mammals, birds, and reptiles), we know very little about most species (especially invertebrates and plants). Of the world's species, many are threatened or endangered. We worry most about threats to mammals, birds, and a few other groups, but a far greater percentage of plants, lichens, mollusks, and other groups are threatened.

Biodiversity is important to us because it can aid ecosystem stability and because we rely on many different organisms for foods, medicines, and other products. We do not know what kinds of undiscovered species may provide medicines and foods in the future. Biodiversity also has important cultural and aesthetic benefits. Pinning a dollar value on these amenities is difficult, but efforts to do so show the wide range of values in conservation areas.

Many factors threaten biodiversity, including habitat loss, invasive species, pollution, population growth, and overharvesting. The acronym HIPPO has been used to describe these threats together as a whole. You can reduce these threats by avoiding exotic pets and unsustainable fisheries or wood products that are harvested unsustainably. Laws to protect biodiversity, including the Endangered Species Act, have been controversial. They have also protected many species, including the bald eagle and gray wolf. Often such laws protect umbrella species, whose habitat also protects many other species. Despite their imperfections, these laws are our only mechanism for preserving biodiversity for future generations.

PRACTICE QUIZ

1. Why are foresters concerned about the effects of bark beetles and fire in western forests?
2. Describe nine major types of terrestrial biomes.
3. Explain how climate graphs (as in fig. 5.6) should be read.
4. Describe conditions under which coral reefs, mangroves, estuaries, and tide pools occur.
5. Throughout the central portion of North America is a large biome once dominated by grasses. Describe how physical conditions and other factors control this biome.
6. Explain the difference among swamps, marshes, and bogs.
7. How do elevation (on mountains) and depth (in water) affect environmental conditions and life-forms?

8. Figure 5.15 shows chlorophyll (plant growth) in oceans and on land. Explain why green, photosynthesizing organisms occur in bands at the equator and along the edges of continents. Explain the very dark green areas and yellow/orange areas on the continents.
9. Define *biodiversity,* and give three types of biodiversity essential in preserving ecological systems and functions.
10. What is a biodiversity hot spot? List several of them (see fig. 5.22).
11. How do humans benefit from biodiversity?
12. What does the acronym HIPPO refer to?
13. Have extinctions occurred in the past? Is there anything unusual about current extinctions?
14. Why are exotic or invasive species a threat to biodiversity? Give several examples of exotic invasive species (see fig. 5.27).
15. What does the Endangered Species Act do?

CRITICAL THINKING AND DISCUSSION

Apply the principles you have learned in this chapter to discuss these questions with other students.

1. Many poor tropical countries point out that a hectare of shrimp ponds can provide 1,000 times as much annual income for a decade or so as the same area in an intact mangrove forest. Debate this point with a friend or classmate. What are the arguments for and against saving mangroves?
2. Genetic diversity, or diversity of genetic types, is believed to enhance stability in a population. Most agricultural crops are genetically very uniform. Why might the usual importance of genetic diversity *not* apply to food crops? Why *might* it apply?
3. Scientists need to be cautious about their theories and assumptions. What arguments could you make for *and* against the statement that humans are causing extinctions unlike any in the history of the earth?
4. A conservation organization has hired you to lead efforts to reduce the loss of biodiversity in a tropical country. Which of the following problems would you focus on first, and why: habitat destruction and fragmentation, hunting and fishing activity, harvesting of wild species for commercial sale, or introduction of exotic organisms?
5. Many ecologists and resource scientists work for government agencies to study resources and resource management. Do these scientists serve the public best if they try to do pure science, or if they try to support the political positions of democratically elected representatives, who, after all, represent the positions of their constituents?
6. You are a forest ecologist living and working in a logging community. An endangered salamander has recently been discovered in your area. What arguments would you make for and against adding the salamander to the official endangered species list?

DATA ANALYSIS Confidence Limits in the Breeding Bird Survey

A central principle of science is the recognition that all knowledge involves uncertainty. No study can observe every possible event in the universe, so there is always missing information. Scientists try to define the limits of their uncertainty, in order to allow a realistic assessment of their results. A corollary of this principle is that the more data we have, the less uncertainty we have. More data increase our confidence that our observations represent the range of possible observations.

One of the most detailed records of wildlife population trends in North America is the Breeding Bird Survey (BBS). Every June, volunteers survey more than 4,000 established 25-mile routes. The accumulated data from thousands of routes, over more than 40 years, indicate population trends, telling which populations are increasing, decreasing, or expanding into new territory. To examine a sample of BBS data, go to Connect, where you can explore the data and explain the importance of uncertainty in data.

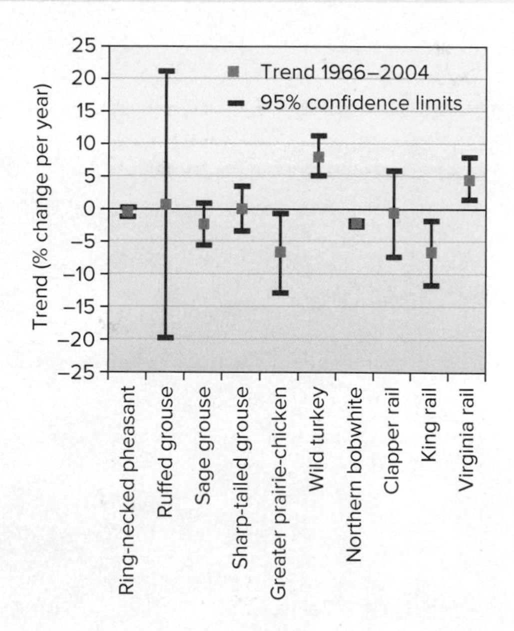

Design Elements: Active Learning (Toad): ©Gaertner/Alamy Stock Photo; Case Study (Globe): ©McGraw-Hill Education; Google Earth: ©McGraw-Hill Education; Abstract Background: ©Martin Kubat/Shutterstock; What do you think (Students using tablets): ©McGraw-Hill Education/Richard Hutchings, photographer; What can you do (Hand holding Globe): ©Christoph Weihs/Shutterstock

CHAPTER

6 Environmental Conservation: Forests, Grasslands, Parks, and Nature Preserves

LEARNING OUTCOMES

After studying this chapter, you should be able to answer the following questions:

- What portion of the world's original forests remains?
- What activities threaten global forests? What steps can be taken to preserve them?
- Why is road construction a challenge to forest conservation?
- Where are the world's most extensive grasslands?
- How are the world's grasslands distributed, and what activities degrade grasslands?
- What are the original purposes of parks and nature preserves in North America?
- What are some steps to help restore natural areas?

Orangutans are among the most critically endangered of all the great apes. Over the past 20 years, about 90 percent of their rainforest habitat in Borneo and Sumatra has been destroyed by logging and conversion to palm oil plantations.

CASE STUDY

Palm Oil and Endangered Species

Are your donuts, toothpaste, and shampoo killing critically endangered orangutans in Sumatra and Borneo? It seems remote, but they might be. Palm oil, a key ingredient in at least half of the packaged foods, cosmetics, and soaps in the supermarket, is almost entirely sourced from plantations that just 20 years ago were moist tropical forest. In Indonesia and Malaysia these forests were the habitat of orangutans, Sumatran tigers and rhinos, and other endangered species. As palm oil has become the world's most widely used vegetable oil, expanding palm oil plantations have become one of the greatest causes of tropical deforestation.

▲ **FIGURE 6.1** Over the past 15 years, palm plantation area in Southeast Asia has grown to more than 14 million hectares (34 million acres), replacing some of the world's richest primary forest. This rapid growth has destroyed habitat and displaced many critically endangered species. ©KhunJompol/Getty Images

A 2017 study of orangutan populations in Borneo, an island owned partly by Malaysia and partly by Indonesia, estimated that at least 100,000 of these rare and reclusive forest primates were killed in just 15 years, between 1999 and 2015. This represents over half of the region's orangutans. By 2050 the population is expected to be only around 50,000, many of them in tiny, dispersed, and nonviable populations. The main reasons for this decline are the rapid conversion of primary forest to palm plantations, deforestation for wood products, and the increasing density of human populations as settlements expand to serve these industries. Habitat loss is a driving factor, but actual mortality in this study was attributed mainly to hunting in the forests around plantations, made possible by the expansion of the plantations and logging roads deep into the primary forest.

In Indonesian, *orang utan* means person of the forest. Orangutans are among our closest primate relatives, sharing at least 97 percent of our genes. Traditional cultures in Borneo may recognize this relationship, because taboos have discouraged hunting and eating them. These taboos seem to be diminishing, however, with the expansion of populations into once-forested regions.

Indonesia and Malaysia produce over 80 percent of the global palm oil supply. In 1960 the two countries together had about 100,000 ha (247,000 acres) of palm oil plantations. That number is now nearly 14 million hectares (34 million acres), according to the UN Food and Agriculture Organization. Expansion of palm plantations usually accompanies other deforestation practices. Often logging companies harvest the valuable hardwoods first; then logging debris is burned to clear the land for planting (and often to cover up illegal logging). Finally, a monoculture of palm trees is planted (fig. 6.1).

These thirsty trees need moist soil and a wet climate, so plantations are often established in lowland peat swamps. Peat is composed mainly of ancient, undecomposed plant material, so draining and burning of a hectare of peatland can release 15,000 tons of CO_2. More than 70 percent of the carbon released from Sumatran forests is from burning peat. Indonesia, which has the third largest area of rainforest in the world as well as the highest rate of deforestation, is now the world's third highest emitter of greenhouse gases. Smoke from burning peat often blankets Singapore, Malaysia, and surrounding regions.

At the 2014 UN Climate Summit in New York, 150 companies, including McDonald's, Nestlé, General Mills, Kraft, and Procter & Gamble, promised to stop using palm oil from recently cleared rainforest and to protect human rights in forest regions. Several logging companies, including the giant Asia Pulp and Paper, pledged to stop draining peat lands and to reduce deforestation by 50 percent by 2020.

Will these be effective promises or empty ones? It is difficult to trace oil origins or to monitor remote areas, but at least this movement sets a baseline for acceptable practices. In 2017 two of the world's largest palm oil traders, Wilmar International and Cargill, announced they would no longer do business with a Guatemalan company, Reforestadora de Palmas del Petén S.A. (REPSA), because of environmental and human rights abuses. REPSA was implicated in the murder of Rigoberto Lilma Choc, a 28-year-old Guatemalan schoolteacher who had protested when effluent from a REPSA palm oil operation poisoned the Pasión River, killing millions of fish. When a Guatemalan judge ordered REPSA to stop operations for 6 months, the ruling was quickly followed by the kidnappings of three human rights activists and by Choc's murder. Cargill then cut ties with REPSA, citing its failure to meet critical criteria for sustainability and ethics.

While the death of 100,000 orangutans has not had the impact of a human murder on global palm oil production and trade, growing awareness can help defend forests, along with forest-dwelling species and people. Throughout the world, monitoring and defending forests is key to protecting biodiversity, climate, water, and cultural diversity. In this chapter we look at the state of forest and grassland reserves, and at efforts to conserve them for future generations. To see Google Earth placemarks that will help you explore these landscapes via satellite images, visit www.connect.mheducation.com.

To read more, see Voigt et al., 2018, Global demand for natural resources eliminated more than 100,000 Bornean orangutans. *Current Biology* 28, 1–9. https://doi.org/10.1016/j.cub.2018.01.053 ■

> *What a country chooses to save is what a country chooses to say about itself.*
>
> —MOLLIE BEATTY, FORMER DIRECTOR, U.S. FISH AND WILDLIFE SERVICE

▲ **FIGURE 6.3** A tropical rainforest in Queensland, Australia. Primary, or old-growth, forests such as this aren't necessarily composed entirely of huge, old trees. Instead, they have trees of many sizes and species that contribute to complex ecological cycles and relationships. ©Digital Vision/PunchStock

6.1 WORLD FORESTS

- *Forests provide habitat, resources, and essential ecological services.*
- *Tropical forests are especially diverse and vulnerable.*
- *Forests help stabilize climate by storing carbon.*

Forests, woodlands, pastures, and rangelands together occupy almost 60 percent of global land cover (fig. 6.2). These ecosystems provide many of our essential resources, such as lumber, paper pulp, and grazing for livestock. They also provide essential ecological services, including regulating climate, controlling water runoff, providing wildlife habitat, purifying air and water, and supporting rainfall. Forests and grasslands also have scenic, cultural, and historic values that deserve protection. But they are also among the most heavily disturbed ecosystems (see chapter 5).

As the opening case study for this chapter shows, balancing competing land uses and needs can be complicated. Many conservation debates have concerned protection or use of forests, prairies, and rangelands. This chapter examines the ways we use and abuse these biological communities, as well as some of the ways we can protect them and conserve their resources. We discuss forests first, followed by grasslands and then strategies for conservation, restoration, and preservation.

Boreal and tropical forests are most abundant

Forests are widely distributed, but most remaining forests are in the cold boreal ("northern") or taiga regions and the humid tropics (fig. 6.3). Assessing forest distribution is tricky, because forests vary in density and height, and many are inaccessible. The UN Food and Agriculture Organization (FAO) defines *forest* as any area where trees cover more than 10 percent of the land. This definition includes woodlands ranging from open **savannas,** whose trees occupy less than 20 percent of the area, to **closed-canopy forests,** in which tree crowns cover most of the ground. The largest tropical forest is in the Amazon River basin. The highest rates of forest loss also are in South America (fig. 6.4). Some of the world's most biologically diverse regions are undergoing rapid deforestation, including Southeast Asia and Central America.

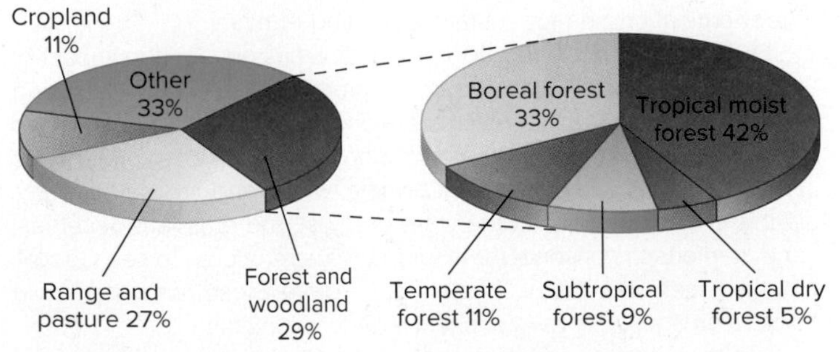

▲ **FIGURE 6.2** World land use and forest types. The "other" category includes tundra, desert, wetlands, and urban areas. Source: UN Food and Agriculture Organization (FAO).

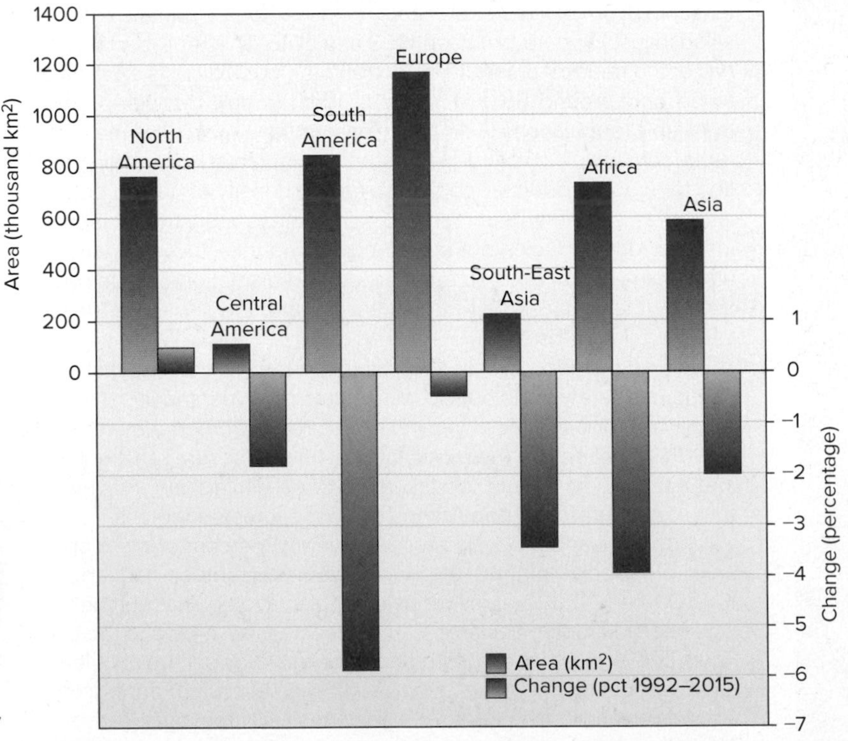

▲ **FIGURE 6.4** Forest area and percentage change 1992–2015. Forest clearing for agriculture accounts for most recent changes. (Note that statistics on forest change can depend on varying definitions of "forest.") Source: Data from FAO, 2018.

Active LEARNING

Calculating Forest Area

Examine figure 6.4, which shows forest losses between 1992 and 2015. Deforestation has been going on much longer than that, though. Use the table below to calculate change from original forest area (before human impacts). To calculate percentage lost, subtract the recent value from the original value. Then divide the difference by the original value.

Changes in Forest Area

Region	Original forest (millions of ha)	2015 (millions of ha)
Africa	2,200	674
Asia	1,900	592
Europe	2,200	1,005
South America	1,400	864
Oceania	400	191
North & Cent. America	700	705
World	8,800	4,031

1. Which region has lost the greatest percentage of its original forest, and what percentage was that?
2. Which region has gained the most forest, and what percentage was that?
3. Africa and Europe (including Siberia) are listed as having the greatest amount of original forest. Among the forest types in figure 6.2, which types are these most likely to be dominant in these two regions? (If in doubt, see chapter 5.) Which region would be capable of supporting palm oil production?
4. How much forest has been lost worldwide since the beginnings of human habitation?
5. Working with round numbers makes comparison easy. How important are the details that are lost when you read approximate numbers from the graph? What kinds of generalization might have gone into producing the FAO's original data? What kinds of generalization might have been unavoidable?

ANSWERS: 1. Africa lost 1,526 million ha, or nearly 70 percent; 2. North America gained 5 million ha; 3. Tropical moist forest (Africa) and boreal forest (Europe); Africa could support oil palms; 4. the world lost about 4,769 million ha of forest, or about 54 percent of its original forest; 5. usually, approximate numbers provide a quick, useful comparison, and additional detail isn't needed until further analysis is done. Defining forests and forest types, or determining extent of either "original" or "current" forest, all involves considerable, usually unavoidable generalization.

Among the forests of greatest ecological importance are the remnants of primeval forests, such as the Indonesian forests described in the opening case study for this chapter, that are home to much of the world's biodiversity, endangered species, and indigenous human cultures. Sometimes called frontier forests, **old-growth forests** are those that cover a relatively large area and have been undisturbed by human activities long enough that trees can live out a natural life cycle and ecological processes can occur in fairly normal fashion. That doesn't mean that all trees need be enormous or thousands of years old. In some forest systems, such as lodgepole pine forests of the Rocky Mountains, most trees live less than a century before being killed by disease or some natural disturbance, such as a fire. Nor does it mean that humans have never been present. Where human occupation entails relatively little impact, an old-growth forest may have been inhabited by people for a very long time. Even forests that have been logged or converted to cropland often can revert to old-growth characteristics if allowed to undergo normal successional processes.

Even though forests still cover about half the area they once did worldwide, only one-quarter of those forests retain old-growth features. The largest remaining areas of old-growth forest are in Russia, Canada, Brazil, Indonesia, and Papua New Guinea. Together, these five countries account for more than three-quarters of all relatively undisturbed forests in the world. In general, remoteness rather than laws protects those forests. Although official data describe only about one-fifth of Russian old-growth forest as threatened, rapid deforestation–both legal and illegal–especially in the Russian Far East, probably puts a much greater area at risk.

Benchmark Data	
Among the ideas and values in this chapter, these are a few worth remembering.	
29%	Percentage of land surface in forest and woodlands
18%	Percentage of anthropogenic climate impacts from forest clearing
17%	Area in preserves
11%	Percentage of land surface in cropland
11:47	Revenue:cost of managing Western grazing lands, in millions of dollars
3:58	Ratio of global protected areas, 1950:2017, in km^2

Forests provide essential products

Wood plays a part in more activities of the modern economy than does any other commodity. There is hardly any industry that doesn't use wood or wood products somewhere in its manufacturing and marketing processes. Think about the amount of mail, newspapers, photocopies, packaging, and other paper products that each of us in developed countries handles, stores, and disposes of in a single day. According to the FAO, total annual world wood consumption is about 5.5 billion m^3. This is more than steel and plastic consumption combined. International trade in wood and wood products amounts to more than $200 billion each year. Developed countries produce less than half of all industrial wood but consume about 80 percent of it. Less-developed countries, mainly in the tropics, produce more than half of all industrial wood but use only 20 percent.

Paper pulp, the fastest-growing forest product, accounts for nearly a fifth of all wood consumption. Most of the world's paper is used in the wealthier countries of North America, Europe, and Asia. Global demand for paper is increasing rapidly, however, as other countries develop. The United States, Russia, and Canada are the largest producers of both paper pulp and industrial wood (lumber and panels). Much industrial logging in Europe and North America occurs on managed plantations, rather than in untouched old-growth forest. However, paper production is increasingly blamed for deforestation in Southeast Asia, West Africa, and other regions.

Fuelwood accounts for nearly half of global wood use. At least 2 billion people depend on firewood or charcoal as a principal source of heating and cooking fuel (fig. 6.5). The average amount of fuelwood used in less-developed countries is about 1 m^3 per person per year, roughly equal to the amount that each American consumes each year as paper products alone. Demand for fuelwood, which is increasing at slightly less than the global population growth rate, is causing severe fuelwood shortages and depleting forests in some developing areas, especially around growing cities. About 1.5 billion people have less fuelwood than they need, and many experts expect shortages to worsen as poor urban areas grow. In some countries, firewood harvesting is a major cause of deforestation, but foresters argue that biomass energy could be produced sustainably in most developing countries, with careful management.

▲ **FIGURE 6.5** Firewood accounts for almost half of all wood harvested worldwide and is the main energy source for nearly half of all humans.
©William P. Cunningham

▲ **FIGURE 6.6** Monoculture forestry, such as this Wisconsin tree farm, produces valuable timber and pulpwood but has little biodiversity.
©William P. Cunningham

Approximately one-quarter of the world's forests are managed for wood production. Ideally, forest management involves scientific planning for sustainable harvests, with particular attention paid to forest regeneration. The FAO reports that in temperate regions, more land is being replanted or allowed to regenerate naturally than is being permanently deforested. Much of this reforestation, however, is in large plantations of single-species, single-use, intensive cropping called **monoculture forestry.** Although this produces rapid growth and easier harvesting than a more diverse forest, a dense, single-species stand often supports little biodiversity and does poorly in providing the ecological services, such as soil erosion control and clean water production, that may be the greatest value of native forests (fig. 6.6).

Some of the countries with the most successful reforestation programs are in Asia. China, for instance, cut down most of its forests 1,000 years ago and has suffered centuries of erosion and terrible floods as a consequence. Recently, however, timber cutting in the headwaters of major rivers has been outlawed, and massive reforestation projects have begun. In the past 20 years, China planted some 50 billion trees, mainly in Xinjiang Province, to stop the spread of deserts. Korea and Japan also have had very successful forest restoration programs. After being almost totally denuded during World War II, both countries are now about 70 percent forested. Many question, though, whether these countries are clearing forests elsewhere as they replant at home.

Tropical forests are being cleared rapidly

Tropical forests are among the richest and most diverse terrestrial systems. Although they now occupy less than 10 percent of the earth's land surface, these forests are thought to contain more than

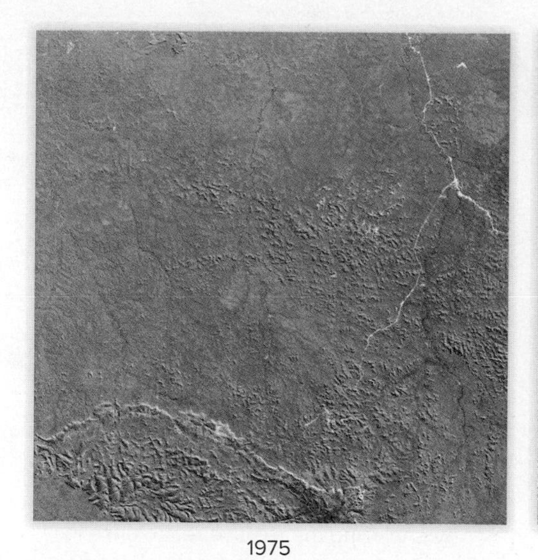
1975

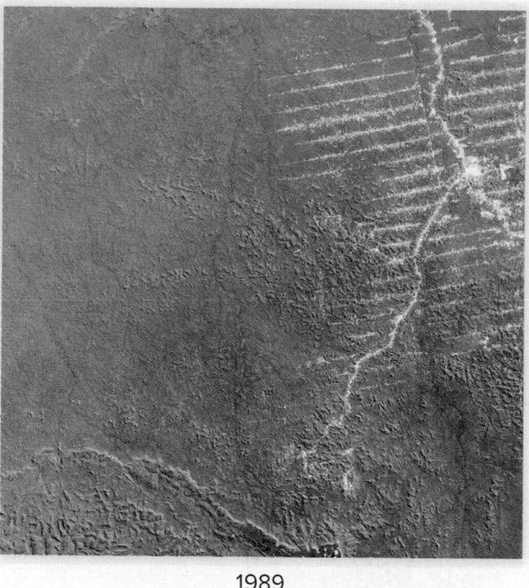
1989

2015

▲ **FIGURE 6.7** Forest destruction in Rondomia, Brazil, between 1975 and 2015. Construction of logging roads creates a featherlike pattern that opens forests to settlement by farmers. Source: USGS EROS Data Center

two-thirds of all higher plant biomass and at least half of all the plant, animal, and microbial species in the world.

A century ago, an estimated 12.5 million km^2 of the tropics (an area larger than the entire United States) were covered with closed-canopy forest. The FAO estimates that only about 40 percent of that forest remains in its original condition and that about 10 million ha, or about 0.6 percent, of existing tropical forests are cleared each year (fig. 6.7).

As the opening case study for this chapter notes, Indonesia is now thought to have the highest rate of **deforestation** (forest loss) in the world. At the beginning of the twentieth century, at least 84 percent of Indonesia's total land was forested. Between 1990 and 2010, the country lost at least 24 million ha of forest (59 million acres), much of it for illegal palm oil plantations, and only 52 percent of the island nation was forested. Despite a 2014 pledge to reduce forest clearing, Indonesia continues to lose an estimated 800,000 ha per year, rivaling deforestation in Brazil.

Estimates for total tropical forest losses range from about 5 million to more than 20 million ha per year. The FAO estimates of 10 million ha deforested per year are generally the most widely accepted. To put that figure in perspective, it means that about 1 acre—or the area of a football field—is cleared every second, on average, around the clock.

In 2004 Brazil was reported to have lost 28,000 km^2 of forest to clearing and fires. Much of this destruction was to make room for soybean production and cattle ranching (fig. 6.8). The rate of deforestation has declined in some years, but recently it has climbed again, reaching 16,000 km^2 in 2017. How are these estimates derived? Remote sensing usually provides our best information on forest cover (Exploring Science, p. 134).

In Africa, the coastal forests of Senegal, Sierra Leone, Ghana, Madagascar, Cameroon, and Liberia already have been mostly demolished. Haiti was once 80 percent forested; today essentially all that forest has been destroyed, and the land lies barren and eroded. India, Myanmar, Cambodia, Thailand, and Vietnam all have little old-growth lowland forest left. In Central America, nearly two-thirds of the original moist tropical forest has been destroyed, mostly within the past 30 years and primarily due to logging and conversion of forest to cattle range. (See related story "Disappearing Butterfly Forests" at www.connect.mheducation.com.)

A variety of factors contribute to deforestation, and different forces predominate in various parts of the world. Logging for valuable tropical hardwoods, such as teak and mahogany, is generally the first step. Although loggers might take only one or two of the largest trees per hectare, the canopy of tropical forests is usually so strongly linked by vines and interlocking branches

▲ **FIGURE 6.8** Forest clearance for cattle ranching is an important driver of deforestation, especially in South America. ©William P. Cunningham

EXPLORING Science — How Can We Know About Forest Loss?

Brazil's Amazon River Basin has the world's largest intact forest. This forest supports some of the world's greatest areas of biodiversity, captures 2 billion tons of carbon dioxide annually, and evaporates moisture that provides rainfall to Brazil's cities and farms. This forest has also been declining continuously, with farming, logging, mining, and hydroelectric dam development. Today soybean farming, expanding to feed the fast-growing pork industry of China, has become the main driver of deforestation. Soy producers, and their bankers, are among the most powerful political players in Brazil, and they are credited with driving new waves of deforestation and wetland destruction. Brazil has cleared between 5,000 and 30,000 km^2 per year every year since the 1970s. In that time, some 400,000 km^2 of primary forest have been lost, an area the size of California.

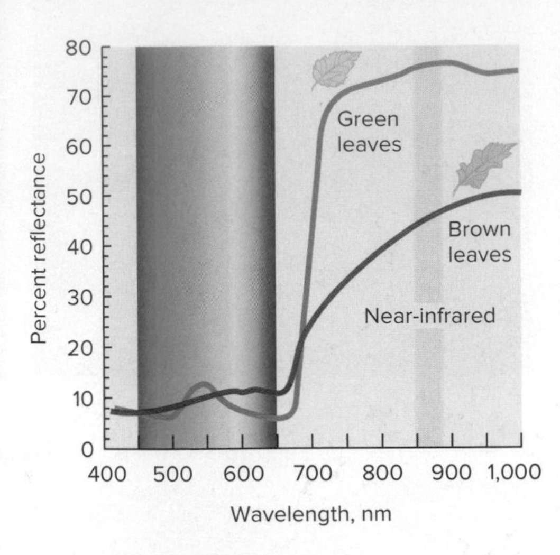

▲ **FIGURE 1** Green leaves reflect more in the green part of the spectrum than in blue or red, but they reflect 70 percent or more in the infrared range, outside the visible range of light. Brown leaves reflect similarly across the visible spectrum and 50 percent or less in the infrared. These distinct patterns let us differentiate land surfaces from satellite images.

How do we know this? How do we know how many acres of forest are lost and what replaced them? Measuring on foot would be impossible. The region is too vast, the forests impassable. Counting and measuring thousands of new soy fields, each year, would be inconceivably time consuming.

Remote sensing is the answer. Anyone familiar with online maps has seen land imagery taken by satellites (see fig. 6.7). These give a picture of the landscape. Satellite images also let us calculate the area forest and farmland.

Remote sensing involves capture and analysis of images from satellites (or from airborne sensors). Hundreds of satellites orbit the earth, transmitting GPS signals, monitoring weather, and capturing earth images. While your eyes detect light energy, in wavelengths that you interpret as blue, green, or red, sensors on a satellite can detect very short (ultraviolet) and long (infrared) wavelengths, which you cannot see.

These wavelengths are the key to differentiating land cover. Green vegetation looks green because it reflects more in the green wavelengths than in red or blue (fig. 1). But green vegetation reflects infrared wavelengths (which you cannot see) even more strongly. The "spectral signatures" (characteristic reflectance patterns in different wavelengths) in figure 1 show that *percent reflectance* for green leaves is about 15 percent in the green wavelengths (about 500–550 nanometers, nm), twice that for blue or red wavelengths. But in the infrared range, reflectance jumps to 70 percent or more. This means that plants are absorbing most visible energy from the sun (blue through red) but reflecting most infrared energy. If you want to "see" where green vegetation is, infrared wavelengths are very useful.

Brown leaves, in contrast, reflect similar amounts of visible wavelengths (slightly more yellow-red colors) and much less infrared than green leaves do.

By analyzing the relative reflectance in different wavelengths, remote sensing software can identify dominant land cover, pixel by pixel, for vast areas of land (fig. 2). It can count the number of pixels filled by green leaves, pavement and buildings, water, even different crop types from soybeans to sugar cane to corn. And analysts can repeat this calculation frequently, every year, every month, or less.

One of the best-known earth-imaging satellites, Landsat, produces images 185 km wide, and each pixel represents an area of just 30 × 30 m on the ground. Landsat orbits approximately from pole to pole, so as the earth spins below the satellite, it captures images of the entire surface every 16 days. The Landsat program began in 1972, so it also provides the best available historical record of land change.

While a photograph gives a good impression of land cover, remote sensing allows us to quantify changes in critical values such as the amount of forest, agriculture, wetlands, or urban development and to answer countless other questions.

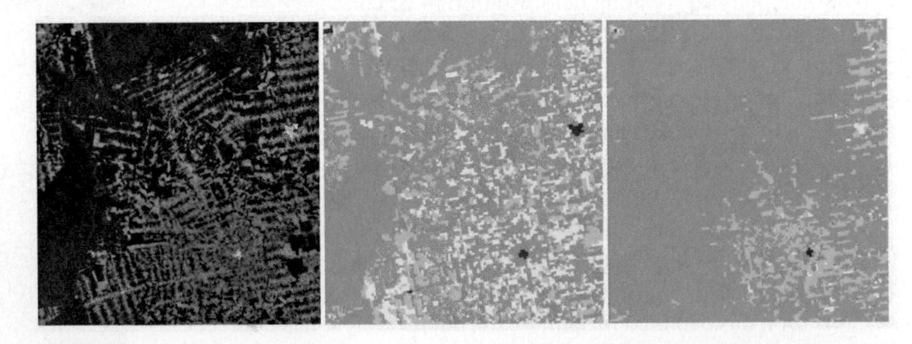

▲ **FIGURE 2** Spectral signatures can classify a visible image (*left*) to forest (green), agriculture (yellow-brown), and cities (red). Cell counts show change from 2014 (*center*) and 2000 (*right*). Source: Data from Brazilian Institute of Geography and Statistics, Monitoramento da Cobertura e Uso da Terra do Brasil 2018.

that felling one tree can bring down a dozen others. Building roads to remove logs kills more trees, but even more important, it allows entry to the forest by farmers, miners, hunters, and others who cause further damage.

Shifting cultivation (sometimes called "slash and burn" or milpa farming) is often blamed for forest destruction. But in many countries, indigenous people have discovered sustainable ways to use complex cycles of mixed polyculture and soil amendment practices to improve soil fertility. Growing nonindigenous populations and logging have destabilized these practices in many areas.

Saving forests stabilizes our climate

Forests provide important regional climate regulation. Trees take up and transpire (release) vast amounts of moisture, and the conversion of liquid water to vapor consumes a great deal of heat from the atmosphere. Evaporated moisture becomes clouds that provide rainfall. In the region around the Amazon rainforest, much of the rain that falls comes directly from transpiration. Amazon forests have even been shown to produce rain 2–3 months before winds deliver moisture from the ocean.

This precipitation is vulnerable to deforestation. As forests are cleared, rainfall can decrease, landscapes dry, and regional temperatures rise. Drought and wildfire become more frequent, and it becomes harder for forests to regrow. Even the Amazon rainforests have been suffering from drought, which causes tree mortality and further reduces rainfall in surrounding farmlands, as forests decline. Ecologists are worried that this positive feedback is producing a self-amplifying decline—with reduced rainfall causing tree mortality, which then further reduces rainfall.

Moisture released from forests contributes to global moisture circulation—recent climate studies suggest that deforestation of the Amazon could reduce precipitation in the American Midwest. Forests are also a huge carbon sink, storing some 422 billion metric tons of carbon in standing biomass. Clearing and burning of forests is responsible for about 18 percent of all the carbon released by human actions every year—more than all vehicles combined—and is a major factor in global climate change (see chapter 10).

Preventing deforestation can be an affordable way to protect climate and rainfall. A study in Uganda, for example, found that paying village landowners $27 per hectare was enough to encourage them to preserve forests, and deforestation was reduced by half compared to control villages. This works out to about 46 cents per ton of CO_2 not emitted, many times cheaper than the cost of carbon control in wealthy countries. For comparison, the cost of carbon capture at U.S. power plants is estimated to be around $130 per ton of CO_2. Clearly, preventing carbon pollution is cheaper than cleaning it up.

REDD schemes can pay for ecosystem services

It is increasingly clear that forests influence the global climate system, so wealthy countries are exploring ways to pay developing countries to keep their forests standing. Reducing Emissions from Deforestation and Forest Degradation (REDD) is a global effort to do this. The hope is that payments can help low-income countries develop economic opportunities that don't require clearing and selling their forests. Agreements by wealthy countries to finance REDD initiatives have been among the notable successes at recent global climate conferences.

The idea, first proposed by Papua New Guinea and Costa Rica at climate talks in 2005, is to recognize the ways wealthy countries depend on climate stablization and other benefits of forests in developing areas. Part of this recognition is that wealthy countries should pay for some of those services by helping to finance forest protection in poorer countries (Key Concepts, p. 136). It is calculated that replanting 300 million ha of degraded forest should capture about 1 billion tons of CO_2 over the next 50 years.

Part of the reason REDD has gained traction is that it protects an array of ecological functions that serve a wide variety of interest groups. More than 1.2 billion people depend on forests for their livelihoods. Governments are interested in political stability in these communities. Forests protect water resources that supply cities, as well as biodiversity that supports ecotourism. Often indigenous and tribal groups have specialized knowledge about forest systems, and REDD promotes efforts to involve these people in conservation strategies. Supplementing their income may also allow them to avoid intensive use of forest resources and yet to remain on the land, where their traditional knowledge and stewardship are valuable resources.

Protection of indigenous rights is an important aim of REDD efforts, but these protections have always been a contentious, and often a dangerous, proposition. In 2017 nearly 200 forest protectors were killed in Latin America and elsewhere in the tropics by the military or private "death squads" over land ownership and pollution issues.

Administering a program as large as REDD will not be easy. The United Nations estimates that fully funding forest protection will cost between $20 billion and $30 billion per year globally. This is a large number but it is just 0.0003 percent of world GDP, or less than the annual net profit of large companies such as ExxonMobil ($45 billion in 2014) or Apple ($100 billion in 2015). Careful monitoring and good governance (often lacking in developing countries) will be needed to make sure this money is spent wisely and that projects are sustainable. With such large amounts of money flowing, it is difficult to control corruption and ensure fair management. Nevertheless, this represents the largest experiment in tropical conservation in world history.

Temperate forests also are at risk

Tropical countries aren't unique in harvesting forests at an unsustainable rate. Northern countries, such as the United States and Canada, also have allowed controversial forest management practices in many areas. For many years the official policy of the U.S. Forest Service was "multiple use," which implied that the forests could be used for everything that we might want to do there simultaneously. Some uses are incompatible, however. Bird-watching, for example, isn't very enjoyable in an open-pit mine. And protecting species that need unbroken old-growth forest isn't easy when you cut down the forest.

Save a tree, save the climate?

Forest destruction and land conversion produce about 17 percent of all human-caused CO_2 emissions—more than all global transportation emissions. REDD (Reducing Emissions from Deforestation and Forest Degradation) aims to reduce those emissions and help avert a climate catastrophe. Reducing deforestation could accomplish about half of global emission reduction goals. Billions of dollars' worth of ecosystem services, and precious biological diversity, can be saved at the same time. Every day over 30,000 hectares of tropical forest are destroyed by logging and burning; another 30,000 ha are degraded. Each year this adds up to an area twice the size of Alabama.

KC 6.1

©Stockbyte/Getty Images

How do deforestation and degradation release carbon?

- Trees are burned, releasing carbon (C) stored in wood and leaves.
- Fallen vegetation decays, releasing stored C (see chapter 2).
- Accumulation of C in soil litter declines; exposed soils dry, and C in soil oxidizes to CO_2.
- The forest ecosystem is no longer available to store C.

What drives deforestation?

- Industrial-scale agriculture (soy and palm oil production, cattle ranching)
- Industrial logging driven by international demand for timber
- Poverty and population pressure as people seek farmland and fuelwood
- Road development, oil development, mining, and dams

KC 6.2

5 mi (11 km)

NASA Landsat image reveals parallel clearings on either side of a road near the Amazon River.

Source: NASA

KC 6.3

Products from deforested lands

- Oil and gasoline
- Food, cosmetics containing palm oil
- Paper products
- Aluminum (from bauxite ore)
- Metals, gems, electronic components
- Many, many others

Lost value from deforestation

Losses in human welfare are estimated at **\$2 trillion to \$4 trillion each year**.[1] Losses in ecosystem services and the value of carbon storage may be still greater.

[1]Sukdev, P. 2010. Putting a price on nature. *Solutions* 1(6):34–43. www.thesolutionsjournal.org.

KC 6.10

What ecological services would be protected under REDD?

We rely on forests for countless goods and services; here are some primary examples:

- Water supplies are maintained by forested areas, which store moisture and release it slowly during a dry season.
- Biodiversity provides for wild foods, medicines, building materials, migratory species, and tourism.
- Climate and weather regulation: Forested areas have less volatile temperature and humidity changes than do cleared areas.

©non15/Shutterstock

KC 6.5a

KC 6.5b

©Stockbyte/Getty Images

©Comstock Images/Alamy Stock Photo

KC 6.4

KC 6.6

©Mary Ann Cunningham

▼ The world's remaining forest area is about 4 billion hectares. Nearly half of these forests are boreal (northern) forest (purple); about half are tropical forest (green).

KC 6.7

Would REDD cost money?

Yes, but it also represents payment for products and services. Many developing countries rely on exporting tropical timber, or conversion to oil palm and soy farms, for most of their income. To cooperate with REDD, they would want this income replaced to some extent.

Wealthier countries rely on resources and ecosystem services from developing areas. Paying for the timber, oil, paper, and food products is easy, but REDD suggests that now we should also pay to protect some ecosystem services we rely on, including global climate stabilization, biodiversity, and water resources.

The United Nations REDD program estimates it will take $20–30 billion annually from developed countries to pay for forest protection, carbon offsets, and alternative development strategies.

What about human rights?

Some 1.2 billion people rely on forests for their livelihoods. More than 2 billion—a third of the world's population—use firewood to cook and to heat their homes. REDD efforts must recognize the rights of native people and local communities. Channeling money to urban central governments could worsen threats to these communities.

©Amazon Conservation Team

KC 6.8

How can we be sure that REDD projects are sustainable and enduring?

Monitoring, good government, and working at the local level are essential for REDD to succeed. A fascinating and successful example of local involvement is that of the Amazon Conservation Team (ACT), which has been partnering with indigenous peoples to map, monitor, and protect their ancestral lands using Google Earth and GPS.

KC 6.9

©Amazon Conservation Team

CAN YOU EXPLAIN?

1. **How does deforestation contribute to carbon emissions?**
2. **What are some tropical forest (or formerly forest) resources you use?**

▲ **FIGURE 6.9** The huge trees of the old-growth temperate rainforest accumulate more total biomass in standing vegetation per unit area than any other ecosystem on earth. They provide habitat to many rare and endangered species, but they also are converted by loggers who can sell a single tree for thousands of dollars. ©William P. Cunningham

Some of the most contentious forestry issues in the United States and Canada in recent years have centered on logging in old-growth temperate rainforests in the Pacific Northwest. As you learned in the opening case study for this chapter, such forests have incredibly high levels of biodiversity, and they can accumulate five times as much standing biomass per hectare as a tropical rainforest (fig. 6.9). Many endemic species, such as the northern spotted owl, Vaux's swift, and marbled murrelet, are so highly adapted to the unique conditions of these ancient forests that they live nowhere else.

The U.S. Northwest forest management plan established in 1994 is a model for integrating scientific study, local needs, and best practices in land use. This plan attempts to integrate human and economic dimensions of issues while protecting the long-term health of forests, wildlife, and waterways. It focuses on scientifically sound, ecologically credible, and legally responsible strategies and implementation. It aims to produce a predictable and sustainable level of timber sales and nontimber resources. And it tries to ensure that federal agencies work together. It may not be enough protection, however, to ensure survival of endangered salmon and trout populations in some rivers. Several salmon and steelhead trout populations have been listed as endangered and more are under consideration. What do you think? How would you balance logging, farming, and cheap hydropower against fishing, native rights, and wildlife protection?

▲ **FIGURE 6.10** Large clear-cuts, such as this, threaten species dependent on old-growth forest and expose steep slopes to soil erosion. Restoring something like the original forest will take hundreds of years. ©Gary Braasch/Getty Images

There are multiple harvesting methods. Most lumber and pulpwood in the United States and Canada currently are harvested by **clear-cutting,** in which every tree in a given area is cut, regardless of size (fig. 6.10). This method is effective for producing even-age stands of sun-loving species, such as aspen or some pines, but often increases soil erosion and eliminates habitat for many forest species when carried out on large blocks. It was once thought that good forest management required immediate removal of all dead trees and logging residue. Research has shown, however, that standing snags and coarse, woody debris play important ecological roles, including soil protection, habitat for a variety of organisms, and nutrient recycling.

Some alternatives to clear-cutting include **shelterwood harvesting,** in which mature trees are removed in a series of two or more cuts, and **strip-cutting,** in which all the trees in a narrow corridor are harvested. For many forest types, the least disruptive harvest method is **selective cutting,** in which only a small percentage of the mature trees are taken in each 10- or 20-year rotation. Ponderosa pine, for example, are usually selectively cut to thin stands and improve growth of the remaining trees. A forest managed by selective cutting can retain many of the characteristics of age distribution and groundcover of a mature old-growth forest. (See related story "Forestry for the Seventh Generation" at www.connect.mheducation.com.)

Many people, concerned about ecosystem services, have called for an end to logging on federal lands in the United States. Just 4 percent of the nation's timber comes from national forests, and this harvest adds only about $4 billion to the American

What Can YOU DO?

Lowering Your Forest Impacts

Forests might seem remote from everyday life. But there are things each of us can do to protect them:

- Recycle paper and re-use it as scratch paper. Make double-sided copies or avoid printing altogether.
- Buy products made from certified sustainably harvested wood. If you build, use composites, particle board, or laminated beams, rather than plywood and timber from old-growth trees.
- Avoid tropical hardwood products, such as ebony, mahogany, rosewood, or teak, unless the manufacturer can guarantee agroforestry or sustainable-harvest sourcing.
- Avoid foods from deforested rainforest. These may include fast-food hamburgers, coffee, bananas, and most products with palm oil.
- Do buy nontimber forest products, such as Brazil nuts, cashews, mushrooms, or shade-grown coffee.
- Visit forested areas and spend your ecotourism money there—but pay attention to minimizing your impacts.
- Contact your representatives, and ask them to support forest protection policies. Stay informed about emerging policies and their impacts.

economy per year. In contrast, recreation, fish and wildlife, clean water, and other ecological services provided by the forest, by their calculations, are worth at least $224 billion each year. Mike Dombeck, former chief of the U.S. Forest Service, has stated that water resources are the most valuable commodity from the nation's forests.

Many remote communities depend on logging jobs to survive. But these jobs rely on abundant subsidies. The federal government builds roads, manages forests, fights fires, and sells timber for less than the administrative costs of the sales. The persistent question is, what would these communities do without federal and state subsidies? How should we weigh these different costs and benefits? Are there alternative ways to support communities dependent on timber harvesting?

Roads on public lands are another controversy. Over the past 40 years, the Forest Service has expanded its system of logging roads more than tenfold, to a current total of nearly 550,000 km (343,000 mi), or more than ten times the length of the interstate highway system. Government economists regard road building as a benefit because it opens up the country to motorized recreation and industrial uses. Wilderness enthusiasts and wildlife supporters, however, see this as an expensive and disruptive program. In 2001 President Clinton established a plan to protect 23.7 million ha (58.5 million acres) of de facto wilderness from roads. Land developers, logging, mining, and energy companies protested this "roadless rule." President G. W. Bush, supported by industry-friendly western judges, overturned the rule and ordered resource managers to expedite logging, mining, and motorized recreation. In 2009 President Obama ordered the rule reinstated. He noted that this measure protects habitat for 1,600 endangered species (including bears and owls) and watersheds for 60 million people. In 2017 President Trump once again overturned the roadless rule and opened up forests to mining and timber extraction.

Fire management is a growing cost

U.S. policy toward forest management was strongly shaped by a series of disastrous fire years during the droughts of the 1930s. Hundreds of millions of hectares of forest were destroyed, whole towns burned to the ground, and hundreds of people died. The Forest Service adopted a policy of aggressive fire control–the stated aim was that every blaze on public land was to be out before 10 A.M. Smokey Bear became the forest mascot and warned us that "only you can prevent forest fires." However, recent studies of fire's ecological role suggest that aggressive fire suppression may have been misguided. Many ecosystems are fire-adapted and require periodic burning for regeneration. And eliminating fire from these forests has allowed woody debris to accumulate, greatly increasing the likelihood of a very big fire (fig. 6.11).

After decades of fire suppression, many forests once characterized by 50–100 mature trees per hectare and open understory now have a thick tangle of up to 2,000 small, spindly, dying saplings. The U.S. Forest Service estimates that 33 million ha (73 million acres), or about 40 percent of all federal forest lands, are at risk of severe fires.

▲ **FIGURE 6.11** By suppressing fires and allowing fuel to accumulate, we make major fires such as this more likely. The safest and most ecologically sound management policy for some forests may be to allow natural or prescribed fires, which don't threaten property or human life, to burn periodically.
Source: John McColgan, Alaska Fires Service/Bureau of Land Management

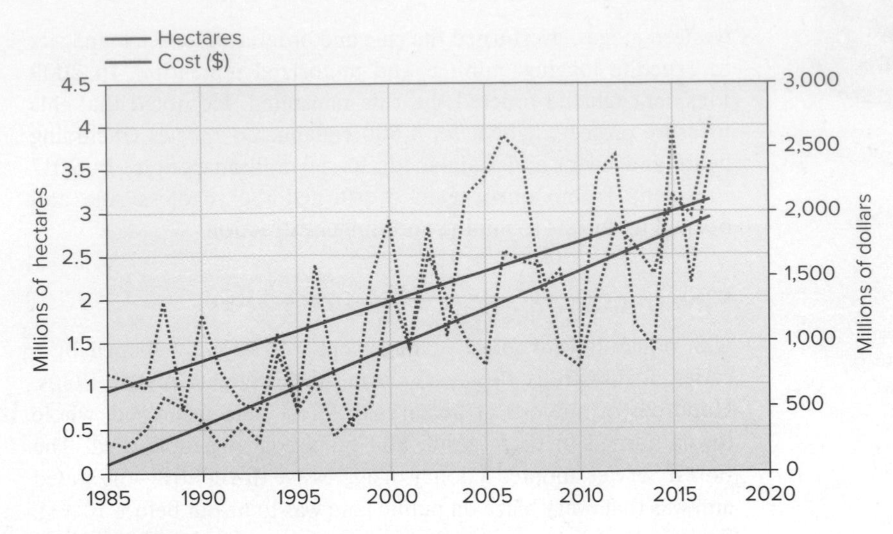

▲ **FIGURE 6.12** Changes in extent of U.S. forest fires and in cost of fire suppression efforts, 1985–2017. Dotted lines show data, and solid lines show trends. By 2025, fire suppression is expected to consume two-thirds of the Forest Service budget, up from 16 percent in 1980. (Costs are not adjusted for inflation.) Source: Data from National Interagency Fire Center.

To make matters worse, Americans increasingly live at the "urban-wildland interface," once-forested areas now scattered with houses, or in remote areas where wildfires are highly likely. With no experience of fire in living memory, many of these residents assume fire is unlikely, but by some estimates 40 million U.S. residents now live in areas with high wildfire risk.

Recent drought in the western United States has heightened fire danger, but awareness is also growing. As long predicted by climatologists, longer, hotter, and drier summers in the West are increasingly producing higher likelihood of fires over time. In much of the American West, the fire season has doubled in length, extending into November or December, and fire severity has increased (fig. 6.12). The fire suppression budget is projected to reach two-thirds of the Forest Service budget in the near future, overwhelming other needs such as road and trail management, ecosystem management, or watershed protection.

Faced with increasing fire risk and a long record of fire suppression, most fire ecologists favor small, prescribed burns. The logging industry prefers to thin forests by selective logging, as well as salvaging burned forests. These activities would require massive extensions of road networks into de facto wilderness and would cost an estimated $12 billion. Whether budgets allow for that remains unclear, but it is certain that the debate will continue.

Ecosystem management is part of forest management

In the 1990s many federal agencies began to shift their policies from a strictly economic focus to **ecosystem management,** which is very similar to the Northwest Forest Plan in its unified, systems approach. Some of its principles include

- Manage across whole landscapes, watersheds, or regions over ecological timescales.
- Depend on scientifically sound, ecologically credible data for decision making.
- Consider human needs and promote sustainable economic development and communities.
- Maintain biological diversity and essential ecosystem processes.
- Utilize cooperative institutional arrangements.
- Generate meaningful stakeholder and public involvement and facilitate collective decision making.
- Adapt management over time, based on conscious experimentation and routine monitoring.

Elements of ecosystem management appear in the *National Report on Sustainable Forests,* which suggests goals for sustainable forest management. Similarly, in 2011, President Obama signed an executive order charging agencies to create a strategic plan for ecosystem-based management of oceans, coasts, and the Great Lakes. Whether these plans for holistic management proceed with subsequent administrations remains to be seen.

6.2 GRASSLANDS

- *Grasslands occupy more than a quarter of the world's land surface and contribute crucially to economies and biodiversity.*
- *Annual grassland disturbance is three times that of tropical forests.*
- *Rotational grazing, as with wild herds, can benefit grasslands.*

After forests, grasslands are among the biomes most heavily used by humans. Prairies, savannas, steppes, open woodlands, and other grasslands occupy about one-quarter of the world's land surface. Much of the U.S. Great Plains and the Prairie Provinces of Canada fall in this category (fig. 6.13). The 3.8 billion ha (12 million mi^2) of pastures and grazing lands in this biome make up about twice the area of all agricultural crops. When you add to this about 4 billion ha of other lands (forest, desert, tundra, marsh, and thorn scrub) used for raising livestock, more than half of all land is used at least occasionally for grazing. At least 3 billion cattle, sheep, goats, camels, buffalo, and other domestic animals on these lands make a valuable contribution to human nutrition. Sustainable pastoralism can increase productivity while maintaining biodiversity in a grassland ecosystem.

Because grasslands, chaparral, and open woodlands are attractive for human occupation, they frequently are converted to cropland, urban areas, or other human-dominated landscapes. Worldwide the rate of grassland disturbance each year is three times that of tropical forest. Although they may appear to be uniform and monotonous to the untrained eye, native prairies can be highly productive and species-rich. According to the U.S. Department of Agriculture, more threatened plant species occur in rangelands than in any other major American biome.

Grazing can be sustainable or damaging

By carefully monitoring the numbers of animals and the condition of the range, ranchers and **pastoralists** (people who live by herding animals) can adjust to variations in rainfall, seasonal plant

▲ **FIGURE 6.13** This short-grass prairie in northern Montana is too dry for trees but nevertheless supports a diverse biological community. ©William P. Cunningham

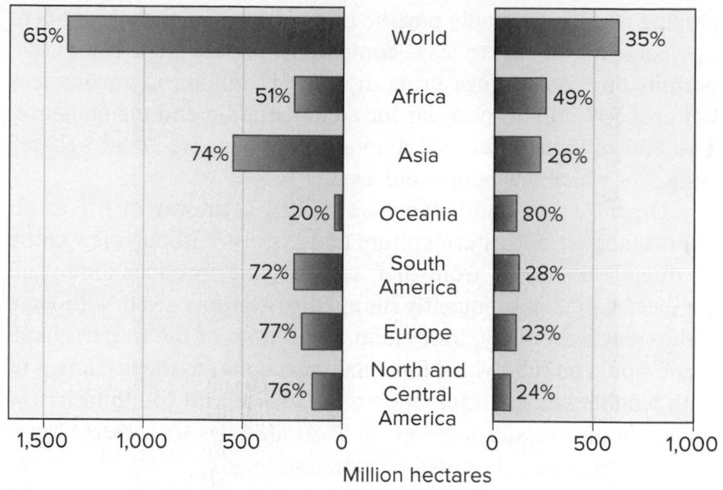

▲ **FIGURE 6.14** Percentage of rangeland soil degradation (red bars) due to overgrazing. Notice that, in Europe, Asia, and the Americas, farming, logging, mining, urbanization, and other causes are responsible for about three-quarters of all soil degradation. In Africa and Oceania, where more grazing occurs and desert and semiarid scrub make up much of the range, grazing damage is higher.

conditions, and the nutritional quality of forage to keep livestock healthy and avoid overusing any particular area. Conscientious management can actually improve the quality of the range.

When land is abused by overgrazing—especially in arid areas—rain runs off quickly before it can soak into the soil to nourish plants or replenish groundwater. Springs and wells dry up. Seeds can't germinate in the dry, overheated soil. The barren ground reflects more of the sun's heat, changing wind patterns, driving away moisture-laden clouds, and leading to further desiccation. This process of conversion of once-fertile land to desert is called **desertification.**

This process is ancient, but in recent years it has been accelerated by expanding populations and the political conditions that lead people to overuse fragile lands. According to the International Soil Reference and Information Centre in the Netherlands, nearly three-quarters of all rangelands in the world show signs of either degraded vegetation or soil erosion. Overgrazing is responsible for about one-third of that degradation (fig. 6.14). The highest percentage of moderate, severe, and extreme land degradation is in Mexico and Central America, while the largest total area is in Asia, where the world's most extensive grasslands occur. It is possible to repair degraded grasslands, but this requires reduced rates of grazing and careful management.

Overgrazing threatens many rangelands

As is the case in many countries, most public grazing lands in the United States are not in good health. Political and economic pressures encourage managers to increase grazing allotments beyond the carrying capacity of the range. Lack of enforcement of existing regulations and limited funds for range improvement have resulted in **overgrazing,** damage to vegetation and soil, including loss of native forage species and erosion. The Natural Resources Defense Council claims that only 30 percent of public rangelands are in fair condition and 55 percent are poor or very poor (fig. 6.15).

Overgrazing has allowed populations of unpalatable or inedible species, such as sage, mesquite, cheatgrass, and cactus, to build up on both public and private rangelands. Wildlife conservation groups regard cattle grazing as the most ubiquitous form of ecosystem degradation and the greatest threat to endangered species in the southwestern United States. They call for a ban on cattle and sheep grazing on all public lands, noting that it provides only 2 percent of the total forage consumed by beef cattle and supports only 2 percent of all livestock producers.

Like federal timber management policy, grazing fees charged for use of public lands often are far below market value and represent an enormous hidden subsidy to western ranchers. Holders of

▲ **FIGURE 6.15** More than half of all publicly owned grazing land in the United States is in poor or very poor condition. Overgrazing and invasive weeds are the biggest problems. ©William H. Mullins/Science Source

grazing permits generally pay the government less than 25 percent of what it would cost to lease comparable private land. The 31,000 permits on federal range bring in only $11 million in grazing fees but cost $47 million per year for administration and maintenance. The $36 million difference amounts to a massive "cow welfare" system of which few people are aware.

On the other hand, ranchers defend their way of life as an important part of western culture and history. Although few cattle go directly to market from their ranches, they produce almost all the beef calves subsequently shipped to feedlots. And without a viable ranch economy, they claim, even more of the western landscape would be subdivided into small ranchettes to the detriment of both wildlife and environmental quality. What do you think? How much should we subsidize extractive industries to preserve rural communities and traditional occupations?

Ranchers are experimenting with new methods

Where a small number of livestock are free to roam a large area, they generally eat the tender, best-tasting grasses and forbs first, leaving the tough, unpalatable species to flourish and gradually dominate the vegetation. In some places, farmers and ranchers find that short-term, intensive grazing helps maintain forage quality. As South African range specialist Allan Savory observed, wild ungulates (hoofed animals), such as wildebeest or zebras in Africa or bison (buffalo) in America, often tend to form dense herds that graze briefly but intensively in a particular location before moving on to the next area. Rest alone doesn't necessarily improve pastures and rangelands. Short-duration, **rotational grazing**–confining animals to a small area for a short time (often only a day or two) before shifting them to a new location–simulates the effects of wild herds (fig. 6.16). Forcing livestock to eat everything equally, to trample the ground thoroughly, and to fertilize heavily with manure before moving on helps keep weeds in check and encourages the growth of more desirable forage species. This approach doesn't work everywhere, however. Many plant communities in the U.S. desert Southwest, for example, apparently evolved in the absence of large, hoofed animals and can't withstand intensive grazing.

▲ **FIGURE 6.16** Intensive, rotational grazing encloses livestock in a small area for a short time (often only 1 day) within a movable electric fence to force them to eat vegetation evenly and fertilize the area heavily. ©William P. Cunningham

▲ **FIGURE 6.17** Red deer (*Cervus elaphus*) are raised in New Zealand for antlers and venison. ©William P. Cunningham

Restoring fire can be as beneficial to grasslands as it is to forests. In some cases ranchers are cooperating with environmental groups in range management and preservation of a ranching economy.

Another approach to ranching in some areas is to raise wild species, such as impala, wildebeest, oryx, or elk (fig. 6.17). These animals forage more efficiently, tolerate harsh climates, often are more pest- and disease-resistant, and fend off predators better than usual domestic livestock. Native species also may have different feeding preferences and needs for water and shelter than cows, goats, or sheep. The African Sahel, for instance, can provide only enough grass to raise about 20 to 30 kg (44 to 66 lb) of beef per hectare. Ranchers can produce three times as much meat with wild native species in the same area because these animals browse on a wider variety of plant materials.

In the United States, ranchers find that elk, American bison, and a variety of African species take less care and supplemental feeding than cattle or sheep and result in a better financial return because their lean meat can bring a better market price than beef or mutton. Some Native American tribes are raising bison both to restore the prairie and to recover cultural heritage.

6.3 PARKS AND PRESERVES

- *Parks and preserves serve ecological, cultural, and recreational purposes.*
- *Protected areas are not all equally protected.*
- *Native people can be important in environmental protection.*

While most forests and grasslands serve useful, or utilitarian, purposes, most societies also set aside some natural areas for aesthetic or recreational purposes. Natural preserves have existed for thousands of years. Ancient Greeks protected sacred groves for religious

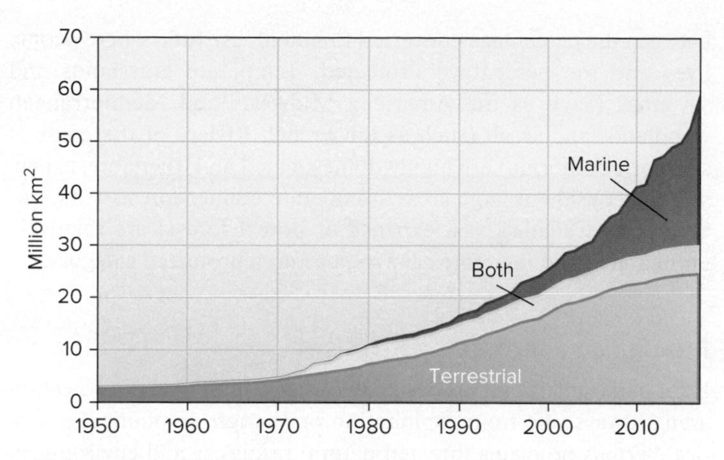

▲ **FIGURE 6.18** Growth of protected areas, 1950–2017. The total area has grown from 3 million to 58 million km^2 in this interval. Source: Data from World Database on Protected Areas

TABLE 6.1 | IUCN Categories of Protected Areas

CATEGORY	ALLOWED HUMAN IMPACT OR INTERVENTION
1. Ecological reserves and wilderness areas	Little or none
2. National parks	Low
3. Natural monuments and archaeological sites	Low to medium
4. Habitat and wildlife management areas	Medium
5. Cultural or scenic landscapes, recreation areas	Medium to high

Source: Data from World Conservation Union, 1990.

purposes. Royal hunting grounds have preserved forests in Europe for centuries. Although these areas were usually reserved for elite classes in society, they have maintained biodiversity and natural landscapes in regions where most lands are heavily used.

The first public parks open to ordinary citizens may have been the tree-sheltered agoras in planned Greek cities. But the idea of providing natural space for recreation, and to preserve natural environments, has really developed in the past 50 years (fig. 6.18). While the first parks were intended mainly for the recreation of growing urban populations, parks have taken on many additional purposes. Today we see our national parks as playgrounds for rest and recreation, as havens for wildlife, as places to experiment with ecological management, and as opportunities to restore ecosystems.

Currently, nearly 17 percent of the land area of the earth is protected in some sort of park, preserve, or wildlife management area. This represents about 25 million km^2 (9.6 million mi^2) in 236,204 different preserves. This is an encouraging environmental success story.

Many countries have created nature preserves

Different levels of protection are found in nature preserves. The World Conservation Union divides protected areas into five categories depending on the intended level of allowed human use (table 6.1). In the most stringent category (ecological reserves and wilderness areas), few human impacts are allowed. In some strict nature preserves, where particularly sensitive wildlife or natural features are located, human entry may be limited only to scientific research groups that visit on rare occasions. Restricting visitor numbers helps protect the area against invasive species or other disruptions. The least restrictive categories (national forests and other natural resource management areas), on the other hand, often allow mining, logging, or grazing.

Venezuela claims to have the highest proportion of its land area protected (66 percent) of any country in the world. About half this land is designated as preserves for indigenous people or for sustainable resource harvesting. With little formal management, however, there is little protection from poaching by hunters, loggers, and illegal gold hunters. It's not uncommon in the developing world to have "paper parks" that exist on maps but have no budget for staff, management, or infrastructure. The United States, by contrast, has only about 22 percent of its land area in protected status, and less than one-third of that amount is in IUCN category I or II (nature reserves, wilderness areas, national parks). The rest is in national forests or wildlife management zones that are designated for multiple use. With hundreds of thousands of state and federal employees, and a high level of public interest and visibility, U.S. public lands are generally well managed. Repeated cuts to conservation and recreation budgets, however put protections at increasing risk.

Brazil has an extensive area in protected status. About 2 million km^2, or 25 percent of the nation's land–mostly in the Amazon basin–is in some protected status. In 2006 the northern Brazilian state of Para, in collaboration with Conservation International (CI) and other nongovernmental organizations, announced the establishment of nine protected areas along the border with Suriname and Guyana. These areas, about half of which was to be strictly protected nature preserves, linked together several existing indigenous areas and nature preserves to create the largest tropical forest reserve in the world. More than 90 percent of the 15-million-ha (58,000 mi^2, or about the size of Illinois) Guyana Shield Corridor is in a pristine natural state. Unfortunately, in 2017, President Temer issued a proclamation abolishing this protected status and opening up the forest to logging and mining.

It is always easiest to preserve lands that are hard for most people to occupy. Greenland has designated 980,000 km^2 of its northern region a national park. Saudi Arabia has established an 825,000 km^2 wildlife management area in its Empty Quarter. These remote areas vary in how much biodiversity they support. Canada's Quttinirpaaq National Park on Ellesmere Island is an example of a preserve with high wilderness values but little biodiversity. Only 800 km (500 miles) from the North Pole, this remote park gets fewer than 100 human visitors per year during its brief, 3-week summer season (fig. 6.19). With little evidence of human occupation, it has abundant solitude and stark beauty but very little wildlife and almost no vegetation.

Figure 6.20 compares the percentages of each major biome in protected status. Not surprisingly, there's an inverse relationship

▲ **FIGURE 6.19** Canada's Quttinirpaaq National Park at the north end of Ellesmere Island has plenty of solitude and pristine landscapes but little biodiversity. ©William P. Cunningham

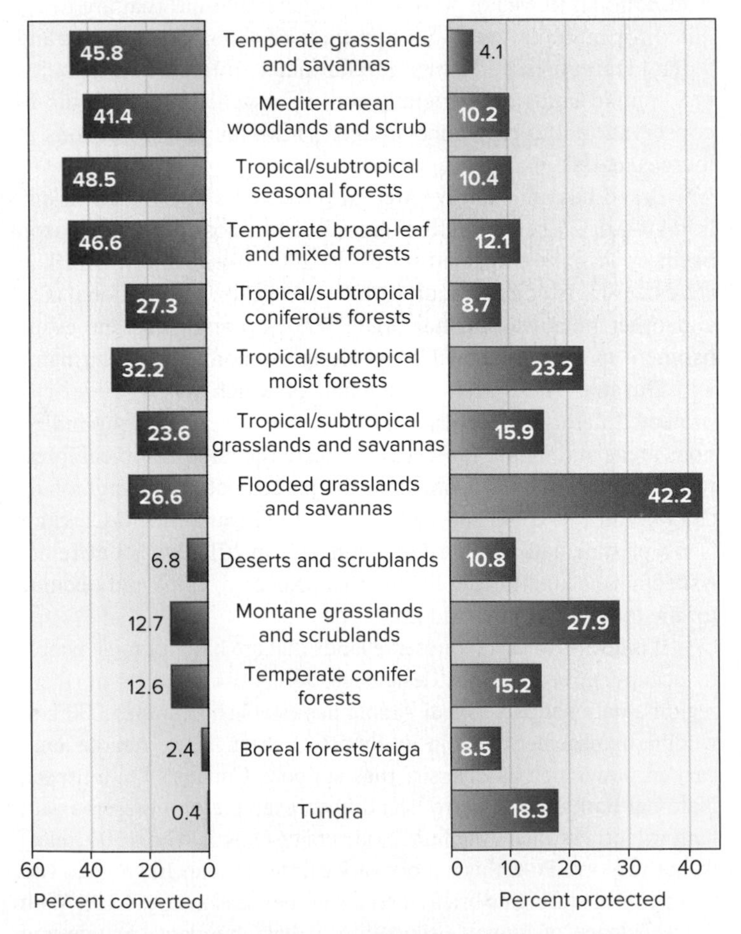

▲ **FIGURE 6.20** With few exceptions, the percentage of each biome converted to human use is roughly inverse to the percentage protected in parks and preserves. Rock and ice, lakes, and Antarctic ecoregions are excluded. Source: World Database on Protected Areas, 2009.

between the percentage converted to human use (and where people live) and the percentage protected. Temperate grasslands and savannas (such as the American Midwest) and Mediterranean woodlands and scrub (such as the French Riviera or the coast of southern California) are highly domesticated and, therefore, expensive to set aside in large areas. Temperate conifer forests (think of Siberia or Canada's vast expanse of boreal forest) are relatively uninhabited and therefore easy to put into a protected category.

Not all preserves are preserved

Even parks and preserves designated with a high level of protection aren't always safe from exploitation or changes in political priorities. Serious problems threaten natural resources and environmental quality in many countries. In Greece, the Pindus National Park is threatened by plans to build a hydroelectric dam in the center of the park, and excessive stock grazing and forestry exploitation in the peripheral zone are causing erosion and loss of wildlife habitat. In Colombia, dam building also threatens the Paramillo National Park. Ecuador's largest nature preserve, Yasuni National Park, which contains one of the world's most megadiverse regions of lowland Amazonian forest, has been opened to oil drilling, while miners and loggers in Peru have invaded portions of Huascaran National Park. In Alaska, oil companies want to drill in the Arctic National Wildlife Refuge (What Do You Think?, p. 145). These are just a few of the many problems faced by parks and preserves around the world. Often countries with the most important biomes lack funds, trained personnel, and experience to manage the areas under their control.

Even in rich countries, such as the United States, some of the crown jewels of the National Park System suffer from overuse and degradation. Yellowstone and Grand Canyon National Parks, for example, have large budgets and are highly regulated but are being "loved to death" because they are so popular. When the U.S. National Park Service was established in 1916, Stephen Mather, the first director, reasoned that he needed to make the parks comfortable and entertaining for tourists as a way of building public support. He created an extensive network of roads in the largest parks, so that visitors could view famous sights from the windows of their automobiles, and he encouraged construction of grand lodges in which guests could stay in luxury.

His plan was successful; the National Park System is cherished and supported by many American citizens. But sometimes entertainment has superseded nature protection. Visitors were allowed—in some cases even encouraged—to feed wildlife. Bears lost their fear of humans and became dependent on an unhealthy diet of garbage and handouts (fig. 6.21). In Yellowstone and Grand Teton National Parks, the elk herd was allowed to grow to 25,000 animals, or about twice the carrying capacity of the habitat. The excess population overgrazed the vegetation to the detriment of many smaller species and the biological community in general. As we discussed earlier in this chapter, 70 years of fire suppression resulted in changes of forest composition and fuel buildup that made huge fires all but inevitable. In Yosemite, you can stay in a world-class hotel, buy a pizza, play video games, do laundry, play golf or tennis, and shop for curios, but you may find it difficult to experience the solitude or enjoy the

What Do YOU THINK?

Wildlife or Oil?

Beyond the rugged Brooks Range in Alaska's far northeastern corner lies one of the world's largest nature preserves, the 7.7-million-ha (19-million-acre) Arctic National Wildlife Refuge (ANWR, pronounced "an-war"). When President Jimmy Carter signed the bill in 1980 to establish the refuge, he said, "Alaska's wilderness areas are truly this country's crown jewels." A small part of the refuge, the 600,000-ha (1.5-million-acre) coastal plains along the Arctic Ocean shore are particularly wildlife-rich. Providing the summer calving grounds of 200,000 caribou, as well as habitat for musk ox, Alaskan brown bears, polar bears, and arctic wolves and millions of migratory birds, this area has been called America's Serengeti.

But oil companies have long wanted to drill on the refuge—particularly in the so-called 1002 study area (named after a section in the founding bill) on the coastal plain, where they believe oil and gas deposits may rival those in the adjacent Prudhoe Bay complex. Not only do corporations hope to find a fossil fuel bonanza in ANWR, but they could ship fuels to market through the already existing Trans-Alaska pipeline, thus extending the life of this multibillion-dollar investment. Over the past 40 years, Republicans in Congress have tried more than 50 times to pass laws to open up the refuge for oil exploration. In 2017 legislators tacked a provision to open the refuge for oil drilling onto an omnibus federal tax bill. Called the Murkowski amendment, the provision ensured that Alaska senator Lisa Murkowski would support the tax bill.

Ecologists warn that oil development could irreversibly damage coastal and marine habitat, potentially decimating wildlife and bird populations. Oil and chemical spills are almost inevitable with drilling, seismic exploration destroys delicate frozen tundra, and traffic drives wildlife away from breeding grounds. Oil and gas development could threaten the existence of the porcupine caribou herd, which migrates annually from its winter range in Canada to calving grounds on the coastal plain. It also jeopardizes the Gwich'in people, an indigenous nation that has relied on caribou for subsistence for centuries and whose traditional lands straddle the U.S.-Canada border north of the Yukon River. Caribou mean more to the Gwich'in than just food. Their name translates to "caribou people," and they claim a spiritual and cultural connection with the herd and the land that dates back at least 20,000 years.

There are debates about how much oil ANWR may hold. Oil companies claim that it could be 9 billion barrels. Senator Murkowski said that oil revenues could raise billions of dollars to help pay for tax cuts passed by Republicans while "creating jobs and strengthening our nation's long-term energy security." On the other hand, during the 1995 ANWR debates, the U.S. Department of the Interior estimated that there was only a 19 percent chance of finding any oil in ANWR, which would amount to less than 200 days' worth of oil for the country. President Bill Clinton vetoed the effort to allow drilling at that time.

Climate scientists also point out that besides the effects of drilling and spills on wildlife and native people, pumping more oil from ANWR is exactly what the world doesn't need. Fossil fuel burning is chiefly responsible for ongoing melting of the Arctic, erosion of habitat as sea levels rise, and increasing storm intensity. Midlatitude regions, the destination market for this oil and gas, are under threat of long-term, Dust Bowl–like drought because of climate warming. Whatever oil is in ANWR, they say, should stay in the ground.

Some energy experts calculate that if everyone in the United States who drives alone to school or work were to share the ride with one other person, we would save more oil in 6 months than we would ever pump out of ANWR, even at the most optimistic estimates. What do you think, will we destroy one of the world's last great wildernesses for the profits of multinational oil companies? How would you resolve this debate?

natural beauty extolled by John Muir as a prime reason for creating the park.

In many of the most famous parks, traffic congestion and crowds of people stress park resources and detract from the experience of unspoiled nature (fig. 6.22). Some national parks, such as Yosemite and Zion, have banned private automobiles from the most congested areas. Visitors park in remote lots and take clean, quiet electric or natural-gas-burning buses to popular sites. Other parks are considering limits on the number of visitors admitted each day. How would you feel about a lottery system that might allow you to visit some famous parks only once in your lifetime but to have an uncrowded, peaceful experience on your one allowed visit? Or would you prefer to be able to visit whenever you wish, even if it meant fighting crowds and congestion?

FIGURE 6.21 Wild animals have always been one of the main attractions in national parks. Many people lose all common sense when interacting with big, dangerous animals. This is not a petting zoo.

▲ **FIGURE 6.22** Thousands of people wait for an eruption of Old Faithful geyser in Yellowstone National Park. Can you find the ranger who's giving a geology lecture? ©William P. Cunningham

Originally, the great wilderness parks of Canada and the United States were distant from development and isolated from most human impacts. This has changed in many cases. Forests are clear-cut right up to some park boundaries. Mine drainage contaminates streams and groundwater. At least 13 U.S. National Monuments are open to oil and gas drilling, including Texas's Padre Island, the only U.S. breeding ground for endangered Kemps Ridley sea turtles. Even in the dry desert air of the Grand Canyon, where visibility was once up to 150 km, it's often too smoggy now to see across the canyon due to air pollution from power plants just outside the park. Snowmobiles and off-road vehicles create pollution and noise and cause erosion while disrupting wildlife in many parks (fig. 6.23).

Chronically underfunded, the U.S. National Park System now has a maintenance backlog estimated to be at least $5 billion. During election campaigns, politicians from both major political parties vow to repair park facilities, but then find other uses for public funds once in office. Ironically, a recent study found that, on average, parks generate $4 in user fees for every $1 they receive in federal subsidies. In other words, they more than pay their own way and should have a healthy surplus if allowed to retain all the money they generate.

▲ **FIGURE 6.23** Off-road vehicles cause severe, long-lasting environmental damage when driven in sensitive areas. ©Carl Lyttle/Getty Images

In recent years, the U.S. National Park System has begun to emphasize nature protection and environmental education over entertainment. This new agenda is being adopted by other countries as well. The IUCN has developed a **world conservation strategy** for protecting natural resources that includes the following three objectives: (1) to maintain essential ecological processes and life-support systems (such as soil regeneration and protection, nutrient recycling, and water purification) on which human survival and development depend; (2) to preserve genetic diversity essential for breeding programs to improve cultivated plants and domestic animals; and (3) to ensure that any utilization of wild species and ecosystems is sustainable.

Marine ecosystems need greater protection

As ocean fish stocks become increasingly depleted globally, biologists are calling for protected areas that can shelter marine organisms from destructive harvest methods. Although about 14 percent of land area is in some conservation status, only about 5 percent of near-shore marine biomes are protected. Limiting the amount and kind of fishing in marine reserves can quickly replenish fish stocks in surrounding areas. In a study of 100 marine refuges around the world, researchers found that, on average, the number of organisms inside no-take preserves was twice as high as surrounding areas where fishing was allowed. In addition, the biomass of organisms was three times as great and individual animals were, on average, 30 percent larger inside the refuge compared to outside. Recent research has shown that closing reserves to fishing even for a few months can have beneficial results in restoring marine populations. The size necessary for a safe haven to protect flora and fauna depends on the species involved, but some marine biologists call on nations to protect at least 20 percent of their nearshore territory as marine refuges.

Coral reefs are among the most threatened marine ecosystems in the world. Surveys show that, worldwide, living coral reefs have declined by about half in the past century, and 90 percent of all reefs face threats from rising sea temperatures, destructive fishing methods, coral mining, sediment runoff, and other human disturbance. In many ways, coral reefs are the old-growth rainforests of the ocean (fig. 6.24). Biologically rich, these sensitive communities can take a century or more to recover from damage. If current trends continue, some researchers predict that in 50 years there will be no viable coral reefs anywhere in the world.

What can be done to reverse this trend? Some countries are establishing large marine reserves specifically to protect coral reefs. Australia's Great Barrier Reef is the largest coral reef complex in the world. Most of it is protected in a marine reserve, but, unfortunately, coral bleaching (caused by high seawater temperatures), pollution, and invasive species are killing much of the reef.

The first major marine protected area in the United States, Papahānaumokuākea Marine National Monument, was established by President George W. Bush in 2006 to protect fish, sea birds, turtles, and other wildlife. Expanded by Barack Obama in 2014, Papahānaumokuākea is the largest contiguous fully protected

▲ FIGURE 6.24 Coral reefs are among both the most biologically rich and most endangered ecosystems in the world. Marine reserves are being established in many places to preserve and protect these irreplaceable resources. ©Comstock Images/PictureQuest

conservation area in U.S. territory and one of the largest marine conservation areas in the world. Extending about 2,300 km (1,440 mi) northwest from Hawaii, the refuge encompasses 1.5 million km^2 (580,000 mi^2) of the Pacific Ocean—an area larger than all the country's national parks combined.

The Pacific Remote Islands National Monument is a group of small islands and coral reefs stretching from Rose Atoll near Somoa to Wake Island, 2,400 km east of Guam. Altogether, these preserves cover about 2 million km^2, including some of the healthiest coral reef systems in the world. They're closed to commercial fishing but allow some sport fishing and sustainable harvest of marine life.

Commercial fishing companies complained, however, about being denied access to these rich fishing grounds. In 2017, despite abundant evidence that marine preserves increase fish populations in surrounding oceans, President Trump reduced the size of both of these national monuments and opened them to commercial fishing. This was part of his campaign to reduce protected areas throughout the United States (see Science and Citizenship, p. 148).

Conservation and economic development can work together

Many of the most biologically rich communities in the world are in developing countries, especially in the tropics. These countries are the guardians of biological resources important to all of us. Unfortunately, where political and economic systems fail to provide residents with land, jobs, food, and other necessities of life, people do whatever is necessary to meet their own needs. Immediate survival takes precedence over long-term environmental goals. Clearly, the struggle to save species and ecosystems can't be divorced from the broader struggle to meet human needs.

People in some developing countries are beginning to realize that their biological resources may be their most valuable assets, and that their preservation is vital for sustainable development. **Ecotourism** (tourism that is ecologically and socially sustainable) can be more beneficial in many places over the long term than extractive industries, such as logging and mining. See What Can You Do? below for some ways to ensure that your vacations are ecologically responsible.

Native people can play important roles in nature protection

The American ideal of wilderness parks untouched by humans is unrealistic in many parts of the world. As we mentioned earlier, some biological communities are so fragile that human intrusions have to be strictly limited to protect delicate natural features or particularly sensitive wildlife. In many important biomes, however, aboriginal people have been present for thousands of years and have a legitimate right to pursue traditional ways of life. Furthermore, many of the approximately 5,000 indigenous or native cultures that remain today possess ecological knowledge about their ancestral homelands that can be valuable in ecosystem management. According to author Alan Durning, "encoded in indigenous languages, customs, and practices may be as much understanding of nature as is stored in the libraries of modern science."

What Can YOU DO?

Being a Responsible Ecotourist

1. *Pretrip preparation.* Learn about the history, geography, ecology, and culture of the area you will visit. Understand the do's and don'ts that will keep you from violating local customs and sensibilities.
2. *Environmental impact.* Stay on designated trails and camp in established sites, if available. Take only photographs and memories and leave only goodwill wherever you go.
3. *Resource impact.* Minimize your use of scarce fuels, food, and water resources. Do you know where your wastes and garbage go?
4. *Cultural impact.* Respect the privacy and dignity of those you meet and try to understand how you would feel in their place. Don't take photos without asking first. Be considerate of religious and cultural sites and practices. Be as aware of cultural pollution as you are of environmental pollution.
5. *Wildlife impact.* Don't harass wildlife or disturb plant life. Modern cameras make it possible to get good photos from a respectful, safe distance. Don't buy ivory, tortoise shell, animal skins, feathers, or other products taken from endangered species.
6. *Environmental benefit.* Is your trip strictly for pleasure, or will it contribute to protecting the local environment? Can you combine ecotourism with work on cleanup campaigns or delivery of educational materials or equipment to local schools or nature clubs?
7. *Advocacy and education.* Get involved in letter writing, lobbying, or educational campaigns to help protect the lands and cultures you have visited. Give talks at schools or to local clubs after you get home to inform your friends and neighbors about what you have learned.

CITIZENSHIP Science Monuments Under Attack

Land preservation has long been a focus of policy wars in the United States. A favorite ground for skirmishes has been the Antiquities Act of 1906, which authorizes the president to designate historic landmarks, structures, and "other objects of historic or scientific interest" as national monuments. President Theodore Roosevelt was the first to use this authority, establishing 17 monuments, many of which, such as Arizona's Grand Canyon, California's Lassen Peak, and Washington State's Olympic Mountains, have become flagship national parks.

The Antiquities Act frequently protects sites of archaeological and scientific value, as well as recreational resources. The original aim of this legislation was to protect prehistoric Indian ruins and antiquities on western federal lands that were being vandalized by visitors. That use has continued today with the recent designation of Bears Ears National Monument and the Grand Staircase–Escalante Monument in 2016. This area contains at least 100,000 archaeological sites associated with early Clovis, Fremont, and Anasazi cultures. The land was long occupied by native people, who consider it sacred. Protecting it was the culmination of years of campaigning by the neighboring Navajo, Hopi, Zuni, Ute, and Ute Mountain tribes.

▲ Bluffs and pinnacles of Navajo sandstone stand along the Escalante River near its junction with the Colorado River in Glenn Canyon. ©William P. Cunningham

The Antiquities Act has been used to protect a wide variety of natural wonders and resources. President Jimmy Carter set aside 56 million acres in Alaska in 17 national monuments, some of which were later promoted to national parks. George W. Bush established the world's first major marine preserve in Hawaii, protecting reproductive areas for fish and other marine life, as well as pristine environments for scientific research and ancient cultural sites. Barack Obama created or expanded 34 national monuments, bringing the total to 129 monuments encompassing 64,750 km^2 (25,000 mi^2) of land plus 3.1 million km^2 (1.2 million mi^2) of marine reserves.

Not everyone appreciates these monuments. Congress has tried more than 11 times over the past century to abolish national monuments, motivated by mining interests or by those who resent Washington's jurisdiction over western lands. These opponents found a new champion in President Trump. Among his first actions on taking office, Trump directed his secretary of the interior, Ryan Zinke, to review the national monuments and marine reserves established by Presidents Clinton and Obama and to recommend which ones could be shrunk or abolished.

The first and most draconian reductions proclaimed by Trump were to Bears Ears and the Grand Staircase–Escalante National Monuments in Utah. This rugged, rolling landscape, dissected by spectacular canyons, lies just west of the Glen Canyon Recreation Area in central Utah. It originally encompassed 1.9 million acres (0.77 million ha) of rarely visited wilderness. It's a hiker's paradise with a maze of narrow slot canyons hiding ephemeral streams and temporary pools in the red slickrock. But mining companies are eager to access mineral rights there. In the Grand Staircase–Escalante, the Kaiparowits Plateau is thought to contain 4 billion tons of coal. Bears Ears has uranium that could feed a now-closed mill just outside the monument.

Trump and Zinke eliminated 85 percent of Bears Ears and half of the Grand Staircase–Escalante. In announcing his plan, Trump said, "We're returning this land to its rightful owners . . . the local residents." Critics responded that the rightful owners are all the citizens of the United States, because it is federal land.

Native American tribes have another perspective. Russell Begaye, president of the Navajo Nation, said, "If Trump wants to return the land to its rightful owners, he should give it back to indigenous people. This was our land for thousands of years before it was taken by the U.S. Government."

The last word hasn't been heard about these or other purged monuments. Both native tribes and environmental groups have vowed to sue to prevent the destruction of treasured public lands. This case represents one of the most high-profile cases in which environmental groups and Native Americans have joined forces to protect natural resources. It represents a welcome expansion of diversity in public discourse about our environment, as well as growing empowerment of native voices.

These policy debates may not affect most of us immediately, but ultimately this is a question of how we think about and manage public lands and resources. Who should make decisions about these lands? Voters? Industry advocates? If locals have a profit incentive in designating uses of public land, does that increase or decrease their right to determine use of the resources? As scientists and citizens, these are questions that concern all of us.

▲ **FIGURE 6.25** Some parks take draconian measures to expel residents and prohibit trespassing. How can we reconcile the rights of local or indigenous people with the need to protect nature? ©William P. Cunningham

Land protection has sometimes displaced people, sometimes aggressively, and this is highly contentious (fig. 6.25). Historically, colonial governments often ignored the rights of tribal groups and evicted them from newly established parks. Sometimes park authorities have evicted local residents in the context of combatting poachers, often well-armed global traffickers in elephant ivory or tiger parts. Many parks struggle with steadily encroaching human populations, which gradually degrade resources. Kenya, for example, has sought to evict residents of the Mau Forest, whose water catchments supply some of the country's critical water resources. Decades of illicit village and farm expansion have dramatically reduced the forest, threatening nationally critical resources. But some residents have now lived there for a generation or more, and they understandably resist eviction. Observers also point out that while poor settlers are evicted, large-scale deforestation and land clearing continue in other areas. Can rights of residents be protected while also protecting parks from expanding human impacts?

Increasingly, conservationists recognize that finding ways to integrate local human needs with those of nature is essential for successful conservation. In 1986, UNESCO (United Nations Educational, Scientific, and Cultural Organization) initiated its **Man and Biosphere (MAB) program,** which encourages the designation of **biosphere reserves,** protected areas divided into zones with different purposes. Critical ecosystem functions and endangered wildlife are protected in a central core region, where limited scientific study is the only human access allowed. Ecotourism and research facilities are located in a relatively pristine buffer zone around the core, while sustainable resource harvesting and permanent habitation are allowed in multiple-use peripheral regions (fig. 6.26).

A well-established example of a biosphere reserve is Mexico's 545,000-ha (2,100-mi^2) Sian Ka'an Reserve on the Tulum Coast of the Yucatán. The core area includes 97 percent of the reserve, with coral reefs, bays, wetlands, and lowland tropical forest. More than 335 bird species have been observed within the reserve, along with endangered manatees, five types of jungle cats, spider and howler monkeys, and four species of increasingly rare sea turtles. Approximately 25,000 people reside in communities and the countryside around Sian Ka'an. In addition to tourism, the economic base of the area includes lobster fishing, small-scale farming, and coconut cultivation.

The Amigos de Sian Ka'an, a local community organization, played a central role in establishing the reserve and is working to protect natural resources while improving living standards for local people. New intensive farming techniques and sustainable harvesting of forest products enable residents to make a living without harming their ecological base. Better lobster harvesting techniques developed at the reserve have improved the catch without depleting native stocks. Local people now see the reserve as a benefit rather than an imposition from the outside. Similar success stories from many parts of the world show how we can support local people and recognize indigenous rights while still protecting important environmental features.

Species survival can depend on preserve size and shape

Many natural parks and preserves are increasingly isolated, remnant fragments of ecosystems that once extended over large areas. As park ecosystems are shrinking, however, they are also becoming more and more important for maintaining biological diversity. Principles of landscape design and landscape structure become important in managing and restoring these shrinking islands of habitat.

For years, conservation biologists have disputed whether it is better to have a *s*ingle *l*arge *o*r *s*everal *s*mall reserves (the SLOSS debate). Ideally, a reserve should be large enough to support viable populations of endangered species, keep ecosystems intact, and isolate critical core areas from damaging external forces. For some species with small territories, several small, isolated refuges can

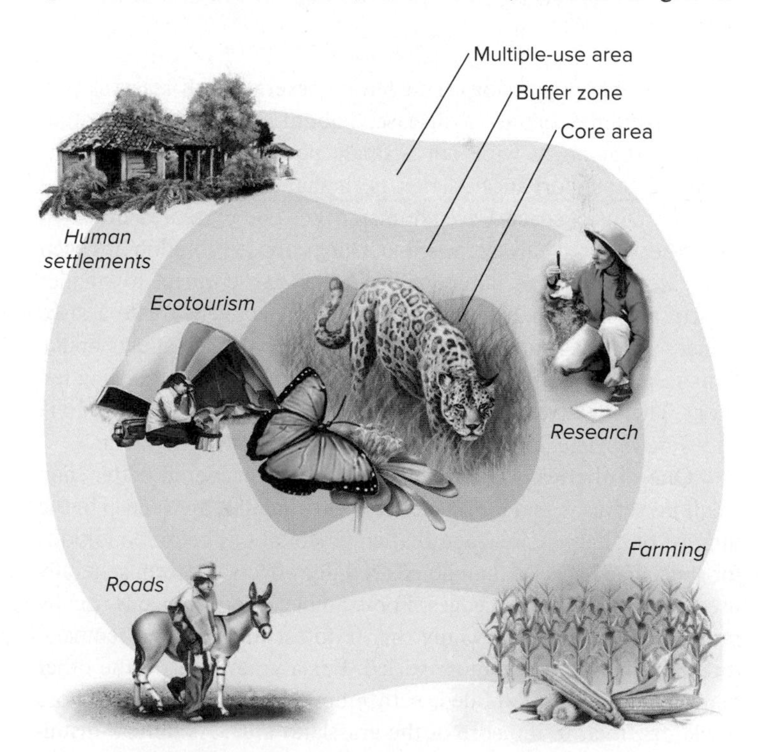

▲ **FIGURE 6.26** A model biosphere reserve. Traditional parks and wildlife refuges have well-defined boundaries to keep wildlife in and people out. Biosphere reserves, by contrast, recognize the need for people to have access to resources. Critical ecosystem is preserved in the core. Research and tourism are allowed in the buffer zone, while sustainable resource harvesting and permanent habitations are situated in the multiple-use area around the perimeter.

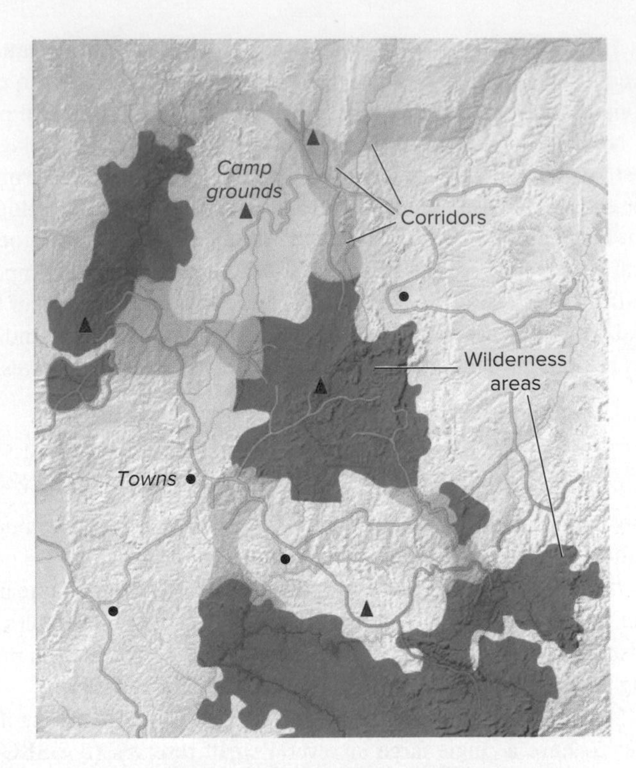

▲ **FIGURE 6.27** Corridors serve as routes of migration, linking isolated populations of plants and animals in scattered nature preserves. Although individual preserves may be too small to sustain viable populations, connecting them through river valleys and coastal corridors can facilitate interbreeding and provide an escape route if local conditions become unfavorable.

support viable populations, and having several small reserves provides insurance against a disease, habitat destruction, or other calamity that might wipe out a single population. But small preserves can't support species such as elephants or tigers, which need large amounts of space. Given human needs and pressures, however, big preserves aren't always possible. One proposed solution has been to create **corridors** of natural habitat that can connect to smaller habitat areas (fig. 6.27). Corridors could effectively create a large preserve from several small ones. Corridors could also allow populations to maintain genetic diversity or expand into new breeding territory. The effectiveness of corridors probably depends on how long and wide they are, and on how readily a species will use them.

One of the reasons large preserves are considered better than small preserves is that they have more **core habitat,** areas deep in the interior of a habitat area, and that core habitat has better conditions for specialized species than do edges. **Edge effects** is a term generally used to describe habitat edges: For example, a forest edge is usually more open, bright, and windy than a forest interior, and temperatures and humidity are more varied. For a grassland, on the other hand, edges may be wooded, with more shade, and perhaps more predators, than in the core of the grassland area. As human disturbance fragments an ecosystem, habitat is broken into increasingly isolated islands, with less core and more edge. Small, isolated fragments of habitat often support fewer species, especially fewer rare species, than do extensive, uninterrupted ecosystems. The size and isolation of a wildlife preserve, then, may be critical to the survival of rare species.

▲ **FIGURE 6.28** How small can a nature preserve be? In an ambitious research project, scientists in the Brazilian rainforest are carefully tracking wildlife in plots of various sizes, either connected to existing forests or surrounded by clear-cuts. As you might expect, the largest and most highly specialized species are the first to disappear.
Courtesy of R.O. Bierregaard

A dramatic experiment in reserve size, shape, and isolation is being carried out in the Brazilian rainforest. In a project funded by the World Wildlife Fund and the Smithsonian Institution, loggers left 23 test sites when they clear-cut a forest. Test sites range from 1 ha (2.47 acres) to 10,000 ha. Clear-cuts surround some, and newly created pasture surrounds others (fig. 6.28); others remain connected to the surrounding forest. Selected species are regularly inventoried to monitor their survival after disturbance. As expected, some species disappear very quickly, especially from small areas. Sun-loving species flourish in the newly created forest edges, but deep-forest, shade-loving species disappear, particularly when the size or shape of a reserve reduces the availability of core habitat. This experiment demonstrates the importance of maintaining core habitat in preserves.

CONCLUSION

Forests and grasslands cover nearly 60 percent of global land area. The vast majority of humans live in these biomes, and we obtain many valuable materials from them. And yet these biomes also are the source of much of the world's biodiversity on which we depend for life-supporting ecological services. How we can live sustainably on our natural resources while preserving enough nature so those resources can be replenished represents one of the most important questions in environmental science.

There is some good news in our search for a balance between exploitation and preservation. Although deforestation and land degradation are continuing at unacceptable rates—particularly in some developing countries—many places are more densely forested now than they were two centuries ago. Protection of indigenous rights and biodiversity always have to struggle with forces that want to exploit natural resources. Overall, at least 17 percent of the earth's land area is now in some sort of protected status, although the level of protection in these preserves varies.

We haven't settled the debate between focusing on individual endangered species and setting aside representative samples of habitat, but pursuing both strategies seems to be working. Protecting charismatic umbrella organisms, such as orangutans in Indonesia or caribou in Alaska, can result in preservation of innumerable unseen species. At the same time, protecting whole landscapes for aesthetic or recreational purposes can also achieve the same end.

PRACTICE QUIZ

1. What do we mean by *closed-canopy* forest and *old-growth* forest?
2. What continent is experiencing the greatest forest losses?
3. What is REDD, and how might it work?
4. Why is fire suppression a controversial strategy?
5. What portion of the U.S. public rangelands are in poor or very poor condition due to overgrazing? Why do some groups say grazing fees amount to a "hidden subsidy"?
6. What is *rotational grazing,* and how does it mimic natural processes?
7. How do the size and design of nature preserves influence their effectiveness? What do landscape ecologists mean by *interior habitat* and *edge effects*?
8. What percentage of the earth's land area has some sort of protected status? How has the amount of protected areas changed globally?
9. What is *ecotourism,* and why is it important?
10. What is a *biosphere reserve,* and how does it differ from a wilderness area or wildlife preserve?

CRITICAL THINKING AND DISCUSSION

Apply the principles you have learned in this chapter to discuss these questions with other students.

1. Conservationists argue that watershed protection and other ecological functions of forests are more economically valuable than timber. Timber companies argue that continued production supports stable jobs and local economies. If you were a judge attempting to decide which group was right, what evidence would you need on both sides? How would you gather this evidence?
2. Divide your class into a ranching group, a conservation group, and a suburban home-builders group, and debate the merits of subsidized grazing in the American West. What is the best use of the land? What landscapes are most desirable? Why? How do you propose to maintain these landscapes?
3. Calculating forest area and forest losses is complicated by the difficulty of defining exactly what constitutes a forest. Outline a definition for what counts as forest in your area, in terms of size, density, height, or other characteristics. Compare your definition to those of your colleagues. Is it easy to agree? Would your definition change if you lived in a different region?
4. There is considerable uncertainty about the extent of degradation on grazing lands. Suppose you were a range management scientist, and it was your job to evaluate degradation for the state of Montana. What data would you need? With an infinite budget, how would you gather the data you need? How would you proceed if you had a very small budget?
5. Why do you suppose dry tropical forest and tundra are well represented in protected areas, while grasslands and wetlands are protected relatively rarely? Consider social, cultural, geographic, and economic reasons in your answer.
6. Oil and gas companies want to drill in several parks, monuments, and wildlife refuges. Do you think this should be allowed? Why or why not? Under what conditions would drilling be allowable?

DATA ANALYSIS Detecting Edge Effects

Edge effects are a fundamental consideration in nature preserves. We usually expect to find dramatic edge effects in pristine habitat with many specialized species. But you may be able to find interior-edge differences on your own college campus, or in a park or other unbuilt area near you. Here are three testable questions you can examine using your own local patch of habitat: (1) Can an edge effect be detected or not? (2) Which species will indicate the difference between edge and interior conditions? (3) At what distance can you detect a difference between edge and interior conditions? Go to Connect to find a field exercise that you can do to form and test a hypothesis regarding edge effects in your own area.

CHAPTER

7 Food and Agriculture

LEARNING OUTCOMES

After studying this chapter, you should be able to answer the following questions:

- How many people are chronically hungry, and why does hunger persist in a world of surpluses?
- What are some health risks of undernourishment, poor diet, and overeating?
- What are our primary food crops?
- Describe five components of soil.
- What was the Green Revolution?
- What are GMOs, and what traits are most commonly introduced with GMOs?
- Describe some environmental costs of farming and ways we can minimize these costs.

Illinois farm fields produce the world's largest crop of corn and soybeans. Increasingly complex pesticide mixes are necessary to support this kind of production.

CASE STUDY

A New Pesticide Cocktail

Every spring the farmers in McLean County, Illinois, set out to grow some of the biggest corn and soy crops the world has ever seen. The county with the largest crops of both soybeans and corn in the state with the largest soy crop and the second-largest corn crop (after Iowa), McLean County is 95 percent cropland, and the two crops make up nearly 90 percent of crop acres. These fields are the economic engine of the U.S. agricultural economy, the world's largest producer and exporter of corn and soybeans.

To an ecologist, this is a very simplified ecosystem. It has just two dominant producer species, soy and corn, and (ideally) one consumer, the farmer. But an ecologist sees a simplified system as a world of niches waiting to be filled. Consumers, especially insect pests, are eager to move in to exploit the burgeoning productivity of crop fields. Plant competitors, especially fast-growing weeds like pigweed and ragweed, are poised to invade. To protect their crops, farmers depend on chemical pesticides. Scores of these have been developed—some broadly toxic, some narrowly targeted to certain pests.

A new stage in this eternal struggle was the introduction of seeds genetically modified to tolerate herbicides. Since the 1990s, joint chemical- and seed-producing conglomerates, especially Bayer-Monsanto and Dow AgroSciences, have produced "systems" of seeds tailored to tolerate a specific herbicide. These allow farm operators to spray herbicides liberally without harming crops. Seed-chemical systems have driven up profits and production at unprecedented rates. They have also driven up costs: Chemicals, seeds, and licenses to use them are expensive, and small producers, unable to compete, have largely disappeared. Since 2010, 94 percent of U.S. soy acres have been herbicide tolerant. Increasingly, these crops have "stacked traits," or combined tolerance of multiple pesticides, to compensate for the rapid evolution of weeds and pests.

As an ecologist would point out, constant disturbance promotes evolution in a species. Constant exposure to glyphosate, the single most abundantly used herbicide in the United States (often known by its trade name, Roundup®), or to atrazine, glufosinate, or other ubiquitous chemicals, destroys most members of a weed population. But the few tolerant individuals that survive will proliferate, free of competition. Tolerant-trait weeds and pests come to dominate the population, making a pesticide ineffectual. For most pesticides, it takes about 5 years before resistant weeds or insects being to appear.

The rapid evolution of pests and weeds, and the resulting need for constantly better pesticides, is known as the pesticide treadmill. After the 1996 introduction of Monsanto's "Roundup®-ready" seeds, it took less than 5 years for resistant pigweeds to appear. By 2017, at least 38 resistant weed species were known, affecting about 40 percent of the 69 million hectares (170 million acres) planted in corn, soybeans, and cotton in the United States (fig. 7.1).

As herbicides become ineffectual, complex "cocktails" of chemicals become necessary. Starting in 2018, the EPA has allowed farmers in Illinois and other major farming states to use Enlist Duo®, a combination of glyphosate and 2,4-D, produced by the chemical

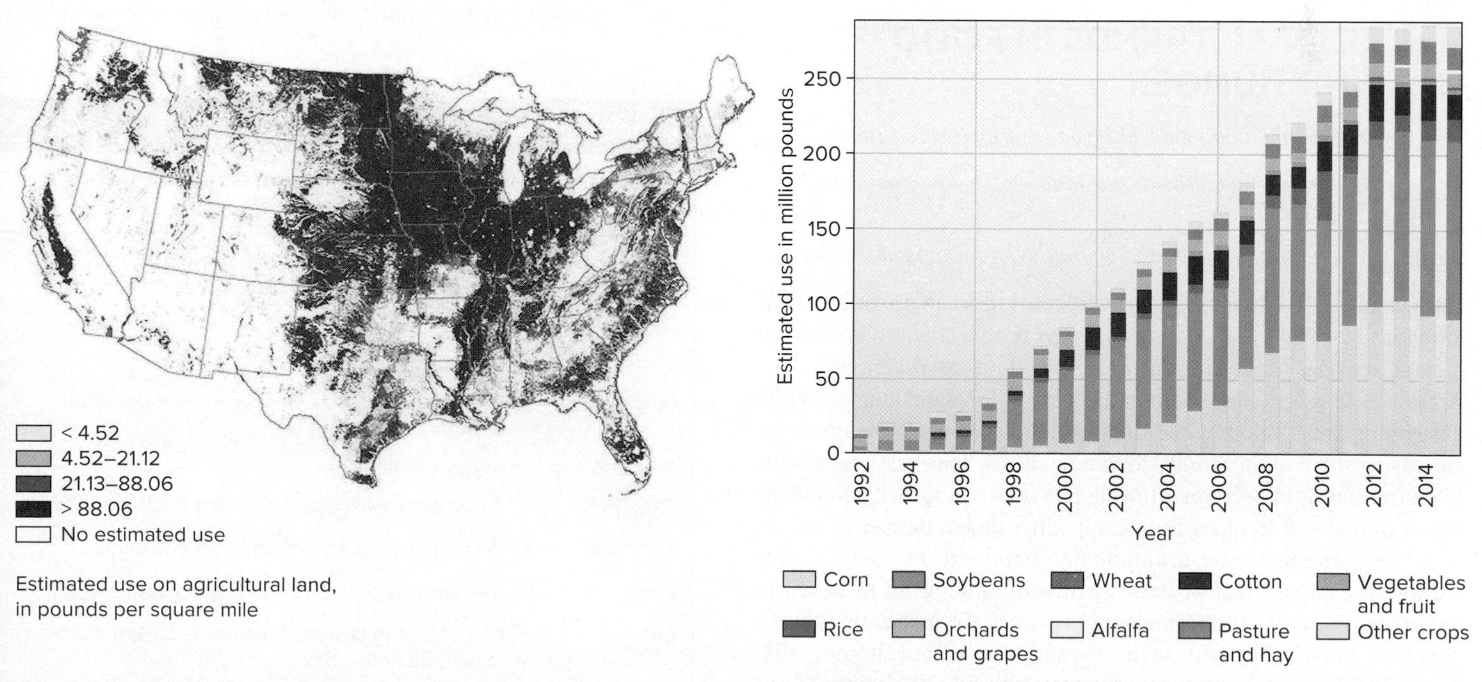

▲ **FIGURE 7.1** Distribution and growth of glyphosate (Roundup) applications. Crops genetically modified to tolerate this herbicide were introduced in the early 1990s. Source: National Water Quality Assessment (NAWQA) project.

(continued)

CASE STUDY continued

manufacturer Dow Agrosciences. Both glyphosate and 2,4-D are probably carcinogenic to humans, according to the World Health Organization (WHO). Neither compound is a general biocide like DDT, but glyphosate is by far the dominant herbicide, used in major crop-producing countries on dozens of crops. And 2,4-D, is one of the oldest and most widely used pesticides. Used since 1945, 2,4-D was an ingredient of the defoliant Agent Orange used in the Viet Nam War. In 2018, a California jury awarded $289 million in a lawsuit charging that Monsanto had deliberately misled the public on the health risks of glyphosate.

The EPA's approval of Enlist Duo® was controversial because nobody knows the health or environmental consequences of accelerated and combined chemical use across the landscape. And nobody knows how long it will be effective in the race against weed evolution.

As more of us move to cities, it becomes harder for us to see and understand the complexities, and the implications, of our agricultural production systems. This is how we produce the food and the trade products we depend on. Soy and corn support the inexpensive meat products we are accustomed to eating. What are the long-term effects on our ecosystems and waterways? What are the effects on farmers and farming communities? Those are also important questions. In this chapter we examine agricultural production, food, and the ways we use these resources. ■

We can't solve problems by using the same kind of thinking we used when we created them.

—ALBERT EINSTEIN

7.1 GLOBAL TRENDS IN FOOD AND HUNGER

- *We produce abundant food, but food security is still a global issue.*
- *Average calorie and protein consumption has increased steadily in recent decades.*
- *Famines are usually triggered by conflict or misguided policies.*

Ensuring that the earth's soil and water resources can provide enough food for the world's population has always been a concern in environmental science. Modern industrial agriculture, as practiced in the American Midwest, has rewritten the story of food and hunger. While developing areas still practice small-scale subsistence agriculture, an increasing share of the world's food now comes from vast operations, often thousands of hectares, growing one or two crops with abundant inputs of fuel and fertilizer for a competitive global market.

These changes have dramatically increased production, providing affordable meat protein worldwide, including in developing areas. Food production has increased so dramatically that we now use corn, soy, and sugar to run our cars (chapter 13). According to the International Monetary Fund, 2005 global food costs (in inflation-adjusted dollars) were the lowest ever recorded, less than one-quarter of the cost in the mid-1970s. Because of overproduction in the United States and Europe, we pay farmers billions of dollars each year to take land out of production.

Now we have new questions in environmental science and food production: What are the environmental costs of intensified production? Which innovations are sustainable? Can we curb the growing health impacts of calorie-rich diets? And can we improve distribution so as to eliminate hunger that persists in so many impoverished areas?

Benchmark Data

Among the ideas and values in this chapter, these are a few worth remembering.

2,200 kcal	Average human dietary energy requrements, according to the World Health Organization
> 3,500 kcal	Average available per person in North America
40–60 g	Daily protein requirements
60 g	Average protein consumption in least-developed countries
110 g	Average in North America
70%; 5%	Percentage overweight; underweight, United States
20%; 10%	Percentage overweight; underweight, China
8 kg	Amount of feed to produce 1 kg of beef
94%	Percentage of U.S. soybeans genetically modified to tolerate herbicides
1 mm	Amount of soil accumulated, per year, in ideal conditions

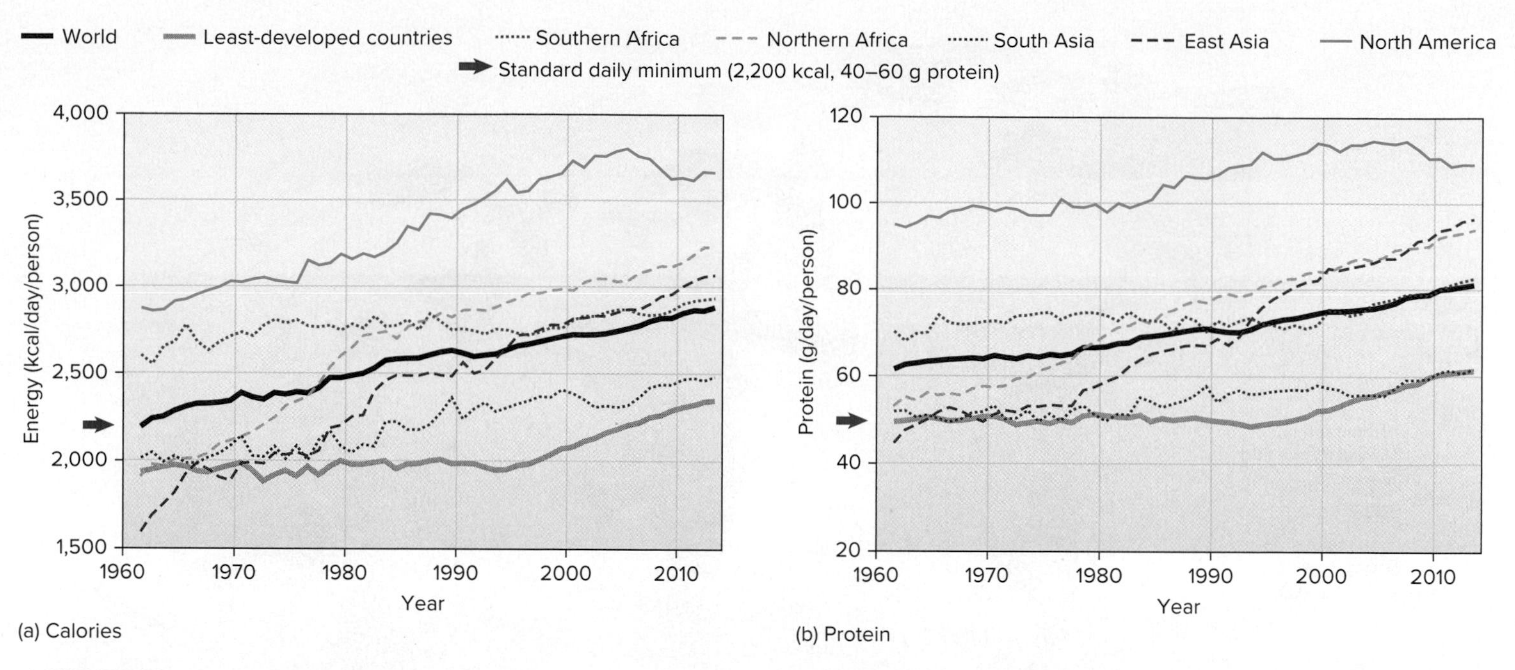

▲ **FIGURE 7.2** Changes in dietary energy (kcal) and protein consumption in selected regions. North America and other developed regions consume more calories and protein than are needed. Source: Data from Food and Agriculture Organization (FAO) 2018.

In this chapter we examine these questions, as well as the ways farmers have managed to feed more and more of the world's growing population. We also consider some of the strategies needed for long-term sustainability in food production.

Food security is unevenly distributed

Fifty years ago, hunger was one of the world's most prominent, persistent problems. In 1960 nearly 60 percent of people in developing countries were chronically undernourished, and the world's population was increasing by more than 2 percent every year. Today some conditions have changed dramatically, while others have changed very little. The world's population has more than doubled, from 3 billion to over 7 billion, but food production has risen even faster. While the average population growth in the past 50 years has been 1.7 percent per year, food production has increased by an average of 2.2 percent per year. Food availability has increased in most countries to well over 2,200 kilocalories, the amount generally considered necessary for a healthy and productive life (fig. 7.2a). Protein intake has also increased in most countries, including China and India, the two most populous countries (fig. 7.2b). Less than 20 percent of people in developing countries now face chronic food shortages, as compared to 60 percent just 50 years ago.

But hunger is still with us. An estimated 900 million people—almost one in every eight people on earth—suffer chronic hunger (fig. 7.3). This number is up slightly from a few years ago, but because of population growth, the *percentage* of malnourished people is still falling (fig. 7.4).

About 95 percent of hungry people are in developing countries. Hunger is especially serious in sub-Saharan Africa, a region plagued by political instability (see figs. 7.2, 7.3). Increasingly, we are coming to understand that **food security,** or the ability to obtain sufficient, healthy food on a day-to-day basis, is a combined problem of economic, environmental, and social conditions. Even in wealthy countries such as the United States, millions lack a sufficient, healthy diet. Poverty, job losses, lack of social services, and other factors lead to persistent hunger—and even more to persistent poor nutrition—despite the fact that we have more, cheaper food (in terms of the work needed to acquire it) than almost any society in history.

Food security is important at multiple scales. In the poorest countries, a severe drought, flood, or insect outbreak can affect vast regions. Individual villages also suffer from a lack of food security: A bad crop year can devastate a family or a village, and local economies can collapse if farmers cannot produce enough to eat and sell. Even within families there can be unequal food security. Males often get both the largest share and the most nutritious food, while women and children—who need food most—all too often get the poorest diet. At least 6 million children under 5 years old die every year of diseases exacerbated by hunger and malnutrition. Providing a healthy diet might eliminate as much as 60 percent of all child deaths worldwide. Because high infant mortality is strongly associated with high population growth rates, reducing childhood hunger is an important strategy for slowing global population growth.

Hungry people can't work their way out of poverty. Nobel Prize-winning economist Robert Fogel has estimated that in 1790, 20 percent of the population of England and France were effectively excluded from the labor force because they were too weak and hungry to work. Fogel calculates that improved nutrition can account for about half of all European economic growth during the nineteenth century. This analysis suggests that reducing hunger in today's poor countries could yield more than $120 billion (U.S.) in economic growth, by producing a healthier, longer-lived, and more productive workforce.

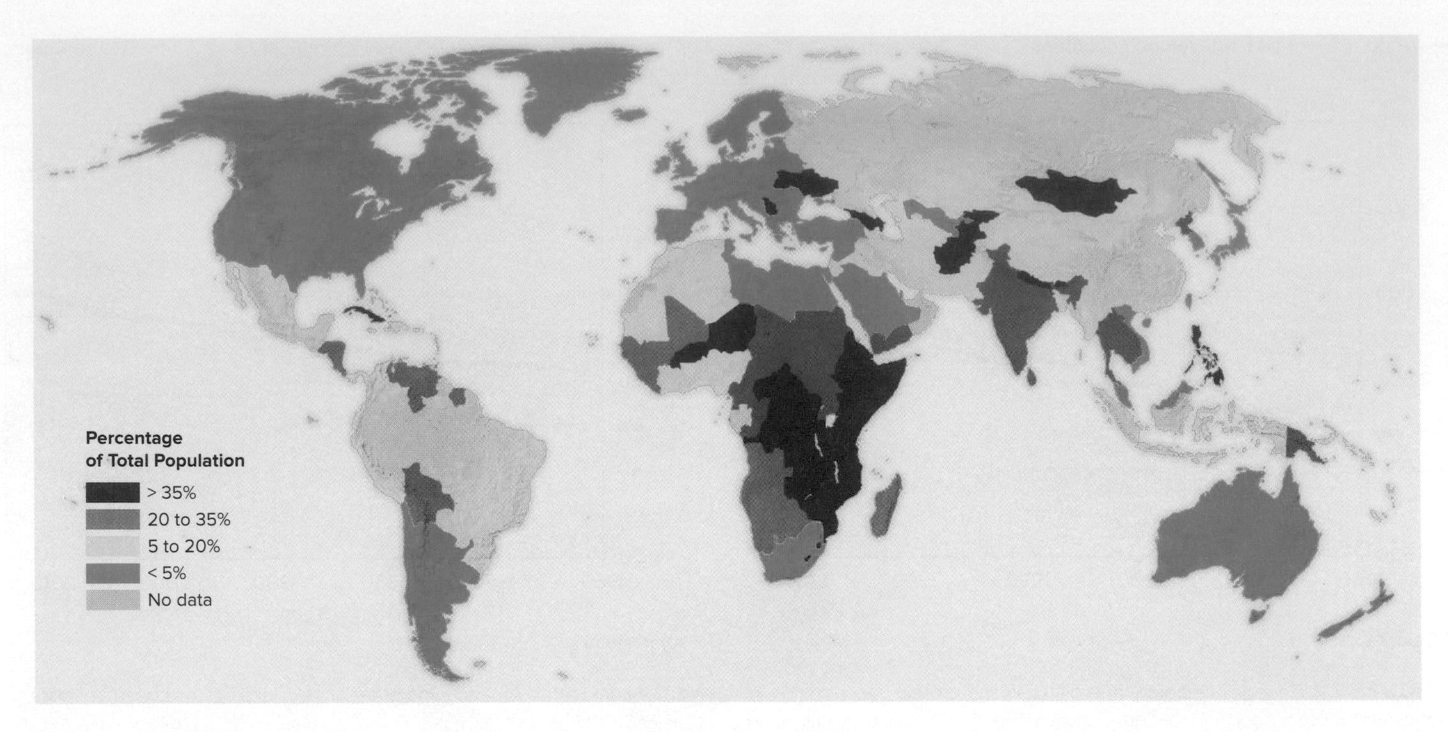

▲ **FIGURE 7.3** Hunger rates worldwide. Most severe and chronic hunger is in developing regions, especially sub-Saharan Africa. Source: United Nations Food and Agriculture Organization.

Famines have political and social roots

Famines are large-scale food shortages, with widespread starvation and social disruption. Famines are often triggered by drought or floods, but root causes of severe food insecurity usually include political instability, such as wars that displace populations, removing villagers from their farms or making farming too dangerous to work in the fields. Economic disparities also drive peasants from the land. In Brazil, for example, wealthy landowners have displaced hundreds of thousands of peasant farmers in recent decades, first to create large cattle ranches and more recently to expand production of soybeans. Displaced farmers often have no choice but to migrate to the already overcrowded slums of major cities when they lose their land (see chapter 14).

World
Developing regions
Sub-Saharan Africa
Northern Africa
South Asia
Eastern Asia
Latin America, Caribbean

Percentage: 0, 5, 10, 15, 20, 25, 30, 35, 40, 45, 50
Year: 1970, 1980, 1990, 2000, 2010

▲ **FIGURE 7.4** Changes in numbers and rates of malnourishment, by region. Source: Data from UN Food and Agriculture Organization, 2015.

Active LEARNING

Mapping Poverty and Plenty

Examine the map in figure 7.3. Using the map of political boundaries at the end of your book, identify ten of the hungriest countries. Then identify ten of the countries with less than 5 percent of people facing chronic undernourishment (green areas). The world's five most populous countries are China, India, the United States, Indonesia, and Brazil. Which classes do these five belong to?

ANSWERS: China, Indonesia, and Brazil have 5–20 percent malnourished; India has 20–35 percent; the United States has < 5 percent.

▲ **FIGURE 7.5** Children wait for their daily ration of porridge at a feeding station in Somalia. When people are driven from their homes by hunger or war, social systems collapse, diseases spread rapidly, and the situation quickly becomes desperate. ©Norbert Schiller/The Image Works

More recently, international "land grabs" have displaced peasants in countries across Africa and parts of Asia. International land speculators, agricultural corporations, and developers have contracted to lease lands occupied by traditional communities but legally owned by governments.

Economist Amartya K. Sen, of Harvard, has shown that while natural disasters often precipitate famines, farmers have almost always managed to survive these events if they aren't thwarted by inept or corrupt governments or greedy elites. Professor Sen points out that armed conflict and political oppression are almost always at the root of famine. No democratic country with a relatively free press, he says, has ever had a major famine.

Natural disasters or drought can contribute to violent conflict and food shortages. Some analysts have argued that recent conflict in Syria results in part from climate change that drove farmers from the land, together with a repressive government that failed to provide timely assistance. Famines often trigger mass migrations to relief camps, where people survive but cannot maintain a healthy and productive life (fig. 7.5). War-torn Sudan and Somalia have experienced this upheaval. In 2011 an estimated 12 to 15 million people in the region faced starvation that was triggered by drought but was rooted in years of conflict.

China's recovery from the famines of the 1960s is a dramatic example of the relationship between politics and famine (see trends in fig. 7.2). Misguided policies from the central government destabilized farming economies across China. Then 2 years of bad crops in 1959–1960 precipitated famines that may have killed 30 million people. In recent years, new political and economic policies have transformed access to food. Even though the population doubled from 650 million in 1960 to 1.3 billion in 2015, hunger is much less common. China now consumes almost twice as much meat (pork, chicken, and beef) as the United States—and much of this meat comes from livestock fed on soybeans bought from Brazil or from the American Midwest.

7.2 HOW MUCH FOOD DO WE NEED?

- *Malnourishment involves vitamin deficiencies as well as insufficient food.*
- *Weight-related issues are outpacing hunger-related ones in developing areas.*
- *We use much of our farmland for producing biofuels, not food.*

A good diet is essential to keep you healthy. You need a balance of foods to provide the right nutrients, as well as enough calories for a productive and energetic lifestyle. The United Nations Food and Agriculture Organization (FAO) estimates that nearly 3 billion people (almost half the world's population) suffer from vitamin, mineral, or protein deficiencies. These shortages result in devastating illnesses and death, as well as reduced mental capacity, developmental abnormalities, and stunted growth.

A healthy diet includes the right nutrients

Malnourishment is a general term for nutritional imbalances caused by a lack of specific nutrients. In conditions of extreme food shortages, a lack of protein in young children can cause kwashiorkor, which is characterized by a bloated belly and discolored hair and skin. *Kwashiorkor* is a West African word meaning displaced child. (A young child is displaced—and deprived of nutritious breast milk—when a new baby is born.) Marasmus (from Greek, to waste away) is another severe condition in children who lack both protein and calories. A child suffering from severe marasmus is generally thin and shriveled (fig. 7.6a). These conditions lower resistance to

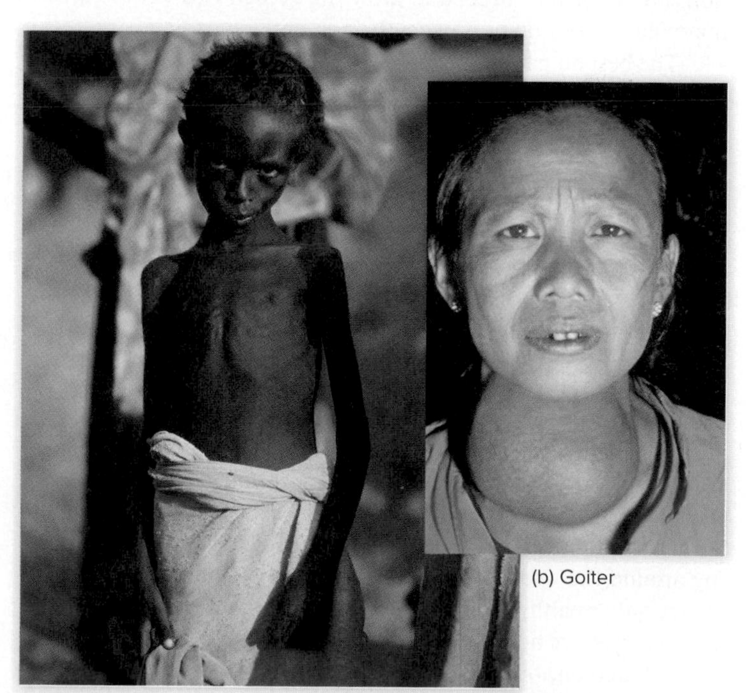

(a) Marasmus

(b) Goiter

▲ **FIGURE 7.6** Dietary deficiencies can cause serious illness. (a) Marasmus results from protein and calorie deficiency and gives children a wizened look and dry, flaky skin. (b) Goiter, a swelling of the thyroid gland, results from an iodine deficiency. (a) ©Scott Daniel Peterson; (b) ©Lester V. Bergman/Getty Images

disease and infections, and children may suffer permanent debilities in mental, as well as physical, development.

Deficiencies in vitamin A, folic acid, and iodine are more widespread problems. Vitamin A and folic acid are found in vegetables, especially dark green, leafy vegetables. Deficiencies in folic acid have been linked to neurological problems in babies. Shortages of vitamin A cause an estimated 350,000 people to go blind every year. Dr. Alfred Sommer, an ophthalmologist from Johns Hopkins University, has shown that giving children just 2 cents' worth of vitamin A twice a year could prevent almost all cases of childhood blindness and premature death associated with shortages of vitamin A. Vitamin supplements also reduced maternal mortality by nearly 40 percent in one study in Nepal.

Iodine deficiencies can cause goiter (fig. 7.6b), a swelling of the thyroid gland. Iodine is essential for synthesis of thyroxin, an endocrine hormone that regulates metabolism and brain development, among other things. The FAO estimates that 740 million people, mostly in Southeast Asia, suffer from iodine deficiency, including 177 million children, whose development and growth have been stunted. Developed countries have largely eliminated this problem by adding iodine to table salt.

Starchy foods, such as maize, polished rice, and manioc (tapioca), form the bulk of the diet for many poor people, but these foods are low in several essential vitamins and minerals. One celebrated effort to deliver crucial nutrients has been through genetic engineering of common foods, such as "golden rice," developed by Monsanto to include a gene for producing vitamin A. This strategy has shown promise, but it also has critics, who argue that genetically modified rice is too expensive for poor populations. In addition, the herbicides needed to grow the golden rice kill the greens that villagers rely on to provide their essential nutrients.

The best human diet is mainly vegetables and grains, with moderate amounts of eggs and dairy products and sparing amounts of meat and oils. A solid base of regular exercise underpins an ideal diet outlined by Harvard dieticians (fig. 7.7) Modest amounts of fats are essential for healthy skin, cell function, and metabolism. But your body is not designed to process excessive amounts of fats or sugars, and these should be consumed sparingly.

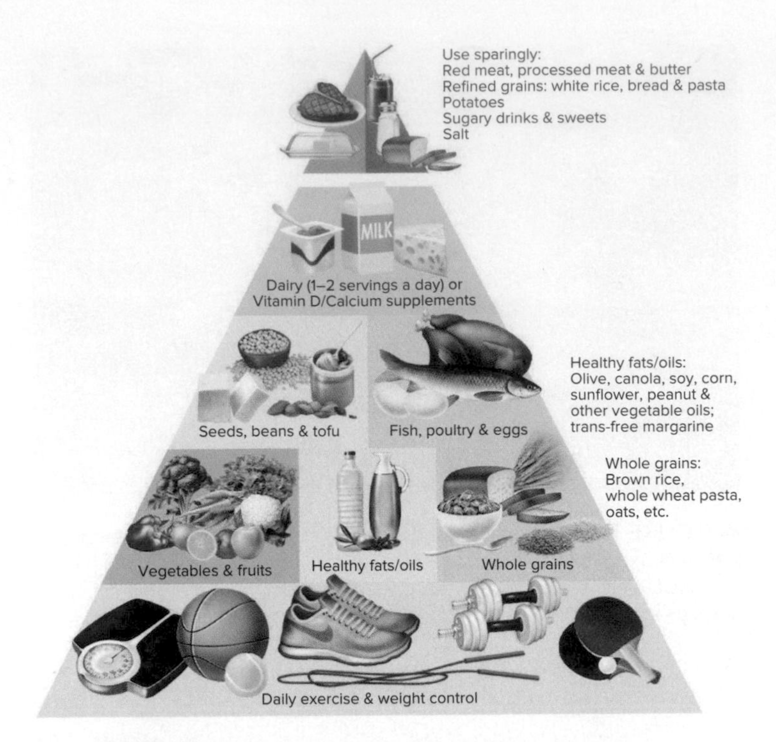

▲ **FIGURE 7.7** The Harvard food pyramid emphasizes fruits, vegetables, and whole grains as the basis of a healthy diet. Unlike most other representations of a healthy diet, this one rests on a foundation of exercise, and it distinguishes the value of whole grains from that of white bread and starches.

Overeating is a growing world problem

For the first time in history, there are probably more overweight people (more than 2 billion) than underweight people (about 900 million). In wealthy countries, more than half of adults are overweight or obese. This trend isn't limited to rich countries, though. Obesity is spreading around the world (fig. 7.8). Diseases once thought to afflict only wealthy nations, such as heart attack, stroke, and diabetes, are now becoming the most prevalent causes of death and disability everywhere (see chapter 8).

In the United States, and increasingly in Europe, China, and developing countries, highly processed, sugary, fatty foods have become a large part of our diet. Over 70 percent of adult Americans are overweight. About one-third of us are seriously overweight, or **obese**—generally considered to mean more than 20 percent over the ideal weight for a person's height and sex. Being overweight increases your risk of hypertension, diabetes, heart attacks, stroke, gallbladder disease, osteoarthritis, respiratory problems, and some cancers. About as

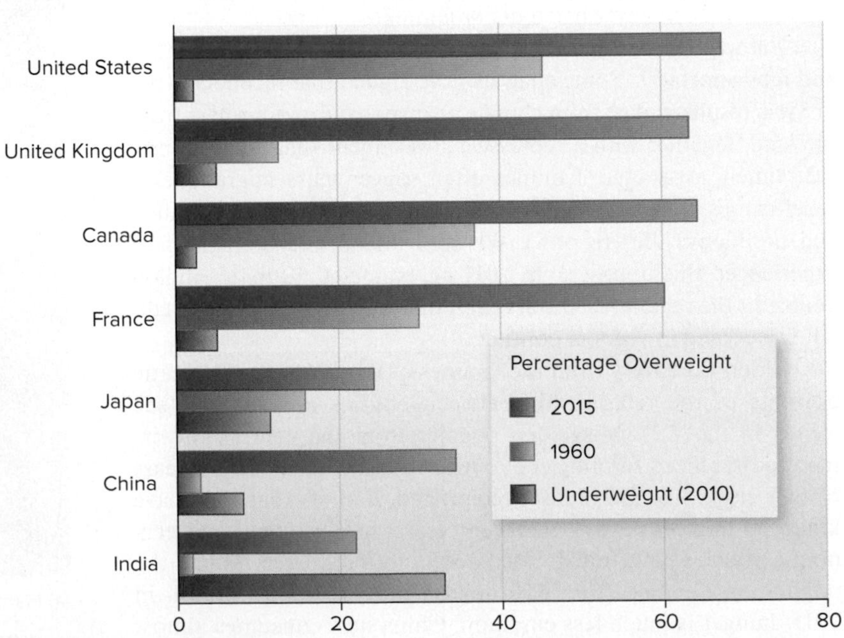

▲ **FIGURE 7.8** While about 900 million people are chronically undernourished, more than twice as many are at risk from eating too much. (Latest available data are shown.) Source: Data from World Health Organization, 2010.

many deaths result from obesity-related illnesses as from smoking every year in the United States.

The obesity epidemic is not a matter of bad behavior; it is a symptom of poverty and food insecurity. High-calorie, low-nutrient foods are cheaper than fresh food, which has a short shelf life. Oil- and sugar-rich prepared foods save time and effort for people too busy to cook regularly or who lack cooking facilities. Healthy food can also be expensive or hard to get, and processed food is marketed aggressively to food-insecure and low-income people.

More production doesn't necessarily reduce hunger

Most strategies for reducing world hunger have to do with increasing efficiency of farm production, expanding use of fertilizers and improved seeds, and converting more unused land or forest to agriculture. But the prevalence of obesity, and unstable farm economies, suggests that lack of supply is not necessarily the principal cause of world hunger. An overabundance of food supplies in much of the world suggests that answers to global hunger may lie in better use and distribution of food resources.

For most farmers in the developed world, overproduction constantly threatens prices for farm products. To reduce food supplies and stabilize prices in the United States, Canada, and Europe, we send millions of tons of food aid to developing areas every year. Often, however, these shipments of free food destabilize farm economies in receiving areas. Prices for local farm products collapse, and political corruption can expand if military groups control distribution. Even in developing areas, lack of food production is not always the cause of hunger.

There are also inefficiencies in food use. Global food waste amounts to some 30 percent of all food production—1.3 billion tons annually—as food is spoiled in storage and transit, used inefficiently, or thrown away after preparation. We also prefer inefficient foods—particularly meats that require abundant feed inputs for every pound of meat we eat.

In the past decade, global hunger has had two newer causes: international financial speculation on food commodities and biofuel production. Food has long been a globally traded commodity, but a law passed by the U.S. Congress in 2000, the Commodities Futures Modernization Act, eliminated long-standing rules restricting risky speculation on a wide variety of commodities. Global food products were among these commodities. Freed from restraints, traders gambled on the value of future crops, which pushed up the price of current crops. This deregulation had far-reaching effects for ordinary people. (In the United States, a widely felt effect of the rule change was the acceleration of trade in home mortgages. This led to a financial "bubble" in home prices, followed by a collapse in which millions of Americans lost their homes.) In 2008, speculative trading in food commodities led to a crisis in food, as a frenzy of speculative trading drove prices of agricultural commodities well beyond their actual value. In response to this bubble in global prices, local food prices rose sharply in towns and villages worldwide. The cost of some basic staples, such as cooking oil and rice, quadrupled. Farm land prices rose as well, often driving struggling peasant farmers from their land and livelihoods. Thus, the effects of frenzied Wall Street trading in wheat, food oils, and other commodities were felt in households worldwide. These changes were inconvenient in wealthy countries, where prices rose on grocery store shelves. In poorer countries, food costs can make up as much as 80 percent of household expenses, and there the effect was far worse. Food riots followed in the Philippines, India, Indonesia, and many other countries.

Biofuels have boosted commodity prices

Food prices declined again after 2008, but they were shored up by an additional change: new policies in the United States and Europe promoting biofuels. Using crops such as soy, corn, palm oil, or sugarcane to drive our cars is an important strategy for supporting farm economies in wealthy countries, as the increased demand keeps prices high. These new policies also led to global increases in production of these crops (opening case study; table 7.1). In the United States, federal ethanol subsidies led to a doubling of corn prices in 2007. But in developing countries, production of soy, palm oil, and other products for export often displaces food production. When biofuel policies compete with food supplies in Asia and Africa, the efficiency of this market becomes less clear.

There has been much debate about the environmental and economic costs and benefits of biofuel production. Some studies have found that biofuels represent a net energy loss, taking more energy to produce than they provide in fuel. In the United States and Europe, the production of biofuels, especially ethanol, has not been economically viable without heavy subsidies for growing and processing crops. On the other hand, plant oils from sunflowers, soybeans, corn, and other oil seed crops can be burned directly in most diesel engines and may be closer than ethanol to a net energy gain (see chapter 12). Brazil produces ethanol more efficiently from its tropical sugarcane crops—although the net energy balance and environmental impacts remain serious questions.

TABLE 7.1 Key Global Food Sources*

CROP	1965	1990	2016
Sugarcane	531	1,053	1,891
Maize	227	483	1,060
Rice	254	519	741
Wheat	264	592	749
Milk	358	524	798
Potatoes	271	267	377
Vegetables	66	140	290
Cassava	86	152	277
Soyabeans	32	108	335
Meat	72	158	297
Barley	93	144	177
Sweet potatoes	108	123	105
Dry beans, pulses	31	41	44

*Production in million metric tons.
Source: Data from UN FAO, 2017.

Do we have enough farmland?

The problem of farmland availability has always been a central question in debates about how to feed the world. Because we currently produce more than enough to feed the world, we probably do have enough farmland to feed more people than currently live on earth—if we ate according to the recommendations in figure 7.7 and if we used all farm products for food, rather than fuel and livestock feed. Could we expand production further by clearing new farmlands? That is harder. Tropical soils often are deeply weathered and infertile, so they make poor farmland unless expensive inputs of lime and fertilizer are added. Much of the world's uncultivated land is too steep, sandy, waterlogged, salty, acidic, cold, dry, or nutrient-poor for most crops.

About 11 percent of the earth's land area, some 1,400 million ha, is used for agricultural production. This land area amounts to about 0.2 ha per person (1 ha = 100 m × 100 m). Arable land per person has fallen from about 0.5 ha per person in 1960, mainly because of population growth. Population projections suggest we will have about 0.15 ha by 2050. In Asia, cropland will be even more scarce—0.09 ha per person by 2050.

Much of the world's uncultivated land could be converted to cropland, but this land currently provides essential ecological services on which farmers depend. Forested watersheds help maintain stream flow, regulate climate, and provide refuge for biological and cultural diversity. Wetlands also regulate water supplies, and forests and grasslands support insect pollinators that ensure productive crops.

Despite the importance of these services, the conversion of tropical forests and savannas to farmland continues, spurred by global trade in export crops, such as soy, palm oil, sugarcane (for ethanol), and corn (maize, largely for animal feed), as well as beef and other livestock. The FAO reports that 13 to 16 million ha of forest land are cleared each year, about half of that in tropical Africa and South America. Brazil, for example, has cleared cleared vast expanses of Amazon and Atlantic rainforest in order to become a global leader in exports of soy, sugar, beef, poultry, and orange juice, and it is emerging as a leading producer of rice and corn.

Most of these export crops serve those who already have enough to eat, however: According to FAO data, 85 percent of global crop exports are bound for Europe, North America, China, Japan, and other wealthy countries. Meanwhile, human rights groups protest that land conversions have displaced traditional communities and subsistence farmers across South America, Africa, and much of Asia.

7.3 WHAT DO WE EAT?

- *Producing meat takes more resources than producing vegetables and grain.*
- *Farmed seafood depends mainly on wild-caught feed.*
- *Antibiotic overuse is making our antibiotics ineffective.*

Of the thousands of edible plants and animals in the world, only a few provide almost all our food. About a dozen types of grasses, 3 root crops, 20 or so fruits and vegetables, 6 mammals, 2 domestic fowl, and a few fish species make up almost all the food we eat (table 7.1). Two grasses, wheat and rice, are especially important because they are the staple foods for most of the 5 billion people in developing countries.

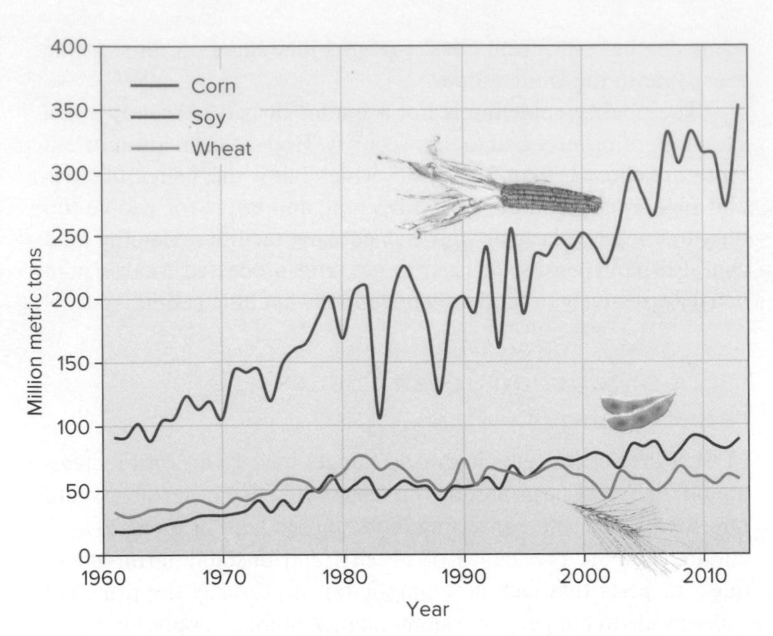

▲ **FIGURE 7.9** U.S. production of our three dominant crops—corn, soybeans, and wheat. Source: Data from USDA, 2015.

In the United States, corn (a grass, also known as maize) and soybeans have become our primary products. We rarely eat either corn or soybeans directly, but corn provides sweeteners; corn oil; the livestock feed for producing our beef, chicken, and pork; industrial starches; and many synthetic vitamins. Soybeans are also fed to livestock, and soy provides protein and oils for processed foods. Because we have developed so many uses for corn, it now accounts for nearly two-thirds of our bulk commodity crops (which include corn, soy, wheat, rice, and other commodities). Together, corn and soy make up about 85 percent of commodity crops, with an annual production of 268 million tons of corn and 88 million tons of soy (fig. 7.9).

Rising meat production is a sign of wealth

Largely because of dramatic increases in corn and soy production, meat consumption has grown worldwide. In developing countries, meat consumption has risen from just 10 kg per person per year in the 1960s to over 26 kg today (fig. 7.10). In the United States, our meat consumption has risen from 90 kg to 136 kg per person per year in the same 50 years. Meat is a concentrated, high-value source of protein, iron, fats, and other nutrients that give us the energy to lead productive lives. Dairy products are also a key protein source: Globally, we consume more than twice as much dairy as meat. But dairy production per capita has declined slightly, while global meat production has more than doubled in the past 50 years.

Meat is a good indicator of wealth because it is expensive to produce in terms of the resources needed to grow an animal (fig. 7.11). As discussed in chapter 2, herbivores use most of the energy they consume in growing muscle and bone, moving around, staying warm, and metabolizing (digesting) food. Only a little food energy is stored

▲ **FIGURE 7.10** Meat and dairy consumption has quadrupled in the past 40 years, and China represents about 40 percent of that increased demand. ©William P. Cunningham

for consumption by carnivores, at the next level of the food pyramid. For every 1 kg of beef, a steer consumes over 8 kg of grain. Pigs are more efficient: Producing 1 kg of pork takes just 3 kg of pig feed. Chickens and herbivorous fish (such as catfish) are still more efficient. Globally, some 660 million metric tons of cereals are used as livestock feed each year. This is just over a third of the world cereal use. As figure 7.11 suggests, we could feed at least eight times as many people by eating those cereals directly.

A number of technological and breeding innovations have made this increased production possible. One of the most important is the **confined animal feeding operation (CAFO),** where animals are housed and fed—mainly soy and corn—for rapid growth (fig. 7.12). These operations dominate livestock raising in the United States, Europe, and increasingly in China and other countries. Animals are housed in giant enclosures, with up to 10,000 hogs or a million chickens in an enormous barn complex (fig. 7.13), or 100,000 cattle in a feed lot. Operators feed the animals specially prepared mixes of corn, soy, and animal protein (parts that can't be sold for human food) that maximize their growth rate. New breeds of livestock have been developed that produce muscle (meat) rapidly, rather than simply getting fat. The turnaround time is getting shorter, too. A U.S. chicken producer can turn baby chicks into chicken nuggets after just 8 weeks of growth. Steers reach full size by just 18 months of age. Increased use of antibiotics, which are mixed in daily feed, make it possible to raise large numbers of animals in close quarters. About 90 percent of U.S. hogs receive antibiotics in their feed.

▲ **FIGURE 7.11** Number of kilograms of grain needed to produce 1 kg of bread or 1 kg live weight gain.

(a)

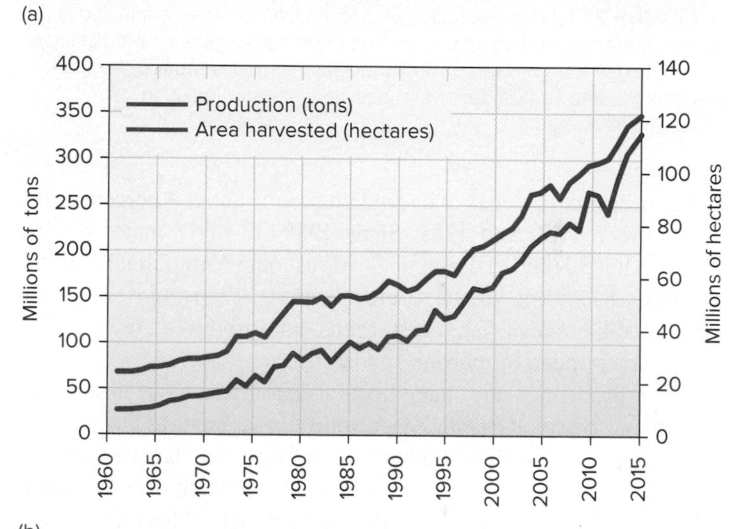

(b)

▲ **FIGURE 7.12** (a) Feedlots have expanded the global market for corn and soybeans. (b) Brazil's soy production, which supports feedlots globally, has grown from near zero to being the country's dominant agricultural product. (a) Source: Photo by Jeff Vanuga, USDA Natural Resources Conservation Service; (b) Source: Data from UN Food and Agriculture Organization, 2015.

Seafood, both wild and farmed, depends on wild-source inputs

The 140 million metric tons of seafood we eat every year is an important part of our diet. Seafood provides about 15 percent of all animal protein eaten by humans, and it is the main animal protein source for about 1 billion people in developing countries. Overharvesting and habitat destruction threaten most of the world's wild fisheries,

▲ **FIGURE 7.13** Most livestock grown in the United States are raised in large-scale, concentrated animal-feeding operations. Up to a million animals can be held in a single facility. High population densities require heavy use of antibiotics and can cause severe local air and water pollution.

▲ **FIGURE 7.14** Fish farms produce much of our seafood, but most rely on wild-caught fish for feed, and pollution and disease are common concerns.

however. Annual catches of ocean fish rose by about 4 percent annually between 1950 and 1988. Since 1989, 13 of 17 major marine fisheries have declined dramatically or become commercially unsustainable. According to the United Nations, three-quarters of the world's edible ocean fish, crustaceans, and mollusks are declining and in urgent need of managed conservation.

The problem is too many boats using efficient but destructive technology to exploit a dwindling resource base. Boats as big as ocean liners travel thousands of kilometers and drag nets large enough to scoop up a dozen jumbo jets, sweeping a large patch of ocean clean of fish in a few hours. Longline fishing boats set cables up to 10 km long with hooks every 2 meters that catch birds, turtles, and other unwanted "by-catch" along with targeted species. Trawlers drag heavy nets across the bottom, scooping up everything indiscriminately and reducing broad swaths of habitat to rubble. One marine biologist compared the technique to harvesting forest mushrooms with a bulldozer. In some operations, up to 15 kg of dead by-catch (nontarget species) are dumped back into the ocean for every kilogram of marketable food. Nearly all countries heavily subsidize their fishing fleets, to preserve jobs and to ensure access in the unregulated open ocean. The FAO estimates that operating costs for the 4 million boats now harvesting wild fish exceed sales by $50 billion (U.S.) per year.

The best solution, according to a United Nations study on ecosystem services, is to establish better international agreements on fisheries. Instead of a free-for-all race to exploit fish as fast as possible, nations could manage fisheries for long-term, sustained production. Just as with hunting laws within a single country, agreement to international fishery rules could improve total food production, fishery employment, and ecosystem stability.

Aquaculture is providing an increasing share of the world's seafood. Fish can be grown in farm ponds that take relatively little space but are highly productive. For cultivation of plant-eating fish such as tilapia, these systems can be very sustainable. Cultivation of high-value carnivorous species such as salmon, however, threatens wild fish populations, which are caught to feed captive fish. A 2012 study of fish catches for fish farms has shown that global declines of seabird populations, such as puffins and albatrosses, closely track the increase in farm-raised carnivorous fish.

In tropical areas, coastal fish and shrimp-rearing ponds have replaced hundreds of thousands of hectares of mangrove forests and wetlands, which serve as irreplaceable nurseries for marine species. Net pens anchored in near-shore areas encourage spread of diseases, as they release feces, uneaten food, antibiotics, and other pollutants into surrounding ecosystems (fig. 7.14).

Biohazards arise in industrial production

Increasingly efficient production has a variety of externalized (unaccounted for) costs. Land conversion from forest or grassland to crop fields increases soil erosion, which degrades water quality. Bacteria in the manure in the feedlots, or liquid wastes in manure storage lagoons (holding tanks) around hog farms, escape into the environment–from airborne fecal dust around feedlots or from breaches in the walls of a manure tank. When Hurricane Floyd hit North Carolina's coastal hog production region in 1999, some 10 million m^3 of hog and poultry waste overflowed into local rivers, creating a dead zone in Pamlico Sound. In 2018, Hurricane Florence caused similar overflows. These disasters were especially severe, but smaller spills and leaks are widespread.

Constant use of antibiotics has been associated with antibiotic-resistant infections in hog-producing areas. Eighty percent of all antibiotics used in the United States are administered to livestock. This massive and constant exposure kills off most bacteria, but occasional strains that resist the effects of the antibiotics can survive. These survivors then reproduce freely, creating new strains that are untreatable in human hosts. Thus, overuse is slowly rendering our standard antibiotics ineffective for human health care, and thousands of people have died from antibiotic-resistant infections. The next time you are prescribed an antibiotic by your doctor, you might ask whether she or he worries about antibiotic-resistant bacteria. What would you do if your prescription were ineffective against an illness?

Although the public is increasingly aware of these environmental and health risks of concentrated meat production, we seem to be willing to accept these risks because this production system has made our favorite foods cheaper, bigger, and more available. A fast-food hamburger today is more than twice the size it was in 1960, especially if you buy the kind with multiple patties and special sauce, and Americans love to eat them. At the same time, this larger burger costs less per pound, in inflation-adjusted dollars, than the 1960 version. As a consequence, for much of the world, consumption of protein and calories has climbed beyond what we really need to be healthy (see figs. 7.2, 7.8).

As environmental scientists, we are faced with a conundrum: Improved efficiency has great environmental costs; it has also given us the abundant, inexpensive foods that we love. We have more protein, but also more obesity, heart disease, and diabetes, than we ever had before. What do you think? Do the environmental risks balance a globally improved quality of life? Should we consider reducing our consumption to lower environmental and health costs? How might we do that?

▲ **FIGURE 7.15** Terracing, as in these Balinese rice paddies, can control erosion and make steep hillsides productive. These terraced rice paddies have produced two or three crops a year for centuries because the soils are carefully managed and organic nutrients are maintained.
©William P. Cunningham

7.4 LIVING SOIL IS A PRECIOUS RESOURCE

- *Soil consists of mineral, organic, and living components.*
- *Soil fauna capture, process, and store nutrients for plants.*
- *Erosion and degradation make millions of hectares of farmland useless each year.*

Understanding the limits and opportunities for feeding the world requires an understanding of the soil that supports us. In this section we examine the nature of soils and the ways we use them. Many of us think of soil as just dirt. But healthy soil is a marvelous substance with astonishing complexity. Soil contains mineral grains weathered from rocks, partially decomposed organic molecules, and a host of living organisms. The complex community of bacteria and fungi is primarily responsible for providing nutrients that plants need to grow. Soil bacteria also help filter and purify the water we drink. Soil can be considered a living ecosystem by itself. How can we manage our soils to maintain and build these precious systems?

Building a few millimeters of soil can take anything from a few years (in a healthy grassland) to a few thousand years (in a desert or tundra). Under the best circumstances, topsoil accumulates at about 1 mm per year. With careful husbandry that prevents erosion and adds organic material, soil can be replenished and renewed indefinitely (fig. 7.15). But many farming techniques deplete soil. Crops consume the nutrients; plowing exposes the soil to erosion by wind or water. Severe erosion can carry away 25 mm or more of soil per year, far more than can accumulate under the best of conditions.

What is soil?

Soil is a complex mixture of six components:

1. *Sand and gravel* (mineral particles from bedrock, either in place or moved from elsewhere, as in windblown sand)
2. *Silts and clays* (extremely small mineral particles; many clays are sticky and hold water because of their flat surfaces and ionic charges; others give red color to soil)
3. *Dead organic material* (decaying plant matter stores nutrients and gives soils a black or brown color)
4. *Soil fauna and flora* (living organisms, including soil bacteria, worms, fungi, roots of plants, and insects, recycle organic compounds and nutrients)
5. *Water* (moisture from rainfall or groundwater, essential for soil fauna and plants)
6. *Air* (tiny pockets of air help soil bacteria and other organisms survive)

Variations in these six components produce almost infinite variety in the world's soils. Abundant clays make soil sticky and wet. Abundant

Active LEARNING

Where in the World Did You Eat Today?

Make a list of every food you ate today or yesterday. From this list, make a graph of the number of items in the following categories: grains, vegetables, dairy, meat, other. Which food type was most abundant? With other students, try to identify the location or region where each food was grown. How many come from a region more than halfway across the country? from another country?

▲ **FIGURE 7.16** Soil ecosystems include countless organisms that consume and decompose organic material, aerate soil, and distribute nutrients through the soil.

organic material and sand make the soil soft and easy to dig. Sandy soils drain quickly, often depriving plants of moisture. Silt particles are larger than clays and smaller than sand, so they aren't sticky and soggy, and they don't drain too quickly. Thus, silty soils are ideal for growing crops, but they are also light and blow away easily when exposed to wind. Soils with abundant soil fauna quickly decay dead leaves and roots, making nutrients available for new plant growth. Compacted soils have few air spaces, making soil fauna and plants grow poorly. You can see some of these differences just looking at soil. Reddish soils are colored by iron-rich, rust-colored clays, the kind that store few nutrients for plants. Deep black soils are rich in decayed organic material and thus are generally rich in nutrients.

Healthy soil fauna can determine soil fertility

Soil bacteria, algae, and fungi decompose and recycle leaf litter into plant-available nutrients, as well as helping to give soils structure and loose texture (fig. 7.16). Microscopic worms and nematodes process organic matter and create air spaces as they burrow through soil. These organisms mostly stay near the surface, often within the top few centimeters. The sweet aroma of freshly turned soil is caused by actinomycetes, bacteria that grow in funguslike strands and give us the antibiotics streptomycin and tetracycline.

The health of the soil ecosystem depends on environmental conditions, including climate, topography, and parent material (the mineral grains or bedrock on which soil is built), and frequency of disturbance. Too much rain washes away nutrients and organic matter, but soil fauna cannot survive with too little rain. In extreme cold, soil fauna recycle nutrients slowly; in extreme heat, they may work so fast that leaf litter on the forest floor is taken up by plants in just weeks or months–so that the soil retains little organic matter. Frequent disturbance prevents the development of a healthy soil ecosystem, as does steep topography that allows rain to wash away soils. In the United States, the best farming soils tend to occur where the climate is not too wet or dry, on glacial silt deposits such as those in the upper Midwest, and on silt- and clay-rich flood deposits, like those along the Mississippi River.

Most soil fauna occur in the uppermost layers of a soil, where they consume leaf litter. This layer is known as the "O" (organic) horizon. Just below the O horizon is a layer of mixed organic and mineral soil material, called the A horizon (fig. 7.17), also known as **topsoil** or surface soil.

The B horizon, or **subsoil,** lies below most organic activity, and it tends to have more clays than the A layer. The B layer accumulates clays that seep downward from the A horizon with rainwater that percolates through the soil. If you dig a hole, you may be able to tell where the B horizon begins, because the soil tends to become slightly more cohesive. If you squeeze a handful of B soil, it should hold its shape better than a handful of A soil.

Sometimes an E (eluviated, or washed-out) layer lies between the A and B horizons. The E layer is loose and light-colored because most of its clays and organic material have been washed down to

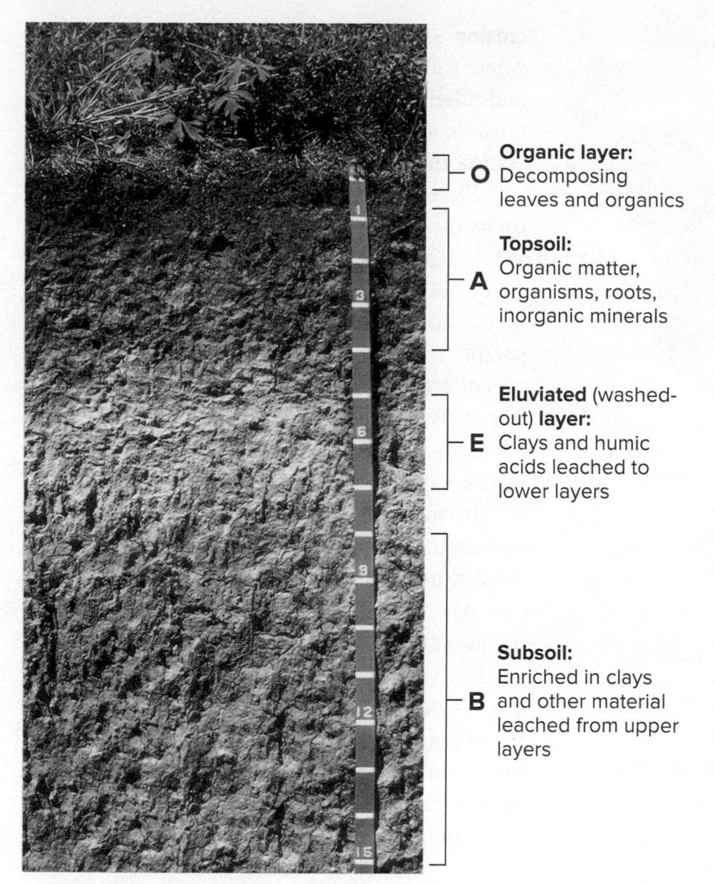

▲ **FIGURE 7.17** An idealized soil profile showing common horizons. Soils vary greatly in composition and thickness of layers. Below the B horizon, there is generally a C horizon of weathered rock, sand, or other parent material. Depth is marked in 10-cm increments. ©Mary Ann Cunningham; Courtesy Soil & Land Resources Division, University of Idaho

the B horizon. The C horizon, below the subsoil, is mainly decomposed rock fragments. Parent materials underlie the C layer. Parent material is the sand, windblown silt, bedrock, or other mineral material on which the soil is built. About 70 percent of the parent material in the United States was transported to its present site by glaciers, wind, and water and is not related to the bedrock formations below it.

Your food comes mostly from the A horizon

Ideal farming soils have a thick, organic-rich A horizon. The soils that support the Corn Belt farm states of the U.S. Midwest have rich, black A horizon that can be more than 2 meters thick, although a century of farming has washed away much of this soil. Most soils have less than half a meter of A horizon. Desert soils, with slow rates of organic activity, might have almost no O or A horizon (fig. 7.18).

Because topsoil is so important to our survival, we differentiate soils largely according to the thickness and composition of their upper layers. The U.S. Department of Agriculture classifies the thousands of different soil types into 11 soil orders (http://soils.usda.gov/technical/classification/orders/). Mollisols (*mollic* = soft, *sol* = soil), for example, have a thick, organic-rich A horizon that develops from the deep, dense roots of prairie grasses. Alfisols have a slightly thinner A horizon, with slightly less organic matter. Alfisols develop in deciduous forests, where leaf litter is abundant. In contrast, the aridisols (*arid* = dry) of the desert Southwest have little organic matter, and they often have accumulations of mineral salts. Organic-rich mollisols and alfisols dominate most of the farming regions of the United States (fig. 7.19).

How do we use and abuse soil?

Agriculture both causes and suffers from environmental degradation. The International Soil Reference and Information Centre in the Netherlands estimates that, every year, 3 million ha of cropland are made useless by erosion, 4 million ha are turned into deserts, and 8 million ha are converted to nonagricultural uses, such as homes, highways, shopping centers, factories, and reservoirs. Over the past 50 years, some 1,900 million ha of agricultural land (an area greater than that now in

(a)

(b)

▲ **FIGURE 7.18** Soils vary dramatically among different climate areas. Temperate grassland soils tend to have a thick, soft, organic-rich A horizon (a). Arid land soils may have little or no workable A horizon, as in this Libyan valley (b). (a) ©Ingram Publishing/SuperStock; (b) ©imageBROKER/Alamy Stock Photo

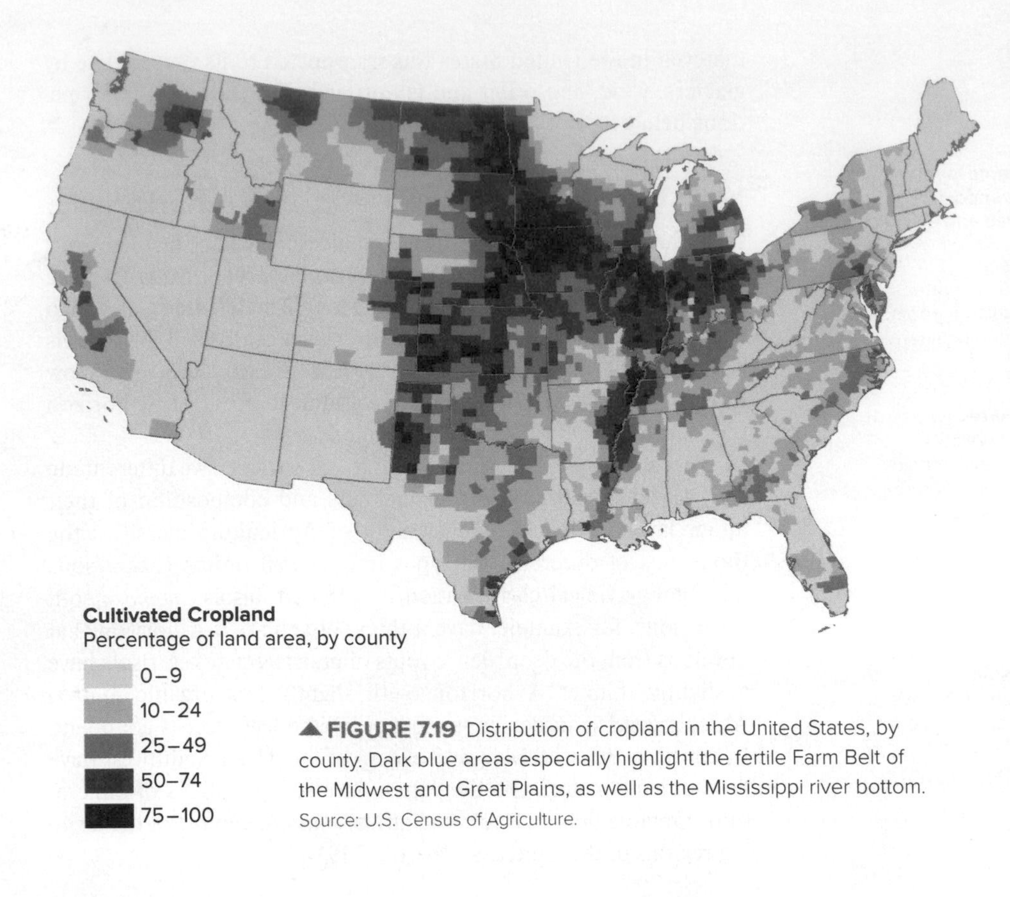

▲ **FIGURE 7.19** Distribution of cropland in the United States, by county. Dark blue areas especially highlight the fertile Farm Belt of the Midwest and Great Plains, as well as the Mississippi river bottom. Source: U.S. Census of Agriculture.

production) have been degraded to some extent. About 300 million ha of this land are strongly degraded, with deep gullies, severe nutrient depletion, or poor crop growth, making restoration difficult and expensive.

Water and wind erosion cause a vast majority of global soil degradation. Chemical degradation, a secondary cause of soil degradation, includes nutrient depletion, salt accumulation, acidification, and pollution. Physical degradation includes factors such as compaction or loss of soil structure after cultivation by heavy machinery, waterlogging from excess irrigation, and laterization—rock-hard solidification of tropical soils exposed to sun and rain.

Water is the leading cause of soil erosion

Erosion is an important natural process that redistributes the products of weathering. Erosion contributes to soil formation, creating regions of windblown silt that form much of the U.S. Farm Belt, and to soil loss, as well as rich river bottom soils on which early civilizations were built. Erosion also depletes farmland soils when they are left bare or worked at the wrong time.

Some erosion occurs so rapidly that you can watch it happen. Running water can scour deep gullies, leaving fence posts and trees standing on pedestals as the land erodes away around them. But most erosion is more subtle. **Sheet erosion** removes a thin layer of soil as a sheet of water flows across a nearly level or gently sloping field. When small rivulets of running water form and cut small channels in the soil, the process is called **rill erosion** (fig. 7.20a). Where rills expand to form bigger channels or ravines, too large to be removed by cultivation, we call the process **gully erosion** (fig. 7.20b). **Streambank erosion** also is a problem in farmland streams, where soil washes away from streambanks. Cattle and other livestock often accelerate streambank erosion, as they trample and graze down the vegetation that holds a bank in place.

Sheet and rill erosion cause most soil loss on farmland, even though this process is often too gradual to notice. Studies in Iowa farm fields have found areas with as much as 50 metric tons lost per hectare during winter and spring runoff, more than 12 times the rate considered tolerable or officially acknowledged by state soil scientists. That loss represents just a few millimeters of soil over the whole surface of the field, but repeated years of storms and spring runoff are gradually sending Iowa's productive capacity downstream to the Gulf of Mexico.

All this waterborne sediment damages aquatic systems, too. Increased sediment loads erode rivers and reduce water quality in lakes. Excess nutrients create eutrophic "dead zones" in estuaries and coastal waterways, most famously in the Gulf of Mexico. Infrastructure such as reservoirs and harbors also can be badly impacted by sediment deposits.

Wind is a close second in erosion

Wind can equal or exceed water in erosive force, especially in dry regions, and where vegetation is sparse (fig. 7.20c). When plant cover and surface litter are removed from the land by agriculture or grazing, wind lifts loose soil particles and sweeps them away. In extreme conditions, windblown dunes encroach on useful land and cover roads and buildings. Over the past 30 years, China has lost 93,000 km^2 (about the size of Indiana) to desertification, or conversion of productive land to desert. Advancing dunes from the Gobi Desert are now only 160 km (100 mi) from Beijing. Every year more than 1 million tons of sand and dust blow from Chinese drylands, often traveling across the Pacific Ocean to the west coast of North America. Since 1985, China has planted more than 40 billion trees to try to stabilize the soil and hold back deserts.

Intensive farming practices are largely responsible for this situation. Row crops, such as corn and soybeans, leave soil exposed for much of the growing season. Deep plowing and heavy herbicide applications create weed-free fields that look tidy but are subject to erosion. Farmers, under pressure to maximize yields and meet loan obligations, often plow through grass-lined watercourses (low areas where water runs off after a rain) and pull out windbreaks and fencerows to accommodate the large machines and to get every last square meter into production.

An estimated 25 billion metric tons of soil are lost from croplands every year due to wind and water erosion. The net effect, worldwide, of this widespread topsoil erosion is a reduction in crop production equivalent to removing about 1 percent of the world's cropland each year.

(a) Sheet erosion

(b) Gully erosion

(c) Wind erosion

▲ **FIGURE 7.20** Land degradation affects more than 1 billion ha yearly, or about two-thirds of all global cropland. Globally, erosion by water accounts for about 56 percent of soil loss from fields (a, b). Wind erosion accounts for another 28 percent (c). (a) Source: Photo by Lynn Betts, courtesy of USDA Natural Resources Conservation Service; (b) Source: Photo by Jeff Vanuga, courtesy of USDA Natural Resources Conservation Center; (c) Source: Natural Resource Conservation Service

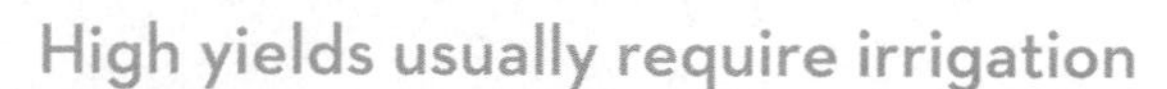

7.5 AGRICULTURAL INPUTS

- *Irrigation is required for most high-yielding agriculture.*
- *Fertilizers, especially N, P, and K, are widely used and overused.*
- *Pesticides vary greatly from general biocides to narrowly targeted compounds.*

Soil is the foundation of food production, but there are many other critical factors. Reliable water resources, nutrients, favorable temperatures and rainfall, productive crop varieties, and the mechanical energy to tend and harvest the crops are also essential. Strategies for applying these different inputs vary greatly among regions and among contrasting farming strategies (Key Concepts, p. 168).

High yields usually require irrigation

Agriculture uses at least two-thirds of all fresh water withdrawn from rivers, lakes, and groundwater. Although estimates vary widely, about 15 percent of all cropland, worldwide, is irrigated. Some countries are water rich and can readily afford to irrigate farmland, while other countries are water poor and must use water very carefully. The efficiency of irrigation water use varies greatly. In some places, high evaporation and seepage losses from unlined and uncovered canals can mean that as much as 80 percent of water withdrawn for irrigation never reaches its intended destination. Poor farmers may overirrigate because they lack the technology to meter water and distribute just the amount needed. In wealthier countries, farmers can afford abundant uses of water, including center-pivot irrigation systems (fig. 7.21). They can also afford water-saving methods such as drip irrigation, which waters only the base of a crop, reducing evaporative losses.

Excessive use not only wastes water but also often results in **waterlogging.** Waterlogged soil is saturated with water, and plant roots die from lack of oxygen. **Salinization,** in which mineral salts accumulate in the soil, is often a problem when irrigation water dissolves and mobilizes salts in the soil. As the water evaporates, it leaves behind a salty crust on the soil surface that is lethal to most plants. Flushing with excess water can wash away this salt accumulation, but the result is even more saline water for downstream users.

(a)

(b)

▲ **FIGURE 7.21** Pivot irrigation systems (a) deliver water to many U.S. farm fields. These systems create green circles visible in aerial images of the landscape (b). (a) ©Cecilia Lim/123RF; (b) ©Kris Hanke/Getty Images

How can we feed the world?

The world's population has climbed from 3 billion to 7 billion in about two generations (since 1960). Despite this growth, the proportion of chronically hungry people has declined from 60 percent to about 20 percent in developing countries, where most population growth has occurred. How have we managed to increase food production so rapidly? What are the pros and cons of these strategies? What additional choices do we have? Presented here are three main strategies in food production.

The Green Revolution involved development of high responders—crops that grow and yield well with increased use of fertilizer, irrigation, and pesticides.

Benefits

- Yields have grown dramatically with increased inputs.
- Efficiency is high for large-scale production, which aids in feeding billions.
- Development of pesticides has improved yields by eliminating competition with other plants or predation by insects.
- Labor costs are low: One farmer can work a huge area.

Problems

- Dependence on pesticides and fertilizers has grown; pesticides lose effectiveness through overuse.
- Agricultural chemicals have unintended ecological consequences, including lost biodiversity, probable loss of pollinating insects, and contamination of drinking water.
- New varieties can be hard for poor farmers to afford, so that wealthier farmers and wealthier regions gain a relative advantage.
- Increased use of nitrogen fertilizer is a major source of greenhouse gases and consumer of fossil fuels.

KC 7.1

Circular green fields, irrigated by center-pivot sprinklers, produce the corn, soybeans, and other crops that are the backbone of our food system.

©Cindy Shaw (created with NAIP USDA with data from USGS).

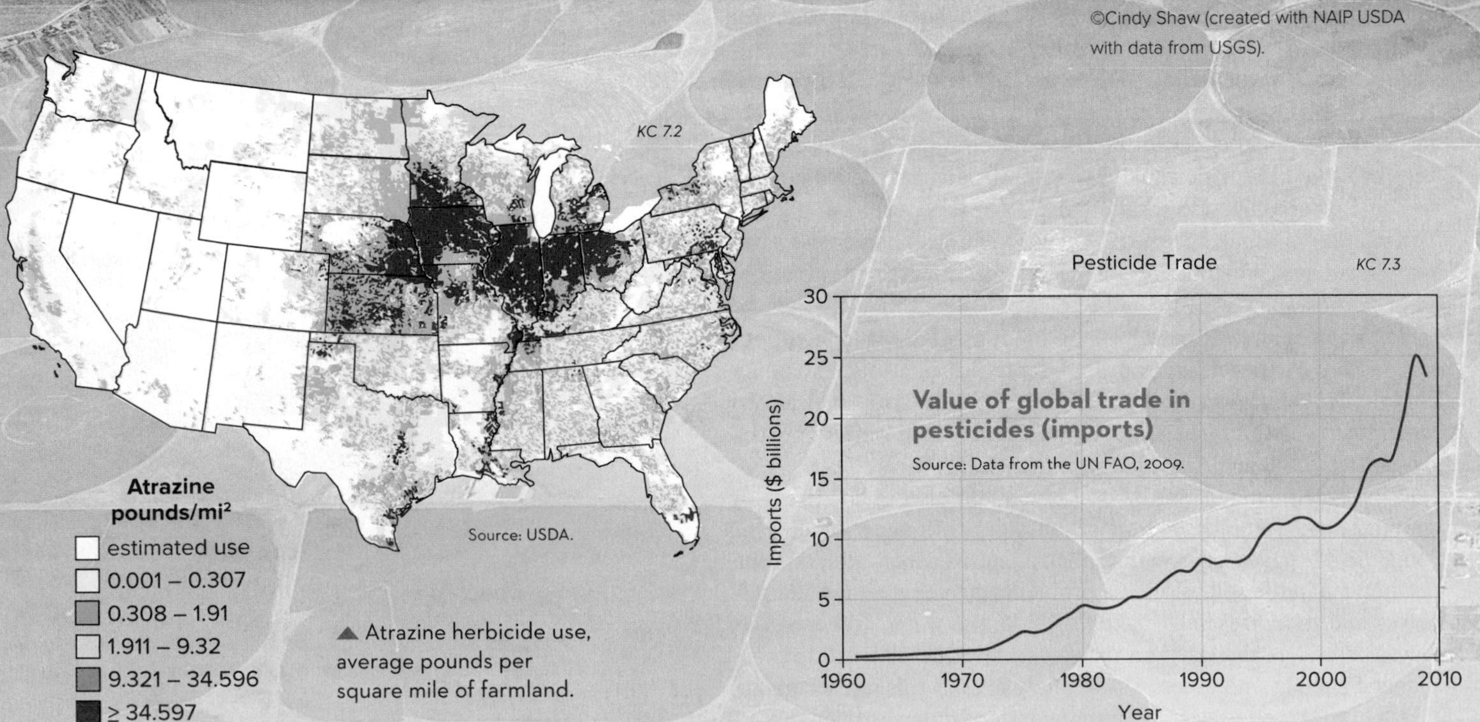

▲ Atrazine herbicide use, average pounds per square mile of farmland.

Source: National Agricultural Imagery Program, USDA

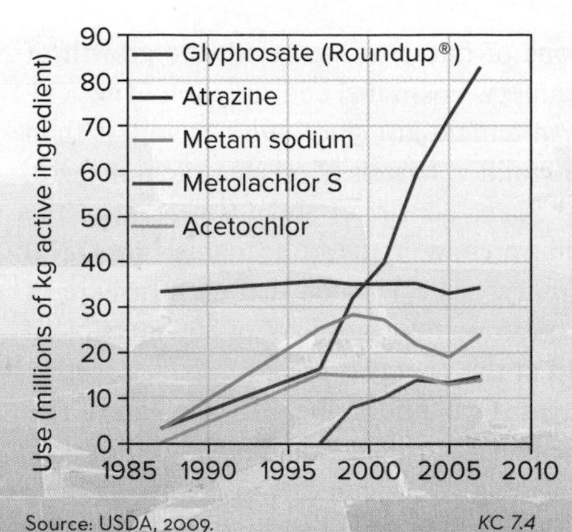

Source: USDA, 2009.

KC 7.4

◀ The majority of GM crops are designed to tolerate high doses of herbicides such as glyphosate (Roundup®), which is now the United States' dominant herbicide. Other GM crops produce their own pesticides.

KC 7.5

©Brent Hofacker/Shutterstock

◀ Conflicting evidence

GM "golden rice" is engineered to provide vitamin A. Indian activist Vandana Shiva charges that golden rice provides less nutrition than traditional greens, which are more affordable for the poor.

Genetically modified (GM) crops have borrowed genes inserted into their DNA, allowing them to produce or tolerate new kinds of organic substances. GM seeds have increased the efficiency of many types of farming.

Benefits

- Yields have increased with GM crops, largely because herbicides help them compete with weeds.
- Bt (*Bacillus thuringiensis*) genes originating from soil bacteria provide a natural insecticide to further protect crops. Many GM crops are designed to produce this insecticide.
- GM crops have allowed expansion of agriculture into formerly unfarmed lands, including Brazil's rainforests and Cerrado region.
- Increased global soy production has raised protein consumption rates in many regions, including China.

Problems

- New varieties are expensive, forcing poor farmers into debt. Wealthier farmers and wealthy regions gain economic advantage.
- In some cases growth hormones (as in GM milk products) are suspected of causing premature puberty and other growth anomalies in humans. These effects remain unclear, however.
- Poorer populations in developing areas often rely on leafy greens to provide much of the nutrition in their diet. Increased herbicide use destroys nontarget crops, increasing malnutrition in some areas.
- Herbicides in drinking water have unknown health effects.

©Sapsiwai/Shutterstock

KC 7.6

KC 7.7

©Corbis/VCG/Getty Images

Organic production involves mixed strategies: Crop rotation retains soil fertility; mixed cropping reduces pest risk; organic fertilizers and pesticides reduce costs of commercial inputs. Organic methods are more sustainable but don't lend themselves to the industrial-scale production of conventional farming, which involves vast expanses of a single crop, such as soy or corn.

Benefits

- Input costs (fertilizers, pesticides, fuel) are minimal.
- Organic production can be sustainable as it preserves or even improves soil, water quality, and biodiversity.
- Crop varieties are usually more mixed, which contributes to healthier diets and more stable farm ecosystems.
- Organic methods are traditional in most areas, and low costs make this approach appropriate for poor farmers, especially in developing regions.
- Integrated pest management (small amounts of pesticides together with other strategies) can protect yields.

Problems

- Labor costs can be high for weed and pest control.
- Careful planning and management may be necessary to ensure good crop management. Creative and innovative problem solving is needed, not simple solutions such as spraying fields.
- Organically farmed food can be more expensive in the United States because it lacks the large-scale distribution networks of conventional products and because conventional methods receive abundant subsidies in tax breaks and price-support payments.
- These methods are unfamiliar to many farmers and can be less efficient in terms of volume of production per unit labor.

©Corbis/VCG/Getty Images

KC 7.8

CAN YOU EXPLAIN?

1. **What was the Green Revolution?**
2. **Which are our top three pesticides? Which of these have you heard of previously?**
3. **Think of what you've eaten today. Which of the three agricultural strategies on this page contributed to your food?**

The FAO estimates that 20 percent of all irrigated land is damaged to some extent by waterlogging or salinity. Water conservation techniques can greatly reduce problems arising from excess water use.

Fertilizers boost production

Plants require small amounts of inorganic nutrients from soil. In large-scale farming, fertilizers are used to ensure a sufficient supply of these nutrients. The major elements required by most plants are nitrogen, phosphorus, potassium, and calcium, with lesser amounts of magnesium and sulfur. Nitrogen, a component of all living cells, is the most common limiting factor for plant growth. Phosphorus and potassium can also be limited in supply, so nitrogen, phosphorus, and potassium are our primary fertilizers. Much of the doubling in worldwide crop production since 1950 has involved increased inorganic fertilizer use. In 1950 the average amount of fertilizer used was 20 kg per hectare. In 2000 the global average was 90 kg per hectare.

Overfertilizing is also a common problem. Growing plants can use only limited amounts of nutrients, and the nutrients not captured by crops run off of fields or seep into groundwater. Excess fertilizers contaminate drinking water and destabilize aquatic ecosystems. Phosphates and nitrates from farm fields and cattle feedlots are a major cause of aquatic ecosystem pollution. Nitrate levels in groundwater have risen to dangerous levels in many areas where intensive farming is practiced. Young children are especially sensitive to the presence of nitrates. Using nitrate-contaminated water to mix infant formula can be fatal for newborns.

There is considerable potential for expanding the world's food supply by increasing fertilizer use in low-income countries, although this would threaten water supplies. African farmers, for instance, use an average of only 19 kg of fertilizer per hectare, less than one-fourth of the world average. The developing world could triple its crop production by raising fertilizer use to the world average, according to some estimates. Finite global supplies of phosphorus would deplete rapidly with this increase in use, however.

Enriching organic material in soils is an alternative way to fertilize crops. Manure is an important natural source of soil nutrients. Cover crops also can be grown and then plowed into the soil. Nitrogen-fixing bacteria living symbiotically in root nodules of legumes are valuable for making nitrogen available as a plant nutrient (see chapter 2). Interplanting and rotating beans or some other leguminous crop with such crops as corn and wheat are traditional ways of increasing nitrogen availability.

Modern agriculture runs on oil

All industrialized agriculture depends on energy. Oil runs tractors and and it delivers crops to market. Fertilizer derives from natural gas, which provides both the energy for fertilizer production and the hydrogen in ammonia fertilizer (NH_3), the most common component. Chemical pesticides are oil derivatives. Irrigation pumps run on oil. Drying and processing crops use gas, oil, and electricity.

Reliance on fossil fuels began in the 1920s, with the adoption of tractors in place of draft animals to pull plows and harvesters. Energy use has risen sharply since the 1940s, with the invention of industrially produced ammonia fertilizer, with pesticides that make it possible to grow vast regions of one crop, and with the growth of tractor and harvesting machinery. Fossil fuel consumption in agriculture has increased more than tenfold just since 1960, from less than 24 million to more than 260 million barrels of oil equivalent in 2016.

Farming in the United States consumes about 9 percent of the total energy we use, depending on how inputs are accounted for. David Pimentel, of Cornell University, has calculated that each hectare of corn in the United States consumes the equivalent of 800 liters of oil (5 barrels' worth). One-third of this energy is used in producing nitrogen fertilizer from natural gas. One-third is fuel for machinery. The final third is for irrigation, chemical pesticides, and other fertilizers.

The average food item in an American diet also travels about 2,000 km (1,250 mi) from the farm. These food miles account for only a small part of the energy involved in production. Still, many consumers try to reduce food miles in their purchases as they try to support the local farming economy.

The net effect of this energy use is growing greenhouse gas emissions. Agriculture contributes about 25 percent of global greenhouse gases. Carbon dioxide emissions account for the bulk of these emissions (fig. 7.22), but an additional factor is methane from the digestion of high-fiber diets among cattle and sheep as well as from anaerobic bacteria in wet rice paddies. Nitrous oxide (N_2O), mainly from ammonia fertilizers, is also an important trace factor.

Pesticide use continues to rise

Biological pests reduce crop yields and spoil as much as half of the crops harvested every year in some areas. Modern agriculture depends on chemicals to kill these pests. There are concerns about the types and amounts of pesticides we use, but our reliance on them has grown steadily over the years (fig. 7.22).

Traditional strategies for evading pest damage often involved mixed crops and crop rotation. With many small patches of multiple crops, pest populations tend to remain relatively small. Reliance on a suite of different food sources can also reduce people's vulnerability to a particular crop pest. Rotating crops from one year to another reduces pests' ability to build up populations over time, and traditional crops often include varieties selected for pest resistance.

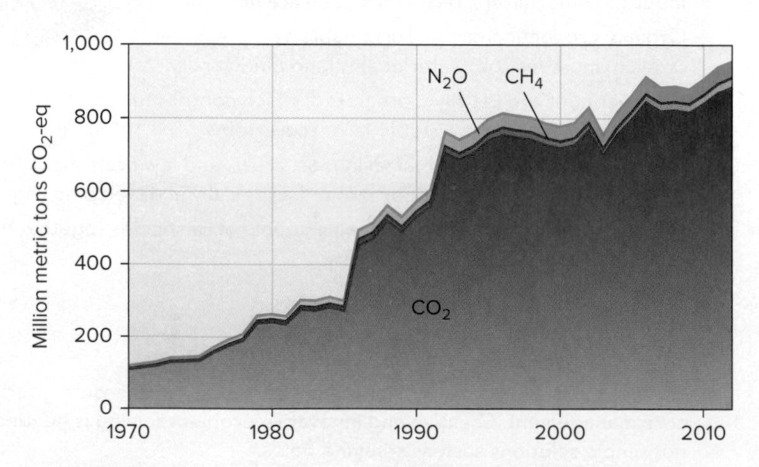

▲ **FIGURE 7.22** Greenhouse gas emissions from energy use in agriculture: CO_2, CO_2-equivalent (for CH_4 and N_2O). Non-energy emissions, such as methane from enteric fermentation and rice paddy cultivation, amount to an additional 5–6 million metric tons CO_2-equivalent. Source: Data from FAO 2018

▲ **FIGURE 7.23** Spraying pesticides by air is quick and cheap, but toxins often drift to nearby fields. ©Philip Wallick/Corbis

Modern agriculture, however, involves vast expanses of a single crop, often with little genetic variation, which increases the need for new methods of pest control. The invention of synthetic organic chemicals, such as DDT (dichlorodiphenyltrichloroethane), transformed our approach to pest control. These chemicals have been an important part of our increased crop production and have helped control many disease-causing organisms.

Organophosphates are the most abundantly used synthetic pesticides. These can be designed to prevent growth of broad-leaf weeds or to attack the nervous systems of insect pests. Glyphosate, the single most heavily used herbicide in the United States, is also known by a variety of trade names, including Roundup®. Glyphosate is applied to 94 percent of U.S. soybeans. "Roundup Ready®" soybeans are one of the most commonly used type of genetically modified crop (see section 7.6), and they dominate the dozens of major crops on which glyphosate is used (see fig. 7.1). In the United States about twice as much glyphosate is used as atrazine, the second most heavily used herbicide (see Key Concepts, p. 169). **Chlorinated hydrocarbons,** also called organochlorines, are persistent and highly toxic to sensitive organisms. Atrazine, which is applied to 96 percent of all U.S. corn, was the most abundantly used pesticide until about 1998, when Roundup Ready® soy became available.

Other important pesticides include fumigants; highly toxic gases such as methylene bromide, which is used to kill fungus on strawberries and other low-growing crops; inorganic pesticides, such as arsenic and copper; and natural "botanical" pesticides derived from plants, such as nicotinoid alkaloids derived from tobacco.

Pesticides are thus important to us, but indiscriminate use has serious effects. Chemicals drifting or washing away from fields endanger nontarget species (fig. 7.23). Often highly persistent and mobile in the environment, many pesticides move through air, water, and soil and bioaccumulate in food chains. Most famously, bioaccumulation of DDT was responsible for nearly eliminating a variety of predatory birds, including peregrine falcons, brown pelicans, and bald eagles, in the 1960s. Many of our dominant pesticides, including atrazine and glyphosate, are known to cause developmental deformities in aquatic organisms such as frogs, salamanders, and fish. Ecologists and endocrinologists have repeatedly found links between these agricultural chemicals and the disappearance of frogs and salamanders in farming regions.

As discussed in the opening case study, one of the most urgent consequences of heavy reliance on pesticides has been the evolution of pesticide-resistant varieties of weeds and parasites. We know that a species' traits can shift if environmental conditions change (chapter 2). In the presence of pesticides, a few tolerant individuals often survive and reproduce, and tolerance of the pesticide can become dominant in a population. New pesticide-tolerant varieties of weeds or parasites then emerge. Around 400 different cases of pesticide-resistant weed populations have been recorded by the Weed Science Society of America (fig. 7.24). In just a decade of glyphosate use, scores of glyphosate-resistant weeds have appeared worldwide. Use of increasingly complex and toxic cocktails of pesticides to combat these new pests results in a **pesticide treadmill** as new resistant strains arise, causing further increases in pesticide applications, which lead to more resistant strains, which require expanded pesticide use.

Your exposure to pesticides is probably higher than you suspect. A study by the U.S. Department of Agriculture found that 73 percent of conventionally grown foods (out of 94,000 samples assayed) had residues from at least one pesticide. (It is important to note, however, that your greatest risk probably involves household pesticides, the most rapidly growing pesticide market.)

Alternatives for reducing our dependence on chemical pesticides include management changes, such as using cover crops and mechanical cultivation, and planting mixed polycultures

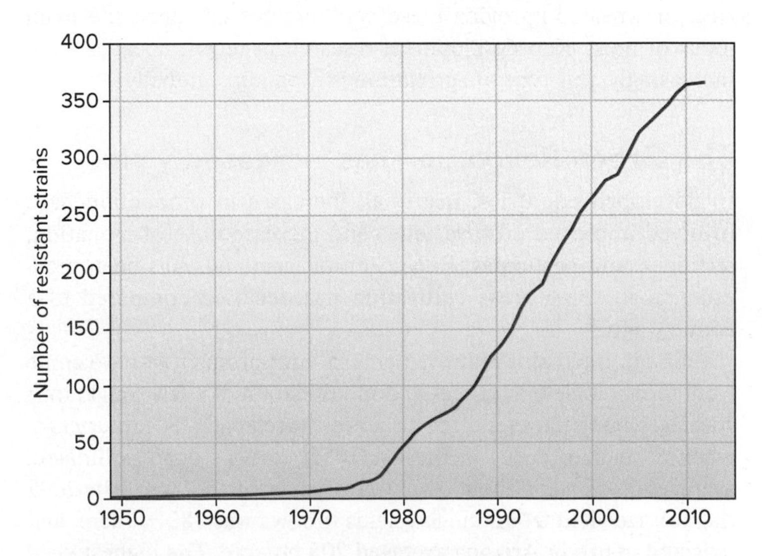

▲ **FIGURE 7.24** Pesticide-resistant varieties have increased, resulting in a "pesticide treadmill," in which farmers must use increasingly expensive combinations of chemicals to suppress weeds, diseases, and pests. Sources: Data from U.S. EPA; Weed Society of America.

rather than vast monoculture fields. Biological controls, such as insect predators or pathogens, can help reduce chemical use. Genetic breeding and biotechnology can produce pest-resistant crop and livestock strains as well. Integrated pest management (IPM) combines all of these alternative methods, together with judicious use of synthetic pesticides under precisely controlled conditions. Consumers can also learn to accept less than perfect fruits and vegetables.

7.6 HOW HAVE WE MANAGED TO FEED BILLIONS?

- *The Green Revolution involved breeding grains for high-yield with fertilizer and irrigation inputs.*
- *Genetic engineering modifies DNA, mainly for herbicide tolerance.*
- *Effects of genetic engineering are often more economic than environmental.*

In most of the world today, as throughout human history, smallholder farmers feed most of the world's population. In wealthy countries, agriculture is largely an industrial-scale process, but in developing regions, small farms of a few hectares or less still employ about one-third of the population. Although these small farms occupy only about half of the world's farmland, the FAO reports that they produce about 80 percent of the world's food.

Commercial-scale agriculture produces the other 20 percent of our food, but much of its production is used for biofuel, for animal feed (which we eat indirectly), for cotton textiles, for food additives, or for other uses.

However, many residents of the developed countries, and of the world's cities, are increasingly accustomed to a rich and varied diet that depends on global trade and large-scale production of commodity crops—especially grains and oil crops, such as corn, soy, rice, or wheat. Improving these commodities has been the main focus of most crop development research in recent decades. And increasingly, this type of agriculture is expanding globally.

The Green Revolution has increased yields

In developed countries, nearly all increases in production have involved improved crop varieties and increased use of irrigation, fertilizer, and pesticides. Expansion of farmlands has been modest, and in some areas cultivation has declined compared to a century ago.

So far, the major improvements in farm production have come from technological advances and modification of a few well-known species. Yield increases often have been spectacular. A century ago, when all maize (corn) in the United States was open pollinated, average yields were about 25 bushels per acre (bu/acre). In 2009 the average yield from rain-fed fields in Iowa was 182 bu/acre, and irrigated maize in Arizona averaged 208 bu/acre. The highest yield ever recorded in field production was 370 bu/acre on an Illinois farm, but theoretical calculations suggest that 500 bu/acre (32 metric tons per hectare) could be possible. Most of this gain was

▲ **FIGURE 7.25** Semidwarf wheat (*right*), bred by Norman Borlaug, has shorter, stiffer stems and is less likely to lodge (fall over) when wet than its conventional cousin (*left*). This "miracle" wheat responds better to water and fertilizer and has played a vital role in feeding a growing human population. ©William P. Cunningham

accomplished by use of synthetic fertilizers along with conventional plant breeding: geneticists laboriously hand-pollinating plants and looking for desired characteristics in the progeny.

Starting about 50 years ago, agricultural research stations began to breed tropical wheat and rice varieties that would provide food for growing populations in developing countries. The first of the "miracle" varieties was a dwarf, high-yielding wheat developed by Norman Borlaug (who received a Nobel Peace Prize for his work) at a research center in Mexico (fig. 7.25). At about the same time, the International Rice Institute in the Philippines developed dwarf rice strains with three or four times the production of varieties in use at the time. The spread of these new high-yield varieties around the world has been called the **Green Revolution.** It is one of the main reasons that world food supplies have more than kept pace with the growing human population over the past few decades.

Most Green Revolution breeds really are "high responders," meaning that they yield well in response to optimal inputs of fertilizer, water, and chemical protection from pests and diseases (fig. 7.26). With fewer inputs, on the other hand, high responders may not produce as well as traditional varieties. Poor farmers who can't afford the expensive seed, fertilizer, and water required to become part of this movement usually are left out of the Green Revolution. In fact, they may be driven out of farming altogether as rising land values and falling commodity prices squeeze them from both sides.

Genetic engineering has benefits and costs

Genetic engineering, splicing a gene from one organism into the chromosome of another, has the potential to greatly increase both the quantity and the quality of our food supply. It is now possible to build entirely new genes, and even new organisms, often called "transgenic" organisms or **genetically modified organisms (GMOs,** or **GM crops),** by taking a bit of DNA from here, a bit from there,

▲ **FIGURE 7.29** Contour plowing and strip cropping help prevent soil erosion on hilly terrain, as well as create a beautiful landscape.
Source: Photo by Lynn Betts, courtesy of USDA Natural Resources Conservation Service

While American agriculture hasn't reached that level of sustainability, federal soil conservation programs have reduced erosion considerably since the disastrous years of the 1920s and 1930s. In a study of one Wisconsin watershed, erosion rates were 90 percent less in 1975–1993 than they were in the 1930s.

Among the most important elements in soil conservation are terracing, groundcover, and reduced tillage. Most erosion happens because water runs downhill. The faster it runs, the more soil it carries off the fields. Comparisons of erosion rates in Africa have shown that a 5 percent slope in a plowed field has three times the water runoff volume and eight times the soil erosion rate of a comparable field with a 1 percent slope. Water runoff can be reduced by grass strips in waterways and by **contour plowing**—that is, plowing across the hill rather than up and down. Contour plowing is often combined with **strip-farming,** the planting of different kinds of crops in alternating strips along the land contours (fig. 7.29). When one crop is harvested, the other is still present to protect the soil and keep water from running straight downhill. The ridges created by cultivation trap water, allowing it to seep into the soil rather than running off.

Terracing involves shaping the land to create level shelves of earth to hold water and soil. The edges of the terrace are planted with soil-anchoring plant species. This is an expensive procedure, requiring either much hand labor or expensive machinery, but it makes farming of very steep hillsides possible. The rice terraces in the Chico River Valley in the Philippines rise as much as 300 m (1,000 ft) above the valley floor. They are considered one of the wonders of the world. In many regions, terrace systems have supported farming communities for centuries.

TABLE 7.2 Soil Cover and Soil Erosion

CROPPING SYSTEM	AVERAGE ANNUAL SOIL LOSS (TONS/HECTARE)	PERCENT RAINFALL RUNOFF
Bare soil (no crop)	41.0	30
Continuous corn	19.7	29
Continuous wheat	10.1	23
Rotation: corn, wheat, clover	2.7	14
Continuous bluegrass	0.3	12

Source: Based on 14 years of data from Missouri Experiment Station, Columbia, Missouri.

Groundcover, reduced tilling protect soil

Annual row crops, such as corn or beans, generally cause the highest erosion rates because they leave soil bare for much of the year (table 7.2). Often the easiest way to provide cover that protects soil from erosion is to leave crop residues on the land after harvest (fig. 7.30). They not only cover the surface to break the erosive effects of wind and water but also reduce evaporation and soil temperature in hot climates and protect ground organisms that help aerate and rebuild soil. In some experiments, crop residue increased water infiltration 99 percent and reduced soil erosion 98 percent.

Leaving crop residues on the field also can increase disease and pest problems, however, and conservation tillage often depends on increased use of pesticides (insecticides, fungicides, and herbicides) to control insects, weeds, and disease. Increased use of agricultural chemicals can diminish the overall environmental benefits of conservation tilling. Increased pesticide use is not always necessary, however. Pesticide use can be minimized through crop rotation and integrated pest management strategies such as trap crops, natural repellents, and biological controls.

Often soil is protected with cover crops such as rye, alfalfa, or clover, which can be planted immediately after harvest to hold and

▲ **FIGURE 7.30** No-till planting involves drilling seeds through debris from last year's crops. Here, soybeans grow through corn mulch. Debris keeps weeds down, reduces wind and water erosion, and keeps moisture in the soil.
©William P. Cunningham

protect the soil. These cover crops can be plowed under at planting time to add organic matter and nutrients to the soil.

In some cases, interplanting of two different crops in the same field not only protects the soil but also is a more efficient use of the land, providing double harvests. Native Americans and pioneer farmers, for instance, planted beans or pumpkins between the corn rows. The beans provided nitrogen needed by the corn, the pumpkins crowded out weeds, and both crops provided foods that nutritionally balance corn. Traditional swidden cultivators in Africa and South America often plant as many as 20 different crops together in small plots. The crops mature at different times, so that there is always something to eat, and the soil is never exposed to erosion for very long. Mixed cultivation can include highly valuable crops, as well as helping to conserve biological diversity (What Do You Think?, p. 177).

Low-input, sustainable agriculture can benefit people and the environment

In contrast to the trend toward industrialization and dependence on chemical fertilizers, pesticides, antibiotics, and artificial growth factors common in conventional agriculture, some farmers are going back to a more natural, agroecological farming style. Finding that they can't—or don't want to—compete with factory farms, these producers are making money and staying in farming by returning to small-scale, low-input agriculture. Brent and Regina Beidler, for example, run one of nearly 200 small, organic dairies in Vermont (fig. 7.31). With just 40 cows, the Biedlers move the cattle every 12 hours, after each milking, to a different paddock. This rotation, with intensive grazing in small paddocks, method mimics the movement of herds of natural grazers. Travelling together in tight herds, the cows are forced to eat unpalatable plants as well as the sweet grasses, so that pastures stay healthy and diverse. With time to regenerate between visits, vegetation in the paddocks can remain healthy and diverse, and soil compaction is minimized.

▲ **FIGURE 7.31** Brent Beidler, of Randolph Center, Vermont, keeps his cows and his pasture healthy with intensive rotational grazing, without chemical pesticides or fertilizers. ©Melanie Stetson Freeman/Getty Images

As on other organic dairies, the Beidlers use no chemical fertilizer or pesticides on pastures. Also like other organic farmers, the Beidlers chose to go organic because it made sense for sustainability of the land, as well as for the health of the cows and the quality of the milk produced. In an organic operation, antibiotics are used only to fight diseases. Cows are kept healthy on a diet of mainly grass and hay, with less grain and antibiotics than in a larger commercial dairy. Studies at Iowa State University have shown that animals on pasture grass rather than grain reduces nitrogen runoff by two-thirds while cutting erosion by more than half.

Low-input farms such as these typically don't turn out the quantity of meat or milk that their intensive agriculture neighbors do, but their production costs are lower, and they get higher prices for their crops, so that the all-important net gain is often higher.

Preserving small-scale family farms also helps preserve rural culture. As Marty Strange of the Center for Rural Affairs in Nebraska asks, "Which is better for the enrollment in rural schools, the membership of rural churches, and the fellowship of rural communities—two farms milking 1,000 cows each or twenty farms milking 100 cows each?" Family farms help keep rural towns alive by purchasing machinery at the local implement dealer, gasoline at the neighborhood filling station, and groceries at the mom-and-pop grocery store.

Consumer choices benefit local farm economies

Since the 1960s, U.S. farm policy and agricultural research have focused on developing large-scale production methods that use fertilizer, pesticides, breeding, and genetic engineering to provide abundant, inexpensive grain, meat, and milk. As a consequence, we can afford to eat more calories and more meat than ever before. Hunger still plagues many regions, but increasing nutrition has reached most of the world's population, not just wealthy countries. We now worry about weight-associated illnesses as a cause of mortality, possibly a first in human history.

Cheap-food policies have raised production dramatically to feed growing populations, both in the United States and abroad. Farm commodity prices have fallen so low that we now spend billions of dollars every year to support prices or repay farmers whose production expenses are greater than the value of their crops. The United States buys millions of tons of surplus food every year, often using it as food aid to famine-stricken regions. This social good is also problematic, as cheap or free food donations undermine small farmers in other countries, who cannot compete with free or nearly free food on the local market.

These policies are deeply entrenched in our way of life, politics, and food systems. But there may be some steps that consumers can take to support the beneficial changes in farming methods and farm policies, while reducing their negative effects.

Supporting local farmers can have a variety of benefits, from keeping money in the local economy to ensuring a fresh and healthy diet. Maintaining a viable farm economy can also help slow the conversion of farmland into expanding suburban subdivisions. As many farmers point out, you don't need to eat 100 percent local. Converting just part of your shopping activity to locally produced goods can make a big difference to farmers.

What Do YOU THINK?

Shade-Grown Coffee and Cocoa

Cocoa pods grow directly on the trunk and large branches of cocoa trees.

©William P. Cunningham

Do your purchases of coffee and chocolate help to protect or destroy tropical forests? Coffee and cocoa are two of the many products grown exclusively in developing countries but consumed almost entirely in the wealthier, developed nations. Coffee grows in cool, mountain areas of the tropics, while cocoa is native to the warm, moist lowlands. What sets these two apart is that both come from small trees adapted to grow in low light, in the shady understory of a mature forest. **Shade-grown** coffee and cocoa (grown beneath an understory of taller trees) allow farmers to produce a crop at the same time as forest habitat remains for birds, butterflies, and other wild species.

Until a few decades ago, most of the world's coffee and cocoa were shade-grown. But new varieties of both crops have been developed that can be grown in full sun. Growing in full sun, trees can be crowded together more closely. With more sunshine, photosynthesis and yields increase.

There are costs, however. Sun-grown trees die earlier from stress and diseases common in crowded growing conditions. Crowding also requires increased use of expensive pesticides and fungicides. Shade-grown coffee and cocoa generally require fewer pesticides (or sometimes none) because the birds and insects residing in the forest canopy eat many of the pests. Ornithologists have found as little as 10 percent as many birds in a full-sun plantation, compared to a shade-grown plantation. The number of bird species in a shaded plantation can be twice that of a full-sun plantation. Shade-grown plantations also need less chemical fertilizer because many of the plants in these complex forests add nutrients to the soil. In addition, shade-grown crops rarely need to be irrigated because heavy leaf fall protects the soil while forest cover reduces evaporation.

Over half the world's coffee and cocoa plantations have been converted to full-sun varieties. Thirteen of the world's 25 biodiversity hot spots occur in coffee or cocoa regions. If all the 20 million ha of coffee and cocoa plantations in these areas are converted to monocultures, an incalculable number of species will be lost.

The Brazilian state of Bahia demonstrates both the ecological importance of these crops and how they might help preserve forest species. At one time, Brazil produced much of the world's cocoa, but in the early 1900s, the crop was introduced into West Africa. Now Côte d'Ivoire alone grows more than 40 percent of the world total. Rapid increases in global supplies have made prices plummet, and the value of Brazil's harvest has dropped by 90 percent. Côte d'Ivoire is aided in this competition by a labor system that reportedly includes widespread child slavery. Even adult workers in Côte d'Ivoire get only about $165 (U.S.) per year (if they get paid at all), compared with a minimum wage of $850 (U.S.) per year in Brazil. As African cocoa production ratchets up, Brazilian landowners are converting their plantations to pastures or other crops.

The area of Bahia where cocoa was once king is part of Brazil's Atlantic Forest, one of the most threatened forest biomes in the world. Only 8 percent of this forest remains undisturbed. Although cocoa plantations don't have the full diversity of intact forests, they do provide an economic rationale for preserving the forest. And Bahia's cocoa plantations protect a surprisingly large sample of the biodiversity that once was there. Brazilian cocoa will probably never be as cheap as that from other areas. There is room in the market, however, for specialty products. If consumers choose to pay a small premium for organic, fair-trade, shade-grown chocolate and coffee, it might provide the incentive needed to preserve biodiversity. Wouldn't you like to know that your chocolate or coffee wasn't grown with child slavery and is helping protect plants and animal species that might otherwise go extinct? What does it take to make that idea spread?

Farmers' markets are usually the easiest way to eat locally (fig. 7.32). The produce is fresh, and profits go directly to the farmer who grows the crop. "Pick your own" farms also let you buy fresh fruit and other products—and they make a fun social outing. Many conventional grocery stores also now offer locally produced, organic, and pesticide-free foods. Buying these products may (or may not) cost a little more than nonorganic and nonlocal produce, but they can be better for you and they can help keep farming and fresh, local food in the community.

Many colleges and universities have adopted policies to buy as much locally grown food as possible. Because schools purchase a lot of vegetables, meat, eggs, and milk, this can mean a large amount of income for local and regional farm economies. Although this policy can take more effort and creativity than ordering from centralized, national distributors, many college food service administrators are happy to try to buy locally, if they see that students are interested. If your school doesn't have such a policy, perhaps you could talk to administrators about starting one.

Many areas also have "community supported agriculture" (CSA) projects, farms supported by local residents who pay ahead of time for shares of the farm's products, which can vary from vegetables to flowers to meat and eggs. CSAs require a lump payment early in the season, but the net cost of food by the end of the season

▲ **FIGURE 7.32** Your local farmers' market is a good source of locally grown organic produce. ©William P. Cunningham

is often less than you would have paid at the grocery store. You also get to meet interesting people and learn more about your local area by participating in a CSA.

You can eat low on the food chain

Because there is less energy involved in producing food from plants than producing it from animals, one way you can reduce your impact on the world's soil and water is to eat a little more grains, vegetables, and dairy and a little less meat. This doesn't mean turning vegetarian—unless you choose to do so. Just returning to the level of protein and fat consumption your grandparents had could make a big difference for the environment and for your health.

Low-input, organic foods also reduce the environmental impact of your food choices. When you buy organic food, you support farmers who use no pesticides or artificial fertilizers. Often these farmers use crop rotations to preserve soil nutrients and manage erosion carefully, to prevent loss of their topsoil. Sometimes these farmers preserve diverse varieties of crops, helping to maintain genetic diversity and pest resistance in crops.

Grass-fed beef and free-range poultry or pork can also be excellent low-input foods. By converting grass to protein, they can be an efficient food source where soils or steep hillsides are unsuitable for cropping. With good management, pastures have minimal soil erosion, because vegetation keeps the soil covered year-round.

CONCLUSION

Food production has grown faster than the human population in recent decades, and the percentage of people facing chronic hunger has declined, although the number has increased. Most of us consume more calories and protein than we need, but some 900 million people still are malnourished. Much, or perhaps most, hunger results from political instability, which displaces farmers, inhibits food distribution, and undermines local farming economies.

Increases in food production result from many innovations in agricultural production. The Green Revolution produced new varieties that yield more crops per hectare, although these crops usually require extra inputs such as fertilizers, irrigation, or pesticides. Genetically modified organisms have also increased yields. Most GMOs are designed to tolerate herbicides, and many produce their own pesticides. Confined animal feeding operations, made possible by large-scale corn and soy production, have greatly increased the efficiency of producing meat. The rise of soy production globally and of meat consumption in the United States, China, and elsewhere results from these innovations.

These changes bring about important environmental effects, such as soil erosion and degradation as well as water contamination from pesticide and fertilizer applications. Health effects are also a concern, including weight-related diseases, such as diabetes, exposure to agricultural chemicals, and antibiotic resistance. There are also many more sustainable approaches, which offer the opportunity to regenerate soils and maintain healthy ecosystems: These systems can offer significant economic advantages to farmers, especially in developing regions, but they are not widely adopted because they are poorly suited to mass production. Consumers can help support sustainable practices by eating locally, eating low on the food chain, buying organic foods or grass-fed meat, and shopping at farmers' markets.

PRACTICE QUIZ

1. What are some of the factors that led to dramatic production increases in Illinois soy fields?
2. Explain how soybeans from Illinois can improve diets in China.
3. What does it mean to be chronically undernourished? How many people in the world currently suffer from this condition?
4. Why do nutritionists worry about food security? Who is most likely to suffer from food insecurity?
5. Describe the conditions that constitute a famine. Why does Amartya Sen say that famines are caused more by politics and economics than by natural disasters?
6. Define *malnutrition* and *obesity*. How many Americans are now considered obese?
7. What three crops provide most human caloric intake?
8. What are confined animal feeding operations, and why are they controversial?
9. What is soil? Why are soil organisms so important?
10. What are four dominant types of soil degradation? What is the primary cause of soil erosion?
11. What do we mean by the Green Revolution?
12. What is genetic engineering, and how can it help or hurt agriculture?
13. What is sustainable agriculture?
14. How could your choices of coffee or cocoa help preserve forests, biodiversity, and local economies in tropical countries?
15. What are the economic advantages of low-input farming?

CRITICAL THINKING AND DISCUSSION

Apply the principles you have learned in this chapter to discuss these questions with other students.

1. Explain the nature of hunger in the world today. How much should we worry about hunger increasing? Why?
2. Review some of the major reasons for global hunger. Which ones are easiest to resolve? How might we approach them?
3. Debate the claim that famines are caused more by human actions (or inactions) than by environmental forces. What evidence would you need to have to settle this question? What hypotheses could you test to help resolve the debate?
4. Should farmers be forced to use ecologically sound techniques that serve their long-term interests, regardless of short-term costs? How might this be done?
5. What aspects of genetically modified food products do you find most beneficial? Which are most worrisome? Are your answers the same as those of your fellow students or family members?
6. Suppose that you were engaged in genetic engineering or pesticide development. What environmental or social safeguards would you impose on your own research, if any? What restrictions would you tolerate from someone else concerned about the effects of your work?

DATA ANALYSIS Mapping Your Food Supply

Understanding where your food comes from helps you understand the environmental questions involved with producing the food you eat. Because this information is so important, culturally and economically, the United States Department of Agriculture (USDA) publishes maps and statistical analysis of our major crops. This rich repository of information lets you explore where your food comes from.

Go to Connect to find a link to some of these maps and to demonstrate your understanding of where we produce some of our major foods.

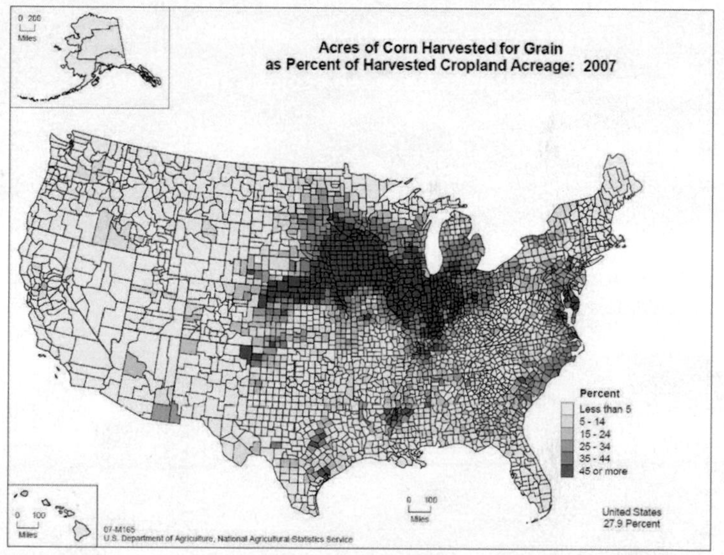

Source: U.S. Department of Agriculture, National Agricultural Statistics Service.

CHAPTER

8 Environmental Health and Toxicology

LEARNING OUTCOMES

After studying this chapter, you should be able to answer the following questions:

- What is environmental health?
- What health risks should worry us most?
- Emergent diseases seem to be more frequent now. What human factors may be involved in this trend?
- Are there connections between ecology and our health?
- When Paracelsus said, "The dose makes the poison," what did he mean?
- What makes some chemicals dangerous and others harmless?
- How much risk is acceptable and to whom?

What hazardous and toxic substances were in the floodwaters that swamped Houston, Texas, after Hurricane Harvey?

©Joe Raedle/Getty Images

CASE STUDY

A Toxic Flood

Houston, Texas, leads the nation in petroleum refining and petrochemical production, with 25 oil refineries, 63 chemical manufacturing plants, and 2,500 other chemical facilities spread across its low-lying coastal plain. Over the years these industries have created huge quantities of harmful by-products. State and federal agencies list 41 Superfund sites in the region, and the EPA's Toxic Release Inventory monitors 495 sites within the Houston metropolitan area, which generate a cumulative 750,000 metric tons (1.5 billion pounds) of toxic waste per year. Much of this material is held in storage tanks or open lagoons until it can be treated.

Even under normal conditions, a significant amount of this waste evaporates into the air or leaks into local waters. On an average day in the neighborhoods around these industrial facilities, the air usually looks bad and smells worse. High levels of nitrogen oxides, benzene, sulfur dioxides, and other harmful pollutants are common. Residents in industrial areas have cancer rates 20 percent higher than in the rest of the city. (Although it's not easy to prove absolutely that this is due to their environment, and not other factors.)

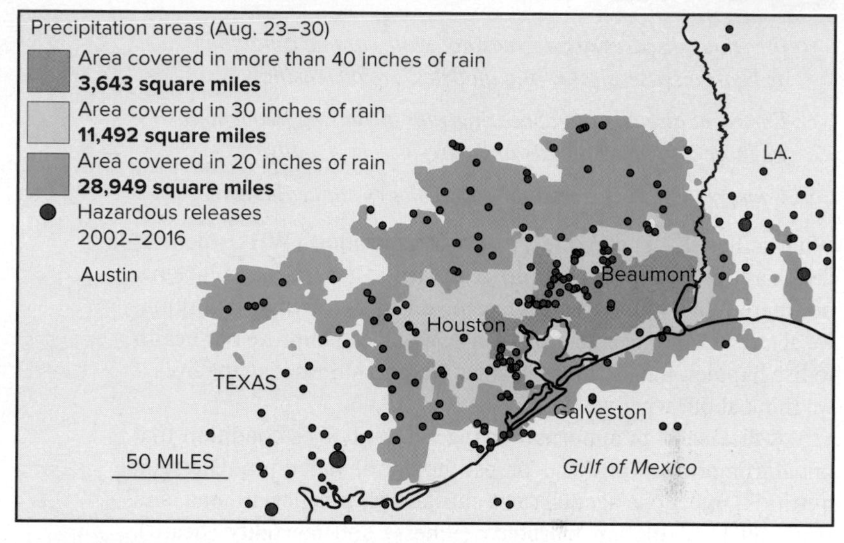

▲ **FIGURE 8.1** Hurricane Harvey produced up to 60 inches of rain in 3 days in Houston, TX (shown by brown outline). The storm flooded hundreds of facilities that produce or use hazardous substances. Source: Dr. Shane Hubbard, Cooperative Institute for Meteorological Satellite Studies

Pollution levels soared in late August 2017 after Hurricane Harvey slammed into the coast as a Category 4 storm with winds of 209 km/hr (130 mph). Some areas of Houston got more than 1.5 m (60 in.) of rain in just 3 days (fig. 8.1). This was over twice the average yearly rainfall in half a week. Combined with a storm surge that pushed Gulf water far inland, storm water flooded an estimated three-quarters of Harris County to more than 0.5 m (1.5 ft).

As the floods spread, waste storage lagoons and sewage treatment plants overflowed, releasing thousands (or perhaps millions) of tons of harmful chemicals and infectious organisms into the water and air. Thirteen of the region's 41 Superfund sites were flooded. Fires flared up as tanks exploded. First responders complained of headaches and vomiting, and subsequent lawsuits claimed that dioxins, furans, and other cancer-causing or toxic compounds were released by ruptured, burning storage tanks.

One of the largest fires was at the Arkema chemical plant in the Houston suburb of Crosby. The plant made organic peroxides, used in plastic manufacturing. Peroxides are highly volatile (they evaporate easily) and highly flammable. Constant refrigeration is required to prevent pressure buildup in storage tanks. So nobody was surprised when, after power lines were blown down and floods shut down emergency backup generators, the refrigeration systems quit and tanks exploded. The factory burned for days. As much as 30 metric tons of sulfur dioxide and other toxic substances may have been released from this factory alone.

Hurricane flooding created a toxic stew of pollutants and untreated sewage that filled the city. Health officials warned Houston residents to avoid all contact with the storm water and to discard any furnishings, food, furniture, or clothing that had come into contact with it. Among the contaminants suspected to have been released into the flood were dioxins, polychlorinated biphenyls (PDBs), phenols, creosote, arsenic, lead, chromium, and pesticides, as well as pathogenic bacteria. Many of these chemicals are no longer used in commercial products, but residues remaining in soil were leached out by the storm.

How dangerous were the Houston floodwaters? That depends on how many people were exposed, for how long, and to what contaminants. Pollutants vary dramatically in effects: Raw sewage can cause diarrhea and, in worst cases, spread waterborne diseases such as cholera or dysentery. Petrochemicals released from industrial sites are toxic, causing rashes, respiratory obstruction, liver damage, and cancer. Effects depend on the solubility, reactivity, and mobility of different compounds. Bioaccumulation, biomagnification, and synergistic effects also can increase risks. Most of us will never experience the crisis conditions of the Houston flood, but we all come into contact in our ordinary life to at least some of the thousands of industrial chemicals found there. What are the effects of a lifetime exposure to low levels of these compounds? These are questions we explore in this chapter. ■

> *To wish to become well is a part of becoming well.*
>
> —SENECA

8.1 ENVIRONMENTAL HEALTH

- *Infectious diseases are decreasing, while chronic conditions, such as heart disease, stroke, and diabetes, are increasing.*
- *Emergent diseases are appearing with increasing frequency as we move into new habitats and carry germs globally.*
- *Conservation medicine combines ecology with health care.*

What is health? The World Health Organization (WHO) defines **health** as a state of complete physical, mental, and social well-being, not merely the absence of disease or infirmity. By that definition, we all are ill to some extent. Likewise, we all can improve our health to live happier, longer, more productive, and more satisfying lives if we think about what we do.

A **disease** is an abnormal change in the body's condition that impairs important physical or psychological functions. Diet and nutrition, infectious agents, toxic substances, genetics, trauma, and stress all play roles in **morbidity** (illness) and **mortality** (death). **Environmental health** focuses on environmental factors that cause disease, including elements of the natural, social, cultural, and technological worlds in which we live. WHO estimates that 24 percent of all global disease burden and 23 percent of premature mortality are due to environmental factors. Among children (0 to 14 years), deaths attributable to environmental factors may be as high as 36 percent. Figure 8.2 shows some of these environmental risk factors as well as the media through which we encounter them.

In ecological terms, your body is an ecosystem. Of the approximately 100 trillion cells that make up each of us, only about 10 percent are actually human (Exploring Science, p. 64). The others are bacteria, fungi, protozoans, arthropods, and other species. Ideally, the various organisms in this complex system maintain a harmonious balance. Beneficial species help regulate the dangerous ones. The health challenge shouldn't be to try to totally eradicate all these other species; we couldn't live without them. Rather, we need to find ways to live in equilibrium with our environment and our fellow travelers.

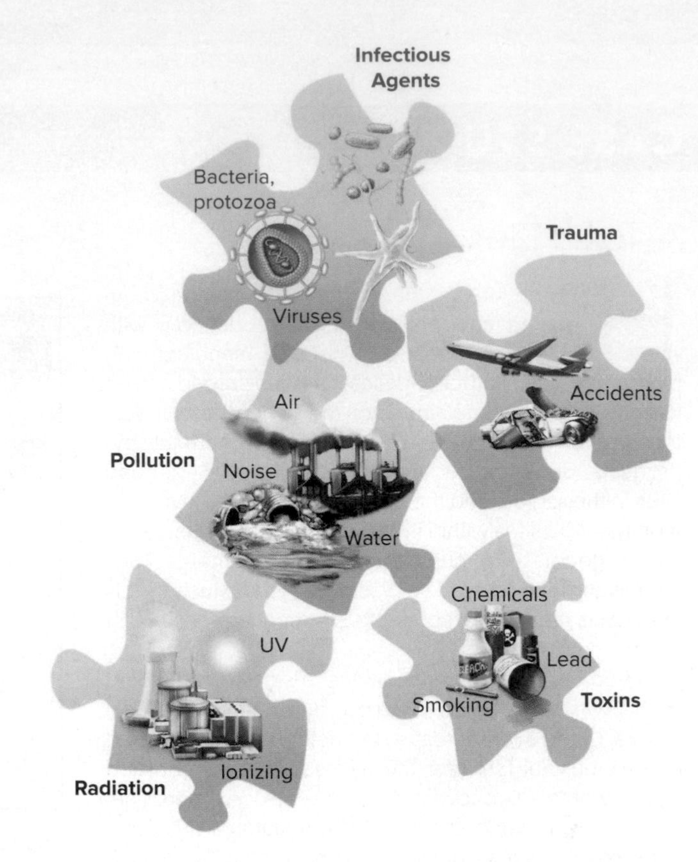

▲ **FIGURE 8.2** Major sources of environmental health risks.

Benchmark Data	
Among the ideas and values in this chapter, these are a few worth remembering for future reference.	
90%	Proportion of cells in your body that are bacteria or other nonhuman cells
1 in 2	Chance of dying from heart disease
1 in 5,000	Chance of dying in airplanes
1 in 4	Number of children in poorest countries who die before age 5
1 in 100	Number of children in richest countries who die before age 5
1	Rank of neonatal conditions in causes of global disease burden
20	Percentage of antibiotics prescribed for inappropriate conditions
70,000	Approximate number of new synthetic chemicals introduced since 1950

Ever since the publication of Rachel Carson's *Silent Spring* in 1962, the discharge, movement, fate, and effects of synthetic chemical toxins have been a special focus of environmental health, but infectious diseases still remain a grave threat. In this chapter we'll study many environmental risk factors. First, however, let's look at some major causes of illness worldwide.

The global disease burden is changing

Although there are many worries about pollution, environmental toxins, emerging diseases, and other health threats, we should take a moment to consider the remarkable progress we've made over the past hundred years in controlling many terrible diseases. Epidemics that once killed millions of people have been reduced or eliminated. Smallpox was completely wiped out in 1977. Polio has been eliminated almost everywhere in the world, except for a few places in Afghanistan, Pakistan, Laos, and Madagascar. Guinea worms, river blindness, and yaws appear to be on their way to elimination. Epidemics of typhoid fever, cholera, and yellow fever that regularly decimated populations a century ago are now rarely encountered in developed countries. AIDS, which once was an immediate death sentence, has become a highly treatable disease. The average HIV-positive person in the United States now lives 24 years after diagnosis if treated faithfully with modern medicines.

One of the best indicators of health gains is a worldwide increase in child survival. Fifty years ago in South Asia, for example, nearly one in four children died before their fifth birthday. Currently, the global 5-year survival rate is over 95 percent. In another couple of decades, if current trends continue, the world under-5 mortality rate should approach that in richer countries of less than 1 percent. Remarkably, East Asia and the Pacific region already have nearly as low a child mortality rate as the United States, despite having a far lower per-capita income. Much of this progress is due simply to better nutrition, improved sanitation, and clean water, which keep children healthy. According to the WHO, half of all gains in child survival are brought about by increases in women's education. Only sub-Saharan Africa has lagged in this health revolution, for a number of reasons that include political instability and poverty.

Improved child survival has important demographic implications, including lowered birth rates and longer life expectancies (see chapter 4). The fact that average global life expectancies have more than doubled over the past century from 30 to 64 years means many more productive years of life for most of us.

The decrease in many infectious illnesses, along with an aging population, is producing a shift in global disease burden. Many people now live for years with chronic illness or disability. According to the WHO, chronic diseases now account for nearly three-fourths of the 56.5 million total deaths worldwide each year and about half of the global disease burden. To account for the costs of chronic illness, health agencies calculate disease burden for a population in terms of **disability-adjusted life years (DALYs).** This statistic is the sum of overall years lost to disability or illness plus years lost to early deaths in a population. For example, a teenager permanently paralyzed by a traffic accident will have many years of suffering, health care costs, and lost potential. Similarly, many years of expected life are lost when a child dies of a neonatal infection. A senior citizen who has a stroke, on the other hand, willl lose far fewer years to disease or disability.

Major causes of death have changed dramatically in recent decades. Chronic conditions, such as cardiovascular disease, cancer, and strokes, were once common only in wealthy countries. These conditions are now among the leading causes of death in lower- and middle-income countries. Although the traditional killers in developing countries–infections, maternal and perinatal (birth) complications, and nutritional deficiencies–still take a terrible toll in the lowest-income areas, diseases such as depression, diabetes, heart attacks, and stroke, which once occurred mainly in rich countries, are rapidly becoming increasingly common everywhere.

Look at the changes in the relative ranking of the major causes of disease burden in table 8.1. Notice how many once-common diseases have been reduced in many cases, while chronic disorders related to lifestyle, age, and affluence are becoming more common. Much of this change in disease burden is occurring in the poorer parts of the world, where people are rapidly adopting the habits and diets of the richer countries. The largest source of unintentional injuries is traffic accidents, while the main sources of intentional injuries are self-harm and interpersonal violence.

TABLE 8.1 Leading Causes of Global Disease Burden

RANK	1990	RANK	2015
1	Respiratory infections	1	Heart disease
2	Diarrhea	2	Respiratory infections
3	Neonatal conditions	3	Stroke
4	Heart disease	4	Preterm birth[2]
5	Stroke	5	Diarrhea
6	COPD[1]	6	Road injury
7	Malaria	7	COPD
8	Tuberculosis	8	Diabetes
9	Malnutrition	9	Birth trauma
10	Neonatal encephalopathy	10	Congenital anomalies
11	Back and neck pain	11	HIV/AIDS
12	Unintentional Injuries	12	Tuberculosis
13	Congenital anomalies	13	Depression
14	Iron anemia	14	Iron anaemia
15	Depression	15	Back and neck pain

[1] Chronic obstructive pulmonary disease (COPD).

[2] Complications associated with preterm birth.

Source: Data from World Health Organization, 2018.

Taking disability as well as death into account in our assessment of disease burden reveals the increasing role of mental health as a worldwide problem. WHO projections suggest that psychiatric and neurological conditions could increase their share of the global burden from the current 10 percent to 15 percent of the total load by 2020. Again, this isn't just a problem of the developed world. Depression is expected to be the second largest cause of all years lived with disability worldwide, as well as the cause of 1.4 percent of all deaths. For women in both developing and developed regions, depression is the leading cause of disease burden, while suicide, which often is the result of untreated depression, is the fourth largest cause of female deaths.

Pollution contributes to many of the health problems in table 8.1 and has become a leading cause of death and disability in many developing countries. The World Health Organization estimates that smog and toxic agents in outdoor air contribute to at least 3.7 million deaths per year from cardiovascular problems and cancer, particularly in urban areas. China, alone, may suffer 40 percent of those deaths.

Another 4.2 million people die from particulates exposure in indoor air from smoky stoves and cooking fires. About 1 million die from contaminants in soil and water. And nearly a million die from diseases linked to poor sanitation. At least 1.1 billion people smoke today, and this number is expected to increase by at least 50 percent, especially in developing countries. If current patterns persist, about 500 million people alive today will eventually be killed by tobacco. Dr. Gro Harlem Brundtland, former director-general of the WHO, observed that reducing tobacco use and air pollution could save billions of lives.

Emergent and infectious diseases still kill millions of people

Although diet- and age-related conditions have become leading killers almost everywhere in the world, communicable diseases still are responsible for about one-third of all disease-related mortality. Diarrhea, acute respiratory illnesses, malaria, measles, tetanus, and a few other infectious diseases kill about 11 million children under age 5 every year in the developing world. Better nutrition, clean water, improved sanitation, and inexpensive inoculations could eliminate most of those deaths (fig. 8.3).

A wide variety of **pathogens** (disease-causing organisms) afflict humans, including viruses, bacteria, protozoans (single-celled animals), parasitic worms, and flukes (fig. 8.4). The greatest loss of life from an individual disease in a single year was the great influenza pandemic of 1918. Epidemiologists now estimate that at least one-third of all humans living at the time were infected and that 50 to 100 million died. Businesses, schools, churches, and sport and entertainment events were shut down for months. Influenza is caused by a family of viruses (fig. 8.4a) that mutate rapidly and move from wild and domestic animals to humans, making control of this disease very difficult. Flu epidemics continue to sweep across the globe every year, but so far none have been as deadly as the 1918 strain.

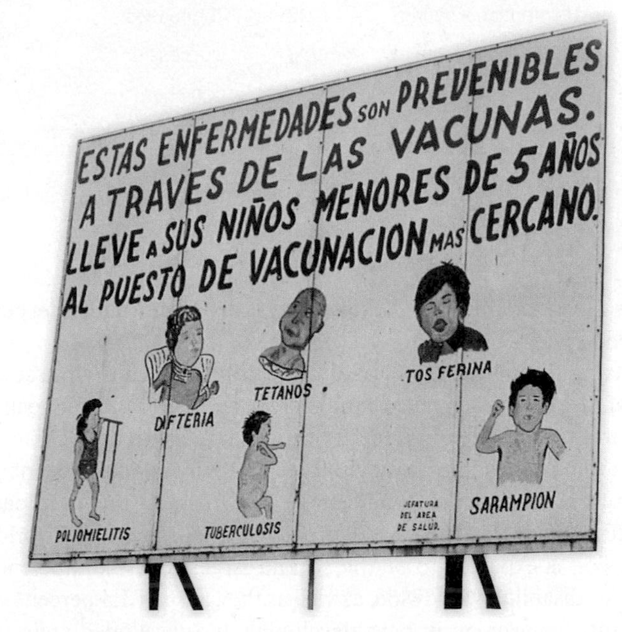

▲ **FIGURE 8.3** At least 3 million children die every year from easily preventable diseases. This billboard in Guatemala encourages parents to have their children vaccinated against polio, diphtheria, TB, tetanus, pertussis (whooping cough), and scarlet fever. ©William P. Cunningham

Every year there are 76 million cases of foodborne illnesses in the United States, resulting in 300,000 hospitalizations and 5,000 deaths. Both bacteria and intestinal protozoa cause these illnesses (fig. 8.4b,c). They are spread from feces through food and water. In 2010 nearly 6 million pounds (about 2,700 metric tons) of ground beef were recalled in the United States because of contamination by *E. coli* strain O157:H7.

At any given time, around 2 billion people—nearly one-third of the world population—suffer from worms, flukes, and other internal parasites. Though parasites rarely kill people, they can be extremely debilitating and can cause poverty that leads to other, more deadly, diseases.

Malaria is one of the most prevalent remaining infectious diseases. Every year about 500 million new cases of this disease occur, and about 1 million people die from it. The territory infected by this disease is expanding as global climate change allows mosquito vectors to move into new territory. Simply providing insecticide-treated bed nets and a few dollars' worth of antimalarial pills could prevent tens of millions of cases of this debilitating disease every year. Tragically, some of the countries where malaria is most widespread tax both bed nets and medicine as luxuries, placing them out of reach for ordinary people.

Emergent diseases are those not previously known or that have been absent for at least 20 years. An increasing number of these epidemics have been popping up in recent years as the global population has grown and become increasingly interconnected (fig. 8.5).

The 2014 Ebola outbreak in West Africa illustrates how rapidly infectious diseases can spread and the dangers of lax oversight. Ebola is a hemorrhagic fever caused by a filovirus. It causes a high fever

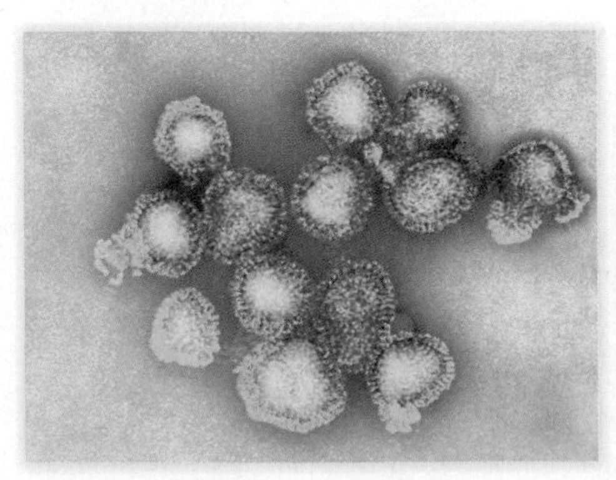

(a) Influenza viruses

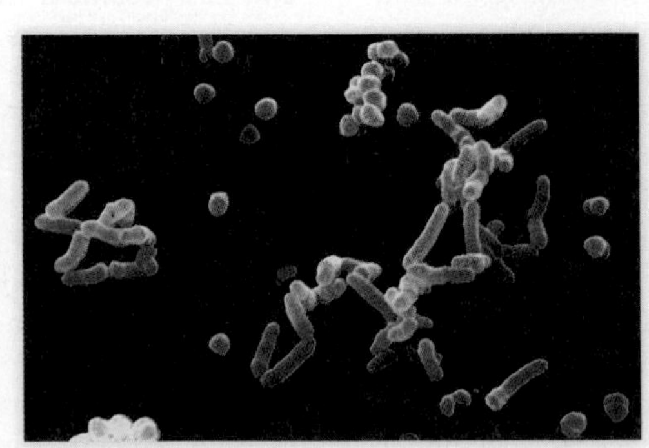

(b) Pathogenic bacteria

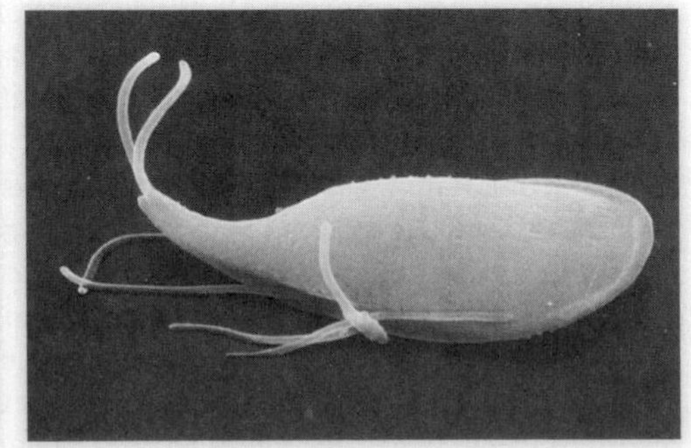

(c) *Giardia*

▲ **FIGURE 8.4** (a) A group of influenza viruses magnified about 300,000 times. (b) Pathogenic bacteria magnified about 50,000 times. (c) *Giardia*, a parasitic intestinal protozoan, magnified about 10,000 times. (a) ©CDC/SCIENCE PHOTO LIBRARY/Getty Images; (b) ©Image Source/Getty Images; (c) Courtesy of Stanley Erlandsen, University of Minnesota

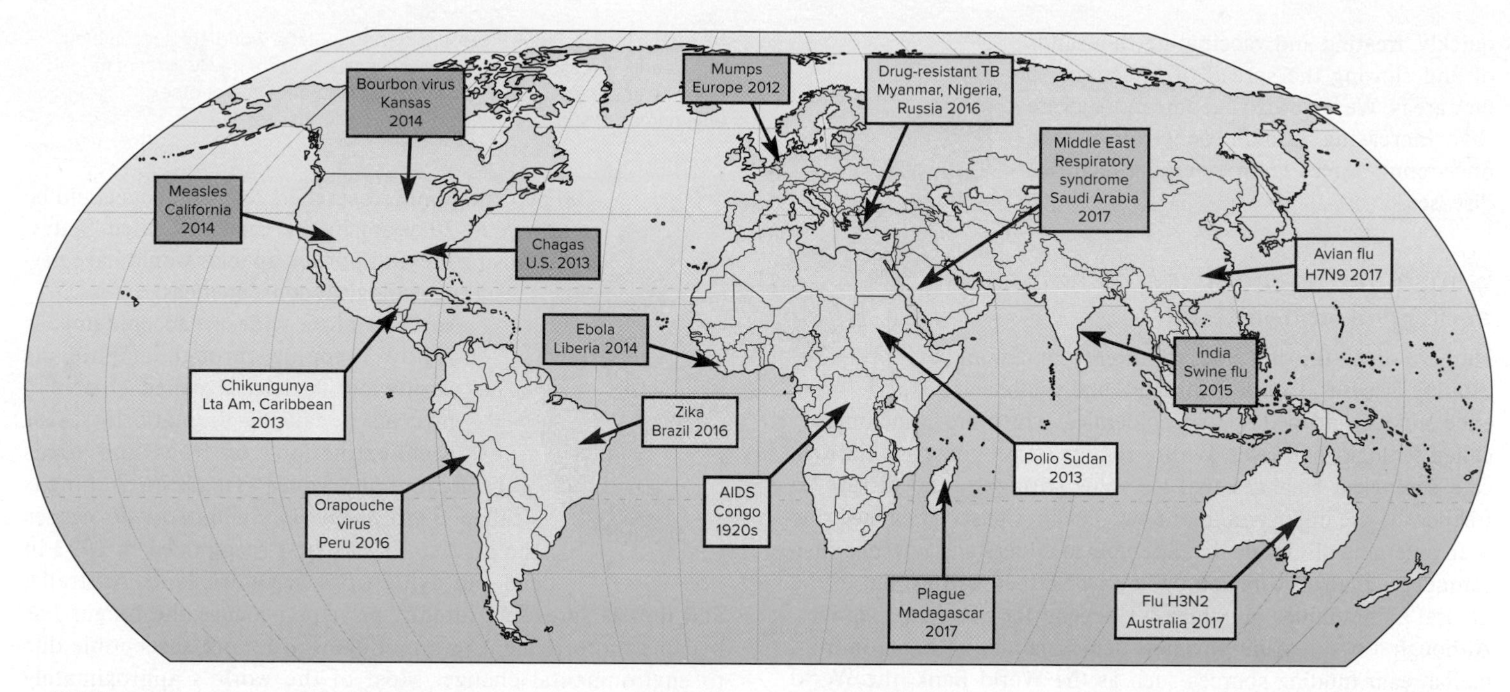

▲ **FIGURE 8.5** Some recent outbreaks of highly lethal infectious diseases. Why are supercontagious organisms emerging in so many different places? Source: Data from U.S. Center for Disease Control and Prevention, 2018.

and diarrhea followed by acute hemorrhage (bleeding). It's highly fatal for apes, monkeys, and humans. In some human outbreaks, 90 percent of patients died. Most outbreaks probably start with someone touching the blood or meat of an animal killed for "bushmeat," especially monkeys and bats, both of which can harbor the Ebola virus. Bushmeat is increasingly in demand as a delicacy in urban areas, and the virus can spread through killing, butchering, and handling of carcasses. It then spreads from person to person. Faster travel by road and air makes it possible for diseases to spread quickly.

At least 24 Ebola outbreaks have occurred across central Africa since it was first described in 1976. Most have been in small, remote villages, where the disease burned out in a few months. Only about 1,600 people are known to have died from Ebola between 1976 and 2013.

The 2014 outbreak started in rural Guinea but quickly spread across the border into neighboring Sierra Leone and Liberia. Among the poorest nations in the world, these countries had little infrastructure and few resources for responding to a public health crisis. The few hospitals were under-equipped, sometimes lacking electricity and running water. Still reeling from devastating civil wars, people were impoverished and often wary of outsiders (fig. 8.6).

It wasn't until the contagion reached the capitals of Freetown and Monrovia, with populations of millions, that officials recognized the severity of the situation and the threat to the whole world if the disease continued to spiral out of control. The total number of infections in 2014 was estimated at 29,000, and more than 11,300 died. The high death rate and highly contagious nature of the disease led to fear and panic, which further hampered control efforts.

What made the 2014 Ebola outbreak so serious? One factor was the increased mobility of people to and from cities, with expanding road networks (especially logging roads deep into forests). Another was the poor, and often corrupt, state of government health agencies. Still another was increasing trade and consumption of bush meat, which had accelerated with economic growth and with increasing availability of firearms. All of these increase the risk of outbreaks.

One lesson from the 2014 Ebola outbreak was the importance of swift and coordinated health care response. Another was that patients often die of dehydration, so providing fluids improves survival. And since 2014, a vaccine has become available. When a new outbreak emerged in 2018 in the Democratic Republic of Congo, government and international health workers responded

▲ **FIGURE 8.6** Ebola workers need complete protection from bodily contact, which is difficult in a hot, humid climate. Protective gear also alienates locals, who tend to be suspicious of strangers and authority. ©KENZO TRIBOUILLARD/Getty Images

quickly, treating and vaccinating thousands of and slowing the spread of infections to new areas. We have also become more aware that increasing human penetration into once-remote areas is an emerging factor in diseases.

Conservation medicine combines ecology and health care

Humans aren't the only ones to suffer from new and devastating diseases. Domestic animals and wildlife also experience sudden and widespread epidemics, which are sometimes called ecological diseases. We are coming to recognize that the delicate ecological balances that we value so highly–and disrupt so frequently–are important to our own health. **Conservation medicine** is an emerging discipline that attempts to understand how our environmental changes threaten our own health as well as that of the natural communities on which we depend for ecological services. Although it is still small, this new field is gaining recognition from mainstream funding sources such as the World Bank, the World Health Organization, and the U.S. National Institutes of Health.

Ebola hemorrhagic fever, for example, is one of the most virulent viruses ever known. In 2002 researchers found that 221 of the 235 western lowland gorillas they had been studying in the Congo disappeared in just a few months. Many chimpanzees also died. Although the study team could find only a few of the dead gorillas, 75 percent of those tested positive for Ebola. Altogether, researchers estimate that 5,000 gorillas died in this small area. Extrapolating to all of central Africa, it's possible that Ebola has killed one-third of all the gorillas in the world.

In 2006, people living near a cave west of Albany, New York, reported something peculiar: Little brown bats (*Myotis lucifugus*) were flying outside during daylight in the middle of the winter. Inspection of the cave by the Department of Conservation found numerous dead bats near the cave mouth. Most had white fuzz on their faces and wings, a condition now known as white-nose syndrome (WNS). Little brown bats are tiny creatures, about the size of your thumb. They depend on about 2 grams of stored fat to get them through the winter. Hibernation is essential to making their energy resources last. Being awakened just once can cost a bat a month's worth of fat.

The white fuzz has now been identified as filamentous fungus (*Geomyces destructans*), which thrives in the cool, moist conditions where bats hibernate. We don't know where the fungus came from, but, true to its name, the pathogen has spread like wildfire, reaching 32 states and five Canadian provinces. Biologists estimate that at least 7 million bats already have died from this disease. It isn't known how the pathogen spreads. Perhaps it moves from animal to animal through physical contact. It's also possible that humans introduce fungal spores on their shoes and clothing when they go from one cave to another.

One mammalogist calls WNS "the chestnut blight of bats" (in reference to a fungus that nearly eliminated the American chestnut in a few decades). So far, six species of bat are known to be susceptible to this plague. In infected colonies, mortality can be 100 percent. Some researchers fear that bats could be extinct in 20 years in the eastern United States. Losing these important species would have devastating ecological consequences.

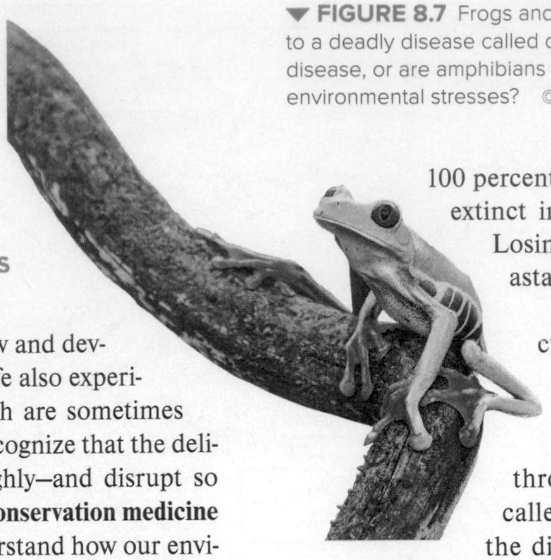

▼ **FIGURE 8.7** Frogs and toads throughout the world are succumbing to a deadly disease called chytridiomycosis. Is this a newly virulent fungal disease, or are amphibians more susceptible because of other environmental stresses? ©Rosalie Kreulen/Shutterstock

An even more widespread epidemic is currently sweeping through amphibians worldwide. A disease called chytridiomycosis is causing dramatic losses or even extinctions of frogs and toads throughout the world (fig. 8.7). A fungus called *Batrachochytrium dendrobatidis* causes the disease. It was first recognized in 1993 in dead and dying frogs in Queensland, Australia, and now is spreading rapidly, perhaps because the fungus has become more virulent or amphibians are more susceptible due to environmental change. Most of the world's approximately 6,000 amphibian species appear to be susceptible to the disease, and at least 2,000 species have declined or even become extinct in their native habitats as part of this global epidemic.

Temperatures above 28°C (82°F) kill the fungus, and treating frogs with warm water can cure the disease in some species. Topical application of the drug chloramphenicol also has successfully cured some frogs. And certain skin bacteria seem to confer immunity to fungal infections. In some places, refuges have been established in which frogs can be maintained under antiseptic conditions until a cure is found. It's hoped that survivors can eventually be reintroduced to their native habitat and species will be preserved.

Climate change also facilitates expansion of parasites and diseases into new territories. Tropical diseases, such as malaria, cholera, yellow fever, and dengue fever, have been moving into areas from which they were formerly absent as mosquitoes, rodents, and other vectors expand into new habitat. This affects other species besides humans. A disease called dermo is spreading northward through oyster populations along the Atlantic coast of North America. This disease is caused by a protozoan parasite (*Perkinsus marinus*) that was first recognized in the Gulf of Mexico about 70 years ago. In the 1950s the disease was found in Chesapeake Bay. Since then the parasite has been moving northward, probably assisted by higher sea temperatures caused by global warming. It is now found as far north as Maine. This disease doesn't appear to be harmful to humans, but it is devastating oyster populations.

Another persistent epidemic, known as colony collapse disorder, is destroying honey bee colonies across North America and parts of Europe. Because bees are essential crop pollinators, at least 70 crops worth at least $15 billion annually are at risk. Many factors have been suggested to explain the mysterious bee die-off, including pesticides, parasites, viruses, fungi, malnutrition, genetically modified crops, and long-distance shipping of bees to pollinate crops. Although many authors claim to have discovered the cause of this disorder, it remains a mystery.

Resistance to antibiotics and pesticides is increasing

Population biologists have long known that constant exposure to a toxic substance produces resistance to it. Constant use of antibiotics, for example, kills off all but the resistant strains of microbes that tolerate the antibiotic. These multiply and come to dominate the population (fig. 8.8a). For bacteria, this problem is especially challenging: Bacteria readily exchange genes, increasing the likelihood of producing resistant populations (fig. 8.8b). Widespread over-use of antibiotics, in people and especially in livestock, is producing a rapidly growing problem of antibiotic-resistant infections.

We raise most of our cattle, hogs, and poultry in densely packed barns and feedlots. Confined animals are dosed constantly with antibiotics to keep them disease-free, and to make them gain weight faster. At least 80 percent of all antibiotics used in the United States are fed to livestock. Monitoring in and around livestock operations consistently finds multi-drug-resistant pathogens in soil, water, and air. These multiple-resistant pathogens are increasingly found in people, a major public health worry. In some countries, such as Denmark, livestock have been raised for decades with only emergency uses of antibiotics. U.S. policy makers, however, have been slow to intervene in these practices.

In humans, at least half of the 100 million antibiotics prescribed every year in the United States are unnecessary or are the wrong ones. In part, that's because at least one-third of Americans incorrectly believe antibiotics are an effective treatment for viral infections. They demand treatments their doctors know will be ineffective, and doctors often find it easier to write a prescription than to argue with a patient. By some estimates, more than 20 percent of antibiotics prescribed for adults are for colds, upper respiratory tract infections, and bronchitis, three conditions for which antibiotics have little or no benefit. Furthermore, many people who start antibiotic treatment fail to carry it out for the time prescribed. Failing to complete an antibiotic treatment just aids the development of resistant strains of a disease.

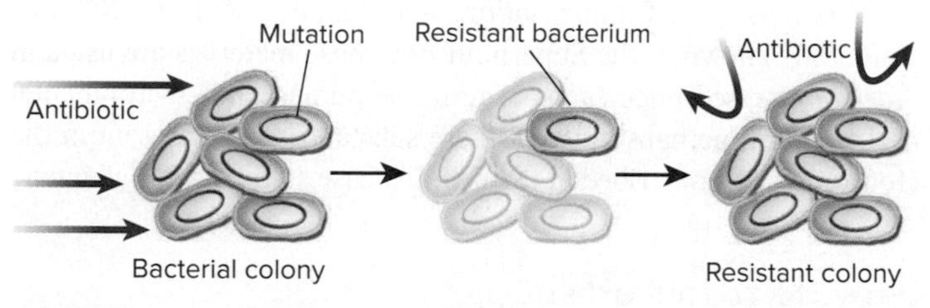

(a) Mutation and selection create drug-resistant strains.

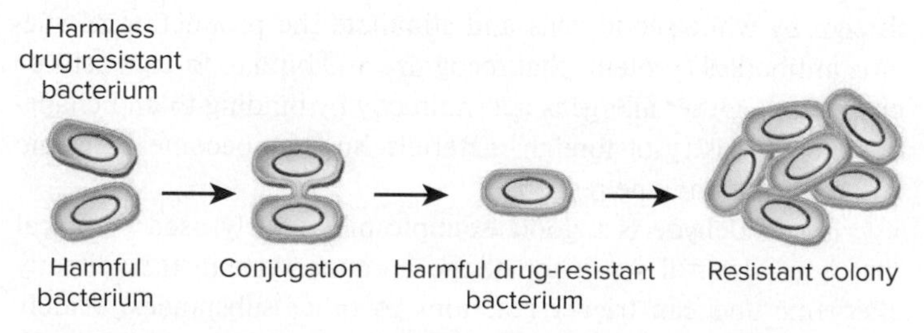

(b) Conjugation transfers drug resistance from one strain to another.

▲ **FIGURE 8.8** How microbes acquire antibiotic resistance. (a) Random mutations make a few cells resistant. When challenged by antibiotics, only those cells survive to give rise to a resistant colony. (b) Sexual reproduction (conjugation) or plasmid transfer moves genes from one strain or species to another.

Besides breeding resistant microbes, the increased use of antibiotics in the first 6 months of a child's life has been linked to a higher incidence of asthma, eczema, and allergic hypersensitivity. Parents should be told that antibiotics can't cure their children's colds and flus, and that symptoms should be treated with home remedies, including rest and fluids.

Hospitals are key centers of antibiotic resistance, but practices have begun to change. They are tracking prescriptions more closely, and they are examining closely how they clean and disinfect rooms, equipment, and the hands of doctors and nurses. Among the most lethal antibiotic-resistant bacteria are CRE (carbapenem-resistant enterobacteriaceae), which have been identified in health care facilities in 42 states. While CRE cause only 1 to 4 percent of all hospital-acquired infections, the CDC has dubbed them the "nightmare bacteria" because they are fatal in half of all cases.

Another serious infection is methicillin-resistant *Staphylococcus aureus* (MRSA). One U.S. study found that 60 percent of the estimated 80,000 MRSA infections in 2011 were related to outpatient hospital procedures, and another 22 percent occurred in the general community.

Clostridium difficile (*C. diff*) is a bacterium that causes colon inflammation. Diarrhea and fever that can lead to other lethal complications are among the most common symptoms. In 2017 this species was estimated to cause about 500,000 infections in the United States and almost 30,000 deaths. Those most at risk were older adults. *C. diff* infections are extremely difficult to eradicate.

Data from the CDC show that although hospital-acquired infections are now declining, about 1 in 25 patients in U.S. hospitals will get some type of infection. This amounts to more than 700,000 illnesses per year. About 75,000 of those patients—most often the elderly—will die. These infections cost U.S. hospitals around $30 billion annually. The average drug-resistant infection costs up to $37,000 to treat.

Dire as these hospital infections are in rich countries, the situation is much worse in poorer places. India may be the epicenter for antibiotic overuse and multi-drug-resistant infections. Bacteria spread easily because at least half the population defecate outdoors, and much of the sewage generated by those who do use toilets is dumped, untreated, into rivers. As a result, Indians have among the highest infection rate in the world, and they collectively take more antibiotics, which are sold over the counter, than any other people. In 2016 at least 60,000 Indian babies died of multi-drug-resistant infections. India's terrible sanitation and uncontrolled antibiotic use, coupled with substandard housing and a complete lack of monitoring of the problem, have created a tsunami of antibiotic resistance that is spreading throughout the world, say epidemiologists. Researchers have recently found "super bugs" carrying a genetic code first identified in India around the world, including to France, Japan, and the United States.

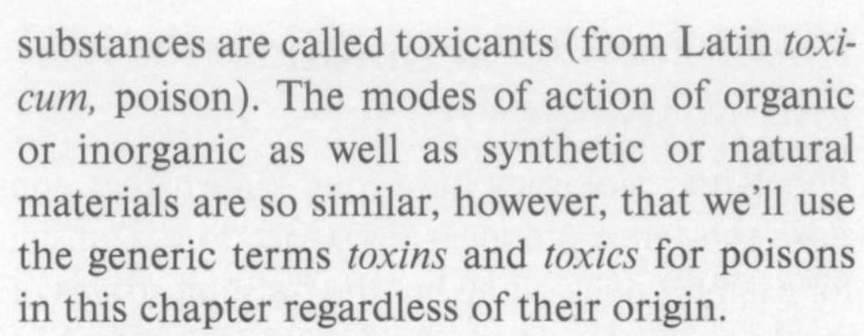

What Can YOU DO?

Tips for Staying Healthy

- Eat a balanced diet with plenty of fresh fruits, vegetables, legumes, and whole grains. Wash fruits and vegetables carefully; they may have come from a country where pesticide and sanitation laws are lax.
- Use unsaturated oils, such as olive or canola, rather than hydrogenated fats, such as margarine. Minimize consumption of sugars and of processed foods.
- Cook meats and other foods at temperatures high enough to kill pathogens; clean utensils and cutting surfaces; store food properly.
- Wash your hands frequently. You transfer more germs from hand to mouth than any other means.
- When you have a cold or flu, don't demand antibiotics from your doctor—they aren't effective against viruses.
- If you're taking antibiotics, continue for the entire time prescribed—quitting as soon as you feel well is an ideal way to select for antibiotic-resistant germs.
- Practice safe sex.
- Don't smoke; avoid smoky places.
- If you drink alcohol, do so in moderation. Never drive when your reflexes or judgment is impaired.
- Exercise regularly: Walk, swim, jog, dance, garden. Do something you enjoy that burns calories and maintains flexibility.
- Get enough sleep. Practice meditation, prayer, or some other form of stress reduction. Get a pet.
- Make a list of friends and family who make you feel more alive and happy. Spend time with one of them at least once a week.

8.2 TOXICOLOGY

- *Many toxic substances are harmful in concentrations as low as billionths, or even trillionths, of a gram.*
- *Allergens, carcinogens, and toxic substances occur in many common products.*
- *Hormone disruptors can interfere with growth, development, and physiology at very low doses.*

Toxic means poisonous. Toxicology is the study of the adverse effects of external factors on an organism or a system. This includes environmental chemicals, drugs, and diet as well as physical factors, such as ionizing radiation, UV light, and electromagnetic forces. Toxicologists also are concerned with movement and fate of poisons in the environment, routes of entry into the body, and effects of exposure to these agents. Toxic substances damage or kill living organisms because they react with cellular components to disrupt metabolic functions. These substances can be harmful even in extremely dilute concentrations. In some cases, billionths, or even trillionths, of a gram can cause irreversible damage.

Many toxicologists limit the term *toxin* to proteins or other molecules synthesized by living organisms. Nonbiological noxious substances are called toxicants (from Latin *toxicum,* poison). The modes of action of organic or inorganic as well as synthetic or natural materials are so similar, however, that we'll use the generic terms *toxins* and *toxics* for poisons in this chapter regardless of their origin.

Hazardous materials aren't necessarily toxic. Some substances are dangerous because they're flammable, explosive, acidic, caustic, irritants, or sensitizers. Many of these materials must be handled carefully in large doses or high concentrations, but they can be rendered relatively innocuous by dilution, incineration, neutralization, or other physical treatment.

Environmental toxicology, or ecotoxicology, specifically deals with the interactions, transformation, fate, and effects of toxic materials in the biosphere, including individual organisms, populations, and whole ecosystems. In aquatic systems the fate of the pollutants is primarily studied in relation to mechanisms and processes at interfaces of the ecosystem components. Special attention is devoted to the sediment/water, water/organisms, and water/air interfaces. In terrestrial environments, the emphasis tends to be on effects of metals on soil fauna community and population characteristics.

Table 8.2 is a list of the top 20 toxic and hazardous substances considered the highest risk by the U.S. Environmental Protection Agency. Compiled from the 275 substances regulated by the Comprehensive Environmental Response, Compensation, and Liability Act (CERCLA), commonly known as the Superfund Act, these materials are listed in order of assessed importance in terms of human and environmental health. Many, perhaps most, of these substances were present in the Houston floods described in the opening case study for this chapter.

How do toxics affect us?

Allergens are substances that activate the immune system. Some allergens act directly as **antigens;** that is, they are recognized as foreign by white blood cells and stimulate the production of specific antibodies (proteins that recognize and bind to foreign cells or chemicals). Other allergens act indirectly by binding to and changing the chemistry of foreign materials, so they become antigenic and cause an immune response.

Formaldehyde is a good example of a widely used chemical that is a powerful sensitizer of the immune system. It is directly allergenic and can trigger reactions to other substances. Widely used in plastics, wood products, insulation, glue, and fabrics, formaldehyde concentrations in indoor air can be thousands of times higher than in normal outdoor air.

Some people suffer from what is called **sick building syndrome:** headaches, allergies, and chronic fatigue caused by poorly vented indoor air contaminated by molds, carbon monoxide, nitrogen

TABLE 8.2 Top 20 Toxic and Hazardous Substances

MATERIAL	MAJOR SOURCES
1. Arsenic	Treated lumber
2. Lead	Paint, gasoline
3. Mercury	Coal combustion
4. Vinyl chloride	Plastics, industrial uses
5. Polychlorinated biphenyls (PCBs)	Electric insulation
6. Benzene	Gasoline, industrial use
7. Cadmium	Batteries
8. Benzo(a)pyrene	Waste incineration
9. Polycyclic aromatic hydrocarbons	Combustion
10. Benzo(b)fluoranthene	Fuels
11. Chloroform	Water purification, industry
12. DDT	Pesticide use
13. Aroclor 1254	Plastics
14. Aroclor 1260	Plastics
15. Trichloroethylene	Solvents
16. Dibenz (a, h)anthracene	Incineration
17. Dieldrin	Pesticides
18. Chromium, hexavalent	Paints, coatings, welding, anticorrosion agents
19. Chlordane	Pesticides
20. Hexachlorobutadiene	Pesticides

Source: Data from U.S. Environmental Protection Agency.

oxides, formaldehyde, and other toxic chemicals released by carpets, insulation, plastics, building materials, and other sources. The Environmental Protection Agency estimates that poor indoor air quality may cost the nation $60 billion a year in absenteeism and reduced productivity (Key Concepts, p. 190).

Neurotoxins are a special class of metabolic poisons that specifically attack nerve cells (neurons). The nervous system is so important in regulating body activities that disruption of its activities is especially fast-acting and devastating. Different types of neurotoxins act in different ways. Heavy metals, such as lead and mercury, kill nerve cells and cause permanent neurological damage. Anesthetics (ether, chloroform, halothane, etc.) and chlorinated hydrocarbons (DDT, Dieldrin, Aldrin) disrupt nerve cell membranes necessary for nerve action. Organophosphates (Malathion, Parathion) and carbamates (carbaryl, zeneb, maneb) inhibit acetylcholinesterase, an enzyme that regulates signal transmission between nerve cells and the tissues or organs they innervate (for example, muscles). Most neurotoxins are both acute and extremely toxic. More than 850 compounds are now recognized as neurotoxins.

In 2011 the Environmental Protection Agency issued standards (which had been due for more than 20 years) on mercury and fine particles from power plants. The agency calculated this new rule would provide about $90 billion per year in public health benefits compared to costs of about $10 billion to power companies. The benefits were based mostly on increased lifetime wages for people who would otherwise suffer from mercury-caused brain damage in childhood. Many other benefits (reduced heart attacks, asthma, and lung diseases) are real but hard to quantify. Opponents in Congress tried to kill the measure because it was "unfriendly to business." What do you think, would you raise business expenses by $10 billion if it saved public costs of $90 billion?

Mutagens are agents, such as chemicals and radiation, that damage or alter genetic material (DNA) in cells. This can lead to birth defects if the damage occurs during embryonic or fetal growth. Later in life, genetic damage may trigger tumor growth. When damage occurs in reproductive cells, the results can be passed on to future generations. Cells have repair mechanisms to detect and restore damaged genetic material, but some changes may be hidden, and the repair process itself can be flawed. It is generally accepted that there is no "safe" threshold for exposure to mutagens. Any exposure has some possibility of causing damage.

Teratogens are chemicals or other factors that specifically cause abnormalities during embryonic growth and development. Some compounds that are not otherwise harmful can cause tragic problems in these sensitive stages of life. Perhaps the most prevalent teratogen in the world is alcohol. Drinking during pregnancy can lead to **fetal alcohol syndrome**—a cluster of symptoms including craniofacial abnormalities, developmental delays, behavioral problems, and mental impairments that last throughout a child's life.

Carcinogens are substances that cause **cancer**—invasive, out-of-control cell growth that results in malignant tumors. Cancer rates rose in most industrialized countries during the twentieth century, and cancer is now the second leading cause of death in the United States, killing about a half a million people per year. Sixteen of the 20 compounds listed by the U.S. EPA as the greatest risk to human health are probable or possible human carcinogens. More than 200 million Americans live in areas where the combined upper limit lifetime cancer risk from these carcinogens exceeds 10 in 1 million, or 10 times the risk normally considered acceptable.

Is your shampoo making you fat?

One of the most recently recognized environmental health threats are **endocrine hormone disrupters,** chemicals that interrupt the normal endocrine hormone functions. Hormones are chemicals released into the bloodstream by glands in one part of the body to regulate the development and function of tissues and organs elsewhere in the body (fig. 8.9). You undoubtedly have heard about sex hormones and their powerful effects on how we look and behave, but these are only one example of the many regulatory hormones that rule our lives.

We now know that some of the most insidious effects of persistent chemicals, such as BPA, DDT and PCBs, are that they interfere with the normal growth, development, and physiology of a variety of animals—presumably including humans—at very low doses. In some cases, picogram concentrations (trillionths of a gram per liter) may be enough to cause developmental abnormalities in sensitive organisms. These chemicals are sometimes called environmental estrogens or androgens, because they often cause sexual dysfunction (reproductive health problems in females or feminization of males, for example). They are just as likely, however, to disrupt

What toxins and hazards are present in your home?

The EPA warns that indoor air can be much more polluted than outdoor air. Many illnesses can be linked to poor air quality and exposure to toxins in the home. Since 1950, at least 70,000 new chemical compounds have been invented and dispersed into our environment. Only a fraction of these have been tested for human toxicity, but it's suspected that many may contribute to allergies, birth defects, cancer, and other disorders. Which of the following materials can be found in your home? What toxins and hazards are present in your home?

Garage
- Antifreeze
- Automotive polishes and waxes
- Batteries
- Insecticides, herbicides, fungicides, and pesticides
- Gasoline and solvents
- Paints and stains
- Pool supplies
- Rust remover
- Wood preservatives

Kitchen/Laundry Area
- Bleach
- Carbon monoxide and fine particulates
- Cleansers and disinfectants
- Laundry detergents
- Drain cleaners
- Floor polishes
- Oven cleaners
- Nonstick by-products
- Window cleaners

Basement
- Carbon monoxide from furnace and water heaters
- Epoxy glues
- Gasoline, kerosene, and other flammable solvents
- Lye and other caustics
- Mold, bacteria, and other pathogens or allergens
- Paint and paint remover
- PVC and other plastics
- Radon gas from subsoil

Attic

- Asbestos
- Fiberglass insulation
- PBDE-treated cellulose

Bathroom

- Chloroform from showers and bath
- Leftover drugs and medications
- Fingernail polish and remover
- Makeup
- Mold
- Mouthwash
- Toilet bowl cleaner

Bedroom

- Aerosols
- Bisphenol A, lead, cadmium in toys and jewelry
- Flame retardants, fungicides, and insecticides in carpets and bedding
- Mothballs

Living Room

- Asbestos from floor or ceiling tiles
- Benzepyrenes from smoking
- Flame retardants
- Freons from air conditioners
- Furniture and metal polishes
- Lead or cadmium from toys
- Paints, fabrics
- Plastics

CAN YOU EXPLAIN?

1. Which space has the largest number of toxic materials?
2. In which room do you spend the most time?
3. Which space has the greatest number of toxins to which you're likely to be exposed on a regular basis?

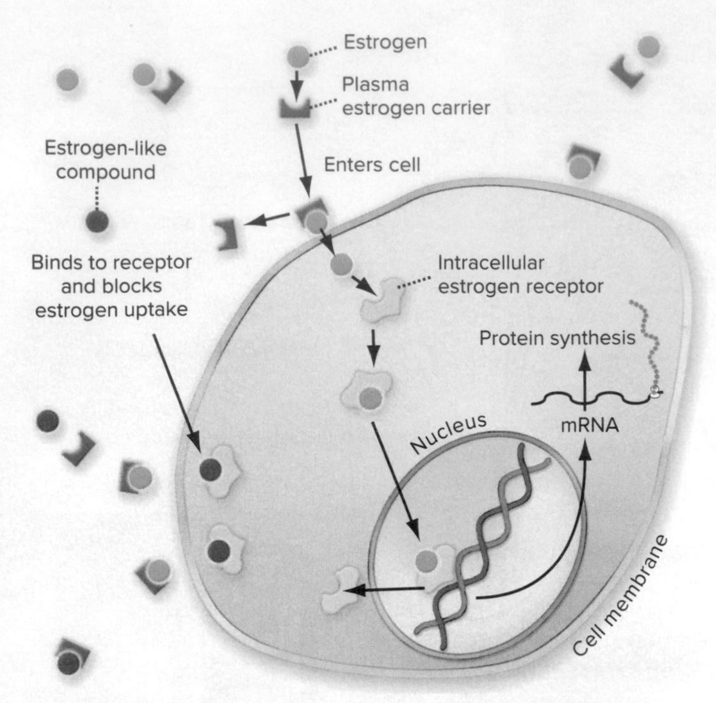

▲ **FIGURE 8.9** Steroid hormone action. Plasma hormone carriers deliver regulatory molecules to the cell surface, where they cross the cell membrane. Intracellular receptors deliver hormones to the nucleus, where they bind to and regulate expression of DNA. Estrogen-like compounds bind to receptors and either block uptake of endogenous hormones or act as a substitute hormone to disrupt gene expression.

TABLE 8.3 | Factors in Environmental Toxicity

FACTORS RELATED TO THE TOXIC AGENT
1. Chemical composition and reactivity
2. Physical characteristics (such as solubility, state)
3. Presence of impurities or contaminants
4. Stability and storage characteristics of toxic agent
5. Availability of vehicle (such as solvent) to carry agent
6. Movement of agent through environment and into cells
FACTORS RELATED TO EXPOSURE
1. Dose (concentration and volume of exposure)
2. Route, rate, and site of exposure
3. Duration and frequency of exposure
4. Time of exposure (time of day, season, year)
FACTORS RELATED TO THE ORGANISM
1. Resistance to uptake, storage, or cell permeability of agent
2. Ability to metabolize, inactivate, sequester, or eliminate agent
3. Tendency to activate or alter nontoxic substances so they become toxic
4. Concurrent infections or physical or chemical stress
5. Species and genetic characteristics of organism
6. Nutritional status of subject
7. Age, sex, body weight, immunological status, and maturity

thyroxin functions or those of other important regulatory molecules as they are to obstruct sex hormones.

Recently, researchers have noticed that many endocrine hormone disrupters also are linked to obesity. Some people call this group **obesogens.** They not only cause diseases, such as diabetes, but also may disrupt regulatory mechanisms for energy storage and metabolism. The Centers for Disease Control and Prevention reports that roughly 70 percent of Americans are overweight and approximately 40 percent are clinically obese. This has a host of adverse health consequences, including cardiovascular disease, certain cancers, bone and joint conditions, and reproductive problems. We used to think that weight problems were simply a matter of too many calories and too little exercise, but there's growing concern that a wide variety of petrochemicals, plasticizers, personal care products, and food additives, such as high-fructose corn syrup, may play a critical role in this epidemic by disrupting hormone pathways.

8.3 MOVEMENT, DISTRIBUTION, AND FATE OF TOXINS

- *Solubility and mobility determine chemical movement in the environment and in our bodies.*
- *Persistent materials, such as heavy metals and some organic compounds, accumulate and concentrate in food webs.*
- *Synergistic (or combined) effects can be far greater than effects of individual substances.*

There are many sources of toxic and hazardous chemicals in the environment. The danger of a chemical is determined by many factors related to the chemical itself, its route or method of exposure, and its persistence in the environment, as well as characteristics of the target organism (table 8.3). We can think of both individuals and an ecosystem as sets of interacting compartments between which chemicals move, based on molecular size, solubility, stability, and reactivity (fig. 8.10). The dose (amount), route of entry, timing of exposure, and sensitivity of the organism all play important roles in determining toxicity. In this section we will consider some of these characteristics and how they affect environmental health.

Solubility and mobility determine when and where chemicals move

Solubility is one of the most important characteristics in determining how, where, and when a toxic material will move through the environment or through the body to its site of action. Chemicals can be divided into two major groups: those that dissolve more readily in water and those that dissolve more readily in oil. Water-soluble compounds move rapidly and widely through the environment because water is ubiquitous. They also tend to have ready access to most cells in the body because aqueous solutions bathe all our cells. Molecules that are oil- or fat-soluble (usually organic molecules) generally need a carrier to move through the environment and into or within the body. Once inside the body, however, oil-soluble toxics penetrate readily into tissues and cells because the membranes that enclose cells are themselves made of similar oil-soluble chemicals.

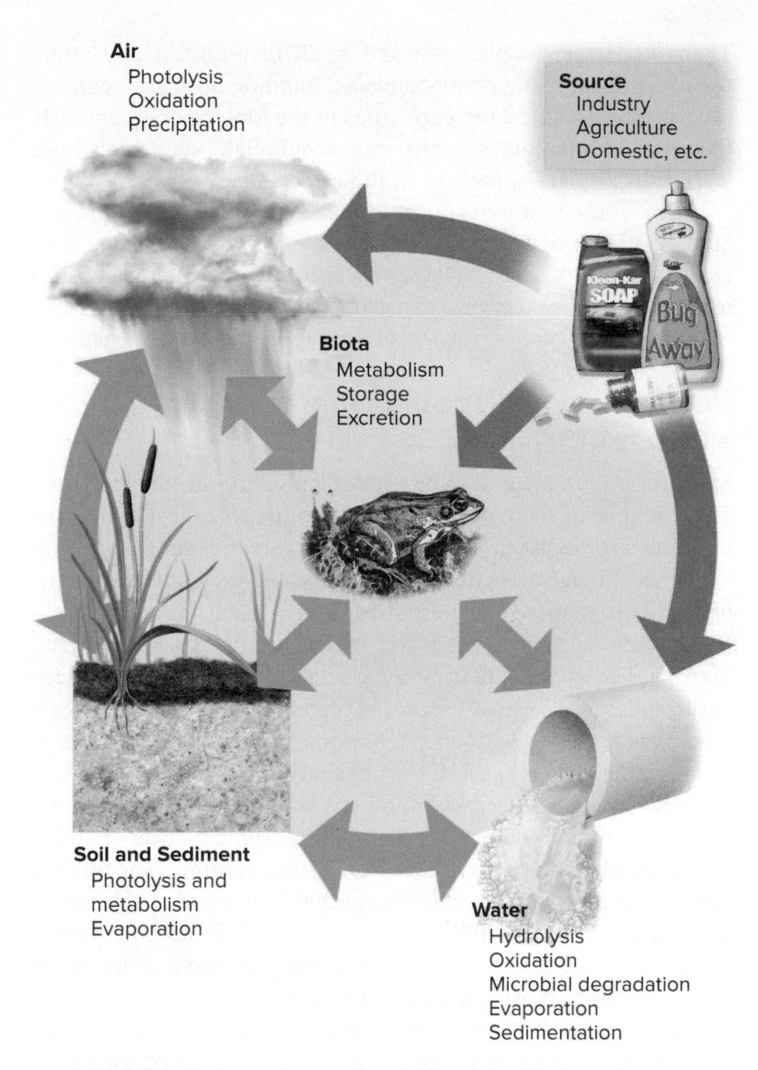

▲ **FIGURE 8.10** Movement and fate of chemicals in the environment. Toxins also move directly from a source to soil and sediment.

Once inside cells, oil-soluble materials are likely to accumulate and to be stored in lipid deposits, where they may be protected from metabolic breakdown and persist for many years.

Exposure and susceptibility determine how we respond

Just as there are many sources of toxic materials in our environment, there are many routes for entry into our bodies (fig. 8.11). Airborne toxics generally cause more ill health than toxics from any other exposure source. We breathe far more air every day than the volume of food we eat or water we drink. Furthermore, the lining of our lungs, which is designed to exchange gases very efficiently, also absorbs toxics very well. Epidemiologists estimate that 3 million people—two-thirds of them children—die each year from diseases caused or exacerbated by air pollution.

But food, water, and skin contact also can expose us to a wide variety of hazards. The largest exposures for many toxics are found in industrial settings, where workers may encounter doses thousands of times higher than would be found anywhere else. The European Agency for Safety and Health at Work warns that 32 million people (20 percent of all employees) in the European Union are exposed to unacceptable levels of carcinogens and other chemicals in their workplaces.

Condition of the organism and timing of exposure also have strong influences on toxicity. Healthy adults, for example, may be relatively insensitive to doses that are very dangerous to young children or to someone already weakened by other diseases. Pound for pound, children drink more water, eat more food, and breathe more air than do adults. Putting fingers, toys, and other objects into their mouths increases children's exposure to toxics in dust or soil.

Furthermore, children generally have less-developed immune systems or processes to degrade or excrete toxics. The developing brain is especially sensitive to damage. Obviously, disrupting the complex and sensitive process of brain growth and development can have tragic long-term consequences.

The best-known example of an environmental risk for children is lead poisoning. Children had high exposure to airborne lead deposited in play-yard dirt and dust, and they sometimes chewed on the sweet-tasting lead paint on window sills. Before lead paint and leaded gasoline were banned in the 1970s, at least 4 million American children had dangerous levels of lead in their blood. Banning these products has been one of the greatest successes in environmental health. Blood lead levels in children have fallen more than 90 percent in the past three decades.

The notorious teratogen thalidomide is a prime example of differences in sensitivity between species and within stages of fetal

▲ **FIGURE 8.11** Routes of exposure to toxic and hazardous environmental factors.

development. A sedative developed in the 1950s, thalidomide was also prescribed to pregnant women to control nausea. But a single dose could cause undeveloped limbs or other deformities in a fetus. Some 10,000 infants were affected before this use was stopped. The drug remains in use in adults, however, for example in some cancers and leprosy.

Bioaccumulation and biomagnification increase chemical concentrations

Cells have mechanisms for **bioaccumulation,** the selective absorption and storage of a great variety of molecules. This allows them to accumulate nutrients and essential minerals, but at the same time they also may absorb and store harmful substances through the same mechanisms. Materials that are rather dilute in the environment can reach dangerous levels inside cells and tissues through this process of bioaccumulation.

Toxic substances also can be magnified through food webs. **Biomagnification** occurs when the toxic burden of a large number of organisms at a lower trophic level is accumulated and concentrated by a predator in a higher trophic level. Phytoplankton and bacteria in aquatic ecosystems, for instance, take up heavy metals or toxic organic molecules from water or sediments (fig. 8.12).

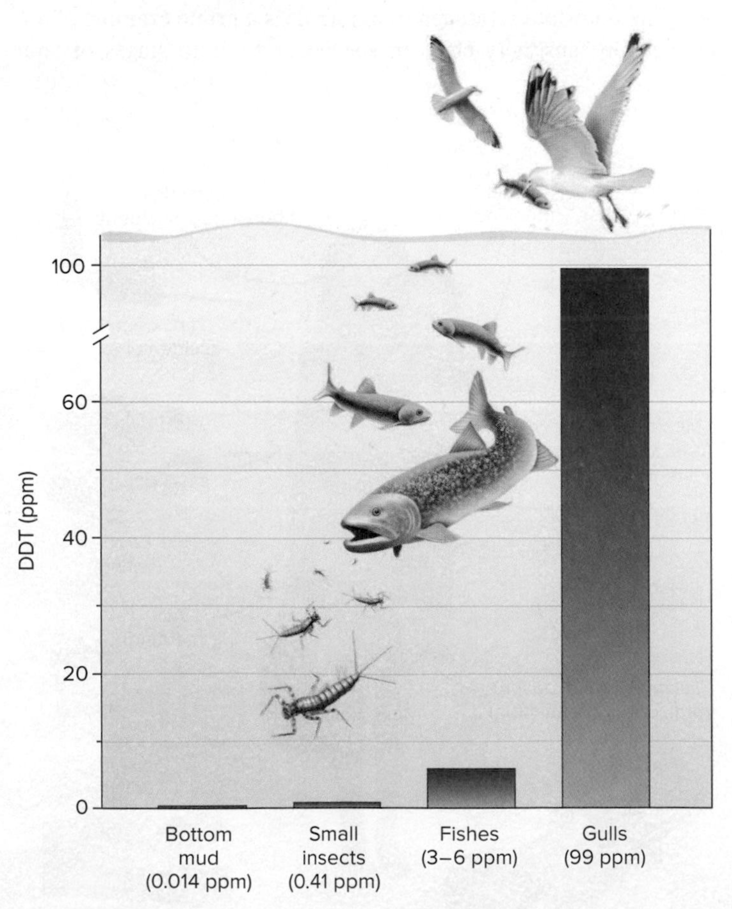

▲ **FIGURE 8.12** Bioaccumulation and biomagnification in a Lake Michigan food chain. The DDT tissue concentration in gulls, a tertiary consumer, was about 240 times that in the small insects sharing the same environment.

Their predators—zooplankton and small fish—collect and retain the toxics from many prey organisms, building up higher concentrations of toxics. The top carnivores in the food chain—game fish, fish-eating birds, and humans—can accumulate such high toxic levels that they suffer adverse health effects.

One of the first well-known examples of bioaccumulation and biomagnification was DDT, which accumulated through food chains, so that by the 1960s it was shown to be interfering with reproduction of bald eagles, peregrine falcons, brown pelicans, and other predatory birds at the top of their food chains.

Persistence makes some materials a greater threat

Many toxic substances degrade when exposed to sun, air, and water. This can destroy them or convert them to inactive forms, but some materials are persistent and can last for years or even centuries as they cycle through ecosystems. Even if released in minute concentrations, they can bioaccumulate in food webs to reach dangerous levels. Heavy metals, such as lead and mercury, are classic examples. Mercury, like lead, can destroy nerve cells and is particularly dangerous to children. The largest source of mercury in the United States is from burning coal. Every year, U.S. power plants release 48 tons of this toxic metal into the air. It works its way through food chains and is concentrated to dangerous levels in fish. Mercury contamination is the most common cause of lakes and rivers failing to meet pollution regulation standards. Forty states have issued warnings that children and pregnant women should not eat local fish. In a nationwide survey of lakes and rivers in 2007, the Environmental Protection Agency found that 55 percent of the 2,700 fish sampled had mercury levels that exceeded dietary recommendations (see chapter 10).

Modern industry has given us many materials that make daily life more comfortable and convenient (fig. 8.13). But some of the 80,000 organic (hydrocarbon-based) chemicals in commercial production can cause long-lasting health problems. Many of these compounds persist a long time in the environment, which can be both a benefit (they do their job for a long time) and a drawback (they also are very hard to eliminate.) We call these **persistent organic pollutants (POPs).** Because of their effectiveness, they have been used abundantly and globally for decades. POPs are found now in the environment and in the tissues and blood of humans and animals everywhere from the tropics to the Arctic. They often accumulate in food webs and reach toxic concentrations in long-living top predators such as humans, sharks, raptors, swordfish, and bears. And they can act, even in low concentrations, as carcinogens, mutagens, neurotoxins, teratogens, and estrogen disrupters. In recent years, global treaties have begun to phase out some of these chemicals, but many remain in widespread use. The POPs of greatest concern include the following:

- **Chloropyrifos** is an organophosphate insecticide used very widely to control insect pests in corn, soybeans, nut trees, and many fruits and vegetables. It is also used for mosquito and termite control, roach and ant traps, and golf courses, green houses, and wood products. It was introduced in 1965 by the Dow Chemical Company, and the EPA estimates that about 9,500 metric tons (21 million pounds) of this compound were used annually in the United States between 1987 and 1998. In

▲ **FIGURE 8.13** Modern industry has given us many materials that make daily life more comfortable and convenient. But we're discovering that some of the 80,000 synthetic chemicals we use can cause long-lasting health problems. ©William P. Cunningham

2007, chloropyrifos, which was the most widely used insecticide in the United States, was banned for most residential use, although commercial use has persisted. The EPA has estimated that 100,000 deaths per year could be traced to chloropyrifos, which interferes with acetylcholine neurotransmitters.

In 2017 an EPA scientific panel recommended that chloripyfos be banned for all uses in the United States because of its high toxicity. But EPA administrator Scott Pruitt rejected the recommendations of his own scientists and squashed efforts to regulate this chemical.

- **Dicamba** is an herbicide in the same family of chlorinated hydrocarbons as 2,4-D. It's a plant hormone that causes broad-leaf weeds to outgrow their nutrient supplies. In 2016 Monsanto introduced genetically modified crop varieties that tolerate dicamba, allowing it to be used liberally on fields without damaging crops. By the next year, 10 million ha (25 million acres) of soy and cotton had been planted in these Monsanto GMO crops and treated with dicamba (also sold by Monsanto). Because this herbicide is effective before weeds sprout, as well as after, much of the spraying was done in late spring or early summer when temperatures and humidity were high, and a significant portion of the herbicide evaporated and drifted into neighboring fields. Neighboring farmers who hadn't paid for the expensive GMO varieties complained that this drift damaged their crops. There are also worries about possible birth defects and reproductive risks from dicamba, especially given the widespread use and volatilization. The agricultural community is bitterly divided about the use of this compound, and there has been violence between neighbors about this dispute.
- **Glyphosate** is an extremely effective herbicide used to kill weeds and grasses marketed by Monsanto under the trade name Roundup®. The Environmental Working Group (EWG) reports that glyphosate is the most widely and heavily used herbicide in the world. The EWG estimates that 8.6 billion kg (18.9 billion pounds) have been used globally since its introduction in 1974. Enough glyphosate was sprayed worldwide in 2014 to apply 0.5 kg per ha (0.75 pound/acre) over all the cropland in the world. The chemical is absorbed through the leaves and works by interfering with the synthesis of essential amino acids. An increasing number of crops have been genetically engineered to be tolerant of glyphosate. Monsanto's Roundup-Ready soybean was the first of these; others include corn, canola, alfalfa, sugar beets, cotton, and wheat. In 2017, 95 percent of soy, 89 percent of corn, and 89 percent of cotton produced in the United States were genetically modified to be herbicide-tolerant (see chapter 7).

In 2015 the European Agency for Research on Cancer listed glyphosate as "probably carcinogenic in humans," on the basis of epidemiological and lab studies. The U.S. EPA, on the other hand, considers glyphosate to be noncarcinogenic and relatively nontoxic. According to the EPA, there is no credible evidence for endocrine disruption by glyphosate. The agency stated that under the worst-case scenario of eating a lifetime of food derived entirely from glyphosate-sprayed fields with residues at maximum permitted levels, no adverse effects would be expected. The State of California disagreed, however, and in 2015 listed the chemical as "known to the State of California to cause cancer."

- **Perfluorinated compounds**, such as perfluorooctane sulfonate (PFOS) and perfluorooctanoic acid (PFOA), are members of a chemical family used to make nonstick, waterproof, and stain-resistant products such as Teflon, Gore-Tex, Scotchguard, and Stainmaster. Industry makes use of their slippery, heat-stable properties to manufacture everything from airplanes and computers to cosmetics and household cleaners. Now these chemicals–which are reported to be infinitely persistent in the environment–are found throughout the world, even in the most remote and seemingly pristine sites.

Almost all Americans have one or more compounds in their blood. Manufacturers of these chemicals have long known that PFOA exposure causes birth defects and cancer in lab animals. They also knew that factory workers have high concentrations of PFOA in their blood and that workers, as well as residents near industrial facilities and waste disposal sites, suffer from a variety of cancers, as well as kidney, liver, and skin ailments. Exposure may be especially dangerous to women and girls, who may be 100 times more sensitive than men to these chemicals. In 2006 the U.S. government and the eight largest fluorochemical companies in America reached an agreement that eliminated PFOA and related compounds by 2015. The European Union has ordered a similar phase-out, but companies in Asia and other regions continue to manufacture and release these compounds. Furthermore, large amounts of these compounds remain in landfills and Superfund sites across America. How we'll clean up this mess remains to be seen.

Chemical interactions can increase toxicity

Some materials produce *antagonistic* reactions. That is, they interfere with the effects or stimulate the breakdown of other chemicals. For instance, vitamins E and A can reduce the response to some carcinogens. Other materials are *additive* when they occur together in exposures. Rats exposed to both lead and arsenic show twice the toxicity of only one of these elements. Perhaps the greatest concern is synergistic effects. **Synergism** is an interaction in which one substance exacerbates the effects of another. For example, occupational asbestos exposure increases lung cancer rates 20-fold. Smoking increases lung cancer rates by the same amount. Asbestos workers who also smoke, however, have a 400-fold increase in cancer rates. How many other toxic chemicals are we exposed to that are below threshold limits individually but combine to give toxic results?

8.4 TOXICITY AND RISK ASSESSMENT

- *The dose makes the poison. Almost everything is toxic at some level.*
- *LD50 and dose-response curves indicate how dangerous a substance is.*
- *Our risk assessment is often selective and nonrational.*

In 1540 the Swiss scientist Paracelsus said, "The dose makes the poison," by which he meant that almost everything is toxic at very high levels but can be safe if diluted enough. This remains the most basic principle of toxicology. Sodium chloride (table salt), for instance, is essential for human life in small doses. If you were forced to eat a kilogram of salt all at once, however, it would make you very sick. A similar amount injected into your bloodstream would be lethal. How a material is delivered–at what rate, through which route of entry, and in what medium–plays a vital role in determining toxicity.

Some toxic substances are so poisonous that a single drop on your skin can kill you. Others require massive amounts injected directly into the blood to be lethal. It is difficult to measure and compare the toxicity of various materials, because species differ in sensitivity and individuals within a species respond differently to a given exposure. In this section we will look at methods of toxicity testing and at how results are analyzed and reported.

We usually test toxic effects on lab animals

Determining toxicity is extremely difficult. The most commonly used and widely accepted toxicity test is to expose a population of laboratory animals to measured doses of a specific substance under controlled conditions. This procedure is expensive, time-consuming, and often painful and debilitating to the animals being tested. It commonly takes hundreds–or even thousands–of animals, several years of hard work, and hundreds of thousands of dollars to thoroughly test toxicity. More humane tests using computer simulations of model reactions, cell cultures, and other substitutes for whole living animals are being developed. However, conventional large-scale animal testing is the method in which scientists have the most confidence and on which most public policies about pollution and environmental or occupational health hazards are based.

Testing toxicity is also difficult because sensitivity varies among the members of a specific population. Figure 8.14 shows a typical dose/response curve for exposure to a hypothetical chemical. Some

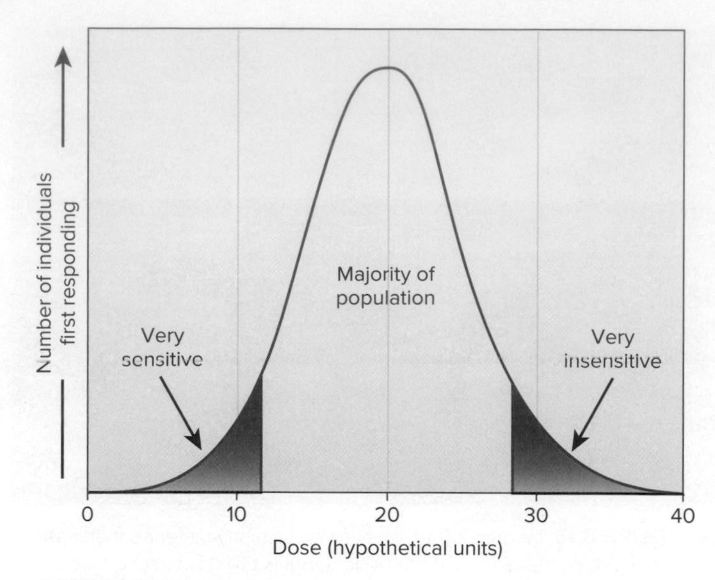

▲ **FIGURE 8.14** Probable variations in sensitivity to a toxin within a population. Some members of a population may be very sensitive to a given toxin, while others are much less sensitive. The majority of the population falls somewhere between the two extremes.

individuals are very sensitive to the material, while others are insensitive. Most, however, fall in a middle category, forming a bell-shaped curve. The question for regulators and politicians is whether we should set pollution levels that will protect everyone, including the most sensitive people, or only aim to protect the average person. It might cost billions of extra dollars to protect a very small number of individuals at the extreme end of the curve. What do you think, is that a good use of resources?

Dose/response curves aren't always symmetrical, making it difficult to compare toxicity of unlike chemicals or different species of organisms. A convenient way to describe toxicity of a chemical is to determine the dose to which 50 percent of the test population is sensitive. In the case of a lethal dose (LD), this is called the **LD50** (fig. 8.15).

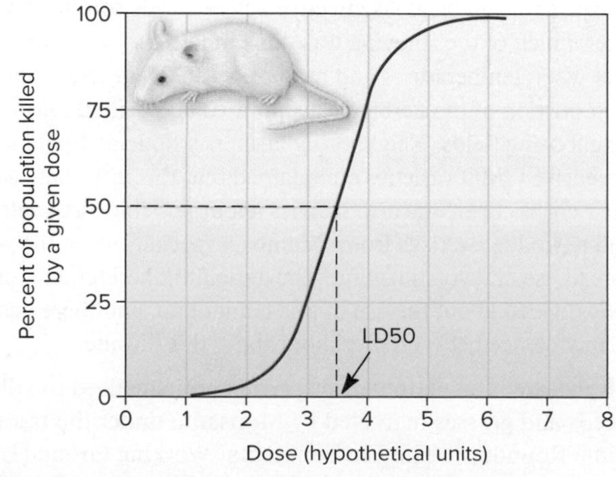

▲ **FIGURE 8.15** Cumulative population response to increasing doses of a toxin. The LD50 is the dose that is lethal to half the population.

Unrelated species can react very differently to the same poison, not only because body sizes vary but also because physiology and metabolism differ. Even closely related species can have very dissimilar reactions to a particular chemical. Hamsters, for instance, are nearly 5,000 times less sensitive to some dioxins than are guinea pigs. Of 226 chemicals found to be carcinogenic in either rats or mice, 95 cause cancer in one species but not the other. These variations make it difficult to estimate the risks for humans, because we don't consider it ethical to perform controlled experiments in which we deliberately expose people to toxins.

Even within a single species, there can be variations in responses between different genetic lines. A current controversy in determining the toxicity of bisphenol A (BPA) concerns the type of rats used for toxicology studies. In most labs, a sturdy strain called the Sprague-Dawley rat is standard. It turns out, however, that these animals, which were bred to grow fast and breed prolifically in lab conditions, are thousands of times less sensitive to endocrine disrupters than ordinary rats. Industry reports that declare BPA to be harmless based on Sprague-Dawley rats are highly suspect.

There is a wide range of toxicity

It's useful to group materials according to their relative toxicity. A moderately harmful substance takes about 1 g per kg of body weight (about 2 oz for an average human) to make a lethal dose. Very toxic materials take about one-tenth that amount, while extremely poisonous materials take one-hundredth as much (only a few drops) to kill most people. Supertoxic chemicals are extremely potent; for some, a few micrograms (millionths of a gram—an amount invisible to the naked eye) make a lethal dose. These materials aren't all synthetic. One of the most toxic chemicals known, for instance, is ricin, a protein found in castor bean seeds. It is so poisonous that 0.3 billionth of a gram given intravenously will kill a mouse. If aspirin were this toxic for humans, a single tablet, divided evenly, could kill 1 million people.

Many carcinogens, mutagens, and teratogens are dangerous at levels far below their direct toxic effect because abnormal cell growth exerts a kind of biological amplification. A single cell, perhaps altered by a molecular event, such as methylation, can multiply into millions of tumor cells or an entire organism. Just as there are different levels of direct toxicity, however, there are different degrees of carcinogenicity, mutagenicity, and teratogenicity. Methanesulfonic acid, for instance, is highly carcinogenic, while the sweetener saccharin is only a possible carcinogen whose effects may be vanishingly small.

Acute versus chronic doses and effects

Most of the toxic effects that we have discussed so far have been **acute effects.** That is, they are caused by a single exposure to the threat and result in an immediate health crisis. Often, if the individual experiencing an acute reaction survives this immediate crisis, the effects are reversible. **Chronic effects,** on the other hand, are long-lasting, perhaps even permanent. A chronic effect can result from a single dose of a very toxic substance, or it can be the result of a continuous or repeated sublethal exposure.

Active LEARNING

Assessing Toxins

The earliest studies of human toxicology came from experiments in which volunteers (usually students or prisoners) were given measured doses of suspected toxins. Today it is considered neither ethical nor humane to deliberately expose individuals to danger, even if they volunteer. Toxicology is now done in either retrospective or prospective studies. In a **retrospective study,** you identify a group of people who have been exposed to a suspected risk factor and then compare their health with that of a control group who are as nearly identical as possible to the experimental group, except for exposure to that particular factor. Unfortunately, people often can't remember where they were or what they were doing many years ago. In a **prospective study,** you identify a study group and a control group and then keep track of everything they do and how it affects their health. Then you watch and wait for years to see if a response appears in the study group but not in the control group. This kind of study is more accurate but expensive because you may need a very large group to study a rare effect, and you have to keep in contact with them for a long time.

Suppose that your class has been chosen for a prospective study of the health risks of a soft drink.

1. The researchers can't afford to keep records of everything that you do or are exposed to over the next 20 or 30 years. Make a list of factors and/or effects to monitor.
2. In a study group of 100 students, how many would have to get sick to convince you that the soft drink was a risk factor? Does the length of your list, above, influence your answer?

We also describe long-lasting exposures as chronic, although their effects may or may not persist after the toxic agent is removed. It usually is difficult to assess the specific health risks of chronic exposures because other factors, such as aging or normal diseases, act simultaneously with the factor under study. It often requires very large populations of experimental animals to obtain statistically significant results for low-level chronic exposures. Toxicologists talk about "megarat" experiments in which it might take a million rats to determine the health risks of some supertoxic chemicals at very low doses. Such an experiment would be terribly expensive for even a single chemical, let alone for the thousands of chemicals and factors suspected of being dangerous.

An alternative to enormous studies involving millions of animals is to give massive amounts—usually the maximum tolerable dose—of a compound being studied to a smaller number of individuals and then to extrapolate what the effects of lower doses might have been. This is a controversial approach because it is not clear that responses to toxics are linear or uniform across a wide range of doses.

Figure 8.16 shows three possible results from low doses of a toxic material. Curve (a) shows a baseline level of response in the population, even at zero dose. This suggests that some other factor in the environment also causes this response. Curve (b) shows a

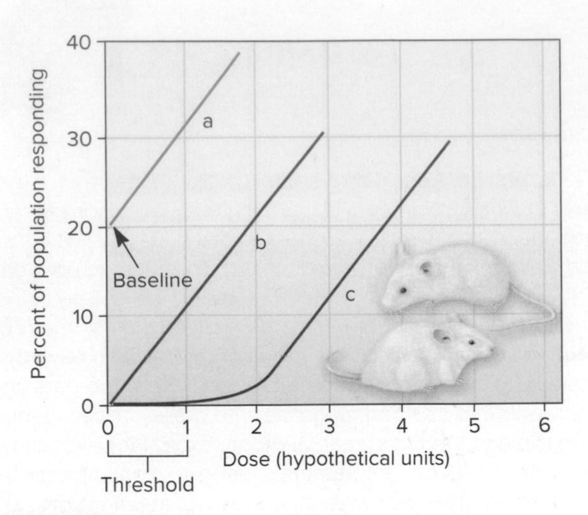

▲ **FIGURE 8.16** Three possible dose-response curves at low doses. (a) Some individuals respond, even at zero dose, indicating that some other factor must be involved. (b) Response is linear down to the lowest possible dose. (c) Threshold must be passed before any response is seen.

straight-line relationship from the highest doses to zero exposure. Many carcinogens and mutagens show this kind of response. Any exposure to such agents, no matter how small, carries some risks. Curve (c) shows a threshold for the response where some minimal dose is necessary before any effect can be observed. This generally suggests the presence of a defense mechanism that prevents the toxin from reaching its target in an active form or repairs the damage that the toxic causes. Low levels of exposure to the material in question may have no deleterious effects, and it might not be necessary to try to keep exposures to zero.

Which, if any, environmental health hazards have thresholds is an important but difficult question. The 1958 Delaney Clause to the U.S. Food and Drug Act forbids the addition of any amount of known carcinogens to food and drugs, based on the assumption that any exposure to these substances represents unacceptable risks. This standard was replaced in 1996 by a "no reasonable harm" requirement, defined as less than one cancer for every million people exposed over a lifetime. This change was supported by a report from the National Academy of Sciences concluding that synthetic chemicals in our diet are unlikely to represent an appreciable cancer risk.

Detectable levels aren't always dangerous

You may have seen or heard dire warnings about toxic materials detected in samples of air, water, or food. A typical headline announced that 23 pesticides were found in 16 food samples. What does that mean? The implication seems to be that any amount of dangerous materials is unacceptable and that counting the numbers of compounds detected is a reliable way to establish danger. We have seen, however, that the dose makes the poison. It matters not only what is there but also how much, where it's located, how accessible it is, and who's exposed. At some level, the mere presence of a substance is insignificant.

Noxious materials may seem to be more widespread now than in the past, and this is surely a valid perception for many substances (fig. 8.17). The daily reports we hear of new chemicals found in new places, however, are also due, in part, to our more sensitive measuring techniques. Twenty years ago, parts per million were generally the limits of detection for most materials. Anything below that amount was often reported as "zero" or "absent," rather than more accurately as "undetected." A decade ago, new machines and techniques were developed to measure parts per billion. Suddenly, substances were found where none had been suspected. Now we can detect parts per trillion or even parts per quadrillion in some cases. Increasingly sophisticated measuring capabilities may lead us to believe that toxic materials have become more prevalent. In fact, our environment may be no more dangerous; we're just better at finding trace amounts.

What accounts for the seemingly enormous increase in conditions, such as asthma, autism, food allergies, and behavioral disorders in American children? Is this evidence for higher levels of environmental pollutants or merely overdiagnosis or increased public paranoia? Perhaps these conditions always existed but weren't so often given labels. In one surprising study, the only factor that appeared to correlate with increased childhood autism was the educational level of the parents. And how can it be that if asthma is caused by air pollution or other environmental contaminants, the disease was, until recently, very rare in China, where conditions are far worse than in the United States? One interesting suggestion is that when we aren't exposed to common pathogens early in life, our immune systems began to attack normal tissues and organs. Clearly, humans are complex organisms and we still have much to learn about our interactions with toxins and our environment.

Low doses can have variable effects

A complication in assessing risk is that relationships of exposure and effects can be nonlinear. That is, if we extrapolate health impacts of large doses from observed effects of small doses, we may dramatically overestimate, or underestimate, actual effects. For example, low doses of the compound DEHP suppress an enzyme essential for rat brain development, but high doses stimulate the enzyme.

▲ **FIGURE 8.17** "Do you want to stop reading those ingredients while we're trying to eat?" Source: Copyright © Richard Guindon. Reprinted with permission.

Similarly, very low amounts of radiation seem to be protective against certain cancers, which is perplexing, because ionizing radiation has long been recognized as a human carcinogen. It's thought now, however, that very low radiation exposure may stimulate DNA repair along with enzymes that destroy free radicals (atoms with unpaired, reactive electrons in their outer shells). Activating these repair mechanisms may defend us from other, unrelated hazards. These nonlinear effects are called **hormesis.**

Another complication is that some substances can have long-lasting effects on genetic expression. For example, researchers found that exposure of pregnant rats to certain chemicals can have effects not only on the exposed rats but also on their daughters and granddaughters. A single dose given on a specific day in pregnancy can be expressed several generations later, even if those offspring have never been exposed to the chemical (Exploring Science, p. 200).

This effect doesn't require a permanent mutation in genes, but it can result in changes, both positive and negative, in expression of whole groups of critical genes over multiple generations. It also can have different outcomes in variants of the same gene. Thus, exposure to a particular toxin could be very harmful to you but have no detectable effects in someone who has slightly different forms of the same genes. This may explain why, in a group of people exposed to the same carcinogen, some will get cancer while others don't. Or it could explain why a particular diet protects the health of some people but not others. As scientists are increasing our knowledge of this intricate network of switches and controls, they're helping explain much about our environment and health.

Our perception of risks isn't always rational

Even if we know with some certainty how toxic a specific chemical is in laboratory tests, it's still difficult to determine **risk** (the probability of harm multiplied by the probability of exposure) if that chemical is released into the environment. As we have seen, many factors complicate the movement and fate of chemicals both around us and within our bodies. Furthermore, public perception of relative dangers from environmental hazards can be skewed so that some risks seem much more important than others.

A number of factors influence how we perceive relative risks associated with different situations.

- People with social, political, or economic interests—including environmentalists—tend to downplay certain risks and emphasize others that suit their own agendas. We do this individually as well, building up the dangers of things that don't benefit us while diminishing or ignoring the negative aspects of activities we enjoy or profit from.
- Most people have difficulty understanding and believing probabilities. We feel that there must be patterns and connections in events, even though statistical theory says otherwise. If the coin turned up heads last time, we feel certain that it will turn up tails next time. In the same way, it is difficult to understand the meaning of a 1-in-10,000 risk of being poisoned by a chemical.
- Our personal experiences often are misleading. When we have not personally experienced a bad outcome, we feel it is more rare and unlikely to occur than it actually may be. Furthermore, the anxieties generated by life's gambles make us want to deny uncertainty and to misjudge many risks (fig. 8.18).
- We have an exaggerated view of our own abilities to control our fate. We generally consider ourselves above-average drivers, safer than most when using appliances or power tools, and less likely than others to suffer medical problems, such as heart attacks. People often feel they can avoid hazards because they are wiser or luckier than others.
- News media give us a biased perspective on the frequency of certain kinds of health hazards, overreporting some accidents or diseases, while downplaying or underreporting others. Sensational, gory, or especially frightful causes of death, such as murders, plane crashes, fires, or terrible accidents, receive a disproportionate amount of attention in the public media. Heart disease, cancer, and stroke kill nearly 15 times as many people in the United States as do accidents and 75 times as many people as do homicides, but the emphasis placed by the media on accidents and homicides is nearly inversely proportional to their relative frequency, compared with either cardiovascular disease or cancer. This gives us an inaccurate picture of the real risks to which we are exposed.
- We tend to have an irrational fear or distrust of certain technologies or activities that leads us to overestimate their dangers. Nuclear power, for instance, is viewed as very risky, while coal-burning power plants seem to be familiar and relatively benign; in fact, coal mining, shipping, and combustion cause an estimated 10,000 deaths each year in the United States, compared with none known so far for nuclear power generation. An old, familiar technology seems safer and more acceptable than does a new, unknown one.
- Alarmist myths and fallacies spread through society, often fueled by xenophobia, politics, or religion. For example, the World Health Organization campaign to eradicate polio worldwide has been thwarted by religious leaders in northern Nigeria—the last country where the disease remains widespread—who claim that oral vaccination is a U.S. plot to spread AIDS or infertility among Muslims.

FIGURE 8.18 How dangerous is trick skating? Many parents regard it as extremely risky, while many students—especially males—believe the risks are acceptable. Perhaps the more important question is whether the benefits outweigh the risks. ©Drpixel/Shutterstock

EXPLORING Science — *The Epigenome*

Could your diet, behavior, or environment affect the lives of your children or grandchildren? For a century or more, scientists assumed that the genes you receive from your parents irreversibly fix your destiny, and that factors such as stress, habits, toxic exposure, or parenting have no effect on future generations.

Now, however, a series of startling discoveries are making us reexamine those ideas. Scientists are finding that a complex set of chemical markers and genetic switches—called the **epigenome**—consisting of DNA and its associated proteins and other small molecules, regulates gene function in ways that can affect numerous functions simultaneously and persist for multiple generations. *Epi* means above, and the epigenome is above ordinary genes in that it regulates their functions. Understanding how this system works helps us see how many environmental factors affect health and may become useful in treating a variety of diseases.

One of the most striking epigenetic experiments was carried out a decade ago by researchers at Duke University. They were studying the effects of diet on a strain of mice carrying a gene called "agouti" that makes them obese, yellow, and prone to cancer and diabetes. Starting just before conception, mother agouti mice were fed a diet rich in B vitamins (folic acid and B_{12}). Amazingly, this simple dietary change resulted in baby mice that were sleek, brown, and healthy. The vitamins somehow had turned off the agouti gene in the offspring.

We know now that B vitamins as well as vegetables, such as onions, garlic, and beets, are methyl donors—that is, they can add a carbon atom and three hydrogens to proteins and nucleic acids. Attaching an extra methyl group can switch genes on or off by changing the way proteins and nucleic acids translate the DNA. Similarly, acetylating DNA (addition of an acetyl group: CH_3CO) can also either stimulate or inhibit gene expression. Both of these reactions are key methods of regulating gene expression.

These reactions involve not only the genes themselves but also a huge set of what we once thought was useless, or junk, DNA in chromosomes as well as a large amount of protein that once seemed to be merely packing material. We now know that both this extra DNA and the histones (proteins around which genes are wrapped) play vital roles in gene expression. And methylating or acetylating these proteins or nucleic acid sequences can have lasting effects on whole families of genes.

More remarkable is that changes in the epigenome can carry through multiple generations. In 2004, Michael Skinner, a geneticist at Washington State University, was studying the effects on rats of exposure to a commonly used fungicide. He found that male rats exposed in utero had lower sperm counts later in life. It took only a single exposure to cause this effect. Amazingly, the effect lasted for at least four generations, even though those subsequent offspring were never exposed to the fungicide. Somehow, the changes in the switching system can be passed from one generation to another along with the DNA it controls.

The way a mother rodent nurtures her young also can cause changes in methylation patterns in her babies' brains that are somewhat like the prenatal vitamins and nutrients that affected the agouti gene. It's thought that licking and grooming activate serotonin receptors that turn on genes to reduce stress responses, resulting in profound brain changes. In another study, rats given extra attention, diet, and mental stimulation (toys) did better at memory tests than did environmentally deprived controls. Altered methylation patterns in the hippocampus—the part of the brain that controls memory—were detected in both these cases. Subsequent generations maintained this methylation pattern.

Epigenetic effects have also been found in humans. One of the most compelling studies involved comparison of two centuries of health records, climate, and food supply in a remote village in northern Sweden. The village of Overkalix was so isolated that when bad weather caused crop failures, famine affected everyone. In good years, on the other hand, there was plenty of food and people stuffed themselves. A remarkable pattern emerged. When other social factors were factored in, grandfathers who were preteens during lean years had grandsons who lived an amazing 32 years longer than those whose grandfathers had gorged themselves as preteens. Similarly, women whose mothers had access to a rich diet while they were pregnant were much more likely to have daughters and granddaughters with health problems and shortened lives.

In an another surprising human health study, researchers found, in a long-term analysis of couples in Bristol, England, that fathers who started smoking before they were 11 years old (just as they were starting puberty and sperm formation was beginning) were much more likely to have sons and grandsons who were overweight and who lived significantly shortened lives than those of nonsmokers. Both these results are attributed to epigenetic effects.

A wide variety of factors can cause epigenetic changes. Smoking, for example, leaves a host of persistent methylation markers in your DNA. So does exposure to a number of pesticides, toxics, drugs, and stressors. At the same time, polyphenols in green tea and deeply colored fruit, B vitamins, and healthy foods, such as garlic, onions, and turmeric, can

▲ Agouti mice have a gene that makes them obese, yellow, and prone to cancer and diabetes. If a mother agouti mouse (*left*) is given B vitamins during pregnancy, the gene is turned off and its baby (*right*) is sleek, brown, and healthy. Amazingly, this genetic repair lasts for several generations before the gene resumes its deleterious effects. (*left*) ©Vasiliy Koval/Shutterstock; (*right*) ©Vasiliy Koval/123RF

Continued ▶

◀ Continued

help prevent deleterious methylations. Not surprisingly, epigenetic changes are implicated in many cancers, including colon, prostate, breast, and blood. This may explain many confusing cases in which our environment seems to have long-lasting effects on health and development that can't be explained by ordinary metabolic effects.

Unlike mutations, epigenetic changes aren't permanent. Eventually, the epigenome returns to normal if the exposure isn't repeated. This makes them candidates for drug therapy. Currently, the U.S. Food and Drug Administration has approved two drugs, Vidaza and Dacogen, that inhibit methylation and are used to treat a precursor to leukemia. Another drug, Zolinza, which enhances acetylation, is approved to treat another form of leukemia. Dozens of other drugs that may treat a variety of diseases, including rheumatoid arthritis, neurodegenerative diseases, and diabetes, are under development.

So your diet, behavior, and environment can have a much stronger impact on both your health and that of your descendants than we previously understood. What you ate, drank, smoked, or did last night may have profound effects on future generations.

How much risk is acceptable?

How much is it worth to minimize and avoid exposure to certain risks? Most people will tolerate a higher probability of occurrence of an event if the harm caused by that event is low. Conversely, harm of greater severity is acceptable only at low levels of frequency. A 1-in-10,000 chance of being killed might be of more concern to you than a 1-in-100 chance of being injured. For most people, a 1-in-100,000 chance of dying from some event or some factor is a threshold for changing what they do. That is, if the chance of death is less than 1 in 100,000, we are not likely to be worried enough to change our ways. If the risk is greater, we will probably do something about it. The Environmental Protection Agency generally assumes that a risk of 1 in 1 million is acceptable for most environmental hazards. Critics of this policy ask, acceptable to whom?

For activities that we enjoy or find profitable, we are often willing to accept far greater risks than this general threshold. Conversely, for risks that benefit someone else, we demand far higher protection. For instance, your chance of dying in a motor vehicle accident in any given year are about 1 in 5,000, but that doesn't deter many people from riding in automobiles. Your chances of dying from lung cancer if you smoke one pack of cigarettes per day are about 1 in 1,000. By comparison, the risk from drinking water with the EPA limit of trichloroethylene is about 2 in 1 billion. Strangely, many people demand water with zero levels of trichloroethylene while continuing to smoke cigarettes.

More than 1 million Americans are diagnosed with skin cancer each year. Some of these cancers are lethal, and most are disfiguring, yet only one-third of teenagers routinely use sunscreen. Tanning beds more than double your chances of cancer, especially if you're young, but about 10 percent of all teenagers admit regularly using these devices.

Table 8.4 lists lifetime odds of dying from some leading causes. These are statistical averages, of course, and there clearly are differences in where one lives and how one behaves that affect the danger level of these activities. Although the average lifetime chance of dying in an automobile accident is 1 in 100, there are clearly things you can do–such as wearing a seat belt, driving defensively, and avoiding risky situations–to improve your odds. Still, it is interesting how we readily accept some risks while shunning others.

Our perception of relative risks is strongly affected by whether risks are known or unknown, whether we feel in control of the outcome, and how dreadful the results are. Risks that are unknown or unpredictable and results that are particularly gruesome or disgusting seem far worse than those that are familiar and socially acceptable.

Studies of public risk perception show that most people react more to emotion than to statistics. We go to great lengths to avoid some dangers while gladly accepting others. Factors that are involuntary, unfamiliar, undetectable to those exposed or catastrophic; those that have delayed effects; and those that are a threat to future generations are especially feared. Factors that are voluntary, familiar, detectable, or immediate cause less anxiety. Even though the actual number of deaths from automobile accidents, smoking, or alcohol, for instance, is thousands of times greater than those from pesticides, nuclear energy, or genetic engineering, the latter group preoccupies us far more than the former.

TABLE 8.4 Lifetime Chances of Dying in the United States

SOURCE	ODDS (1 IN x)
Heart disease	2
Cancer	3
Smoking	9
Lung disease	15
Pneumonia	30
Automobile accident	100
Suicide	100
Falls	200
Firearms	200
Fires	1,000
Airplane accident	5,000
Jumping from high places	6,000
Drowning	10,000
Lightning	56,000
Hornets, wasps, bees	76,000
Dog bite	230,000
Poisonous snakes, spiders	700,000
Botulism	1 million
Falling space debris	5 million
Drinking water with EPA limit of trichloroethylene	10 million

Source: Data from U.S. National Safety Council, 2003.

Active LEARNING

Calculating Probabilities

You can calculate the statistical danger of a risky activity by multiplying the probability of danger by the frequency of the activity. For example, in the United States, 1 person in 3 will be injured in a car accident in his or her lifetime (so the probability of injury is 1 per 3 persons, or ⅓). In a population of 30 car-riding people, the cumulative risk of injury is 30 people × (1 injury/3 people) = 10 injuries over 30 lifetimes.

1. If the average person takes 50,000 trips in a lifetime, and the accident risk is ⅓ per lifetime, what is the probability of an accident per trip?
2. If you have been riding safely for 20 years, what is the probability of an accident during your next trip?

ANSWERS: 1. Probability of injury per trip = (1 injury/3 lifetimes) × (1 lifetime/50,000 trips) = 1 injury/150,000 trips. 2. 1 in 150,000. Statistically, you have the same chance each time.

8.5 ESTABLISHING HEALTH POLICY

- *It's difficult to evaluate multiple risks and benefits simultaneously.*
- *Health standards need to consider combined exposures, differing sensitivities, and differences between chronic and acute exposures.*

Risk management combines principles of environmental health and toxicology with regulatory decisions based on socioeconomic, technical, and political considerations (fig. 8.19). The biggest problem in making regulatory decisions is that we are usually exposed to many sources of harm, often unknowingly. It is difficult to separate the effects of all these different hazards and to evaluate their risks accurately, especially when the exposures are near the threshold of measurement and response. In spite of often vague and contradictory data, public policymakers must make decisions.

The struggle over whether to vaccinate children against common illnesses is a good example of the difficulties in risk assessment. In 1998 a British physician published a paper suggesting that the measles, mumps, and rubella (MMR) vaccine is linked to autism. In 2010 the U.K. General Medical Council found the author of that study guilty of "dishonesty and misleading conduct" for failing to disclose his personal interest in this research and scientific and ethical errors led to retraction of the original paper by *The Lancet.* At least 20 subsequent studies have found no link between vaccines and autism, but none of that scientific evidence is reassuring to thousands of angry and frightened parents who demand answers for why their children are autistic. Many of them remain convinced that vaccines cause this distressing condition and they refuse to allow their children to be vaccinated. Physicians argue that this is a danger not only to the children but also to the population at large, which is at risk from epidemics when there's a large pool of nonimmune children. Lack of immunizations is blamed for the 2014 measles outbreak in California. Nevertheless, the absence of a clear, convincing explanation for what does cause autism simply fuels many people's suspicions. This is a good example of the power of anecdotal evidence and personal bias versus scientific evidence.

In setting standards for environmental toxics, we need to consider (1) combined effects of exposure to many different sources of damage, (2) different sensitivities of members of the population, and (3) effects of chronic as well as acute exposures. Some people argue that pollution levels should be set at the highest amount that does *not* cause measurable effects. Others demand that pollution be reduced to zero if possible, or as low as is technologically feasible. It may not be reasonable to demand that we be protected from every potentially harmful contaminant in our environment, no matter how small the risk. As we have seen, our bodies have mechanisms that enable us to avoid or repair many kinds of damage, so that most of us can withstand a minimal level of exposure without harm.

On the other hand, each challenge to our cells by toxic substances represents stress on our bodies. Although each individual stress may not be life-threatening, the cumulative effects of all the environmental stresses, both natural and human-caused, to which we are exposed may seriously shorten or restrict our lives. Furthermore, some individuals in any population are more susceptible to those stresses than others. Should we set pollution standards so that no one is adversely affected, even the most sensitive individuals, or should the acceptable level of risk be based on the average member of the population?

Finally, policy decisions about hazardous and toxic materials also need to be based on information about how such materials affect the plants, animals, and other organisms that define and maintain our environment. In some cases, pollution can harm or destroy whole ecosystems with devastating effects on the life-supporting cycles on which we depend. In other cases, only the most sensitive species are threatened. Table 8.5 shows the Environmental Protection Agency's assessment of relative risks to human welfare.

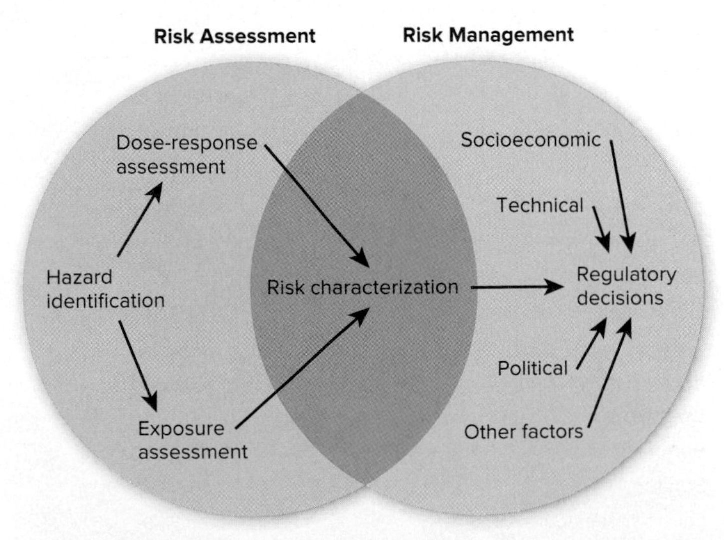

▲ **FIGURE 8.19** Risk assessment organizes and analyzes data to determine relative risk. Risk management sets priorities and evaluates relevant factors to make regulatory decisions.

TABLE 8.5 Relative Risks to Human Welfare

RELATIVELY HIGH-RISK PROBLEMS
Habitat alteration and destruction
Species extinction and loss of biological diversity
Stratospheric ozone depletion
Global climate change
RELATIVELY MEDIUM-RISK PROBLEMS
Herbicides/pesticides
Toxins and pollutants in surface waters
Acid deposition
Airborne toxins
RELATIVELY LOW-RISK PROBLEMS
Oil spills
Groundwater pollution
Radionuclides
Thermal pollution

Source: Data from U.S. Environmental Protection Agency.

This ranking reflects a concern that our exclusive focus on reducing pollution to protect human health has neglected risks to natural ecological systems. While there have been many benefits from a case-by-case approach in which we evaluate the health risks of individual chemicals, we have often missed broader ecological problems that may be of greater ultimate importance.

CONCLUSION

We have made marvelous progress in reducing some of the worst diseases that have long plagued humans. Smallpox is the first major disease to be completely eliminated. Guinea worms and polio are nearly eradicated worldwide; typhoid fever, cholera, yellow fever, tuberculosis, mumps, and other highly communicable diseases are rarely encountered in advanced countries. Childhood mortality has decreased 90 percent globally, and people almost everywhere are living twice as long, on average, as they did a century ago.

But the technological innovations and affluence that have diminished many terrible diseases have also introduced new risks. Traffic accidents and chronic conditions, such as cardiovascular disease, cancer, depression, dementia, and diabetes, that once were confined to richer countries now have become leading health problems nearly everywhere. Part of this change is that we no longer die at an early age of infectious disease, so we live long enough to develop the infirmities of old age. Another factor is that affluent lifestyles, lack of exercise, and unhealthy diets aggravate these chronic conditions.

New, emergent diseases are appearing at an increasing rate. With increased international travel, diseases can spread around the globe in a few days. Epidemiologists warn that the next deadly epidemic may be only a plane ride away. In addition, modern industry is introducing thousands of new chemical substances every year, most of which aren't studied thoroughly for long-term health effects. Endocrine disrupters, neurotoxics, carcinogens, mutagens, teratogens, obesogens, and other toxics can have tragic outcomes. The effects of lead on children's mental development is an example of both how we have introduced materials with unintended consequences and how we have succeeded in controlling a serious health risk. Many other industrial chemicals could be having similar harmful effects.

PRACTICE QUIZ

1. Define the terms *health* and *disease*.
2. Name the five leading causes of global disease burden in 2015.
3. Define *emergent diseases* and give some recent examples.
4. What is *conservation medicine*?
5. What is the difference between toxic and hazardous? Give some examples of materials in each category.
6. What are *endocrine disrupters*, and why are they of concern?
7. What are *bioaccumulation* and *biomagnification*?
8. Why is chloropyrifos a concern?
9. What is an *LD50*?
10. Distinguish between acute and chronic toxicity.

CRITICAL THINKING AND DISCUSSION

Apply the principles you have learned in this chapter to discuss these questions with other students.

1. Is it ever possible to be completely healthy?
2. Why should we be concerned with diseases, such as Ebola, that occur only in tropical countries? What's wrong with simply closing our borders and letting the diseases run their course? Are there ethical, moral, economic, or political implications?
3. Why do we spend more money on heart diseases or cancer than childhood diseases?
4. Why do we tend to assume that natural chemicals are safe while industrial chemicals are always dangerous? Is this accurate?
5. Examine the list of reasons that people have a flawed perception of certain risks. Do you see things that you or someone you know sometimes do?
6. Do you agree that 1 in 1 million risk of death is an acceptable risk? Notice that almost everything in table 8.4 carries a greater risk than this. Does this make you want to change your habits?

DATA ANALYSIS How Do We Evaluate Risk and Fear?

A central question in environmental health is how we perceive different risks around us. When we evaluate environmental hazards, how do we assess known factors, uncertain risks, and the unfamiliarity of new factors we encounter? Which considerations weigh most heavily in our decisions and our actions as we try to avoid environmental risks?

The graph shown here shows one set of answers to this question, using aggregate responses of many people to risk. Go to Connect to find further discussion and a set of questions regarding risk, uncertainty, and fear.

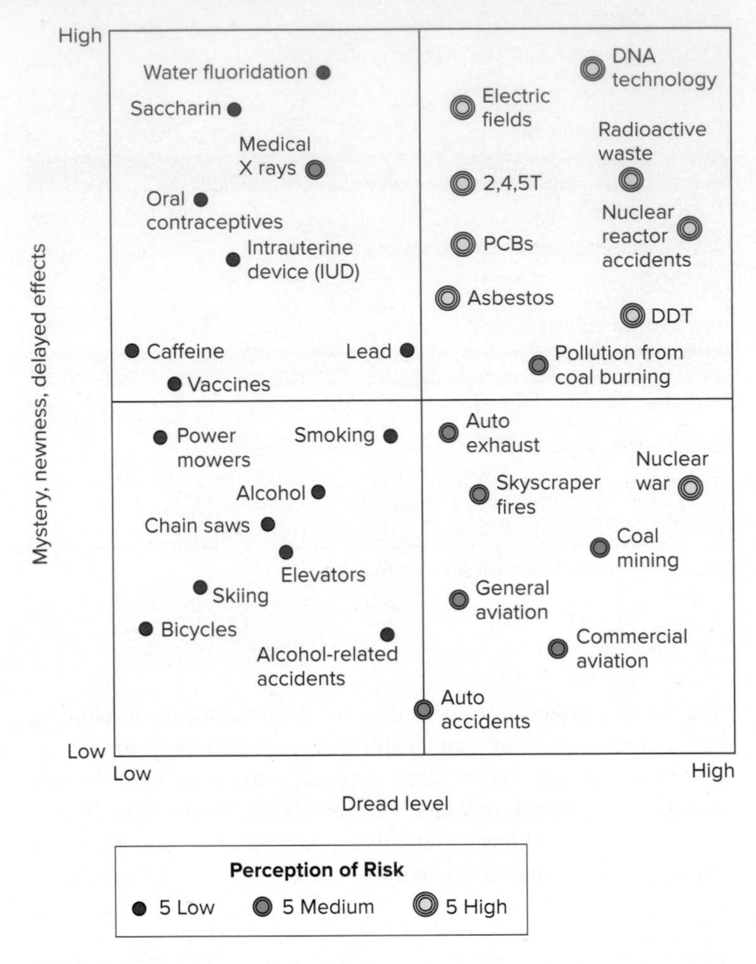

▲ Public perception of risk, depending on familiarity, apparent potential for harm, and personal control over the risk.

CHAPTER

9 Climate

LEARNING OUTCOMES

After studying this chapter, you should be able to answer the following questions:

- How do the troposphere and stratosphere differ?
- What are some factors in natural climate variability?
- Explain the greenhouse effect and how it is changing our climate.
- How do we know the nature and cause of recent climate change?
- List some effects of climate change.
- What are some strategies for minimizing global climate change?

Rising sea levels, accompanied by higher storm surges and greater storm intensity, will affect many cities as the global climate warms.

CASE STUDY

Shrinking Florida

What will Florida be like when its beaches are under water? It's hard to say precisely, but the tourism industry is sure to struggle as beach resorts and coastal cities flood. The interior of the state may become more crowded as residents of coastal cities are forced to move inland, away from rising tides and storm surges. Because drought and hot summers will likely make much of the state's agriculture uneconomical, there should be plenty of land for building inland, although freshwater resources will be scarce. Everglades National Park and much of the Florida Keys will exist only in photographs. Forests will be less extensive, as dry summers produce more frequent, larger wildfires.

All this might seem apocalyptic, but it conforms to what the climate trends indicate. Climate warming has already caused global mean sea level to rise about 20 cm (8 in.) in the past century. That sounds slight, but it brings higher storm waves and tides. By 2050, another 25 cm (10 in.) of sea level rise is likely, and by 2100 sea level is expected to be 1 m higher; some estimates put the change closer to 2 m. A meter of sea level rise is enough to inundate most of Florida's coast, including much of Miami, Boca Raton, Fort Meyers, and Cape Canaveral (fig. 9.1). The National Climate Assessment of 2014 named Miami as one of the U.S. cities most vulnerable to physical and economic damage resulting from human-caused climate warming.

Why is sea level rising? Thermal expansion associated with heat absorbed from the atmosphere explains about half of the change so far, because water expands slightly as it warms. Melting land-based glaciers contribute about one-quarter of the increase, and melting ice sheets contribute the remainder. This balance may change in the coming decades: Antarctica's land-based ice sheets have begun to flow to the sea at an accelerating rate, and Greenland is melting at a rate few climatologists expected a few years ago. Greenland's massive ice cap holds enough water to raise sea level by about 6 m (about 20 ft) if it all melts. That degree of melting is likely to take centuries, but intermediate changes will cause tremendous damage.

South Florida faces special challenges because its porous limestone bedrock allows seawater to infiltrate aquifers. Saltwater intrusion has already affected wells, as farms and cities pump out groundwater and sea water presses in to replace it. This process will accelerate as sea levels rise and shorelines retreat.

The extra heat energy in the atmosphere produces an added risk of increased hurricane frequency and intensity in the Gulf Coast region. Hurricanes occur when intense sunlight heats the ocean surface, warming the air, evaporating water, and causing high-energy atmospheric circulation that evolves into swirling winds of 120 km/hr (74 mi/hr) or more. Warming ocean surfaces increase the heat energy available to these weather systems. Climatologists anticipate that the simultaneous development of a hotter atmosphere and warmer oceans will lead to more and bigger hurricanes over time. Since 40 percent of all U.S. hurricanes hit Florida, according to the National Oceanic and Atmospheric Administration (NOAA), this is a major threat to the state.

Many policymakers don't like to talk about climate change, but it will affect all of us. The pattern of warming has become more and more clear, with increased heat, drought, and storm records piling up in recent years. Climate change will present a different world to our grandchildren. Because many state governors have refused to address climate science (or even to acknowledge its existence) when planning for emergencies like hurricanes and heat waves, the Federal Emergency Management Agency (FEMA) has threatened to withhold disaster funding from states that fail to plan for climate change. The shared economic and social threats are just as real as the environmental costs.

In this chapter we examine how our climate works and what causes climate changes, including anthropogenic (human-caused) factors. We also highlight some of the many steps we can all take to reduce these changes and their impacts. ■

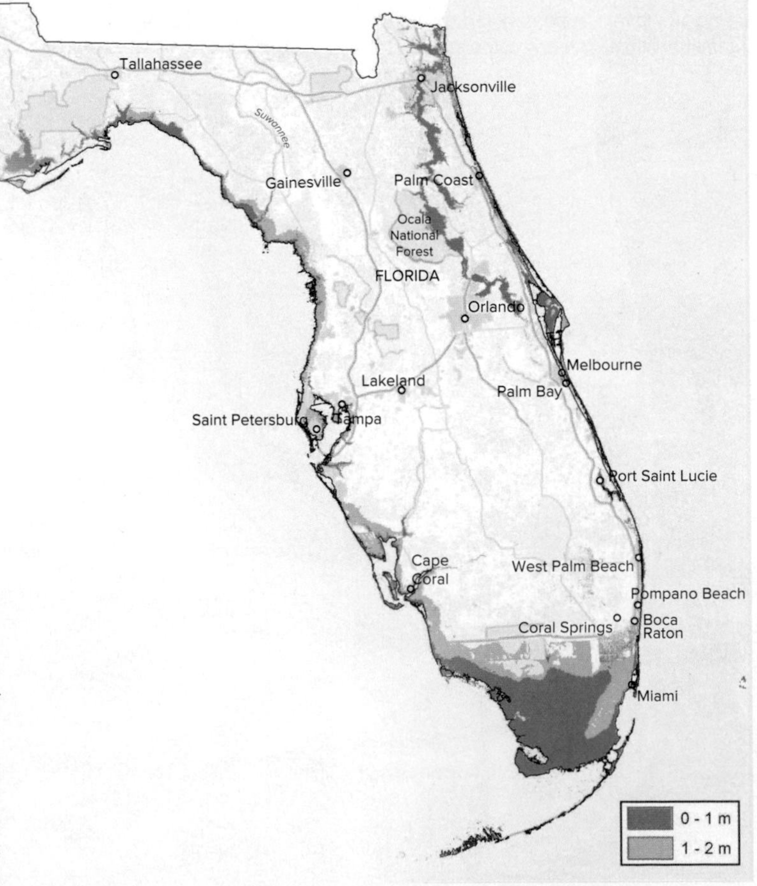

▲ **FIGURE 9.1** Projected sea level rise by 2100.

> *The next decade is critical. If emissions do not peak by around 2020, . . . the needed 50% reduction by 2050 will become much more costly. In fact, the opportunity may be lost completely.*
>
> —INTERNATIONAL ENERGY AGENCY, 2010

9.1 WHAT IS THE ATMOSPHERE?

- *The atmosphere is layered, with more massive molecules near the ground surface.*
- *Circulation in the troposphere redistributes heat and moisture.*
- *The composition of the atmosphere influences how much heat energy is stored.*

Climate refers to the atmosphere, how it behaves, and how it is composed. Climate processes involve the movement and storage of energy, including interactions with the land surface and with oceans. Rising sea levels that threaten coastal Florida are one example of this interaction (fig. 9.2).

Earth's atmosphere consists of gas molecules, relatively densely packed near the surface and thinning gradually to about 500 km (300 mi) above the earth's surface. In the lowest layer of the atmosphere, air moves ceaselessly—flowing, swirling, and continually redistributing heat and moisture from one part of the globe to another. The daily temperatures, wind, and precipitation that we call **weather** occur in the troposphere. Long-term temperatures and precipitation trends we refer to as **climate.**

The earliest atmosphere on earth probably consisted mainly of hydrogen and helium. Over billions of years, most of that hydrogen and helium diffused into space. Volcanic emissions added carbon, nitrogen, oxygen, sulfur, and other elements to the atmosphere. Virtually all of the molecular oxygen (O_2) we breathe was probably produced by photosynthesis in blue-green bacteria, algae, and green plants.

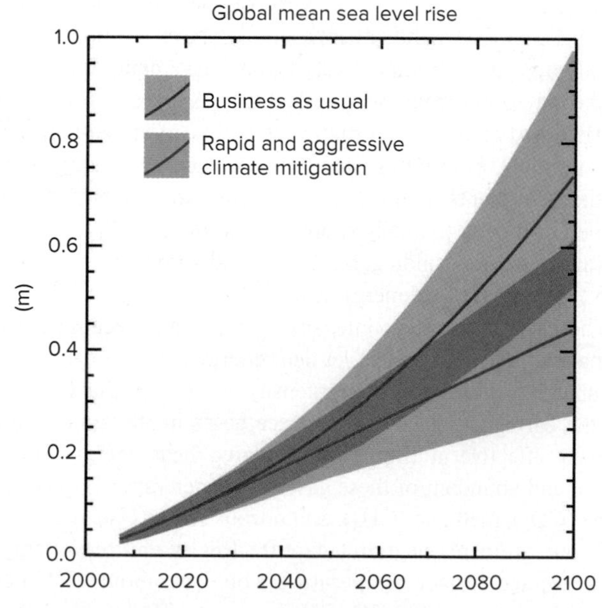

▲ **FIGURE 9.2** Projected change in sea level, with range of modeled values (shaded areas). Blue represents rapid and aggressive climate mitigation reducing to 0 emissions by 2100; orange represents business as usual. Some models show sea level change of 2 m. Source: IPPC 2013.

Clean, dry air is 78 percent nitrogen and almost 21 percent oxygen, with the remaining 1 percent composed of argon, carbon dioxide (CO_2), and a variety of other gases. Water vapor (H_2O in gas form) varies from near 0 to 4 percent, depending on air temperature and available moisture. Minute particles and liquid droplets—collectively called **aerosols**—also are suspended in the air. Atmospheric aerosols and water vapor play important roles in the earth's energy budget and in rain production.

The atmosphere has four distinct zones of contrasting temperature, due to differences in absorption of solar energy (fig. 9.3). The layer immediately adjacent to the earth's surface is called the **troposphere** (*tropein* means to turn or change, in Greek). Within the troposphere, air circulates in great vertical and horizontal **convection currents,** constantly redistributing heat and moisture around the globe. Convection occurs in three general cells (fig. 9.4), which are driven by heating at the sun-warmed earth's surface (which causes air to rise). At altitude, air cools and sinks, or subsides. Intense heating near the equator produces vigorous circulation, so the Hadley cell reaches higher into the atmosphere than the Ferrell and Polar cells. These convection patterns shape many regional weather patterns, including shifting frontal weather in midlatitudes, tropical rainfall, and dry conditions where subsiding currents meet. Jet streams, strong, high-altitude winds, occur where convection cells meet.

The troposphere ranges in depth from about 18 km (11 mi) over the equator to about 8 km (5 mi) over the poles where air is cold and dense. Because gravity holds most air molecules close to the earth's surface, the troposphere is much denser than the other layers: It contains about 75 percent of the total mass of the atmosphere. Air temperature drops rapidly with increasing altitude in this layer, reaching about −60°C (−76°F) at the top of the troposphere. A sudden reversal of this temperature gradient creates a boundary called the tropopause. Above this boundary, **ozone** (O_3) molecules in the stratosphere absorb solar energy. In particular, ozone absorbs ultraviolet (UV) radiation (wavelengths of 290–330 nm; see fig. 2.13). This absorbed energy makes the stratosphere

Benchmark Data

Among the ideas and values in this chapter, these are a few worth remembering.

5–18 km	Approximate depth of the atmosphere, at poles and equator
280 ppm	Pre-industrial (1750) concentration of CO_2
350 ppm	Concentration that limits climate warming to < 2 degrees C
410 ppm	Concentration in 2018
40 GT	Approximate amount of CO_2 released annually
5–20% of GDP	Cost to global economy of current climate change trajectory
1% of GDP	Cost to aggresively control CO_2 emissions

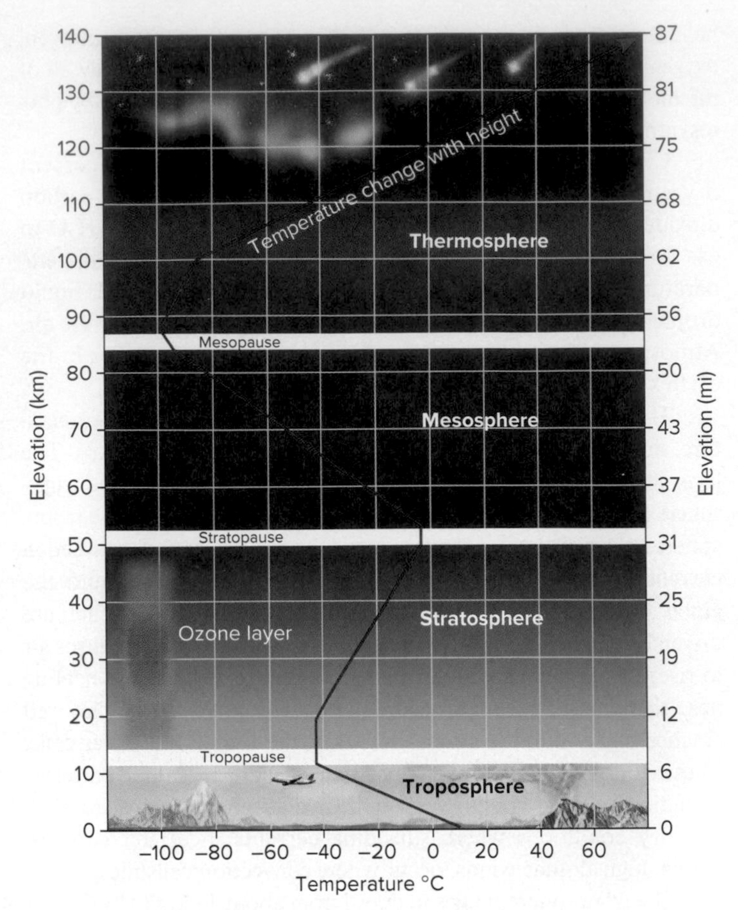

▲ **FIGURE 9.3** Layers of the atmosphere vary in temperature and composition. Our weather happens in the troposphere. Stratospheric ozone is important for blocking ultraviolet solar energy.

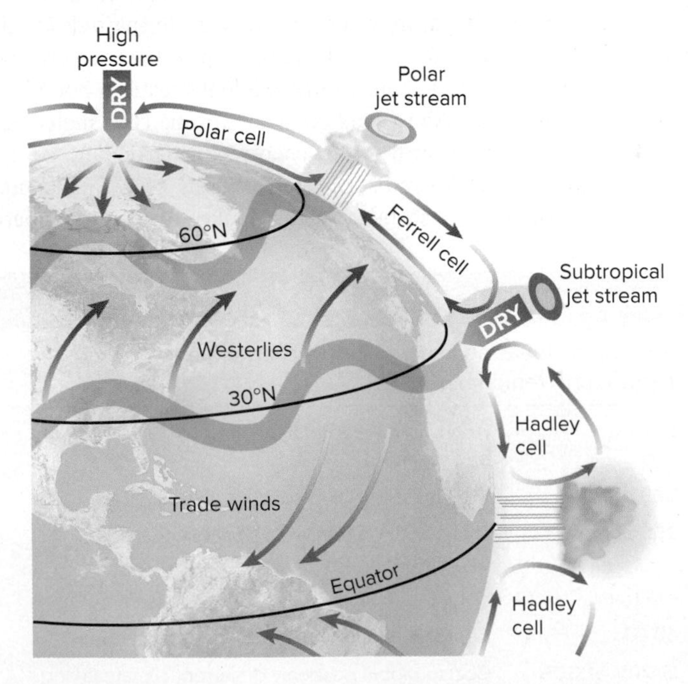

▲ **FIGURE 9.4** Convection cells circulate air, moisture, and heat around the globe. Jet streams develop where cells meet, and surface winds result from convection. Convection cells expand and shift seasonally.

warmer than the upper troposphere. Absorption of UV radiation by ozone also protects living organisms on earth, because UV radiation damages living tissues. Tropospheric air cannot continue to rise when it is cooler than the surrounding air, so there is little mixing across this boundary.

The **stratosphere** extends about 50 km (31 mi) out from the tropopause. It is far more dilute than the troposphere, but it has a similar composition—except that it has almost no water vapor and nearly 1,000 times more ozone.

Unlike the troposphere, the stratosphere is relatively calm. There is so little mixing in the stratosphere that volcanic ash and human-caused contaminants can remain in suspension there for many years.

Above the stratosphere, the temperature diminishes again, creating the mesosphere, or middle layer. The thermosphere (heated layer) begins at about 50 km. This is a region of highly ionized (electrically charged) gases, heated by a steady flow of high-energy solar and cosmic radiation. Sometimes, intense pulses of radiation near the poles cause ionized gases to glow, producing northern (or southern) lights.

The atmosphere captures energy selectively

The sun supplies the earth with abundant energy, especially near the equator. Of the solar energy that reaches the outer atmosphere, about one-quarter is reflected by clouds and atmospheric gases, and another quarter is absorbed by water vapor, carbon dioxide, methane, and other gases (fig. 9.5). This absorbed energy warms the atmosphere slightly. About half of incoming solar radiation (insolation) reaches the earth's surface. Most of this energy is in the form of light or infrared energy (including heat).

Some incoming solar energy is reflected by bright surfaces, such as snow, ice, and sand. The rest is absorbed by the earth's surface and by water. Surfaces that *reflect* energy have a high **albedo** (reflectance, or "whiteness"). Fresh snow and dense clouds, for instance, can reflect as much as 85 to 90 percent of the light falling on them (table 9.1). Surfaces that absorb energy have a low albedo and generally appear dark. Black soil, asphalt pavement, and water, for example, have low albedo, with reflectance as low as 3 to 5 percent.

Absorbed energy heats materials (think of an asphalt parking lot in summer), evaporates water, and provides the energy for photosynthesis in plants. Following the second law of thermodynamics, absorbed energy is gradually re-emitted as lower-quality heat energy. The walls of a brick building, for example, absorb light (high-intensity energy) and re-emit that energy as heat (low-intensity energy).

The change in energy intensity is important because the gases that make up our atmosphere let light energy pass through, but these gases absorb or reflect the lower-intensity heat energy that is re-emitted from the earth (fig. 9.5). Several trace gases in the atmosphere are especially effective at trapping re-radiated heat energy. The most effective and abundant of these gases are water vapor (H_2O), carbon dioxide (CO_2), methane (CH_4), and nitrous oxide (N_2O).

If our atmosphere didn't capture this re-emitted energy, the earth's average surface temperature would be about −6°C (21°F), rather than the current 14°C (57°F) average. Thus, energy capture is necessary for liquid water on earth, and for life as we know it. The **greenhouse effect** is a common term to describe the capture of energy by gases in the atmosphere. Something like the glass of a

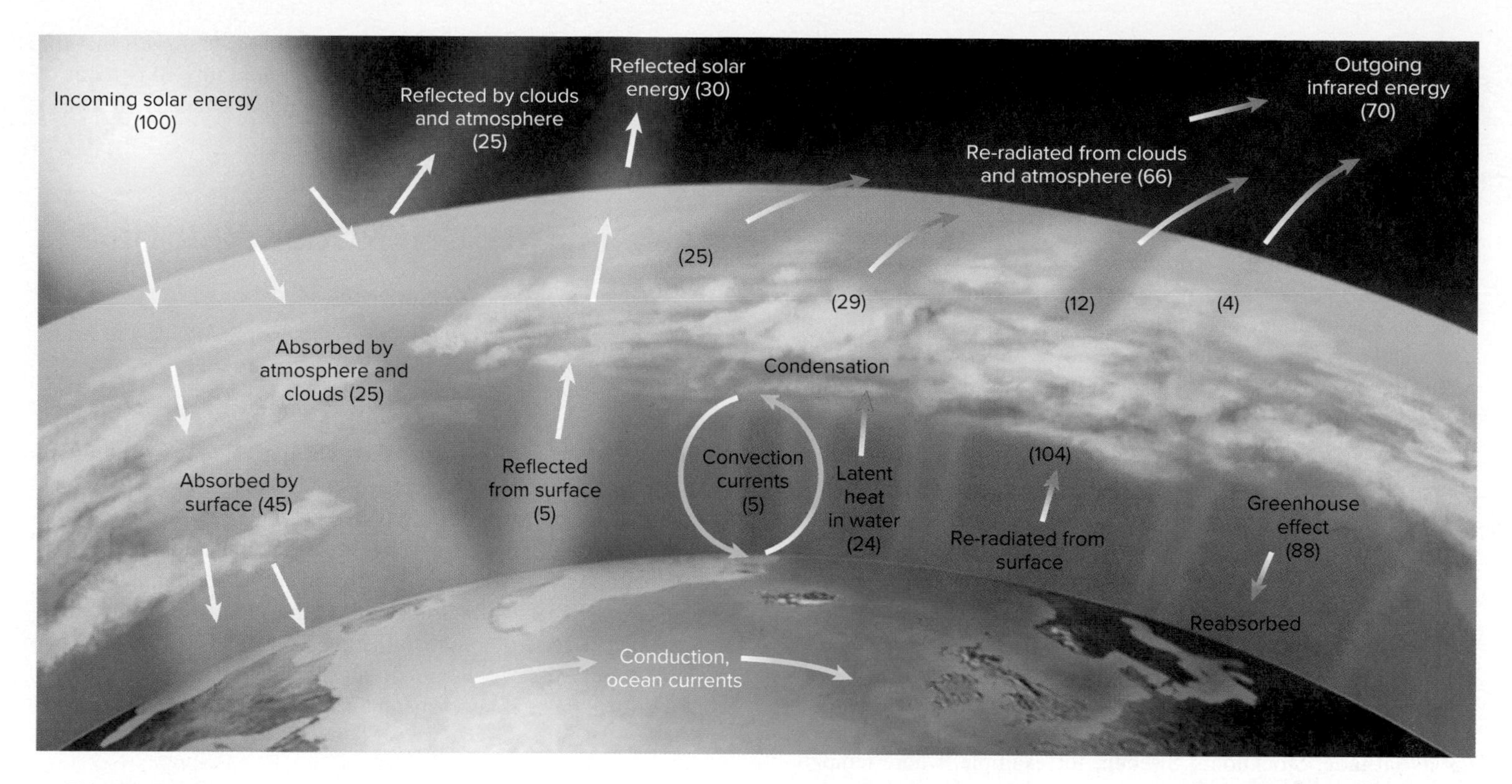

▲ **FIGURE 9.5** Energy balance between incoming and outgoing radiation. The atmosphere absorbs or reflects about half of the solar energy reaching the earth. Most of the energy re-emitted from the earth's surface is long-wave, infrared energy. Gases and aerosols in the atmosphere absorb and re-radiate most of this energy, keeping the surface much warmer than it would otherwise be. This absorption is known as the greenhouse effect.

TABLE 9.1 Albedo (Reflectivity) of Earth Surfaces

SURFACE	ALBEDO (%)
Fresh snow	80–85
Dense clouds	70–90
Water (low sun)	50–80
Sand	20–30
Forest	5–10
Water (sun overhead)	5
Dark soil	3

greenhouse, the atmosphere transmits sunlight but traps some heat inside. Also like a greenhouse, the atmosphere lets that energy dissipate gradually to space (Key Concepts, p. 220).

In a greenhouse, the temperature depends on the rates of incoming and outgoing energy. In the "greenhouse" of our atmosphere, we are currently slowing the rate of outgoing energy by increasing concentrations of heat-trapping gases, and the temperature is rising as a consequence. We have raised the amount of CO_2, CH_4, and N_2O to levels the earth has not seen since before the appearance of humans as a species. The question is whether we will be able to agree on changing this trend.

Evaporated water stores and redistributes heat

Much of the incoming solar energy is used up in evaporating water. Every gram of evaporating water absorbs 580 calories of energy as it transforms from liquid to gas. Water vapor in the air stores that 580 calories per gram, and we call that stored heat **latent heat.** Later, when the water vapor condenses, it releases 580 calories per gram. Globally, latent heat contains huge amounts of energy, enough to power thunderstorms, hurricanes, and tornadoes. Imagine the sun shining on the Gulf of Mexico in the winter. Warm sunshine and plenty of water allow continuous evaporation that converts an immense amount of solar (light) energy into latent heat stored in evaporated water. Now imagine a wind blowing the humid air north from the Gulf toward Canada. The air cools as it rises and moves north. Eventually, cooling causes the water vapor to condense. Rain (or snow) falls as a consequence. Note that it is not only water that has moved from the Gulf to the Midwest: 580 calories of heat have also moved with every gram of moisture. The heat and water have now moved from the sunny Gulf to the colder Midwest. This redistribution of heat and water around the globe is essential to life on earth.

Why does it rain? Understanding this will help you understand the distribution of latent heat, and water resources, around the globe. Rain falls when there are two conditions: (1) a moisture source, such as an ocean, from which water can evaporate into the atmosphere, and (2) a lifting mechanism. Lifting is important because air cools at high elevations. You may have observed this cooling if you have driven over a mountain pass. Lifting can occur when winds push warm, moist air up and over mountains. Lifting can also occur when weather systems collide: When a cold front meets a warm front, for example, the warm air is forced up and over the colder, denser air mass. Lifting also occurs when hot air, warmed near the earth's surface on a sunny day, rises in convection currents. Any of these three mechanisms can cause air to rise and cool. Moisture in the cooling air then condenses. We see the moisture falling as rain or snow.

Next time you watch the weather report, see if you can find references to these processes in predicted rain or snowfall.

Ocean currents also redistribute heat

Warm and cold ocean currents strongly influence climate conditions on land. Surface ocean currents result from wind pushing on the ocean surface. As surface water moves, deep water wells up to replace it, creating deeper ocean currents. Differences in water density–depending on the temperature and saltiness of the water–also drive ocean circulation. Huge cycling currents called gyres carry water north and south, redistributing heat from low latitudes to high latitudes (see appendix 3, p. A-4, global climate map). For example, the Alaska current, flowing from Alaska southward to California, keeps San Francisco cool and foggy during the summer.

The Gulf Stream, one of the best-known currents, carries warm Caribbean water north past Canada's maritime provinces to northern Europe (fig. 9.6). This current is immense, some 800 times the volume of the Amazon, the world's largest river. The heat transported from the Gulf keeps Europe much warmer than it should be for its latitude. Stockholm, Sweden, for example, where temperatures rarely fall much below freezing, is at the same latitude as Churchill, Manitoba, where polar bears live in summer. As the warm Gulf Stream passes Scandinavia and swirls around Iceland, the water cools and evaporates, becomes dense and salty, and plunges downward, creating a strong, deep, southward current.

Together, this surface and deep-water circulation system is called the **thermohaline** (temperature and salinity-related) **circulation,** because both temperature and salt concentrations control the density of water, and contrasts in density drive its movement. Dr. Wallace Broecker of the Lamont Doherty Earth Observatory, who first described this great conveyor system, also found that it can shut down suddenly. About 11,000 years ago, as the earth was warming at the end of the last ice age, cold glacial melt-water surged into the North Atlantic and interrupted the thermohaline circulation cycle. Europe was plunged into a cold period that lasted for 1,300 years. Temperatures may have changed dramatically in just a few years.

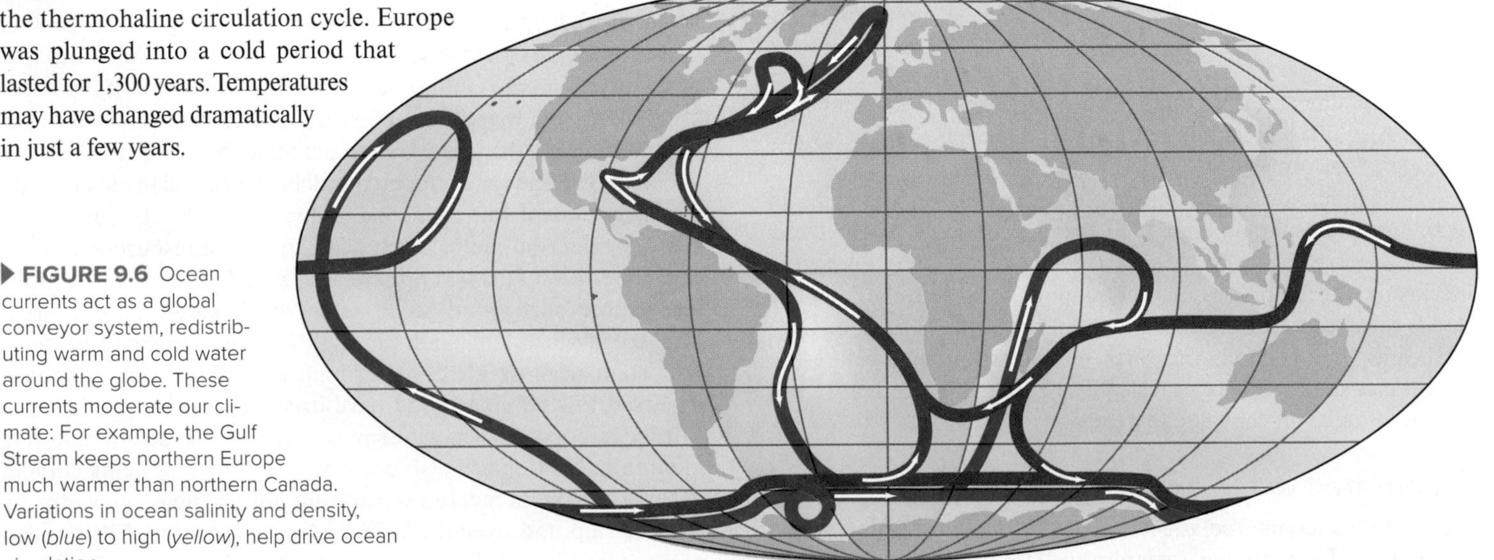

▶ **FIGURE 9.6** Ocean currents act as a global conveyor system, redistributing warm and cold water around the globe. These currents moderate our climate: For example, the Gulf Stream keeps northern Europe much warmer than northern Canada. Variations in ocean salinity and density, low (*blue*) to high (*yellow*), help drive ocean circulation.

Could this happen again? Some climatologists suggest that the melting of the Greenland ice sheet, which contains 10 percent of the globe's glacial ice, could cause sudden changes in ocean circulation.

9.2 CLIMATE CHANGES OVER TIME

- *Milankovitch cycles drive long-term climate changes.*
- *Ice cores contain CO_2 and oxygen isotopes used to reconstruct past temperatures and atmospheric composition.*
- *El Niño is one example of an ocean-atmosphere oscillation.*

Climatologist Wallace Broecker has said that "climate is an angry beast, and we are poking it with sticks." He meant that we assume our climate is stable, but our thoughtless actions may be stirring it to sudden and dramatic changes. How stable is climate? That depends upon the time frame you consider. Over centuries and millennia, we know that climate shifts somewhat, but usually we expect little change on the scale of a human lifetime. But this is no longer a reasonable expectation. If climate does shift, how fast might it change, and what will those changes mean for the environmental systems on which we depend?

Ice cores tell us about climate history

Every time it snows, small amounts of air are trapped in the snow layers. In Greenland and Antarctica and other places where cold is persistent, yearly snows slowly accumulate over the centuries. New layers compress lower layers into ice, but still tiny air bubbles remain, even thousands of meters deep into glacial ice. Each bubble is a tiny sample of the atmosphere at the time that snow fell.

Climatologists have discovered that by drilling deep into an ice sheet, they can extract ice cores, from which they can collect air-bubble samples. Samples taken every few centimeters show how the atmosphere has changed over time. Ice core records have

(a)

(b)

▲ **FIGURE 9.7** (a) Ice cores contain air bubbles that give samples of the atmosphere thousands of years ago, as in this sample of 45,000-year-old ice from Antarctica. (b) Dr. Mark Twickler, of the University of New Hampshire, holds a section of the 3,000-m Greenland ice sheet core, which records 250,000 years of climate history. Source: (a) NASA's Goddard Space Flight Center/Ludovic Brucker; (b) Courtesy Candace Kohl, University of California, San Diego

revolutionized our understanding of climate history (fig. 9.7). We can now see how concentrations of atmospheric CO_2 have varied. We can detect ash layers and spikes in sulfate concentrations that record volcanic eruptions.

Most important, we learn about ancient temperatures by comparing oxygen isotopes (atoms of different mass) in these air samples. In cold years, water molecules containing slightly lighter oxygen atoms evaporate more easily than water with slightly heavier atoms. By comparing the proportions of heavier and lighter oxygen atoms, climatologists can reconstruct temperatures over time and plot temperature changes against CO_2 concentrations and other atmospheric components.

The first very long record was from the Vostok ice core, which reached 3,100 m into the Antarctic ice and which gave us a record of temperatures and atmospheric CO_2 over the past 420,000 years. A team of Russian scientists worked for 37 years at the Vostok site, about 1,000 km from the South Pole, to extract this ice core. A similar core has been drilled from the Greenland ice sheet. More recently, the European Project for Ice Coring in Antarctica (EPICA) has produced a record reaching back over 800,000 years (fig. 9.8). All these cores show that climate has varied dramatically over time but that there is a close correlation between atmospheric temperatures and CO_2 concentrations.

From these ice cores, we know that CO_2 concentrations have varied between 180 and 300 ppm (parts per million) in the past 800,000 years. Therefore, we know that today's concentrations of approximately about 410 ppm (and rising about 2 ppm per year) are one-third higher than the earth has seen in nearly a million years. Concentrations of methane and nitrous oxide, two other important greenhouse gases, are also higher than in any records in the EPICA core. We also know from oxygen isotopes that present temperatures are nearly as warm as any in the ice core records. Further warming in the coming decades is likely to exceed anything ever seen by our species, *Homo sapiens,* which appeared just 200,000 years ago.

What causes natural climatic swings?

Ice core records also show that there have been repeated, cyclical climate changes over time. What causes these periodic (repeated) changes? Modest changes correspond to a cycle in the sun's intensity,

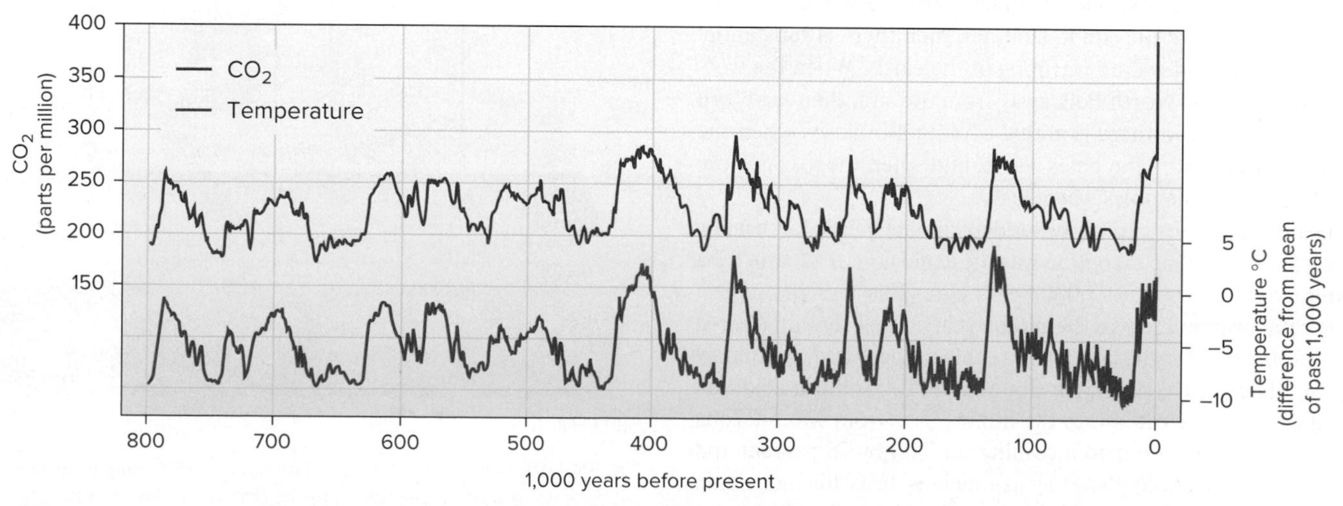

▲ **FIGURE 9.8** Air bubbles in ice cores provide samples of ancient atmospheric composition, going back 800,000 years in this record from the EPICA ice core. Concentrations of CO_2 (*red line*) map closely to temperatures (*blue*, derived from oxygen isotopes). Recent temperatures lag behind rising CO_2, possibly because oceans have been absorbing heat.

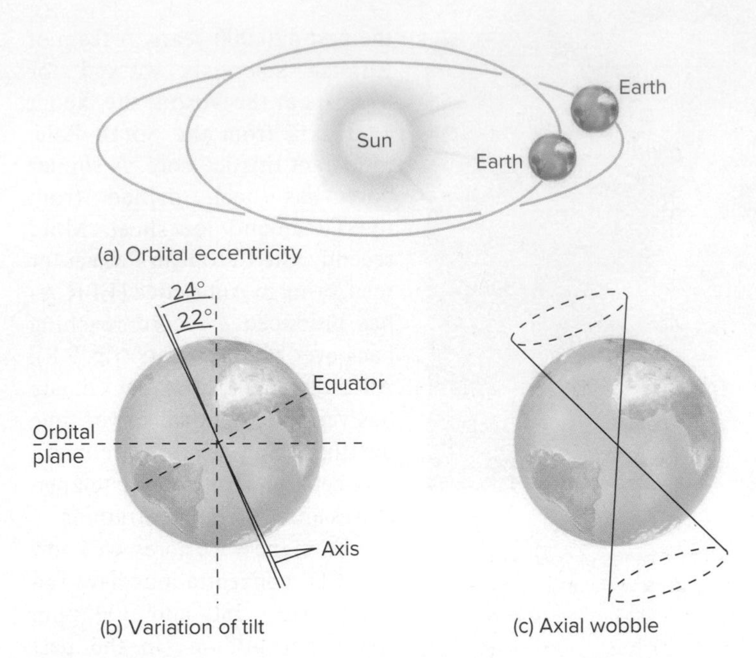

▲ **FIGURE 9.9** Milankovitch cycles include very long, cyclical changes in (a) the eccentricity of the orbit (about 100,000 years); (b) the degree of axis tilt (about 40,000 years); and (c) wobble, or precession of the axis (about 26,000 years).

which peaks about every 11 years. More dramatic changes are associated with periodic shifts in the earth's orbit and tilt (fig. 9.9). These are known as the **Milankovitch cycles,** after the Serbian scientist Milutin Milankovitch, who first described them in the 1920s. There are three of these cycles: (1) The earth's elliptical orbit stretches and shortens in a cycle of about 100,000 years; (2) the earth's axis changes its angle of tilt in a 40,000-year cycle; (3) over a 26,000-year period, the axis wobbles like an out-of-balance spinning top. These cycles cause variation in the intensity of incoming solar energy at different latitudes, and in the intensity of summer heating or winter cooling.

The interplay of these cycles also seems to explain the glacial periods of the last 800,000 years, which you can see in the cold/warm cycles in figure 9.8. For example, when the wobble orients the North Pole toward the sun in summer, then there is substantial summer warming and overall warming of the earth. When the wobble points toward the North Pole away from the sun, then northern summers are cold, and there is global cooling. Similarly, when the axis tilts toward the sun, the poles warm, and when the axis is more parallel to the sun, the poles warm less.

Volcanic eruptions can cause sudden climate shifts, but usually only for a few years. One exception was the explosion of Mount Toba in western Sumatra about 73,000 years ago. This was the largest volcanic cataclysm in the past 28 million years. The eruption ejected at least 2,800 km^3 of material (compare this to the 1 km^3 emitted by Mount St. Helens in Washington State in 1980). Sulfuric acid and particulate material ejected into the atmosphere from Mount Toba are estimated to have dimmed incoming sunlight by 75 percent and to have cooled the whole planet by as much as 16°C for more than 160 years. In climate history, 160 years is a short time, however, and this was the largest eruption in 28 million years—so volcanoes are a notable but not dominant factor in recent climate trends.

El Niño/Southern Oscillation is one of many regional cycles

On the scale of years or decades, the climate also changes according to oscillations in the ocean and atmosphere. These coupled ocean-atmosphere oscillations occur in all the world's oceans, but the **El Niño/Southern Oscillation** (ENSO) is probably the best known. ENSO affects weather across the Pacific and adjacent continents, causing heavy monsoons or serious droughts.

The core of this system is a huge pool of warm surface water in the Pacific Ocean that sloshes slowly back and forth between Indonesia and South America like water in a giant bathtub. Most years, steady equatorial trade winds hold this pool in place in the western Pacific (fig. 9.10). From Southeast Asia to Australia, this concentration of warm equatorial water provides latent heat (water vapor) that drives strong upward convection (low pressure) in the atmosphere. Resulting heavy rains in Indonesia support dense tropical forests.

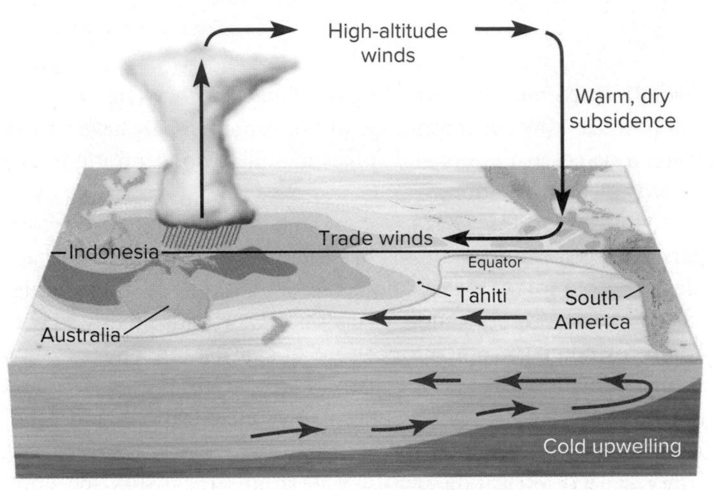

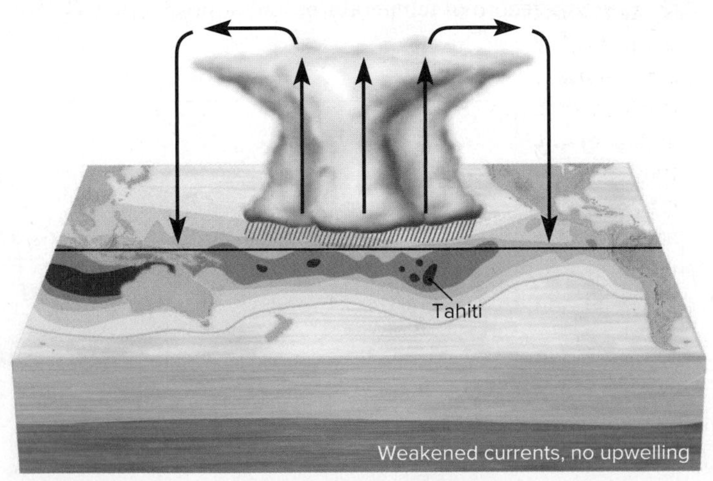

▲ **FIGURE 9.10** The El Niño/La Niña Southern Oscillation. (a) Normally, surface trade winds drive cold currents from South America toward Indonesia, and cold, deep water wells up near Peru. (b) During El Niño years, winds and currents weaken, and warm, low-pressure conditions shift eastward, bringing storms to the Americas.

On the American side of the Pacific, cold upwelling water along the South American coast replaces westward-flowing surface waters. This upwelling deep water is rich in nutrients. It supports dense schools of anchovies and other fish. In the atmosphere, dry, sinking air in Mexico and California replaces the air moving steadily westward in the trade winds. Normally dry conditions in the southwestern United States are a result.

Every 3 to 5 years, for reasons that we don't fully understand, Indonesian convection (rising air currents) weakens, and westward wind and ocean currents fail. Warm surface water surges back east across the Pacific. Rains increase in the western United States and Mexico, and drought occurs in Indonesia. Upwelling currents that support South American fisheries also fail.

Fishermen in Peru were the first to notice irregular cycles of rising ocean temperatures because the fish disappeared when the water warmed. They named this event El Niño (Spanish for "the Christ child") because it often occurs around Christmas time. The counterpart to El Niño, when the eastern tropical Pacific cools, has come to be called La Niña ("little girl").

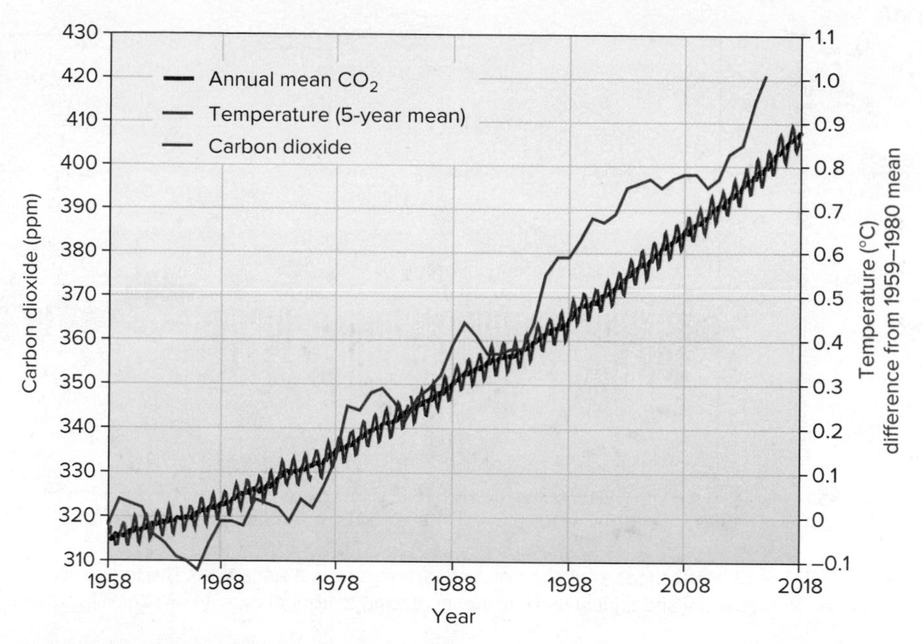

▲ **FIGURE 9.11** Measurements of atmospheric CO_2 taken at the top of Mauna Loa, Hawaii, show an increase of about 2.2 percent per year in recent years. Monthly mean CO_2 (*red*) and annual average CO_2 (*black*) track closely with the general trend in global temperatures (*blue*). Source: Data from National Aeronautics and Space Administration and National Oceanic and Atmospheric Administration.

ENSO cycles have far-reaching effects. During an El Niño year, the northern jet stream—which is normally over Canada—splits and is drawn south over the United States. This pulls moist air from the Pacific and Gulf of Mexico inland, bringing intense storms and heavy rains from California across the midwestern states. The intervening La Niña years bring hot, dry weather to the same areas. Oregon, Washington, and British Columbia, on the other hand, tend to have warm, sunny weather in El Niño years rather than their usual rain. Droughts in Australia and Indonesia during El Niño episodes cause disastrous crop failures and forest fires, including one in Borneo in 1983 that burned 3.3 million ha (8 million acres).

Some climatologists believe that El Niño conditions are becoming stronger or more frequent because of global climate change. There is evidence that warm ocean surface temperatures are spreading, which could contribute to the intensity of El Niño convection patterns.

9.3 HOW DO WE KNOW THE CLIMATE IS CHANGING FASTER THAN USUAL?

- *The Keeling curve documents changing CO_2 concentrations.*
- *There is clear evidence for a human role in current climate change.*
- *Polar regions and continental interiors are warming most rapidly.*

Many scientists consider anthropogenic (human-caused) global climate change to be the most important environmental issue of our time. The possibility that humans might alter world climate is not a new idea. In 1859, John Tyndall measured the infrared absorption of various gases and described the greenhouse effect. In 1895, Svante Arrhenius, who subsequently received a Nobel Prize for his work in chemistry, predicted that CO_2 released by coal burning could cause global warming.

Pivotal evidence that human activities are increasing atmospheric CO_2 came from an observatory on top of the Mauna Loa volcano in Hawaii. The observatory was established in 1957 as part of an International Geophysical Year, to provide atmospheric data in a remote, pristine environment. Measurements taken by atmospheric chemist Charles David Keeling showed CO_2 levels increasing about 0.5 percent per year. The "Keeling curve" (fig. 9.11), which has been continuously updated, shows that CO_2 concentrations have risen from 315 ppm in 1958 to 410 ppm in 2018. The line is jagged because of annual fluctuations: Concentrations decline each May as plants in the Northern Hemisphere (which has far greater continental area than the Southern Hemisphere) begin photosynthesizing and capturing CO_2. During the northern winter, plant decay releases CO_2, causing levels to rise. The effect of increasing CO_2 is a general pattern of rising temperatures worldwide (fig. 9.12).

Scientific consensus is clear

Because the climate is so complex, climate scientists worldwide have collaborated in collecting and sharing data, and in programming models to describe how the climate system works. Evidence shows regional variation in warming and cooling trends, and there are minor differences among models. But among climate scientists who work with the data and models, there is no disagreement about the direction of change. The evidence shows unequivocally, as in the Mauna Loa graph, that climate is changing, and the global average is warming because of increased retention of energy in the lower atmosphere.

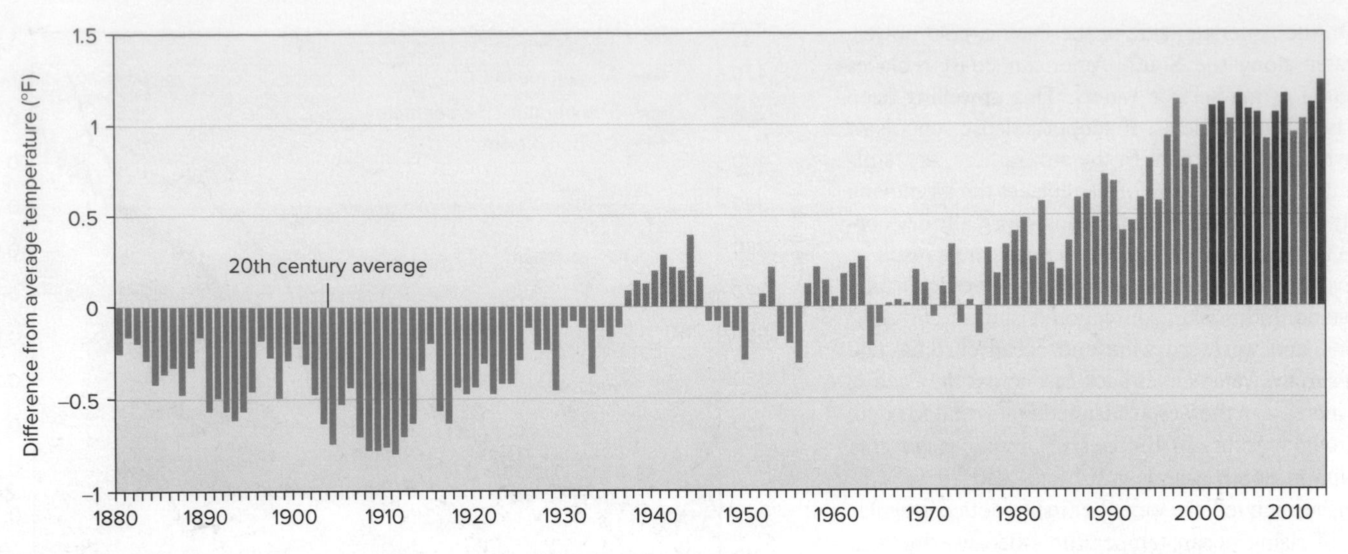

▲ **FIGURE 9.12** Differences from average annual temperature, since 1880. The ten warmest years on record (darkest red) have occurred in the most recent decades. New monthly and annual records are increasing in frequency. Source: Graph by NOAA Climate.gov, based on data from the National Climatic Data Center.

Active LEARNING

Can you explain key evidence on climate change?

Climate scientists present much of their evidence in graphs, so it's worth pausing to make sure you understand what they show. Here are two important cases:

1. In figure 9.11, what is the range of years shown?
2. Why does the red monthly CO_2 value zigzag?
3. The organization 350.org (chapter 16) takes its name from the recommendation that CO_2 concentrations return to less than 350 parts per million. When did we pass 350 ppm?
4. In the Exploring Science figure (p. 222), what is the range of years shown in the graph?
5. What do the pink and blue shaded areas represent?
6. Why are the pink and blue areas not just a simple line to show modeled trends?
7. Explain what it tells us when observations (black line) fit the pink area.

ANSWERS: 1. 1958 to 2018; 2. As noted in the text, CO_2 declines in the Northern Hemisphere summer because of vegetation growth; 3. 1988; 4. 1910–2010; 5. Models run with and without human impacts; 6. Shaded areas show uncertainty, because different models produce different results; 7. Only models including anthropogenic factors explain observed patterns; thus, anthropogenic factors explain observed climate change.

The most comprehensive effort to describe the state of climate knowledge is that of the **Intergovernmental Panel on Climate Change (IPCC)**. As the name indicates, the IPCC is a collaboration among governments, with scientists and government representatives from 130 countries. The aim of the IPCC is to review scientific evidence on the causes and likely effects of human-caused climate change (Exploring Science, p. 222).

In 2013–2014 the IPCC released its Fifth Assessment Report (AR5). The result of 6 years of work by 2,500 scientists, the four volumes of the report represent a consensus by more than 90 percent of all the scientists working on climate change. The conclusion is a 99 percent certainty that observed climate change is caused by human activity. You can view the report, with figures and related documents, at the IPCC's website: http://www.ipcc.ch.

Though scientific consensus has long been clear, in a few places, notably the United States, a public debate has developed about climate evidence, a debate unrelated to science. Popular media commentators and some politicians have recently converted "belief" in climate evidence into a matter of identity and political philosophy, rather than a question of what the evidence shows.

This political positioning is new. In 1997 and 2002, just 5 years apart, both the Clinton administration and President George W. Bush awarded Charles Keeling medals of honor for his studies of atmospheric CO_2 and climate change. At that time, members of both major parties considered Keeling's work to be of national interest and global importance. A decade later, views on climate had become a polarizing political issue between parties. Some commentators disputed evidence for change, while others dismissed the data and the science altogether.

Sociologists have pointed out that this is largely a matter of group identity and worldview, rather than a matter of science and observation. Most people tend to conform to beliefs of the people around them, or the people they admire. We tend to agree with messages delivered repeatedly by trusted figures, including those on TV. There is also growing evidence that uncertainty in science directly reflects misinformation campaigns by fossil fuel industries. This contrast in views reminds us how important it is that voters have some understanding of data and of the scientific process.

Meanwhile, most U.S. businesses are making plans for climate change. When President Obama announced new rules for reducing greenhouse gas emissions from federal agencies, several dozen major U.S. contractors immediately followed suit, presenting plans for evaluating and reducing their climate impacts. Insurance companies in particular have long paid close attention to climate science. Rising costs of claims associated with storms, crop losses, forest fires, and other climate-related events present a real threat to their profits. More recent adminisrative rejection of climate science, and science in general, has created confusion for the business community, however.

Rising heat waves, sea level, and storms are expected

The IPCC's Assessment Report presents a variety of climate scenarios for predicted emissions of greenhouse gases. For each scenario, the IPCC modeled future emissions, starting in 2000. Scenarios differ in expected population growth, economic growth, energy conservation and efficiency, and adoption of greenhouse gas controls (or lack thereof). The different scenarios project a temperature increase by 2100 of 1–6°C (2–11°F) compared to temperatures at the end of the twentieth century (fig. 9.13).

According to the IPCC, the "best estimate" for temperature rise is now about 2–4°C (about 3–8°F). A change of 4°C is just slightly less than the difference in global temperature between now and the middle of the last glacial period, which was about 5°C cooler.

Observations since the Fourth Assessment Report in 2007 show that all the IPCC scenarios were too conservative. Greenhouse gas emissions, temperatures, sea level, and energy use have accelerated faster than even the worst IPCC projections. Sea level rise, estimated in 2007 to be 17–57 cm (7–23 in.) by 2100, is now expected to reach 1–2 m (3–6 feet) by 2100.

If recent rapid melting of polar ice sheets and Greenland glaciers continues, this change will be higher still. Complete melting of Greenland's ice sheet would raise sea level by more than 6 m (nearly 20 ft). This would flood most of Florida, a broad swath of the Gulf Coast, most of Manhattan Island, Shanghai, Hong Kong, Tokyo, Kolkata, Mumbai, and about two-thirds of the other largest cities in the world (fig. 9.14).

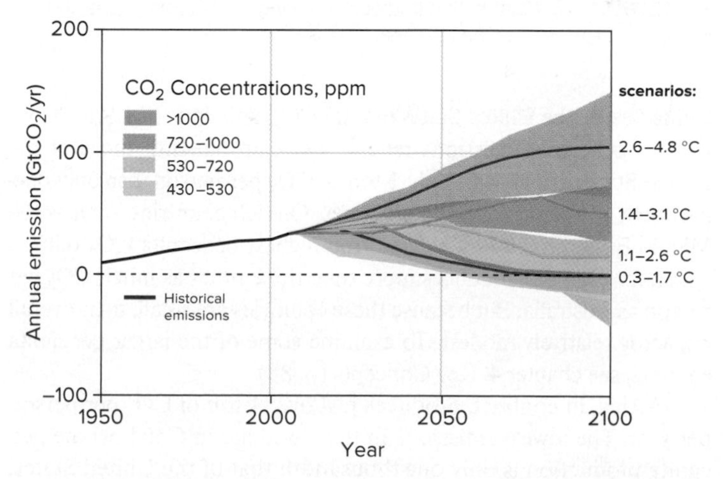

▲ **FIGURE 9.13** Annual anthropogenic CO_2 emissions for four climate change scenarios, with differing temperature ranges by 2100. The 0.3–1.7°C scenario involves aggressive and rapid reduction of GHGs; the 2.6–4.8°C scenario represents business as usual. The likely range of CO_2 concentrations associated with each scenario is shown. Source: IPCC 2013, AR5 Working Group 1 SPM fig. SPM.5.

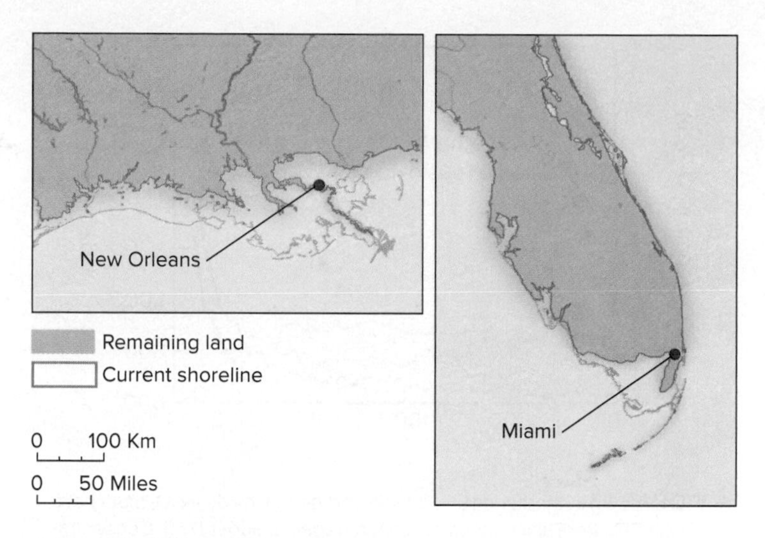

▲ **FIGURE 9.14** Approximate change in land surface with 1 m (3 ft) of sea level rise, a change that is likely by 2100. If no action is taken, sea level change may be 2 m (6 ft) or more.

Tipping points are a concern for many climatologists. Projections indicate that if we don't control emissions in the next few decades we will pass points of no return in melting of permafrost, in the loss of Greenland's ice cap, and other factors.

The United States military is concerned about global warming. In 2007 the U.S. Military Advisory Board said, "Climate change, national security, and energy dependence are a related set of global challenges that will lead to tensions even in stable regions of the world." Some humanitarian aid agencies point to unusual drought as a factor in recent civil wars and famines in southern Sudan in Somalia, and in Syria. These crises are rooted in drought and food shortages caused by changing weather patterns that have led to years of below-normal rainfall and desertification. Global climate change may bring more such conflicts along with the millions of refugees and tragic suffering in many regions.

Policymakers have made little progress in finding solutions. Climate control is a classic free-rider problem, in which nobody wants to take action for fear that someone else might benefit from their sacrifices.

Will sacrifices need be as big as some policymakers suggest? Evidence suggests not. Increasingly, economists point out that shifting our energy strategy from coal (our largest emitter of greenhouse gases and other pollutants) to wind, solar, and greater efficiency could produce millions of new jobs and save billions in health care costs associated with coal burning.

The main greenhouse gases are CO_2, CH_4, and N_2O

Since preindustrial times atmospheric concentrations of CO_2, methane (CH_4), and nitrous oxide (N_2O) have climbed by over 31 percent, 151 percent, and 17 percent, respectively (fig. 9.15). Carbon dioxide is

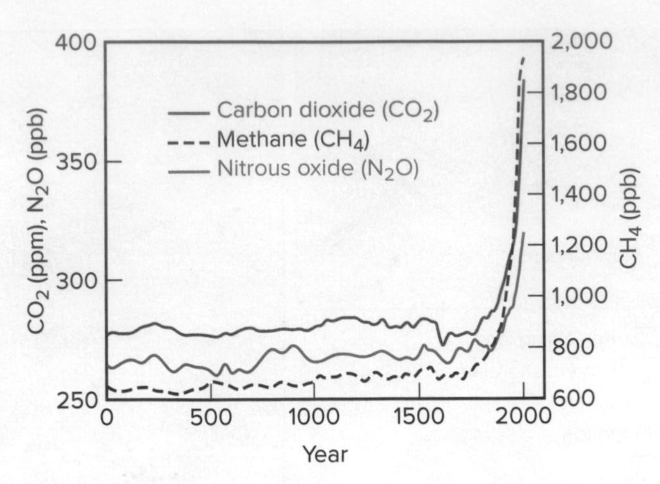

▲ **FIGURE 9.15** Methane (CH_4, *black*) and nitrous oxide (N_2O, *blue*) have risen with CO_2 (*red*) since industrialization began in about 1750. Concentrations are shown in parts per million (ppm) and parts per billion (ppb). Source: USGS 2009.

by far the most important of these because of its abundance and because it lasts for decades or centuries in the atmosphere (fig. 9.16a). Carbon dioxide makes up about 80 percent of greenhouse gases. Fossil fuel burning produces most of the 40 GT (billion tons) of CO_2 released each year. Other key sources are forest burning and other land use impacts. About 4 GT is taken up by plants and 3 GT are absorbed by oceans. If current trends continue, CO_2 concentrations could reach about 500 ppm (approaching twice the preindustrial level of 280 ppm) by the end of the twenty-first century.

Methane (CH_4) is much less abundant than CO_2, but it absorbs far more infrared energy per molecule and is accumulating in the atmosphere about twice as fast as CO_2. Methane can be produced anywhere organic matter decays without oxygen, especially under water. Natural gas, which is mainly CH_4, derives from ancient organic material. Methane also is released by ruminant animals, wet-rice paddies, coal mines, landfills, fracking, and pipeline leaks.

Reservoirs for hydroelectricity, usually promoted as a clean power source, are an important source of methane because they capture submerged, decaying vegetation. Philip Fearnside, an ecologist at Brazil's National Institute for Amazon Research, calculates that rotting vegetation in the reservoir behind the Cura-Una dam in Para Province emits so much carbon dioxide and methane every year that it causes three and a half times as much global warming as would generating the same amount of energy by burning fossil fuels. Tropical dams produce roughly 3 percent of global CH_4 emissions.

Nitrous oxide (N_2O) is produced mainly by chemical reactions between atmospheric N and O, which combine in the presence of heat from internal combustion engines. Other sources are burning of organic material and soil microbial activity.

Chlorofluorocarbons (CFCs) and other gases containing fluorine also capture energy (see fig. 10.10). CFC releases in developed countries have declined since many of their uses were banned, but increasing production in developing countries, such as China and India, remains a problem. Together, fluorine gases and N_2O account for about 17 percent of human-caused global warming (fig. 9.16a).

The United States, with less than 5 percent of the world's population, releases one-quarter or more of the global CO_2 emissions. In 2007

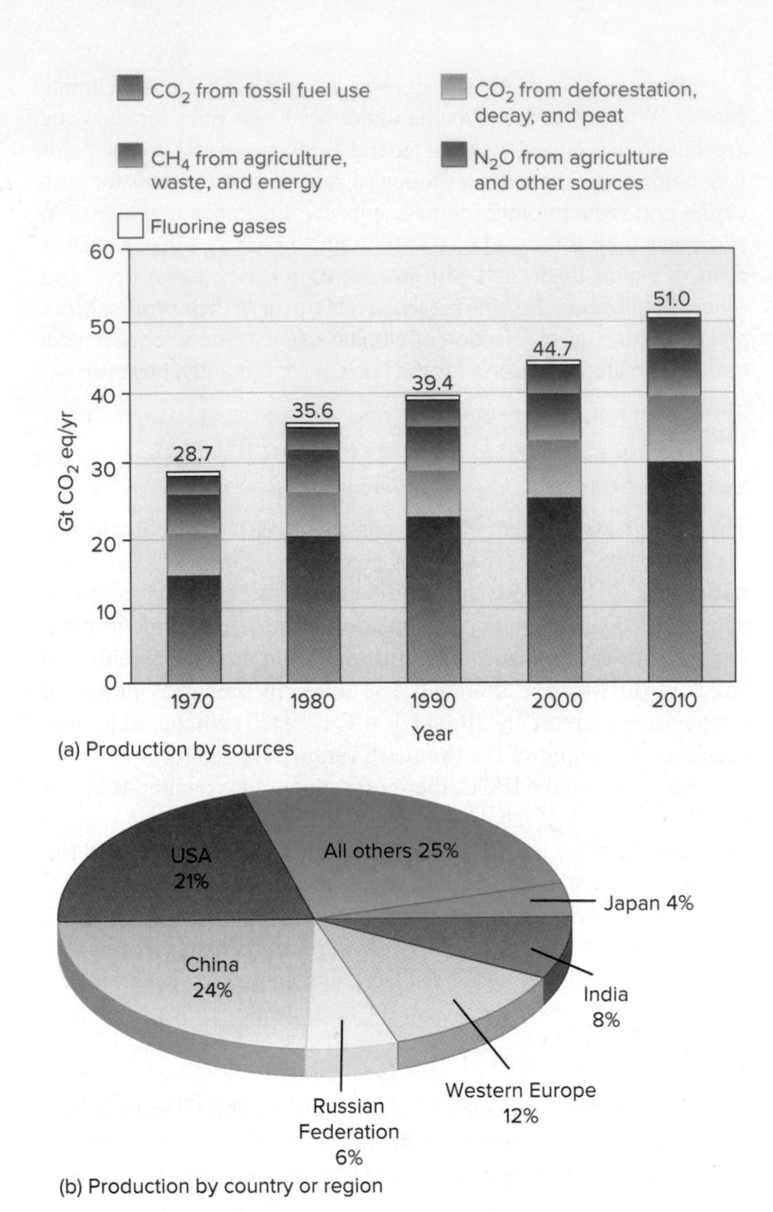

▲ **FIGURE 9.16** Contribution to global warming by different gases and activities (a) and by countries (b). Source: IPCC.

China passed the United States in total CO_2 emissions (fig. 9.16b), but China's per-capita emissions remain less than one-fifth those of the United States. India, with only 1 ton of CO_2 per person, has only one-twentieth as much as the United States. Oil-rich countries, such as the Middle Eastern oil emirates, have the highest per-capita CO_2 output. Qatar, for example, produces more than three times as much CO_2 per person as Australia. But because these countries are small, their overall impact is relatively modest. To examine some of the larger per-capita emitters, see chapter 4, Key Concepts (p. 84).

Africa, in contrast, produces just over 1 ton of CO_2 per person per year. The lowest emissions in the world are in Chad, where per-capita production is only one-thousandth that of the United States.

Some countries with high standards of living release relatively little CO_2. Sweden, for example, produces only 6.5 tons per person per year, or about one-third that of the United States. Remarkably, by adopting renewable energy and conservation measures, Sweden has reduced its carbon emissions by 40 percent over the past 30 years.

At the same time, Sweden has seen dramatic increases in both personal income and quality of life measures.

Perhaps the biggest question in environmental science today is whether China and India, which now have the world's largest populations and are among the fastest-growing economies, will follow the development path of the United States or that of Sweden. Rising affluence in China has fueled a rapidly growing demand for energy, the vast majority of which comes from coal. China continues to invest heavily in coal, at home and abroad. Another large source of CO_2 in China is cement production which both uses fossil fuels and releases CO_2 from the limestone used in production. Worldwide, cement manufacturing accounts for 4 percent of all CO_2 emissions. Chinese cement companies, stimulated by the world's largest building boom, now produce over half of the world's supply, and these plants produce about 12 percent of China's domestic CO_2 emissions.

We greatly underestimate methane emissions

We generally calculate emissions from coal, oil, and natural gas from records of how much fuel is sold and burned. But recent studies have started asking whether that is an accurate measure. What about accidental leakage? Fracking, in particular, involves high-pressure fracturing of bedrock, which also weakens well casings. About 5 percent of fracked well casings leak right away, and the few studies available suggest that most leak within a few years. Gas leaks from valves, pump stations, frost-cracked delivery lines, and other parts of the supply chain—generally called fugitive emissions—are common, but how common they are is not well known.

Estimates are that around 1 to 5 percent of gas escapes as fugitive emissions. This doesn't sound like much, but it would dramatically alter our greenhouse gas calculations. One kg of methane absorbs 86 times as much energy as 1 kg of CO_2 over 20 years (after which most of the methane has broken down to other molecules). Cornell University has estimated that if it accounted for fugitive emissions in its natural gas supply, its calculated GHG footprint would more than triple, from 214,000 tons to 820,000 tons of CO_2 equivalent (fig. 9.17).

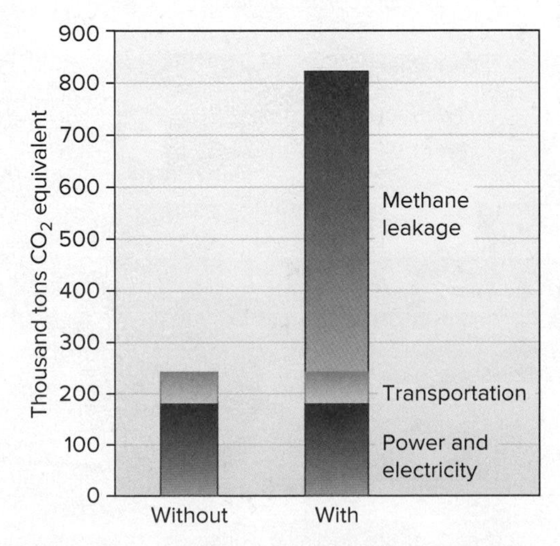

▲ **FIGURE 9.17** Cornell University recalculated its greenhouse gas inventory with and without methane leakage in production of the gas the university purchases. Accounting for upstream fugitive emissions more than tripled Cornell's GHG figures. This accounting is necessary for reducing emissions.

The oil and gas industry is responsible for nearly all fugitive emissions. Fixing buried pipelines and thousands of valves is expensive, but the International Energy Agency calculates that fixing only the cost-effective leaks—the ones that would pay for themselves by selling captured gas—would prevent around 76 million tons of methane leakage per year globally. That leakage is equivalent to nearly 47 times the CO_2 emissions of the entire European Union.

Controlling methane is one of the most cost-effective and quickest ways we have to slow climate change. For this reason, rules requiring controls of methane leakage were a policy of the late Obama administration. The Trump administration eliminated those rules as one of its earliest actions. It remains to be seen whether or how soon methane controls are implemented. Better accounting of fugitive emissions in more GHG calculation projects would add pressure to companies and policymakers on this issue.

What does 2° look like?

Climate scientists report that keeping climate warming to less than 2 degrees C (far preferably, 1.5 degrees) is necessary to avoid the worst effects of climate change. Already we have warmed the climate by nearly 1 degree, bringing temperatures higher than they have been in thousands of years—since the beginning of civilization as we know it. Expected warming, following business-as-usual practices of energy use, land use change, and carbon emissions, is unevenly distributed (fig. 9.18a). Warming will be, and already has been, greatest in the Arctic, but large continental interior areas, such as the American Midwest or central Siberia, will also warm more than average. Currently, we are on a trajectory toward an average of 2.5–5 degrees warming (see fig. 9.13), far more than the "safe" limit policymakers hope to achieve. Even if we keep warming to an average of 2 degrees, dramatic effects are expected. What might those effects be?

Droughts will be more common and more severe. Warmer temperatures and longer warm seasons increase both evaporation and plant transpiration, so soil and vegetation become drier even if precipitation doesn't decline. Already, incidence of severe drought is increasing. In Africa, for example, droughts have increased about 30 percent since 1970 (fig. 9.18b).

Climate refugees will increase. Already uncounted refugees have fled drought-stricken farmlands across Africa and the Middle East. Ongoing conflicts in Syria, and global migration and political crises involving refugee resettlement, have been widely blamed on the collapse of farming economies (made worse by misguided farm policies) across the region. Aid agencies are confident that the number of migrants will increase with further warming.

Rising sea levels will be another source of climate refugees. Around a billion people now live on seacoasts and nearby fertile coastal plains that are at risk. Think about the crisis caused by a few million people trying to flee wars and famines in 2017. What will be the effects if hundreds of millions of people are forced to leave their homes and seek higher ground? Anti-immigrant, nationalist parties already are changing the political landscape in many countries. What are the political and economic consequences if that migration increases several hundred-fold?

Mortality from heat waves will become more common. Death in heat waves is especially likely for ill or elderly people and for

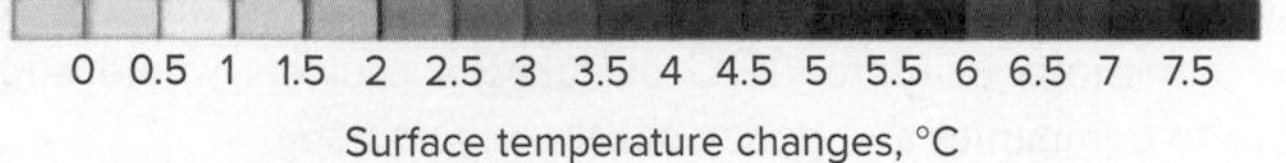

(a) Surface temperature is projected to rise sharply, even under conservative scenarios, with rapid adoption of new energy technology, reduction of fossil fuel use, and a declining population after 2050. Source: IPCC AR4, WG1.

(b) Expanding drought is expected to increase water scarcity and displace far more climate refugees. Source: Bartosz Hadyniak/Getty Images.

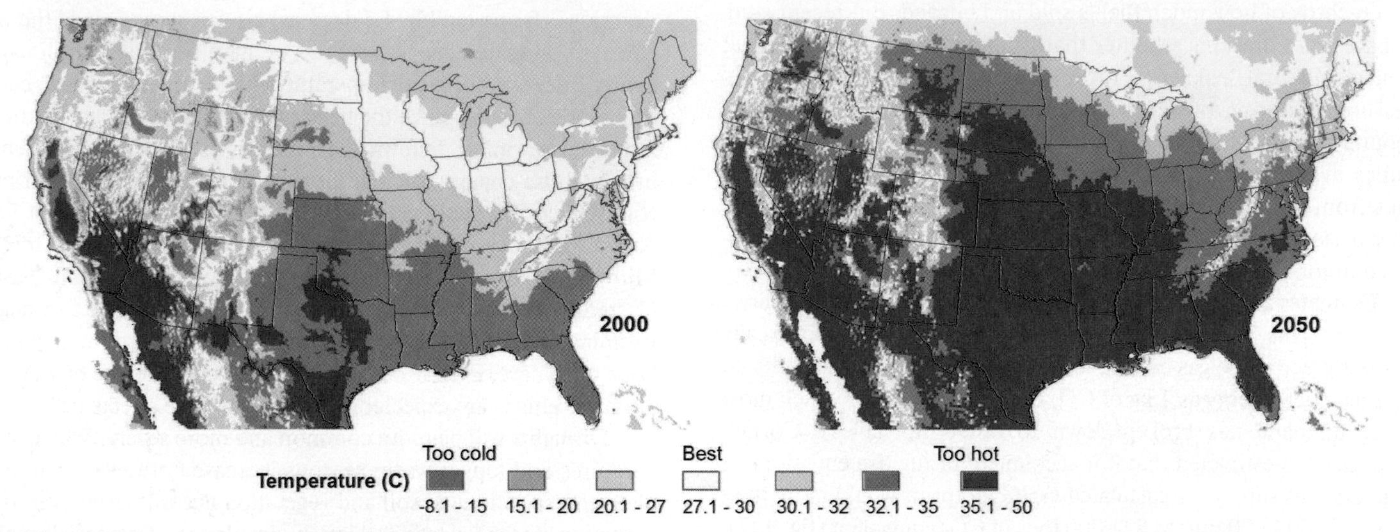

(c) Suitable summer temperatures for growth of major crops, such as corn, will shift northward. July maximum temperature (average for 2000 ± 10 years and 2050 ± 10 years) is shown. Source: Data from CCAFS (Climate Change, Agriculture and Food Security), CMIP5 and WorldClim climate models, RCP 8.5.

(d) The Greenland ice sheet, which contains enough water to raise sea levels 6 m, is melting at an accelerating rate. Source: Bernhard_Staehli/Getty Images.

(e) Ice-dependent species are declining worldwide. ©Frank Krahmer/Photographer's Choice RF/Getty Images

▲ **FIGURE 9.18** Some of the major consequences of climate change.

low-income populations who lack options for escaping a heat wave. A rare heat wave across Europe in 2003 was blamed for 35,000 deaths, and public health officials fear these events will become more frequent.

Forest fires will increase. Already, the U.S. Forest Service reports that the extent of wildfires has more than tripled from 40 years ago (see fig. 6.12), while the length of the fire season lasts months longer. The National Research Council finds that the area burned is likely to quadruple with a 1.8°C rise in temperature. With further warming, much of U.S. forests may transition to shrublands or grasslands.

Food production is threatened by high temperatures, as well as droughts. Traditional agricultural regions mainly occur where average growing-season temperatures are suitable for plant growth and reproduction. Already, crop production is unreliable in many traditional farming regions. The U.S. Corn Belt, which produces by far our largest commodity crop, is especially under threat, because a warming climate will heat the expansive mid-section of the continent more rapidly than the coasts (fig. 9.18c).

Expanding ranges of diseases is likely. Up to now, the spread of most disease vectors, such as mosquitoes that carry Zika virus, yellow fever, or malaria, has involved expansion into new warm-climate ranges. These species may find newly habitable ranges with climate warming.

Hurricanes and extreme weather will be more severe. Already, storms are becoming stronger and more damaging, as greater energy retention in the atmosphere leads to more vigorous circulation, including storms, extreme rainfall, snowfall, and hurricane frequency. Insurance companies are dropping storm coverage and raising premium rates as they plan for more extreme disaster costs.

Coral reefs worldwide are "bleaching" (losing the colorful algae they rely on for survival) as water temperatures rise above 30°C (85°F). Marine ecologists worry that warming oceans could be the final blow for reefs, which are already threatened by pollution and overfishing, and for the fisheries and ecosystems they support.

Oceans are also becoming more acidic as they absorb CO_2 from the atmosphere. This absorption slows warming, but acidity dissolves calcium carbonate, making it harder for mollusks to grow shells and for corals to build reefs. It will take centuries to dissipate the stored heat even if we reduce our greenhouse gas emissions immediately.

The world's ice is disappearing. Greenland is losing ice at a rate of 270 billion tons in some years, compared to an average of around 0 net loss from 1960. Greenland's ice cap holds enough water to raise sea level by about 6 m (20 ft), so climate scientists are watching the accelerating loss closely. Arctic sea ice is less than half as thick and half as extensive as it was 30 years ago—this is both a consequence of more extreme warming near the poles (fig. 9.18a) and a cause of further warming. Populations of ice-dependent species, such as seals, walrus, and polar bears, have declined as ice habitat has disappeared. Ice shelves in Antarctica are breaking up and drifting away. Populations of Emperor and Adélie penguin have declined by half over the past 50 years as the ice shelves melt (fig. 9.18e).

Economic costs of all these changes are expected to amount to 5 to 20 percent of global GDP. This is far greater than the costs of reducing emissions.

These threats are dire, but the good news is that we have most of the technology, and increasingly we understand policy needs to change these trends. The question is whether we will choose to use them.

Ice loss produces positive feedbacks

Climatologists warn that summer sea ice in the Arctic Ocean has thinned and declined in extent by about half. Late-summer arctic ice may be gone entirely by 2030 or sooner. Why does this matter to the rest of us? Because sea ice reflects a great deal of incoming solar energy. Ice, like fresh snow, has an albedo of 80 to 90 percent (table 9.1). This means ice reflects most incoming sunlight, whereas water reflects as little as 5 percent of incoming energy, depending on the angle at which sunlight hits the water surface. Absorbed energy warms the Arctic Ocean. Because water heats and cools slowly, that heat remains in the ocean much longer than it would on land or in the atmosphere.

Melting ice has a strong positive feedback effect. As ice melts, exposed water absorbs energy and warms. Warm water retains its heat long into the fall, reducing the new ice that forms over winter. Thinner, less extensive ice melts rapidly in spring, allowing greater warming the following summer, further enhancing heat absorption (fig. 9.19).

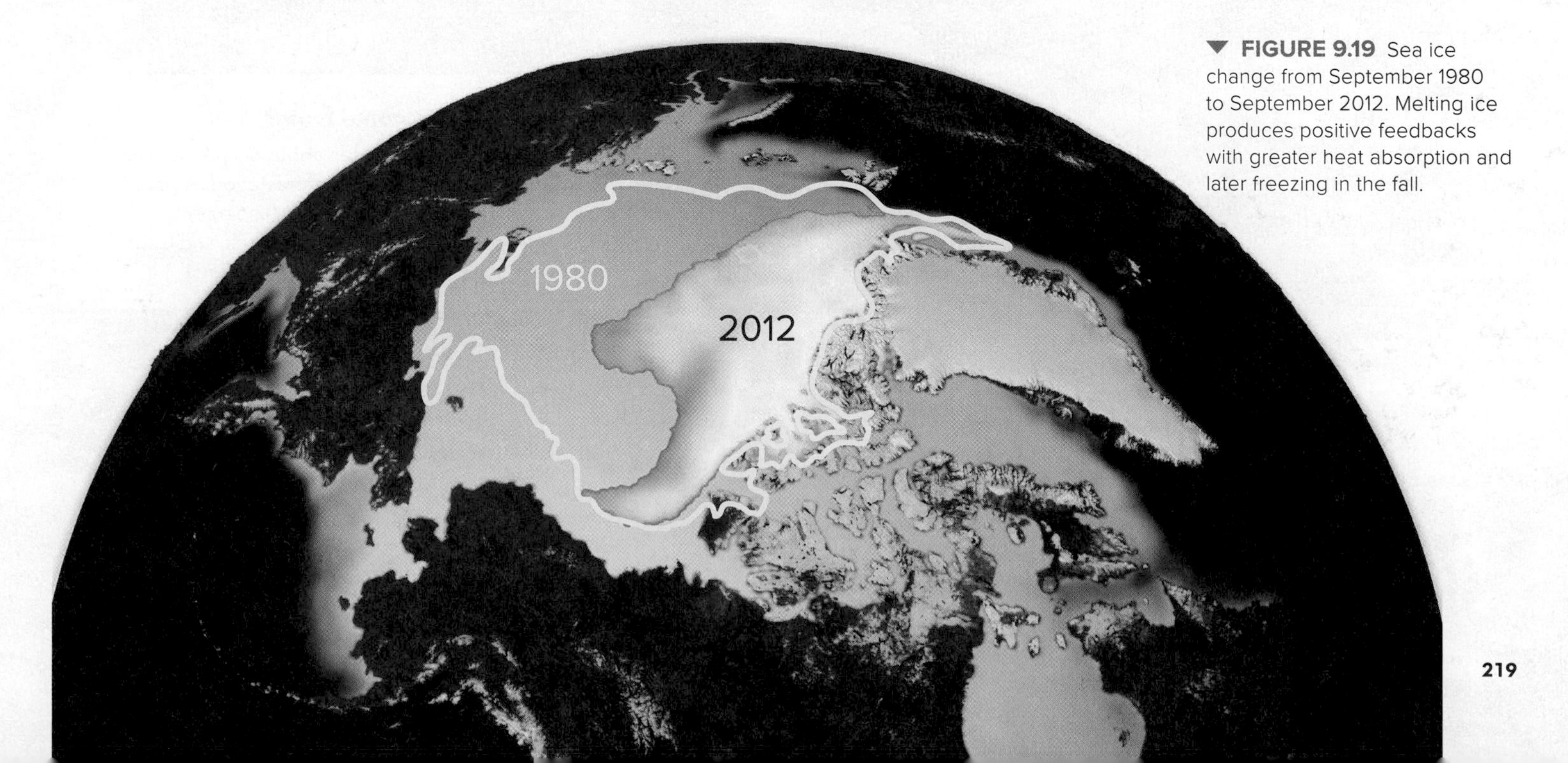

FIGURE 9.19 Sea ice change from September 1980 to September 2012. Melting ice produces positive feedbacks with greater heat absorption and later freezing in the fall.

KEY CONCEPTS

Climate change in a nutshell: How does it work?

The greenhouse effect describes the heating of the earth's atmosphere. Roughly similar to a glass greenhouse, our atmosphere is transparent to visible light energy from the sun, but gases in the atmosphere block the longer-wavelength infrared energy that is re-emitted by the earth. Certain **greenhouse gases,** especially H_2O, CO_2, CH_4, and N_2O, block infrared wavelengths, keeping that energy in the atmosphere (see figure at right). In general, this "greenhouse effect" keeps average earth temperatures warm enough to support life. A higher concentration of these gases increases the amount of heat retained in the atmosphere.

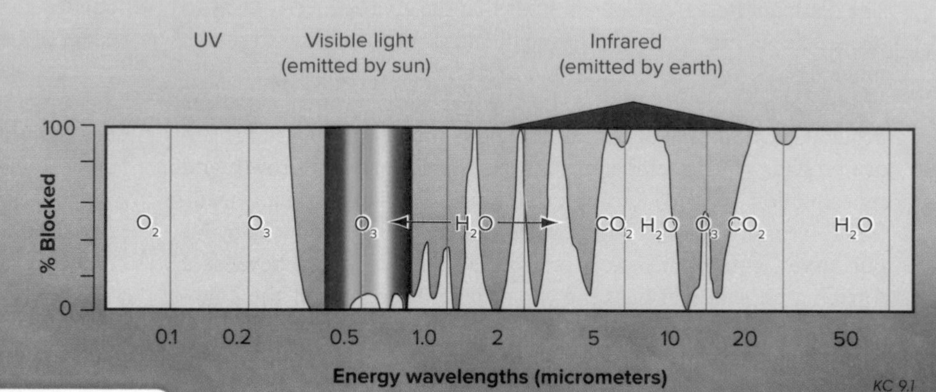

▲ Different molecules block different wavelengths. CO_2 and H_2O especially prevent infrared energy from escaping the atmosphere.

What are GHGs?

Greenhouses gases (GHGs) are molecules in the atmosphere that block long-wave energy from escaping to space. Water vapor is our most abundant GHG, but human activities have not caused as much change in atmospheric water vapor as other GHGs. We have dramatically increased CO_2, CH_4, N_2O, and other gases since industrialization began in about 1800.

These gases naturally keep our planet warm, but recent increases in GHGs are warming the planet enough to destabilize our economies and resource uses.

Gas	% of Climate Forcing*
Carbon dioxide (CO_2)	60%
Methane (CH_4)	20%
Nitrous oxide (N_2O)	10%
Aerosols, other gases	10%

*Percentage of anthropogenic change: depends on (1) amount emitted, (2) energy-capture effectiveness, and (3) persistence in the atmosphere.

Where do GHGs come from?

Fossil fuel burning produces about 60 percent of GHG emissions, followed by deforestation (17 percent) and industrial and agricultural N_2O (14 percent). Fracking, rice paddies, belching livestock, and tropical dams produce CH_4 (9 percent of emissions).

How do we know that recent climate changes are caused by human activity?

IPCC models show that observed temperature trends (black line) do not fit mathematical models built without human-caused factors (blue shaded area shows range of model predictions). Observed trends do fit models built with factors such as fossil fuel use and forest clearing (pink shaded area). ▶

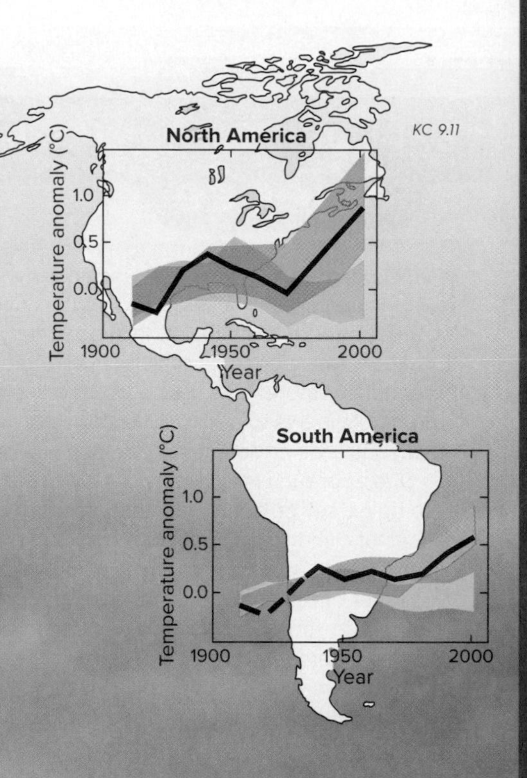

Hasn't the climate changed in geologic time?

Climate conditions have always changed, but never so dramatically since the beginning of civilization, and usually more slowly than now. Our current course is set for higher temperatures by 2100 than have occurred in at least 800,000 years. Current CO_2 concentrations are about 30 percent higher than at any time in the past 800,000 years, according to ice core data. The rate of current climate change is also new: Changes now occurring in 100 years took 800 to 5,000 years at the end of the ice ages. ▼

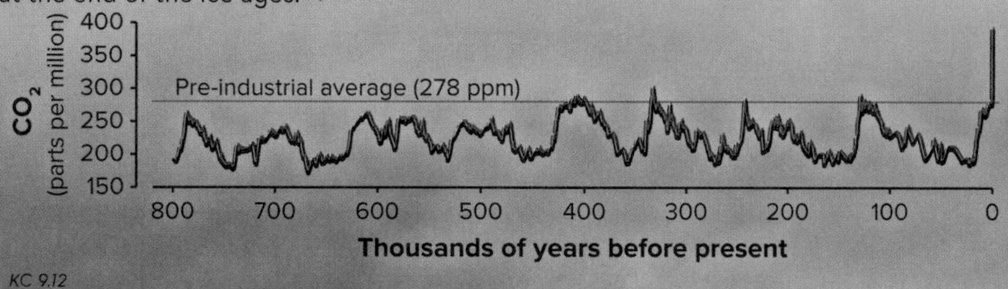

KC 9.12

What effects are observed and expected?

Disappearing ice: Arctic ice, which helps stabilize climate, has declined by nearly half in summer. Mountain glaciers and snow, which provide water to about 75 percent of the western United States and over 1 billion people in Asia, are disappearing worldwide.

Wildfire and pests: Increased fire frequency and severity, aided by expanding parasites, is causing ecosystem change and even human mortality.

Early spring: Early onset of warm weather has led to early flowering, migrations, and hotter summers.

Rising sea level: We are committed to about 0.5 m rise. Without rapid CO_2 reductions, we may soon be committed to 2 m or more.

More storms: More energetic atmospheric circulation is likely to bring more, heavier storms. Heavier rain and snow in the eastern U.S. may already be evident.

Cumulative costs of climate change: $5-90 trillion* by 2100, in damaged infrastructure, lost property values, health costs.

*Pew Environment Group, 2010.

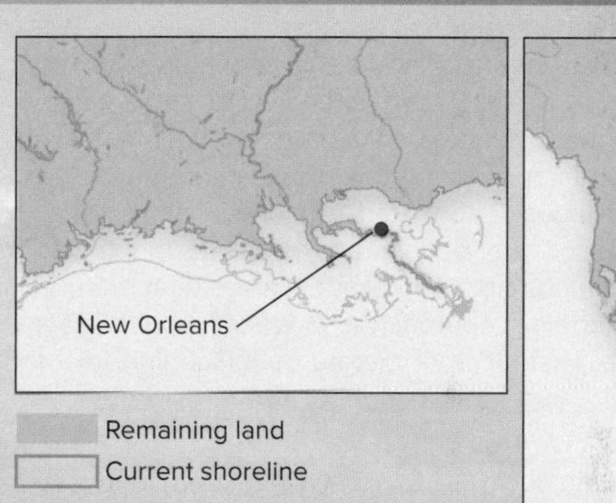

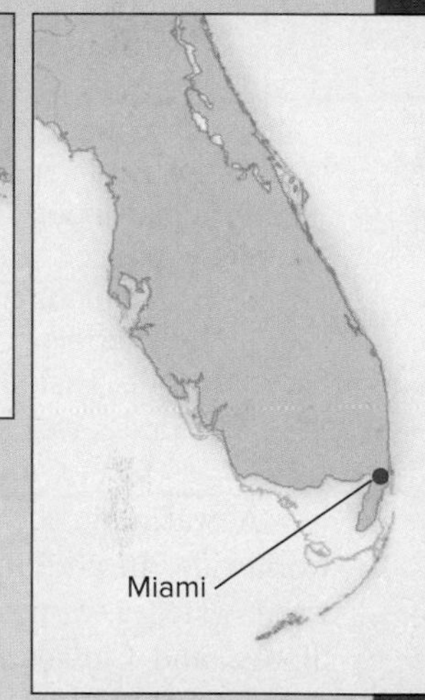

KC 9.13

Observed changes are greater than models anticipated.

Observed GHG emissions (black lines) have accelerated faster than all the IPCC's projected scenarios (colored lines). Changes in temperature and sea level therefore may be greater than anticipated. ▶

KC 9.14

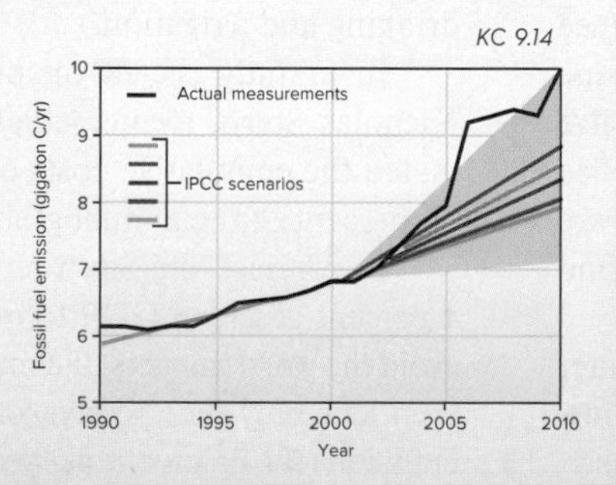

Are we committed to this path?

Not necessarily. If we adopt alternative plans, especially for energy production, transportation, and more efficient use of energy, we could constrain temperature change to only 2–4 degrees globally. We are committed to at least 20–40 cm of sea level change, however, as the ocean continues to absorb excess heat from the atmosphere.

CAN YOU EXPLAIN?

1. What is a greenhouse gas? What are three main anthropogenic gases?
2. Explain the pink and blue bands in the maps at top right. What do the black lines show?
3. Examine the New Orleans and Miami maps. Identify some strategies to protect these cities against rising sea levels and storm frequencies.
4. In this chapter, what are some strategies we have to reduce climate change?

(KC 9.1) Evgeny Bahchev/Shutterstock

EXPLORING Science: How Do We Know That Climate Change Is Human-Caused?

Climate change is one vast, manipulative experiment: We are injecting greenhouse gases into the atmosphere and observing the changes that result. In most manipulative experiments, though, we have controls, which we can compare to treatments to be certain of the effects. Since we have only one earth, we have no controls in this experiment. So how do we test a hypothesis in an uncontrolled experiment?

One approach is to use models. You build a computer model, a complex set of equations, that includes all the known natural causes of climate fluctuation, such as Milankovitch cycles and solar variation. You also include the known human-caused inputs (fossil fuel emissions, methane, aerosols, soot, and so on). Then you run the model and see if it can reproduce observed past changes in temperatures.

If you can accurately "predict" past changes, then your model is a good description of how the system works. You've done a good job of representing how the atmosphere responds to CO_2 inputs, how oceans absorb heat, how changes in snow cover accelerate energy absorption, and so on.

If you can create a model that represents the system quite well, then you can rerun the model, but this time you leave out all the anthropogenic inputs. If the model *without* human inputs is *inconsistent* with observed changes in temperature, and if the model *with* human inputs is *consistent* with observations, then you can be extremely confident, beyond a reasonable doubt, that human inputs had made the difference and caused temperature changes.

This modeling approach is precisely what climate scientists have done. Historical temperatures (black line in the following graph) do not fit the temperatures predicted by models with only natural factors (blue line). Historical temperatures do fit models that include human inputs (red line). Shaded areas represent as you can see in the IPCC's summary of model results, observed temperatures (black line below) do not fit the temperatures predicted by models with only natural factors (blue line). Observed temperatures do fit models that include human inputs (red line). In this way, models let us test hypotheses about complex systems.

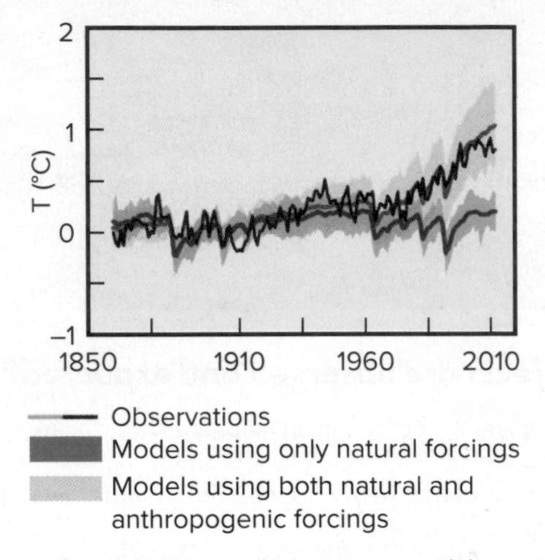

▲ Comparison of modeled historical temperatures with human-caused forcing (*pink*) and without it (*blue*). *Shaded areas* show 95 percent confidence intervals for temperatures predicted by models. Observed temperatures (*black line*) correspond only to the pink models. Source: IPCC 2013.

A warming, ice-free Arctic Ocean contributes to warming worldwide. In addition to providing a vast new thermal mass of warm water, a warm Arctic accelerates warming of land masses in Siberia and Canada. Melting permafrost in Canada and Siberia releases CH_4 and CO_2, creating another positive feedback.

An ice-free Arctic also influences ocean circulation, allowing fresh, low-density water to interrupt thermohaline circulation (see fig. 9.6). A warmer Arctic also alters the ecosystem: Polar bears, walrus, seals, and most other arctic mammals are adapted to live and hunt from the ice. Algae and plankton that grow under the ice are often the foundation of an ecosystem. Cold-water fish are replaced by warmer-water fish, and birds may not be able to find suitable fish in suitable locations for feeding.

Arctic shorelines also erode rapidly when exposed to winter storms, violent waves and storm surges, and salt-water flooding. Some portions of Siberia are eroding at a rate of 40 m per year. Damage to coastal villages and infrastructure has been severe.

Climate change will cost far more than climate protection

The financial costs of climate change will be enormous. According to a 2010 study by the Pew Trust, climate change is likely to cost between \$5 trillion and \$90 trillion by 2100, depending on how economic discount rates and other factors are calculated. The costs assessed included factors such as lost agricultural productivity from drought, damage to infrastructure from flooding and storms, lost biological productivity, health costs from heat stress, and lost water supplies to the billion or so people who depend on snowmelt for drinking and irrigation.

In a study issued on behalf of the British government, Sir Nicholas Stern, former chief economist of the World Bank, estimated the cumulative costs of climate change equal between 5 and 20 percent of the annual global gross domestic product (GDP).

In contrast, the Stern report estimated it would cost only about 1 percent of global GDP to reduce greenhouse gas emissions now to avoid the worst impacts of climate change. The IPCC says it would cost even less, only 0.12 percent of annual global GDP, to reduce carbon emissions the 2 percent per year necessary to stabilize world climate.

For many people, global climate change is a moral and ethical issue as well as a practical one. Religious leaders are joining with scientists and business leaders to campaign for measures to reduce greenhouse gas emissions. Ultimately, the people likely to suffer most from global warming are the poorest in Africa, Asia, and Latin America, who have contributed least to the problem. There is also a question of intergenerational equity. The actions we take–or fail to take—in the next 10 to 20 years will have a profound effect on those living in the second half of this century and in the next. What kind of world are we leaving to our children and grandchildren? What price will they pay if we fail to act?

Why are there disputes over climate evidence?

Scientific studies have long been unanimous about the direction of climate trends, but commentators on television, in newspapers, and on radio continue to fiercely dispute the evidence. Why is this? Part of the reason may be that change is threatening, and on many of us would rather ignore it or dispute it than acknowledge it. Part of the reason may be a lack of information. Another reason is that while scientists tend to look at trends in data, the public might be more impressed by a few recent and memorable events, such as an especially snowy winter in their local area. And on talk radio and TV, colorful opinions sell better than evidence. Climate scientists offer the following responses to some of the claims in the popular media.

Reducing climate change requires abandoning our current way of life. Reducing climate change requires that we use *different* energy. If we replace coal-powered electricity with wind, solar, fossil fuel, and improved efficiency, we can drastically cut our emissions but keep our computers, TVs, cars, and other conveniences. Reducing dependence will also reduce air pollution, health expenditures, and destruction of vegetation and buildings. Increasingly, economies are decoupling economic growth from energy consumption. U.S. emissions of GHGs did not increase from 2005 to 2017, despite economic growth of 14 percent.

There is no alternative to current energy systems. Without investments in alternative energy sources, this would be true, but renewable energy companies are demonstrating that this claim is false. These businesses are showing that alternative energy and improved efficiency can already provide what we need and that there's a great deal of money to be made in new technology. In the coming years there is likely to be more profit in new technologies than in the traditional energy and transportation technologies of the 1940s. Much of the technological innovation in energy is occurring in the United States, although adoption here has been slower than elsewhere.

A comfortable lifestyle requires high CO_2 output. The data don't support this assumption. Most northern European countries have higher standards of living (in terms of education, health care, life span, vacation time, financial security) than residents of the United States and Canada, yet their CO_2 emissions are as low as half those of North Americans. Residents of San Francisco consume about one-sixth as much energy as residents of Kansas City, yet quality of life is not necessarily six times greater in Kansas City than in San Francisco.

Natural changes such as solar variation can explain observed warming. Solar input fluctuates, but changes are slight and do not coincide with the direction of changes in temperatures (fig. 9.20). Milankovitch cycles also cannot explain the rapid changes in the past few decades. Increased GHG emissions, however, do correspond closely with observed temperature and sea level changes (see fig. 9.14).

The climate has changed before, so this is nothing new. Today's CO_2 level of around 410 ppm exceeds by at least 30 percent anything the earth has seen for nearly a million years, and perhaps as

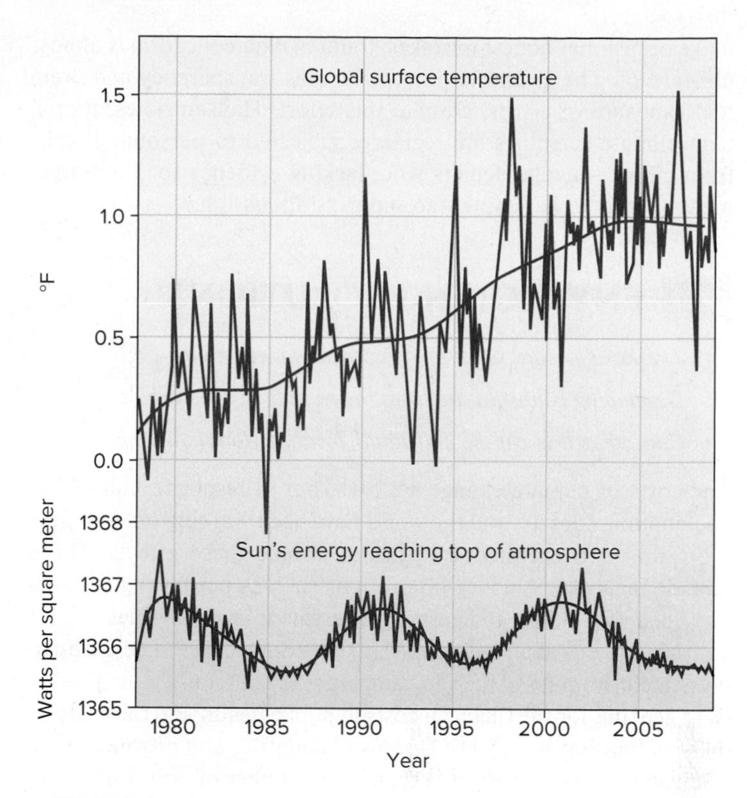

▲ **FIGURE 9.20** Solar energy received at the top of the earth's atmosphere has been measured by satellites since 1978. It has followed a natural 11-year cycle of small shifts but no overall increase (*bottom*). Over the same period, surface temperatures have increased markedly (*top*). Source: Climate Change Compendium, 2009

long as 15 million years. Recent change is also far more rapid than natural fluctuations. Antarctic ice cores indicate that CO_2 concentrations for the past 800,000 years have varied from 180 to 300 ppm (see fig. 9.8). This natural variation in CO_2 appears to be a feedback in glacial cycles, resulting from changes in biotic activity in warm periods. Because temperature has closely tracked CO_2 over time, it is likely that temperatures by 2100 will exceed anything in the past million years. The rate of change is probably also unprecedented. Changes that took 1,000 to 5,000 years at the end of ice ages are now occurring on the scale of a human lifetime.

Temperature changes are leveling off. Short-term variation in trends can always be found if we view the data selectively (fig. 9.20), but over decades the trends in surface air temperatures and in sea level continue to rise (see fig. 9.2).

We had cool temperatures and snowstorms last year, not heat and drought. The climate is complex, and change differs among regions and seasons. For example, a weak and wobbly jet stream can result from the declining north-south contrast in temperatures, as the Arctic warms. Loops in the weakened jet stream can draw arctic air far south to Texas and warm air up to Alaska. Extreme snow might fall in New York while California sees deep drought.

Climate scientists don't know everything, and they have made errors. The gaps and uncertainties in climate data are minute compared to the evident trends. There are many unknowns, such as details of precipitation change, but the trends are unequivocal. Climatologist James Hansen has noted that while most people make occasional honest mistakes, fraud in data collection is almost unheard of. The scientific process ensures transparency and eventual exposure of errors. Despite this effort, Hansen notes, prominent climate scientists are regularly subjected to personal attacks from climate-change deniers who, lacking evidence for their arguments, resort to harassment to suppress discussion.

9.4 ENVISIONING SOLUTIONS

- *The Paris Accord seeks to keep warming well below 2°C.*
- *Developing countries are investing rapidly in renewables.*
- *Climate action can be individual, local, national, or global.*

The costs of climate change are high, but in response individuals and communities around the world have been working on countless promising and exciting strategies to reduce these effects. These include technology, policy innovations such as permitting community solar (fig. 9.21), and growing investment in renewables.

Movements at all scales have arisen, from local climate action to divestment from fossil fuels to campaigns to "put a price on it"—that is, to account for the financial costs to public health and the environment of burning fossil fuels. Dozens of countries and provinces have calculated a "social cost of carbon," a multiplier of, say, $40/ton of CO_2 emissions. Sweden, for example, incorporates the social cost of carbon into taxes on fuel purchases—the revenue is then used to help fund efficiency improvements and renewables for communities.

Current U.S. emissions are about 9 percent higher than in 1990 but have declined slightly since 2008, largely because of a decline in coal use. Each of us can make a contribution in this effort. Simply driving less and buying efficient vehicles could save about 1.5 billion tons of carbon emissions by 2054 (What Can You Do?, p. 226).

The Paris Accord establishes new goals

Until 2012, the benchmark global agreement on controlling climate change was the Kyoto Protocol. Established at a 1997 meeting in Kyoto, Japan, this agreement called for countries to voluntarily set their own targets for reducing emissions of CO_2, CH_4, N_2O, and fluorine gases. Although the Kyoto Protocol did not succeed in reducing overall global emissions, many countries, mostly in Europe, met or exceeded their target reduction of 5 to 10 percent below 1990 emissions by 2012. Developing nations, such as India and China, were exempt, allowing them to expand their economies and improve standards of living. The United States, until recently the world's greatest producer of greenhouse gases, was almost unique among wealthy countries in declining to sign on to the agreement.

▲ **FIGURE 9.21** Community solar, now allowed in many communities, is a way to participate in the renewable energy transition even if you don't own a roof suitable for solar panels. ©Monty Rakusen/Image Source/Getty Images

In 2015 a new global agreement, the Paris Accord, was agreed to. This agreement has much greater urgency, but also greater participation, than the Kyoto Protocol. The Paris Accord required ratification by at least 55 countries, accounting for 55 percent of global emissions, to come into force. Remarkably, this goal was achieved within a year. The 195 countries in attendance at the Paris meeting agreed on a number of major points, including these:

- Holding the global average temperature increase to well below 2°C (above pre-industrial levels) is necessary. Below 1.5°C should be the target.
- Zero carbon emissions is a global goal, to avoid compounding climate effects already committed by past emissions.
- Each participating country establishes voluntary emission reductions goals, and those goals, and progress toward them, must be publicly visible.
- Reduction plans submitted thus far are not sufficient to keep warming below 2°C, so plans must be revised every 5 years.
- Climate finance is necessary: Advanced economies agreed to strive toward donations of $100 billion per year to a "green carbon fund" to support low-carbon development in emerging economies.

Can any country force another to comply with the Paris agreement? No, participation is voluntary, and each country devises and monitors its own "intended nationally determined contributions" to global carbon reductions (INDCs). Can a country withdraw from the agreement? Yes, but only after a 3-year waiting period.

This agreement calls for wealthy countries to contribute to a green fund, to help finance emissions reductions in poor countries. It is not clear if countries will follow through on their pledges, but many developing regions are not waiting for the green carbon fund to start investing in clean energy. In Bangladesh and India, for example, both central banks and private funds are investing in "green bonds" that support renewable energy and that are producing hundreds of thousands of new jobs.

Developing countries are investing rapidly in renewable energy. From 2014 to 2017, investments totaled $413 billion in developed countries and $419 billion in developing countries. Record low costs for solar and wind installations have been reported in Chile and Morocco, Zambia, Mexico, and Peru, according to analysts at Bloomberg Energy Finance and the UNEP. India has pledged to meet 40 percent of its electricity needs from non-fossil sources by 2030. By late 2017, India had installed nearly 10 gigawatts of solar power, with another 10.6 GW in process. Electric vehicles are also a growing initiative in India, which has suffered from disastrous air quality conditions in recent decades.

Perhaps most important, agreement on the need for better policies is almost universally accepted now. Countries have always behaved selfishly, but international agreements make bad behavior harder to hide. For example, President Obama's Clean Power Plan, which protected public health and economy as well as the climate, was abandoned in the first months of the Trump administration, which prioritized shoring up coal and gas. But the Paris agreement makes these policy changes globally visible, and trading partners have threatened to end trade deals with the United States if we fail to pursue our INDCs and remain in the Paris Accord.

We have many drawdown options right now

There have been many proposals for dramatic new inventions that can fix the problem all at once—nuclear fusion, space-based solar energy, or giant mirrors that would reflect solar energy away from the earth's surface. These are intriguing ideas, but all are still in the distant future, and it is clear that we need to act now.

The good news is that we have lots of solutions available. And polls show most people are enthusiastic about reducing climate impacts. The task is to make it easy for people to participate.

A wedge approach is one strategy, proposed in 2004 by Stephen Pacala and Robert Socolow, of Princeton University's Climate Mitigation Initiative. They calculated that currently available technologies—efficient vehicles, buildings, power plants, alternative fuels—could solve our problems quickly, if we invested in them now. Pacala and Socolow's paper described 14 "wedges" (fig. 9.22). Most of the wedges are technologically easy—for example, increasing our vehicle fuel efficiency from the expected 30 miles per gallon in 2058 to 60 mpg, promoting public transportation, installing more efficient lighting, or insulating buildings.

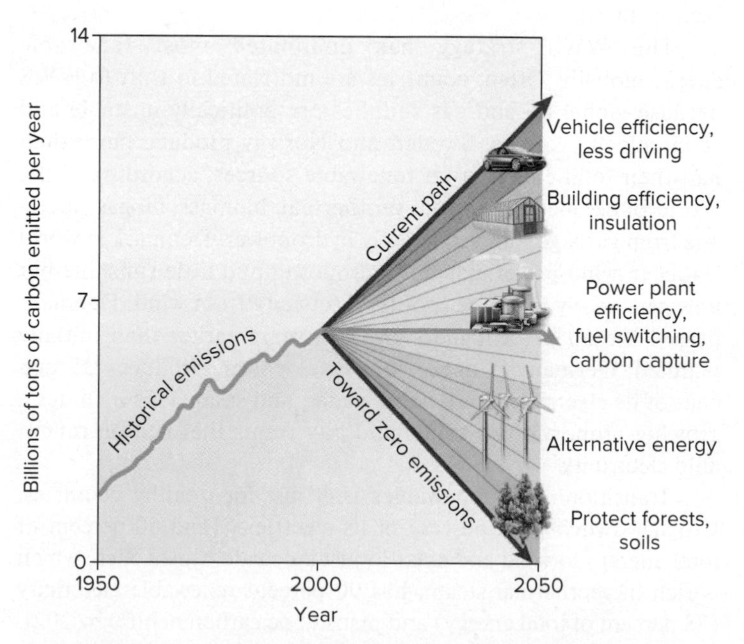

▲ **FIGURE 9.22** A wedge approach could use multiple strategies to reduce or stabilize carbon emissions rapidly and relatively cheaply.

TABLE 9.2 Top 15 Options Identified by the Drawdown Study for Reducing Emissions with Lifetime Net Savings

SOLUTION	GT CO_2	SAVINGS ($BILLION)
Refrigerant management	89.74	−$903
Wind turbines (onshore)	84.6	$7,425
Reduced food waste	70.53	NA
Plant-rich diet	66.11	NA
Tropical forest protection	61.23	NA
Educating girls	59.6	NA
Family planning	59.6	NA
Solar farms	36.9	$5,024
Silvopasture	31.19	$699
Rooftop solar	24.6	$3,548
Regenerative agriculture	23.15	$1,928
Temperate forest protection	22.61	NA
Peatland protection	21.57	NA
Tropical agroforestry	20.19	$627
Forest expansion	18.06	$392
Total for 80 solutions	1,051	$73,874

Each wedge represents 1 GT (1 billion tons) of carbon emissions avoided in 2058, compared to a "business-as-usual" scenario. Each wedge might be small now, but their impacts will grow over time, producing major impacts in 50 years. Accomplishing just half of these wedges could level off our emissions. Accomplishing all of them draw us down below 1990 emissions levels.

A comprehensive update to this approach, *Drawdown*, was published by Paul Hawken and colleagues in 2017. This study emphasized that ultimately we need to draw down emissions to negative levels, because the atmosphere has already exceeded safe concentrations of GHGs.

The Drawdown study was unusually comprehensive, ranking 80 strategies that included a wide array of social practices (such as educating girls, so as to empower them to plan their own family size; better forest management; plant-rich diets) as well as implementing existing technology (table 9.2). The study also found that these changes don't have to cost more overall. It found a net savings of $74 trillion from steps to reduce 1,051 GT CO_2 by 2050.

You can learn much more by examining the Drawdown study online.

Wind, water, and solar could meet all our needs

Access to affordable and feasible technology is no longer the problem it once was. Renewable energy costs have plummeted as the market has grown, and many scenarios have been developed for converting to most or all energy production from renewable sources.

What Do YOU THINK?

Unburnable carbon

Climatologists have shown that the earth will become largely uninhabitable if we continue to burn oil, gas, and coal at the rate we have done in recent decades. Major cities will be under water, and most farmland will be desert. Alternative energy sources are therefore a priority. But fossil fuels are abundant. We have enough natural gas and coal to last for centuries and enough oil to last 40–60 years even at current high-use rates. Because we can burn only a fraction of these fuels without making our planet uninhabitable, analysts often refer to these as **unburnable carbon.** Financial investments in these resources are often called **stranded assets,** because they have no monetary value if they cannot be sold.

Three important obstacles have blocked wider adoption of alternative energy and conservation in recent years. One is inertia: We already have trillions of dollars invested in existing cities, roads, and power infrastructure. Replacing all this infrastructure is not easy to budget, even though most of it is gradually and steadily replaced and upgraded.

A second factor is ready availability of abundant fossil fuels. As long as they are easy to use and comparatively cheap to produce, we will have difficulty weaning ourselves from them.

A third obstacle is that owners of this carbon strongly resist energy and climate policies that would turn their wealth into stranded assets. Understandably enough, they have invested millions of dollars in campaigns to block fossil fuel regulation, to support lawmakers who deny climate change, and to fund research that seeks alternative explanations for climate change, such as sunspots or random variability.

At the same time global subsidies for fossil fuel development, exploration, and marketing exceed $550 billion per year, according to the International Energy Agency (IEA). This is about 4.5 times the $120 billion for all renewable energy. The United States is the world's most generous supporter of fossil fuels, providing about one-fifth of global subsidies. The IEA and other groups have begun to question whether this ratio can be changed.

If you were in charge of energy policy, what would you do? Would you compensate oil and gas companies for their unburnable carbon? Would you try to restrict corporate influence on climate policy, or not? If so, how?

What Can YOU DO?

Climate Action

The main thing we all need to do is reduce energy consumption and work for policies that support low-carbon energy. The most important steps are changing national energy policies. At the same time, millions of individual actions also have powerful positive impacts.

Use your phone. Contact your senators and representatives to tell them to protect your future. Vote for climate-conscious candidates.

Save gas. Drive less, use an efficient vehicle, and make plans to live where you can use public transit or bicycles. Vehicles may represent our biggest individual climate impact.

Buy efficient appliances. Avoiding catastrophic climate change is about more than just changing lightbulbs, but better bulbs and appliances are also important, and the impact multiplies as millions of us follow better practices.

Eat low on the food chain. Meat and dairy production are important sources of methane and CO_2.

Weatherize your home. Improve your comfort and reduce heat loss in winter or heat gain in summer. Turn down the thermostat and dress for the season.

Buy wind and solar energy. Many utilities allow you to purchase alternatives.

Practice energy policy at school and at work. Ask administrators what they are doing to reduce energy use, and help them write plans to do better.

Mark Jacobson and his colleagues, for example, outlined a strategy by which all U.S. energy—for transportation, heating, and industrial uses, as well as electricity—could be 80 percent wind, water, and solar (WWS) by 2030, and 100 percent WWS by 2050, with existing and feasible technologies. They also found savings of $10,000 per person per year in financial benefits from this transition, including energy cost savings, increased employment, reduced health care costs, and avoided storm- and heat-related losses (see section 13.7).

The WWS strategy has dominated most renewable energy globally. Often, countries are motivated to shift to WWS because global oil and gas supplies are politically unstable and economically volatile. Sweden and Norway produce more than half their total energy from renewable sources, according to the World Bank, including wind, geothermal, biomass, biogas (methane from garbage), and especially hydropower. Denmark, a world leader in wind power, has no hydropower and little sunshine but now gets nearly 60 percent of its electricity from wind. Denmark passed the 50 percent mark over 15 years earlier than initially planned. Germany, a major industrial power, produces 35 percent of its electricity from wind, water, and solar. This includes a growing proportion of trains and city trams that run on renewable electricity.

Transitioning to renewables isn't just for wealthy countries. Uruguay switched 95 percent of its electricity (and 60 percent of total energy) to wind and solar in just a decade. Costa Rica, which is rich in geothermal steam, has 90 percent renewable electricity (38 percent of total energy) and plans to be carbon neutral by 2021. China, the world's fastest-growing economy, produces nearly a quarter of its electricity from renewable sources.

Local initiatives are everywhere

Local action is gaining momentum, as renewables become more familiar and more affordable. Nearly 1,000 U.S. cities and 39 states have announced their own plans to combat global warming. And over 600 university and college campuses have pledged to reduce greenhouse emissions. Mayors and governors have committed locally to meeting Paris Accord priorities.

Part of the motivation for these steps is that alternative solutions often have multiple benefits. Building and vehicle efficiency saves money in the long run. Buying better lightbulbs saves money immediately. Planning cities for better transit, biking, and walking saves tax dollars we now spend on far-flung road and service networks. A side effect of these movements is that renewable energy can make control of energy more local, keeping money close to home. People are also learning that efficiency and climate action not only save money and build community but are more engaging than wringing our hands over impending climate disaster.

Colleges and universities are also taking action. Ball State University, in Indiana, was one of the first to transition its buildings from fossil-fuel boilers to electric-powered geothermal, starting in 2012. Skidmore College in northern New York has been 40 percent heated by geothermal heat pumps since 2014, and Carleton College, in Minnesota, transitioned to nearly 100 percent geothermal heating and cooling in 2019 (fig. 9.23). Vermont's Middlebury College, which has access to forest waste products as well as biogas from local dairies, achieved carbon neutrality in 2016.

States are leading the way

Among U.S. states, many are seeing changes in both practices and policies. While some states are digging further into fossil fuels, many are betting on renewables. Texas leads the nation in wind power production, followed by Iowa, Oklahoma, California, and Illinois. New York, California, Massachusetts, and other states are planning for 80 percent reductions in carbon emissions (below 1990 levels) by 2050—the "80 by 50" target called for by the Paris agreement and the IPCC.

▲ **FIGURE 9.23** Carleton College, in Minnesota, is transitioning to geothermal heating, and by sourcing electricity from wind, the college aims to be 100 percent fossil-free by 2050. ©Carleton College

California has often led the way in environmental policy. In 2004, the state passed revolutionary legislation that required automakers to cut tailpipe emissions of carbon dioxide from cars and trucks, which has since been picked up by New York and other states. When car manufacturers failed to comply, California sued the six largest automakers in 2006, charging that they were costing the state billions of dollars in health and environmental damages.

California is motivated to do this partly because its economy relies almost entirely on declining winter snowpack for both urban water use and farm irrigation. Recent years of severe droughts have affected much of the state and worried cities and counties. Californians also have gotten tired of waiting for action in Washington, where the dominant view has been that climate controls will cost jobs. Contrary to this argument, California has seen rapid job growth in clean energy. Clean energy employs over 500,000 people and has brought in over $9 billion in venture capital, or 60 percent of all clean-energy investments nationwide.

Carbon trading is another financial approach to incentivizing carbon reductions. Companies are granted permits for a specified amounts of CO_2 emissions; companies that reduce emissions below permitted levels can sell the extra permits. At least 27 states and four Canadian provinces participate in regional carbon-trading compacts. One of these, the northeastern Regional Greenhouse Gas Initiative (RGGI), began trading carbon credits for 233 plants in 2008. Carbon credit auctions bring in hundreds of millions of dollars in revenue each year, which supports conservation and renewable energy initiatives in participating states. Carbon trading is not perfect: Auction prices are often too low to provide real incentives for some industries, and many question whether a "right to pollute" is appropriate. However, these compacts are widely considered successful—and politically palatable—approaches to reducing emissions.

Carbon capture saves CO_2 but is expensive

It is possible, though expensive, to store CO_2 by injecting it deep into geologic formations. Since 1996, Norway's Statoil has been pumping more than 1 million metric tons of CO_2 per year into an aquifer 1,000 m below the seafloor in the North Sea. The pressurized CO_2 enhances oil recovery. It also saves money because otherwise the company would have to pay a $50 per ton carbon tax on its emissions. Around the world, deep, salty aquifers could store a century's output of CO_2 at current fossil fuel consumption rates. A number of companies have started, or are now planning, similar schemes.

Carbon capture and storage are appealing solutions that would allow us to continue to burn fossil fuels (What Do You Think?, p. 226). However, compressing CO_2 to transport and store it takes about one-quarter of the energy produced at a plant, so it greatly increases the cost of producing a kilowatt hour of electricity. Cost-effective strategies for carbon capture from fossil fuels remain elusive. GHG capture can also be improved in other contexts.

Capturing methane from landfills, oil wells, and coal mines would make an important contribution to both fuel and climate problems. Rice paddies, with submerged, decaying plant matter, are also important methane sources: Changing flooding schedules and fertilization techniques can reduce methane production in paddies.

Ruminant animals (such as cows, camels, and buffalo) create large amounts of methane in their digestive systems. Feeding them more grass and less corn can reduce these emissions. Some analysts have suggested that eating less beef—skipping just one day per week—could reduce our individual contributions to climate change more than driving a hybrid Toyota Prius would.

CONCLUSION

Climate change is the most far-reaching issue in environmental science today. Although the challenge is almost inconceivably large, solutions are possible if we choose to act, as individuals and as a society. Temperatures are now higher than they have been in thousands of years, and climate scientists say that if we don't reduce greenhouse gas emissions soon, drought, flooding of cities, and conflict may be inevitable.

Understanding the climate system is essential to understanding the ways in which changing composition of the atmosphere (more carbon dioxide, methane, and nitrous oxide, in particular) matters to us. Basic concepts to remember about the climate system include how the earth's surfaces absorb solar heat, how atmospheric convection transfers heat, and that different gases in the atmosphere absorb and store heat that is re-emitted from the earth. Increasing heat storage in the lower atmosphere can cause increasingly vigorous convection, more extreme storms and droughts, melting ice caps, and rising sea levels. Changing patterns of monsoons, cyclonic storms, frontal weather, and other precipitation patterns could have extreme consequences for humans and ecosystems.

Despite the importance of natural climate variation, observed trends in temperature and sea level are more rapid and extreme than other changes in the climate record. Climate models show that these changes can only be explained by human activity. Increasing use of fossil fuels is our most important effect, but forest clearing, decomposition of agricultural soils, and increased methane production are also extremely important.

Countless new strategies are being developed, with many wedges involved in resolving climate problems. Renewable energy holds great promise. International accords have also produced some results. Thousands of communities and institutions have also developed local plans to do their part. We have most of the tools we need to save the climate; our challenge is to decide to apply them.

PRACTICE QUIZ

1. What are the dominant gases that make up clean, dry air?
2. Name and describe four layers of the atmosphere.
3. What is the greenhouse effect? What is a greenhouse gas?
4. What are some factors that influence natural climate variation?
5. Atmospheric convection is an important cause of rainfall. What drives Hadley cell convection?
6. What is a monsoon, and why is it seasonal?
7. What is a cyclonic storm?
8. What is the IPCC, and what is its function?
9. What method has the IPCC used to demonstrate a human cause for recent climate changes? Why can't we do a proper manipulative study to prove a human cause?
10. What are five key ideas in the Paris Accord?
11. List 5 to 10 effects of changing climate.
12. What is a climate stabilization wedge? Why is it an important concept?
13. What are several of the "Drawdown" solutions, and which ones are most effective?

CRITICAL THINKING AND DISCUSSION

Apply the principles you have learned in this chapter to discuss these questions with other students.

1. Weather patterns change constantly over time. From your own memory, what weather events can you recall? Can you find evidence in your own experience of climate change? What does your ability to recall climate changes tell you about the importance of data collection?
2. One of the problems with forming climate policies such as the Paris Accord is that economists and scientists define problems differently and have contrasting priorities. How would an economist and an ecologist explain disputes over the Paris Accord differently?
3. Economists and scientists often have difficulty reaching common terms for defining and solving issues such as the Clean Air Act renewal. How might their conflicting definitions be reshaped to make the discussion more successful?
4. Why do you think controlling greenhouse gases is such a difficult problem? List some of the technological, economic, political, emotional, and other factors involved. Whose responsibility is it to reduce our impacts on climate?
5. How does the decades-long, global-scale nature of climate change make it hard for new policies to be enacted? What factors might be influential in people's perception of the severity of the problem?
6. Would you favor building more nuclear power plants to reduce CO_2 emissions? Why or why not?
7. Of the climate solutions listed in table 9.2, which ones seem most important to you, in terms of benefits beyond carbon emissions?

DATA ANALYSIS Examining the IPCC Fifth Assessment Report (AR5)

The Intergovernmental Panel on Climate Change (IPCC) has a rich repository of figures and data, and because these data are likely to influence some policy actions in your future, it's worthwhile to examine the IPCC reports. The data and conclusions, as well as the points of uncertainty, are presented there to help the public understand the issues with the best available data.

An excellent overview is in the Summary for Policy Makers (SPM) that accompanies the Fifth Assessment Report. Find a copy of this report on Connect. Examine the figures in the report and answer questions in Connect to demonstrate your understanding of the ideas and the issues.

Change in temperature, sea level, and Northern Hemisphere snow cover

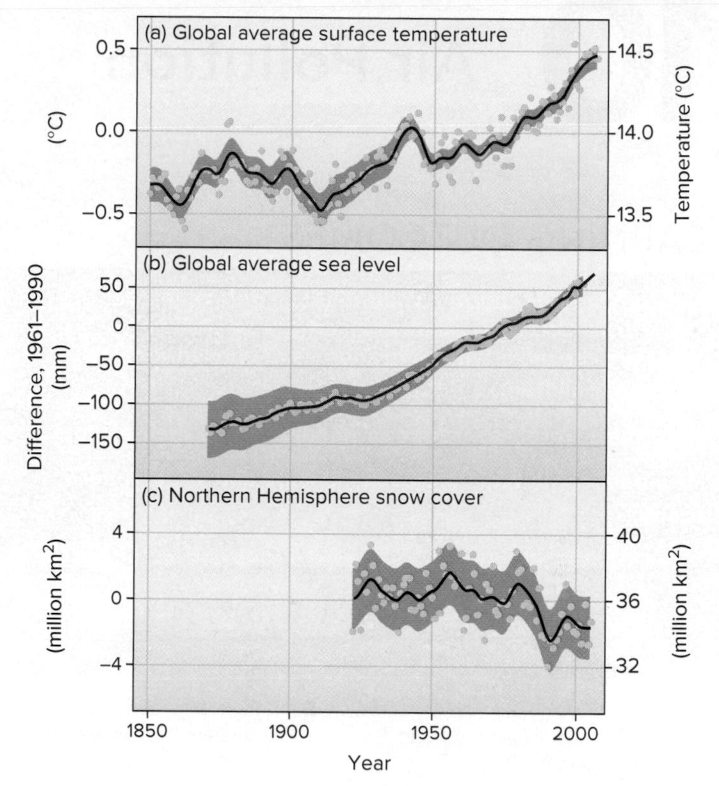

▶ See the evidence: View the IPCC report at **http://www.ipcc.ch/ipccreports/ar4-syr.htm**.

CHAPTER

10 Air Pollution

LEARNING OUTCOMES

After studying this chapter, you should be able to answer the following questions:

- What are the main types and sources of conventional or "criteria" pollutants?
- Describe several hazardous air pollutants and their effects.
- How do air pollutants affect the climate and stratospheric ozone?
- In what ways can air pollution affect human health?
- What policies and strategies do we have for reducing air pollution?
- Has world air quality been getting better or worse? Why?

According to the World Health Organization, Delhi, India, has the worst air quality of any major city in the world. Breathing the air there is equivalent to smoking 50 cigarettes per day.

©Saurav022/Shutterstock

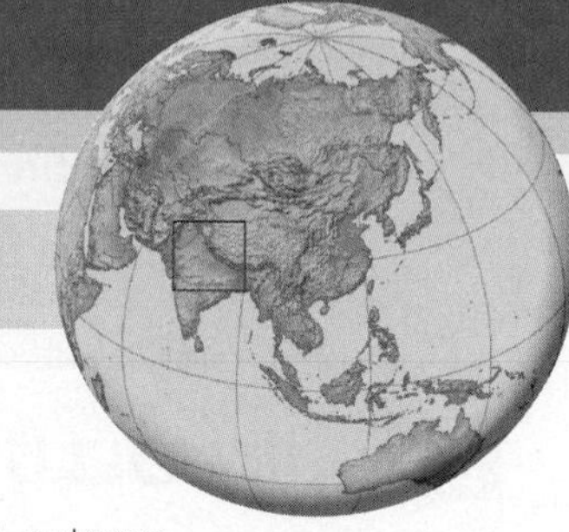

CASE STUDY

Delhi's Air Quality Crisis

Delhi, India, has the worst air quality of any major city on the planet. In a survey of 1,600 world cities, the World Health Organization (WHO) found that Delhi had far more polluted air than any other large urban area. You may have heard about the dangers of breathing the polluted air in Beijing, China, but, on average, Delhi's air is twice as bad as Beijing's. India has the world's highest death rate from chronic respiratory diseases and asthma, according to the WHO. It's estimated that 2.5 million people die every year in India from diseases related to air pollution.

Air pollution comes in many forms. Smoke, haze, dust, odors, corrosive gases, and toxic compounds are among our most widespread pollutants. Depending on which ones, when, how much, and for what length of time we're exposed, these pollutants can result in asthma, bronchitis, emphysema, heart disease, strokes, neurological damage, and other health problems. Of the many air pollutants to which we're exposed, particulate material (dust, ash, soot, smoke, and other suspended solids), are often the most visible and among the most dangerous. Fine particles less than 2.5 micrometers in diameter (PM2.5) can penetrate deep into your lungs and damage fragile lung tissues, leading to scarring and decreased lung function.

Polluted air is especially dangerous for children. About 2.2 million children in Delhi (roughly half of all children in the city) are thought to have lung damage that can lead to asthma, emphysema, or obstructive pulmonary disease. Chronic exposure to air pollutants may lower Intelligence Quotient (IQ), and increase risks for autism, epilepsy, diabetes, and perhaps adult-onset diseases, such as multiple sclerosis and cancer.

In November 2017, Delhi experienced a frightening smog episode in which the air quality index spiked off the charts (fig. 10.1). This index is based on five major air pollutants, ground-level ozone, particulates, carbon monoxide, sulfur dioxide, and nitrogen dioxide. The scale runs from 0 to 500, but on November 8, 2017, the index in Delhi reached 1,010, more than 20 times the level considered safe. According to some experts, breathing the air in Delhi was equivalent to smoking 50 cigarettes per day.

There are many sources of air pollutants in Delhi. Motor vehicle exhaust is a major problem. The number of registered motor vehicles in India increased from 0.3 million in 1951 to more than 220 million in 2017. Delhi, a megacity with about 20 million residents with millions of cars, trucks, motor bikes, tractors, and three-wheelers, has the highest density of motor vehicles in the country. According to some studies, motor vehicles are responsible for as much as 80 percent of nitrogen oxides that create acid rain and smog. In 2002, Delhi passed a law requiring thousands of three-wheeled motorized rickshaws, called "tuk tuks," that clog urban streets to change to clean-burning natural gas. Air quality did improve temporarily, but trucks and automobiles have proliferated, making pollution levels higher than ever.

Burning crop residues is another huge cause of pollution in India. Traditionally, farmers have burned stubble left in the field after harvest. This helps control plant pests and diseases as well as returning mineral nutrients back to the soil. But the burning usually is done in winter months when light winds and lack of rain let smoke linger in the air. As the smoke drifts into the city it mixes with diesel fumes, dust from construction sites, burning garbage, and other pollutants to create a dense smog that cuts visibility to a few hundred meters. Airports have been closed, trains delayed, and roads clogged with traffic accidents as drivers can't see where they are going.

Industrial facilities, including coal-burning power plants, oil refineries, chemical industries, and brick kilns, also contribute to Delhi's unhealthy air. Politics plays a role as well. Although Delhi's chief minister described the city as a "gas chamber," other governmental officials are reluctant to regulate farmers, industries, or transportation sectors, all of which have powerful political constituencies. In this chapter we examine the sources and types of air pollutants that affect all of us as well as the policies and technology that can help protect air quality. ■

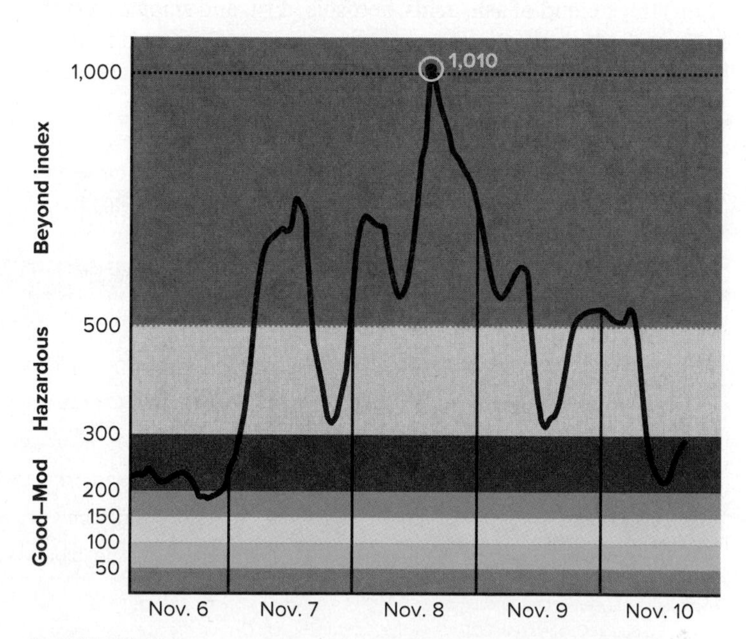

▲ **FIGURE 10.1** Air quality index in Delhi, India, in November 2017 as measured by the American Embassy. The peak pollution on November 8 was 20 times higher than levels considered safe for human health.

If you think education is expensive, try ignorance.

—DEREK BOK

Benchmark Data

Among the ideas and values in this chapter, these are a few worth remembering.

2.5 million	Estimated early deaths from air pollution in India, per year
2.5 μm	Size of particulate matter that causes most serious lung damage
$2,000 billion	Economic benefits of controlling U.S. air pollution
$65 billion	Cost of controlling U.S. air pollution
25:1	Ratio of benefits:cost

10.1 AIR POLLUTION AND HEALTH

- *Air pollutants affect health and environmental quality.*
- *The EPA regulates six major criteria pollutants as well as other contaminants.*
- *Mercury, lead, and organic compounds are examples of hazardous air pollutants.*

As the example of Delhi, India, shows, poor air quality in the burgeoning cities of the developing world is a serious threat to human health. In many Indian cities airborne dust, smoke, and soot often are 20 times higher than levels considered safe for human health. According to the WHO, 10 of the world's 20 smoggiest cities are in India. India's urban residents are six times more likely than rural people to die of lung cancer. Respiratory ailments, cardiovascular diseases, lung cancer, infant mortality, and miscarriages are as much as 50 percent higher in countries with high pollution levels than in those with cleaner air. In 2018, a group of 40 distinguished health experts concluded that air pollution kills some 6.5 million people worldwide. About 92 percent of those deaths occur in low-income countries, where pollution control policies are weak and health care is uncertain.

Air pollution studies over southern Asia reveal that a 3 km (2 mi) thick cloud of ash, acids, aerosols, dust, and smog covers the entire Indian subcontinent for much of the year. Produced by forest fires, the burning of agricultural wastes, and dramatic increases in the use of fossil fuels, this smog layer cuts the amount of solar energy reaching the earth's surface beneath it by up to 15 percent. Nobel laureate Paul Crutzen predicts that this smog layer—80 percent of which is human-made—could disrupt monsoon weather patterns and cut rainfall over northern Pakistan, Afghanistan, western China, and central Asia by up to 40 percent.

When this "Asian Brown Cloud" drifts out over the Indian Ocean at the end of the monsoon season, it cools sea temperatures and may be changing regional climate patterns in the Pacific Ocean as well. This plume of soot and gases can travel halfway around the globe in a week, with unknown impacts on the world's climate and environmental quality.

Worldwide, air pollution emissions add up to billions of metric tons per year (table 10.1). Air pollution impairs human health, damages crops and ecosystems, and corrodes buildings and infrastructure. Greenhouse gases are altering our climate. Aesthetic degradation, such as odors and lost visibility, are also important

TABLE 10.1 Estimated Fluxes of Pollutants and Trace Gases to the Atmosphere

		APPROXIMATE ANNUAL FLUX (MILLIONS OF METRIC TONS/YR)	
SPECIES	MAJOR SOURCES	NATURAL	ANTHROPOGENIC
CO_2 (carbon dioxide)	Respiration, fossil fuel burning, land clearing, industry	370,000	36,800*
CH_4 (methane)	Rice paddies, wetlands, gas drilling, landfills, cattle, termites	155	558
CO (carbon monoxide)	Incomplete combustion, CH_4 oxidation, plant metabolism	1,580	562
Non-methane hydrocarbons	Fossil fuels, industrial uses, plant isoprenes, other biogenics	860	170
NO_x (nitrogen oxides)	Fossil fuel burning, lightning, biomass burning, soil microbes	90	122
SO_x (sulfur oxides)	Fossil fuel burning, industry, biomass burning, volcanoes, oceans	35	103
SPM (suspended particulate matter)	Fossil fuels, industry, mining, biomass burning, dust, sea salt	583	362

*Fossil fuels and cement production only.

Source: European Commission EDGAR, Global Carbon Project, Carbon Dioxide Information and Analysis Center.

consequences of air pollution. These factors rarely threaten life or health directly, but they can strongly impact our quality of life. They also increase stress, which affects health.

Many people don't remember that only a few decades ago, many American and European cities also endured conditions similar to those currently experienced today in Delhi, Beijing, and other large cities in the developing world. Chronic bad air, and occasional severe smog events, gradually led to the adoption of pollution controls. Legal enforcement has improved, and many students today don't realize how bad air quality once was in their hometowns. Many American and European cities still have bad air: Major port cities, oil and gas extraction areas, and industrial cities are particularly bad. But in the past 40 years, air quality protections have increased in number and in effectiveness, greatly improving public health. In most developed economies, there are established, legally enforceable rules to protect the air we all breathe. Industry has also relocated to hungry regions of the developing world, where environmental and health protections are poorly enforced.

The Clean Air Act regulates major pollutants

Air pollution control has evolved gradually. The U.S. Clean Air Act of 1963 was the first national legislation in the United States aimed at air quality. The act provided federal grants to aid states in pollution control but was careful to preserve states' rights to set or enforce air quality regulations. It soon became obvious that piecemeal, local standards did not resolve the problem, because neither pollutants nor the markets for energy and industrial products are contained within state boundaries.

Amendments to the law in 1970 designated new standards, to be applied equally across the country, for six major pollutants: sulfur dioxide, nitrogen oxides, carbon monoxide, ozone (and its precursor volatile organic compounds), lead, and particulate matter. These six are referred to as **conventional** or **criteria pollutants**, and they were addressed first because they contributed the largest volume of air quality degradation and are considered the most serious threat to human health and welfare. Transportation and power plants are the dominant sources of most criteria pollutants (fig. 10.2). National ambient air quality standards (NAAQS) designated allowable levels for these pollutants in the **ambient air** (the air around us). Primary standards (table 10.2) are intended to protect human health. Secondary standards are also set to protect crops, materials, climate, visibility, and personal comfort.

In addition to the six conventional pollutants, the Clean Air Act regulates an array of unconventional pollutants, compounds that are

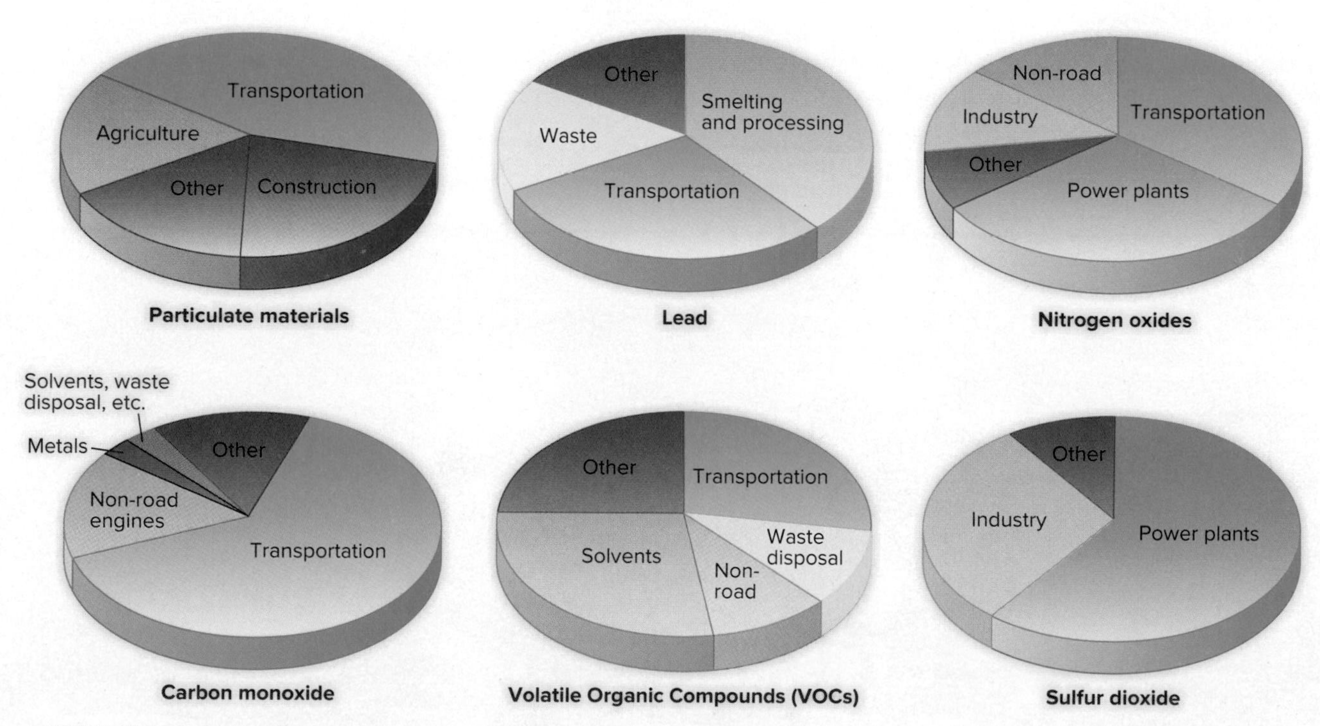

▲ **FIGURE 10.2** Anthropogenic sources of six of the primary "criteria" air pollutants in the United States. Source: UNEP.

TABLE 10.2 National Ambient Air Quality Standards (NAAQS)

POLLUTANT	PRIMARY (HEALTH-BASED) AVERAGING TIME	STANDARDS (ALLOWABLE CONCENTRATIONS)
PM[a]	Annual geometric mean[b]	50 μg/m^3
	24 hours	150 μg/m^3
SO_2	Annual arithmetic mean[c]	80 μg/m^3 (0.03 ppm)
	24 hours	120 μg/m^3 (0.14 ppm)
CO	8 hours	10 mg/m^3 (9 ppm)
	1 hour	40 mg/m^3 (35 ppm)
NO_2	Annual arithmetic mean	80 μg/m^3 (0.05 ppm)
O_3	Daily max 8 hour avg.	157 μg/m^3 (0.08 ppm)
Lead	Maximum quarterly avg.	1.5 μg/m^3

[a] Total suspended particulate material, PM2.5 and PM10.

[b] The geometric mean is obtained by taking the *n*th root of the product of *n* numbers. This tends to reduce the impact of a few very large numbers in a set.

[c] An arithmetic mean is the average determined by dividing the sum of a group of data points by the number of points.

Source: UNEP.

produced in less volume than conventional pollutants but that are especially toxic or hazardous, such as asbestos, benzene, mercury, polychlorinated biphenyls (PCBs), and vinyl chloride. Most of these are uncommon in nature or have no natural sources (fig. 10.3).

Many pollutants come from a **point source**, such as a smokestack. **Fugitive**, or **nonpoint-source**, **emissions** are those that do not go through a smokestack. Leaking valves and pipe joints contribute as much as 90 percent of the hydrocarbons and volatile organic chemicals emitted from oil refineries and chemical plants, and increasingly from natural gas wells. Dust, as from mining, agriculture, and building construction and demolition, is also considered fugitive emissions.

Primary pollutants are substances that are harmful when released. Secondary pollutants, by contrast, become harmful after they react with other gases or substances in the air. In particular, **photochemical oxidants** (compounds created by reactions driven by solar energy) and atmospheric acids are probably the most important secondary pollutants.

Conventional pollutants are abundant and serious

Most conventional pollutants are produced primarily by burning fossil fuels, especially in coal-powered electric plants and in cars and trucks, as well as in processing natural gas and oil. Others, especially sulfur and metals, are by-products of mining and manufacturing processes. Of the 188 air toxics listed in the Clean Air Act, about two-thirds are volatile organic compounds, and most of the rest are metal compounds. In this section we will discuss the characteristics and origin of the major pollutants.

▲ **FIGURE 10.3** Many of our most serious pollutants are fugitive emissions from petrochemical facilities such as this one in Baton Rouge, Louisiana. ©Mary Ann Cunningham

Active LEARNING

Compare Sources of Pollutants

Getting a handle on the nature of air pollution can be difficult because there are many pollutants, all originating from a variety of sources. A good place to start is to remember the identity and sources of the six major "criteria" pollutants first targeted by the Clean Air Act. These pollutants aren't as dangerous in minute doses as mercury or other hazardous pollutants, but they are both serious and abundant.

Examine closely the pollutant sources shown in figure 10.2, to answer these questions:

1. Close your eyes and list the six pollutants shown.
2. Volatile organic compounds (VOCs) are often evaluated together with ozone (O_3), because together these contribute to a variety of photochemical oxidants and other pollutants. In this context, what does "volatile" mean? What does "organic compound" mean?
3. Examine the chart for sulfur dioxide. What is its single largest source? What is another pollutant produced in abundance from that source?
4. Which pollutants have transportation as a major contributor? How many of the six are these? As you read this chapter, what are some of our strategies for minimizing these pollutants?
5. What is the main source of nitrogen in nitrogen oxides from your car engine? (Examine your text and figures for the answer.)

ANSWERS: 1. particulate material, carbon monoxide, lead, volatile organic compounds, nitrogen oxides, sulfur dioxide; 2. "volatile" means easily evaporated, and organic compound means a base of carbon atoms or rings; 3. industry, nitrogen oxides; 4. five of the six pollutants: all but sulfur dioxide—pollution control strategies include better combustion, cleaner fuels, and other efforts; 5. atmospheric nitrogen, which binds to atmospheric oxygen in the heat of fuel combustion.

Sulfur dioxide (SO_2) is a colorless, corrosive gas that damages both plants and animals. Once in the atmosphere, it can be further oxidized to sulfur trioxide (SO_3), which reacts with water vapor or dissolves in water droplets to form sulfuric acid (H_2SO_4), a major component of acid rain. Sulfur dioxide and sulfate ions are probably second only to smoking as causes of air pollution–related health damage. Sulfate particles and droplets also reduce visibility in the United States by as much as 80 percent. Minute droplets of sulfuric acid can penetrate deep into lungs, causing permanent damage, as well as irritating eyes and corroding buildings. In plants, sulfur dioxide destroys chlorophyll, eventually killing tissues (fig. 10.4).

Nitrogen oxides (NO_x) are highly reactive gases formed when the heat of combustion initiates reactions between atmospheric nitrogen (N_2) and oxygen (O_2). The initial product, nitric oxide (NO), oxidizes further in the atmosphere to nitrogen dioxide (NO_2), a reddish-brown gas that gives photochemical smog its distinctive color. Because these gases convert readily from one form to the other, the general term NO_x (with x indicating an unspecified number) is used to describe these gases. Nitrogen oxides combine with water to form nitric acid (HNO_3), which is also a major component of acid precipitation (fig. 10.5). Excess nitrogen in water is also causing eutrophication of inland waters and coastal seas.

▲ **FIGURE 10.4** This soybean leaf was exposed to 2.1 mg/m^3 sulfur dioxide for 24 hours. White patches show where chlorophyll has been destroyed.
©William P. Cunningham

Carbon monoxide (CO) is less common but more dangerous than the principal form of atmospheric carbon, carbon dioxide (CO_2). CO is a colorless, odorless, but highly toxic gas produced mainly by incomplete combustion of fuel (coal, oil, charcoal, wood, or gas). CO inhibits respiration in animals by binding irreversibly to hemoglobin in blood. In the United States, two-thirds of the CO emissions are created by internal combustion engines in transportation. Land-clearing fires and cooking fires also are major sources. About 90 percent of the CO in the air is consumed in photochemical reactions that produce ozone.

Ozone (O_3) is important in the upper atmosphere, where it shields us against ultraviolet radiation from the sun (see chapter 9), but at the ground level ozone is a highly reactive oxidizing agent that damages eyes, lungs, and plant tissues, as well as paint, rubber, and plastics. Ground-level ozone is a secondary pollutant, created by chemical reactions that are initiated by solar energy (fig. 10.5). In general, pollutants created by these light-initiated reactions are known as photochemical oxidants (that is, oxidizing agents that irritate tissues and damage materials). One of the most important of these reactions involves formation of single atoms of oxygen by splitting nitrogen dioxide (NO_2). This atomic oxygen then binds to a molecule of O_2 to make ozone (O_3). The acrid, biting odor of ozone is a distinctive characteristic of photochemical smog.

A variety of **volatile organic compounds (VOCs)** interact with ozone to produce photochemical oxidants in smog. A wide array of these organic (carbon-based), volatile (easily evaporated) chemicals derive from industrial processes such as refining of oil and gas, or plastics and chemical manufacturing. Some of the most dangerous

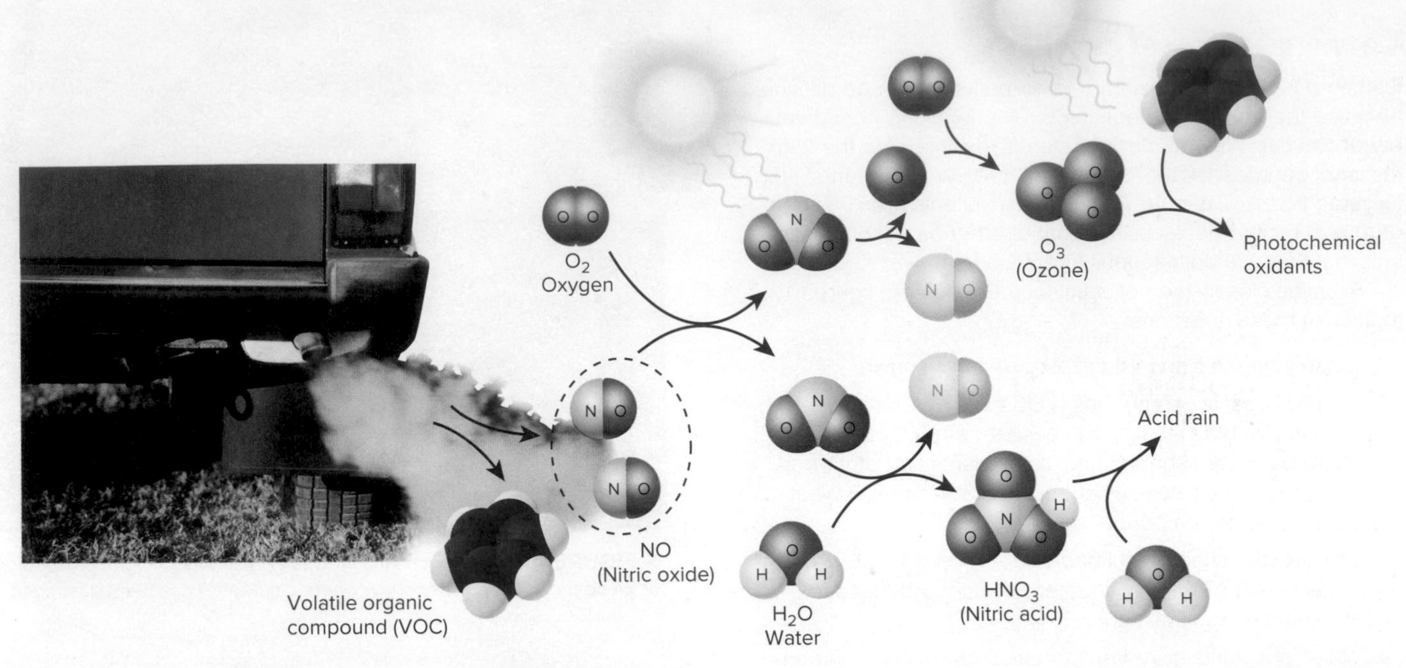

▲ **FIGURE 10.5** The heat of fuel combustion causes nitrogen oxides to form from atmospheric N_2 and O_2. NO_2 interacts with water (H_2O) to form HNO_3, a component of acid rain. In addition, solar radiation can force NO_2 to release a free oxygen atom, which joins to atmospheric O_2, creating ozone (O_3). Fuel combustion also produces incompletely burned hydrocarbons (including volatile organic compounds). Both O_3 and VOCs contribute to photochemical oxidants, in reactions activated by sunlight. The VOC shown here is benzene, a ring of six carbon atoms with a hydrogen atom attached to each carbon. ©Houghton Mifflin Harcourt/Photodisc/Getty Images

and common of these industrial VOCs are benzene, toluene, formaldehyde, vinyl chloride, phenols, chloroform, and trichloroethylene. Principal sources are incompletely burned fuels from vehicles, power plants, chemical plants, and petroleum refineries.

Lead, our most abundantly produced metal air pollutant, impairs nerve and brain functions. It does this by binding to essential enzymes and cellular components and disabling them. A wide range of industrial and mining processes produce lead, especially smelting of metal ores, mining, and burning of coal and municipal waste, in which lead is a trace element, and burning of gasoline to which lead has been added. Historically, leaded gasoline was the main source of lead in the United States, but leaded gas was phased out in the 1980s. Since 1986, when the ban was enforced, children's average blood lead levels have dropped 90 percent and average IQs have risen three points. Banning leaded gasoline in the United States was one of the most successful pollution–control measures in American history. Now, 50 nations have renounced leaded gasoline. The global economic benefit of this step is estimated to be more than $200 billion per year.

Particulate material includes dust, ash, soot, lint, smoke, pollen, spores, algal cells, and many other suspended materials. **Aerosols**, or extremely minute particles or liquid droplets suspended in the air, are included in this class. Particulates often are the most apparent form of air pollution, since they reduce visibility and leave dirty deposits on windows, painted surfaces, and textiles. Breathable particles smaller than 2.5 micrometers are among the most dangerous of this group because they can damage lung tissues. Asbestos fibers and cigarette smoke are among the most dangerous respirable particles in urban and indoor air because they are carcinogenic.

Hazardous air pollutants can cause cancer and nerve damage

A special category of toxins is monitored by the U.S. EPA because they are particularly dangerous even in low concentrations. Called **hazardous air pollutants (HAPs)**, these chemicals cause cancer and nerve damage as well as disrupt hormone function and fetal development. These persistent substances remain in ecosystems for long periods of time and accumulate in animal and human tissues. Most of these chemicals are either metal compounds, chlorinated hydrocarbons, or volatile organic compounds. Gasoline vapors, solvents, and components of plastics are all HAPs that you may encounter on a daily basis.

Many HAPs are emitted by chemical-processing factories that produce gasoline, plastics, solvents, pharmaceuticals, and other organic compounds. Benzene, toluene, xylene, and other volatile organic compounds are among these. Dioxins, carbon-based compounds containing chlorine, are released mainly by burning plastics and medical waste containing chlorine. The EPA reports that 100 million Americans (one-third of us) live in areas where the cancer rate from HAPs is ten times the normally accepted standard for action (1 in 1 million). Benzene, formaldehyde, acetaldehyde, and 1,3 butadiene are responsible for most of this HAP cancer risk. To help the public track local air quality levels, the EPA recently estimated the concentration of HAPs in localities across the continental United States. You can check pollutant levels and types in your own community by looking online for the Environmental Defense Fund HAP scorecard.

To help inform communities about toxic substances produced and handled in their area, Congress established the **Toxic Release**

▲ **FIGURE 10.6** A variety of hazardous emissions from some 23,000 facilities in the United States are monitored by the Toxic Release Inventory (TRI). ©PhotoDisc/Getty Images

Inventory (TRI) in 1986. This inventory collects self-reported statistics from 23,000 factories, refineries, hard rock mines, power plants, and chemical manufacturers to report on toxin releases (above certain minimum amounts) and waste management methods for 667 toxic chemicals. Although this total is less than 1 percent of all chemicals registered for use, and represents a limited range of sources, the TRI is widely considered the most comprehensive source of information about toxic pollution in the United States (fig. 10.6).

Mercury is a key neurotoxin

Airborne metals originate mainly from combustion of fuel, especially coal, which contains traces of mercury, arsenic, cadmium, and other metals, as well as sulfur and other trace elements. Airborne mercury has received special attention because it is a widespread and persistent neurotoxin (a substance that damages the brain and nervous system). Minute doses can cause nerve damage and other impairments, especially in young children and developing fetuses. Some 70 percent of airborne mercury is released by coal-burning power plants. Metal ore smelting and waste combustion also produce airborne mercury and other metals.

About 75 percent of human exposure to mercury comes from eating fish. This is because aquatic bacteria are mainly responsible for converting airborne mercury into methyl mercury, a form that accumulates in living animal tissues. Once methyl mercury enters the food web, it bioaccumulates in the flesh and bloodstream of predators. As a consequence, large, long-lived, predatory fish contain especially high levels of mercury in their tissues. Contaminated tuna fish alone is responsible for about 40 percent of all U.S. exposure to mercury (fig. 10.7). Swordfish, shrimp, and other seafood are also important mercury sources in our diet.

A 2009 report by the United States Geological Survey found that mercury levels in Pacific Ocean tuna have risen 30 percent in 20 years, with another 50 percent rise projected by 2050. Increased coal burning in which continues to build new coal-burning power plants, is understood to be the main cause of growing global mercury emissions in the Pacific. However, U.S. coal plants also produce mercury that is deposited across North America, in the Atlantic, and across Europe. Long-range transport of mercury through the air is even causing bioaccumulation in aquatic ecosystems in remote, high-Arctic areas. There, mercury poisoning can be a serious risk for people and wildlife in whose food chain is based on fish.

Freshwater fish also carry risks. Mercury contamination is the most common cause of impairment of U.S. rivers and lakes, and 45 states have issued warnings against frequent consumption of fresh-caught fish. A 2007 study tested more than 2,700 fish from 636 rivers and streams in 12 western states, and mercury was found in every one of them.

The U.S. National Institutes of Health (NIH) estimates that between 300,000 and 600,000 of the 4 million children born each year in the United States are exposed in the womb to mercury levels that could cause diminished intelligence or developmental impairments. According to the NIH, elevated mercury levels cost the U.S. economy $8.7 billion each year in higher medical and educational costs and in lost workforce productivity.

Mercury became fully regulated by the Clean Air Act in 2000, after decades of debate. Since then emissions have declined in many areas, as the metal is captured before it leaves the smokestack, but globally mercury is still a growing problem.

▲ **FIGURE 10.7** Canned tuna is our dominant source of mercury, because tuna is a top predator fish that we eat in abundance. ©Charles Brutlag/Shutterstock

Indoor air can be worse than outdoor air

We have spent a considerable amount of effort and money to control the major outdoor air pollutants, but indoor air pollution is a growing concern. With increasing use of volatile organic compounds in carpets, furniture, cleaning products, paints, and other substances, the U.S. EPA often finds high indoor concentrations of toxic air pollutants. Compounds such as chloroform, benzene, carbon tetrachloride, formaldehyde, and styrene can be 70 times greater in indoor air than outdoor air. Many homes have concentrations that would be illegal in the workplace. Because most of us now spend more time inside than outside, we also have higher exposure.

Cigarette smoke is without doubt the most important air contaminant in developed countries in terms of human health. The U.S. surgeon general has estimated that 400,000 people die each year in the United States from emphysema, heart attacks, strokes, lung cancer, or other diseases caused by smoking. These diseases are responsible for 20 percent of all mortality in the United States, or four times as much as infectious agents. Total costs for early deaths and smoking-related illnesses are estimated to be $100 billion per year. Eliminating smoking probably would save more lives than any other pollution-control measure.

In the less-developed countries of Africa, Asia, and Latin America, where such organic fuels as firewood, charcoal, dried dung, and agricultural wastes make up the majority of household energy, smoky, poorly ventilated heating and cooking fires are the worst sources of indoor air pollution (fig. 10.8). The World Health Organization estimates that 3 billion people are affected, especially women and small children, who spend long hours each day around open fires or unventilated stoves in enclosed spaces. More than half of all air pollution-related deaths worldwide are thought to be caused by bad indoor air.

▲ **FIGURE 10.8** Indoor air pollution affects some 3 billion people, mainly women and children, who spend days in poorly ventilated kitchens where carbon monoxide, particulates, and cancer-causing hydrocarbons often reach dangerous levels. ©China Tourism Press/Getty Images

10.2 AIR POLLUTION AND CLIMATE

- *Atmospheric circulation distributes pollutants around the globe.*
- *Black carbon (soot) causes climate change by melting reflective ice and snow.*
- *Chlorofluorocarbons, which destroy stratospheric ozone, declined with 1987 Montreal Protocol.*

Air pollutants have important interactions with our climate. Pollutants cause climate change: Carbon dioxide, methane, nitrous oxide, and halogen gases (such as chlorofluorocarbons used in refrigerants) are primary culprits (see chapter 9). At the same time, warming temperatures increase ozone and photochemical oxidant concentrations in the air we breathe (see fig. 10.5). And normal atmospheric circulation distributes pollutants around the globe, exposing remote populations to industrial-source pollutants.

Air pollutants travel the globe

Dust and fine aerosols can be carried great distances by the wind. Pollution from the industrial belt between the Great Lakes and the Ohio River Valley regularly contaminates the Canadian Maritime Provinces and sometimes can be traced as far as Ireland. Similarly, dust storms from China's Gobi and Takla Makan Deserts routinely close schools, factories, and airports in Japan and Korea, and often reach western North America. In one particularly severe dust storm in 1998, chemical analysis showed that 75 percent of the particulate pollution in Seattle, Washington, air came from China. Similarly, dust from North Africa regularly crosses the Atlantic and contaminates the air in Florida and the Caribbean Islands (fig. 10.9). This dust can carry pathogens and is thought to be the source of diseases attacking Caribbean corals.

▲ **FIGURE 10.9** A massive dust storm extends more than 1,600 km from the coast of Western Sahara and Morocco. Long-distance transport of dust and pollutants is an important source of contaminants worldwide. Source: SeaWiFS Project, GSFC, NASA

Soil scientists estimate that 3 billion tons of sand and dust are blown around the world every year.

Industrial contaminants are increasingly common in places usually considered among the cleanest in the world. Samoa, Greenland, and even Antarctica and the North Pole all have heavy metals, pesticides, and radioactive elements in their air. Since the 1950s, pilots flying in the high Arctic have reported dense layers of reddish-brown haze clouding the arctic atmosphere. Aerosols of sulfates, soot, dust, and toxic heavy metals, such as vanadium, manganese, and lead, travel to the pole from the industrialized parts of Europe and Russia. Soot and particulates that settle out on arctic snowfields also absorb the sun's heat, rather than reflecting it as snow does. This "black carbon" is an important factor in global climate change (Exploring Science, p. 240).

Circulation of the atmosphere tends to transport contaminants toward the poles. Like mercury (discussed above), volatile compounds (VOCs) evaporate from warm areas, travel through the atmosphere, then condense and precipitate in cooler regions. Over several years, contaminants migrate to the coldest places, generally at high latitudes, where they bioaccumulate in food chains. Whales, polar bears, sharks, and other top carnivores in polar regions have been shown to have dangerously high levels of pesticides, metals, and other hazardous air pollutants in their bodies. A study of Inuit people of Broughton Island, well above the Arctic Circle, found higher levels of polychlorinated biphenyls (PCBs) in their blood than any other known population except victims of industrial accidents. Far from any source of this industrial by-product, these people accumulate PCBs from the flesh of fish, caribou, and other animals they eat.

CO_2 and halogens are key greenhouse gases

Until recently CO_2 emissions were not regulated, even though it has been clear since the 1990s that reducing CO_2 emissions is necessary for slowing climate change. At normal concentrations, CO_2 is harmless, but emissions are rising steadily, at about 0.5 percent per year, due to human activities. Warming temperatures directly affect human health in several ways: Rates of heat stroke and dehydration soar during heat waves, as do deaths of the elderly and ill. Pollutant levels rise with increasing temperatures. Water resources decline as temperatures rise, and particulate matter in the atmosphere increases with drought.

For health reasons such as these, the EPA has been charged with defining allowable limits for CO_2 emissions. Rules were first proposed in the 1990s, but industry opposition, mainly from coal and oil producers, has blocked the development and implementation of rules. In 2011 Congress threatened to slash EPA funding by one-third, in part to prevent pollution monitoring and regulation.

Part of the debate has involved the belief that economic growth requires constant increases in fossil fuel use. That assumption no longer holds: In recent years, emissions in many regions have stabilized while GDP has continued to rise. Germany, an industrial producer, reduced its greenhouse gas emissions 28 percent between 1991 and 2016, while its GDP grew nearly 50 percent. Objections also arise in states and industries whose economies depend on fossil fuels. Transitioning away from a lucrative industry is always difficult, even when the industry endangers public health.

The Supreme Court has charged the EPA with controlling greenhouse gases

The question of whether the EPA should regulate greenhouse gases was so contentious that it went to the Supreme Court in 2007. The Court ruled that it was the EPA's responsibility to limit these gases, on the grounds that greenhouse gases endanger public health and welfare within the meaning of the Clean Air Act. The Court, and subsequent EPA documents, noted that these risks include increased drought, more frequent and intense heat waves and wildfires, sea level rise, and harm to water resources, agriculture, wildlife, and ecosystems.

In addition to these risks, the U.S. military has cited climate change as a growing security threat. A coalition of generals and admirals signed a report from the Center for Naval Analyses stating that climate change "presents significant national security challenges," including violence resulting from scarcity of water and migration from sea level rise and crop failure.

Since the Supreme Court ruling, the EPA is charged with regulating six greenhouse gases: carbon dioxide, methane, nitrous oxide, hydrofluorocarbons, perfluorocarbons, and sulfur hexafluoride. These are gases whose emissions have grown dramatically in recent decades.

Three of these six gases contain halogens, a group of lightweight, highly reactive elements (fluorine, chlorine, bromine, and iodine). These gases are far more potent greenhouse gases per molecule than CO_2 (fig. 10.10). Because they are generally toxic in their elemental form, they are commonly used as fumigants and disinfectants, but they also have hundreds of uses in industrial and commercial products. Chlorofluorocarbons (CFCs) have been banned for most uses in industrialized countries, but about 600 million tons of these compounds are used annually worldwide in spray propellants, in refrigeration compressors, and for foam blowing. They diffuse into the stratosphere, where they release chlorine and fluorine atoms, which destroy the ozone shield that protects the earth from ultraviolet radiation.

How do we reduce emissions of greenhouse gases? Reducing fuel use through conservation and alternative energy is a first step.

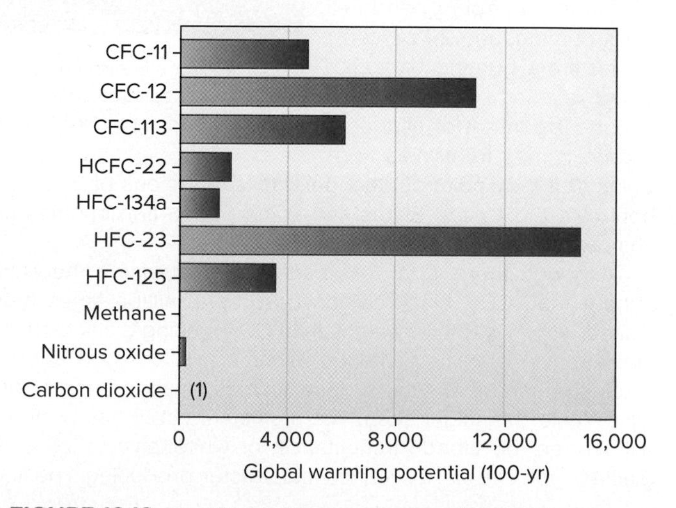

▲ **FIGURE 10.10** The Montreal Protocol has contributed to reducing greenhouse gases, as well as preserving stratospheric ozone, because CFCs have high global warming potential and longevity compared to CO_2.

EXPLORING Science

Black Carbon

Most of the air pollution in Delhi, India, described in the case study for this chapter, was particulate matter: that is, tiny, airborne fragments of matter mostly produced by combustion of organic material. A major source is fossil fuel burning, in power plants, factories, and vehicles, but other key sources include biofuels (such as wood or dung), and burning of forests and crop residue. All these types of combustion produces smoke, soot, or ash that can linger in the atmosphere for days or weeks.

As we've already discussed, fine particulates are hazardous for human health. The most dangerous particles are less than 2.5 micrometers, or 0.0025 mm, small enough to penetrate deep into the lungs. Called PM2.5, these tiny fragments damage sensitive epithelial linings and obstruct breathing, leading to asthma, heart disease, and other conditions. Also important are slightly larger but still breathable particulates 10 micrometers (0.01 mm) or less in diameter called PM10.

Still larger particles of soot, smoke, and ash, known as black carbon, have important environmental effects. They block sunlight, affecting plant growth, but even more important, they are heavy enough to fall to the earth's surface, where they are very effective at absorbing solar energy. Human activities release an estimated 6 to 8 million metric tons of black carbon per year. Approximately 40 percent is from fossil fuel combustion, 40 percent is from open biomass burning, and 20 percent comes from burning biofuels. Some researchers report that about 60 percent of all black carbon comes from India and China. But they point out that per-capita emissions of black carbon from the United States and some European countries are comparable to those in South and East Asia.

▲ Black carbon in soot and smoke from fires, power plants, and diesel engines contributes to both health problems and global warming.

Although there is far less particulate material in the atmosphere than CO_2, black carbon absorbs a million times more solar energy, weight for weight, than CO_2, making black carbon a critical driver of climate change. When it falls on snow and ice, black carbon darkens the surface, increasing absorption of sunlight. While this might seem like a small effect, it has profound impacts on the albedo (reflectance, or whiteness) of ice caps, glaciers, and sea ice. Darker ice melts faster, producing a positive freedback: As the snow and ice melt, the carbon on the surface is concentrated, absorbing more energy and causing further melting. Eventually, bare rock or open water is exposed, which absorbs even more sunlight and warms the climate further.

A recent study found that Greenland's ice and snow have been darkening significantly over the past 20 years. Albedo could decrease 10 percent or more by the end of this century. The black carbon causing melting isn't all from recent combustion. Smoke, soot, and ash from ancient forest fires and volcanic activity are stored in glaciers and snowfields. As ice and snow melt, some of this stored carbon is washed away, but as much as one-third remains on the surface and gradually builds up to enhance further melting. Research by James Hansen and coauthors at the National Oceanic and Atmospheric Administration (NOAA) suggests that black carbon could be responsible for more than 30 percent of Arctic warming.

What can we do about this problem? Since 40 percent of black carbon comes from biomass burning, a first step would be to reduce burning forests to clear land, and burning crop residue. Another 40 percent comes from fossil fuel combustion, so targeting brick kilns and coke ovens, inefficient stoves, and diesel vehicles everywhere can help reduce black carbon emissions. Black carbon releases also can be reduced by requiring cleaner fuel and new standards for engines and boilers. The U.S. EPA estimates that a switch to fuel-efficient automobiles and trucks (especially electric ones) could cut U.S. transportation emissions by 86 percent. Ships burning heavy bunker fuel are especially polluting, and more of them sailing through the Arctic as sea lanes open up will deposit heavy soot more directly on Arctic ice cover. Requiring these ships to burn cleaner fuel is imperative. Similarly, switching from coal-fired power plants to solar and wind energy would slash emissions. Because particulates remain in the atmosphere for a relatively short time, cutting emissions will immediately reduce the rate of warming, particularly the rapidly changing Arctic, in addition to improving health outcomes around the world. Isn't it time to do something about this problem?

Changing subsidy systems that support coal burning is another. A cap-and-trade system, involving a market for trading in emission rights, or "credits," has been the most popular strategy (see section 10.4), but markets have generally failed to set a price that effectively discourages CO_2 production. Carbon taxes, or fees charged on the sale of fossil fuels, are widely supported. Because Congress refused to consider cap-and-trade systems or carbon taxes for reducing greenhouse gases, President Obama turned to regulatory limits. Under the Clean Power Rule the EPA set CO_2 limits on power plants, with a target of 30 percent lower emissions by 2030 than in 2005 (see chapter 9). However, administrative actions can easily be overturned. The Trump administration eliminated this rule immediately upon taking office in 2017.

CFCs also destroy ozone in the stratosphere

Controlling chlorofluorocarbon emissions is critical for the climate, but concern about CFCs initially focused on an entirely different issue: stopping the growth of an ozone "hole" (a region of reduced concentration) in the stratosphere (fig. 10.11). This phenomenon was discovered in 1985 but has probably been developing since at least the 1960s. Chlorine-based aerosols, such as chlorofluorocarbons (CFCs) and hydrochlorofluorocarbons (HCFCs), are the principal agents of ozone depletion. Nontoxic, nonflammable, chemically inert, long-lasting, and cheaply produced, these compounds were extremely useful as industrial gases and in refrigerators, air conditioners, Styrofoam insulation, and aerosol spray cans for many years. From the 1930s until the 1980s, CFCs were used all over the world and widely dispersed through the atmosphere.

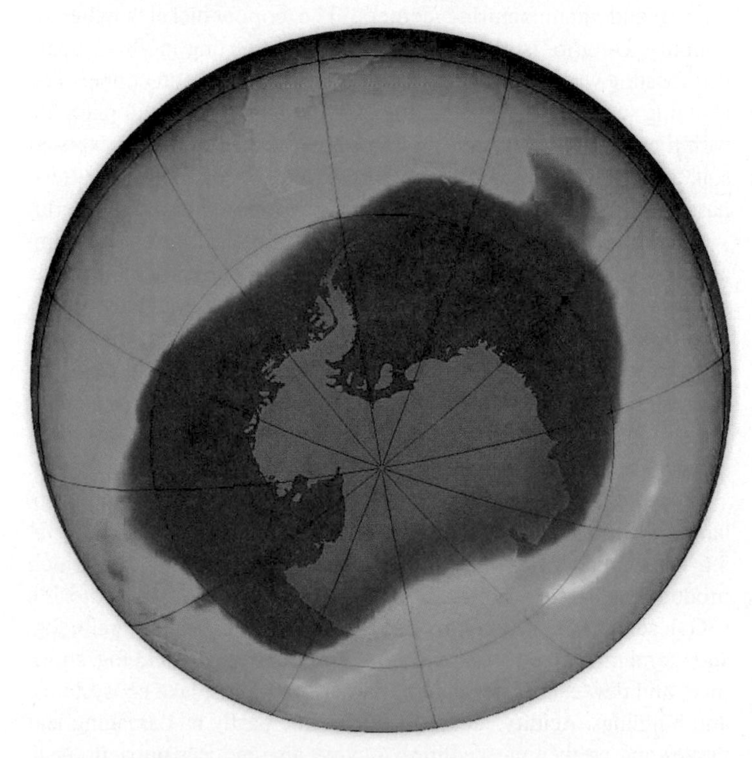

▲ **FIGURE 10.11** Reduced concentrations of stratospheric ozone, which occur in the southern springtime, are shown by purple shading. This 2006 ozone "hole" was the largest recorded, although CFC production has been declining since 1990. Source: NASA.

TABLE 10.3 Stratospheric Ozone Destruction by Chlorine Atoms and UV Radiation

STEPS	PRODUCT
1. $CFCl_3$ (chlorofluorocarbon) + UV energy	$CFCl_2 + Cl$
2. $Cl + O_3$	$ClO + O_2$
3. O_2 + UV energy	$2O$
4. $ClO + 2O$	$O_2 + Cl$
5. Return to step 2	

Source: UNEP

Ozone (O_3) is a pollutant near the ground because it irritates skin and plant tissues, but ozone in the stratosphere is valuable. The O_3 molecule is especially effective at absorbing ultraviolet (UV) radiation as it enters the atmosphere from space. UV radiation damages plant and animal cells, potentially causing mutations that produce cancer. A 1 percent loss of ozone could result in about a million extra human skin cancers per year worldwide. Excessive UV exposure could reduce agricultural production and disrupt ecosystems. Scientists worry, for example, that high UV levels in Antarctica could reduce populations of plankton, the tiny floating organisms that form the base of a food chain that includes fish, seals, penguins, and whales in Antarctic seas.

Antarctica's exceptionally cold winter temperatures (−85° to −90°C) help break down ozone. During the long, dark winter months, strong winds known as the circumpolar vortex circle the pole. These winds isolate Antarctic air and allow stratospheric temperatures to drop low enough to create ice crystals at high altitudes—something that rarely happens elsewhere in the world. Ozone and chlorine-containing molecules are absorbed on the surfaces of these ice particles. When the sun returns in the spring, it provides energy to liberate chlorine ions, which readily bond with ozone, breaking it down to molecular oxygen (table 10.3). It is only during the Antarctic spring (September through December) that conditions are ideal for rapid ozone destruction. During that season, temperatures are still cold enough for high-altitude ice crystals, but the sun gradually becomes strong enough to drive photochemical reactions.

As the Antarctic summer arrives, temperatures warm slightly. The circumpolar vortex weakens, and air from warmer latitudes mixes with Antarctic air, replenishing ozone concentrations in the ozone hole. Slight decreases worldwide result from this mixing, however. Ozone re-forms naturally, but not nearly as fast as it is destroyed. Since the chlorine atoms are not themselves consumed in reactions with ozone, they continue to destroy ozone for years, until they finally precipitate or are washed out of the air.

CFC control has had remarkable success

The discovery of stratospheric ozone losses brought about a remarkably quick international response. In 1987 an international meeting in Montreal, Canada, produced the Montreal Protocol, the first of several major international agreements on phasing out most use of CFCs by 2000. As evidence accumulated, showing that losses were larger and more widespread than previously thought, the deadline for the elimination of all CFCs (halons, carbon tetrachloride, and methyl chloroform) was moved up to 1996, and a $500 million fund

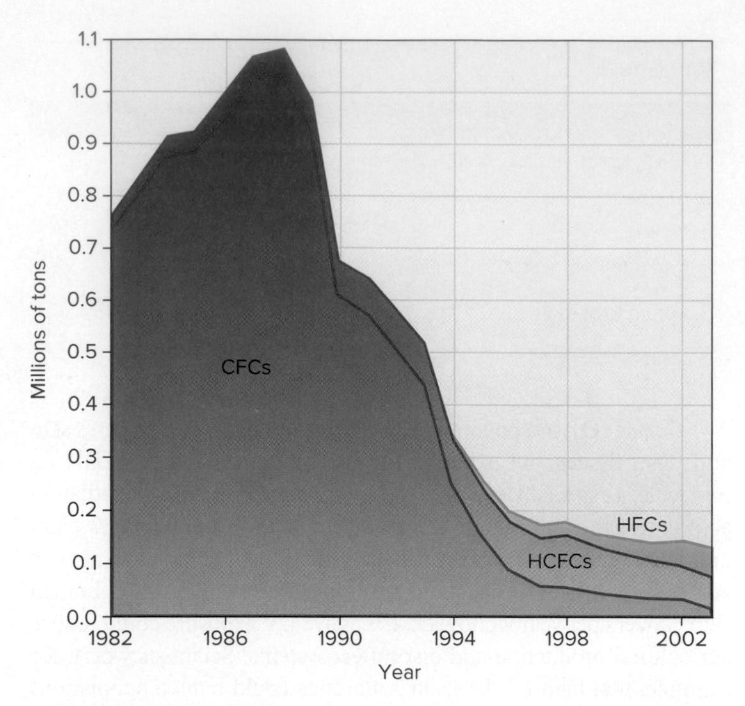

▲ **FIGURE 10.12** The Montreal Protocol has been remarkably successful in eliminating CFC production. The remaining HFC and HCFC use is primarily in recently industrialized countries, such as China and India.

was established to assist poorer countries in switching to non-CFC technologies. Fortunately, alternatives to CFCs for most uses already exist. The first substitutes are hydrochlorofluorocarbons (HCFCs), which release much less chlorine per molecule. Many refrigerant and air conditioning systems also use ammonia (NH_3), which is much safer for the climate.

There is evidence that the CFC ban is already having an effect. CFC production in most industrialized countries has fallen sharply since 1988 (fig. 10.12), and CFCs are now being removed from the atmosphere more rapidly than they are being added. In 50 years or so, stratospheric ozone levels are expected to be back to normal.

The Montreal Protocol has also helped reduce climate change, because CFCs and related compounds are such powerful and long-lasting greenhouse gases.

10.3 ENVIRONMENTAL AND HEALTH EFFECTS

- *Acid deposition results from sulfur or nitrogen emissions.*
- *Inversions concentrate pollutants near the ground.*
- *Heat islands and smog directly affect human health.*

Consequences of breathing dirty air include increased probability of heart attacks, respiratory diseases, and lung cancer. This can mean as much as a 5- to 10-year decrease in life expectancy if you live in the worst parts of Los Angeles or Baltimore, for example. Of course, the intensity and duration of your exposure, as well as your age and general health, are extremely important: You are much more likely to be at risk if you are very young, very old, or already suffering from some respiratory or cardiovascular disease. Bronchitis and emphysema are common chronic conditions resulting from air pollution. The U.S. Office of Technology Assessment estimates that 250,000 people suffer from pollution-related bronchitis and emphysema in the United States, and some 50,000 excess deaths each year are attributable to complications of these diseases, which are probably second only to heart attack as a cause of death.

How does air pollution cause these health effects? Because they are strong oxidizing agents, sulfates, SO_2, NO_x, and O_3 irritate and damage delicate tissues in the eyes and lungs. Fine, suspended particulate materials penetrate deep into the lungs, causing irritation, scarring, and even tumor growth. Heart stress results from impaired lung functions. Carbon monoxide binds to hemoglobin, reducing oxygen flow to the brain. Headaches, dizziness, and heart stress result. Lead also binds to hemoglobin, damaging critical neurons in the brain and resulting in mental and physical impairment and developmental retardation.

Environmental effects of air pollution can involve similarly complex, and often gradually accumulating, effects.

Acid deposition results from SO_4 and NO_x

Deposition of acidic droplets or particles, from rain, fog, snow, or aerosols in the atmosphere, became recognized as a widespread pollution problem only since the 1980s. But the effects of air pollution on plants, especially sulfuric acid deposition, have been known in industrial areas since at least the 1850s. In the early days of industrialization, fumes from furnaces, smelters, refineries, and chemical plants destroyed vegetation and created desolate, barren landscapes around mining and manufacturing centers. The copper-nickel smelter at Sudbury, Ontario, is a spectacular example. Starting in 1886, open-bed roasting was used to purify sulfide ores of nickel and copper. The resulting sulfur dioxide and sulfuric acid destroyed nearly all plant life within about 30 km of the smelter. Rains washed away the exposed soil, leaving a barren moonscape of blackened bedrock. This pattern has been widespread in mining and smelting regions around the world.

Pollutant levels too low to produce visible symptoms of damage may still have important effects. Field studies show that yields in some crops, such as soybeans, may be reduced as much as 50 percent by currently existing levels of oxidants in ambient air. Some plant pathologists suggest that ozone and photochemical oxidants are responsible for as much as 90 percent of agricultural, ornamental, and forest losses from air pollution. The total costs of this damage may be as much as $10 billion per year in North America alone.

Acidic deposition is now understood to affect forests and croplands far from industrial centers. Rain is normally slightly acidic (pH 5.6: see chapter 2), owing to reactions of CO_2 and rainwater, which produce a mild carbonic acid. Industrial emissions of sulfur dioxide (SO_2), sulfate (SO_4), and nitrogen oxides (NO_x) can acidify rain, fog, snow, and mist to pH 4 or lower. Ongoing exposure to acid fog, snow, mist, and dew causes permanent damage to plants, lake ecosystems, and buildings. Acidity causes forest decline partly by damaging leaf tissues and weakening seedlings. Acidity also reduces nutrient availability in forest soils, and it mobilizes toxic concentrations of metals in soils, especially aluminum. Weakened trees become susceptible to other stressors such as diseases and insect pests.

▲ **FIGURE 10.13** A Fraser fir forest on Mount Mitchell, North Carolina, killed by acid rain, insect pests, and other stressors. ©William P. Cunningham

Lakes in Scandinavia were among the first aquatic ecosystems discovered to be damaged by acid precipitation. Prevailing winds from Germany, Poland, and other parts of Europe deliver acids generated by industrial and automobile emissions–principally H_2SO_4 and HNO_3. The thin, acidic soils and nutrient-poor lakes and streams in the mountains of southern Norway and Sweden have been severely affected by this acid deposition. Most noticeable is the reduction of trout, salmon, and other game fish, whose eggs and young die below pH 5. Aquatic plants, insects, and invertebrates also suffer. Many lakes in Sweden are now so acidic that they will no longer support game fish or other sensitive aquatic organisms. Large parts of Europe and eastern North America have also been damaged by acid precipitation.

High-elevation forests are most severely affected. Mountain tops often have thin, often acidic soils under normal conditions, with little ability to neutralize acidic rain, snow, and mist. On Mount Mitchell in North Carolina, nearly all the trees above 2,000 m (6,000 ft) have lost needles, and about half are dead (fig. 10.13). Damage has been reported throughout Europe, from the Netherlands to Switzerland, as well as in China and the states of the former Soviet Union. In 1985 West German foresters estimated that about half the total forest area in West Germany (more than 4 million ha) was declining. The loss to the forest industry is estimated to be about 1 billion Euros per year.

A vigorous program of pollution control has been undertaken by Canada, the United States, and several European countries since the widespread recognition of acid rain. SO_2 and NO_x emissions from power plants have decreased dramatically over the past three decades over much of Europe and eastern North America as a result of pollution-control measures (fig. 10.14).

Although acid precipitation has decreased dramatically in the United States, it remains high in other places. In cities throughout the world, air pollution is destroying some of the oldest and most glorious buildings and works of art. Smoke and soot coat buildings, paintings, and textiles. Acids dissolve limestone and marble, destroying features and structures of historic buildings (fig. 10.15). The Parthenon in Athens, the Taj Mahal in Agra, the Coliseum in Rome, and medieval cathedrals in Europe are slowly dissolving and flaking away because of acidic fumes in the air. Acid deposition also speeds corrosion of steel in reinforced concrete, weakening buildings, roads, and bridges. Limestone, marble, and some kinds of sandstone flake and crumble.

Urban areas endure inversions and heat islands

In urban areas, pollution is most extreme when temperature inversions develop, concentrating dangerous levels of pollutants within cities (as in the opening case study). A temperature inversion is a situation in which stable, cold air rests near the ground, with warm layers above (fig. 10.16). This situation reverses the normal conditions: Usually, air is warmed by heat re-emitted from the ground surface, and it cools with elevation above the earth's surface. Warming air rises, mixing the atmospheric layers and helping pollution to disperse from its sources. When cool, dense air lies below a warmer, lighter layer, air remains stable and still, and pollutants accumulate near the ground, where they irritate

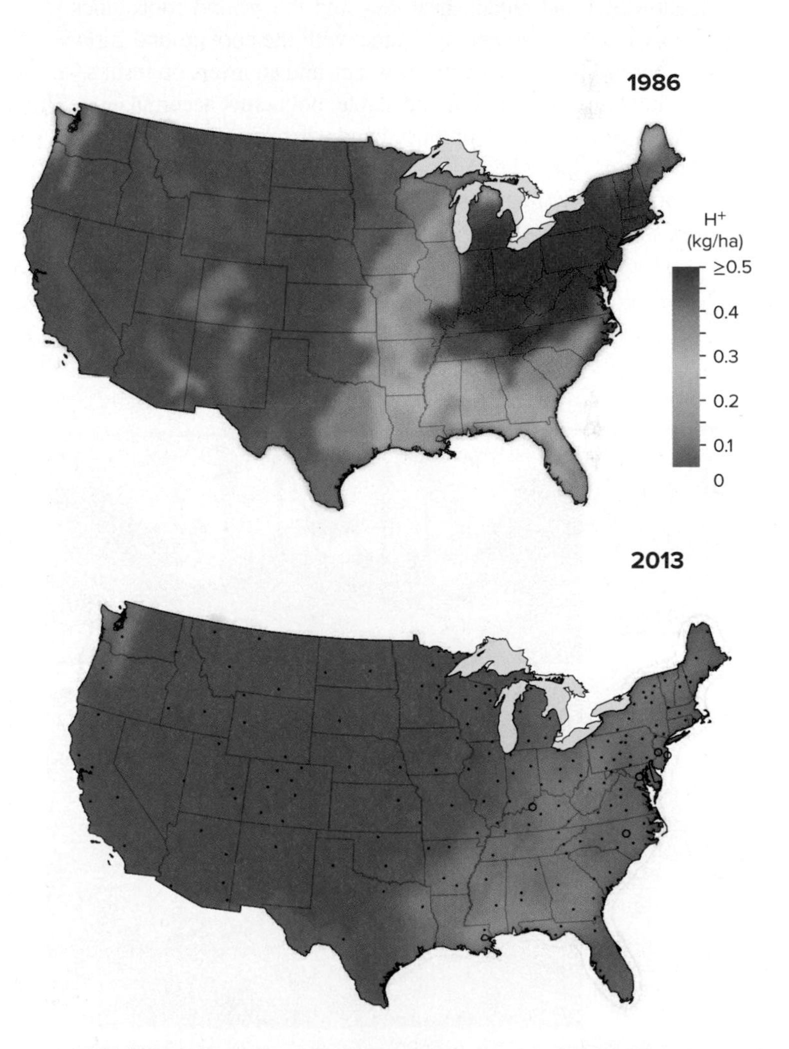

▲ **FIGURE 10.14** Acid precipitation over the United States in 1986 (a) and 2013 (b). Controls on sulfur dioxide emissions have been one of the greatest successes in pollution control. Source: National Atmospheric Deposition Program/National Trends Network http://nadp.isws.illinois.edu.

▶ **FIGURE 10.15** Atmospheric acids, especially sulfuric and nitric acids, have almost eaten away the face of this medieval statue. Annual losses from air pollution damage to buildings and materials amount to billions of dollars. ©Ryan McGinnis/Alamy Stock Photo

our lungs and eyes. Stable inversion conditions are usually created by rapid nighttime cooling, especially in a valley, where air movement is restricted.

Los Angeles has ideal conditions for inversions. Mountains surround the city on three sides, reducing wind movement; heavy traffic and industry create a supply of pollutants; skies are generally clear at night, allowing rapid radiant heat loss, and the ground cools quickly. Surface air layers are cooled by contact with the cool ground surface, while upper layers remain relatively warm, and an inversion results. As long as the atmosphere is still and stable, pollutants accumulate near the ground, where they are produced and where we breathe them.

Abundant sunlight in Los Angeles initiates photochemical oxidation in the concentrated aerosols and gaseous chemicals in the inversion layer. A brown haze of ozone and nitrogen dioxide quickly develops. Although recent air quality regulations have helped tremendously, on summer days, ozone concentrations in the Los Angeles basin still can reach unhealthy levels.

Heat islands and dust domes occur in cities even without inversion conditions. With their low albedo, concrete and brick surfaces in cities absorb large amounts of solar energy. A lack of vegetation or water results in very slight evaporation (latent heat production); instead, available solar energy is turned into heat. As a result, temperatures in cities are frequently 3° to 5°C (5° to 9°F) warmer than in the surrounding countryside, a condition known as an urban heat island. Tall buildings create convective updrafts that sweep pollutants into the air. Stable air masses created by this heat island over the city concentrate pollutants in a dust dome.

Smog and haze reduce visibility

Smog doesn't just afflict cities, such as Delhi, India. We have only recently realized that pollution also affects rural areas. Even supposedly pristine places such as our national parks are suffering from air pollution. Grand Canyon National Park, where maximum visibility used to be 300 km (185 mi), is now so smoggy on some winter days that visitors can't see the opposite rim only 20 km (12.5 mi) across the canyon. Mining operations, smelters, and power plants (some of which were moved to the desert to improve air quality in cities such as Los Angeles) are the main culprits (fig. 10.17).

Huge regions are affected by pollution. A gigantic "haze blob" as much as 3,000 km (about 2,000 mi) across covers much of the eastern United States in the summer, cutting visibility as much as 80 percent.

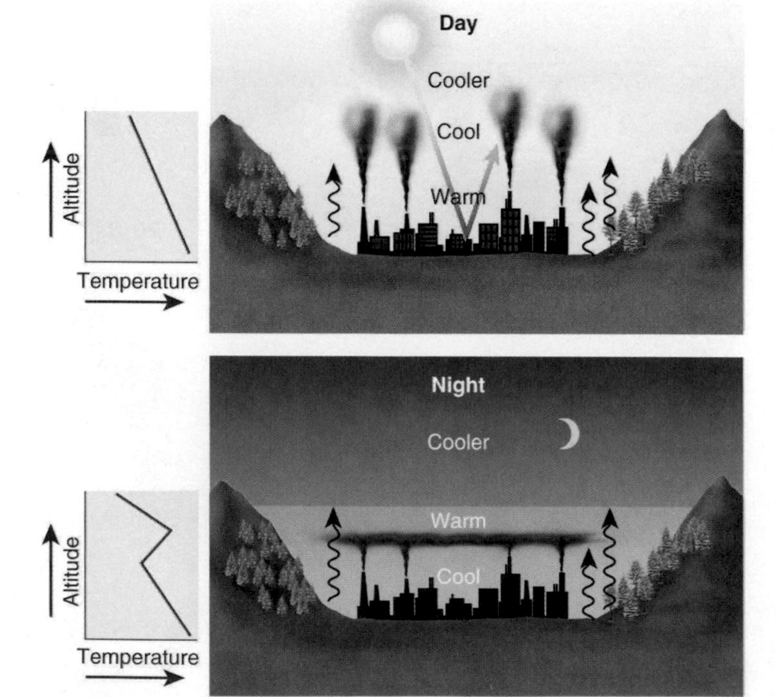

▲ **FIGURE 10.16** Atmospheric temperature inversions occur where ground-level air cools more quickly than upper air. With cold air resting below the warmer air, there is little mixing, and pollutants are trapped near the ground.

▲ **FIGURE 10.17** Reduced visibility is a widespread impact of air pollutants, as on this Los Angeles afternoon. ©Jacobs Stock Photography/Getty Images

People become accustomed to these conditions and don't realize that the air once was clear. Studies indicate, however, that if all human-made sources of air pollution were shut down, the air would clear up in a few days, and there would be about 150-km (90-mi) visibility nearly everywhere, rather than the 15 km to which we have become accustomed.

Recent studies have found that rural sources also impact urban areas. In some parts of Los Angeles, for example, ammonia (NH_3) drifting from nearby dairy feedlots causes as much smog as cars do. Dust, pulverized manure, and methane are additional airborne pollutants that drift from dairy feedlots into the city. Windblown dust from cultivated fields is also a dominant cause of impaired visibility in some regions, especially during spring and summer.

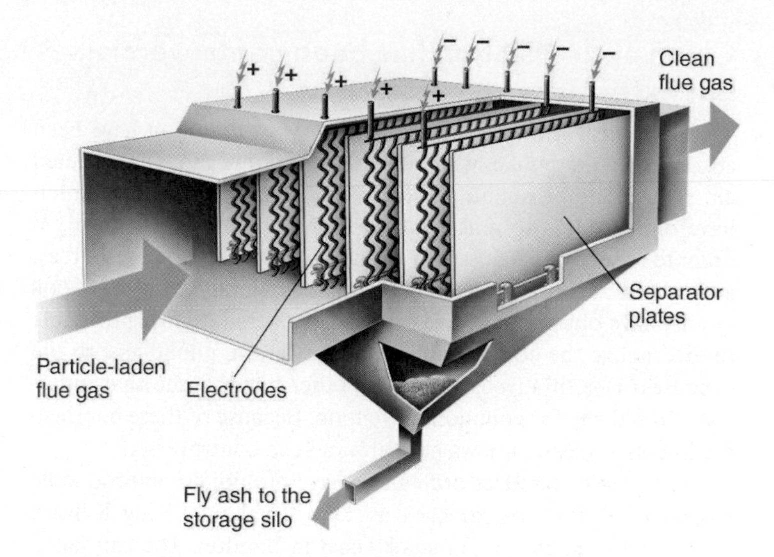

▲ **FIGURE 10.18** An electrostatic precipitator can remove 99 percent of unburned particulates in the effluent (smoke) from power plants. Electrodes transfer a static electric charge to dust and smoke particles, which then adhere to collector plates. Particles are then shaken off of the plates and collected for re-use or disposal.

10.4 AIR POLLUTION CONTROL

- *Pollutants can be captured after production or avoided.*
- *Clean air legislation has reduced CO and SO_4 but not NO_x.*
- *Data show that the CAA has saved money and has not diminished economic growth.*

"Dilution is the solution to pollution": This catch phrase has long characterized our main approach to air pollution control. Tall smokestacks were built to send emissions far from the source, where they became difficult to detect or trace to their source. With increasing global industrialization, though, dilution is no longer an effective strategy. We have needed to find different methods for pollution control.

The best strategy is reducing production

Since most air pollution in the developed world is associated with transportation and energy production, the most effective strategy would be conservation: Reducing electricity consumption, insulating homes and offices, and developing better public transportation, and more alternative energy, all greatly reduce air pollution at the source. Pollutants can also be captured from effluent after burning.

Particulate removal involves filtering air emissions. Filters trap particulates in a mesh of cotton cloth, spun glass fibers, or asbestos-cellulose. Industrial air filters are generally giant bags through which effluent gas is blown, much like the bag on a vacuum cleaner. Bags can be huge, 10 to 15 m long and 2 to 3 m wide. In power plants, electrostatic precipitators are the most common particulate controls (fig. 10.18). Ash particles pick up an electrostatic surface charge as they pass between large electrodes. The electrically charged particles then precipitate (collect) on an oppositely charged collecting plate. These precipitators consume a large amount of electricity, but maintenance is relatively simple, and collection efficiency can be as high as 99 percent. The ash collected by both of these techniques is sometimes reusable as construction material, but often it is hazardous waste because it contains mercury, lead, or arsenic captured from coal smoke. This waste must be buried in landfills.

Sulfur removal is important because sulfur oxides are among the most damaging of all air pollutants for human health, infrastructure, and ecosystems. Switching from soft coal with a high sulfur content to low-sulfur coal is the surest way to reduce sulfur emissions. But high-sulfur coal is often used for political reasons or because it is cheap. In the United States, high-sulfur coal comes mainly from Appalachia, a region with political leverage because of its chronic poverty and its powerful coal interests (which now use the controversial process of mountaintop removal to mine coal: see discussion in chapter 12). In China, much domestic coal is rich in sulfur, and companies use it because it is both cheap and convenient. Coal can also be cleaned: It can be crushed, washed, and gasified to remove sulfur and metals before combustion. These measures improve heat efficiency, but they also replace air pollution with solid waste and water pollution, and they are expensive.

Sulfur can be extracted after combustion with catalytic converters, which oxidize or reduce sulfur in effluent gas. Residues, including elemental sulfur, sulfuric acid, and ammonium sulfate can be marketable products. This approach can help companies make money instead of waste, but markets must be reasonably close, and the product must be pure enough for easy reuse.

Nitrogen oxides (NO_x) can be reduced in both internal combustion engines and industrial boilers by as much as 50 percent by carefully controlling the flow of air and fuel. Staged burners, for example, control burning temperatures and oxygen flow to prevent formation of NO_x. The **catalytic converter** on your car uses platinum-palladium and rhodium catalysts to remove up to 90 percent of NO_x, hydrocarbons, and carbon monoxide at the same time.

Hydrocarbon controls mainly involve complete combustion or the control of evaporation. Hydrocarbons and volatile organic compounds are produced by incomplete combustion of fuels or by solvent evaporation from chemical factories, paints, dry cleaning, plastic manufacturing, printing, and other industrial processes. Closed systems that prevent escape of fugitive gases can reduce many of these emissions. Controlling leaks from industrial valves, pipes, and storage tanks can have a significant impact on air quality. Afterburners are often the best method for destroying volatile organic chemicals in industrial exhaust stacks.

Clean air legislation has been controversial but extremely successful

Through most of human history, the costs of pollution have been absorbed by the public, which breathes or grows crops in polluted air, rather than by polluters themselves. Rules to control pollution have often sought to make polluters pay for pollution control, in order to reduce public losses in sickness, death, degraded resources, and other costs associated with pollution. Naturally, emitters of pollutants have objected to these rules. It is easier and more profitable to externalize the costs of pollution, and let the public absorb the expenses. Health advocates, on the other hand, argue that industries should pay for pollution prevention. Because of these contrasting interests, clean air laws have always been controversial.

Over time countless ordinances have prohibited objectionable smoke and odors. As far back as 1306, England's King Edward tried to ban the burning of smoky coal in London. The ban failed but was attempted repeatedly over the centuries. It wasn't until 1956 that better enforcement–triggered by an episode of unusually lethal air quality in 1952–and alternative fuels finally made clean air rules effective in London.

In the United States, the Clean Air Act of 1963 was the first national law for air pollution control. The act provided federal grants to states to combat pollution but was careful to preserve states' rights to set and enforce their own standards for air quality and enforcement. Because this approach was uneven and difficult to enforce, the act was largely rewritten in the 1970 Clean Air Act amendments, standardizing policies nationally. These amendments identified the "criteria" pollutants discussed earlier. The 1970 rules also established primary standards, intended to protect human health, and secondary standards, to protect materials, crops, climate, visibility, and personal comfort.

One of the most contested aspects of the act is the "new source review," which was established in 1977. This provision was originally adopted because industry argued that it would be intolerably expensive to install new pollution-control equipment on old power plants and factories that were about to close down anyway. Congress agreed to "grandfather," or exempt, existing equipment from new pollution limits with the stipulation that when they were upgraded or replaced, more stringent rules would apply. The result has been that owners have kept old facilities operating precisely because they were exempted from pollution control. In fact, corporations have poured millions into aging power plants and factories, expanding their capacity rather than build new ones. Decades later, many of those grandfathered plants are still going strong and continue to be among the biggest contributors to smog and acid rain.

The 1990 amendments included major changes in incentives as well as rules for additional pollutants. Among the major provisions were establishment of new controls for ozone-depleting CFCs, new rules for controlling emissions of benzene, chloroform, and other hazardous air pollutants, and a requirement that comprehensive federal and state standards be set for both industrial and transportation-based sources of common pollutants. The 1990 amendments provided incentives and rules to support development of alternative fuels and technology. These amendments also established the EPA's right to fine violators of air pollution standards.

Trading pollution credits is one approach

The 1990 revisions also created new incentives for pollution control. One of these is a market-based cap-and-trade system. In this approach, the EPA sets maximum emission levels for pollutants. Facilities can then buy and sell emission "credits," or permitted allotments of pollutants. Companies can decide if it's cheaper to install pollution control equipment or to simply buy someone else's credits.

Cap-and-trade has worked well for sulfur dioxide. When trading began in 1990, economists estimated that eliminating 10 million tons of sulfur dioxide would cost $15 billion per year. Left to find the most economical ways to reduce emissions, however, utilities have been able to reach clean air goals for one-tenth that price. A serious shortcoming of this approach is that while trading has resulted in overall pollution reduction, some local "hot spots" remain where owners have found it cheaper to pay someone else to reduce pollution than to do it themselves. This has been the approach adopted for CO_2 and for mercury, among other pollutants.

Presidents and EPA administrators have varied greatly in their enthusiasm for enforcing rules. Business-friendly administrations tend to rely on voluntary emissions controls and a trading program for air pollution allowances. Administrations more focused on public health have sought enforcement of rules, because voluntary action alone rarely reduces pollution.

Amendments have involved acrimonious debate, with bills sometimes languishing in Congress for years because of disputes over burdens of responsibility and cost and definitions of risk. A 2002 report concluded that simply by enforcing existing clean air legislation, the United States could save at least 6,000 lives per year and prevent 140,000 asthma attacks.

Despite controversies, the Clean Air Act has been tremendously successful (fig 10.19). Measured only in economic terms, a comparison of the costs of regulation (about $50 billion) and the economic benefits of reduced illness, property damage, and

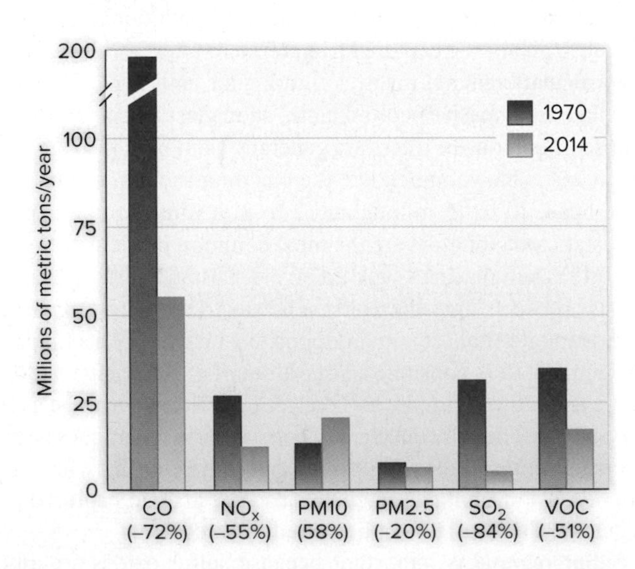

▲ **FIGURE 10.19** Air pollution trends in the United States, 1970 to 2014. Since Congress passed the Clean Air Act , most criteria air pollutants decreased significantly, even while population has increased. Pollution protections also aid the economy by reducing illness and creating jobs in pollution prevention. Source: Environmental Protection Agency, 2016.

increased productivity (about $1,300 billion), the economic benefits had outweighed costs by more than 25 to 1 by 2010 (Key Concepts, p. 248–249). Like automobile industry officials who argued that seat belts in cars were technologically infeasible, unnecessary, and costly, opponents of pollution regulations have often argued that regulations would be technologically infeasible and would hinder economic growth. The evidence does not support these arguments when cumulative costs and benefits are taken into account.

10.5 THE ONGOING CHALLENGE

- *Some highly successful air pollution control efforts in recent years have been lead, acid rain, and chlorofluorocarbons.*
- *Air quality in developing countries, such as India and China, remains problematic.*
- *Progress is possible. Some highly polluted cities in the United States and elsewhere have cleaned their air significantly.*

Although the United States has not yet achieved the Clean Air Act goals in many parts of the country, air quality has improved dramatically in the last decade in terms of the major large-volume pollutants. For 23 of the largest U.S. cities, the number of days each year in which air quality reached the hazardous level is down 93 percent from a decade ago. Of 97 metropolitan areas that failed to meet clean air standards in the 1980s, 41 are now in compliance. Eighty percent of the United States now meets the National Ambient Air Quality Standards.

Most pollutants have declined sharply since the introduction of Clean Air Act rules. Lead has been reduced by 98 percent, SO_2 by 35 percent, and CO by 32 percent. Filters, scrubbers, and precipitators on power plants and other large, stationary sources are responsible for most of the particulate and SO_2 reductions. Catalytic converters on automobiles are responsible for most of the CO and O_3 reductions. The only conventional "criteria" pollutants that have not dropped significantly are particulates and NO_x and particulate matter. Because automobiles are the main source of NO_x, this pollutant has remained high or grown in many areas. Particulate matter, mostly dust and soot produced by agriculture, fuel combustion, metal smelting, concrete manufacturing, and other activities, has also grown in some areas.

Among the areas of sharply increasing air pollutants are rural natural gas producing regions, such as central Wyoming and Colorado. Tens of thousands of gas-producing wells, many of them leaking small amounts of gas, and unregulated waste storage tanks produce ozone, volatile organic compounds, and other contaminants that can exceed some of the worst urban levels in the United States.

Pollution persists in developing areas

The major metropolitan areas of many developing countries are growing at explosive rates, and many have abysmal environmental quality. Cities in India and China have been especially notorious. China continues to have crippling smog on cold winter days and in the heat of summer, but India has overtaken China in recent years. According to the World Health Organization, 15 of the world's 30 worst cities in terms of PM2.5 levels were in India in 2017. Another five were in Pakistan or Bangladesh. Industry, agriculture, and road traffic all contribute soot, dust, metals, NO_x, and other pollutants. Airborne heavy metals drifting over farm fields contaminate food supplies. Delhi had the world's highest annual average levels of PM2.5 in 2017.

▲ **FIGURE 10.20** Growing numbers of vehicles, together with heavy industry and field burning, have increased air pollution in many parts of Asia. ©William P. Cunningham

Both India and China have sought to introduce pollution controls, but urban and rural opposition has remained strong. Factories and power plants burn low-quality coal, have few pollution controls, and continue to produce tens of millions of tons of soot and sulfur dioxide annually. At the same time, farmers continue to burn crop residues, and the number of vehicles on the roads, another dominant source of particulates and other pollutants, has climbed steadily (fig. 10.20).

Globally, the World Health Organization estimates that there are 3.12 million premature deaths per year from outdoor air pollution. Most of these are in large cities of newly industrialized regions. Another 3.38 million premature deaths are attributed to indoor air pollution. Deaths result from stroke, heart disease, and cancer as well as respiratory illnesses. South Asia and East Asia are thought to account for more than half those deaths.

Change is possible

In China, public attention to air pollution was suddenly awakened in 2015 when journalist Chai Jing released a self-produced exposé on YouTube called "Under the Dome." The movie provoked global dialog and was viewed 150 million times in just a few days before it was taken down in China by authorities. It has remained online outside of China. Many observers called the video China's "Silent Spring moment" after Rachel Carson's ground-breaking book about pesticides in the United States (see chapter 1). Carson's writing changed U.S. attitudes

KEY CONCEPTS

Can we afford clean air?

Designed to protect human health, crops, and buildings, the Clean Air Act requires the control of conventional (criteria) pollutants, metals, organic compounds, and other substances.

As part of the CAA, Congress directed the EPA to evaluate the economic costs and benefits of enforcing the act's provisions. Emitters of air pollutants have charged that emission controls are expensive, and that those costs reduce economic productivity and threaten job creation.

In the most recent of these reports, published in 2011, the EPA calculated the economic costs and benefits of the 1990 revisions to the CAA. The report is available on the EPA website. The EPA compared the economic costs that would have been incurred (costs of health care, lost productivity, infrastructure degradation, and other factors) *if we had not implemented the 1990 air pollution controls.* These costs were compared to *observed costs of implementation* of those rules.

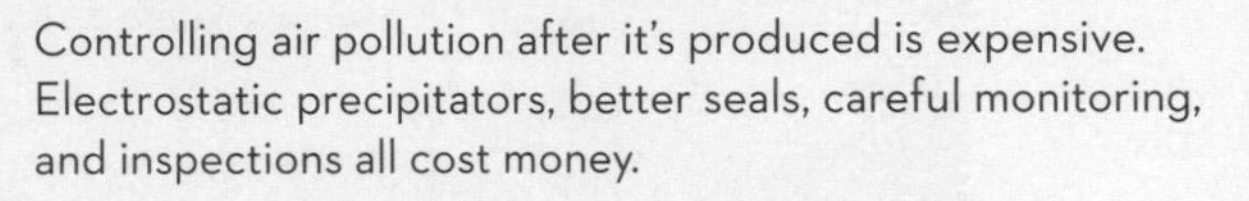

Controlling air pollution after it's produced is expensive. Electrostatic precipitators, better seals, careful monitoring, and inspections all cost money.

The EPA study found that by 2020, cumulative public and private costs of the 1990 rules amounted to **$65 billion**.

Cumulative savings amounted to **$2,000 billion**.

At this rate, can we afford not to keep our air clean?

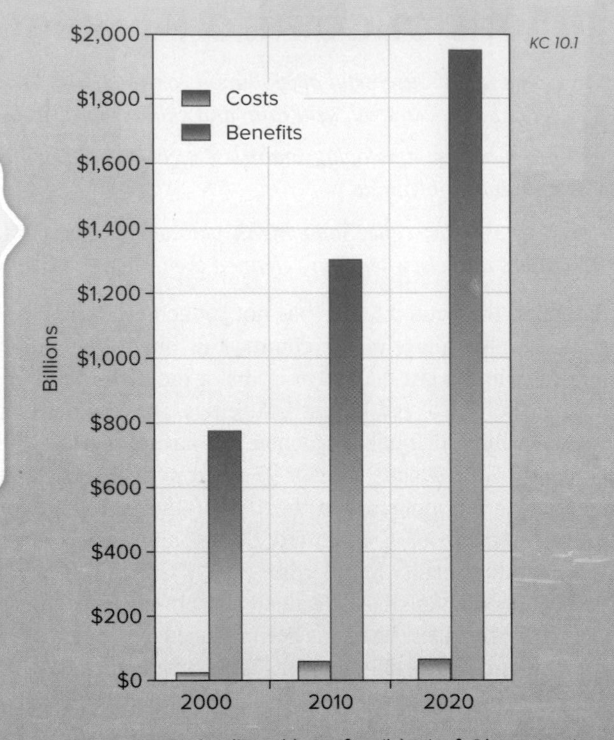

▲ Direct costs (red) and benefits (blue) of Clean Air Act provisions by 2000, 2010, and 2020, in billions of 2006 dollars. Source: EPA 2011 Clean Air Impacts Summary Report.

What are the 1990 rules?

Among the major provisions were establishment of

- controls for ozone-depleting CFCs
- a market-based "cap-and-trade" system, with emissions permits that can be sold, to reduce acid rain-producing SO_2 and NO_x
- requirement of federal and state rules for both industrial and transportation sources
- new rules for controlling emissions of benzene, chloroform, and other hazardous air pollutants
- "new source review" for newly constructed power plants and other major emitters
- the EPA's right to fine violators of air pollution standards
- a framework for developing alternative fuels and technology

What are the costs?

Direct costs of compliance by 2020 for five major source categories (red bars) and for additional minor categories (purple). Vehicles and electricity generation, our dominant sources, bear most costs. ▶

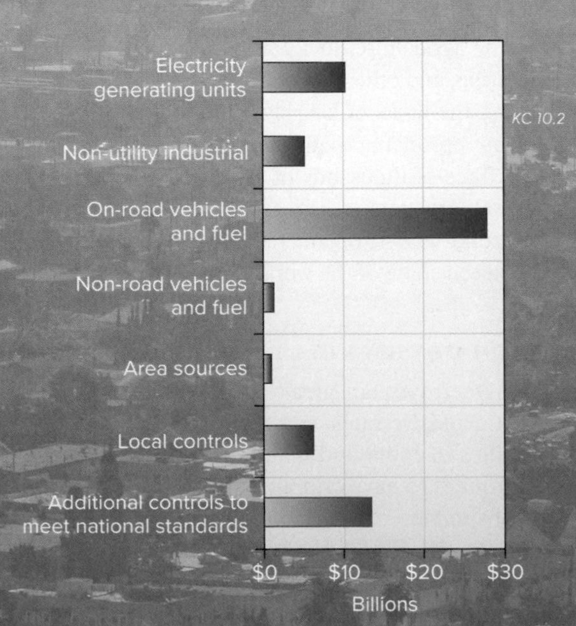

Source: 2011 Clean Air Impacts Summary Report.

©Paul Briden/Shutterstock

What are the benefits of the 1990 Clean Air Act amendments?

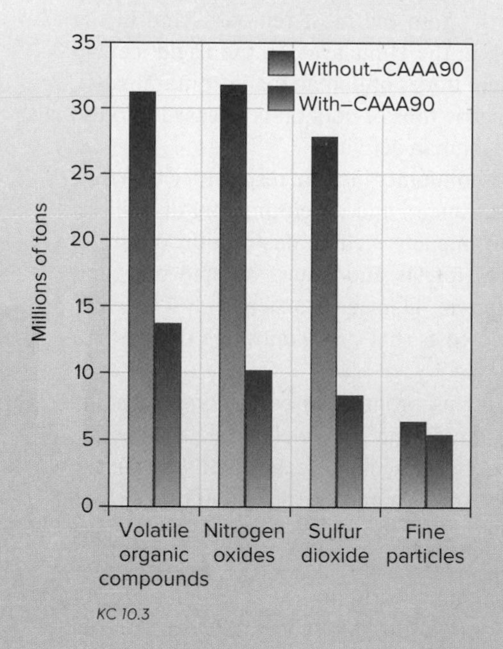

KC 10.3

Without the Clean Air Act amendments (CAAA) of 1990:

◀ Production of key pollutants would be greater in 2020.

Mortality, bronchitis, heart disease, and lost work days would all be greater. ▶

Visibility would be worse. ▼

Health Effect Reductions (PM2.5 & Ozone Only)	Pollutant(s)	Year 2010	Year 2020
PM2.5 Adult Mortality	PM	160,000	230,000
PM2.5 Infant Mortality	PM	230	280
Ozone Mortality	Ozone	4,300	7,100
Chronic Bronchitis	PM	54,000	75,000
Acute Bronchitis	PM	130,000	180,000
Heart Disease	PM	130,000	200,000
Asthma Exacerbation	PM	1,700,000	2,400,000
Hospital Admissions	PM, Ozone	86,000	135,000
Emergency Room Visits	PM, Ozone	86,000	120,000
Restricted Activity Days	PM, Ozone	84,000,000	110,000,000
School Loss Days	Ozone	3,200,000	5,400,000
Lost Work Days	PM	13,000,000	17,000,000

Without CAAA

KC 10.4

With CAAA

Visibility in Deciviews, 2020

0 3 6 9 12 15 18 21 24 27 30

Best — Worst

Impacts on the economy?

Since 1970, the six commonly found air pollutants have decreased by more than 50 percent. Hazardous air pollutants from large industrial sources, such as chemical plants, petroleum refineries, and paper mills, have declined by nearly 70 percent. Production of most ozone-depleting chemicals has ceased. New cars are more than 90 percent cleaner and will be even cleaner in the future.

At the same time, the U.S. gross domestic product (GDP) has tripled, vehicle use has doubled, and energy consumption has risen 50 percent.

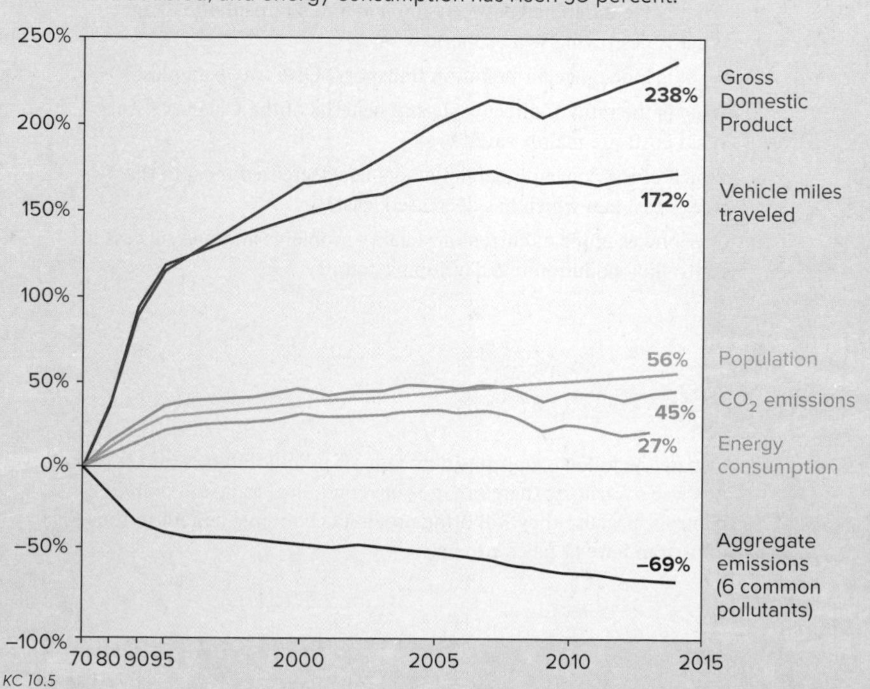

KC 10.5

CAN YOU EXPLAIN?

1. **What are several key provisions of the 1990 Clean Air Act amendments?**
2. **What is the idea of "cap-and-trade" market mechanisms?**
3. **In what ways have the amendments saved money?**

KC 10.6

about pollution, galvanized public opinion, and helped lead to bedrock environmental protections in the United States. Citizen protest of this sort often has been the catalyst that has led to action on pollution issues. Political leaders need to know that the population cares in order to push back against industry. Time will tell whether Chai Jing will be China's Rachel Carson.

Despite its current challenges, Delhi has also shown it can make progress. In the 1990s the city required catalytic converters on vehicles, and in 2002 buses, taxis, and motor rickshaws were ordered to switch to compressed natural gas. Sulfur dioxide and carbon monoxide levels dropped 80 percent and 70 percent, respectively. Particulate emissions dropped by 50 percent. Conditions have since worsened, but this history shows how effective pollution control policies can be when they are enforced.

Political change can also lead to better air quality. Decades ago, Cubatao, Brazil, was described as the "Valley of Death," one of the most dangerously polluted places in the world. A steel plant, a huge oil refinery, and fertilizer and chemical factories churned out thousands of tons of air pollutants every year that were trapped between onshore winds and the uplifted plateau on which São Paulo sits. Trees died on the surrounding hills. Birth defects and respiratory diseases were alarmingly high. Since then, however, the citizens of Cubatao have made remarkable progress in cleaning up their environment. The end of military rule and restoration of democracy allowed residents to publicize their complaints. The environment became an important political issue. The state of São Paulo invested about $100 million and the private sector spent twice as much to clean up most pollution sources in the valley. Particulate pollution was reduced 75 percent, ammonia emissions were reduced 97 percent, hydrocarbons that cause ozone and smog were cut 86 percent, and sulfur dioxide production fell 84 percent. Fish are returning to the rivers, and forests are regrowing on the mountains.

CONCLUSION

Air pollution is often the most obvious and widespread type of pollution. Everywhere on earth, from the most remote island in the Pacific, to the highest peak in the Himalayas, to the frigid ice cap over the North Pole, there are traces of human-made contaminants, remnants of the 2 billion metric tons of pollutants released into the air worldwide every year by human activities.

Health effects of these pollutants include respiratory diseases, birth defects, heart attacks, cancer, and developmental disabilities in children. Environmental impacts include destruction of stratospheric ozone, poisoning of forests and waters by acid rain, and corrosion of building materials. Damages to health and environment in turn have economic costs that vastly outweigh the costs of pollution prevention.

We have made encouraging progress in controlling air pollution, progress that has economic benefits as well as health benefits. Many people aren't aware of how much worse air quality was in the industrial centers of North America and Europe a century or two ago than they are now. Cities such as London, Pittsburgh, Chicago, Baltimore, and New York had air quality as bad as or worse than most megacities of the developing world now.

The success of the Montreal Protocol and the Clean Air Act show that real progress is achievable if we choose to act. Though the stratospheric ozone hole persists because of the residual chlorine in the air released decades ago, we expect the ozone depletion to end in about 50 years. The Clean Air Act, similarly, has dramatically improved air quality and reduced health costs. Local pollution reductions in developing countries, such as Brazil and India, show that better air quality is not just for the wealthy. What is needed is a decision to act. As the Chinese philosopher Lao-tzu wrote, a journey of a thousand miles must begin with a single step.

PRACTICE QUIZ

1. Define *primary* and *secondary air pollutants.*
2. What are the six "criteria" pollutants in the original Clean Air Act? Why were they chosen? What are some additional hazardous air toxins have been added to the list regulated by the Clean Air Act?
3. What pollutants in indoor air may be hazardous to your health? What is the greatest indoor air problem globally?
4. What is acid deposition? What causes it?
5. What is an atmospheric inversion, and how does it trap air pollutants?
6. What is the difference between ambient and stratospheric ozone? What is destroying stratospheric ozone?
7. What is long-range air pollution transport? Give two examples.
8. What is the ratio of direct costs and benefits of the Clean Air Act? What costs are mainly saved?
9. Which of the conventional pollutants has decreased most in the recent past and which has decreased least?
10. Give one example of current air quality problems and one success in controlling pollution in a developing country.

CRITICAL THINKING AND DISCUSSION

Apply the principles you have learned in this chapter to discuss these questions with other students.

1. How would you choose between government regulations and market-based trading programs for air pollution control? Are there situations where one approach would work better than the other?
2. Debate the following proposition: Our air pollution blows into someone else's territory; therefore, it is uneconomical to install pollution controls, because they will bring no direct economic benefit to those of us who have to pay for them.

3. Utility managers once claimed that it would cost \$1,000 per fish to control acid precipitation in the Adirondack lakes and that it would be cheaper to buy fish for anglers than to put scrubbers on power plants. Does that justify continuing pollution? Why or why not?
4. Economists and scientists often have difficulty reaching common terms for defining and solving issues such as the Clean Air Act renewal. How might their conflicting definitions be reshaped to make the discussion more successful?
5. Why do you think controlling pollutants like mercury is such a difficult problem? List some of the technological, economic, political, emotional, and other factors involved. Whose responsibility is it to reduce these emissions?

DATA ANALYSIS How Polluted Is Your Hometown?

How does air quality in your area compare to that in other places? You can examine trends in major air pollutants, both national and local trends in your area, on the EPA's website. The EPA is the principal agency in charge of protecting air quality and informing the public about the air we breathe and how healthy it is.

Go to Connect to find a link to data and maps showing trends in SO_2 emissions since 1980. At the same site you can see trends in NO_x, CO, lead, and other criteria pollutants. Examine national trends; then look at your local area on the map on the same page to answer questions about trends in your area, and to compare your area to others.

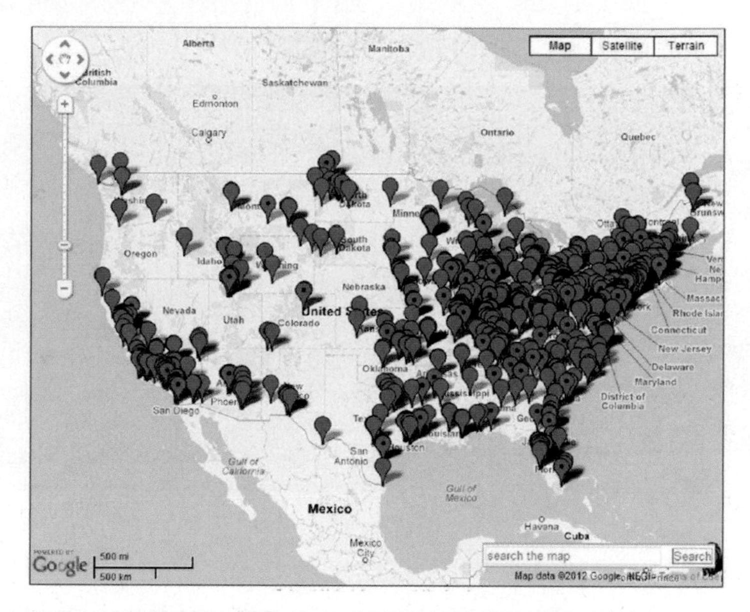

▲ Examine pollutant trends in your area on the EPA website.

Design Elements: Active Learning (Toad): ©Gaertner/Alamy Stock Photo; Case Study (Globe): ©McGraw-Hill Education; Google Earth: ©McGraw-Hill Education; Abstract Background: ©Martin Kubat/Shutterstock; What do you think (Students using tablets): ©McGraw-Hill Education/Richard Hutchings, photographer; What can you do (Hand holding Globe): ©Christoph Weihs/Shutterstock

CHAPTER

11 Water: Resources and Pollution

LEARNING OUTCOMES

After studying this chapter, you should be able to answer the following questions:

- Where does our water come from? How do we use it?
- Where and why do water shortages occur?
- How can we increase water supplies? What are some costs of these methods?
- How can you conserve water?
- What is *water pollution*? What are its sources and effects?
- Why are sewage treatment and clean water important in developing countries?
- How can we control water pollution?

Years of historic drought and declining snowpack have threatened water resources that support California's cities, farms, and hydropower.

CASE STUDY

A Water State of Emergency

It's often said in the American West that whisky is for drinking, and water is for fighting. This old line has new resonance lately in California, where many of the state's 39 million people, and the country's most productive farmland, have struggled with persistent drought. What the future holds is uncertain, but climatologists say it's unlikely that the coming decades will bring much relief to the drought.

In spring 2015, California's governor, Jerry Brown, declared the state's first ever mandatory water use restrictions. Following 4 years of the worst drought in 120 years of record keeping, and after a year of unsuccessful efforts to encourage voluntary reductions, Governor Brown announced a mandatory 25 percent cut in water consumption. The cuts could be enforced with fines if necessary. At the time of the declaration, snowpack in the nearby Sierra Nevada Mountains, which normally provide one-third of the state's water, stood at less than 6 percent of normal.

Both cities and farms in California have grown and prospered on imported water, delivered from the Sierra Nevada or even farther afield. In 1913, when the Los Angeles Aqueduct opened, Los Angeles was a city of 350,000. The aqueduct delivered water from the Owens Valley 200 miles away, and it was the beginning of a vast network of pipes, pumps, and canals that has allowed Los Angeles to mushroom to a city of 10 million, with another 8 million in the surrounding region. With almost no surface water in the city, every Angeleno depends on imported water, or on water pumped from underground, to drink, bathe, wash cars, and water lawns (fig. 11.1).

Agriculture in arid southern California also was a modest affair before the aqueducts. Ranchers could graze livestock in the hills, but thirsty crops like almonds, oranges, and tomatoes couldn't survive in most of this dry country. Then irrigation turned the desert into a garden. Agriculture now uses 80 percent of California's water and produces a quarter of the food Americans eat.

Most of this water originates as snowpack in the Sierra Nevada, the "snowy range" in Spanish. The Sierras catch and store moisture, in the form of snow and rain, coming in from the Pacific Ocean. Normally, snow has accumulated to depths exceeding 4 m (13 ft) at high elevations. Snow melts well into the summer, feeding the lakes and rivers that keep the region's aqueducts full. Snowmelt also powers the state's hydroelectric dams. When rivers no longer have enough water to run hydropower stations, electricity may become more expensive.

Snowmelt in the Sierra has decreased for three related reasons. First is drought: Precipitation always varies over decades, but the drought leading up to the water-saving mandate was the worst ever recorded in the state. It may have been the worst in 1,000 years or more. Second is short winters: Global patterns of climate warming have produced warmer winters here, with shorter snow seasons, more precipitation that falls as rain instead of snow, and earlier melting in the spring. Third is rising temperatures and increased evapotranspiration: In a warmer climate, water evaporates faster and plants use more moisture than in a cool climate. Hotter temperatures raise water demands in valley farm fields, but they also reduce water supplies coming down from the mountains.

How can water use be cut? Governor Brown's rule included restrictions on watering golf courses and cemeteries, bans on using potable water for landscape watering, and incentives to replace 50 million square feet of irrigated lawns with dry landscaping. Financial assistance was provided to help residents purchase water-saving washing machines, toilets, and shower heads. Agricultural water use restrictions were not specified, but better reporting of water use was required. A city that failed to comply could be fined $10,000 per day.

By the time of the new rules, most Californians were already well aware of the crisis. Some cities had already started removing lawns and giving incentives for water-saving appliances. Farmers without legal water rights had been forced to idle half a million acres of land in 2014. The main disagreements voiced publicly about the water restrictions was that they didn't go far enough, or they did too little to address the vast amount of agricultural irrigation.

In this chapter we will explore the world's water resources, how they are distributed, and how they are used. In the long run it remains to be seen whether water is really for fighting over or if the world's governments, cities, and farmers can accommodate a warming climate and its impacts on the water of life. ■

▲ **FIGURE 11.1** California's water system uses hundreds of miles of aqueducts to deliver water over the mountains and across deserts from the Sierras and the Colorado Rockies. Source: Library of Congress Prints and Photographs Division [HAER CA-298-AH-4 (CT)]

I tell you gentlemen; you are piling up a heritage of conflict and litigation of water rights, for there is not sufficient water to supply the land.

—JOHN WESLEY POWELL

11.1 WATER RESOURCES

- *Water shortages affect economies, food production, and security.*
- *Evapotranspiration, the release of water vapor by plants + evaporation, depends on temperature.*
- *Less than 1 percent of water is fresh surface water.*

Water is essential for life. It is the medium in which all living processes occur. Water dissolves nutrients and distributes them to cells, regulates body temperature, supports cells, and removes waste products. You are 60 percent water. You could survive for weeks without food but only a few days without water.

Worsening water shortages worry cities and farmers worldwide. In India's Ganges River plain, home to half a billion people, groundwater levels have declined some 30 cm in just a decade (fig. 11.2). China, Syria, Iraq, and other countries face increasing water shortages. Globally, military analysts expect that water shortages will increasingly become the focus of wars and refugee crises. Understanding how these resources are distributed, and why they are changing, is the first step toward finding better strategies to protect this precious resource.

Conserving water resources, as California is trying to do, doesn't need to mean disaster. As in California, it may mean fewer green lawns in the desert. It means better management of resources, and it means adjusting expectations to what the land can support.

How does the hydrologic cycle redistribute water?

Solar energy constantly evaporates water, and evaporated water condenses to liquid (rain, fog, dew) or solid snow and ice when it cools. The general term for falling liquid or solid water is **precipitation**, and most water precipitates somewhere distant from where it evaporated (see fig. 2.17). Most snowfall on the Sierra Nevada, for example, comes from evaporation over the Pacific Ocean.

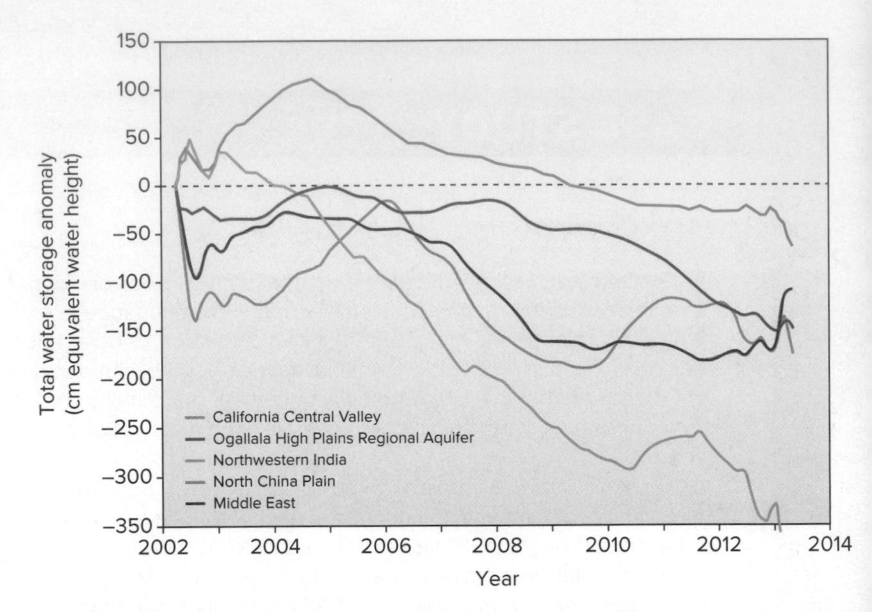

▲ **FIGURE 11.2** Withdrawals are reducing water storage in many of the world's most important aquifers, including those in politically tense regions. Calculated variation in groundwater is derived form NASA's Gravity Recovery and Climate Experiment (GRACE) satellites, which measure changes in the mass (water content) of the land surface. Source: Famiglietti. J. S., "The Global Groundwater Crisis," *Nature Climate Change*, vol. 49, 2014, 945–948.

These familiar processes supply nearly all the water that keeps us and our ecosystems alive. Some 500,000 km^3 evaporates from the oceans every year; of this 90 percent falls back into the ocean, but some 47,000 km^3 of moisture drifts over land, where it falls, along with some 72,000 km^3 of water evaporated from lakes, rivers, and plants (table 11.1). Plants play a major role in this process. They take up moisture and release water vapor from leaf pores, in a process known as **transpiration**. The term **evapotranspiration** refers to the combined processes of evaporation and transpiration. Both evaporation and transpiration are much more active in hot climates than in cool climates: More solar energy causes more evaporation, and plants must transpire faster to avoid baking in the heat.

Solar energy drives the cycle of water evaporation, transport, and precipitation. But both solar energy and water that can be evaporated are unevenly distributed around the globe. Consequently, water resources are unevenly distributed. At Iquique in the Chilean desert, for instance, no rain has fallen in recorded history. At the other end of the scale, the Himalayan town of Cherrapunji, India, reported 26.5 m of rain (86.8 ft) in 1860.

In general, water availability is very high near the equator, as in South America, Africa, and Southeast Asia, where solar intensity

Benchmark Data

Among the ideas and values in this chapter, these are a few worth remembering.

2.4%	Percentage of all water on earth that is fresh
0.8%	Surface water percentage of fresh water
>13,000	Liters water needed to produce 1 kg of beef (average)
60 million	Number of the 315 million Americans who get water from national forests
2 ppm	Oxygen concentration in water needed for most aquatic life

TABLE 11.1 Units of Water Measurement

Units of Water Measurement
1 m^3 = 1,000 liters = 264 gal
1 km^3 = 1,000,000,000 m^3
1 acre-foot = 1 acre 1 ft deep = 1,234 m^3
1 m^3/sec (for river flow) = 264 gal/sec

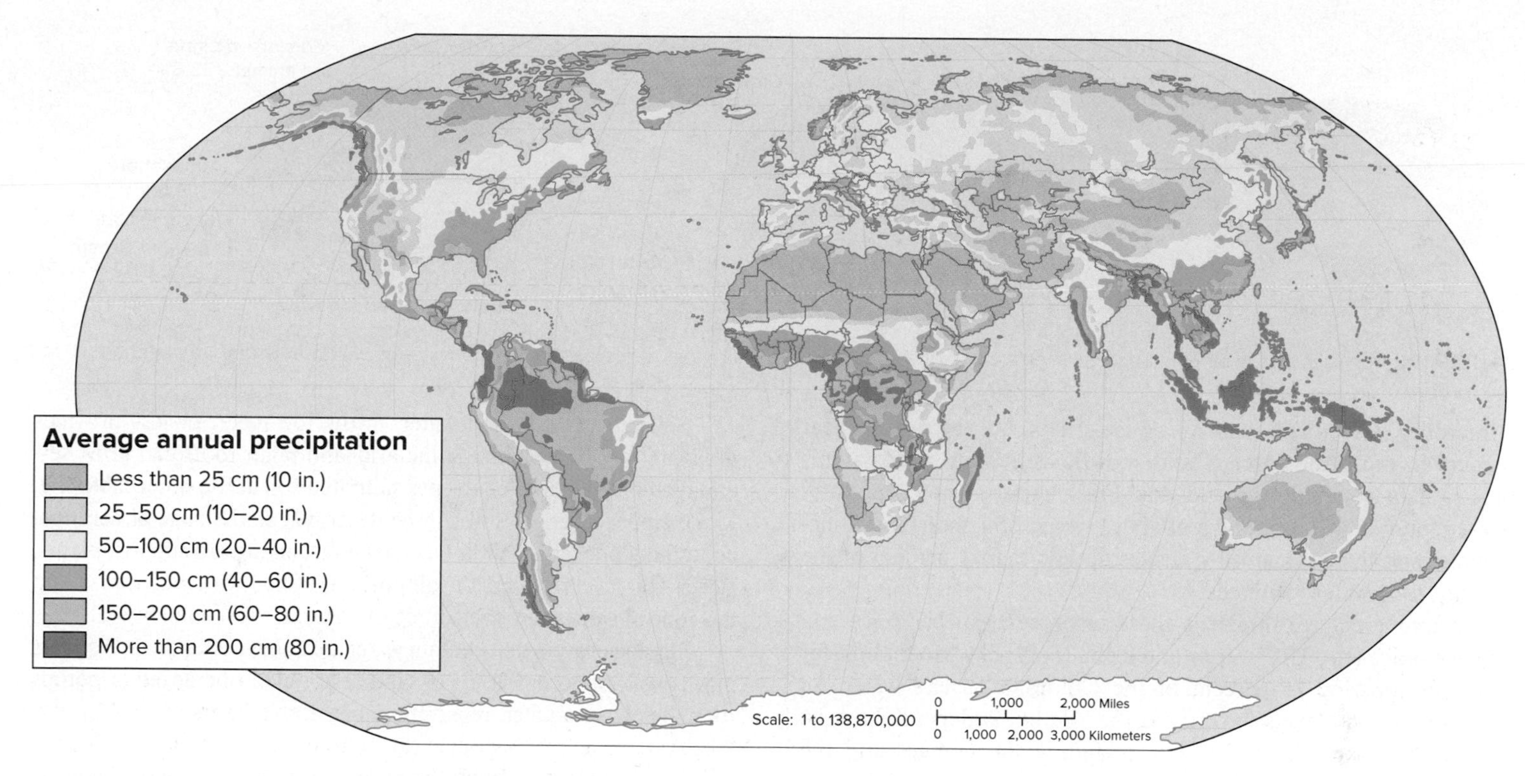

▲ **FIGURE 11.3** Average annual precipitation. Note wet areas that support tropical rainforests occur along the equator, while the major world deserts occur in zones of dry, descending air between 20° and 40° north and south.

and evapotranspiration rates are high (fig. 11.3). Mountainous regions near coasts, such as the Pacific coast of North America, also capture moisture, as in the Sierra Nevada. The leeward (downwind) side of mountains tends to be depleted of moisture. In Hawaii, the windward side of Mount Waialeale has an annual rainfall around 1,200 cm (460 in.). The leeward side, only a few kilometers away, has an average yearly rainfall of only 46 cm (18 in.).

Deserts occur on every continent just outside the tropics (the Sahara, the Namib, the Gobi, the Sonoran, and others). Here, high-pressure weather systems (see chapter 9) tend to maintain clear, dry climates. Rainfall is also slight near the poles, another high-pressure region.

Water availability often varies seasonally. India's monsoonal climate, for example, has intense drought most of the year and intense rainfall at the end of summer. Unusually heavy monsoons can cause floods and devastation; unusually light monsoons can spell disaster for farmers.

Major water compartments vary in residence time

The distribution of water often is described in terms of interacting compartments in which water resides, sometimes briefly and other times for eons (table 11.2). The length of time water typically stays in a compartment is its **residence time**. On average, a water molecule stays in the ocean for about 3,000 years before it evaporates and starts through the hydrologic cycle again. A water molecule stays in the atmosphere only about a week. Nearly all the world's water is in the oceans (fig. 11.4). Oceans play a crucial role in moderating the earth's temperature, and over 90 percent of the world's living biomass is contained in the oceans.

Glaciers, ice caps, and perennial snowfields contain nearly 90 percent of the world's fresh water. Most of this ice is in

TABLE 11.2 Earth's Water Compartments

COMPARTMENT	VOLUME (1,000 km^3)	PERCENT OF TOTAL WATER	AVERAGE RESIDENCE TIME
Total	1,386,000	100	2,800 years
Oceans	1,338,000	96.5	3,000 to 30,000 years*
Ice and snow	24,364	1.76	1 to 100,000 years*
Saline groundwater	12,870	0.93	Days to thousands of years*
Fresh groundwater	10,530	0.76	Days to thousands of years*
Fresh lakes	91	0.007	1 to 500 years*
Saline lakes	85	0.006	1 to 1,000 years*
Soil moisture	16.5	0.001	2 weeks to 1 year*
Atmosphere	12.9	0.001	1 week
Marshes, wetlands	11.5	0.001	Months to years
Rivers, streams	2.12	0.0002	1 week to 1 month
Living organisms	1.12	0.0001	1 week

*Depends on depth and other factors.
Source: Data from UNEP, 2002.

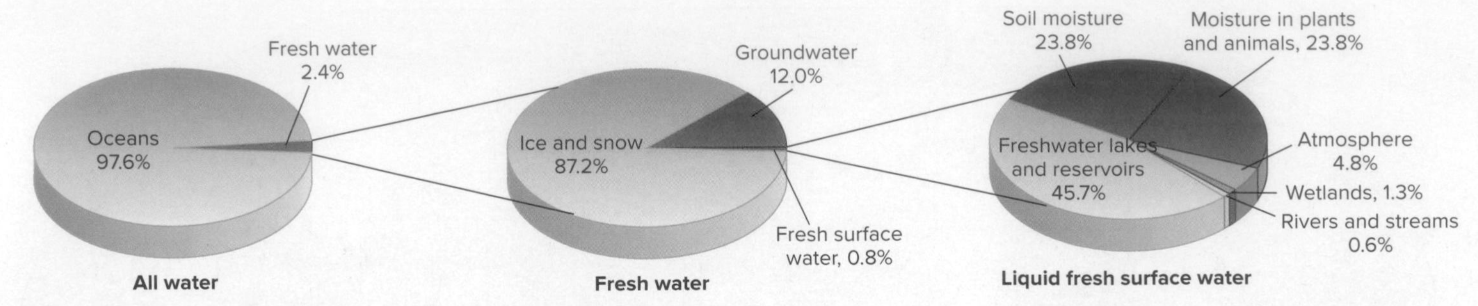

▲ **FIGURE 11.4** Less than 1 percent of fresh water, and less than 0.02 percent of all water, is fresh, liquid surface water on which terrestrial life depends.

Antarctica, Greenland, and Arctic ice sheets, but the portion that occurs in mountain glaciers and snowfields supports ecosystems and human populations worldwide. Fresh surface water, the water we encounter in lakes and wetlands, rivers, soil moisture, atmosphere, and living organisms, makes up less than 1 percent of the world's freshwater resources.

Climate change threatens these resources, especially ice and snow storage (fig. 11.5). Snow and ice in the Rocky Mountains, for example, provide 75 percent of the Colorado River's water, on which cities and farms throughout the Southwest depend. In Asia, six of the world's largest rivers, including the Ganges and the Yangtze, originate in Tibetan glaciers. There are warnings that these glaciers could vanish in a few decades. This loss could eliminate the drinking water and irrigation for 3 billion people. The human and economic effects of this are hard to imagine.

Groundwater storage is vast and cycles slowly

Originating as precipitation that percolates into layers of soil and rock, groundwater makes up the largest compartment of liquid, fresh water. The groundwater within 1 km of the surface is more than 100 times the volume of all the freshwater lakes, rivers, and reservoirs combined. In general, we can only dig wells a few tens or hundreds of meters deep, however, so much of this water is inaccessible.

Soil moisture, or soil water, in the top meter or less provides moisture for most plants. Some arid-land plant roots can grow several meters down to seek deeper saturated soil at the **water table** (fig. 11.6). Upper layers of soil can be described as the **zone of aeration**, containing pockets of air but usually moist enough to support plants. Lower soil layers, where all soil pores are filled with water, make up the **zone of saturation**, the source of water in most shallow wells.

Geologic layers that contain water are known as **aquifers**. Aquifers may consist of porous layers of sand or gravel or of cracked or porous rock. Below an aquifer, relatively impermeable layers of rock or clay keep water from seeping out at the bottom. Instead, water seeps more or less horizontally through the porous layer. Depending on geology, it can take from a few hours to many years for water to move a few hundred meters through an aquifer. If impermeable layers lie above an aquifer, pressure can develop within the water-bearing layer. Pressure in the aquifer can make a well flow freely at the surface. These free-flowing wells and springs are known as *artesian* wells or springs.

Areas where surface water filters into an aquifer are **recharge zones** (fig. 11.7). Most aquifers recharge extremely slowly, and road

▼ **FIGURE 11.5** Glaciers and snowfields provide much of the water on which billions of people rely. The snowpack in the western Rocky Mountains, for example, supplies about 75 percent of the annual flow of the Colorado River. Global climate change is shrinking glaciers and causing snowmelt to come earlier in the year, disrupting this vital water source. ©William P. Cunningham

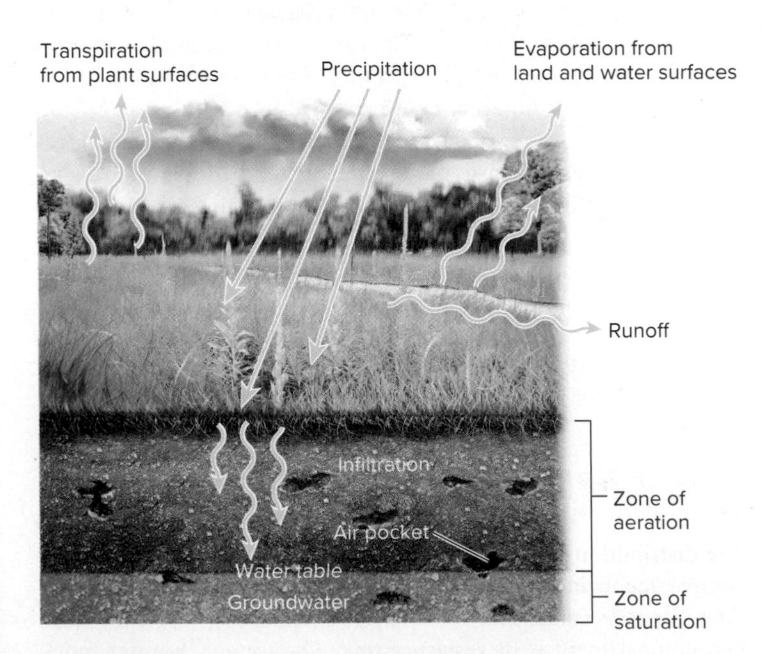

▲ **FIGURE 11.6** Precipitation that does not evaporate or run off over the surface percolates through the soil in a process called infiltration. The upper layers of soil hold droplets of moisture between air-filled spaces (usually much smaller than shown here). Lower layers, where all spaces are filled with water, make up the zone of saturation, or groundwater.

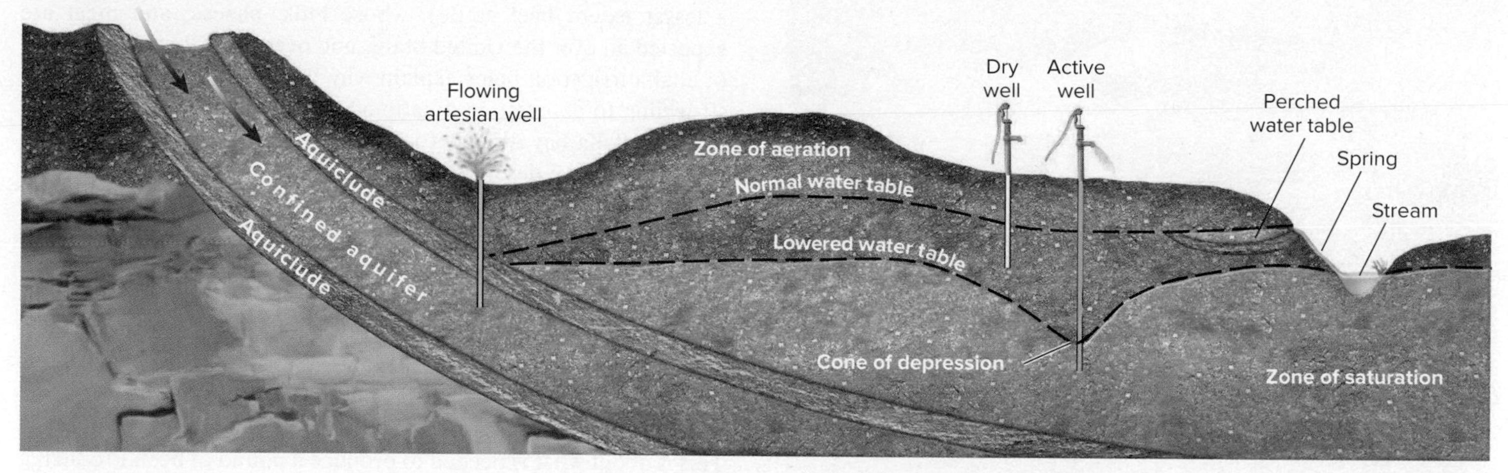

▲ **FIGURE 11.7** An aquifer is a porous or cracked layer of rock. Impervious rock layers (aquicludes) keep water within a confined aquifer. Pressure from uphill makes an artesian well flow freely. Pumping can create a cone of depression, which leaves shallower wells dry.

and house construction or water use at the surface can further slow recharge rates. Contaminants can also enter aquifers through recharge zones. Urban or agricultural runoff in recharge zones can permanently pollute a groundwater source.

Surface water and atmospheric moisture cycle quickly

Fresh, flowing surface water is one of our most precious resources. Rivers contain a relatively small amount of water at any one time. Most rivers would begin to dry up in weeks or days if they were not constantly replenished by precipitation, snowmelt, or groundwater seepage.

We compare the size of rivers in terms of **discharge**, or the amount of water that passes a fixed point in a given amount of time. This is usually expressed as liters or cubic feet of water per second. The 16 largest rivers in the world carry nearly half of all surface runoff on the earth, and a large fraction of that occurs in a single river, the Amazon, which carries nearly as much water as the next seven biggest rivers together.

Lakes contain nearly 100 times as much water as all rivers and streams combined, but much of this water is in a few of the world's largest lakes. Lake Baikal in Siberia, the Great Lakes of North America, the Great Rift Lakes of Africa, and a few other lakes contain vast amounts of water. Worldwide, lakes are almost as important as rivers in terms of water supplies, food, transportation, and settlement.

Wetlands—bogs, swamps, wet meadows, and marshes—play a vital and often unappreciated role in the hydrologic cycle. Their lush plant growth stabilizes soil and holds back surface runoff, allowing time for infiltration into aquifers and producing even, year-long stream flow. When wetlands are disturbed, their natural water-absorbing capacity is reduced, and surface waters run off quickly, resulting in floods and erosion during the rainy season and low stream flow the rest of the year.

The atmosphere contains only 0.001 percent of the total water supply, but it is the most important mechanism for redistributing water around the world. An individual water molecule resides in the atmosphere for about a week, on average. Some water evaporates and falls within hours. Water can also travel halfway around the world before it falls, replenishing streams and aquifers on land.

Active LEARNING

Mapping the Water-Rich and Water-Poor Countries

The top ten water-rich countries, in terms of water availability per capita, and the ten most water-poor countries are listed below. Locate these countries on the fold-out map at the end of your book. Describe the patterns. Where are the water-rich countries concentrated? (Hint: Does latitude matter?) Where are the water-poor countries most concentrated?

Water-rich countries: Iceland, Surinam, Guyana, Papua New Guinea, Gabon, Solomon Islands, Canada, Norway, Panama, Brazil

Water-poor countries: Kuwait, Egypt, United Arab Emirates, Malta, Jordan, Saudi Arabia, Singapore, Moldavia, Israel, Oman

ANSWER: Water-rich countries (per capita) are either in the far north, where populations and evaporation are low, or in the tropics. Water-poor countries are in the desert belt at about 15° to 25° latitude or are densely populated island nations (e.g., Malta, Singapore).

11.2 HOW MUCH WATER DO WE USE?

- *Food consumption patterns influence our "water footprint."*
- *Virtual water is exported in products.*
- *Access to clean water is one of the most critical health care factors.*

Perhaps more than any other environmental factor, water availability determines where we live and what we do (fig. 11.8). Some places like Iceland have small populations and abundant rainfall. Others like Egypt have large populations and limited water supplies. Almost all of Egypt's population lives adjacent to the Nile

▲ **FIGURE 11.8** Water has always been the key to survival. Who has access to this precious resource and who doesn't have long been a source of tension and conflict. ©Ray Ellis/Science Source

River. Technology has allowed exceptions to this rule: Bahrain, which receives almost no rainfall, uses expensive, energy–intensive desalination systems to produce fresh water from the sea. Fast-growing cities in the United States, such as Tampa, Florida, and San Diego, California, also depend on desalination for part of their water resources.

The amount of water we use depends on a number of factors, especially how much agriculture a region has, how we manage landscapes, and personal consumption practices. Direct household usage varies dramatically within a region or state. On average, in the United States we consume about 1600 m^2 per person per year. This high rate results largely from agricultural water use, as well as a warm climate and a rich diet. About 30 percent of that "water footprint" is from beef consumption. Mexicans, in contrast, consume about 750 m^3/year. In northern Europe, where climates are cool and people consume modest amounts of meat, consumption ranges from about 100 to 700 m^2 per year. Developing countries tend to use much less, often less than 50 m^2 per person per year.

Worldwide, agriculture consumes about 75 percent of water resources. Industry consumes another 20 percent, and domestic uses make up the remaining 5 percent.

We export "virtual water"

Evaluating water consumption can be tricky, because much of it is exported in a variety of forms. In California, for example, about 15 percent of all water use is consumed in producing alfalfa. Alfalfa hay feeds dairy cows (and to a lesser extent beef cattle), whose milk, cheese, and meat are exported all over the United States and overseas. (Rapid expansion of alfalfa irrigation helps explain why Wisconsin and Vermont are struggling to compete in a national dairy market.) California also exports alfalfa hay directly to China, Japan, and elsewhere. These exports are often described as **virtual water** exports.

Similarly, California's almond crop consumes about 10 percent of California's water. Almond growth has exploded in recent years: From about 60,000 tons produced in 1960, the industry expanded 35-fold to over 2 million tons in 2018. That represents 99 percent of all U.S. almonds and 80 percent of the *world's* almond supply. To keep trees alive year round as well as to produce the nuts requires nearly 4 l of water (more than 1 gallon) for each almond nut. On a per-kg basis, this amounts to 16,000 l/kg (1,900 gal/lb) of almonds. This is about what is needed to produce a pound of beef. Producing a head of lettuce takes about 800 l/kg (100 gal/lb).

By some estimates, the water content in exported alfalfa amounts to 380 million m^3 (100 billion gallons) per year, enough to supply about 1 million families with drinking water for a year. Conservation proponents argue that reducing water exports could dramatically reduce the water crisis in the American West.

Some products are thirstier than others

Our water footprint depends a great deal on what we eat. Raising a kilogram of beef in a concentrated feeding operation, for example, takes more than 15,000 liters of water (fig. 11.9) because cattle are fed corn, which we grow with water-intensive irrigation systems. The result is that the average hamburger takes 2,400 liters (630 gal) to produce. Raising goats is a much more water-efficient way to produce meat, because goats eat a more varied diet, and they metabolize their food more efficiently than cattle.

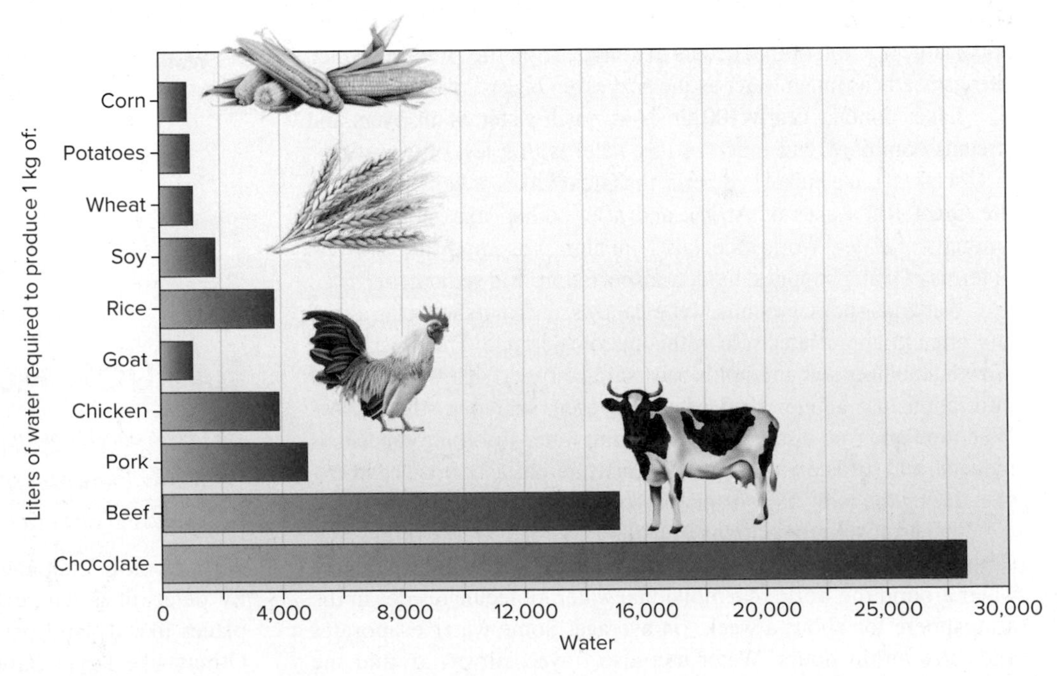

▲ **FIGURE 11.9** Water required to produce 1 kg of some important foods.

Rice cultivation (which generally occurs in wet paddies) takes about three times as much water as raising potatoes or wheat. If rice is grown in a rainy region, providing this water is rarely a major problem. But some countries, including the United States, depend on abundant irrigation for rice growing. For some foods, the greatest water use isn't in growing the crop but in preparing it for our consumption. Chocolate, for example, is highly processed before we eat it.

Industrial uses include energy production

Industrial uses represent about one-fifth of all water withdrawals worldwide. Some European countries use 70 percent of water for industry; less-industrialized countries use as little as 5 percent. Cooling water for power plants is by far the largest single industrial use of water, typically accounting for 50 to 75 percent of industrial withdrawal. If treated properly, much of the cooling water can be recycled for other uses.

Biofuel production, such as ethanol from corn, also demands extreme amounts of water. It takes 4 to 7 liters of water to produce 1 liter of ethanol, depending on how much irrigation was required to produce the corn feedstock.

Since 2008, another dominant industrial use of water has been hydraulic fracturing ("fracking") for natural gas and oil production. As the name suggests, water is pumped under pressure into oil and gas wells, to help fracture geologic formations and release hydrocarbons locked in the rocks. Fracking one well uses 15 to 23 million liters of water (4 to 6 million gallons). Sometimes wells are fracked more than once in order to release additional gas.

Most of this water returns to the surface, but it is too contaminated to be reused without a great deal of treatment. Contaminants include carcinogenic hydrocarbons as well as detergents, salts, and radioactive particles from deep underground. In areas of intensive oil and gas development, as in western North Dakota, northern Texas, and Oklahoma, water resources can be severely stressed by fracking demands. Drought-stressed California used 900 million liters (237 million gal) of water for fracking between 2005 and 2015. The U.S. total was nearly 1 trillion liters (240 billion gal).

Domestic water supplies protect health

Household water use, for drinking, washing, and cooking, accounts for only about 5 percent of world water use. Amounts of use vary with wealth, however. The United Nations reports that people in developed countries consume, on average, about ten times more water per day than those in developing nations. Poorer countries often can't afford the infrastructure to obtain and deliver water to citizens. When a city cannot pipe safe public water to neighborhoods, residents have to buy high-priced water delivered by truck. The cost of this water can help keep families in poverty, but they have little choice. Providing clean water to poor populations is one of the primary aims of international development organizations.

Clean drinking water and basic sanitation are necessary to prevent communicable diseases and to maintain a healthy life. The United Nations estimates that at least a billion people lack access to safe drinking water and 2.5 billion don't have adequate sanitation. As populations grow, more people move into cities, agriculture and industry compete for increasingly scarce water supplies, and water shortages are expected to become even worse.

▲ **FIGURE 11.10** Women and children often spend hours every day collecting water—which often is unsafe for drinking—from local water sources. ©Layton Thompson

Rural people often have less access to good water than do city dwellers. Water-related diseases such as diarrhea are common in areas with poor sanitation. Every year about 1.6 million people, 90 percent of them children under 5 years old, die from these diseases. More children die from diarrhea than malaria, measles, and HIV/AIDS combined.

More than two-thirds of the world's households have to fetch water from outside the home (fig. 11.10). This is heavy work, done mainly by women and children, and can take several hours a day, time that could otherwise be spent going to school, tending children, or earning an income.

11.3 DEALING WITH WATER SCARCITY

- *Drought stress is expected to increase with climate change.*
- *Subsidence collapses and eliminates groundwater storage.*
- *Water diversions redistribute vast amounts of water.*

The World Health Organization considers an average of 1,000 m^3 (264,000 gal) per person per year to be a necessary amount of water for modern domestic, industrial, and agricultural uses. It is fairly common in some areas to use 70 percent of annual river flows and, in a few cases, to use as much as 120 percent of annual renewable resources by drawing on fossil groundwater. This obviously isn't sustainable in the long run.

About a third of the world's current population lives in countries where water supplies don't meet everyone's minimum essential water needs (fig. 11.11). Compare figure 11.11 with figure 11.3. Notice that the scarcity map shows average conditions for each country. Local conditions can be much worse than the country-wide mean. Thus, China is vulnerable to water shortages due to drought in its northern

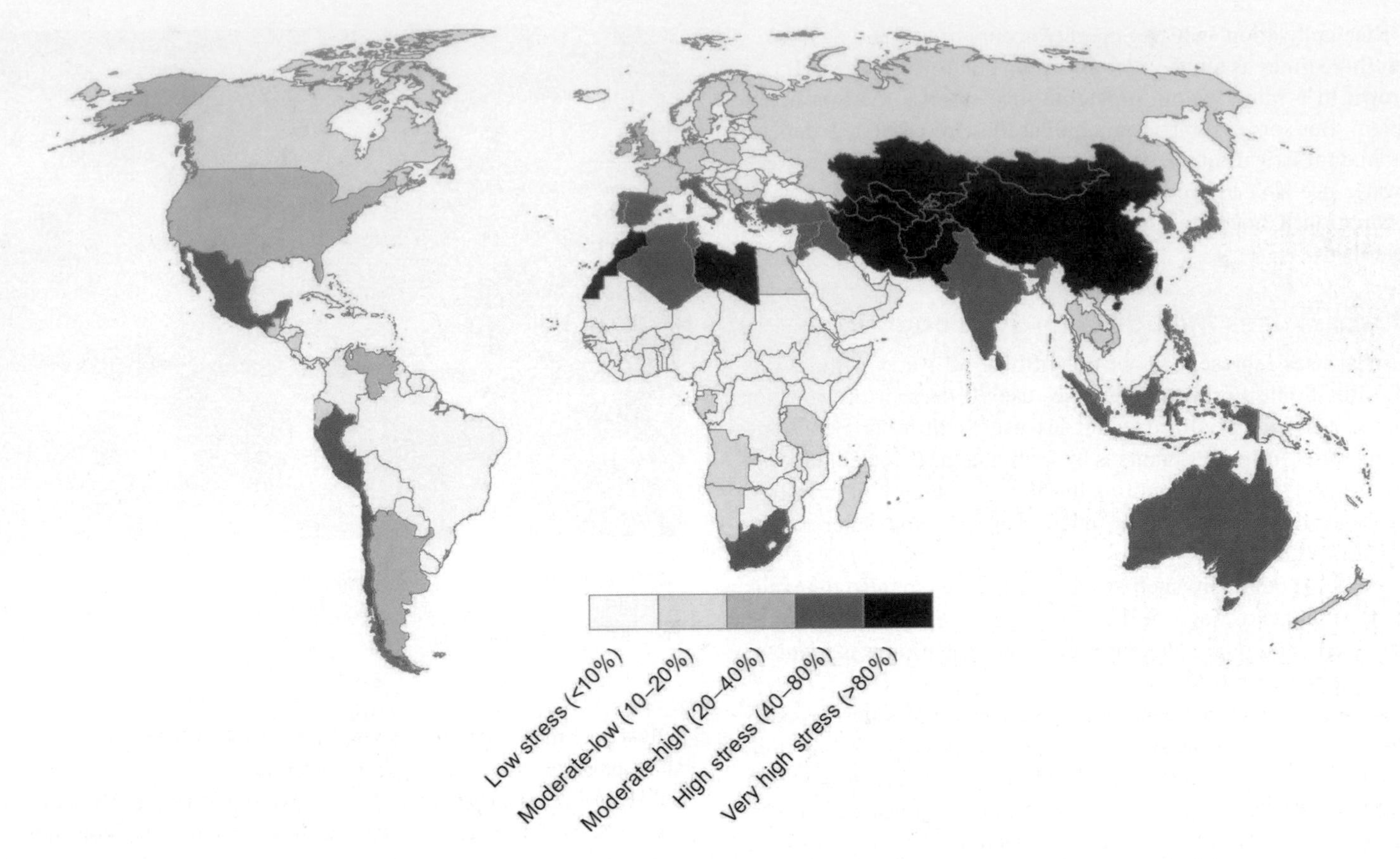

▲ **FIGURE 11.11** Water stress can be measured as the ratio of water withdrawals to supply. Source: Data from the World Resources Institute.

and western regions even though its southeast has been suffering catastrophic floods. Not surprisingly, the biggest problems in water stress and scarcity at a human level occur in Africa and Asia where rainfall is low and poor countries can't afford to adapt. Some affluent countries are very dry but have the money and political capacity to adjust to drought (What Do You Think?, p. 261).

Water use has been increasing about twice as fast as population growth over the past century. And water withdrawals are expected to continue to grow as more land is irrigated to feed an expanding population. The United Nations Food and Agriculture Organization predicts that, by 2025, 1.8 billion people will be living with severe water scarcity and as many as two-thirds of all humans will experience water-stressed conditions.

Climate change is expected to exacerbate water scarcity. A general rule is that with global warming, dry areas will generally get drier, while wet places will be wetter. One of the United Nations Millennium Development Goals (discussed in chapter 16) has been to reduce by one-half the proportion of people without reliable access to clean water and improved sanitation, but that will be difficult if water supplies become more stressed.

Drought, climate, and water shortages

In general, a drought is an extended period of consistently below average precipitation that has a substantial impact on ecosystems, agriculture, and economies. The worst drought in recent U.S. history, in economic and social terms, was in the 1930s. Poor soil conservation practices and a series of dry years in the Great Plains combined to create the "dust bowl." Wind stripped topsoil from millions of hectares of land, and billowing dust clouds turned day into night. Thousands of families were forced to leave farms and migrate to other areas. As the opening case study shows, much of the western United States has been exceptionally dry over the past decade. In many states, conditions are now worse than during the 1930s dust bowl.

Droughts in the American Southwest aren't a new phenomenon. In fact, the dry spells in recent years have been of relatively short duration compared to events in the region's climate history. The dust bowl of the 1930s, for example, only lasted about 6 years. By contrast, the megadroughts that destroyed the Anasazi (or Ancient Pueblo) cultures in the twelfth and thirteenth centuries (fig. 11.12) lasted between 25 and 50 years.

Recent drought in the western United States is no surprise. Major John Wesley Powell, who surveyed western states for the farming resources and led the first boat trip through the Grand Canyon in 1869, reported that there was not enough water in the West. "I tell you gentlemen," he wrote on his return to Washington, "you are piling up a heritage of conflict and litigation of water rights, for there is not sufficient water to supply the land."

Powell recommended that the political organization of the West be based on watersheds so that everyone in a given

What Do YOU THINK?

Water and Power

In the American West it's often said that water flows uphill toward money. But what happens when the water just stops flowing? The Colorado River is the source of life, wealth, and power for much of the West. More than 30 million people and a $1.2 trillion regional economy that includes Phoenix, Las Vegas, Los Angeles, and Denver, depend on its water. Two hydroelectric dams, the Hoover Dam and the Glen Canyon Dam, produce 3 gigawatts (GW) of electricity, enough for 3 million households. But the future of these dams is in doubt. Withdrawals for crop irrigation, rapid urban growth, and now long-term drought and climate change are lowering water supplies to levels not seen since the dams were built.

In 2008, Tim Barnett and David Pierce from the Scripps Institute in California published a provocative article, "When Will Lake Mead Go Dry?" Drawing on historic rates of water consumption, evaporation from the reservoirs behind the two dams, Lake Powell and Lake Mead, and precipitation trends, the paper estimated when the two reservoirs would reach "dead pool" status—too little water to drain by gravity. Barnett and Pierce projected a 50 percent chance that the two dams will reach this stage by 2021. Water deliveries could fall short by 60 to 90 percent in coming decades. Electric generation could stop well before that. To create power at the Hoover Dam, Lake Mead's water level has to be at least 320 m (1,050 ft) above sea level; in 2015 the low water level fell below 329 m (1,078 ft). Releases from upstream dams have helped maintain roughly this level since then, but the Hoover Dam has been downgraded from 2 GW to 1.6 GW capacity because of low water, and it runs mainly at peak demand hours. Electricity generation will continue to decline, and water deliveries to Los Angeles, Phoenix, Las Vegas, and Tucson will become more unreliable, unless changes in water management are made.

Negotiating a new water management plan is difficult when water is power. Water allocation in the Colorado was supposed to be settled back in 1922 by the Colorado Compact, an agreement that designated specific amounts of the river's water for seven western states and Mexico. Altogether the Compact allocates 16.5 million acre feet (af) from the river per year. But normal flow in the river is only 14–15 million af. The compact that was supposed to resolve conflict set the stage for more disputes.

To distribute this water to farms and cities, the federal government has spent billions of dollars building massive water diversion projects, such as the Colorado River Aqueduct, which provides water for Los Angeles; the All-American Canal, which irrigates California's Imperial Valley; and the Central Arizona Project, which pumps water over the mountains and across the desert to Phoenix and Tucson. Much of the water irrigates crops—cotton, alfalfa, lettuce, citrus fruit, catfish, and other products—especially in California's Imperial Valley. The rest provides water for lawns, golf courses, and other urban uses. Las Vegas and other western cities are working to remove lawns and establish conservation rules. But if we're going to grow crops in the desert, we can't get by without irrigation.

All these projects did not anticipate climate change. Most of the water in the Colorado comes from snowmelt in the Rocky Mountains, which is decreasing as the climate warms. Evaporation from the Lake Mead, Lake Powell, and other reservoirs—which are effectively giant evaporation pans in the desert sun—is already responsible for about 1.5–2 million af of water loss per year. In the coming years, temperatures in the Colorado basin are expected to rise 2–4 °C, accelerating evaporation as well as reducing snowmelt.

▲ The Colorado River flows 2,330 km (1,450 mi) through seven western states. Its water supports 30 million people and a $1.2 trillion regional economy, but drought, climate change, and rapid urban growth threaten the sustainability of this resource.

The future does not look good for regions relying on the Colorado River. Lake Powell is less than half full, and Lake Mead 37 percent full, and both are dropping. The shores of both lakes now display a wide "bath-tub ring" of deposited minerals left by the receding water.

What would you do if you could control water policy in the Colorado basin? Are there other places where crops could grow if arid lands of the West lost irrigation? What would the costs be if Los Angeles, Phoenix, Las Vegas, and other major cities were to run out of water and power?

For further reading, see:

Barnett, T. P., and D. W. Pierce. 2008. When will Lake Mead go dry? *Journal of Water Resources Research*, vol. 44, W03201.

▲ **FIGURE 11.12** The stunning cliff dwellings of Mesa Verde National Park were abandoned by the Anasazi, or ancient Puebloan people, during a severe megadrought between 1275 and 1299. Will some of our modern cities face the same fate? ©William P. Cunningham

jurisdiction would be bound together by the available water. He thought that farms should be limited to local surface-water supplies, and that cities should be small, oasis settlements. Instead, we've built huge metropolitan areas, such as Los Angeles, Phoenix, Las Vegas, and Denver, in places where there is little or no natural water supply.

Groundwater supplies are being depleted

Groundwater provides nearly 40 percent of the fresh water for agricultural and domestic use in the United States. Nearly half of all Americans and about 95 percent of the rural population depend on groundwater for drinking and other domestic purposes. Overuse of these supplies dries up wells, natural springs, and even groundwater-fed wetlands, rivers, and lakes. Pollution of aquifers through dumping of contaminants on recharge zones, leaks through abandoned wells, or deliberate injection of toxic wastes can make this valuable resource unfit for use.

In California's agricultural Central Valley (opening case study) groundwater pumping has accelerated dramatically as Sierra snowpack and other surface waters have disappeared. Farmers have been drilling wells ever deeper, sometimes more than 500 m (1600 ft) down, in efforts to capture diminishing groundwater before their neighbors do. Drilling such a deep well can cost more than a quarter million dollars, but without them, many farmers would be out of business.

Farmers, and the state, know this is a short-sighted strategy. Shallower aquifers will take years or decades to recharge—if there is sufficient precipitation. Deeper aquifers will take centuries to refill.

Many of the depleted aquifers may not be able to refill, however. With the water removed, the loss of water pressure in the aquifer allows the depleted porous rock layers to compress. **Subsidence**, the gradual sinking of the ground surface, follows. In some parts of California's Central Valley, the ground has subsided a meter in just a few years. Roads buckle and canals shift as the land sinks. Subsidence is nothing new to this region: The San Joaquin valley sank more than 10 m (33 ft) in the earlier days of pumping (fig. 11.13).

▲ **FIGURE 11.13** A photo taken in 1977 showing subsidence caused by groundwater pumping in California's San Joaquin Valley. The 1925 sign indicates the ground level in that year. Source: Dick Ireland, USGS

Groundwater depletion is an ongoing process in many of the world's great aquifers (see fig. 11.2). The Ogallala High Plains Regional Aquifer, which underlies a vast area from Nebraska to Texas, is a well-known case of long, slow collapse. A porous bed of sand, gravel, and sandstone, the Ogallala once held more water than all the freshwater lakes, streams, and rivers on earth. Much of the water has probably been there for thousands of years. Decades of pumping to irrigated cotton, corn, soybeans, and wheat through much of the arid Great Plains have steadily lowered the water table. In many areas the aquifer is permanently depleted—at least on a human time scale. This "fossil" water is essentially a non-renewable resource.

Another consequence of aquifer depletion is saltwater intrusion. Along coastlines and in areas where saltwater deposits are left from ancient oceans, overuse of freshwater reservoirs often allows saltwater to intrude into aquifers used for domestic and agricultural purposes.

Diversion projects redistribute water

Dams and canals are a foundation of civilization because they store and redistribute water for farms and cities. Many great civilizations have been organized around large-scale canal systems, including ancient empires of Sumeria, Egypt, and India. As modern dams and water diversion projects have grown in scale and number, though, their environmental costs have raised serious questions about efficiency, costs, and the loss of river ecosystems.

More than half of the world's 227 largest rivers have been dammed or diverted. Of the 50,000 large dams in the world, nearly half are in China, and China continues to build and plan dams on its remaining rivers. Dams are justified in terms of flood control, water storage, and electricity production. However, the costs are tremendous: Villages must be relocated; fertile farmlands and fishing grounds are flooded; evaporation reduces water resources. Downstream regions, often in other countries, may be deprived of water altogether. Loss of water leads to overpumping of groundwater, farm failures, and economic strains. China's planned dams on the Brahmaputra River in Tibet, for example, is likely to starve Cambodia and Laos of vital water supplies.

Similarly, Turkey has put large dams on the Tigris and Euphrates. Downstream, Syria and Iraq are overdrawing their groundwater to compensate. In the United States, water diversions from the Colorado River reduced river flow so dramatically that the United States was forced to build a $150 million desalination plant to restore the river's depleted flow to usable quality as it entered Mexico. For users of the water, this is a helpful subsidy. Since the cost is distributed, most U.S. taxpayers know little about it. Does that mean the project makes economic sense? Studies asking this question of the world's largest dams have found that at least one-third of them never made economic sense to the public, although they often have enriched those who built them.

The largest ever river diversion is the South-North Water Transfer Project, which brings water from the Yangtze River in the south to Beijing, a city of 20 million in the north. Beijing receives little rain and has long been short of water. The south-to-north water diversion may be the most expensive engineering project ever built, with a cost of about $80 billion by 2014, when the early sections of it went into operation.

Las Vegas is also digging a $3.5 billion tunnel that will burrow into Lake Mead, 100 m (300 ft) below the normal outlet (fig. 11.14). Lake Mead is formed by the Hoover Dam, one of the largest dams in the United States. The hydropower station in the dam provides electricity to Las Vegas and the surrounding region. Like other reservoirs, the lake is going dry. Even if, or when, the water becomes too low to run the hydropower dam, Las Vegas hopes the tunnel will at least maintain a water supply for the city.

One of the most disastrous diversions in world history is that of the Aral Sea. Situated in arid Central Asia, on the border of Kazakhstan and Uzbekistan, the Aral Sea is a shallow inland sea fed by rivers from distant mountains. Starting in the 1950s, the Soviet Union began diverting these rivers to water cotton fields and rice fields. Gradually, the Aral Sea has evaporated, leaving vast, toxic salt flats (fig. 11.15). The economic value of the cotton and rice has probably never equaled the cost of lost fisheries, villages, and health.

Recently, some river flow has been restored to the "Small Aral," or northern lobe of the once-great sea. Water levels have risen 8 m and native fish are being reintroduced. It's hoped that one day commercial fishing may be resumed. The fate of the larger, southern remnant is more uncertain. There may never be enough water to refill it, and if there were, the toxins left in the lake bed could still make the water badly degraded.

▲ **FIGURE 11.14** Hoover Dam powers Las Vegas, Nevada. Lake Mead, behind the dam, loses about 1.3 billion m^3 per year to evaporation. Reduced inflows now threaten the viability of this system. Source: USDA, NRCS, photo by Lynn Betts

Questions of justice often surround dam projects

While dams provide hydroelectric power and water to distant cities, local residents often suffer economic and cultural losses. In some cases, dam builders have been charged with using public money to increase the value of privately held farmlands, as well as encouraging inappropriate farming and urban growth in arid lands.

International Rivers, an environmental and human rights organization, reports that dam projects have forced more than 23 million people from their homes and land, and many are still suffering the impacts of dislocation years after it occurred. Often the people being displaced are ethnic minorities. Currently, at least 144 dams on eight rivers in Southeast Asia have been proposed or are under construction. This includes the Lancang (Upper Mekong), the Nu (Upper Salween), and the Jinsha (Upper Yangtze). Several of these projects are in or adjacent to the Three Rivers World Heritage Site, threatening the ecological and cultural integrity of one of the most spectacular and biologically rich areas in the world.

There's increasing concern that big dams in seismically active areas can trigger earthquakes. In more than 70 cases worldwide, large dams have been linked with increased seismic activity. Geologists suggest that filling the reservoir behind the nearby Zipingpu Dam on the Min River caused the devastating 7.9-magnitude Sichuan earthquake that killed an estimated 90,000 people in 2008. If true, it would be the world's deadliest dam-induced earthquake ever. But it pales in comparison to the potential catastrophe if the Three Gorges Dam on the Yangtze were to collapse. As one engineer says, "It would be a flood of biblical proportions for the 100 million people who live downstream."

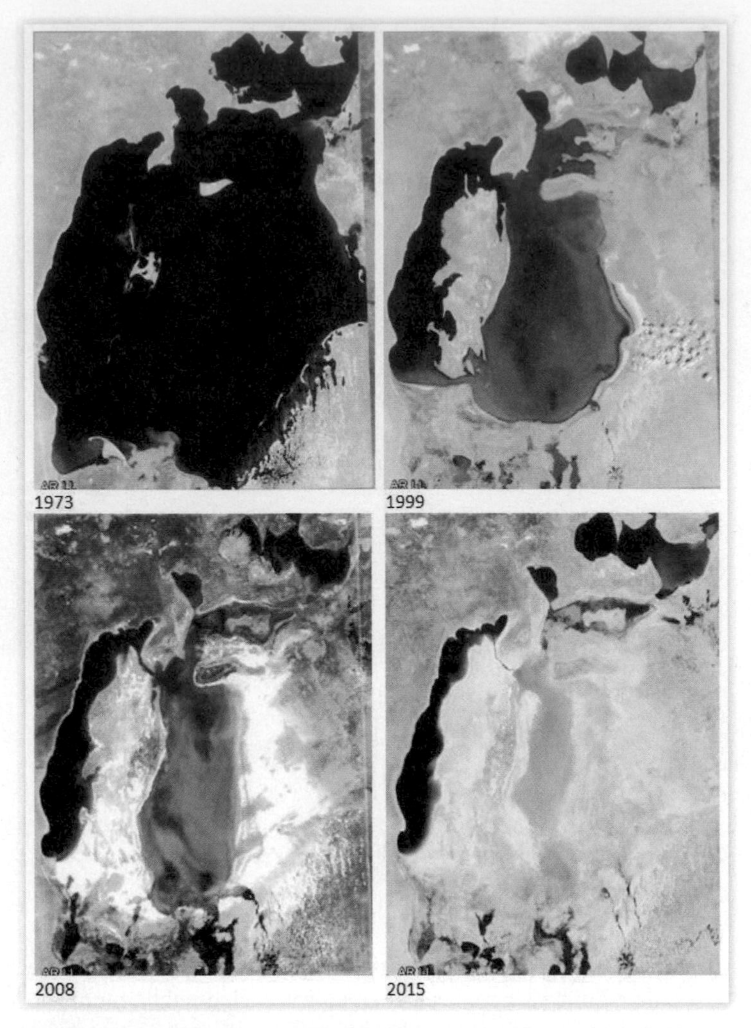

▲ **FIGURE 11.15** For 30 years, rivers feeding the Aral Sea were diverted to irrigate cotton and rice fields. The sea has lost more than 90 percent of its water. The "Small Aral" (upper right lobe) has separated from the main lake and is now being refilled. Source: USGS Earth Resources Observation and Science (EROS) Center.

Would you fight for water?

Many environmental scientists have warned that water shortages could lead to wars between nations. *Fortune* magazine wrote, "Water will be to the 21st century what oil was to the 20th." With one-third of all humans living in areas with water stress now, the situation may become much worse as the population grows and climate change dries up some areas and brings more severe storms to others. Already, we've seen skirmishes—if not outright warfare—over insufficient water. Think of recent global conflicts: how many occurred in the water-stressed regions shown in fig. 11.11?

Often water is an underlying cause of conflict. Violence in Syria has been connected with displaced rural populations whose farms failed in a drying climate. Many conflicts in sub-Saharan Africa also have involved populations driven from drying farming and fishing regions. An underlying cause of the genocide now occurring in the Darfur region of Sudan is water scarcity. When rain was plentiful, Arab pastoralists and African farmers coexisted peacefully. Drought—perhaps caused by global warming—has upset that truce. The hundreds of thousands who have fled to Chad could be considered climate refugees as well as war victims.

Although they haven't usually risen to the level of war, there have been at least 37 military confrontations in the past 50 years in which water has been at least one of the motivating factors. Thirty of those conflicts have been between Israel and its neighbors. India, Pakistan, and Bangladesh also have confronted each other over water rights, and Turkey and Iraq threatened to send their armies to protect access to the water in the Tigris and Euphrates Rivers. As water resources decline and populations become more urban, many companies are working to cash in on selling water. When a municipality privatizes water rights—sells controlling rights to a private company—the public loses control over its water, and prices rise.

Public anger over privatization of the public water supply in Bolivia sparked a revolution that overthrew the government in 2000. Water sales are already a \$400-billion-a-year business. Multinational corporations are moving to take control of water systems in many countries. Who owns water and how much they are able to charge for it could become the question of the century. Investors are now betting on scarce water resources by buying future water rights. One Canadian water company, Global Water Corporation, puts it best: "Water has moved from being an endless commodity that may be taken for granted to a rationed necessity that may be taken by force."

Freshwater shortages may become much worse in the future because of global climate change. Figure 11.16 shows the best estimate from the Intergovernmental Panel on Climate Change (IPCC) on likely changes in global precipitation. In combination with higher temperatures and increased evapotranspiration, these models suggest decreasing water resources in many already stressed areas.

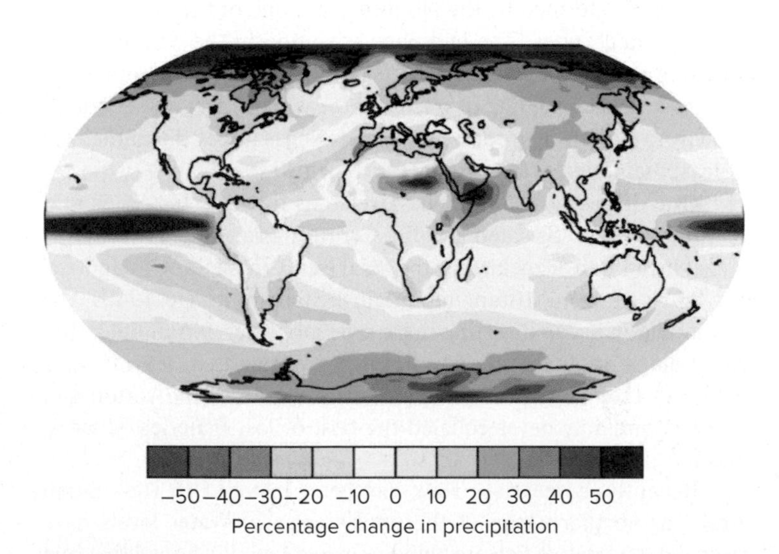

▲ **FIGURE 11.16** Change in precipitation in 2081–2100 (relative to 1986–2005). This map represents the average of 35 different models, with representative radiative forcing of 8.5 watts/m^2. With rising temperatures and evapotranspiration, modest increases in precipitation may not mean increased availability of water resources. Source: IPCC Climate Change 2014 Synthesis Report, Summary for Policymakers, Figure SPM.7, 12.

Land and water conservation protect resources

As with most resource questions, better management of water is usually cheaper and more effective than dealing with the consequences of poor management. Water conservation practices can happen at any step in the water use process: Protecting landcover, healthy soils, and vegetation in farmlands can slow runoff, so that water resources last longer into a dry season. Conserving wetlands provides water storage in upper reaches of a watershed. Multiple small dams on tributary streams can increase water storage without the extreme economic, social, and environmental costs of a large dam. Protecting rivers and floodplains also helps retain water on the land and slow its return to the sea.

Healthy forests are especially important for water conservation. They help protect moisture and keep sediment from clogging streams. Intact forests are especially important on steep slopes, high in the mountains, and in aquifer recharge zones (see fig. 11.7).

The former U.S. Forest Service chief, Mike Dombeck, was especially clear on this as he tried to push the agency toward new measures to protect the nation's forests. "Water," he said, "is the most valuable and least appreciated resource the national forests provide. More than 60 million people in 33 states obtain their drinking water from national forest lands. Protecting watersheds is far more economically important than logging or mining, and will be given the highest priority in forest planning."

What Can YOU DO?

Saving Water and Preventing Pollution

Each of us can conserve much of the water we use and avoid water pollution in many simple ways.

- Refer to figure 11.9, and reduce your consumption of water-intensive foods. Avoid food waste, to save money as well as water.
- Don't flush every time you use the toilet. Take shorter showers, and shower instead of taking baths.
- Always turn off the faucet while brushing your teeth or washing dishes. Draw a basin of water for washing and another for rinsing dishes. Don't run the dishwasher when it's half full.
- Use water-conserving appliances: low-flow showers, low-flush toilets, and aerated faucets.
- Fix leaking faucets, tubs, and toilets. A leaky toilet can waste 50 gal per day. To check your toilet, add a few drops of dark food coloring to the tank and wait 15 minutes. If the tank is leaking, the water in the bowl will change color.
- Put a brick or full water bottle in your toilet tank to reduce the volume of water in each flush.
- Avoid using toxic or hazardous chemicals for simple cleaning or plumbing jobs. A plunger or plumber's snake will often unclog a drain just as well as caustic acids or lye. Hot water and soap can accomplish most cleaning tasks.
- If you have a lawn, use water, fertilizer, and pesticides sparingly. Plant native, low-maintenance plants that have low water needs.
- If possible, use recycled (gray) water for lawns, house plants, and car washing.

Everyone can help conserve water

We could probably save as much as half of the water we now use for domestic purposes without great sacrifice or serious changes in our lifestyles. Simple steps, such as taking shorter showers, fixing leaks, and washing cars, dishes, and clothes as efficiently as possible, can go a long way toward forestalling the water shortages that many authorities predict. Isn't it better to adapt to more conservative uses now when we have a choice than to be forced to do it by scarcity in the future?

Water-conserving appliances, such as low-volume showerheads and efficient dishwashers, can reduce water consumption greatly (What Can You Do?, above). If you live in an arid part of the country, you might consider whether you really need a lush, green lawn that requires constant watering, feeding, and care. Planting native vegetation can be both ecologically sound and aesthetically pleasing (fig. 11.17). As part of its water conservation efforts, Las Vegas is paying residents to replace turf with natural vegetation, has asked golf courses to rip up fairways, and is encouraging hotels to use recycled water in fountains. Also, water cops patrol the streets to identify illegal watering or car washing.

Toilets are our greatest domestic water user (fig. 11.18). Usually, each flush uses several gallons of water to dispose of a few ounces of waste. On average, each person in the United States uses about 50,000 L (13,000 gal) of drinking-quality water annually to flush toilets. Low-flush toilets can drastically reduce this water use. Gray water (recycled from other uses) could be used for flushing, but installing separated plumbing systems is expensive.

▲ **FIGURE 11.17** By using native plants in a natural setting, residents of Phoenix save water and fit into the surrounding landscape.
©William P. Cunningham

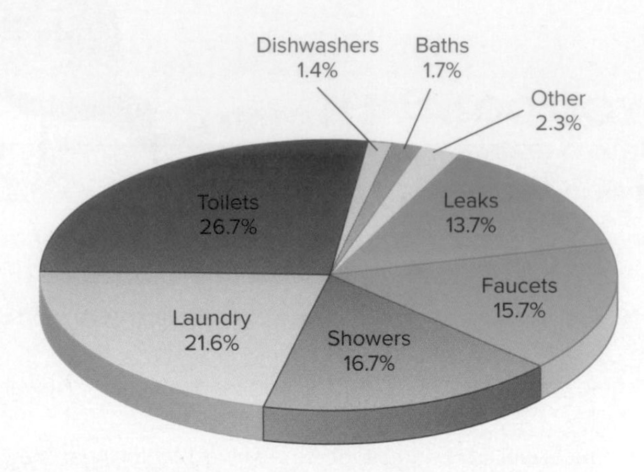

▲ **FIGURE 11.18** Typical household water use in the United States.
Source: Data from the American Water Works Association, 2010.

Communities are starting to recycle water

Reclaimed stormwater, and even water from sewage treatment plants, can be filtered, purified, and re-used. It's expensive, and usually we rely on soil bacteria and ecosystems to do this work for free. But where water shortages are immediate, it's worth the cost to recycle unclean water.

California, for example, has always shunted its stormwater and wastewater directly to the Pacific. Most treated wastewater–some almost clean enough to drink–is sent to the sea. But the state has been ramping up its water recycling programs recently. At present about 860 million m^3 (700,000 acre feet) of water is recycled each year, mostly for irrigation.

Some countries with desperate water shortages are using recycled water for drinking. These include Singapore, Australia, and parts of the United States (see What Do You Think?, p. 261). Would this be difficulty for you?

Growing recognition that water is a precious and finite resource has changed policies and encouraged conservation across the United States. Despite a growing population, the United States is now saving some 144 million liters (38 million gal) per day–a tenth the volume of Lake Erie–compared with per-capita consumption rates of 30 years ago. With nearly 90 million more people in the United States now than in 1980, we get by with 10 percent less water. New requirements for water-efficient fixtures in many cities help conserve water on the home front.

Pricing has an effect on our water usage. Ironically, water from Lake Mead, which is facing a supply crisis, currently costs Las Vegas residents 33 cents per m^3. By comparison, the same amount costs $3 in Atlanta and $7 in Copenhagen, where water is abundant. What do you think those prices do to motivate conservation?

Charging a higher proportion of real costs to users of public water projects can rationalize use patterns as will water marketing policies that allow prospective users to bid on water rights. Some countries already have effective water pricing and allocation policies that encourage the most socially beneficial uses and discourage wasteful water uses. It will be important, as water markets develop, to be sure that environmental, recreational, and wildlife values are not sacrificed to the lure of high-bidding industrial and domestic users.

11.4 WATER POLLUTANTS

- *Point source and nonpoint sources differ in distribution and controls.*
- *Different kinds of pollutants–pathogens, nutrients, metals–have very different impacts.*
- *Bottled water often contains plasticizers and other contaminants.*

Any physical, biological, or chemical change in water quality that adversely affects living organisms or makes water unsuitable for desired uses can be considered pollution. There are natural sources of water contamination, such as poison springs, oil seeps, and sedimentation from erosion, but here we'll focus primarily on human-caused changes that affect water quality or usability.

Pollution includes point sources and nonpoint sources

Pollution-control standards and regulations usually distinguish between point and nonpoint pollution sources. Factories, power plants, sewage treatment plants, underground coal mines, and oil wells are classified as **point sources** because they discharge pollution from specific locations, such as drainpipes, ditches, or sewer outfalls (fig. 11.19). These sources are discrete and identifiable, so they are relatively easy to monitor and regulate. It is generally possible to divert effluent from the waste streams of these sources and treat it before it enters the environment.

In contrast, **nonpoint sources** of water pollution are diffuse, having no specific location where they discharge into a particular

▲ **FIGURE 11.19** Sewer outfalls, industrial effluent pipes, acid draining out of abandoned mines, and other point sources of pollution are generally easy to recognize. Pollution-control laws have made this sight less common today than it once was. ©Simon Fraser/SPL/Science Source

body of water. They are much harder to monitor and regulate than point sources because their origins are hard to identify. Nonpoint sources include runoff from farm fields and feedlots, golf courses, lawns and gardens, construction sites, logging areas, roads, streets, and parking lots. While point sources may be fairly uniform and predictable throughout the year, nonpoint sources are often highly episodic. The first heavy rainfall after a dry period may flush high concentrations of gasoline, lead, oil, and rubber residues off city streets, for instance, while subsequent runoff may be much cleaner.

Perhaps the ultimate in diffuse, nonpoint pollution is atmospheric deposition of contaminants carried by air currents and precipitated into watersheds or directly onto surface waters as rain, snow, or dry particles. The Great Lakes, for example, have been found to be accumulating industrial chemicals, such as PCBs (polychlorinated biphenyls) and dioxins, as well as agricultural toxins, such as the insecticide toxaphene, that cannot be accounted for by local sources alone. The nearest sources for many of these chemicals are sometimes thousands of kilometers away.

Biological pollution includes pathogens and waste

Although the types, sources, and effects of water pollutants are often interrelated, it is convenient to divide them into major categories for discussion (table 11.3). Here, we look at some of the important sources and effects of different pollutants.

Pathogens The most serious water pollutants in terms of human health worldwide are pathogenic (disease-causing) organisms (see chapter 8). Among the most important waterborne diseases are typhoid, cholera, bacterial and amoebic dysentery, enteritis, polio, infectious hepatitis, and schistosomiasis. Malaria, yellow fever, and filariasis are transmitted by insects that have aquatic larvae. Altogether, at least 25 million deaths each year are blamed on water-related diseases. Nearly two-thirds of the mortalities of children under 5 years old in poorer countries are linked to these diseases.

The main source of these pathogens is untreated or improperly treated human wastes. Animal wastes from feedlots or fields near waterways and food-processing factories with inadequate waste treatment facilities also are sources of disease-causing organisms.

In developed countries, sewage treatment plants and other pollution-control techniques have reduced or eliminated most of the worst sources of pathogens in inland surface waters. Furthermore, drinking water is generally disinfected by chlorination, so epidemics of waterborne diseases are rare in these countries. The United Nations estimates that 90 percent of the people in developed countries have adequate (safe) sewage disposal, and 95 percent have clean drinking water.

The situation is quite different in less-developed countries, where billions of people lack adequate sanitation and access to clean drinking water. Conditions are especially bad in remote, rural areas, where sewage treatment is usually primitive or nonexistent and purified water is either unavailable or too expensive to obtain. The World Health Organization estimates that 80 percent of all sickness and disease in less-developed countries can be attributed to waterborne infectious agents and inadequate sanitation.

Detecting specific pathogens in water is difficult, time-consuming, and costly, so water quality is usually described in terms of concentrations of **coliform bacteria**–any of the many types that commonly live in the colon, or intestines, of humans and other animals. The most common of these is *Escherichia coli* (or *E. coli*). Other bacteria, such as *Shigella*, *Salmonella*, or *Listeria*, can also cause serious, even fatal, illness. If any coliform bacteria are present in a water sample, infectious pathogens are assumed to be present as well, and the Environmental Protection Agency (EPA) considers the water unsafe for drinking.

Biological Oxygen Demand The amount of oxygen dissolved in water is a good indicator of water quality and of the kinds of life it will support. An oxygen content above 6 parts per million (ppm) will support game fish and other desirable forms of aquatic life. At oxygen levels below 2 ppm, water will support mainly worms, bacteria, fungi, and other detritus feeders and decomposers. Oxygen is

TABLE 11.3 | Major Categories of Water Pollutants

CATEGORY	EXAMPLES	SOURCES
CAUSE OF ECOSYSTEM DISRUPTION		
1. Oxygen-demanding wastes	Animal manure, plant residues	Sewage, agricultural runoff, paper mills, food processing
2. Plant nutrients	Nitrates, phosphates, ammonium	Agricultural and urban fertilizers, sewage, manure
3. Sediment	Soil, silt	Land erosion
4. Thermal changes	Heat	Power plants, industrial cooling
CAUSE OF HEALTH PROBLEMS		
1. Pathogens	Bacteria, viruses, parasites	Human and animal excreta
2. Inorganic chemicals	Salts, acids, caustics, metals	Industrial effluents, household cleansers, surface runoff
3. Organic chemicals	Pesticides, plastics, detergents, oil, gasoline	Industrial, household, and farm use
4. Radioactive materials	Uranium, thorium, cesium, iodine, radon	Mining and processing of ores, power plants, weapons production, natural sources

added to water by diffusion from the air, especially when turbulence and mixing rates are high, and by photosynthesis of green plants, algae, and cyanobacteria. Turbulent, rapidly flowing water is constantly aerated, so it often recovers quickly from oxygen-depleting processes. Oxygen is removed from water by respiration and chemical processes that consume oxygen. Because oxygen is so important in water, **dissolved oxygen (DO)** levels are often measured to compare water quality in different places.

Adding organic materials, such as sewage or paper pulp, to water stimulates activity and oxygen consumption by decomposers. Consequently, **biochemical oxygen demand (BOD)**, or the amount of dissolved oxygen consumed by aquatic microorganisms, is another standard measure of water contamination. Alternatively, chemical oxygen demand (COD) is a measure of all organic matter in water.

Downstream from a point source, such as a municipal sewage plant discharge, a characteristic decline and restoration of water quality can be detected either by measuring DO content or by observing the types of flora and fauna that live in successive sections of the river. The oxygen decline downstream is called the **oxygen sag** (fig. 11.20). Upstream from the pollution source, oxygen levels support normal populations of clean-water organisms. Immediately below the source of pollution, oxygen levels begin to fall as decomposers metabolize waste materials. Rough fish, such as carp, bullheads, and gar, are able to survive in this oxygen-poor environment, where they eat both decomposer organisms and the waste itself.

Farther downstream, the water may become so oxygen depleted that only the most resistant microorganisms and invertebrates can survive. Eventually, most of the nutrients are used up, decomposer populations are smaller, and the water becomes oxygenated once again. Depending on the volumes and flow rates of the effluent plume and the river receiving it, normal communities may not appear for several miles downstream.

Nutrients cause eutrophication

Water clarity (transparency) is affected by sediments, chemicals, and the abundance of plankton organisms; clarity is a useful measure of water quality and water pollution. Rivers and lakes that have clear water and low biological productivity are said to be **oligotrophic** (*oligo* = little + *trophic* = nutrition). By contrast, **eutrophic** (*eu* + *trophic* = well-nourished) waters are rich in organisms and organic materials. Eutrophication, an increase in nutrient levels and biological productivity, often accompanies successional changes (see chapter 3) in lakes. Tributary streams bring in sediments and nutrients that stimulate plant growth. Over time, ponds and lakes often fill in, becoming marshes or even terrestrial biomes. The rate of eutrophication depends on water chemistry and depth, volume of inflow, mineral content of the surrounding watershed, and biota of the lake itself.

Human activities can greatly accelerate eutrophication, an effect called **cultural eutrophication**. Cultural eutrophication is caused mainly by increased nutrient input into a water body.

Increased productivity in an aquatic system sometimes can be beneficial. Fish and other desirable species may grow faster, providing a welcome food source. Often, however, eutrophication produces "blooms" of algae or thick growths of aquatic plants stimulated by elevated phosphorus or nitrogen levels (fig. 11.21). Bacterial populations then increase, fed by larger amounts of

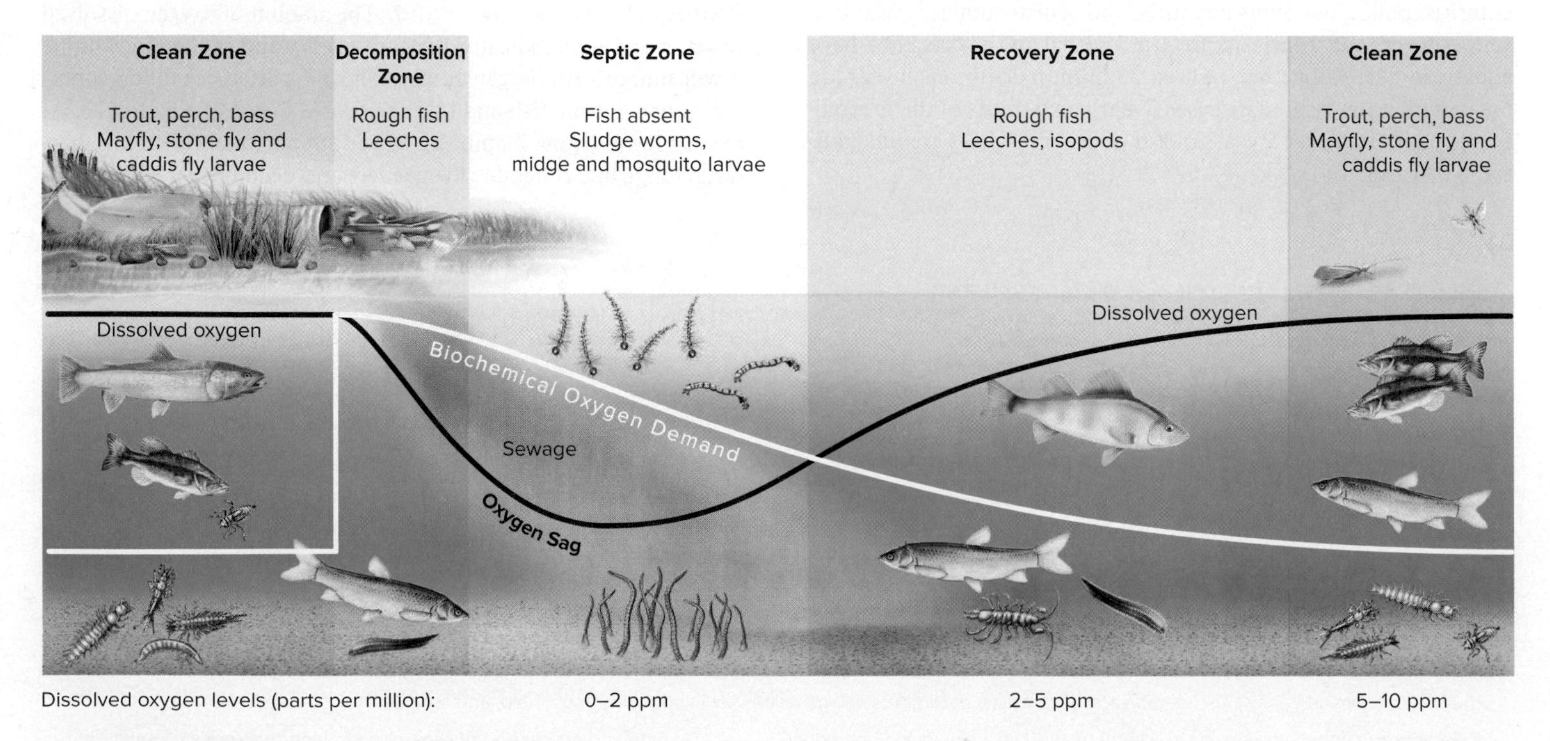

▲ **FIGURE 11.20** Oxygen sag downstream of an organic source. A great deal of time and distance may be required for the stream and its inhabitants to recover.

▲ **FIGURE 11.21** Eutrophic lake. Nutrients from agriculture and domestic sources have stimulated growth of algae and aquatic plants. This reduces water quality, alters species composition, and lowers the lake's recreational and aesthetic values. ©William P. Cunningham

organic matter. The water often becomes cloudy, or turbid, and has unpleasant tastes and odors. Cultural eutrophication can accelerate the "aging" of a water body enormously over natural rates. Lakes and reservoirs that normally might exist for hundreds or thousands of years can be filled in a matter of decades.

Eutrophication also occurs in marine ecosystems, especially in nearshore waters and partially enclosed bays or estuaries. Partially enclosed seas, such as the Black, Baltic, and Mediterranean Seas, tend to be in especially critical condition. During the tourist season, the coastal population of the Mediterranean, for example, swells to 200 million people. Eighty-five percent of the effluents from nearby large cities go untreated into the sea. Beach pollution, fish kills, and contaminated shellfish result. Extensive "dead zones" often form where rivers dump nutrients into estuaries and shallow seas. A federal study of the condition of U.S. coastal waters found that 28 percent of estuaries are impaired for aquatic life, and 80 percent of all coastal water is in fair to poor condition.

Inorganic pollutants include metals, salts, and acids

Some toxic inorganic chemicals are naturally released into water from rocks by weathering processes (see chapter 12). Humans accelerate the transfer rates in these cycles thousands of times above natural background levels by mining, processing, using, and discarding minerals.

Among the chemicals of greatest concern are heavy metals, such as mercury, lead, tin, and cadmium. Supertoxic elements, such as selenium and arsenic, also have reached hazardous levels in some waters. Other inorganic materials, such as acids, salts, nitrates, and chlorine, that are nontoxic at low concentrations may become concentrated enough to lower water quality and adversely affect biological communities.

Metals Many metals, such as mercury, lead, cadmium, and nickel, are highly toxic in minute concentrations. Because metals are highly persistent, they accumulate in food chains and have a cumulative effect in humans.

Currently, the most widespread toxic metal contamination in North America is mercury released from incinerators and coal-burning power plants. Transported through the air, mercury precipitates in water supplies, where it bioconcentrates in food webs to reach dangerous levels in top predators. As a general rule, Americans are warned not to eat more than one meal of wild-caught fish per week (fig. 11.22). Top marine predators, such as shark, swordfish, bluefin tuna, and king mackerel, tend to have especially high mercury content. Pregnant women and small children should avoid these species entirely. Public health officials estimate that 600,000 American children now have mercury levels in their bodies high enough to cause mental and developmental problems, while one woman in six in the United States has blood-mercury concentrations that would endanger a fetus.

Mine drainage and leaching of mining wastes are serious sources of metal pollution in water. A survey of water quality in eastern Tennessee found that 43 percent of all surface streams and lakes and more than half of all groundwater used for drinking supplies were contaminated by acids and metals from mine drainage. In some cases, metal levels were 200 times higher than what is considered safe for drinking water.

Nonmetallic Salts Some soils contain high concentrations of soluble salts, including toxic selenium and arsenic (see related story "Arsenic in Drinking Water" at www.connect.mheducation.com). Tens of millions of people are at risk in India and Bangladesh where groundwater is polluted with arsenic. Irrigation and drainage of desert soils can mobilize toxic salts and result in serious pollution problems, as in Kesterson Marsh in California, where selenium poisoning killed thousands of migratory birds in the 1980s.

▲ **FIGURE 11.22** Mercury contamination is the most common cause of impairment of U.S. rivers and lakes. Forty states have issued warnings about eating locally caught freshwater fish. Long-lived, top predators are especially likely to bioaccumulate toxic concentrations of mercury.

EXPLORING Science

Inexpensive Water Purification

When Ashok Gadgil was a child in Bombay, India, five of his cousins died in infancy from diarrhea spread by contaminated water. Although he didn't understand the implications of those deaths at the time, as an adult he realized how heartbreaking and preventable those deaths were. After earning a degree in physics from the University of Bombay, Gadgil moved to the University of California at Berkeley, where he was awarded a PhD in 1979. He's now senior staff scientist in the Environmental Energy Technology Division, where he works on solar energy and indoor air pollution.

▲ A woman fills her jug with clean water from the village WaterHealth kiosk. More than 6 million people's lives have been improved by this innovative system of water purification. ©Waterhealth International

But Dr. Gadgil wanted to do something about the problem of waterborne diseases in India and other developing countries. Although progress has been made in bringing clean water to poor people in many countries, about a billion people still lack access to safe drinking water. After studying ways to sterilize water, he decided that UV light treatment had the greatest potential for poor countries. It requires far less energy than boiling, and it takes less sophisticated chemical monitoring than chlorination.

There are many existing UV water treatment systems, but they generally involve water flowing around an unshielded fluorescent lamp. However, minerals in the water collect on the glass lamp and must be removed regularly to maintain effectiveness. Regular disassembly, cleaning, and reassembly of the apparatus are difficult in primitive conditions. The solution, Gadgil realized, was to mount the UV source above the water where it couldn't develop mineral deposits. He designed a system in which water flows through a shallow, stainless steel trough. The apparatus can be gravity fed and requires only a car battery as an energy source.

The system can disinfect 15 liters (4 gallons) of water per minute, killing more than 99.9 percent of all bacteria and viruses. This produces enough clean water for a village of 1,000 people. This simple system costs only about 5 cents per ton (950 liter). Of course, removing pathogens doesn't do anything about minerals, such as arsenic, or dangerous organic chemicals, so UV sterilization is often combined with filtering systems to remove those contaminants.

WaterHealth International, the company founded to bring this technology to market, now makes several versions of Gadgil's disinfection apparatus for different applications. A popular version provides a complete water purification system, including a small kiosk, jugs for water distribution, and training on how to operate everything.

A village-size system costs about $5,000. Grants and loans are available for construction, but villagers own and run the facility to ensure there's local responsibility. Each family in the cooperative pays about $1 per month for pure water. These systems have been installed in thousands of villages in India, Bangladesh, Africa, and the Philippines. Currently, about 6.6 million people are getting clean, healthy water at an easily affordable price from the simple system Dr. Gadgil invented.

Ordinarily nontoxic salts, such as sodium chloride (table salt), that are harmless at low concentrations also can be mobilized by irrigation and concentrated by evaporation, reaching levels that are dangerous for plants and animals. Salinity levels in the Colorado River and surrounding farm fields have become so high in recent years that millions of hectares of valuable croplands have had to be abandoned. In northern states, millions of tons of sodium chloride and calcium chloride are used to melt road ice in the winter. Leaching of road salts into surface waters has deleterious effects on aquatic ecosystems.

Acids and Bases Acids are released as by-products of industrial processes, such as leather tanning, metal smelting and plating, petroleum distillation, and organic chemical synthesis. Coal mining is an especially important source of acid water pollution. Sulfur compounds in coal react with oxygen and water to make sulfuric acid. Thousands of kilometers of streams in the United States have been acidified by acid mine drainage, some so severely that they are essentially lifeless.

Acid precipitation (see chapter 10) also acidifies surface-water systems. In addition to damaging living organisms directly, these acids leach aluminum and other elements from soil and rock, further destabilizing ecosystems.

Organic chemicals include pesticides and industrial substances

Thousands of different natural and synthetic organic chemicals are used in the chemical industry to make pesticides, plastics, pharmaceuticals, pigments, and other products that we use in everyday life. Many of these chemicals are highly toxic (see chapter 8). Exposure to very low concentrations (perhaps even parts per quadrillion, in the case of dioxins) can cause birth defects, genetic disorders, and cancer. Some can persist in the environment because they are resistant to degradation and toxic to organisms that ingest them.

The two principal sources of toxic organic chemicals in water are (1) improper disposal of industrial and household wastes and (2) pesticide runoff from farm fields, forests, roadsides, golf courses, and private lawns. The EPA estimates that about 500,000 metric tons of pesticides are used in the United States each year. Much of this washes into the nearest waterway, where it passes through ecosystems and may accumulate in high levels in nontarget organisms. The bioaccumulation of DDT in aquatic ecosystems was one of the first of these pathways to be understood (see chapter 8). Dioxins and other chlorinated hydrocarbons (hydrocarbon molecules that contain chlorine atoms) have been shown to accumulate to dangerous levels in the fat of salmon, fish-eating birds, and humans and to cause health problems similar to those resulting from toxic metal compounds.

Hundreds of millions of tons of hazardous organic wastes are thought to be stored in dumps, landfills, lagoons, and underground tanks in the United States (see chapter 14). Many, perhaps most, of these sites have leaked toxic chemicals into surface waters, groundwater, or both. The EPA estimates that about 26,000 hazardous waste sites will require cleanup because they pose an imminent threat to public health, mostly through water pollution.

Is bottled water safer?

It has become trendy to drink bottled water. Every year, Americans buy about 28 billion bottles of water at a cost of about $15 billion with the mistaken belief that it's safer than tap water. Worldwide, some 160 billion liters (42 billion gallons) of bottled water are consumed annually. Public health experts say that municipal water is often safer than bottled water because most large cities test their water supplies every hour for up to 25 different chemicals and pathogens, while the requirements for bottled water are much less rigorous. About one-quarter of all bottled water in the United States is simply reprocessed municipal water, and much of the rest is drawn from groundwater aquifers, which may or may not be safe. A recent survey of bottled water in China found that two-thirds of the samples tested had dangerous levels of pathogens and toxins.

Though the plastics used for bottling water are easily recycled, 80 percent of the bottles purchased in the United States end up in a landfill (the recycling rate is even poorer in most other countries). Overall, the average energy cost to make the plastic, fill the bottle, transport it to market, and then deal with the waste would be "like filling up a quarter of every bottle with oil," says water-expert Peter Gleick. Furthermore, it takes three to five times as much water to make the bottles as they hold. In blind tasting tests, most adults either can't tell the difference between municipal and bottled water or actually prefer municipal water. Furthermore, if water is held in a plastic bottle for weeks or months, plasticizers and other toxic chemicals can leach from the bottle into the water.

In most cases, bottled water is expensive, wasteful, and often less safe than most municipal water. Drink tap water and do a favor for your environment, your budget, and, possibly, your health.

Sediment is one of our most abundant pollutants

Sediment is a natural and necessary part of river systems. Sediment fertilizes floodplains and creates fertile deltas. But human activities, chiefly farming and urbanization, greatly accelerate erosion and increase sediment loads in rivers. Silt and sediment are considered the largest source of water pollution in the United States, being responsible for 40 percent of the impaired river miles in EPA water quality surveys. Cropland erosion contributes about 25 billion metric tons of soil, sediment, and suspended solids to world surface waters each year. Forest disturbance, road building, urban construction sites, and other sources add at least 50 billion additional tons.

This sediment fills lakes and reservoirs, obstructs shipping channels, clogs hydroelectric turbines, and makes purification of drinking water more costly. Sediments smother gravel beds in which insects take refuge and fish lay their eggs. Sunlight is blocked, so that plants cannot carry out photosynthesis, and oxygen levels decline. Murky, cloudy water also is less attractive for swimming, boating, fishing, and other recreational uses (fig. 11.23). Sediment washed into the ocean clogs estuaries and coral reefs.

Thermal pollution, usually effluent from cooling systems of power plants or other industries, alters water temperature. Raising or lowering water temperatures from normal levels can adversely affect water quality and aquatic life. Water temperatures are usually much more stable than air temperatures, so aquatic organisms tend to be poorly adapted to rapid temperature changes. Lowering the temperature of tropical oceans by even 1°C can be lethal to some corals and other reef species. Raising water temperatures can have similar devastating effects on sensitive organisms. Oxygen solubility in water decreases as temperatures increase, so species requiring high oxygen levels are adversely affected by warming water.

Humans also cause thermal pollution by altering vegetation cover and runoff patterns. Reducing water flow, clearing streamside trees, and adding sediment all make water warmer and alter the ecosystems in a lake or stream.

Warm-water plumes from power plants often attract fish and birds, which find food and refuge there, especially in cold weather. This

▲ **FIGURE 11.23** Sediment and industrial waste flow from this drainage canal into Lake Erie. ©Laurence Lowry/Science Source

artificial environment can be a fatal trap, however. Florida's manatees, an endangered mammal, are attracted to the abundant food supply and warm water in power plant thermal plumes. Often they are enticed into spending the winter much farther north than they normally would. On several occasions, a midwinter power plant breakdown has exposed a dozen or more of these rare animals to a sudden, deadly thermal shock.

11.5 PERSISTENT CHALLENGES

- *Pollution is especially common in low-income areas.*
- *Pathogens, metals, nutrients, and sediment are major impairments.*
- *Wastewater treatment removes solids.*

The greatest impediments to improving water quality in most places are sediment, nutrients, and pathogens, especially from nonpoint discharges of pollutants (fig. 11.24, graph of pollution types). These sources are harder to identify and to reduce or treat than are specific point sources. About three-fourths of the water pollution in the United States comes from soil erosion, fallout of air pollutants, and surface runoff from urban areas, farm fields, and feedlots. In the United States, as much as 25 percent of the 46,800,000 metric tons (52 million tons) of fertilizer spread on farmland each year is carried away by runoff.

Cattle in feedlots produce some 129,600,000 metric tons (144 million tons) of manure each year, and the runoff from these sites is rich in viruses, bacteria, nitrates, phosphates, and other contaminants. A single cow produces about 30 kg (66 lb) of manure per day. Some feedlots have 100,000 animals with no provision for capturing or treating runoff water. Imagine drawing your drinking water downstream from such a facility. Pets also can be a problem. It is estimated that the wastes from about a half million dogs in New York City are disposed of primarily through storm sewers and therefore do not go through sewage treatment.

Loading of both nitrates and phosphates in surface water has decreased from point sources but has increased about fourfold since 1972 from nonpoint sources. Fossil fuel combustion has become a major source of nitrates, sulfates, arsenic, cadmium, mercury, and other toxic pollutants that find their way into water. Carried to remote areas by atmospheric transport, these combustion products now are found nearly everywhere in the world. Toxic organic compounds, such as DDT, PCBs, and dioxins, also are transported long distances by wind currents.

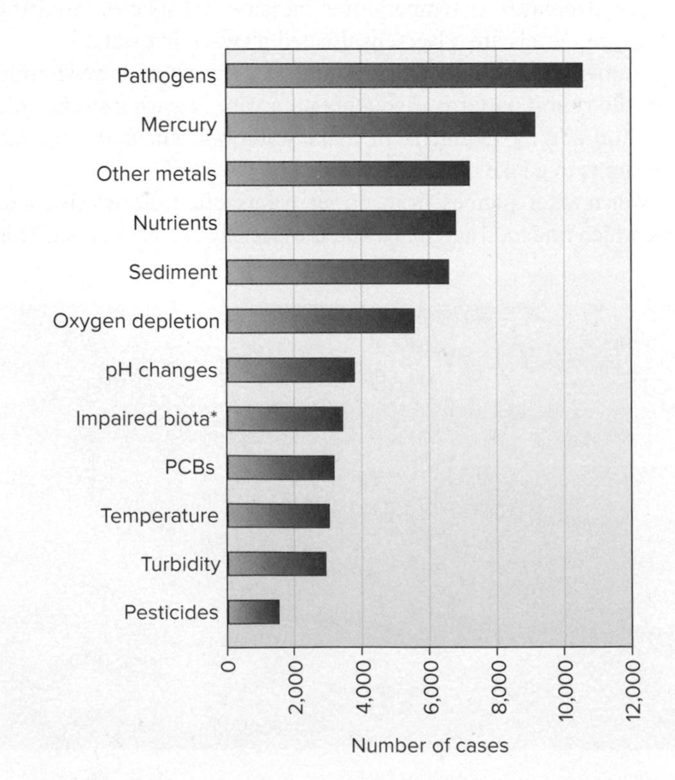

▲ **FIGURE 11.24** Twelve leading causes of surface-water impairment in the United States. *Undetermined causes. Source: Data from EPA, 2009.

Developing countries often have serious water pollution

Installing water treatment and pollution control technologies can take decades and cost billions of dollars. And it requires established, professional agencies to oversee, enforce, and support pollution control measures. Even among developed countries, water quality is uneven. Sweden provides at least secondary sewage treatment to 98 percent of its population. The United States provides treatment for 70 percent of the population, Spain only 18 percent, and Greece less than 2 percent. Russia and its neighbors have enormous water quality challenges left over from the Soviet era.

Although parts of China are as wealthy as anywhere, much of the country remains rural and poor. It's estimated that 70 percent of China's surface water is unsafe for use (fig. 11.25). Shanxi Province illustrates China's challenges. An industrial powerhouse,

▼ **FIGURE 11.25** Half of the water in China's major rivers is too polluted to be suitable for any human use. Although the government has spent billions of yuan in recent years, dumping of industrial and domestic waste continues at dangerous levels. ©William P. Cunningham

in the north-central part of the country, Shanxi has about one-third of China's known coal resources and produces about two-thirds of the country's energy. In addition to power plants, major industries include steel mills, tar factories, and chemical plants.

Economic growth has been pursued in recent decades at the expense of environmental quality. According to the Chinese Environmental Protection Agency, the country's ten worst polluted cities are all in Shanxi. Factories have been allowed to exceed pollution discharges with impunity. For example, 3 million tons of wastewater is produced every day in the province, with two-thirds of it discharged directly into local rivers without any treatment. Locals complain that the rivers, which once were clean and fresh, now run black with industrial waste. Among the 26 rivers in the province, 80 percent were rated Grade V (unfit for any human use) or higher in 2006. More than half the wells in Shanxi are reported to have dangerously high arsenic levels. Many of the 85,000 reported public protests in China in 2006 involved complaints about air and water pollution.

The less-developed countries of South America, Africa, and Asia have even worse water quality than do the poorer countries of Europe. Sewage treatment is usually either totally lacking or woefully inadequate. In urban areas, 95 percent of all sewage is discharged untreated into rivers, lakes, or the ocean. Low technological capabilities and little money for pollution control are made even worse by burgeoning populations, rapid urbanization, and the shift of much heavy industry (especially the dirtier ones) from developed countries where pollution laws are strict to less-developed countries where regulations are more lenient.

Appalling environmental conditions often result from these combined factors. In many countries, open sewers run through urban areas (fig. 11.26). Two-thirds of India's surface waters are contaminated sufficiently to be considered dangerous to human health. The Yamuna River in New Delhi has 7,500 coliform bacteria per 100 ml (37 times the level considered safe for swimming in the United States) *before* entering the city. The coliform count increases to an incredible 24 *million* cells per 100 ml as the river leaves the city! At the same time, the river picks up some 20 million liters of industrial effluents every day from New Delhi. It's no wonder that disease rates are high and life expectancy is low in this area. Only 1 percent of India's towns and cities have any sewage treatment, and only eight cities have anything beyond primary treatment.

▲ **FIGURE 11.26** Ditches in this Haitian slum serve as open sewers into which all manner of refuse and waste are dumped. The health risks of living under these conditions are severe. ©Robert Nickelsberg/Getty Images

Groundwater is especially hard to clean up

About half the people in the United States, including 95 percent of those in rural areas, depend on underground aquifers for their drinking water. This vital resource is threatened in many areas by overuse and pollution and by a wide variety of industrial, agricultural, and domestic contaminants. For decades it was widely assumed that groundwater was impervious to pollution because soil would bind chemicals and cleanse water as it percolated through. Springwater or artesian well water was considered to be the definitive standard of water purity, but that is no longer true in many areas.

A recent source of aquifer pollution in many areas comes from hydraulic fracturing (or fracking) to release gas and oil from tight shale formations (see chapter 12 for more discussion). The U.S. EPA has found high levels of methane and hydraulic fracturing fluids in water wells near fracking operations. From 2005 to 2015, drilling companies used nearly 11 million tons of toxic diesel fuel, hydrochloric acid, benzene, formaldehyde, methanol, and hundreds of unidentified compounds in fracking.

The EPA estimates that every day some 4.5 trillion liters (1.2 trillion gal) of contaminated water seep into the ground in the United States from septic tanks, leaking underground storage tanks at gas stations, municipal and industrial landfills and waste disposal sites, surface impoundments, agricultural fields, forests, and abandoned wells (fig. 11.27). The most toxic of these are probably waste disposal sites. Agricultural chemicals and wastes are responsible for the largest total volume of pollutants and area affected. Because deep underground aquifers often have residence times of thousands of years, many contaminants are extremely stable once underground. It is possible, but expensive, to pump water out of aquifers, clean it, and then pump it back.

In farm country, especially in the Midwest's Corn Belt, fertilizers and pesticides commonly contaminate aquifers and wells. Herbicides such as atrazine and alachlor are widely used on corn and soybeans and show up in about half of all wells in Iowa, for example. Nitrates from fertilizers often exceed safety standards in rural drinking water. These high nitrate levels are dangerous to infants (nitrates combine with hemoglobin in the blood and result in "blue-baby" syndrome).

Every year, epidemiologists estimate that around 1.5 million Americans fall ill from infections caused by fecal contamination. In 1993, for instance, a pathogen called cryptosporidium got into the Milwaukee public water system, making 400,000 people sick and killing at least 100 people. The total costs of these diseases amount to billions of dollars per year. Preventative measures, such as protecting water sources and aquifer recharge zones and updating treatment and distribution systems, would cost far less.

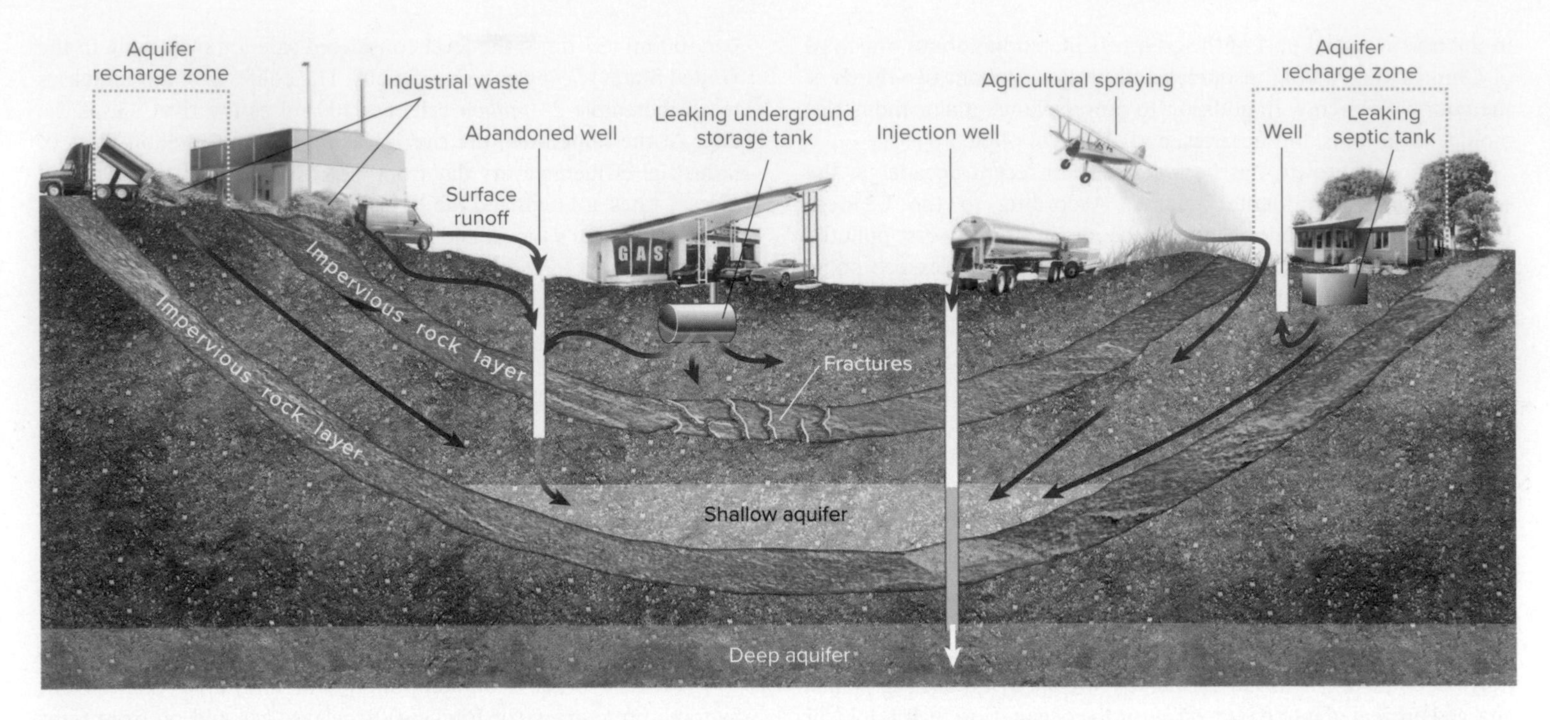

▲ **FIGURE 11.27** Sources of groundwater pollution. Septic systems, landfills, and industrial activities on aquifer recharge zones leach contaminants into aquifers. Wells provide a direct route for injection of pollutants into aquifers.

Ocean pollution has few controls

Although we don't use ocean waters directly, ocean pollution is serious and one of the fastest-growing water pollution problems. Coastal bays, estuaries, shoals, and reefs are often overwhelmed by pollution. Dead zones and poisonous algal blooms are increasingly widespread. Toxic chemicals, heavy metals, oil, sediment, and plastic refuse affect some of the most attractive and productive ocean regions. The potential losses caused by this pollution amount to billions of dollars each year. In terms of quality of life, the costs are incalculable.

Discarded plastic flotsam and jetsam are becoming a ubiquitous mark of human impact on the oceans. Even the most remote beaches of distant islands are likely to have mounds of trash. Every year we introduce over 8 million metric tons of plastic bottles, packaging, and other debris into the oceans. Originating as urban litter or waste dumped at sea, it chokes and ensnares seabirds, fish, turtles, and mammals (fig. 11.28).

Researchers have discovered a vast swath of the Pacific Ocean filled with a soup of plastic refuse. Dubbed the "Great Pacific Garbage Patch," this slowly swirling vortex fills two "convergence

▲ **FIGURE 11.28** A deadly necklace. Marine biologists estimate that castoff nets, plastic beverage yokes, and other packing residue kill hundreds of thousands of birds, mammals, and fish each year. Source: (*left*) Joe Lucas/Marine Entanglement Research Program/National Marine Fisheries Service NOAA; Source: (*right*) NOAA and Georgia Department of Natural Resources

zones" that collect trash from all over the world. One of these gyres occurs between Hawaii and California; the other one is closer to Japan. Each is larger than Texas. Currents sweep all sorts of refuse into these huge vortices that have been called the world's biggest garbage dumps. Much of this plastic consists of tiny particles suspended at or just below the water surface. The smallest particles may be more dangerous biologically than bigger pieces. Plankton and small fish ingest the plastic bits along with the contaminants to their surface, and introduce them into the marine food chain. See chapter 14 for further discussion of this topic.

Oil pollution affects beaches and open seas around the world. Oceanographers estimate that between 3 million and 6 million metric tons of oil are discharged into the world's oceans each year from oil tankers, fuel leaks, intentional discharges of fuel oil, and coastal industries. About half of this amount is due to maritime transport. Of this portion, most is not from dramatic, headline-making accidents such as the 2010 *Deepwater Horizon* spill in the Gulf of Mexico. Rather, routine open-sea bilge pumping and tank cleaning are the primary source. These activities are illegal but very common.

Fortunately, awareness of ocean pollution is growing. Oil spill cleanup technologies and response teams are improving, although most oil is eventually decomposed by natural bacteria. Efforts are growing to control waste plastic. Criminals who ship toxic waste to poor countries are being prosecuted. Volunteer efforts also are helping to reduce beach pollution locally: In one day, volunteers in Texas gathered more than 300 tons of plastic refuse from Gulf Coast beaches.

11.6 WATER TREATMENT AND REMEDIATION

- *Nonpoint pollution requires controlling upstream sources.*
- *Wastewater treatment involves removal of bacteria, sediments, and nutrients.*
- *Pollutants of emerging concern include drugs, steroids, and disinfectants.*

The cheapest and most effective way to reduce pollution is to avoid producing it or releasing it in the first place. Eliminating lead from gasoline has resulted in a dramatic decrease in the amount of lead in U.S. surface waters. Studies have shown that as much as 90 percent less road deicing salt can be used in many areas without significantly affecting the safety of winter roads. Careful handling of oil and petroleum products can greatly reduce the amount of water pollution caused by these materials. Although we still have problems with persistent chlorinated hydrocarbons spread widely in the environment, the banning of DDT and PCBs in the 1970s has resulted in significant reductions in levels in wildlife.

Industry can reduce pollution by recycling or reclaiming materials that otherwise might be discarded in the waste stream. These approaches usually have economic as well as environmental benefits. Companies can extract valuable metals and chemicals and sell them, instead of releasing them as toxic contaminants into the water system. Both markets and reclamation technologies are improving as awareness of these opportunities grows. In addition, modifying land use is an important component of reducing pollution.

Impaired water can be restored

The good news is that there are many examples of improved water quality. Consider, for example, the case of Minamata, Japan. Mercury contamination in the bay poisoned residents who ate local fish, and effects were so notorious that "Minamata Disease" became synonymous with mercury poisoning in the 1950s. The mercury source was local industries, which for years discarded waste into the bay. A neurotoxin, mercury accumulated in the food chain, causing nerve damage in adults and severe developmental defects in children. But when mercury discharges were banned, mercury concentrations eventually became diluted, and in 1997 the bay was declared clean.

Similarly, the Rhine River, which connects most of the major industrial centers of Europe, has long been an environmental disaster, flowing with urban, industrial, and agricultural waste. More than 50 million people live in the Rhine basin, and some 20 million get drinking water from the river. After a disastrous fire at a chemical warehouse near Basel, Switzerland, created a toxic spill that killed millions of fish and made the water undrinkable, the countries along the river finally committed to controlling waste, runoff, and industrial dumping.

Oxygen concentrations in the Rhine have gone up fivefold since 1970 (from less than 2 mg/l to nearly 10 mg/l) in long stretches of the river. Chemical oxygen demand has fallen fivefold, and organochlorine levels have decreased as much as tenfold. Fish and aquatic invertebrates have returned to the river. In 1992, for the first time in decades, mature salmon were caught in the Rhine.

Nonpoint sources require prevention

Farmers have long contributed a huge share of water pollution, including sediment, fertilizers, and pesticides that flow from fields. Soil conservation practices on farmlands (see chapter 7) aim to keep soil and contaminants on fields, where they are needed. Precise application of fertilizer, irrigation water, and pesticides saves money and reduces water contamination. Preserving wetlands, which help capture sediment and contaminants, also helps protect surface and groundwaters.

In urban areas, reducing waste that enters storm sewers is essential. It is getting easier for city residents to recycle waste oil and to properly dispose of paint and other household chemicals that they once dumped into storm sewers or the garbage. Urbanites can also minimize use of fertilizers and pesticides. Regular street sweeping greatly reduces nutrient loads (from decomposing leaves and debris) in rivers and lakes.

The tremendous challenge of managing these sources is seen in Chesapeake Bay, America's largest estuary. Once fabled for its abundant oysters, crabs, shad, striped bass, and other valuable fisheries, the bay had deteriorated seriously by the early 1970s. Citizens' groups, local communities, state legislatures, and the federal government together established an innovative pollution-control program that made the bay the first estuary in America targeted for protection and restoration.

Among the principal objectives of this plan is reducing nutrient loading through land use regulations in the bay's six watershed states to control agricultural and urban runoff. Pollution–prevention measures, such as banning phosphate detergents, also are important, as are upgrading wastewater treatment plants and improving compliance with discharge and filling permits. Efforts are under way to replant thousands of hectares of sea grasses and to restore wetlands that filter out pollutants.

Since the 1980s, annual phosphorus discharges into Chesapeake Bay have dropped 40 percent. Nitrogen levels, however, have remained constant or have even risen in some tributaries. Although progress has been made, the goals of reducing both nitrogen and phosphate levels by 40 percent and restoring viable fish and shellfish populations are still decades away. Still, as former EPA administrator Carol Browner said, it demonstrates the power of cooperation in environmental protection.

How do we treat municipal waste?

Under natural conditions, water purification occurs constantly in soils and water. Bacteria take up and transform nutrients or break down oils. Sand and soil filter water; plant roots and fungi use nutrients in the water and simultaneously capture metals and other components. When water is cool and moving, oxygen from the air mixes in, eliminating stagnant conditions where harmful organisms can grow.

The high population densities of cities, however, produce much more waste than natural systems can process. As we have already seen, human and animal wastes usually create the most serious health-related water pollution problems. More than 500 types of disease-causing (pathogenic) bacteria, viruses, and parasites can travel from human or animal excrement through water.

Most developed countries require that cities and towns build municipal water treatment systems to purify the human and household waste. Most rural households use septic systems, which allow solids to settle in a tank, where bacteria decompose them. Liquids percolate through soil, where soil bacteria presumably purify them. Where population densities are not too high, this can be an effective method of waste disposal. With urban sprawl, however, groundwater pollution often becomes a problem.

Municipal treatment has three levels of quality

Over the past 100 years, sanitary engineers have developed ingenious and effective municipal wastewater treatment systems to protect human health, ecosystem stability, and water quality (fig. 11.29). This topic is an important part of pollution control and is a principal responsibility of every municipal government.

Primary treatment physically separates large solids from the waste stream with screens and settling tanks. Settling tanks allow grit and some dissolved (suspended) organic solids to fall out as sludge. Water drained from the top of settling tanks still carries up to 75 percent of the organic matter, including many pathogens. These pathogens and organics are removed by **secondary treatment**, in which aerobic bacteria break down dissolved organic compounds. In secondary treatment, effluent is aerated, often with sprayers or in an aeration tank, in which air is pumped through the microorganism-rich slurry. Fluids can also be stored in a sewage lagoon, where sunlight, algae, and air process waste more cheaply but more slowly. Effluent from secondary treatment processes is usually disinfected with chlorine, UV light, or ozone to kill harmful bacteria before it is released to a nearby waterway.

Tertiary treatment removes dissolved metals and nutrients, especially nitrates and phosphates, from the secondary effluent. Although wastewater is usually free of pathogens and organic material after secondary treatment, it still contains high levels of these inorganic nutrients. If discharged into surface waters, these nutrients stimulate algal blooms and eutrophication. Allowing effluent to flow through a wetland or lagoon can remove nitrates and phosphates. Alternatively, chemicals often are used to bind and precipitate nutrients.

Sewage sludge can be a valuable fertilizer, but it can be unsafe if it contains metals and toxic chemicals. Some cities spread sludge on farms and forest lands, while others convert it to methane (natural gas). Many cities, however, incinerate or landfill sludge, both expensive options. Often, sanitary sewers are connected to storm sewers, which carry contaminated runoff from streets, parking lots, and yards. This allows treatment to remove oil, gasoline, fertilizers, and pesticides. Heavy storms, however, often overload municipal systems, resulting in large volumes of raw sewage and toxic surface runoff being dumped directly into rivers or lakes.

Conventional Treatment Misses New Pollutants In 2002 the USGS released the first-ever study of pharmaceuticals and hormones in streams. Scientists sampled 130 streams, looking for 95 contaminants, including antibiotics, natural and synthetic hormones, detergents, plasticizers, insecticides, and fire retardants (fig. 11.30). All these substances were found, usually in low concentrations. One stream had 38 of the compounds tested. Drinking-water standards exist for only 14 of the 95 substances. A similar study found the same substances in groundwater, which is much harder to clean than surface waters. What are the effects of these widely used chemicals on our environment or on people consuming the water? Nobody knows. This study is a first step toward filling huge gaps in our knowledge about their distribution, though.

Natural wastewater treatment can be an answer

Natural wastewater treatment systems offer a promising alternative to remote locations, developing countries, and small factories that can't afford

FIGURE 11.29 In conventional sewage treatment, aerobic bacteria digest organic materials in high-pressure aeration tanks. This is described as secondary treatment.
©Steve Allen/Brand X Pictures/Alamy Stock Photo

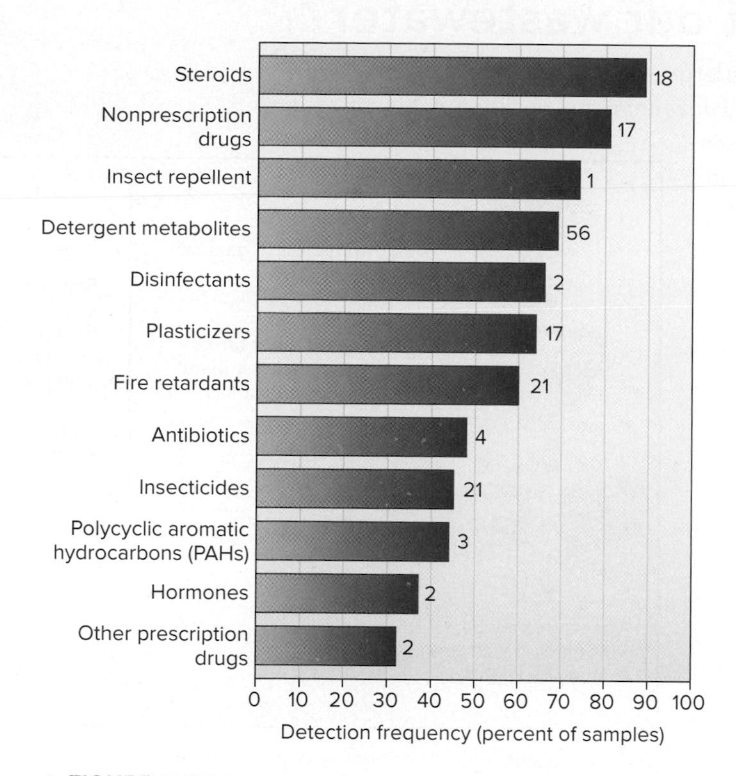

▲ **FIGURE 11.30** Detection frequency of organic, wastewater contaminants in a recent USGS survey. Maximum concentrations in water samples are shown above the bars in micrograms per liter. Dominant substances included DEET insect repellent, caffeine, and triclosan, which comes from antibacterial soaps.

conventional treatment. These systems are still unfamiliar and unconventional, so they are relatively uncommon, but they offer many advantages. Natural wastewater treatment systems are normally cheaper to build and operate than conventional systems. They use less energy and less chlorine or other purifiers, because gravity moves water, and plants and bacteria do most (or all) disinfection. With fewer pumps and filters to manage, less staff time is needed. Plants remove nutrients, metals, and other contaminants that are not captured by most conventional systems.

Constructed wetlands are a complex of artificial marshes designed to filter and decompose waste. One of the best known of these is in Arcata, California, which was required to build a new and very expensive sewer system upgrade 30 years ago. As an alternative, the city transformed a 65-ha garbage dump into a series of ponds and marshes that serve as a simple, low-cost, waste treatment facility. Arcata saved millions of dollars and improved its environment simultaneously. The marsh is a haven for wildlife and has become a prized recreation area for the city.

Similar wetland waste treatment systems are now operating in many developing countries. Effluent from these operations can be used to irrigate crops or raise fish for human consumption if care is taken first to destroy pathogens. Usually 20 to 30 days of exposure to sun, air, and aquatic plants is enough to make the water safe. These systems make an important contribution to human food supplies. A 2,500-ha waste-fed aquaculture facility in Kolkata, for example, supplies about 7,000 metric tons of fish annually to local markets.

Many institutions don't have the space for a constructed wetland. One of these is the Cedar Grove Cheese Factory in southern Wisconsin, which has built a "**Living Machine**®," a sequence of tanks, bacteria, algae, and small artificial wetlands (fig. 11.31). This system converts factory effluent to nearly pure water and vegetation. It removes 99 percent of the biological oxygen demand, 98 percent of the suspended solids, 93 percent of the nitrogen, and 57 percent of the phosphorus.

Systems like this can be built adjacent to, or even inside of, buildings. Combinations of plants and animals, including algae, rooted aquatic plants, clams, snails, and fish are present, each chosen to provide a particular service in a contained environment. The water leaving such a system is of drinkable quality, and it's cleaner than the water received by the facility. This approach can save resources and money, and it can serve as a valuable educational tool.

Remediation can involve containment, extraction, or biological treatment

Just as there are many sources of water contamination, there are many ways to clean it up. New developments in environmental engineering are providing promising solutions to many water pollution problems. Containment methods keep dirty water from spreading. Many pollutants can be destroyed or detoxified by chemical reactions that oxidize, reduce, neutralize, hydrolyze, precipitate, or otherwise change their chemical composition. Where chemical techniques are ineffective, physical methods may work. Solvents and other volatile organic compounds, for instance, can be stripped from solution by aeration and then burned in an incinerator.

▲ **FIGURE 11.31** This "Living Machine," installed at a cheese factory in Wisconsin to purify wastewater naturally, is a sequence of tanks containing plants, algae, bacteria, and other aquatic organisms that filter water and remove organic waste and nutrients from factory effluent. ©Cedar Grove Cheese Inc.

Could natural systems treat our wastewater?

Conventional sewage treatment systems are designed to treat large volumes of effluent quickly and efficiently. Water treatment is necessary for public health and environmental quality, but it is expensive. Industrial-scale installations, high energy inputs, and caustic chemicals are needed. Huge quantities of sludge must be incinerated or trucked off-site for disposal.

▲ An aeration tank helps aerobic (oxygen-using) bacteria digest organic compounds. *(left)* ©Thinkstock Images/Getty Images; *(right)* ©Steve Allen/Brand X Pictures/Alamy Stock Photo

KC 11.3

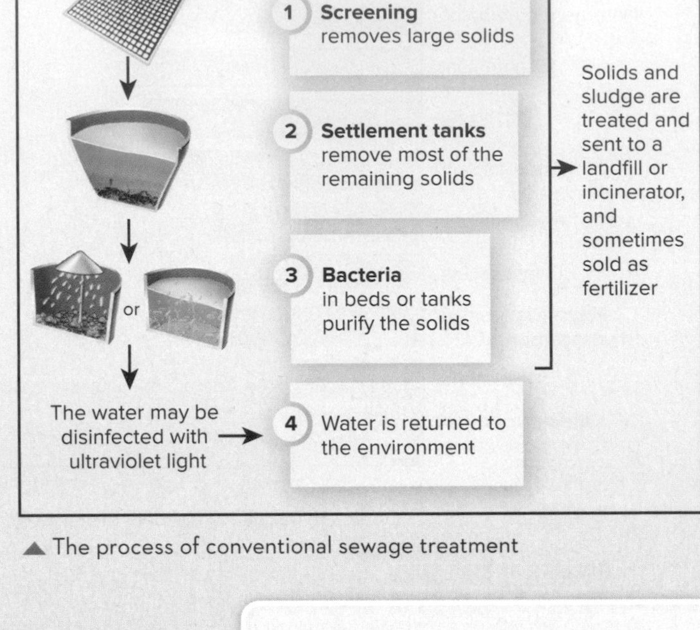

▲ The process of conventional sewage treatment

Conventional treatment misses new pollutants. Pharmaceuticals and hormones, detergents, plasticizers, insecticides, and fire retardants are released freely into surface waters, because these systems are not designed for those contaminants.

KC 11.6

◀ A constructed wetland outside can be an attractive landscaping feature that further purifies water. ©William P. Cunningham

◀ Where space is available, a larger constructed wetland can serve as recreational space, a wildlife refuge, a living ecosystem, and a recharge area for groundwater or streamflow. ©William P. Cunningham

KC 11.7

Constructed wetland systems can be designed with endless varieties, but all filter water through a combination of beneficial microorganisms and plants. Here are common components:

- **Anaerobic (oxygen-free) tanks:** here anaerobic bacteria convert nitrate (NO_3) to nitrogen gas (N_2), and organic molecules to methane (CH_4). In some systems, methane can be captured for fuel.
- **Aerobic (oxygen-available) tanks:** aerobic bacteria convert ammonium (NH_4) to nitrate (NO_3); green plants and algae take up nutrients.
- **Gravel-bedded wetland:** beneficial microorganisms and plants growing in a gravel bed capture nutrients and organic material. In some systems, the wetland provides wildlife habitat and recreational space.
- **Presumable disinfection:** water is clean leaving the system, but rules usually require that chlorine be added to ensure disinfection. Ozone or ultraviolet light can also be used.

In this system, after passing through the growing tanks, the effluent water runs over a waterfall and into a small fish pond for additional oxygenation and nutrient removal. This verdant greenhouse is open to the public and adds an appealing indoor space in a cold, dry climate. ©Mary Ann Cunningham ▶

The growing tanks need to be in a greenhouse or other sunny space to provide light for plants. ©Mary Ann Cunningham ▶

KC 11.8

KC 11.9

Natural wastewater treatment is unfamiliar but usually cheaper

We depend on ecological systems—natural bacteria and plants in water and soil—to finish off conventional treatment. Can we use these systems for the entire treatment process? Although they remain unfamiliar to most cities and towns, **wetland-based treatment systems** have operated successfully for decades—at least as long as the lifetime of a conventional plant. Because they incorporate healthy bacteria and plant communities, there is potential for uptake of novel contaminants and metals as well as organic contaminants. These systems also remove nutrients better than most conventional systems do. These systems can be half as expensive as conventional systems because they have

- few sprayers, electrical systems, and pumps ⟶ cheaper installation
- gravity water movement ⟶ low energy consumption
- few moving parts or chemicals ⟶ low maintenance
- biotic treatment ⟶ little or no chlorine use
- nutrient uptake ⟶ more complete removal of nutrients, metals, and possibly organic compounds

▲ Drinkable quality water is produced by a well-designed natural system. This photo shows before and after treatment. Most people are squeamish about the prospect of drinking treated wastewater, so recycled water is generally used for other purposes such as toilets, washing, or irrigation. Since these uses make up about 95 percent of many municipal water supplies, they can represent a significant savings. ©Peter Essick/Getty Images

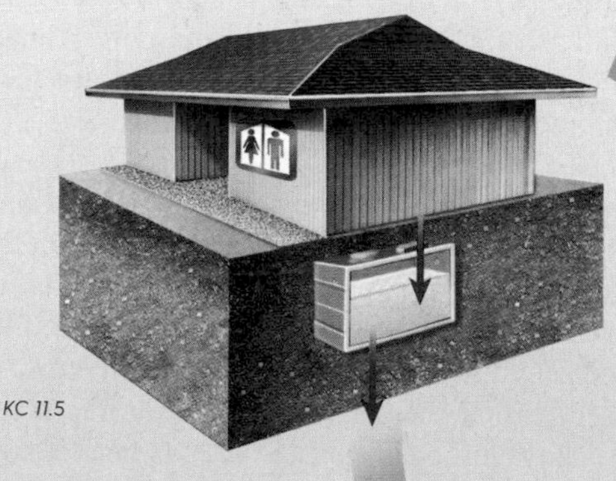

CAN YOU EXPLAIN?

1. **Based on your reading of this chapter, what are the primary contaminants for which water is treated?**
2. **What is the role of bacteria in a system like this?**
3. **What factors make conventional treatment expensive?**
4. **Why is conventional treatment more widely used?**

Often, living organisms can clean contaminated water effectively and inexpensively. We call this **bioremediation.** Restored wetlands, for instance, along stream banks or lake margins can effectively filter out sediment and remove pollutants. Some plants are very efficient at taking up heavy metals and organic contaminants. Bioremediation offers exciting and inexpensive alternatives to conventional cleanup.

11.7 LEGAL PROTECTIONS FOR WATER

- *The Clean Water Act had great impacts, but we have far to go.*
- *The CWA funded badly needed sewage treatment plants.*
- *Permits allow for monitoring and limiting pollutants.*

Water pollution control has been among the most broadly popular and effective of all environmental legislation in the United States. Surface-water pollution is often both highly visible and one of the most common threats to environmental quality. In more-developed countries, reducing water pollution has been a high priority over the past few decades (fig. 11.32). Billions of dollars have been spent on control programs, and considerable progress has been made. Still, much remains to be done.

Clean water protection has also been complicated. Numerous major laws have been necessary, with amendments over the years (table 11.4). Most laws and rules have faced objections, because they generally seek to restrict individual actions that threaten the public interest. For example, unregulated waste disposal saves money and time for the waste producer, but the public bears the burden of increased health costs and impaired water resources. Opponents of federal regulation have tried repeatedly to weaken or eliminate the Clean Water Act. They object to the cost and to putting environmental interests ahead of economic interests. Despite this, water protections have been widely supported.

▲ **FIGURE 11.32** Pollution control laws have sought to make water safe for fishing, recreation, and other uses. These policies protect resources for everyone's benefit. Source: Pos, Robert H, U.S. Fish and Wildlife Service

TABLE 11.4 Some Important Water Quality Legislation

1. *Federal Water Pollution Control Act (1972).* Establishes uniform nationwide controls for each category of major polluting industries.
2. *Safe Drinking Water Act (1974).* Requires minimum safety standards for every community water supply. Among the contaminants regulated are bacteria, nitrates, and pesticides; radioactivity and turbidity also are regulated.
3. *Resource Conservation and Recovery Act (RCRA) (1976).* Regulates the storage, shipping, processing, and disposal of hazardous wastes and sets limits on the sewering of toxic chemicals.
4. *Comprehensive Environmental Response, Compensation, and Liability Act (CERCLA) (1980)* and *Superfund Amendments and Reauthorization Act (SARA) (1984).* Provide for sealing, excavation, or remediation of toxic and hazardous waste dumps.
5. *Clean Water Act (1985) (amending the 1972 Water Pollution Control Act).* Sets as a national goal the attainment of "fishable and swimmable" quality for all surface waters in the United States.

The Clean Water Act was ambitious, popular, and largely successful

Passage of the U.S. Clean Water Act of 1972 was a bold, bipartisan step that made clean water a national priority. Along with the Endangered Species Act and the Clean Air Act, this is one of the most significant and effective pieces of environmental legislation ever passed by the U.S. Congress. It also is an immense and complex law, with more than 500 sections regulating everything from urban runoff, industrial discharges, and municipal sewage treatment to land use practices and wetland drainage.

The ambitious goal of the Clean Water Act was to return all U.S. surface waters to "fishable and swimmable" conditions. For point sources, the act requires discharge permits and use of the best practicable control technology (BPT). For toxic substances, the act sets national goals of best available, economically achievable technology (BAT) and zero discharge goals for 126 priority toxic pollutants. As discussed earlier, these regulations have had a positive effect on water quality. Although surface water is not yet swimmable or fishable everywhere, its quality in the United States has significantly improved on average over the past quarter century. Perhaps the most important result of the act has been investment of $54 billion in federal funds and more than $128 billion in state and local funds for municipal sewage treatment facilities.

The CWA helped fund infrastructure

Since the Clean Water Act was passed in 1972, the United States has spent more than $180 billion in public funds and perhaps ten times as much in private investments on water pollution control. Most of that effort has been aimed at point sources, especially to build or upgrade thousands of municipal sewage treatment plants. As a result, nearly everyone in urban areas is now served by municipal sewage systems.

The first generation of sewage treatment infrastructure is aging now, however, and it was built for cities much smaller than we have today. Increasingly, even modest rainstorms are causing raw sewage discharges into rivers and lakes. Finding the funds to upgrade aging plants is an important priority.

The CWA established permitting systems

Only about 10 percent of our water pollution now comes from industrial and municipal point sources. An important reason for this is that the Clean Water Act established a National Pollution Discharge Elimination System (NPDES), which issues a permit for any industry, municipality, or other entity discharging waste in surface waters. While this might seem like giving permission to pollute, the permits recognize that surface waters can dilute some level of pollutants, such as nutrient discharge from small-town wastewater treatment, and the permits ensure some degree of disclosure and monitoring of what is entering our waterways. If there are too many violations of permit limits, regulators can impose a fine, or even sue a polluter.

In 1998 the EPA began shifting the focus to pollutant loads in the water body. This approach aids watershed-level monitoring and protection, and it gives the public better information about the health of their watersheds. In addition, states can have greater flexibility as they identify impaired water bodies and set priorities. For water bodies that don't meet water quality goals, states must develop *total maximum daily load* (**TMDL**) standards—the amount of a particular pollutant that a water body can receive from both point and nonpoint sources—for each pollutant and each listed water body. You can look up TMDLs for your local watersheds on the EPA's website. Some 4,000 watersheds are now monitored for water quality.

Of the 5.6 million km of rivers monitored for TMDLs, only 480,000 km fail to meet their clean water goals. Only 12.5 percent of total lake area (in about 20,000 lakes) fail to meet their goal.

The CWA has made real but incomplete progress

Surface waters have seen real improvements in many places. One of the best-known cases, Lake Erie, was widely regarded as "dead," an ecological disaster in the 1960s. Following control of industrial and urban discharge, the lake became, according to local promoters, the "walleye capital of the world." Bacteria counts and algae blooms decreased more than 90 percent from 1962. Water became clear and algae levels declined.

Lake Erie also shows how much easier it is to control point sources than nonpoint sources. In recent years, water quality has deteriorated again, as farm runoff—from newly intensified corn and soybean production in Great Lakes states—has raised nitrogen and phosphorus levels in the lake. Agricultural pesticides have likewise increased in the lake, presenting new challenges to regulators and to the public who are interested in protecting the nation's waters.

The EPA reports that 21,000 water bodies still do not meet their designated uses. An overwhelming majority of Americans—almost 218 million—live within 16 km (10 mi) of an impaired water body.

On the other hand, fish and aquatic insects have returned to many waters that formerly were too oxygen-deprived for most life. Swimming and other water-contact sports are again permitted in rivers, in lakes, and at ocean beaches that once were closed by health officials. The Clean Water Act goal of making all U.S. surface waters "fishable and swimmable" has not been fully met, but the EPA reports that 91 percent of all monitored river miles and 88 percent of all assessed lake acres are suitable for their designated uses (fig. 11.33).

▲ **FIGURE 11.33** Not all rivers and lakes are "fishable or swimmable," but we've made substantial progress since the Clean Water Act was passed in 1972. ©William P. Cunningham

CONCLUSION

Water is a precious resource. As human populations grow and climate change affects rainfall patterns, water is likely to become even more scarce in the future. Already, about 2 billion people live in water-stressed countries (where there are inadequate supplies to meet all demands), and at least half those people don't have access to clean drinking water. Depending on population growth rates and climate change, it's possible that by 2050 there could be 7 billion people (about 60 percent of the world population) living in areas with water stress or scarcity. Conflicts involving water rights, directly or indirectly, are becoming more common both among countries and within countries. This is made more likely by the fact that most major rivers cross two or more countries before reaching the sea, and droughts, such as the one in the southeastern United States, may become more frequent and severe with global warming. Many experts agree with *Fortune* magazine that "water will be to the 21st century what oil was to the 20th."

Forty years ago, rivers in the United States were so polluted that some caught fire, while others ran red, black, orange, or other unnatural colors with toxic industrial wastes. Many cities still dumped raw sewage into local rivers and lakes, so that warnings had to be posted to avoid any bodily contact. We've made huge progress since that time. Not all rivers and lakes are "fishable or swimmable," but federal, state, and local pollution controls have greatly improved our water quality in most places.

In rapidly developing countries, such as China and India, water pollution remains a serious threat to human health and ecosystem well-being. It will take a massive investment to correct this growing problem. But there are relatively low-cost solutions to many pollution issues. Constructed wetlands for ecological sewage treatment provide an example of the kind of creative strategies we need to resolve widespread challenges. These and other solutions—some of which you might help develop some day—point the direction to smarter, more sustainable management of water resources.

PRACTICE QUIZ

1. Describe the path a molecule of water might follow through the hydrologic cycle from the ocean to land and back again.
2. About what percent of the world's water is liquid, fresh, surface water that supports most terrestrial life?
3. What is an *aquifer*? How does water get into an aquifer? Explain the idea of an *artesian well* and a *cone of depression*.
4. Which sector of water use consumes most globally? Overall, has water use increased in the past century? Has efficiency increased or decreased in the three main use sectors?
5. Describe at least one example of the environmental costs of water diversion from rivers to farms or cities.
6. Explain the difference between point and nonpoint pollution. Which is harder to control? Why?
7. Why are nutrients considered pollution? Explain the ideas of *eutrophication* and an *oxygen sag*.
8. Describe primary, secondary, and tertiary water treatment.
9. What are some sources of groundwater contamination? Why is groundwater pollution such a difficult problem?
10. What is natural wastewater treatment, and how does it work?

CRITICAL THINKING AND DISCUSSION

Apply the principles you have learned in this chapter to discuss these questions with other students.

1. What changes could occur in the hydrologic cycle if our climate were to warm or cool significantly?
2. Why does it take so long for deep ocean waters to circulate through the hydrologic cycle? What happens to substances that contaminate deep ocean water or deep aquifers in the ground?
3. If you were a judge responsible for allocating the dwindling water supply in the Colorado River among the various stakeholders, how would you assign water rights? What would be your criteria for needs and rights?
4. Do you think that water pollution is worse now than it was in the past? What considerations go into a judgment such as this? How do your personal experiences influence your opinion?
5. What additional information would you need to make a judgment about whether conditions are getting better or worse? How would you weigh different sources, types, and effects of water pollution?
6. Under what conditions might sediment in water or cultural eutrophication be beneficial? How should we balance positive and negative effects?

DATA ANALYSIS Graphing Global Water Stress and Scarcity

According to the United Nations, **water stress** is when annual water supplies drop below 1,700 m^3 per person. **Water scarcity** is defined as annual water supplies below 1,000 m^3 per person. More than 2.8 billion people in 48 countries will face either water stress or scarcity conditions by 2025. Of these 48 countries, 40 are expected to be in West Asia or Africa. By 2050, far more people could be facing water shortages, depending both on population projections and scenarios for water supplies based on global warming and consumption patterns.

To explore some of the issues and questions about future water scarcity, go to Connect, and answer questions about this figure and others from this chapter.

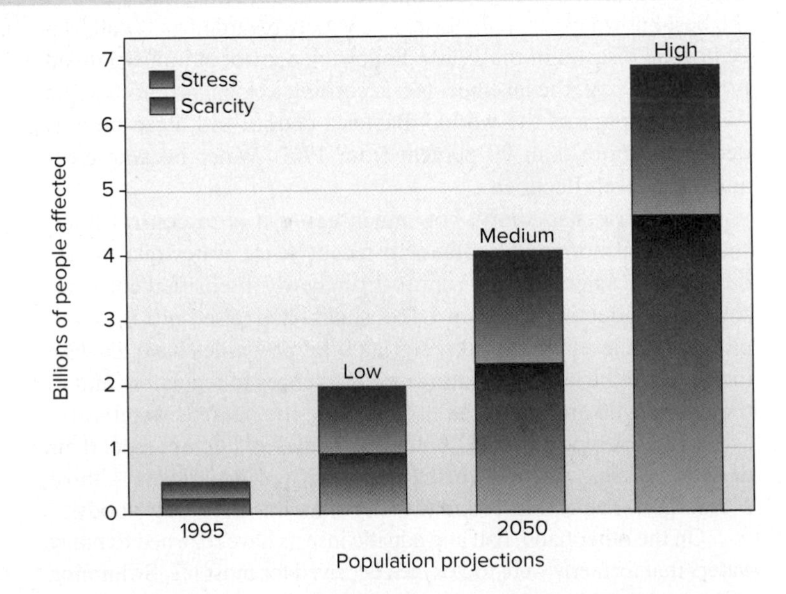

Design Elements: Active Learning (Toad): ©Gaertner/Alamy Stock Photo; Case Study (Globe): ©McGraw-Hill Education; Google Earth: ©McGraw-Hill Education; Abstract Background: ©Martin Kubat/Shutterstock; What do you think (Students using tablets): ©McGraw-Hill Education/Richard Hutchings, photographer; What can you do (Hand holding Globe): ©Christoph Weihs/Shutterstock

CHAPTER

12 Environmental Geology and Earth Resources

LEARNING OUTCOMES

After studying this chapter, you should be able to answer the following questions:

- What are tectonic plates, and how does their movement shape our world?
- Where and why do volcanoes and earthquakes occur?
- What are some of the environmental and social costs of mining and oil- and gas-drilling?
- How can we reduce our consumption of geologic resources?
- Explain why floods and mass wasting are problems.

The Bristol Bay watershed supports the largest remaining sockeye salmon fishery on the planet, producing nearly half the world's wild sockeye harvest, but mining companies want to mine copper ore here.

©Felt Soul Media

CASE STUDY

Salmon or Copper?

Alaska has many natural wonders, but among the most spectacular is the migration of millions of salmon every year from the dark waters of Bristol Bay on the state's southwest coast up the Mulchatna and Nushagak Rivers into shallow headwater streams near Lake Clark and Katmai National Parks. It's one of the last, great wild salmon runs in the world (fig. 12.1).

The Bristol Bay watershed is a sprawling wilderness of winding streams and rivers, vast marshy wetlands, tundra, and forests of alder and spruce, filled with salmon, bears, caribou, moose, and wolves. This breathtaking place is one of the most productive marine ecosystems in the world. All five species of Pacific salmon—sockeye, Chinook, coho, chum, and pink—spawn there and have sustained native salmon-based cultures for thousands of years.

The watershed supports the largest sockeye, or King, salmon fishery on the planet, producing nearly half the world's wild sockeye harvest. This fishery generates $1.5 billion a year and supports nearly 20,000 jobs throughout the United States. Over 4,000 Alaskans, including many native Yup'ik and Dena'ina, rely on fish as the mainstay of their diet.

But the Bristol Bay watershed contains other riches besides the salmon bonanza. Geologists estimate that the Pebble ore deposit in the uplands between the two largest river systems in the watershed may contain 57 billion metric tons of copper, 15 million tons of molybdenum, 5,000 tons of silver, and 2,000 tons of gold. Palladium and rhenium also occur in lower concentrations. Mining engineers have claimed that the deposit contains over $300 billion worth of recoverable metals.

Since 2001, Northern Dynasty, a Canadian mining company, has sought to develop an open pit mine in the area. This pit could be deeper than the Grand Canyon, impact an area larger than Manhattan, and fill a major football stadium up to 3,900 times with mine waste. The liquid waste would be held behind earthen dams, some taller than a 50-story skyscraper. The EPA review of the project concluded that even without a collapse of those tailings ponds (as happened at similar mines in British Columbia in 2014 and Brazil in 2015), the Pebble mine would destroy 150 km (94 mi) of salmon streams and 2,200 ha (5,400 acres) of wetlands, lakes, and ponds and could imperil not only the salmon fishery but the whole ecosystem.

The mine would also require building of a new deepwater port on Cook Inlet, a natural gas pipeline running across the ocean floor from the Kenai Peninsula to the mine site, and 100 km (60 mi) of private roads that would skirt the boundary of Lake Clark National Park and cross approximately 60 salmon streams. Routine release of silt, pollutants, and runoff into headwater streams, the

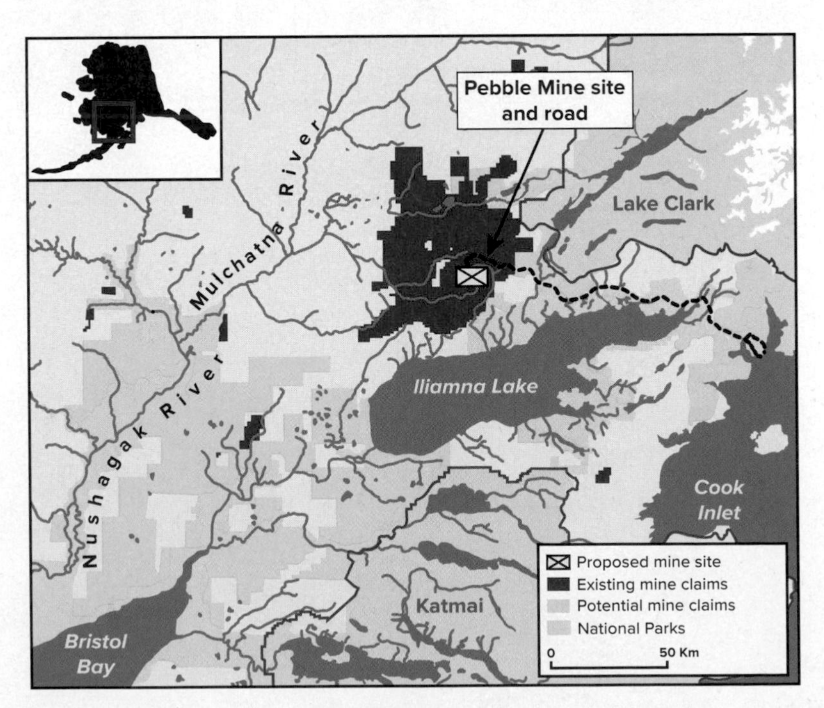

▲ **FIGURE 12.1** The Bristol Bay watershed lies in the headwaters of the Mulchatna and Nushagak Rivers near the borders of Lake Clark and Katmai National Parks. It provides the spawning grounds for the world's last great sockeye salmon run, but it also contains valuable metal resources.

CASE STUDY continued

EPA ruled, would irreparably harm the Bristol Bay fishery, wildlife, and people.

In 2014, the Obama administration, responding to concerns of a coalition of fishers, native Alaskans, environmental groups, and local businesses, denied mining permits for the Pebble mine, saying that it could have "significant and potentially catastrophic impacts" on the region. But in 2017, then-EPA administrator Scott Pruitt directed his staff to reverse previous protections and to expedite permitting for the mine.

What do you think about the trade-off of metals for salmon? We all use metals in our daily life. Every American uses about 20 pounds per year, on average, in various manufactured goods. Our demand for new metals is driven partly by the fact that we recycle very little. We recycle only about one-third of commercial copper—and less than 9 percent of post-consumer copper—and we recycle far less of most other metals. So we require a constant stream of new materials. At the same time, many of us enjoy fish, particularly salmon, too. Large mining companies, including Mitsubishi, Rio Tinto, and Anglo American, invested in Pebble years ago but pulled out, discouraged by local opposition and the dangers of operating in this cold, remote location.

Pebble Partnership claims that the mine will employ about 2,000 construction workers and provide about 850 permanent jobs over the 20-year life of the mine. Environmental groups point out that a healthy salmon fishery supports about 15 times that many permanent jobs. The metals will provide billions in profits every year during the mine's life, but little of those profits will go to Alaska or even the United States. Polling by mine opponents found that 58 percent of Alaskans overall, and 80 percent of Bristol Bay residents, oppose the project. Given our low rate of metal re-use, could we do better and preserve the salmon of Bristol Bay?

In this chapter, we'll look at the principles of geology, earth processes, and earth resources. We examine the ways our resource use affects our environment, and we also consider the ways in which we humans have become geologic agents. For related resources, including Google Earth placemarks that show locations discussed in this chapter, visit www.mcgrawhillconnect.com. ■

When we heal the earth, we heal ourselves.

—DAVID ORR

12.1 EARTH PROCESSES SHAPE OUR RESOURCES

- *Convection in the mantle causes tectonic plate movement, earthquakes, and volcanoes.*
- *When oceanic and continental plates collide, the lighter continental plate rides up, forming mountains; the oceanic plate subsides, forming trenches.*
- *The volcanic "ring of fire" results from these plate collisions.*

All of us benefit from the earth's geological resources. Right now you are undoubtedly using products made from these resources: plastics of many types are made from petroleum; iron, copper, and aluminum mines produce the materials for the electrical wiring that brings you power; rare earth metals are essential in your cell phone, and your computer. All of us also share responsibility for

Benchmark Data

Among the ideas and values in this chapter, these are a few worth remembering.

33%	Percentage of earth that is iron
30%	Percentage that is oxygen
1 cm/year	Approximate average rate of tectonic plate movement
250 MJ/kg	Energy needed to refine aluminum from ore
8 MJ/kg	Energy to recycle aluminum
14%	Percentage of U.S. household aluminum recycled

the environmental and social devastation that often results from mining, drilling, and processing materials.

Fortunately, there are many promising solutions to reduce these costs, including recycling and alternative materials. But why are these risks and resources distributed as they are? To understand how and where geological risks and resources are created, we need to learn about the earth's structure and the processes that shape it.

Earth is a dynamic planet

Although we think of the ground under our feet as solid and stable, the earth is a dynamic and constantly changing structure. Titanic forces inside the earth cause continents to split, move apart, and then crash into each other in slow but inexorable collisions.

The earth is a layered sphere. The **core,** or interior, is composed of a dense, intensely hot mass of metal—mostly iron—thousands of kilometers in diameter (fig. 12.2). Solid in the center but more fluid in the outer core, this immense mass generates the magnetic field that envelops the earth.

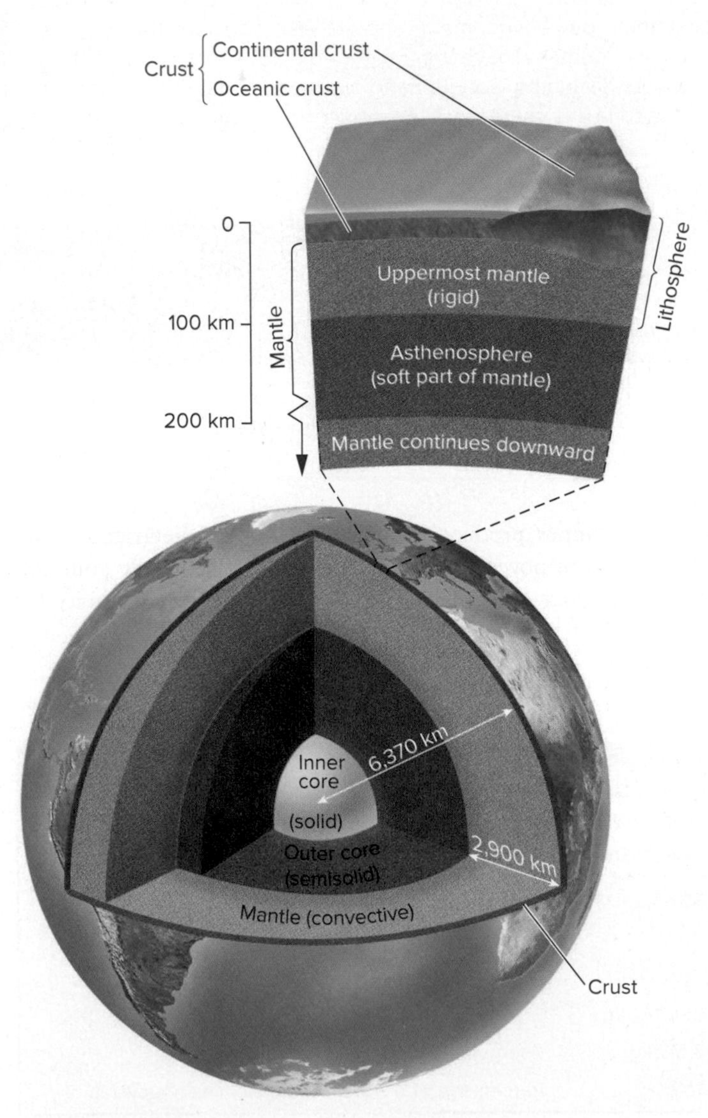

▲ **FIGURE 12.2** Earth's cross section. Slow convection in the mantle causes the thin, brittle crust to move.

TABLE 12.1 Eight Most Common Chemical Elements (Percent) in Whole Earth and Crust

WHOLE EARTH		CRUST	
Iron	33.3	Oxygen	45.2
Oxygen	29.8	Silicon	27.2
Silicon	15.6	Aluminum	8.2
Magnesium	13.9	Iron	5.8
Nickel	2.0	Calcium	5.1
Calcium	1.8	Magnesium	2.8
Aluminum	1.5	Sodium	2.3
Sodium	0.2	Potassium	1.7

Surrounding the molten outer core is a hot, pliable layer of rock called the **mantle.** The mantle is much less dense than the core because it contains a high concentration of lighter elements, such as oxygen, silicon, and magnesium.

The outermost layer of the earth is the cool, lightweight, brittle rock **crust.** The crust below oceans is relatively thin (8–15 km), dense, and young (less than 200 million years old) because of constant recycling. Crust under continents is relatively thick (25–75 km) and light, and as old as 3.8 billion years, with new material being added continually. It also is predominantly granitic, whereas oceanic crust is mainly dense basaltic rock. Table 12.1 compares the composition of the whole earth (dominated by the dense core) and the crust.

Tectonic processes reshape continents and cause earthquakes

The huge convection currents in the mantle are thought to break the overlying crust into a mosaic of huge blocks called **tectonic plates** (fig. 12.3). These plates slide slowly across the earth's surface like wind-driven ice floes on water, in some places breaking up into smaller pieces, in other places crashing ponderously into each other to create new, larger landmasses. Ocean basins form where continents crack and pull apart. The Atlantic Ocean, for example, is growing slowly as Europe and Africa move away from the Americas. **Magma** (molten rock) forced up through the cracks forms new oceanic crust that piles up under water in **mid-ocean ridges.** Creating the largest mountain range in the world, these ridges wind around the earth for 74,000 km (46,000 mi). Although concealed from our view, this jagged range boasts higher peaks, deeper canyons, and sheerer cliffs than any continental mountains. Slowly spreading from these fracture zones, ocean plates push against continental plates.

Earthquakes are caused by jerking as plates grind past each other. Mountain ranges like those on the west coasts of North and South America are pushed up at the margins of colliding continental plates. The Himalayas are still rising as the Indian subcontinent collides inexorably with Asia. Southern California is sailing very slowly north toward Alaska. In about 30 million years, Los Angeles will pass San Francisco, if both still exist by then.

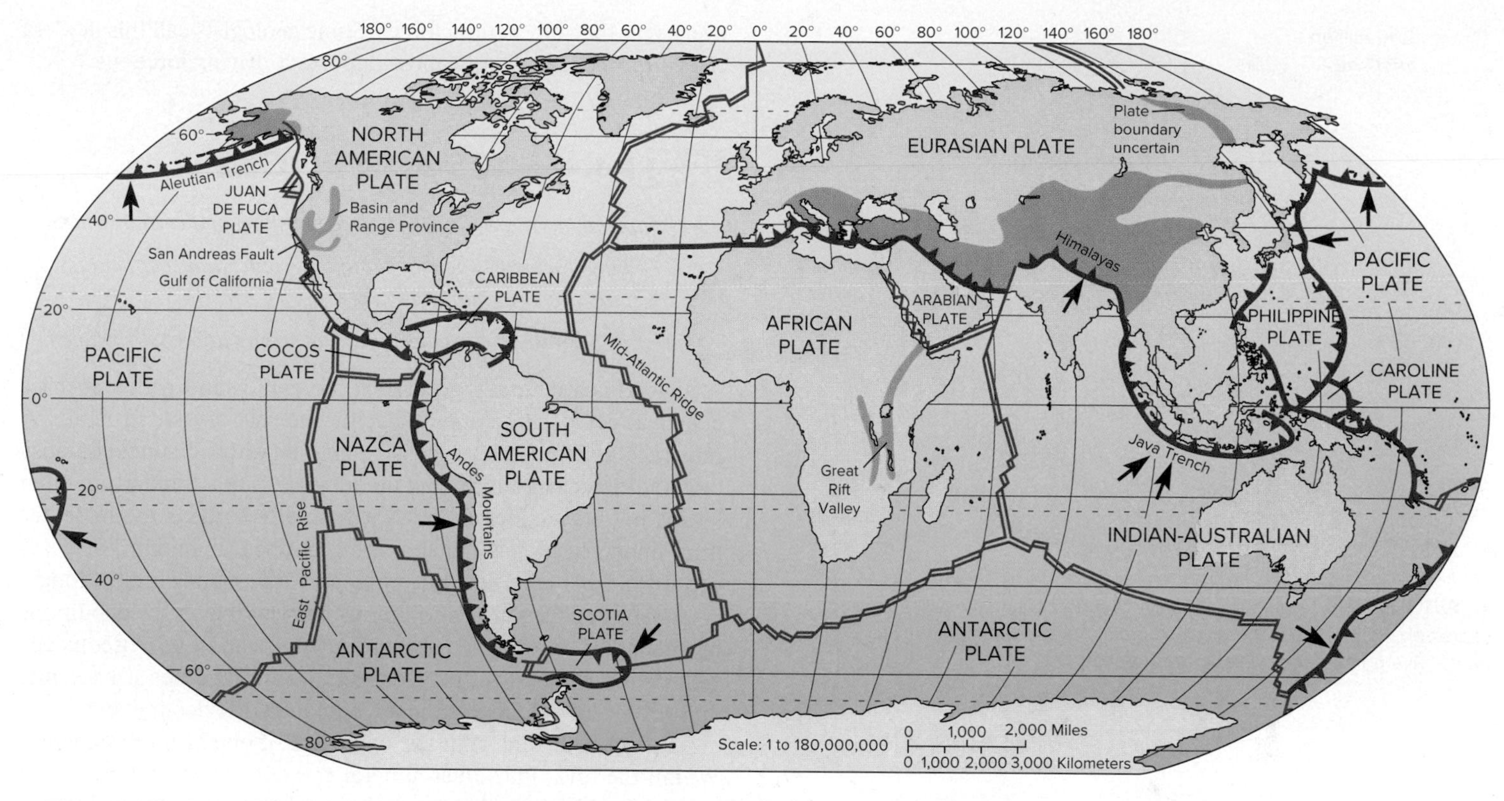

▲ **FIGURE 12.3** Map of tectonic plates. Plate boundaries are dynamic zones, characterized by earthquakes, volcanism, and the formation of great rifts and mountain ranges (shaded areas). Arrows indicate the direction of subduction, where one plate is diving beneath another. These zones are sites of deep trenches in the ocean floor and high levels of seismic and volcanic activity.

When an oceanic plate collides with a continental landmass, the continental plate usually rides up over the seafloor, while the oceanic plate is **subducted**, or pushed down into the mantle, where it melts and rises back to the surface as magma (fig. 12.4). Deep ocean trenches mark these subduction zones, and volcanoes form where the magma erupts through vents and fissures in the overlying crust. Trenches and volcanic mountains ring the Pacific Ocean rim from Indonesia to Japan to Alaska and down the west coast of the Americas, forming a so-called ring of fire where oceanic plates are being subducted under the continental plates. This ring is the source of more earthquakes and volcanic activity than any other region on the earth.

Over millions of years, continents can drift long distances. Antarctica and Australia once were connected to Africa, for instance, somewhere near the equator and supported luxuriant forests. Geologists suggest that several times in the earth's history, most or all of the continents have gathered to form supercontinents, which have broken up and re-formed over hundreds of millions of years (fig. 12.5). The redistribution of continents has profound effects on the earth's climate and may help explain the periodic mass extinctions of organisms marking the divisions between many major geologic periods (fig. 12.6).

We humans have now become a major geological force. We may move more dirt and rock and cause more species extinction

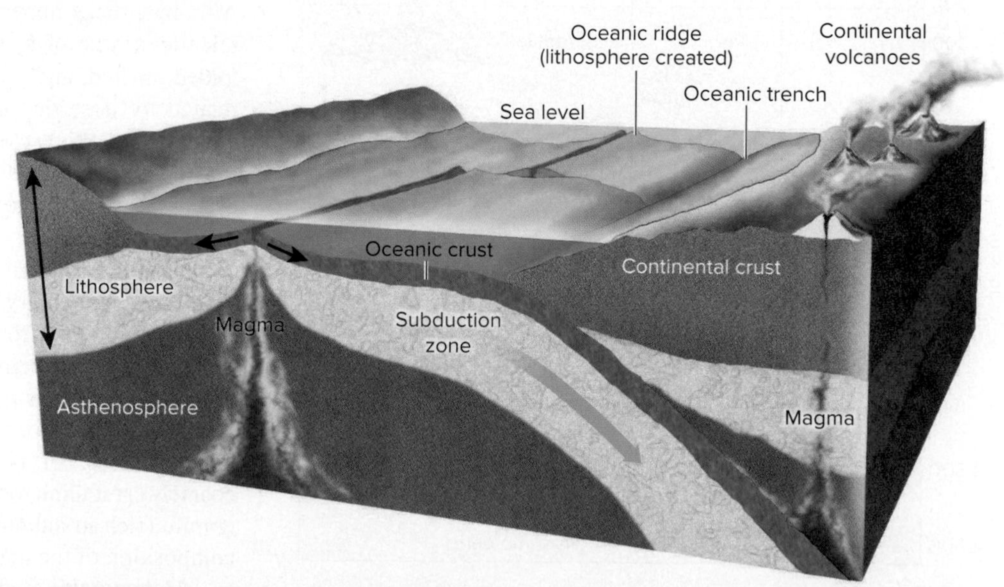

▲ **FIGURE 12.4** Tectonic plate movement. Where thin, oceanic plates diverge, upwelling magma forms mid-ocean ridges. A chain of volcanoes, like the Hawaiian Islands, may form as plates pass over a "hot spot." Where plates converge, melting can cause volcanoes, such as the Cascades.

every year than any natural forces. Some geologists call this new era the Anthropocene after its most important driving force—us.

▲ **FIGURE 12.5** Pangaea, an ancient supercontinent of 200 million years ago, combined all the world's continents in a single landmass. Continents have combined and separated repeatedly.

Millions of years before present (ma)	Era	Periods	Life on Earth
3	**Cenozoic**	Quaternary	First humans
		Tertiary	First important mammals
65	**Mesozoic**	Cretaceous	Extinction of dinosaurs
		Jurassic	
		Triassic	First dinosaurs
245	**Paleozoic**		
300		Permian	First reptiles
		Pennsylvanian	
		Mississippian	
400		Devonian	Fish become abundant
		Silurian	
		Ordovician	
		Cambrian	First abundant fossils
545	**Precambrian**		Earliest single-celled fossils
3,500			(The Precambrian accounts for the vast majority of geologic time)
4,500			Origin of the earth

▲ **FIGURE 12.6** Periods and eras in geological time, and major life-forms that mark some periods. Some authors claim we've now entered a new era, the Anthropocene, dominated by humans.

12.2 MINERALS AND ROCKS

- *A mineral has a specific chemical composition and crystal structure.*
- *The three rock types, igneous, metamorphic, and sedimentary, constantly re-form into each other.*
- *Metals are valuable because they are strong and easily shaped.*

A **mineral** is a naturally occurring, inorganic solid with a specific chemical composition and a specific internal crystal structure. A mineral is solid; therefore, ice is a mineral (with a distinct composition and crystal structure), but liquid water is not. Similarly, molten lava is not crystalline, although it generally hardens to create distinct minerals. Metals (such as iron, copper, aluminum, or gold) come from mineral ores, but once purified, metals are no longer crystalline and thus are not minerals. Depending on the conditions in which they were formed, mineral crystals can be microscopically small, such as asbestos fibers, or huge, such as the tree-size selenite crystals recently discovered in a Chihuahua, Mexico, mine.

A **rock** is a solid, cohesive aggregate of one or more minerals. Within the rock, individual mineral crystals (or grains) are mixed together and held firmly in a solid mass. The grains may be large or small, depending on how the rock was formed, but each grain retains its own unique mineral qualities. Each rock type has a characteristic mixture of minerals, grain sizes, and ways in which the grains are mixed and held together. Granite, for example, is a mixture of quartz, feldspar, and mica crystals. Rocks with a granite-like mineral content but much finer crystals are called rhyolite; chemically similar rocks with large crystals are called pegmatite.

The rock cycle creates and recycles rocks

Although rocks appear hard and permanent, they are part of a relentless cycle of formation and destruction. They are crushed, folded, melted, and recrystallized over time by dynamic processes related to those that shape the large-scale features of the earth's crust. We call this cycle of creation, destruction, and metamorphosis the **rock cycle** (fig. 12.7). Understanding something of how this cycle works helps explain the origin and characteristics of different types of rocks.

There are three major rock classifications: igneous, metamorphic, and sedimentary. **Igneous rocks** (from *igni*, the Latin word for fire) are solidified from hot, molten magma or lava. Most rock in the earth's crust is igneous. Magma extruded to the surface from volcanic vents cools quickly to make finely crystalline rocks, such as basalt, rhyolite, or andesite. Magma that cools slowly in subsurface chambers or is intruded between overlying layers makes coarsely crystalline rocks, such as gabbro (rich in iron and silica) or granite (rich in aluminum and silica), depending on the chemical composition of the magma.

Metamorphic rocks form from the melting, contorting, and recrystallizing of other rocks. Deep in the ground, tectonic forces squeeze, fold, heat, and transform solid rock. Under these

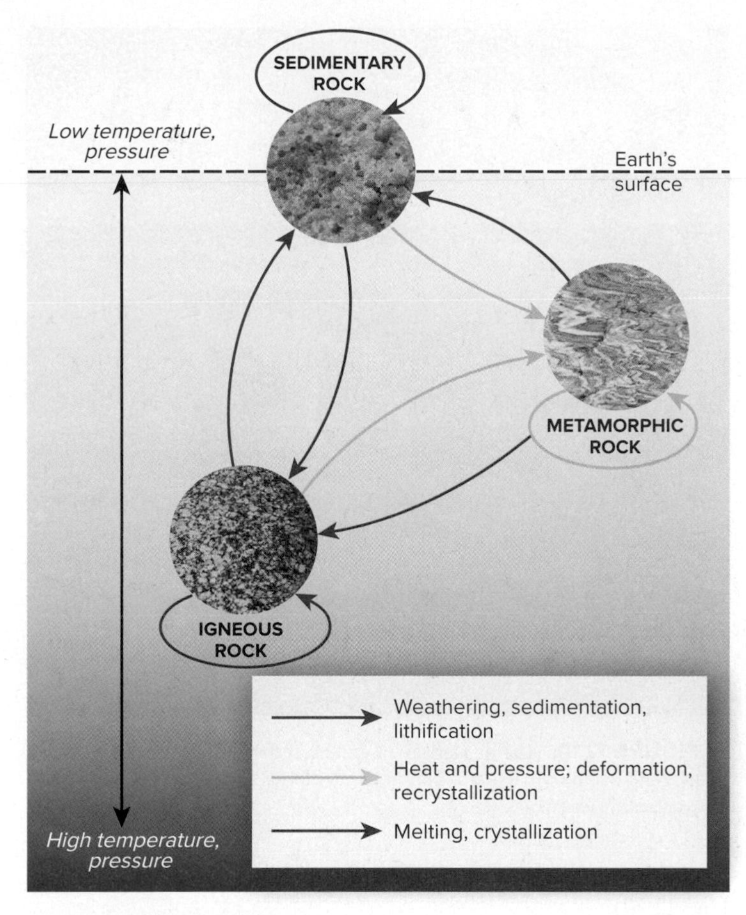

▲ **FIGURE 12.7** The rock cycle includes a variety of geologic processes that can transform any rock.

conditions, chemical reactions can alter both the composition and the structure of the component minerals. Metamorphic rocks are classified by their chemical composition and by the degree of recrystallization: Some minerals form only under extreme pressure and heat (diamonds or jade, for example); others form under more moderate conditions (graphite or talc). Some common metamorphic rocks are marble (from limestone), quartzite (from sandstone), and slate (from mudstone and shale). Metamorphic rocks often have beautiful colors and patterns left by the twisting and folding that created them.

Sedimentary rocks are formed when loose grains of other rocks are consolidated by time and pressure. Sandstone, for example, is solidified from layers of sand, and mudstone consists of extremely hardened mud and clay. Tuff is formed from volcanic ash, and conglomerates are aggregates of sand and gravel. Some sedimentary rocks develop from crystals that precipitate out of extremely salty water. Rock salt, made of the mineral halite, is ground up to produce ordinary table salt (sodium chloride). Salt deposits often form when a body of saltwater dries up, leaving salt crystals behind. Limestone is a rock composed of cemented remains of marine organisms. You can often see the shapes of shells and corals in a piece of limestone. Sedimentary formations often have distinctive layers that show different conditions when they were laid down. Erosion can reveal these layers and inform us of their history (fig. 12.8).

▲ **FIGURE 12.8** Different colors of soft sedimentary rocks deposited in ancient seas during the Tertiary period 63 to 40 million years ago have been carved by erosion into the fluted spires and hoodoos of the Pink Cliffs of Bryce Canyon National Park. ©Natphotos/Getty Images

Weathering and sedimentation

Most crystalline rocks are extremely hard and durable, but exposure to air, water, changing temperatures, and reactive chemical agents slowly breaks them down in a process called **weathering** (fig. 12.9). Mechanical weathering is the physical breakup of rocks into smaller particles without a change in chemical composition of the constituent minerals. You have probably seen the results of mechanical weathering in rounded rocks in rivers or on shorelines, smoothed by constant tumbling in waves or currents. On a larger scale, mountain valleys are carved by rivers and glaciers.

Chemical weathering is the selective removal or alteration of specific minerals in rocks. This alteration leads to weakening and disintegration of rock. Among the more important chemical weathering processes are oxidation (combination of oxygen with an element to form an oxide or a hydroxide mineral) and hydrolysis (hydrogen atoms from water molecules combine with other chemicals to form acids). The products of these reactions are more

▲ **FIGURE 12.9** Weathering slowly reduces an igneous rock to loose sediment. Here, exposure to moisture expands minerals in the rock, and frost may also force the rock apart. Courtesy of David McGeary

susceptible to both mechanical weathering and dissolving in water. For instance, when carbonic acid (formed when rainwater absorbs CO_2) percolates through porous limestone layers in the ground, it dissolves the rock and creates caves.

Particles of rock loosened by wind, water, ice, and other weathering forces are carried downhill, downwind, or downstream until they come to rest again in a new location. The deposition of these materials is called **sedimentation.** Water, wind, and glaciers deposit particles of sand, clay, and silt far from their source. Much of the American Midwest, for instance, is covered with hundreds of meters of sedimentary material left by glaciers (till, or rock debris deposited by glacial ice), wind (loess, or fine dust deposits), river deposits of sand and gravel, and ocean deposits of sand, silt, clay, and limestone.

Economic Geology and Mineralogy

The earth is unusually rich in mineral variety. Mineralogists have identified some 4,400 different mineral species, far more, we believe, than any of our neighboring planets. What makes the difference? The processes of plate tectonics and the rock cycle on this planet have gradually concentrated uncommon elements and allowed them to crystallize into new minerals. But this accounts for only about one-third of our geologic legacy. The biggest difference is life. Most of our minerals are oxides, but there was little free oxygen in the atmosphere until it was released by photosynthetic organisms, thus triggering evolution of our great variety of minerals.

Economic mineralogy is the study of resources that are valuable for manufacturing and trade. Most economic minerals are metal ores, minerals with unusually high concentrations of metals. Lead, for example, generally comes from the mineral galena (PbS), and copper comes from sulfide ores, such as bornite (Cu_5FeS_4). Nonmetallic geologic resources include graphite, feldspar, quartz crystals, diamonds, and other crystals that are valued for their usefulness or beauty. Metals have been so important in human affairs that major epochs of human history are commonly known by their dominant materials and the technology involved in using those materials (Stone Age, Bronze Age, Iron Age, etc.). The mining, processing, and distribution of these materials have broad implications for both our culture and our environment. Most economically valuable crustal resources exist everywhere in small amounts; the important thing is to find them concentrated in economically recoverable levels.

The U.S. mining law passed in 1872 encourages mining on public lands as a way of boosting the economy and utilizing natural resources. There have been repeated efforts to update this law and to recover public revenue from publicly owned resources, but powerful friends in Congress together with a tradition of supporting extractive industries in many states have continually blocked these reforms.

Metals are essential to our economy

Metals are malleable substances that are useful and valuable because they are strong, relatively light, and can be reshaped for many purposes. The availability of metals and the methods to extract and use them have determined technological developments, as well as economic and political power for individuals and nations (fig. 12.10).

▲ **FIGURE 12.10** The availability of metals and minerals and the ways we extract and use them have profound effects on our society and environment. ©Digital Vision/PunchStock

The metals consumed in greatest quantity by world industry include iron (740 million metric tons annually), aluminum (40 million metric tons), manganese (22.4 million metric tons), copper and chromium (8 million metric tons each), and nickel (0.7 million metric tons). The complex, worldwide network that extracts, processes, and distributes these metals has become crucially important to the economic and social stability of all nations. Table 12.2 shows some important uses of these metals.

TABLE 12.2 | Primary Uses of Some Major Metals

METAL	USE
Aluminum	Packaging foods and beverages (38%), transportation, electronics
Chromium	High-strength steel alloys
Copper	Building construction, electric and electronic industries
Iron	Heavy machinery, steel production
Lead	Leaded gasoline, car batteries, paints, ammunition
Manganese	High-strength, heat-resistant steel alloys
Nickel	Chemical industry, steel alloys
Platinum group	Automobile catalytic converters, electronics, medical uses
Gold	Medical, aerospace, electronic uses; accumulation as monetary standard
Silver	Photography, electronics, jewelry

The largest countries (in surface area) have the greatest likelihood of mountain building (orogeny), volcanism, or other events that create economic mineral deposits. Russia, China, Canada, the United States, and Australia, for example, are especially rich in metal resources. Africa has relatively few geologic resources, except for South Africa, which is unusually rich in diamonds, gold, and other valuable minerals. In 2011, geologists announced the discovery of an estimated $1 trillion (U.S.) in metals and other valuable minerals in Afghanistan. Critics claimed this announcement was merely an excuse for continued military occupation of this troubled country. In any case, difficult terrain and political instability will make these deposits difficult to access.

The rapid growth of green technologies, such as renewable energy and electric vehicles, has made a group of rare earth metals especially important. Worries about impending shortages of these minerals complicate future developments in this sector (Key Concepts, pp. 292–293).

Nonmetal mineral resources include gravel, clay, glass, and salts

Nonmetal minerals constitute a broad class that covers resources from gemstones to sand, gravel, salts, limestone, and soils. Sand and gravel production for road and building construction comprise by far the greatest volume and dollar value of all nonmetal mineral resources and a far greater volume than all metal ores. Sand and gravel are used mainly in brick and concrete construction, in paving, as loose road filler, and for sandblasting. High-purity silica sand is our source of glass. These materials usually are retrieved from surface pit mines and quarries, where they were deposited by glaciers, winds, or ancient oceans. Some industry sources predict a building crisis in the United States as supplies of easily accessible sand and gravel are exhausted.

Limestone, like sand and gravel, is mined and quarried for concrete and crushed for road rock. It also is cut for building stone, pulverized for use as an agricultural soil additive that neutralizes acidic soil, and roasted in lime kilns and cement plants to make plaster (hydrated lime) and cement.

Evaporites (materials deposited by evaporation of chemical solutions) are mined for halite, gypsum, and potash. These are often found at or above 97 percent purity. Halite, or rock salt, is used for water softening and ice melting on winter roads in some northern areas. When refined, it is a source of table salt. Gypsum (calcium sulfate) now makes our plaster wallboard, but it has been used to cover walls ever since the Egyptians plastered their frescoed tombs along the Nile River some 5,000 years ago. Potash is an evaporite composed of a variety of potassium chlorides and potassium sulfates. These highly soluble potassium salts have long been used as a soil fertilizer. Sulfur deposits are mined mainly for sulfuric acid production. In the United States, sulfuric acid use amounts to more than 200 lb per person per year, mostly because of its use in industry, car batteries, and some medicinal products. Interestingly, the world's largest supply of lithium, which is essential for lightweight batteries for electric cars, cell phones, and other consumer goods, is a vast salt flat, called the Salar de Uyuni, in southwestern Bolivia. The lithium, which was deposited when a huge salt lake evaporated eons ago, could make this impoverished country the Saudi Arabia of the electronic age.

Currently, the earth provides almost all our fuel

At present, modern society functions largely on energy produced from geologic deposits of oil, coal, and natural gas. Nuclear energy, which runs on uranium, provides about 20 percent of our electricity. But, as chapter 13 shows, renewable sources, such as sun, wind, hydropower, biomass, and geothermal energy, could replace all the fossil fuels and nuclear power we now use while reducing pollution and reducing global climate change.

Oil, coal, and gas are organic, created over millions of years as extreme heat and pressure transformed the remains of ancient organisms. They are not minerals, because they have no crystalline structure, but they can be considered part of economic mineralogy because they are such important geologic resources. In addition to providing energy, oil is the source material for plastics, and natural gas is used to make agricultural fertilizers. We'll discuss fossil fuels further in chapter 13.

12.3 ENVIRONMENTAL EFFECTS OF RESOURCE EXTRACTION

- *Mining produces dust, acids, and toxic metal pollutants.*
- *Metal is extracted by smelting or chemical extraction.*
- *The value of minerals and the cost of cleanup lead to conflict over these resources.*

Each of us depends daily on geologic resources mined or pumped from sites around the world. We use scores of metals and minerals, many of which we've never even heard of, in our lights, computers, watches, fertilizers, and cars. Extracting and purifying these resources can have severe environmental and social consequences. The most obvious effect of mining and well drilling is often the disturbance or removal of the land surface. Farther-reaching effects, though, include air and water pollution. The EPA lists more than 100 toxic air pollutants, from acetone to xylene, released from U.S. mines and wells every year. Nearly 80,000 metric tons of particulate matter (dust) and 11,000 tons of sulfur dioxide are released from nonmetal mining alone. Pollution from chemical and sediment runoff is a major

Active LEARNING

What Geologic Resources Are You Using Right Now?

Make a list of the geologic materials that are found in some of the objects you are using right now. For example, the computer used to write this chapter is made largely of plastic (from oil), silicon chips (sand), and copper wire, and it runs on energy from coal and uranium-powered electric plants.

Start your list with some of the following items: glasses, chair, table, pencil, lightbulb, window, building, wristwatch, coffee cup, tooth fillings.

KEY CONCEPTS

Where does your cell phone come from?

Mobile phones, computers, and other electronic gadgets have transformed our lives, but few of us think about the geologic resources that went into making those devices. We enjoy these items for their useful lives, but we're always eager to trade up to the next newer and better models. While each individual gadget may be small and contain only tiny amounts of precious metals, rare earths, fossil fuels, and other materials, collectively they make a big impact.

Currently, there are at least 1 billion personal computers and 5 billion mobile phones in use around the world, and the numbers are climbing rapidly. In the United States, most people change their cell phone every 18 to 24 months. Computers last only 2 to 3 years, contributing to the mountains of eWaste discarded in the United States every year.

Here are a few examples of the many source areas that contribute to our cell phones.

We depend increasingly on electronics, but where do they originate?

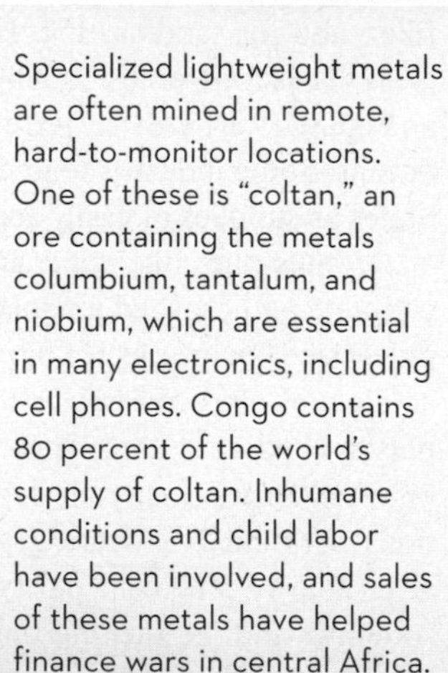

KC 12.1

©McGraw-Hill Education/Marker Dierker, photographer

KC 12.2

©Arthur Tilley/Getty Images

Although they may be tiny, cell phones contain a surprising amount of metal that could be recycled, and yet hundreds of millions are thrown away every year or lie unused in drawers and cupboards. **A typical cell phone can contain gold, silver, copper, palladium, platinum, lead, zinc, mercury, chromium, cadmium, rhodium, beryllium, arsenic, and lithium** among other metals and chemical compounds.

Specialized lightweight metals are often mined in remote, hard-to-monitor locations. One of these is "coltan," an ore containing the metals columbium, tantalum, and niobium, which are essential in many electronics, including cell phones. Congo contains 80 percent of the world's supply of coltan. Inhumane conditions and child labor have been involved, and sales of these metals have helped finance wars in central Africa.

Depending on the ore quality, a metric ton of ore may yield only 0.3 g of gold, and the miners may have to move 2 to 5 tons of overburden (unwanted rock) to get to that ore.

Abandoned mines often leak strong acids laced with arsenic, mercury, and other toxic metals into local groundwater and surface waters. By some calculations, humans now move more earth than the glaciers did.

©GeoStock/Getty Images

Open-pit copper mine

KC 12.4

KC 12.3

©Tom Stoddart Archive/Getty Images

Workers mine coltan.

As we've seen elsewhere in this chapter, supplies of rare earth metals, such as neodymium, dysprosium, lanthanum, and yttrium are also critical for modern electronics. Monopolies in the sources of some of these materials, together with adverse environmental impacts in their extraction and purification, could constrain some technologies.

©PhotoDisc/Getty Images

KC 12.5

Oil refineries produce our plastics from oil.

Every year, billions of cell phones and computers are discarded around the world. Added to the piles of refrigerators, air conditioners, TV sets, and other unwanted appliances, disposal of electronic waste is a huge problem. In the United States, where 3 million tons of electronics are tossed out every year, 70 percent of the heavy metals in landfills are from eWaste. Increasingly, this electronic garbage is shipped to developing countries, where scavengers disassemble it under dangerous conditions in an effort to recover valuable metals. Modern recycling facilities can recover 99 percent of the contents much more safely and efficiently.

An offshore oil well provides energy and plastics.

KC 12.7

©Keith Wood/Getty Images

The plastic case for your phone or computer is made from petroleum. Energy is also used to manufacture and ship all the gadgets we use. Extracting, shipping, and refining fossil fuels are among our biggest geologic impacts. Obtaining the 20 billion barrels of petroleum or the 7 billion short tons of coal we use globally every year has uncalculated environmental, social, and economic impacts.

KC 12.6

©William P. Cunningham

eWaste collection and monitoring are increasingly important.

©Ingram Publishing/Getty Images

KC 12.8

Gold smelter

Smelting (extracting metals from ore by baking it) takes a huge amount of energy and often releases large amounts of air and water pollution, especially in countries where environmental regulation is weak.

CAN YOU EXPLAIN?

1. **If each cell phone contains 0.03 g of gold, how much would be in 5 billion phones?**
2. **List 15 of the elements or earth materials found in cell phones.**
3. **Do we have an ethical responsibility for what happens to our waste once we're through with it?**

▲ **FIGURE 12.11** Thousands of abandoned mines on public lands poison streams and groundwater with acid, metal-laced drainage. ©Bryan F. Peterson

problem in many local watersheds. Acidic mine runoff has damaged or destroyed aquatic ecosystems in many places (fig. 12.11).

Gold and other metals are often found in sulfide ores that produce sulfuric acid when exposed to air and water. In addition, metal elements often occur in very low concentrations—10 to 20 parts per billion may be economically extractable for gold, platinum, and other metals. Consequently, vast quantities of ore must be crushed and washed to extract these metals. Cyanide, mercury, and other toxic substances are used to chemically separate metals from the minerals that contain them, and these substances can easily contaminate lakes and streams. Furthermore, a great deal of water is used in washing crushed ore with cyanide and other solutions. The USGS estimates that in arid Nevada, mining consumes about 230,000 m^3 (60 million gal) per day. After use in ore processing, much of this water contains sulfuric acid, arsenic, heavy metals, and other contaminants and is unsuitable for any other use.

Mining and drilling can degrade water quality

There are many techniques for extracting geologic materials. The most common methods are open-pit mining, strip-mining, and underground mining. As the opening case study for this chapter shows, these mining techniques can yield valuable metals, but they can also do irreversible damage to ecosystems.

An ancient method of accumulating gold, diamonds, and coal is placer mining, in which pure flakes or nuggets are washed from stream sediments. Since the California gold rush of 1849, placer miners have used water cannons to blast away hillsides. This method, which chokes stream ecosystems with sediment, is still used in Alaska, Canada, and many other regions. Another ancient and much more dangerous method is underground mining. Ancient Roman, European, and Chinese miners tunneled deep into tin, lead, copper, coal, and other mineral seams. Mine tunnels occasionally collapse, and natural gas in coal mines can explode.

Water pollution often is a serious consequence of mining. When water seeps into mine pits or underground shafts, it can dissolve toxic minerals. This is especially true where ores are high in sulfur, which produces sulfuric acid when exposed to air and water. Similarly, when water seeps through waste rock piles, it can become highly acidic and laced with heavy metals. Thousands of kilometers of American streams have been poisoned by mine drainage. Oil and gas well drilling also often produces salty, toxic water that must be treated or disposed of. Pumping it down deep disposal wells is a relatively cheap way to get rid of unwanted waste but can pollute underground aquifers and cause induced seismicity (Exploring Science, p. 295).

An ongoing controversy in the United States involves extraction of methane gas from coal deposits that are too deep or too dispersed for mining. Vast deposits of coal-bearing shale underlie both the Rocky Mountains and the Appalachian Mountains. Because the gas doesn't migrate easily through tight shales, it often takes many closely spaced wells to extract this methane (see fig. 13.12). In Wyoming's Powder River basin, for example, 140,000 wells were drilled for methane extraction. Together with the vast network of roads, pipelines, pumping stations, and service facilities, this industry is having serious impacts on ranching, wildlife, and recreation in formerly remote areas.

Perhaps even worse is the effect on water supplies. Each well produces up to 75,000 liters of salty water per day. Dumping this toxic

▲ **FIGURE 12.12** In the past decade, gas and oil drilling have expanded dramatically in the northeastern United States and the Great Plains. Often, these wells require a controversial procedure called hydraulic fracturing, or fracking, to release gas from sedimentary formations. ©Jim West/Alamy Stock Photo

EXPLORING Science *Induced Seismicity*

We used to think that humans couldn't cause earthquakes, but surges in seismic activity associated with well drilling and disposal of wastewater are changing our assumptions. Between 2000 and 2007, Oklahoma, for example, averaged about 20 measurable earthquakes per year. Starting about 2008, that number began shooting up, reaching more than 6,000 tremors by 2015. This was more than all the other 48 continental states combined. Oklahoma has become the earthquake capital of the United States. Some towns that had never experienced a quake had as many as 50 in a single day. In 2016, Pawnee, Oklahoma, had a 5.8-magnitude quake, which was the strongest in state history and the most powerful quake in the eastern United States in 70 years. This swarm of quakes correlates with a rush of oil and gas well drilling, and in particular with a proliferation of deep wastewater wells. Maps of well location and earthquake occurrence match well (fig. 1).

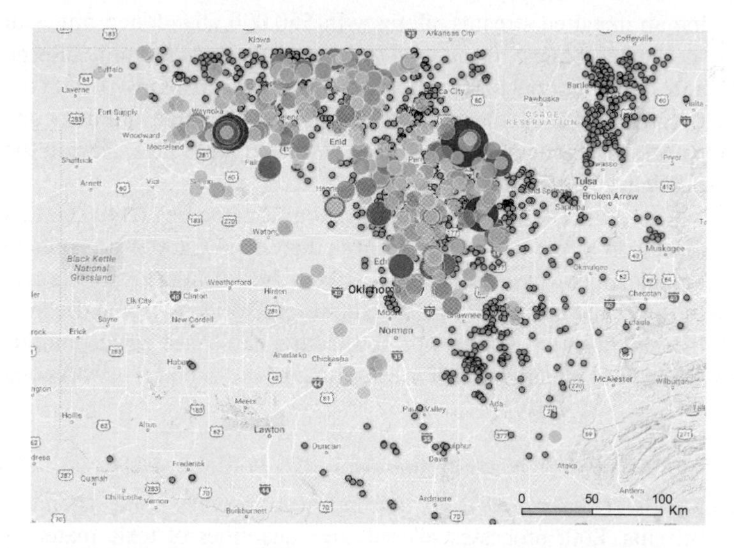

▲ **FIGURE 1** Earthquakes, historically uncommon in Oklahoma, have become frequent and often large around injection wells (blue) used for disposing of fracking wastewater. Earthquakes (orange) greater than magnitude 3.0 for 2016–17 are shown. Source: Oklahoma Geological Survey.

The issue is that Oklahoma has a wastewater disposal problem. Some oil wells there deliver up to 65 barrels of salty, contaminated groundwater for every barrel of oil. The "produced water" is much too toxic to dump into surface ponds or streams, and it would cost more than the oil is worth to treat it. So most of it is pumped back down into deep geological formations. In north central Oklahoma alone, about 1,000 wells injected some 17 billion liters (850 million barrels) of wastewater into deep strata in 2015. Many in the oil and gas industry, however, denied any connection to seismic activity, pointing out that they had been pumping liquids into the ground for decades without causing any noticeable tremors.

Establishing a link to injection wells took a combination of geologic map data and statistics. Geologists used detailed 3-dimensional maps of fault lines and geological formations, together with detailed data on earthquakes and aftershocks, to trace the origin of a magnitude 5.7 earthquake to within 200 m of an injection well. From long records of earthquake events, they also showed that earthquake activity in 2015 was about 600 times the normal rate. That is hard to ascribe to coincidence. They also showed that injection increased water pressure on faults, probably reducing friction and causing slippage. After publication of these findings, and after public attention was captured by dozens of additional major shocks, the state of Oklahoma agreed to address injection well activity.

Arkansas, Kansas, and Colorado also have experienced dramatic increases in seismic activity associated with disposal wells. Sometimes only a few wells, or well zones, are dangerous. Injection wells in the northeast part of figure 1, for example, are not associated with major quakes. In Arkansas, injection was halted for just four wells, and earthquake activity declined sharply. Oklahoma is now developing a rating system to designate safe and unsafe areas for injection.

The question now is whether seismicity will return to normal if we reduce injection wells. We will need them as long as we continue fracking. Is this geological hazard now the new normal?

---

waste into streams causes widespread pollution. To boost well output, mining companies rely on hydraulic fracturing (or "fracking"). A mixture of water, sand, and toxic chemicals is pumped into the ground and rock formations at extremely high pressure. The pressurized fluid cracks sediments and releases the gas. This often disrupts aquifers, however, and contaminates wells. And in some areas, deep injection of fracking liquids has been correlated with increased earthquake frequency.

For decades, coal-bed methane extraction was a problem only in western states, but this controversial technology is now moving to the East Coast as well (fig. 12.12). The Marcellus and Devonian Shales underlie parts of ten eastern states, ranging from northern Georgia to Upstate New York. It has long been recognized that methane can be extracted from these formations, but estimates of economically recoverable amounts were relatively small. New developments in horizontal drilling and hydraulic fracturing along with increased exploratory drilling have now made this deposit a potentially "supergiant" gas field. The U.S. Geological Survey now estimates that the Marcellus/Devonian formation may contain 500 trillion ft^3 (13 trillion m^3) of methane. If all of it were accessible, it would make a 100-year supply for the United States at current consumption rates. But the same issues, concerning a multitude of wells, water pollution, and threats to water supplies on which millions of people depend, raise thorny problems.

▲ **FIGURE 12.13** Some giant mining machines stand as tall as a 20-story building and can scoop up thousands of cubic meters of rock per hour. ©James P. Blair/National Geographic Creative

Surface mining destroys landscapes

Open-pit mines are used to extract massive beds of metal ores and other minerals. The size of modern open pits can be hard to comprehend. The Bingham Canyon mine, near Salt Lake City, Utah, is 800 m (2,640 ft) deep and nearly 4 km (2.5 mi) wide at the top. More than 5 billion tons of copper ore and waste material have been removed from the hole since 1906. A chief environmental challenge of open-pit mining is that groundwater accumulates in the pit. In metal mines, a toxic soup results. No one yet knows how to detoxify these lakes, which endanger wildlife and nearby watersheds.

About two-thirds of the coal mined in the United States comes from surface mines. Because coal is often found in expansive, horizontal beds, the entire land surface can be stripped away to cheaply and quickly expose the coal (fig. 12.13). Up to 100 m (328 ft) of overburden, or surface material, may be removed in relatively flat areas. The waste rock is generally placed back into the mine, but usually in long ridges called spoil banks. Spoil banks are very susceptible to erosion and chemical weathering. Because the spoil banks have no topsoil (the complex organic mixture that supports vegetation—see chapter 7), revegetation occurs very slowly.

A particularly damaging mining practice called mountaintop removal is controversial because of the destruction it wreaks on the environment as well as local communities. Forests are clear-cut and thousands of tons of high-powered explosives rip open the mountain. Giant machines scrape off the mountaintop to get to underlying coal seams (fig. 12.14). Sometimes the overburden can be piled back on top of the mountain or ridge, but often it's simply shoved into the valley below. Nearly 1.2 million acres (0.5 million ha) of land in Appalachia on more than 500 mountain peaks and ridgelines have been turned into mining wastelands, and about 2,000 mi (3,200 km) of headwater streams have been buried or severely polluted by the "valley fills" created by this process. The broken rock left from mining has high concentrations of sulfur, arsenic, mercury, and other toxic metals. As rainwater percolates through the rock rubble, it leaches toxins and becomes highly acidic as sulfur compounds are oxidized to sulfuric acid.

Coal must be crushed and washed before being sent to market. Millions of gallons of toxic wastewater and sludge are stored in huge ponds behind earthen dams in coal-mining country. A number of these dams have failed, sending floods of toxic sludge onto communities below. As coal demand has declined in the United States, coal companies have turned to shipping their product to foreign countries. We're destroying our landscape to sell coal to China; is that right?

The 1977 federal Surface Mining Control and Reclamation Act (SMCRA) requires better restoration of strip-mined lands, especially where mines replaced prime farmland. Since then, the record of strip-mine reclamation has improved substantially. Complete mine restoration is expensive, often more than $10,000 per hectare. Restoration is also difficult because the developing soil is usually acidic and compacted by the heavy machinery used to reshape the land surface.

The Mineral Policy Center in Washington, D.C., estimates that 19,000 km (12,000 mi) of rivers and streams in the United States are contaminated by mine drainage. The EPA estimates that cleaning up impaired streams, along with 550,000 abandoned mines in the United States, may cost $70 billion. Worldwide, mine closing and rehabilitation costs are estimated in the trillions of dollars. Because of the volatile prices of metals and coal, many mining companies have gone bankrupt before restoring mine sites, leaving the public responsible for cleanup.

In 2012, the Obama administration declared nearly 400,000 ha (1 million acres) of public land near the Grand Canyon off-limits to uranium mining because of worries about the potential for pollution in the nearby national park. In 2017, however, President Trump declared the area open for mining. For now, courts have ruled against mining and in favor of park protection. Stay tuned to learn how this works out.

Processing contaminates air, water, and soil

Metals are extracted from ores by heating or by using chemical solvents. Both processes release large quantities of toxic materials that can be even more environmentally hazardous than mining.

▲ **FIGURE 12.14** The tops of at least 500 mountains in the southern Appalachians have been sheared off to access buried coal. ©Vivian Stockman, Oct. 19, 2003

Smelting—roasting ore to release metals—is a major source of air pollution. One of the most notorious examples of ecological devastation from smelting in the United States is a wasteland near Ducktown, Tennessee. In the mid-1800s, mining companies began excavating the rich copper deposits in the area. To extract copper from the ore, they built huge, open-air wood fires, using timber from the surrounding forest. Dense clouds of sulfur dioxide released from sulfide ores poisoned the vegetation and acidified the soil over a 50 mi^2 (13,000 ha) area. Rains washed the soil off the denuded land, creating a barren moonscape.

Sulfur emissions from Ducktown smelters were reduced in 1907 after Georgia sued Tennessee over air pollution. In the 1930s the Tennessee Valley Authority (TVA) began treating the soil and replanting trees to cut down on erosion. Recently, upward of $250,000 per year has been spent on this effort. The trees and other plants are still spindly and feeble, but more than two-thirds of the area is considered "adequately" covered with vegetation. Similarly, smelting of copper-nickel ore in Sudbury, Ontario, a century ago caused widespread ecological destruction that is slowly being repaired following pollution-control measures.

Chemical extraction is used to dissolve or mobilize pulverized ore, but it uses and pollutes a great deal of water. A widely used method is **heap-leach extraction,** which involves piling crushed ore in huge heaps and spraying it with a dilute alkaline-cyanide solution. The solution percolates through the pile and dissolves gold. The gold-containing solution is then pumped to a processing plant that removes the gold by electrolysis. A thick clay pad and plastic liner beneath the ore heap are supposed to keep the poisonous cyanide solution from contaminating surface or groundwater, but leaks are common.

Once all the gold is recovered, mine operators have simply walked away from the operation, leaving vast amounts of toxic effluent in open ponds behind earthen dams. A case in point is the Summitville Mine near Alamosa, Colorado. After extracting $98 million in gold, the absentee owners declared bankruptcy in 1992, abandoning millions of tons of mine waste and huge, leaking ponds of cyanide. The Environmental Protection Agency may spend more than $100 million trying to clean up the mess and keep the cyanide pool from spilling into the Alamosa River.

Recycling saves energy as well as materials

Conservation offers great potential for extending our supplies of economic minerals and reducing the effects of mining and processing. The advantages of conservation are significant: less waste to dispose of, less land lost to mining, and less consumption of money, energy, and water resources.

Some waste products already are being exploited, especially for scarce or valuable metals. Aluminum, for instance, must be extracted from bauxite by electrolysis, an expensive, energy-intensive process. Recycling waste aluminum, such as beverage cans, on the other hand, consumes only *3 percent* of the energy needed to extract new aluminum (table 12.3).

Aluminum is also easy to recycle: much of it is beverage cans or other easily separated products. And aluminum has a far higher dollar value than iron, glass, or other common materials. Recycling is so rapid and effective that half of all the aluminum cans now on a grocer's shelf will be made into another can within 2 months.

TABLE 12.3 Energy Requirements in Producing Various Materials from Ore and Raw Source Materials

	ENERGY REQUIREMENT (MJ/KG)[1]	
PRODUCT	NEW	FROM SCRAP
Glass	25	25
Steel	50	26
Plastics	162	n.a.[2]
Aluminum	250	8
Titanium	400	n.a.[2]
Copper	60	7
Paper	24	15

[1] Megajoules per kilogram.
[2] Not available.
Source: Data from E. T. Hayes, *Implications of Materials Processing*, 1997.

Despite these efficiencies, U.S. aluminum recycling rates remain low. We now recycle only 36 percent of our aluminum food packaging, and only about 14 percent of all household aluminum, according to the EPA. (This is still better than approximately 0 percent of plastic cups and packaging). Recycling aluminum saves so much energy and natural resources, it only makes sense to improve on this record.

Platinum, the catalyst in automobile catalytic exhaust converters, is valuable enough to be regularly retrieved and recycled from used cars. Gold and silver are valuable enough to warrant recovery, even when not found in concentrated form. Copper, lead, iron, and steel are widely recycled because they are readily available in a pure and massive form, including copper pipes, lead batteries, and steel and iron auto parts. Nearly all scrapped automobiles and car batteries (which are legally required to be recycled) are now recycled in the United States (fig. 12.15).

FIGURE 12.15 The richest metal source we have—our mountains of scrapped cars—offers a rich, inexpensive, and ecologically beneficial resource that can be "mined" for a number of metals. ©Lisa S./Shutterstock

Although total U.S. steel production has fallen in recent decades—largely because of inexpensive supplies from newer and more efficient competitors—a new type of mill subsisting entirely on a readily available supply of scrap/waste steel and iron is a growing industry. **Minimills,** which remelt and reshape scrap iron and steel, are smaller and cheaper to operate than traditional integrated mills that perform every process from preparing raw ore to finishing iron and steel products. Minimills use less than half as much energy per ton of steel produced, compared to integrated mill furnaces. Minimills now produce about half of all U.S. steel. Some minimills use as much as 90 percent recycled iron and steel. Now that all new steel made in North America must contain at least 28 percent recycled contents, iron and steel recycling in the United States reached 83 percent in 2010.

New materials can replace mined resources

Mineral and metal consumption can be reduced by new materials or new technologies developed to replace traditional uses. This is a long-standing tradition; for example, bronze replaced stone technology, and iron replaced bronze. More recently, the introduction of plastic pipe has decreased our consumption of copper, lead, and steel pipes. In the same way, the development of fiber-optic technology and satellite communication reduces the need for copper telephone wires.

Iron and steel have been the backbone of heavy industry, but we are now moving toward other materials. One of our primary uses for iron and steel has been machinery and vehicle parts. In automobile production, steel is being replaced by polymers (long-chain organic molecules similar to plastics), aluminum, ceramics, and new high-technology alloys. All of these reduce vehicle weight and cost while increasing fuel efficiency. Some of the newer alloys that combine steel with titanium, vanadium, or other metals wear much better than traditional steel. Ceramic engine parts provide heat insulation around pistons, bearings, and cylinders, keeping the rest of the engine cool and operating efficiently. Plastics and glass fiber–reinforced polymers are used in body parts and some engine components.

Electronics and communications (telephone) technology, once major consumers of copper and aluminum, now use ultrahigh-purity glass cables to transmit pulses of light, instead of metal wires carrying electron pulses. Once again, this technology has been developed for its greater efficiency and lower cost, but it also reduces consumption of our most basic metals.

12.4 GEOLOGIC HAZARDS

- *Earthquakes result from sudden shifts of tectonic plates.*
- *Earthquakes and floods cause the most damage and mortality of geologic hazards.*
- *Erosion and landslides frequently cause property damage.*

Earthquakes, along with volcanoes, floods, and landslides, are normal earth processes, events that have made our earth what it is today. However, when they affect human populations, their consequences can be among the worst and most feared disasters that befall us.

Earthquakes are frequent and deadly hazards

Earthquakes are among our planet's most destructive geological disasters. In 2004 an earthquake and tsunami just off the coast of Banda Aceh, Indonesia, for example, killed over 230,000 people and caused damage as far away as Africa. A far larger toll is thought to have been caused by a 1976 earthquake in Tangshan, China. Government officials reported 655,000 deaths, although some geologists doubt it was that high.

Earthquakes are sudden movements in the earth's crust that occur along faults (planes of weakness), where one rock mass slides past another one. When movement along faults occurs gradually and relatively smoothly, it is called creep, or seismic slip, and may be undetectable to the casual observer. When friction prevents rocks from slipping easily, stress builds up until it is finally released with a sudden jerk. The point on a fault at which the first movement occurs during an earthquake is called the epicenter.

Earthquakes are especially catastrophic for cities built on soft alluvial soil, or water-saturated ground, which can liquefy when shaken. Death rates can be in the thousands when dense cities in developing areas are hit. The Kathmandu, Nepal, earthquake in 2015, with a magnitude of 7.8, collapsed thousands of unreinforced brick and timber buildings and killed over 7,000 people. This region experiences earthquakes often, but Kathmandu is much larger and more crowded now than it was a decade or two ago. These conditions make earthquakes in developing areas especially serious.

Earthquakes frequently occur along the edges of tectonic plates, especially where one plate is being subducted, or pushed down, beneath another. Earthquakes also occur in the centers of continents, however. In fact, the largest earthquake in recorded history in North America was one of an estimated magnitude 8.8 that struck the area around New Madrid, Missouri, in 1812 (fig. 12.16). Fortunately, few people lived there at the time, and the damage was minimal.

Modern building codes in earthquake zones attempt to prevent damage and casualties by constructing buildings that can withstand tremors. The primary methods used are heavily reinforced structures,

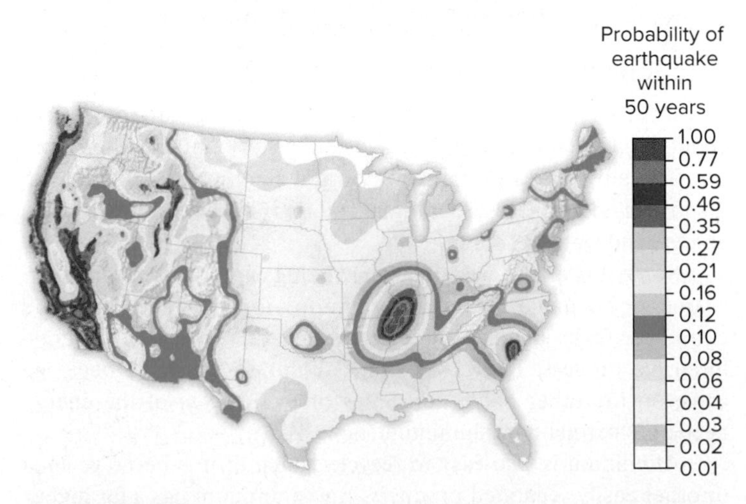

▲ **FIGURE 12.16** A seismic map of the lower 48 states shows the risk of earthquakes over the next 50 years (1.00 = a 2 percent chance). You might be surprised to learn that some of the highest risk is along the Mississippi River near New Madrid, Missouri. Source: USGS, 2010.

▲ **FIGURE 12.17** In 2011 a magnitude 9.0 earthquake just off the coast of Japan created a massive tsunami that destroyed homes, killed at least 25,000 people, and destroyed four nuclear power plants. ©JIJI PRESS/Getty Images

▲ **FIGURE 12.18** Volcanic ash covers damaged houses and dead vegetation after the November 2010 eruption of Mt. Merapi (background) in central Java. More than 300,000 residents were displaced and at least 325 were killed by this multi-day eruption. ©BAY ISMOYO/Getty Images

strategically placed weak spots in the building that can absorb vibration from the rest of the structure, and pads or floats beneath the building on which it can shift harmlessly with ground motion.

There's evidence that human activities can trigger earthquakes. Increased seismic activity has been triggered in many places by pumping huge amounts of contaminated water from oil and gas wells and other industrial activities into deep disposal wells (Exploring Science, p. 295). The tremendous mass of water contained behind large dams can also cause seismic activity. Chinese geologists, for example, suspect that the 2008 earthquake that killed 69,000 people in Sichuan Province may have been triggered by recent building of the Zipingpu dam on the Min River only a few kilometers from the earthquake epicenter.

One of the most notorious effects of earthquakes is the **tsunami** (Japanese for "harbor wave.") These giant sea swells (sometimes called tidal waves), such as the one that struck Japan in 2011, can move at 1,000 kph (600 mph), or faster, away from the center of an earthquake. About 25,000 people were killed and four nuclear reactors were destroyed by the 2011 tsunami (fig. 12.17). When these swells approach the shore, they can create breakers as high as 65 m (nearly 200 ft). Tsunamis also can be caused by underwater volcanic explosions or massive seafloor slumping.

Volcanoes eject deadly gases and ash

Volcanoes and undersea magma vents are the sources of most of the earth's crust. Over hundreds of millions of years, gaseous emissions from these sources formed the earth's earliest oceans and atmosphere. Many of the world's fertile soils are weathered volcanic materials. Volcanoes have also been an ever-present threat to human populations (fig. 12.18). One of the most famous historic volcanic eruptions was that of Mount Vesuvius in southern Italy, which buried the cities of Herculaneum and Pompeii in A.D. 79. The mountain had been showing signs of activity before it erupted, but many citizens chose to stay and take a chance on survival. On August 24, the mountain buried the two towns in ash. Thousands were killed by the dense, hot, toxic gases that accompanied the ash flowing down from the volcano's mouth.

Nuees ardentes (French for "glowing clouds") are deadly, denser-than-air mixtures of hot gases and ash like those that inundated Pompeii and Herculaneum. Temperatures in these clouds may exceed 1,000°C, and they move at more than 100 kph (60 mph). *Nuees ardentes* destroyed the town of St. Pierre on the Caribbean island of Martinique on May 8, 1902. Mount Pelee released a cloud of *nuees ardentes* that rolled down through the town, killing between 25,000 and 40,000 people within a few minutes. All of the town's residents died except for a single prisoner being held in the town dungeon.

Disastrous mudslides are also associated with volcanoes. The 1985 eruption of Nevado del Ruíz, 130 km (85 mi) northwest of Bogotá, Colombia, caused mudslides that buried most of the town of Armero and devastated the town of Chinchina. An estimated 25,000 people were killed. Heavy mudslides also accompanied the eruption of Mount St. Helens in Washington in 1980. Sediments mixed with melted snow destroyed roads, bridges, and property, but because of sufficient advance warning, there were few casualties. Geologists worry that similar mudflows from an eruption at Mount Rainier would threaten much larger populations (fig. 12.19).

Volcanic eruptions often release large volumes of ash and dust into the air. Mount St. Helens expelled 1 km^3 of dust and ash, causing ash fall across much of North America. This was only a minor eruption. An eruption in a bigger class of volcanoes was that of Tambora in Indonesia in 1815, which expelled 175 km^3 of dust and ash, more than 58 times that of Mount St. Helens. These dust clouds circled the globe and reduced sunlight and air temperatures enough so that 1815 was known as the year without a summer.

It is not just a volcano's dust that blocks sunlight. Sulfur emissions from volcanic eruptions combine with rain and atmospheric moisture to produce sulfuric acid (H_2SO_4). Droplets of H_2SO_4

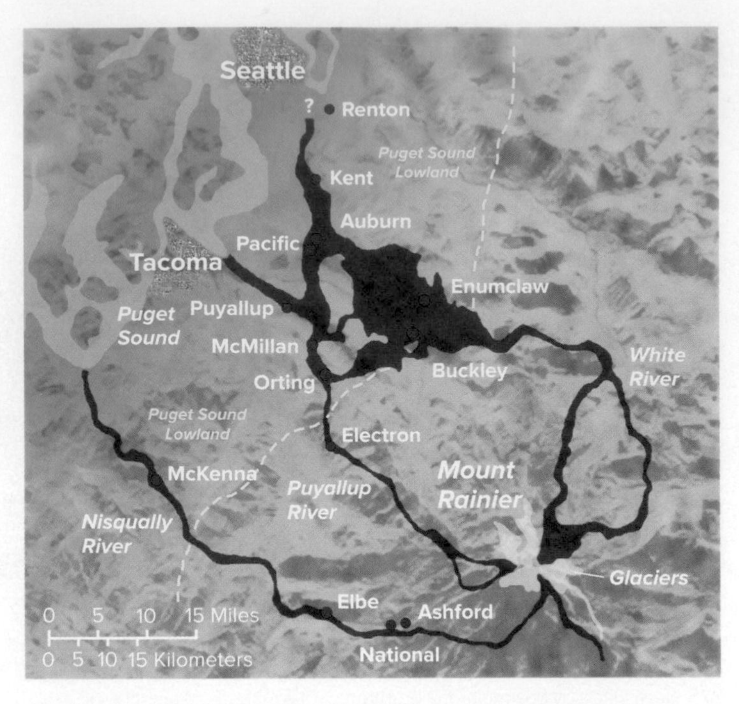

▲ **FIGURE 12.19** The highest risk for earthquakes, tsunamis, and volcanic activity in the contiguous 48 states is in the Pacific Northwest, where the Juan de Fuca plate is diving under the North American plate. Historic mudflows from Mount Rainier (brown areas) cover large areas that are now settled near Seattle and Tacoma, Washington.

▲ **FIGURE 12.20** In 2010, unusually heavy monsoon rains flooded about one-quarter of Pakistan, forcing about 20 million people from their homes and killing at least 2,000. ©Digital Vision/PunchStock

interfere with solar radiation and can significantly cool the world climate. In 1991 Mount Pinatubo in the Philippines emitted 20 million tons of sulfur dioxide aerosols, which remained in the stratosphere for 2 years. This thin haze cooled the entire earth by 1°C for about 2 years. It also caused a 10 to 15 percent reduction in stratospheric ozone, allowing increased ultraviolet light to reach the earth's surface.

Floods are part of a river's land-shaping processes

Like earthquakes and volcanoes, **floods** are normal events that cause damage when people get in the way. As rivers carve and shape the landscape, they build broad **floodplains,** level expanses that are periodically inundated. Large rivers often have huge floodplains. Many cities have been built on these flat, fertile plains, to be convenient to the river. Floodplains may appear safe for many years, but eventually they do flood. The severity of floods can be described by the height of water above the normal stream banks or by how frequently a similar event occurs—on average—for a given area. Note that these are statistical averages. A "10-year flood" would be expected to occur once in every 10 years; a "100-year flood" would be expected to occur once every century. But two 100-year floods can occur in successive years or even in the same year.

Among direct natural disasters, floods take the largest number of human lives and cause the most property damage (fig. 12.20). A flood on the Yangtze River in China in 1931, for example, killed 3.7 million people, making it the most deadly natural disaster in recorded history.

The biggest flood in U.S. history occurred in 1927 when the Mississippi River and its tributaries overflowed their banks from Minnesota to the Gulf of Mexico. Exceptional rain and snowfalls across the central United States in 1926 set the stage for this catastrophe. In April 1927, Mississippi levees broke in more than 145 places. Below St. Louis, the river spread out up to 80 miles (128 km) from side to side. More than 200,000 people were displaced. Some languished in refugee camps with little food, shelter, or medical help for months. White authorities' refusals to help African-Americans during this time played a role in the diaspora to northern cities—particularly Chicago—in the 1930s.

Flooding associated with hurricanes and other major storms are a growing risk as oceans and the climate warm. Unprecedented rainfall and deadly flooding have been seen repeatedly, as with Hurricane Harvey, which dropped 1.5 m of rain on Houston, TX (2017), Tropical Storm Lane, which produced 1.3 m on the island of Hawai'i (2018), and Hurricane Florence (up to 1 m of rain in North and South Carolina, 2018). Extreme storms like these occur around the globe, often endangering life and destroying infrastructure. Vulnerability to these events is made far worse as we build more communities in floodplains and along coastlines and barrier islands.

We have also made floods worse by eliminating places where rainfall can go. Paving the land surface prevents infiltration of stormwater, a factor that was blamed for much of Houston's flooding in Hurricane Harvey. Clearing forests for agriculture and destroying natural wetlands also increase both the volume and the rate of water discharge after a storm. In Iowa, for example, at least 99 percent of the natural wetlands that existed before settlement have been filled for farmland and urban development.

Land use changes and extreme monsoon rains, intensified by climate change, also caused huge floods across South Asia in 2017. More than 40 million people were displaced, and at least 1,200 died, especially in the Gangetic plains across India, Nepal, and Bangladesh. This may only be the beginning of a new normal.

Flood control

After the 1927 floods the U.S. Army Corps of Engineers was ordered to control the Mississippi River. The world's longest system of levees and flood walls was built to contain water within riverbanks, and river channels were dredged and deepened to allow water to recede faster. But every levee simply transfers the problem downstream. The water has to go somewhere. If it doesn't soak into the ground upstream, it will exacerbate floods somewhere else–leading to more levee development, and then to more flooding farther downstream, and so on.

More than $25 billion of river-control systems have been built on the Mississippi and its tributaries since 1928. These systems have protected many communities over the past century. In recent years, however, this elaborate system helped turn large floods into major disasters. Deprived of the ability to spill out over floodplains, the river is pushed downstream to create faster currents and deeper floods until eventually a levee gives way somewhere. Hydrologists calculate that Mississippi floods in 1993 were about 3 m (10 ft) higher than they would have been, given the same rainfall in 1900, before so many wetlands were filled in and flood-control structures were built.

Under current rules, the government is obligated to finance most levees and flood-control structures. Many people think that it would be much better to spend this money to restore wetlands, revegetate water courses, and remove flood-vulnerable structures. Floodplains could provide wildlife habitat, parks, recreation areas, and other uses not susceptible to flood damage.

Mass wasting includes slides and slumps

Gravity constantly pulls downward on every material everywhere on earth. Hillsides, beaches, even relatively flat farm fields can lose material to erosion. Often water helps mobilize loose material, and catastrophic slumping, beach erosion, and gully development can occur in a storm. A general term for downhill slides of earth is "mass wasting."

Landslides are sudden collapses of hillsides. In the United States alone, landslides and related mass wasting cause over $1 billion in property damage every year. When unconsolidated sediments on a hillside are saturated by a storm or exposed by logging, road building, or house construction, slopes are especially susceptible to sudden landslides (fig. 12.21).

◀ **FIGURE 12.21** Damaged homes sit on a hillside after a landslide in Laguna Beach, California. ©NICK UT/AP Images

Often people are unaware of the risks they face by locating on or under unstable hillsides. Sometimes they simply ignore clear and obvious danger. In southern California, where land prices are high, people often build expensive houses on steep hills and in narrow canyons. Most of the time, this dry environment appears quite stable, but in fact, steep hillsides slip and slump frequently. Especially when soil is exposed or rainfall is heavy, mudslides and debris flows can destroy whole neighborhoods. In developing countries, mudslides have buried entire villages in minutes. **Soil creep,** on the other hand, moves material inexorably downhill at an imperceptibly slow pace.

Erosion destroys fields and undermines buildings

Erosion constantly reworks the land surface. This is normal and natural, but it can be inconvenient when infrastructure is involved (see fig. 12.21). We describe erosion in different ways, depending on context and the severity. Gullying is the development of deep trenches on relatively flat ground. Especially on farm fields, which have a great deal of loose soil unprotected by plant roots, rainwater running across the surface can dig deep channels. Sometimes land becomes useless for farming because gullying is so severe and because erosion has removed the fertile topsoil. Agricultural soil erosion has been described as an invisible crisis. Erosion has reduced the fertility of millions of acres of prime farmland in the United States alone.

Beach erosion occurs on all sandy shorelines because the motion of the waves is constantly redistributing sand and other sediments. One of the world's longest and most spectacular beach complexes runs down the Atlantic coast of North America from New England to Florida and around the Gulf of Mexico. Much of this beach lies on some 350 long, thin **barrier islands** that stand between the mainland and the open sea. Behind these barrier islands lie shallow bays or brackish lagoons fringed by marshes or swamps.

Early inhabitants recognized that exposed, sandy shores were hazardous places to live, and they settled on the bay side of barrier islands or as far upstream on coastal rivers as was practical. Modern residents, however, place a high value on living where they have an ocean view and ready access to the beach. And they assume that modern technology makes them immune to natural forces. The most valuable and prestigious property is closest to the shore. Over the past 50 years, more than 1 million acres (400,000 ha) of estuaries and coastal marshes have been filled to make way for housing or recreational developments.

Construction directly on beaches and barrier islands can cause irreparable damage to the whole ecosystem. Under normal circumstances, fragile vegetative cover holds the shifting sand in place. Damaging this vegetation with construction, building roads, and breaching dunes with roads can destabilize barrier islands. Storms then wash away beaches or even whole islands. Hurricane Katrina in 2005 caused $100 billion in property damage along the Gulf Coast of the United States, mostly from the storm washing over barrier islands and coastlines (fig. 12.22). FEMA estimates that 25 percent of all coastal homes in the United States could have the ground washed out from under them by 2060 as intensified storms and rising sea levels caused by global warming make barrier islands and low-lying areas even riskier places to live.

Cities and individual property owners often spend millions of dollars to protect beaches from erosion and repair damage after storms. Sand is dredged from the ocean floor or hauled in by the truckload, only to wash away again in the next storm. Building artificial barriers, such as groins or jetties, can trap migrating sand and build beaches in one area, but they often starve downstream beaches and make erosion there even worse.

As is the case for inland floodplains, government policies often encourage people to build where they probably shouldn't. Subsidies for road building and bridges, support for water and sewer projects, tax exemptions for second homes, flood insurance, and disaster relief are all good for the real estate and construction businesses but invite people to build in risky places. Flood insurance typically costs $400 per year for $100,000 of coverage. In 2017 FEMA paid out $24 billion in disaster claims, 80 percent of which were flood- or storm-related. Settlement usually requires that structures be rebuilt exactly where and as they were before. There is no restriction on how many claims can be made, and policies are rarely canceled, no matter what the risk. Some beach houses have been rebuilt—at public expense—five times in two decades. The General Accounting Office reports that 2 percent of federal flood policies are responsible for 30 percent of all claims.

▲ **FIGURE 12.22** The aftermath of Hurricane Katrina on Dauphin Island, Alabama. Since 1970, this barrier island at the mouth of Mobile Bay has been overwashed at least five times by storms. More than 20 million yd^3 (15 million m^3) of sand have been dredged or trucked in to restore the island. Some beach houses have been rebuilt, mostly at public expense, five times. Does it make sense to keep rebuilding in such an exposed place? Source: USGS

The Coastal Barrier Resources Act of 1982 prohibited federal support, including flood insurance, for development on sensitive islands and beaches. In 1992, however, the U.S. Supreme Court ruled that ordinances forbidding floodplain development amount to an unconstitutional "taking," or confiscation, of private property.

CONCLUSION

Geologic hazards, including earthquakes, volcanic eruptions, tsunamis, floods, and landslides, represent major threats. Devastating events have altered human history many times in the past, sending geopolitical, economic, genetic, and even artistic repercussions around the planet. But the same processes that cause threats to humans also create resources, such as fossil fuels, metals, and building materials. The study of geology has allowed us to predict where these threats and resources will occur. The extraction of earth resources often carries severe environmental costs, however, including water contamination, habitat destruction, and air pollution.

The earth's surface is shaped by shifting pieces of crust, which split and collide slowly but continuously. Earthquakes, volcanoes, and mountains generally occur on plate margins. Rocks are composed of minerals, and rocks can be described according to their origins in molten material (igneous rocks), eroded or deposited sediments (sedimentary rocks), or materials transformed by heat and pressure deep in the earth (metamorphic rocks).

Earth resources, including oil, gas, and coal, are the foundation of our economy. Metals are expensive to extract, but they are extremely valuable because they are ductile (bendable) yet strong and because they carry electricity (e.g., copper). Mining and refining of metals and other geologic resources can cause severe environmental damage. In some cases, land can be restored to something close to its original state, but when you dig a mile-deep hole in the ground, or lop the whole top off a mountain and dump the waste into a nearby valley, it's unlikely that the damage will ever be undone.

Many materials can be recycled, saving money, energy, and environmental quality. Recycling aluminum consumes one-twentieth of the energy to extract new aluminum, for example, and recycling copper takes about one-eighth as much energy. We can also save energy and resources by replacing traditional materials with newer, more efficient ones. Fiber-optic communication lines have replaced much of our copper wiring, adding speed and efficiency while saving copper use.

Understanding both geologic hazards and resources is essential in environmental science. These factors influence environmental quality, resource provision, and environmental risks that affect all our lives.

PRACTICE QUIZ

1. How does tectonic plate movement create ocean basins, mid-ocean ridges, and volcanoes?
2. What is the "ring of fire"?
3. Describe the processes and components of the rock cycle.
4. What is the difference between metals and nonmetal mineral resources?
5. What is a *mineral* and a *rock*? Why are pure metals not minerals?
6. Which countries are the single greatest producers of our major metals?
7. Describe some of the mining, processing, and drilling methods that can degrade water or air quality.
8. Compare the different mining methods of underground, open-pit, strip, and placer mining, as well as mountaintop removal.
9. What resources, aside from minerals themselves, can be saved by recycling?
10. Describe the most deadly risks of volcanoes.
11. What is *mass wasting*? Give three examples and explain why they are a problem.
12. Why is building on barrier islands risky?
13. What is a *floodplain*? Why is building on floodplains controversial?
14. Describe the processes of chemical weathering and mechanical weathering. How do these processes contribute to the recycling of rocks?
15. The Mesozoic period begins and ends with the appearance and disappearance of dinosaurs. What fossils mark the other geologic eras?

CRITICAL THINKING AND DISCUSSION

Apply the principles you have learned in this chapter to discuss these questions with other students.

1. Understanding and solving the environmental problems of mining are basically geologic problems, but geologists need information from a variety of environmental and scientific fields. What are some of the other sciences (or disciplines) that could contribute to solving mine contamination problems?
2. Geologists are responsible for identifying and mapping mineral resources, but mineral resources are buried below the soil and covered with vegetation. How do you suppose geologists in the field find clues about the distribution of rock types?
3. If you had an igneous rock with very fine crystals and another with very large crystals, which would you expect to have formed deep in the ground, and why?
4. Heat and pressure tend to help concentrate metal ores. Explain why such ores might often occur in mountains such as the Andes in South America.
5. The idea of tectonic plates shifting across the earth's surface is central to explanations of geologic processes. Why is this idea still called the "theory" of plate tectonic movement?
6. Geologic data from fossils and sediments provided important evidence for past climate change. What sorts of evidence in the rocks and landscape around you suggest that the place where you live once looked much different than it does today?
7. We all use metals and fossil fuels obtained by mining. What responsibility do each of us have for the methods that produce the goods and services we enjoy?

DATA ANALYSIS Exploring Recent Earthquakes

The USGS Earthquake Center is a site that gives you access to global earthquake monitoring data. Understanding the distribution of earthquakes will help you understand the patterns of earth movement, volcanoes, and mountain building processes. Earthquakes are also one of our most important geologic hazards worldwide. Go to Connect to find data and questions that let you examine real-time earthquake records and to demonstrate your understanding of this geologic phenomenon.

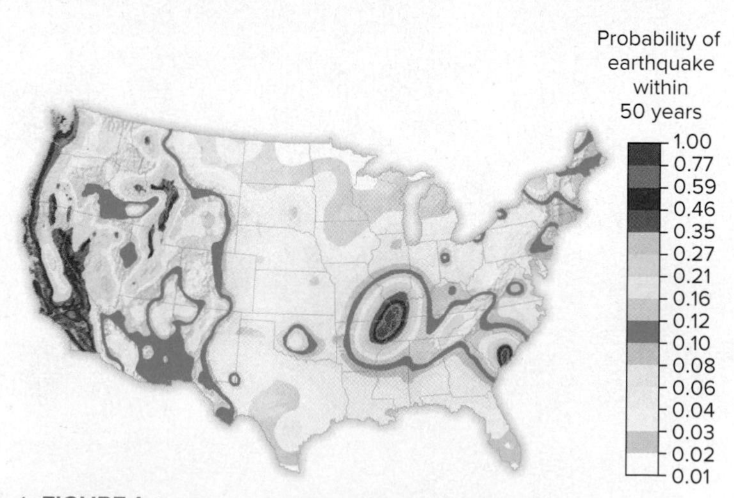

▲ **FIGURE 1** Probability of earthquakes in the coming half-century.

Design Elements: Active Learning (Toad): ©Gaertner/Alamy Stock Photo; Case Study (Globe): ©McGraw-Hill Education; Google Earth: ©McGraw-Hill Education; Abstract Background: ©Martin Kubat/Shutterstock; What do you think (Students using tablets): ©McGraw-Hill Education/Richard Hutchings, photographer; What can you do (Hand holding Globe): ©Christoph Weihs/Shutterstock

CHAPTER

13 Energy

LEARNING OUTCOMES

After studying this chapter, you should be able to answer the following questions:

- What are our dominant sources of energy?
- What is peak oil production? Why have estimates of oil resources been changing?
- What are arguments for and against increasing use of natural gas?
- What are the environmental effects of coal burning?
- How do nuclear reactors work? What are some of their advantages and disadvantages?
- What are our main renewable forms of energy?
- Could solar, wind, hydropower, and other renewables eliminate the need for fossil fuels?
- What are photovoltaic cells, and how do they work?
- What are biofuels? What are arguments for and against their use?

At the 2014 People's Climate March, more than 400,000 people gathered to call for alternatives to fossil fuels. The same weekend, New York City announced a goal of 80 percent CO_2 reduction by 2050.

CASE STUDY

Greening Gotham: Can New York Reach an 80 by 50 Goal?

New York City has a gritty reputation, and it's not known as a world leader in energy conservation. The biggest city in the United States, New York uses more energy than almost any other city, almost all of it derived from oil, natural gas, and nuclear power. As a northern city, New York needs about 6 months of heating and 4 months of air conditioning each year for its aging, often leaky building stock, and it has limited solar potential. At the same time, New York City planners know that climate change threatens the city with rising sea levels, increased hurricane frequency, intense heat waves, and economic instability as a warming climate undermines agriculture, transportation, health, and other underpinnings of the city's economic foundations. A serious heat wave can be especially deadly for low-income or elderly urban residents. If these threats are to be reduced or averted, New York and other cities have to do their share to cut greenhouse gas emissions, and they need to do it soon.

The Intergovernmental Panel on Climate Change (IPCC) reports that to avoid the worst impacts of climate change, the world needs to reduce its greenhouse gas emissions 80 percent by the year 2050. This is a daunting goal for any city. New York's energy infrastructure, like most of its building stock, was established half a century ago, or more. Can the city replace or upgrade its aging infrastructure on a tight budget and a tight timeline?

In 2014 the city's Office of Sustainability published a new policy that New York would aim for the IPCC's goal of 80 percent reductions by 2050, often called an "80 by 50" goal. The goal is ambitious, and the path to get there is not entirely clear, but Mayor Bill de Blasio and his administration agreed that it was necessary to protect the city's long-term interests. They also knew that if New York could lead the way, other cities would take the idea seriously and follow.

New York already uses less energy per person than any other major American city. Densely built apartments and offices share walls and conserve heat. Public transportation uses a minute fraction of the energy private cars do, per rider. Bicycle infrastructure has transformed the city in recent years, providing cheap and efficient transportation options. Many people walk to work, school, and stores because distances are short. People with small kitchens also tend to buy and waste less food. In New York the biggest challenges are leaky, thin-walled buildings and a lack of easy places to plan wind, solar, or other alternative energy sources.

New York is starting with its building stock. Financial incentives are helping landlords upgrade heating and cooling systems. The public housing authority has begun installing geothermal systems, better windows and building sealing, and more efficient heating systems. Upgrades save New Yorkers money as well as energy, and they make buildings more comfortable. City housing even built its first apartment building to *passive house* standards, which prescribe buildings so tightly made that they use almost no energy for heating, with more natural lighting and more comfortable design than the standard housing of a generation ago (fig. 13.1).

The city also works closely with large institutions, helping them reduce electricity consumption at peak hours; helping them finance improved lighting, windows, and insulation; and encouraging people to turn off lights and turn down thermostats. Large institutions are also investing in co-generation—generating electricity with waste heat from heating and cooling systems—which reduces energy demand on public utilities.

New policies are as important as new technology and tighter buildings. Reducing paperwork and bureaucratic obstacles to upgrades saves time and money. New rules at the state level can give utilities new incentives to integrate solar, wind, geothermal, and other renewable energy sources into their energy production system. The city also collaborates with public advocacy organizations to help define new incentives for the public to invest in solar, wind, and energy conservation. The city is also negotiating with the State of New York to develop new rules that promote shared solar energy investments, and easier permitting of offshore wind. Because without some investment in alternatives, all the insulation in the world won't do the job that needs to be done.

In many ways, the story of energy in the future is going to be different from the story in the past. Just as oil and automobiles transformed life a century ago, alternative energy sources, consumption practices, and ownership policies are likely to transform how we live and work. New Yorkers are betting that the future scenario will have exciting co-benefits, including safer, more reliable, and more affordable energy, if we can just commit to doing it right.

▲ **FIGURE 13.1** Knickerbocker Commons, in Brooklyn, is the first mid-sized apartment building designed to passive house standards in the United States. ©Henry Gifford, 803 Knickerbocker Avenue.

> *The stone age didn't end because we ran out of stones.*
>
> —SHEIK YAMANI, FORMER SAUDI OIL MINISTER

13.1 ENERGY RESOURCES

- *Fossil fuels provide most of the world's energy.*
- *Transitioning to renewables is becoming economical as well as important for the climate.*
- *Energy is mainly measured in kilowatt-hours or joules.*

Modern life runs on energy. Electricity and abundant fuel make our lives better by providing heating, cooling, and transportation. We use energy to grow and prepare food, to work at our jobs, and to entertain ourselves. Fossil fuels–oil, gas, and coal–provide the bulk of this energy. Hydropower and nuclear power supplement these, and increasingly wind, solar, and other renewable forms of energy are part of our standard energy supply.

A central energy question in environmental science has long been, when will we run out of oil? Oil is our single most important energy source–in terms of both volume consumed and diversity of uses, from running cars, trains, and ships to powering electric generating stations and heating our homes (fig. 13.2). Globally, oil provides 33 percent of our total energy supply, followed by natural gas (24 percent) and coal (28 percent). Together these fossil fuels make up 86 percent of our energy supply. Hydropower (7 percent) and nuclear (4 percent) make up most of the rest. Wind, solar, and other renewable energy forms are growing fast, but they still make up only about 3 percent of global energy. However, that is triple what they were a few years ago. The energy mix is changing; how it shifts will depend on how China and India, the two most populous, most rapidly developing countries, choose to develop.

One reason for the importance of oil, gas, and coal is the subsidies we provide for producing and consuming them. In 2013 global subsidies for fossil fuels amounted to about $1.9 trillion, or about 600 times the support for wind, solar, and all other alternative energy sources, according to the International Energy Agency.

Benchmark Data

Among the ideas and values in this chapter, these are a few worth remembering.

86%	Global energy from fossil fuels
7%	Global energy from hydropower
100%	Percentage of global energy that could be provided by renewables
2/3	Fraction of energy lost as heat in producing electricity from fossil fuel combustion
4,500 kWh	Amount of residential electricity per person per year in U.S.
5%	Percentage of residential energy that is elctricity

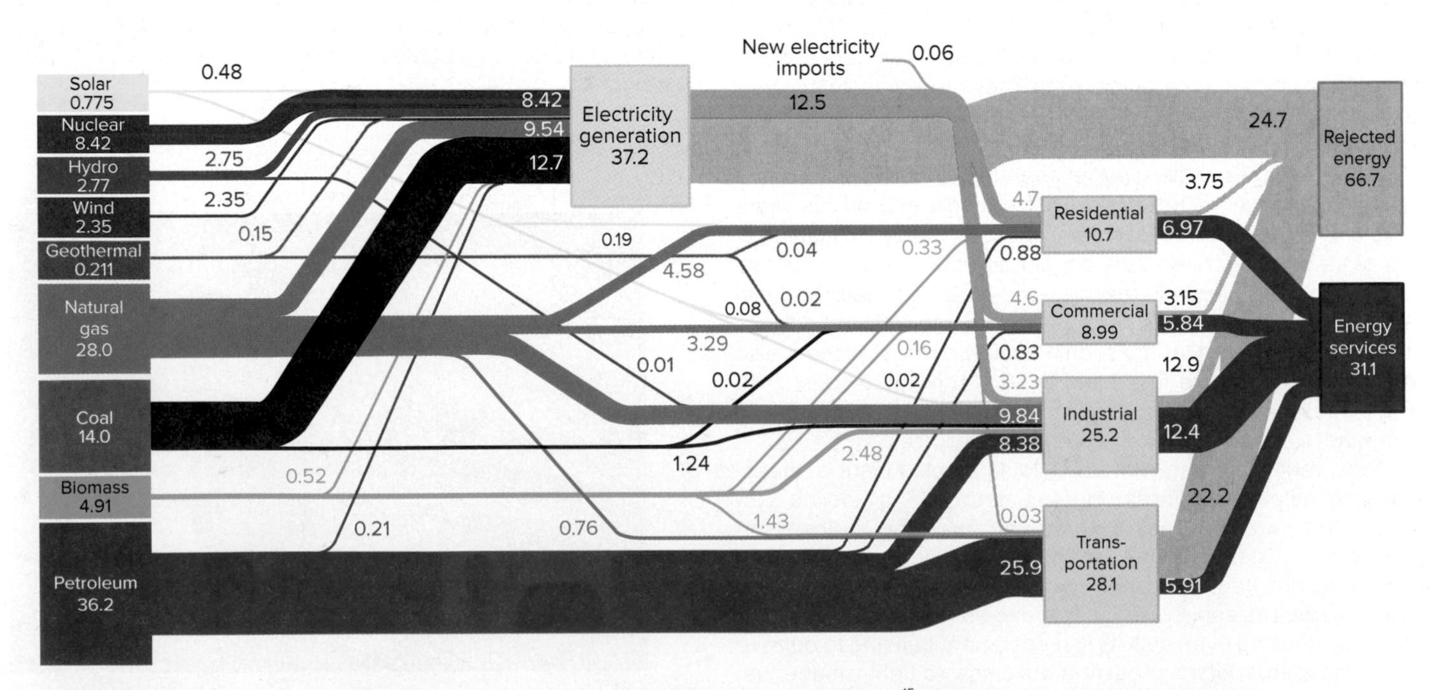

▲ **FIGURE 13.2** Sources and uses of electrical energy in the United States, in quadrillion (10^{15}) British thermal units BTUs. Line widths are proportional to amount of energy. "Rejected energy," lost through leakage and inefficiency accounts for 66.7 percent of energy production. Most coal-powered plants, for example, waste about 65 percent of energy in coal fuel. Source: LLNL March, 2017. Data is based on DOE/EIA MER, 2016. Quad = quadrillion (or 10^{15} BTUs).

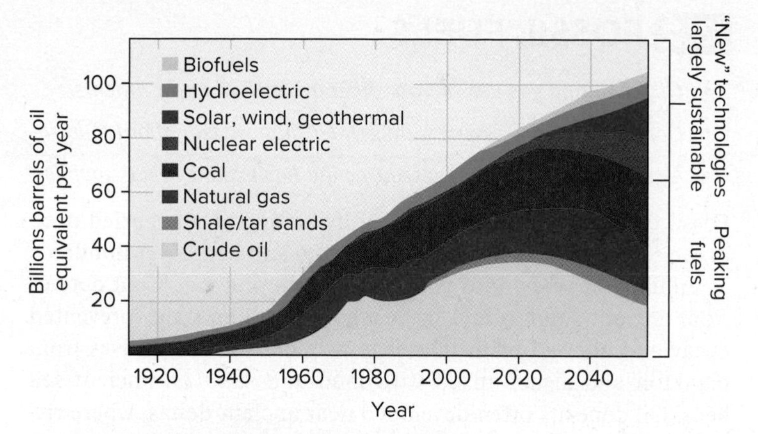

▲ **FIGURE 13.3** Fossil fuels, which now supply about 88 percent of all commercial energy in the world, are likely to decline as changes in technology, policy, and cost make renewables more attractive.

The future of energy is not the past

New techniques for oil and gas extraction have sharply increased our estimates of recoverable gas and oil reserves in the past 20 years. Horizontal drilling can reach several km through oil or gas-rich layers of rock, and hydraulic fracturing ("fracking") of geologic formations releases tightly bound gas and oil thousands of meters deep. Deep sea drilling technology has made vast undersea reserves newly accessible.

While energy analysts have long worried about the end of oil, climate analysts now warn that we have far more oil, gas, and coal than we can safely use. Burning all coal, oil, and gas deposits would produce five times as much carbon as we can afford to release without disrupting most agriculture and most ecosystems, as well as drowning most coastal cities (see chapter 9).

For these reasons, our energy future is likely to be different from our energy past (fig. 13.3). The future of energy will involve shifting transportation, building heating and cooling, and manufacturing to electricity, produced from renewable sources. Efficiency improvements will be necessary to make this possible.

This sounds like a daunting task, but the transition, while still small, is happening faster than analysts have expected (fig. 13.4). Renewable energy analysts find that a transition off of fossil fuels transition is not just necessary but also possible. And this transition brings co-benefits of a healthier environment, better jobs, less air and water pollution, and fewer violent conflicts over oil reserves.

We measure energy in units such as J and W

Energy can be measured in many ways, and understanding some terms will help in this chapter. The application of force over distance, as in pushing a block across a table, is measured in **joules** (J, table 13.1). Electricity is measured in **watts** (1 W = 1 joule per second).

Household energy use is measured in watts or in **kilowatts** (1 kW = 1,000 watts). A laptop computer uses about 100 W; a hair dryer uses about 1,500 W (table 13.2). To measure the amount of electricity used over time, we use watt-hours or **kilowatt-hours (kWh).** You can think of watts and kilowatt-hours a little like a garden hose pouring into a bucket: Watts are like the amount of water coming out at any instant. Watt-hours (or kilowatt-hours) are like the volume of water that collects in the bucket over time.

A typical power plant might supply electricity at a rate of 1,000 megawatts (MW)–that is, 1 gigawatt (GW). This is enough for about 640,000 households in the United States or about 1.3 million households in Europe. Some larger power plants are designed to supply 2,000 MW or more. Wind turbines produce around 1-5 MW (offshore wind turbines can reach 8 MW), so it takes a lot of wind turbines to add up to one conventional power plant. But it takes a

TABLE 13.1 Energy Units

1 joule (J) = work needed to accelerate 1 kg 1 m/sec^2 for 1 m (or 1 amp/sec flowing through 1 ohm resistance)
1 watt (W) = 1 J per second
1 terawatt (TW) = 1 trillion watts
1 kilowatt hour (kWh) = 1,000 W exerted for 1 hour (= 3.6 million J)
1 megawatt (MW) = 1 million (10^6) W
1 gigajoule (GJ) = 1 billion (10^9) J
1 standard barrel (bbl) of oil = 42 gal (160 liters)

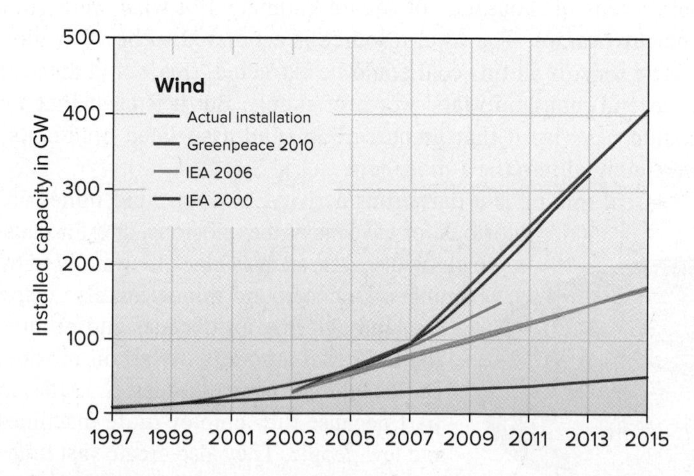

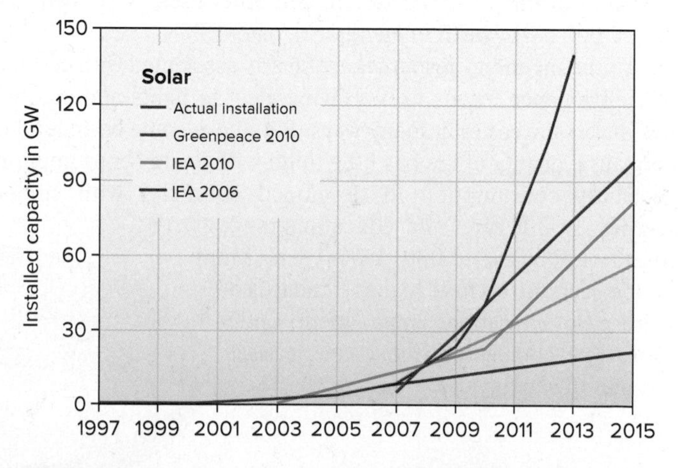

▲ **FIGURE 13.4** Expectations have usually underestimated actual progress. Installations of both wind (a) and solar (b) have accelerated over time. Projections are from the International Energy Agency (IEA) and Greenpeace. Note that the vertical scale for wind is larger than that for solar. Source: Meister Consultant Group, 2015.

TABLE 13.2 Energy Uses (Average)

USES	kWh/YEAR*
Computer	100
Television	125
100 W lightbulb	250
12 W LED bulb	33
Dehumidifier	400
Dishwasher	600
Electric stove/oven	650
Clothes dryer	900
Refrigerator	1,100

*Averages shown; actual rates vary greatly.
Source: U.S. Department of Energy

vast area of coal mines and oil fields, and fuel delivery infrastructure, to supply fuel to a conventional power plant. Most commercial-scale solar arrays produce around 150–500 MW of electricity.

The energy industry in the United States also uses traditional measures such as British thermal unit (Btu), the amount of energy it takes to heat one pound of water one degree F. Oil is measured in **barrels** (bbl). One barrel equals 42 gallons (160 l). Natural gas, which is a gas like air, not liquid like gasoline, is measured in cubic meters or cubic feet (m^3 or ft^3). Coal is measured in tons (2,000 lb) or metric tons (1,000 kg, or 2,200 lb).

How much energy do we use?

Of the four major energy sectors, about 22 percent is residential use (fig. 13.5). On average, Americans use about 4,500 kWh of electricity per person per year for household activities, twice what we used in 1970. In 1960 we used just 1,000 kWh per person per year. Europeans use about 2,000 kWh per year, and Chinese about 500 kWh. In low-income countries, people use less than 100 kWh per year–about what Americans use in a week.

But electricity is only about half of the energy we use. Including household heating, transportation, and other uses, Americans use about 9,000 kWh worth of energy per year.

Abundant energy resources are clearly associated with comfort and convenience in our lives. Having electric lights and automobiles makes life easier in many ways. But above some basic level of energy use, quality of life has little to do with energy consumption. Electricity consumption in developed countries with similar incomes, health care, and education ranges from around 1,500 to 10,000 kWh/year. Many European countries have higher standards of health care, education, employment, and leisure time, compared to most Americans, with half the energy use.

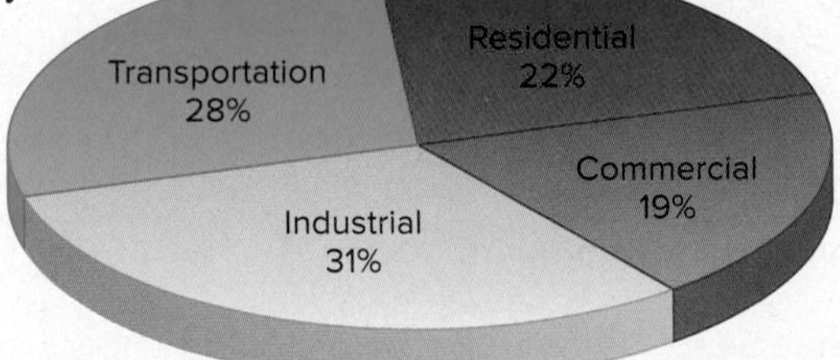

FIGURE 13.5 U.S. energy consumption by sector in 2012. Source: U.S. EIA.

13.2 FOSSIL FUELS

- *Oil, coal, and gas derive from different ancient environments.*
- *Coal, oil, and gas reserves are greater than we can afford to burn.*
- *Unconventional oil and gas extraction has expanded our supplies.*

Fossil fuels are ancient remains of living organisms, buried deep in the earth and transformed by pressure and heat, over hundreds of millions of years, into coal, oil, and natural gas. Coal derives from ancient swamps and bogs, where standing water prevented decay and allowed plant matter to accumulate. Oil derives from plankton and algae, mixed with mud and sand on ancient sea beds. Oil deposits often developed near ancient deltas, where rivers carried nutrients from land into the sea. Oil is generally extracted from the sand and mud (solidified to shale rock formations) with which it accumulated. Most of these resources originate from 280–360 million years ago, when the earth's climate was much warmer and wetter than it is now.

Both oil and coal vary in quality and usefulness. Hard, rock-like black coal is more pure, so it provides more heat and produces less air pollution, than soft brown coal. "Light, sweet" crude oil is fluid and easy to burn with relatively little purification. Tar sands, as the name implies, are thick, tar-like deposits mixed with sediment. Extracting these deposits requires a great deal of energy and money and is only economical when oil prices are high.

Natural gas is the smallest and simplest of these **hydrocarbons** (hydrogen-carbon compounds). Largely composed of **methane** (CH_4) that forms in association with coal and oil, gas tends to seep to the top of a coal seam or an oil deposit. But much gas also remains tightly bound in rock formations, as in the deep shale formations of Texas, Oklahoma, and Pennsylvania.

Coal resources are greater than we can use

World coal deposits are enormous, ten times greater than conventional oil and gas resources combined. Almost all the world's coal is in North America, Europe, and Asia, and just three countries, the United States, Russia, and China, account for two-thirds of all proven reserves. Coal seams can be 100 m thick and can extend across tens of thousands of square kilometers of what were once ancient swamps. The total resource is estimated to be 10 trillion metric tons. If all this coal could be extracted, this would amount to several thousand years' worth of supply. But it is clear that we couldn't survive if that much carbon (and associated pollutants) were emitted into the atmosphere.

Coal mining is a dangerous activity. Underground mines are notorious for cave-ins and explosions, and for causing lung diseases, such as black lung suffered by miners. Underground mines can also catch fire, smouldering for decades and producing unknown amounts of carbon dioxide. Surface mines, or strip mines, are safer, in part because they employ more machines and few people. They also create vast holes in the ground and contaminate surface waters if not managed well.

▲ **FIGURE 13.6** One of the most environmentally destructive methods of coal mining is mountaintop removal. Up to 250 m of the mountain is scraped off and pushed into the valley below, burying forests, streams, farms, and sometimes whole towns. ©David T. Stephenson/Shutterstock

An especially damaging technique employed in Appalachia is called mountaintop removal. Typically, the whole top of a mountain ridge is scraped off to access buried coal (fig. 13.6). In 2010 the EPA announced it would ban "valley fill," in which waste rock is pushed into nearby valleys, but existing operations are "grandfathered in" and allowed to continue.

Coal burning releases huge amounts of air pollution. For each kWh, a coal plant produces roughly twice as much CO_2 as natural gas plant. Coal produces about one-third of U.S. electricity but over two-thirds of the CO_2 and sulfur dioxide (SO_2) and 90 percent of the nitrogen oxides (NO_x) associated with electricity production. Coal also contains toxic impurities, such as mercury, arsenic, chromium, and lead, which are released into the air during combustion. Coal burning emits about 40 tons of mercury each year, as well as trace amounts of uranium and thorium. You would be exposed to more radioactivity living 70 years next to a coal power plant than next to a nuclear plant–assuming no accidents at the nuclear plant.

The Mercury and Air Toxics Standards announced by the EPA in 2012 were intended to dramatically reduce mercury emissions from coal-fired power plants. This action was required by the 1970 Clean Air Act, but five decades later, it continues to be delayed in courts. Coal industry lobbyists, arguing that the health rules were burdensome or that their plants were aging and couldn't be upgraded, have succeeded in preventing mercury controls despite multiple legislative efforts.

Coal use is declining in the United States

Recent improvements in technology have made other choices, from gas and oil to wind and solar, cheaper than coal. Coal has lower energy content per pound than oil or gas, so it is a less efficient fuel. At the same time, most of our coal power plants are reaching the end of their intended life. Aging plants are being retired steadily, because it is expensive to retrofit them to meet new public health standards.

The EPA estimated that mercury controls would cost utilities about $9 billion but would save $90 billion in health care costs by 2016 by reducing our exposure to mercury, arsenic, chromium, and fine particulates that cause mental impairments, cardiovascular diseases, asthma, and other disorders. In 2012 the EPA also proposed limiting CO_2 emissions from power plants, another rule that remains mired in litigation. If this rule goes into effect, new facilities will be allowed to emit no more than 1,000 lb (454 kg) of CO_2 per megawatt hour of electricity produced, less than half the current average. Natural gas plants can easily meet that standard, but for most coal plants, the only way to meet this limit with coal is to install expensive carbon capture and storage equipment.

China and India, both of which have very large coal resources, now burn about half of all coal mined annually in the world. Both of these countries have been increasing coal production greatly in the recent past to fuel their rapidly growing economies. Continuing to do so could cause runaway global climate change, so it's good that these two countries have begun investing in renewable energy (see section 13.7).

When will we run out of oil?

In the 1940s Dr. M. King Hubbert, a Shell Oil geophysicist, predicted that oil production in the United States would peak in 1970, based on estimates of U.S. oil reserves at the time (fig. 13.7). Subsequent studies expanded this analysis to global oil reserves, with similar conclusions. When we will reach peak oil production, and what will happen after that, has been a persistent concern among energy analysts.

Despite these predictions, production has continued to rise as new extraction technologies have greatly increased accessible

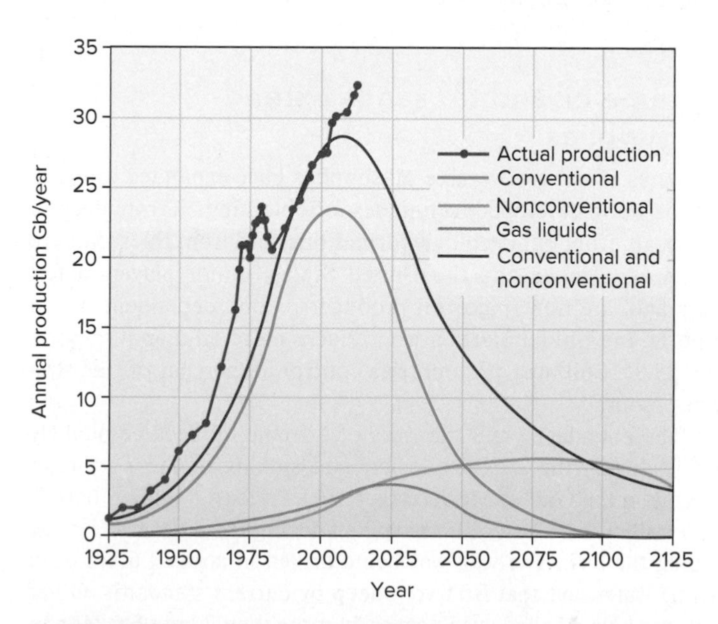

▲ **FIGURE 13.7** Worldwide production of crude oil with predicted Hubbert production. Gb = billion barrels. Source: Jean Laherrère, www.hubbertpeak.org; International Energy Agency, 2011.

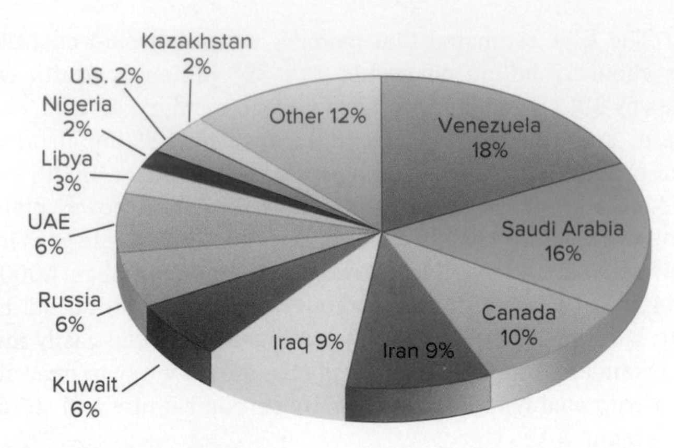

▲ **FIGURE 13.8** Proven oil reserves. The Middle East now accounts for a much smaller percent of the known, economically recoverable oil than in previous years. The numbers here add to more than 100 percent due to rounding. Source: CIA Factbook, 2016.

▲ **FIGURE 13.9** In 2010, the oil drilling rig *Deepwater Horizon* exploded and sank, spilling 5 million barrels (800 million liters) of crude oil into the Gulf of Mexico. It was drilling in water 1 mile (1.6 km) deep, but other wells are now more than twice as deep. ©U.S. Coast Guard/Getty Images

oil supplies. From 1993 to 2017, proven (commercially extractable) reserves grew by over 60 percent, from 1.04 trillion barrels to 1.69 trillion barrels. But consumption also grew, from 25 to 40 billion barrels per year. With oil production and consumption both rising, estimates of how long the resource will last have not changed greatly: Oil reserves are still expected to last 40–60 years, depending on rates of consumption.

Oil has been important because it has countless uses, from gasoline and other fuels to plastics and pesticides. Most of the growth in resources has been in South and Central America, including Venezuela and Brazil, although the majority of reserves remain in the Middle East (fig. 13.8).

Consumption has been climbing rapidly in fast-growing economies such as China, India, and Brazil. Emerging economies in Africa more than doubled their oil consumption between 1993 and 2017.

Extreme oil and tar sands extend our supplies

Estimates of our recoverable oil supplies have expanded dramatically as we've developed techniques for obtaining oil from deep in the ocean, from tight geologic formations, and from thick, impure tar sands. Canada and the United States, minor players a few years ago, are now major oil producers. Long dependent on oil imports, the United States now produces more oil than it imports (fig. 13.8). Still, just 12 countries control 88 percent of this strategic resource.

The abundance and the risks of extreme oil became publicly visible during the 2010 explosion and sinking of the *Deepwater Horizon* in the Gulf of Mexico (fig. 13.9). At least 5 million barrels (800 million liters) of oil were spilled during the 4 months it took to plug the leak. The well was being drilled in about 1.6 km deep (1 mi) water, but that isn't very deep by current standards. Other wells are being drilled even deeper, in more than 3 km of water and then 6 km or more below the sea floor, as in Venezuela and Mexico. Brazil's ultradeep deposits, some 300 km (186 mi) offshore, could hold 50 to 100 billion barrels. This could make Brazil fifth or sixth in the world in oil resources.

Tar sands are another recently exploited form of oil. Composed of sand and shale particles coated with thick, tarlike bitumen, tar sands can be excavated and mixed with steam and hot water to melt off the bitumen. The energy use involved in this process makes it extremely expensive. Alberta, the province with most of Canada's tar sands, has proposed building a nuclear power plant to provide steam for extracting the bitumen.

This system is economical only when oil prices are high, and a global oil glut has kept prices low for some years. However, the infrastructure is extremely costly, so production must continue even when other sources are cheaper.

The Keystone XL pipeline, built to deliver tar sands oil from Alberta to Houston, Texas, for processing and export, has been controversial because carbon emissions associated with tar sands are the

▼ **FIGURE 13.10** Alberta tar sands are now the largest single source of oil for the United States, but water pollution, forest destruction, and the energy used to liquefy the sticky tar are among the many costs of extracting this oil. ©dan_prat/Getty Images

highest of any major oil production process. By some estimates, emissions are higher than those from coal. A typical facility producing 125,000 bbl of tar sands oil per day also releases about 5,000 metric tons of CO_2 per year and contaminates billions of cubic meters of fresh water, producing about 15 million m^3 of toxic sludge (fig. 13.10). Native Cree, Chipewyan, and Métis communities protest that mining threatens their water, wildlife, forest, and traditional ways of life.

Canadian tar sands are estimated to be equivalent to 1.7 trillion bbl of oil. Venezuela has nearly as much. Together, these are nearly three times as much as all conventional liquid oil reserves.

The United States also has large supplies of unconventional oil. **Oil shales** are fine-grained sedimentary rock rich in solid organic material called kerogen. Like tar sands, the kerogen can be heated, liquefied, and pumped out like liquid crude oil. Oil shale beds up to 600 m (1,800 ft) thick—containing several trillion barrels of oil—underlie much of Colorado, Utah, and Wyoming.

Access to markets is a key challenge

A battle over the Keystone XL pipeline, proposed as a means to carry tar sands oil from mines in Alberta to Houston, Texas, brought this fuel source to public attention. Supporters of the Keystone pipeline claimed it would bring energy security to the United States and provide 20,000 jobs. Opponents countered that the pipeline wouldn't help the United States much because the oil would be shipped to Texas for sale abroad. This could raise U.S. oil prices rather than lower them. Critics also have said that the pipeline supporters' estimates of job creation included every possible job involved in supporting actual pipeline workers. A more realistic number, according to opponents, was about 50 permanent jobs on the pipeline. Furthermore, a rupture in the Keystone pipeline could contaminate the Ogallala aquifer, which supplies drinking water and irrigation for much of the Great Plains.

TransCanada, the company behind the pipeline, is also pursuing a northern "Gateway" route that would cross the Canadian Rockies on its way to a terminal in the fjords of British Columbia's Great Bear Rainforest. First Nations people fear that heavy tanker traffic through the narrow, twisting fjords could result in another Exxon *Valdez*-size accident in this pristine wilderness. Pipelines carrying tar sands oil have a much higher rupture rate than those for conventional oil. The residual sand is more abrasive, the oil is more acidic and corrosive, and heavy oil must be heated to higher temperatures to be shipped, all making tar sands pipelines more accident prone.

Natural gas is growing in importance

Natural gas (mostly methane, CH_4) is the world's second-largest commercial fuel, making up about one-quarter of global energy consumption. Gas burns more cleanly than either coal or oil, and it generally produces only half as much CO_2 as an equivalent amount of coal when it is burned. Many people argue that switching from coal to gas will help slow climate change, although many also dispute this.

More than half of all the world's proven natural gas reserves are in the Middle East and Russia (fig. 13.11). Both eastern and western Europe are highly dependent on imported gas. The total ultimately recoverable natural gas resources are thought to be

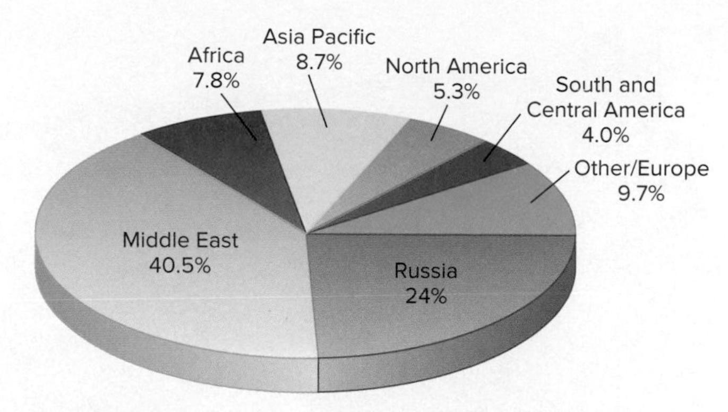

▲ **FIGURE 13.11** Proven natural gas reserves by region.
Source: Data from British Petroleum, 2014.

28 billion m^3, corresponding to about 80 percent as much energy as the estimated recoverable reserves of crude oil.

Large amounts of methane are known to occur in many relatively shallow sedimentary beds. Accessing this gas has involved intense drilling activity in the western United States. It often takes many closely spaced wells and directional drilling to extract methane from these "coal-bed" methane deposits. In Wyoming's Powder River basin, for example, 140,000 wells have been proposed for methane extraction. Together with the vast network of roads, pipelines, pumping stations, and service facilities, this industry is having serious impacts on ranching, wildlife, and recreation in formerly remote areas (fig. 13.12).

Hydraulic fracturing opens up tight gas resources

Deep gas-bearing shale formations underlie vast areas of Appalachia, Texas, and other regions (fig. 13.13). The Marcellus

▲ **FIGURE 13.12** Some of the thousands of gas wells in the Jonah Field in Wyoming's Upper Green River Basin. ©Peter Aengst/Wilderness Society/Lighthawk

FIGURE 13.13 Shale gas plays in the United States. The cost of extracting these resources varies with depth. Source: Based on data from the Energy Information Association.

and Devonian Shales, underlying Pennsylvania and New York, have been the focus of considerable debate. The Barnett and Haynesvile formations in Texas have been less debated and extremely productive. The U.S. Geological Survey estimates that the Marcellus/Devonian formation may contain a 100-year supply for the United States at current consumption rates. Large amounts of gas now available have driven prices down sharply in recent years.

But these shale deposits are generally "tight" formations through which gas doesn't flow easily. To boost well output, drilling companies rely on hydraulic fracturing, or fracking. A mixture of water, sand, and various chemicals is pumped into the ground and rock formations at extremely high pressure. The pressurized fluid cracks sediments and releases the gas. Fracturing rock formations often disrupts aquifers, however, and contaminates water wells.

Drilling companies generally refuse to reveal the chemical composition of the fluids used in fracking. They claim it's a proprietary secret, but it's well known that a number of petroleum distillates, such as diesel fuel, benzene, toluene, xylene, polycyclic aromatic hydrocarbons, glycol ethers, as well as hydrochloric acid or sodium hydroxide, are often used. Many of these chemicals are known to be toxic to humans and wildlife. The U.S. EPA recently forced mining companies to reveal the contents, but not specific fractional composition, of their fracking fluids used on public land.

Natural gas leaks are a growing concern. Cracks in well linings and fractures in rock formations can let gas leak into groundwater. A study by the National Academy of Sciences found that shallow wells near gas wells in Pennsylvania and New York had 17 times as much methane as those from sites far from wells.

Leaky valves and pipelines emit vast amounts of methane, a powerful greenhouse gas, every year. No one knows how much gas escapes this way. Estimates are that 3 to 8 percent of gas production may be released, especially at fracked wells. Some studies argue that as long as these leaks are unmonitored and unrepaired, they make gas 20 times worse than coal for our climate (see section 9.3). Fracked gas is expected to supply about two-thirds of our gas supply in coming decades.

13.3 NUCLEAR POWER AND HYDROPOWER

- *Nuclear and hydropower make electricity by spinning generators.*
- *Nuclear fission breaks up unstable uranium into other elements.*
- *Large hydro produces abundant electricity but has great impacts.*

After fossil fuels, nuclear and hydropower are dominant global sources of energy. Both produce electricity, in contrast to fossil fuels that can also provide transportation and heating fuel (see fig. 13.2). Both nuclear and hydropower are expensive ventures made possible by billions of dollars in government funding over many decades. This funding has provided electricity to billions of consumers at a low price. External, unaccounted for costs, however, can be substantial.

In 1953 President Dwight Eisenhower presented his "Atoms for Peace" speech to the United Nations. He announced that the United States would build nuclear-powered electrical generators to provide clean, abundant energy. He predicted that nuclear energy would fill the deficit caused by predicted shortages of oil and natural gas. It

▲ **FIGURE 13.14** New York's Indian Point nuclear power plant is ranked the riskiest in the country by the U.S. Nuclear Regulatory Commission due to its age and location on the Hudson River just 24 miles (38 km) north of Manhattan Island. What would it cost to evacuate New York City if these reactors melted down? ©William P. Cunningham

would provide power "too cheap to meter." Today there are about 440 reactors in use worldwide, 104 of them in the United States. Half of the U.S. reactors are about 40 years old and are approaching the end of their expected operational life (fig. 13.14). Cracking pipes, leaking valves, and other parts increasingly require repair or replacement as a plant ages. The costs of demolishing a worn-out plant may be ten times as much as building it in the first place.

Many have argued that we must invest in nuclear as a lower-carbon stopgap to slow climate change, because reactors don't release greenhouse gases during ordinary operation of the reactor. Emissions do result from mining and processing of uranium, producing cement for reactors, and decommissioning old reactors. Still, many experts see nuclear as a critical part of controlling GHG emissions. Even though radioactive fuel and waste are risky, they argue, the global risks of climate change are much worse.

But economics are a challenge for nuclear power. Nuclear facilities have always been too costly to produce without heavy subsidies, but now renewables and gas pose new competitive threats. Growing security concerns also reduce enthusiasm for systems that produce vast amounts of radioactive waste. All these issues have made nuclear energy less attractive than promoters expected in the 1950s. Of the 140 reactors on order in 1975, 100 were canceled.

Only two nuclear reactors are under construction in the United States, a pair of Georgia Power reactors. Approved in 2012, they were the first permitted since 1978, with an expected cost of $14 billion, but by 2018 their cost estimate was raised to $25 billion. State utility regulators have guaranteed that customers will pay the costs, regardless of whether the reactors are ever completed.

How do nuclear reactors work?

The most commonly used fuel in nuclear power plants is U^{235}, a naturally occurring radioactive isotope of uranium. Uranium ore must be purified to a concentration of about 3 percent U^{235}, enough to sustain a chain reaction in most reactors. The uranium is then formed into cylindrical pellets slightly thicker than a pencil and about 1.5 cm long. Although small, these pellets pack an amazing amount of energy. Each 8.5 g pellet is equivalent to a ton of coal or 4 bbl of crude oil.

The pellets are stacked in hollow metal rods approximately 4 m (13 ft) long. About 100 of these rods are bundled together to make a **fuel assembly.** Thousands of fuel assemblies containing about 100 tons of uranium are bundled in a heavy steel vessel called the reactor core. Radioactive uranium atoms are unstable—that is, when struck by a high-energy subatomic particle called a neutron, they undergo **nuclear fission** (splitting), releasing energy and more neutrons. When uranium is packed tightly in the reactor core, the neutrons released by one atom will trigger the fission of another uranium atom and the release of still more neutrons (fig. 13.15). Thus, a self-sustaining **chain reaction** is set in motion, and vast amounts of energy are released.

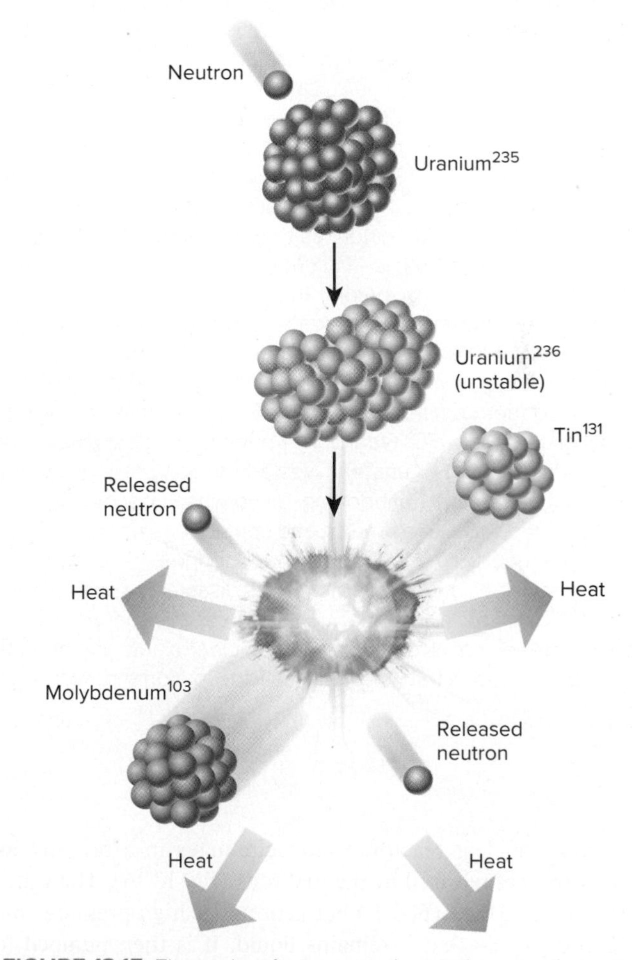

▲ **FIGURE 13.15** The nuclear fission carried out in the core of a nuclear reactor. The unstable isotope uranium-235 absorbs a neutron and splits to form tin-131 and molybdenum-103. Two or three neutrons are released per fission event and continue the chain reaction. A tiny amount of mass is converted to energy (mostly heat).

What Do YOU THINK?

Twilight for Nuclear Power?

On Friday, March 11, 2011, at 2:46 p.m. Tokyo time, a magnitude 9.0 earthquake hit northern Japan. The largest earthquake in Japan's recorded history damaged buildings and roads in its own right, but even worse, it generated tsunami waves up to 30 m (98 ft) high that crushed buildings, toppled power lines, and washed away cars, boats, and millions of tons of debris. Authorities reported 15,846 deaths, 6,011 injuries, 3,320 people missing, and 125,000 buildings damaged or destroyed by waves.

Perhaps the worst result of this catastrophe was the destruction of four of the six nuclear reactors at the Fukushima Daiichi power station 170 mi (273 km) north of Tokyo. The reactors shut down, as they were designed to do, when the earthquake hit, but that eliminated the electricity needed to pump cooling water through the intensely hot reactor core. Backup generators and connections to the regional power grid that would have provided emergency power were destroyed by the tsunami. The reactors quickly overheated, and the fuel rods began to melt in three of the six reactors' cores. Hydrogen explosions in the reactor buildings at the complex destroyed roofs and walls and scattered radioactive debris around the area. In addition, spent fuel rods in storage pools of two units also overheated and caught fire, releasing even more radiation.

The plant operators sprayed seawater onto reactors to cool the reactors and put out fires, but that washed radioactive pollution into the ocean, contaminating marine ecosystems and seafood. In 2012 Japan began tunneling under the nuclear complex to install a giant concrete diaper to try to contain radioactive drainage from the site. High radioactivity caused authorities to order evacuation of 140,000 people living within a 12-mile (20-km) radius around the facility. But the toll could have been worse. If the melting fuel and fires hadn't been contained, the radiation release could have been ten times greater than the 1986 disaster at Chernobyl in Ukraine.

At one point, government officials seriously considered evacuating the Tokyo metropolitan area. That might have meant moving up to 40 million people, which would have been the largest mass relocation in world history. Fortunately, westerly winds blew most of the radiation out to sea and abandoning Tokyo wasn't necessary.

Still, cleanup will take decades, and some areas near the reactors may never be habitable again. Altogether, Japanese officials estimate that losses may be $300 (U.S.) billion. This disaster is causing people in many countries to reconsider nuclear power. In Japan, which once got about one-third of its electricity from nuclear plants and had plans to expand that share to more than half the nation's power supply, more than 80 percent of the population now say they oppose nuclear power. After the disaster, all the nation's 54 reactors were shut down. An intense debate occurred about whether to restart any of them.

▲ Three of the four nuclear reactors at Japan's Fukushima Diiachi power station that were destroyed by fuel melting and hydrogen explosions after the 2011 tsunami knocked out emergency cooling systems. ©DigitalGlobe/Getty Images

After Fukushima, Germany immediately shut down eight reactors and promised to close all the rest of its nuclear plants by 2022. Italy, Switzerland, and Spain voted to keep their countries nonnuclear. And in France, which gets three-quarters of its electricity from nuclear power, 62 percent of the population favored a phase-out of this energy source.

Could this be the death knell for nuclear power? Public opinion has swung against nuclear power after Chernobyl, and again after Fukushima. Many factors weigh in public opinions about nuclear power—fear of accidents, worry about long-term storage of radioactive waste, concern for how we will get the electricity the market demands, and more. What do you think? Which of these factors are most important to you, and why? Or which other factors? If you were making the decision, what would lead you to increase, or to decrease, investments in nuclear power?

In most nuclear reactors, water circulates in a primary loop and absorbs heat released by the fuel rods (fig. 13.16). The water is superheated to 317°C (600°F) but kept under high pressure (more than 2,000 psi) so that it remains liquid. It is then pumped to a secondary loop, where it heats water to produce steam, which spins an electricity-generating turbine.

The chain reaction is moderated (slowed) in a power plant by a neutron-absorbing cooling solution that circulates between the fuel rods. In addition, **control rods** of neutron-absorbing material, such as cadmium or boron, are inserted into spaces between fuel assemblies to shut down the fission reaction or are withdrawn to allow it to proceed. The greatest danger in one of these complex machines is a cooling system failure. If the pumps fail or pipes break during operation, the nuclear fuel quickly overheats, and a "meltdown" can result that releases deadly radioactive material (What Do You Think?, above).

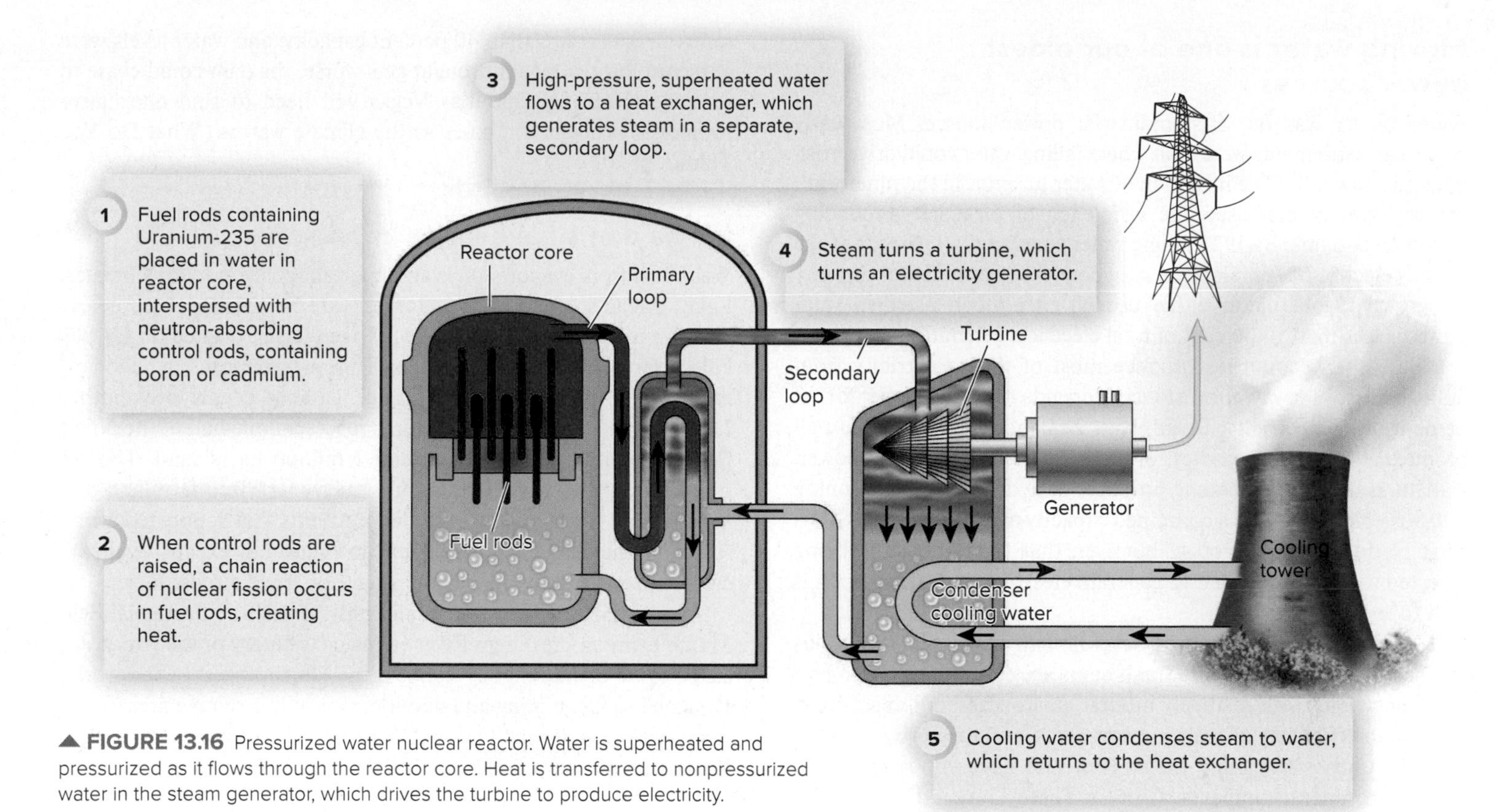

▲ **FIGURE 13.16** Pressurized water nuclear reactor. Water is superheated and pressurized as it flows through the reactor core. Heat is transferred to nonpressurized water in the steam generator, which drives the turbine to produce electricity.

We lack safe storage for radioactive waste

One of the most difficult problems associated with nuclear power is the disposal of wastes produced during mining, fuel production, and reactor operation.

Enormous piles of mine wastes and abandoned mill tailings in uranium-producing countries represent serious waste disposal problem. Production of 1,000 tons of uranium fuel typically generate 100,000 tons of tailings and 3.5 million liters of liquid waste. There now are approximately 200 million tons of radioactive waste in piles around mines and processing plants in the United States. This material is carried by the wind or washes into streams, contaminating areas far from its original source.

In addition to the leftovers from fuel production, the United States has about 77,000 tons of high-level (very radioactive) wastes. The high-level wastes consist mainly of spent fuel rods from commercial nuclear power plants and assorted wastes from nuclear weapons production. While they're still intensely radioactive, spent fuel assemblies are being stored in deep, water-filled pools at the power plants. These pools were originally intended only as temporary storage until the wastes were shipped to reprocessing centers or permanent disposal sites.

In 1987 the U.S. Department of Energy announced plans to build the first high-level waste repository in the desert under Yucca Mountain in Nevada. Waste was to be buried deep in the ground, where it was hoped it would remain unexposed to groundwater and earthquakes for the tens of thousands of years required for the radioactive materials to decay to a safe level. But continuing worries about the stability of the site led the Obama administration to cut off funding for the project in 2009 after 20 years of research and $100 billion in exploratory drilling and development.

For the foreseeable future, the high-level wastes that were to go to Yucca Mountain will be held in large casks in temporary surface storage facilities located at 131 sites in 39 states (fig. 13.17). But local residents living near these sites fear casks will leak.

If the owners of nuclear facilities had to pay the full cost for fuel, waste storage, and insurance against catastrophic accidents, there would be little interest in this energy source. Rather than be too cheap to meter, it would be too expensive to matter.

▲ **FIGURE 13.17** Spent fuel is being stored temporarily in large, above-ground "dry casks" at many nuclear power plants. Source: Office of Civilian Radioactive Waste Management, Department of Energy

Moving water is one of our oldest power sources

Water power was our first industrial power source. Most early American settlements were built where falling water could drive gristmills and sawmills. The invention of water turbines in the nineteenth century greatly increased the efficiency of electricity-producing hydropower dams. By 1925 falling water generated 40 percent of the world's electric power. Since then, hydroelectric production capacity has grown 15-fold, but fossil fuel use has risen much faster, so water power is less than 10 percent of total electrical generation.

Still, many countries produce most of their electricity from falling water. Norway, for instance, depends on hydropower for 99 percent of its electricity; Brazil, New Zealand, and Switzerland all produce at least three-quarters of their electricity with water power. Canada is the world's leading producer of hydroelectricity, running 400 power stations with a combined capacity exceeding 60,000 MW. First Nations people protest, however, that their rivers are being diverted and lands flooded to generate electricity, most of which is sold to the United States.

Much of the hydropower development since the 1930s has focused on enormous dams. Small dams are common and can produce electricity with relatively modest environmental impacts. But there is an efficiency of scale in giant dams, and they bring prestige to the countries that build them. They also have important social and environmental impacts. China's Three Gorges Dam on the Yangtze River, for instance, is the largest in the world. It spans 2.0 km (1.2 mi) and is 185 m (600 ft) tall (fig. 13.18). The reservoir it creates has displaced more than 1 million people and drowned important farmlands.

In arid regions, hydropower generators can be vulnerable to low water levels. Lake Mead, on the Colorado River, provides power and water to Las Vegas, Nevada, and other desert cities. The largest reservoir in the United States, Lake Mead was created when the Hoover Dam was completed in 1935. Since then water levels have fluctuated dramatically, as river flow into Lake Mead has varied. In recent years of drought, engineers have worried that there could be insufficient water to drive the power turbines in the dam. In 2014 the reservoir was at less than 40 percent capacity, and water levels were at record lows. If recent drought gets worse, the dam could cease to produce electricity, and Las Vegas will need to find alternative sources, if drought continues as the climate warms (What Do You Think?, p. 261).

▲ **FIGURE 13.18** Very large dams, such as China's Three Gorges Dam, produce abundant electricity but have tremendous environmental and social costs. ©menabrea/Getty Images

Large dams have large impacts

Water loss from evaporation is an important concern in dry climates. Lake Mead's evaporative losses are estimated at nearly 1 billion m^3 per year. Even worse is Lake Nasser, created on Egypt's Nile River by the Aswan High Dam. This reservoir loses 15 billion m^3 each year to evaporation and seepage. Unlined canals lose another 1.5 billion m^3. Together, these losses represent one-half of the Nile's flow, and enough water to irrigate 2 million ha of land. The silt trapped by the Aswan High Dam formerly fertilized farmland during seasonal flooding and provided nutrients that supported a rich fishery in the delta region. Farmers now must buy expensive chemical fertilizers, and the fish catches have dropped almost to zero.

Large dams also destroy biodiversity. Brazil's controversial Belo Monte Dam on the Xingu River (a major tributary of the Amazon), slated to be operational by 2020, is the fourth largest in the world. Designed to fuel mining and development in this remote area, it will flood around 450 km^2 (175 mi^2) of tropical rainforest. Indigenous Kayapo people have protested that the dam is destroying traditional hunting lands and will eliminate many endemic species. A more global concern is methane emissions. As in all tropical dams, submerged vegetation in the drowned forest will emit methane as it decays. This methane could have a greater global warming impact than burning coal to produce an equivalent amount of electricity.

Despite these concerns, large dams are an important energy source: They produce electricity at a scale that electric utilities can manage and sell for large consumers, including mining, industry, and large cities. Public subsidies are necessary to build these dams, but the electricity produced is abundant and is cheaper than many other sources, when social and environmental costs are excluded.

13.4 ENERGY EFFICIENCY AND CONSERVATION

- *Efficiency is usually the cheapest "source" of energy.*
- *Life-cycle costs account for operating costs.*
- *High-efficiency houses are necessary for energy use reductions.*

It's often said that the cheapest form of energy is conservation. Efficiency is not as visible or exciting as a new power plant, but it saves money and helps avoid new production, with its various costs. This is why New York City (opening case study) is focusing on building efficiency first, even before it starts spending money on new energy sources.

Conservation can save money as well as reducing our energy footprint. Better sealing of doors and windows can be inexpensive and make your house or apartment much more comfortable. Reducing hot water consumption and remembering to turn off lights usually cost nothing and lower energy bills. There are

What Can YOU DO?

Steps to Save Energy and Money

1. Live close to work and school, or near transit routes, so you can minimize driving.
2. Ride a bicycle, walk, and use stairs instead of elevators.
3. Keep your thermostat low in winter and high in summer. Fans are 99% cheaper to run than air conditioners.
4. Buy fewer disposable items: Producing and shipping them costs energy.
5. Turn off lights, televisions, computers, and other appliances when not needed.
6. Line-dry your laundry.
7. Recycle.
8. Cut back on meat consumption: If every American ate 20 percent less meat, we would save as much energy as if everyone used a hybrid car.
9. Buy local food to reduce shipping energy and keep your money in the local economy.

countless steps each of us can take that cost nothing and save money as well as energy (What Can You Do?, above).

The Environmental Protection Agency and other government agencies often help us out in saving energy and money. EPA-approved Energy Star appliances are rated for improved efficiency and savings. There are also government incentives and rebates for energy efficiency measures. Automobile efficiency in the United States has doubled because of gas mileage regulations, from only 13 miles per gallon in 1975 to nearly 30 mpg today. In 2012 the Obama administration announced plans to increase national fuel economy standards to 54.5 miles per gallon by the year 2025. This would reduce U.S. oil consumption by 2.2 million barrels a day. That is 12 percent of the 18.5 million barrels per day we consume. Imagine what we would do for any other energy source that offered 12 percent of our needs basically for free.

But you don't have to wait until 2025 for an efficient vehicle. Low-emission, hybrid gas-electric vehicles already get up to 72 mpg (30.3 km/l), and electric cars get the equivalent of 120 mpg (50.5 km/l) or more (Active Learning, above). Of course, walking, biking, and public transportation lower your personal energy footprint far more.

Costs can depend on how you calculate them

Sometimes the cost and benefit of conservation depend on how many years you're thinking of when you calculate expenses and savings. If we calculate costs on a short time frame, it can be expensive to replace existing appliances, roof insulation, and furnaces with newer, better types. If we calculate over the lifetime of an appliance, roof, or furnace, it is usually cheaper to buy the energy-efficient version, even if the cost is higher in the short term. If we're buying something new anyway, then the more efficient option is almost always the smarter option.

Active LEARNING

Driving Down Gas Costs

Most Americans drive at least 1,000 miles per month in vehicles that get about 20 miles per gallon. Suppose gasoline costs $4.00 per gallon.

1. At these rates, how much does driving cost in a year? You can calculate the annual cost of driving with the following equation. Before multiplying the numbers, cross out the units that appear on both the bottom and the top of the fractions—if the units cancel out and give you $/year, then you know your equation is set up right. Then use a paper and pencil, and multiply the top numbers and divide by the bottom numbers. What is your annual cost?

$$\frac{12 \text{ months}}{\text{year}} \times \frac{1{,}000 \text{ mi}}{\text{month}} \times \frac{1 \text{ gal}}{20 \text{ mi}} \times \frac{\$4.00}{\text{gal}} =$$

2. Driving fast lowers your mileage by about 25 percent. At 75 mph, you get about three-fourths as many miles per gallon as you get driving 60 mph. This would drop your 20 mpg rate to 15 mpg. Recalculate the equation above, but replace the 20 with a 15. What is your annual cost now?

3. Efficiency also declines by about 33 percent if you drive aggressively, because rapid acceleration and braking cost energy. Aggressive driving would drop your 20 mpg mileage to about 13.4 mpg. How much would your yearly gas cost at 13.4 mpg?

4. A U.S. goal has been to reach an average fleet efficiency of 54.5 mpg by 2025. How much would your yearly gas cost at 54.5 mpg? What is the difference between that cost and your cost at 20 mpg?

ANSWERS: 1. $2,400/year; 2. $3,200/year; 3. $3,592/year; 4. $881/year, or $1,519 less than at 20 mpg.

Lightbulb efficiency is a familiar example. The cost of 60 W incandescent bulbs has long been very low, often $1 per bulb. A compact fluorescent bulb that produces the same amount of light might be five or ten times that. A newer LED (light-emitting diode) bulb might be around $10, or possibly much less. But over a year of use the cheapest bulb costs ten times more to operate (fig. 13.19). Over 10 years the cost difference multiplies. Because incandescent bulbs put such a high demand on a community's energy resources, and because they are responsible for very high greenhouse gas emissions, they are being phased out in many places. LEDs and CFLs (compact fluorescent lightbulbs) are also much safer because they are less hot, and less likely to start a fire when unattended, than incandescent bulbs.

If a whole city (or a college or university) switches to LEDs, the savings multiply. When Ann Arbor, Michigan, replaced over 1,000 streetlights with LED lights, the city saved over $80,000 per year, and the new fixtures paid for themselves in just 2 years.

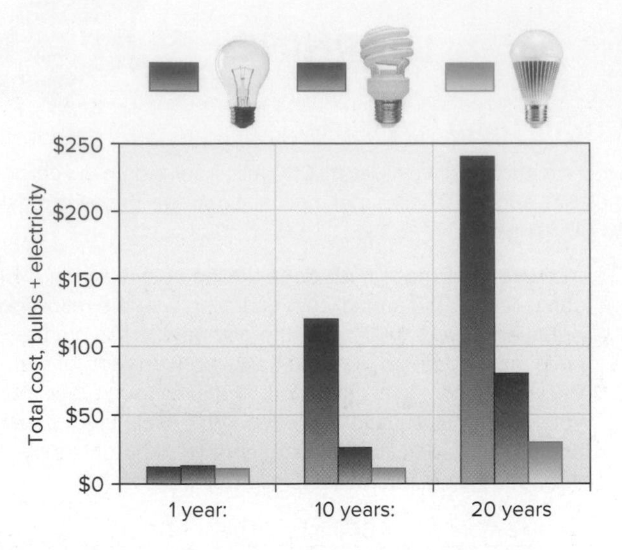

▲ **FIGURE 13.19** The cost effectiveness of efficient technology often depends on the time frame used in calculating costs. This graph shows costs of three light bulbs, assuming an LED bulb costs about $10, lasts 45 years, and costs $1/year to use; a CFL costs $10, lasts 7 years, and costs $2.50 per year; and an incandescent bulb costs $1, lasts about 1 year of steady use, and costs $11 per year to use.

▲ **FIGURE 13.20** Energy loss from buildings is visble in an infrared image. Improving building efficiency is one of the most important ways to reduce energy consumption. ©William P. Cunningham

Tight houses save money

Heating and cooling are the biggest energy uses in most regions (fig. 13.20). Improving energy efficiency in heating and cooling is much more expensive than switching lightbulbs, but builders calculate that these measures can also pay for themselves in a few years in energy savings.

Many efficiency improvements you can do yourself. Caulking around doors, windows, foundation joints, electrical outlets, and other holes in your walls can reduce air leakage, keeping your living space warmer in winter and cooler in summer. Insulating the roof and foundation is harder but even more valuable. Sealing the corners of roofs and foundations—the joints where air sneaks through and around insulation layers—also makes a tremendous difference.

Innovations in "green" building have been stirring interest in both commercial and household construction. Much of the innovation has occurred in large commercial structures that have the largest budgets—and more to gain through efficiency—than most individual households have. Elements of green building are evolving rapidly, but they include extra insulation in walls and roofs, coated windows to keep summer heat out and winter heat in, and recycled materials, which save energy in production. Orienting windows toward the sun and providing roof overhangs for shade are important for comfort as well as for saving money.

Passive housing is becoming standard in some areas

Germany has long been a leader in energy innovation and efficiency. Many regions in Germany have begun to require that new buildings conform to **passive house** standards. These standards are strict energy use limits, along with guidelines for building practices that reduce energy consumption to just 10 percent of what normal buildings use (fig. 13.21). Many cities and states across Europe have begun to follow Germany in either requiring or encouraging passive standards.

◀ **FIGURE 13.21** Energy-efficient buildings lower energy costs dramatically. Passive house standards, in particular, eliminate thin walls and windows, thermal bridges (which conduct heat through a wall), and poorly sealed joints. These standards cut heat loss in cold climates and heat gain in hot climates.

In the United States, incentives, such as faster permitting or tax breaks to building projects that meet passive standards, are increasingly common (see fig. 13.1).

Relatively rare a few years ago, houses and office buildings of this standard are becoming more common in Europe, North America, and Asia. In a place where building standards are already high—as they are in Germany—a passive house can cost only 10 percent more than a standard house, and that extra cost pays for itself in just a few years of reduced energy use. Where building standards are more lax and housing is inexpensive, the cost difference can be much greater, but so can the savings. Residents also say passive houses are more comfortable, with even heating, better natural light, and less noise than conventional housing.

What does passive housing involve? The main principles are these:

- No thermal bridges (wood or metal components conduct heat through walls, roof, or foundation)
- Roof and wall insulation 24–32 cm (10–14 in.) thick
- Well-sealed windows and doors to prevent heat loss or gain
- Heat exchangers, to warm or cool fresh air as it is exchanged with stale air leaving the building
- Triple-pane windows that reduce radiative heat loss or gain
- Windows well positioned to let in daylight

When a passive building produces electricity—for example, with solar panels on the roof—it can become a "net-zero" building that uses no more energy than it produces. Some of these are even "energy-plus" buildings, producing more energy than they consume on an annual basis.

Additional measures are becoming common in many areas: Appliances often have timers so that they can run overnight, when energy use is low. Thermostats can automatically adjust the temperature when you are not home or when you are sleeping. Smart metering gives you information on how much energy is being used and how much it costs. Using one of these systems, you might program your water heater to operate only after midnight, when surplus wind power is available. Often these can be operated from your smart phone, even when you're not home.

Cogeneration makes electricity from waste heat

A fast-growing source of new energy is cogeneration, the simultaneous production of both electricity and steam or hot water in the same plant. Cogeneration is mainly used in institutions that have large boilers producing steam or hot water to heat multiple buildings. The net energy yield in these boilers is normally 30–35 percent. Using waste heat to spin a turbine and produce electricity can produce efficiencies of 80–90 percent.

Colleges, universities, and other large institutions are increasingly installing cogeneration. Cities with district heating (a central heating plant that distributes water or steam) also use cogeneration. These facilities reduce the load on utilities, helping them avoid the expense of building new power plants. Utilities also buy electricity from cogenerators when it is needed.

Generating electricity onsite also avoids losses on long transmission lines, which lose up to 20 percent of the electricity they carry. The EPA estimates that cogeneration could produce almost 20 percent of U.S. electrical use, or the equivalent of 400 coal-fired plants.

13.5 WIND AND SOLAR ENERGY

- *Wind and solar are now less expensive and quicker to install than conventional sources.*
- *Solar PV works by knocking electrons from material in the cell.*
- *China, India, and developing countries are investing rapidly in renewables.*

In his 2011 State of the Union speech, President Barack Obama said, "To truly transform our economy, protect our security, and save our planet from the ravages of climate change, we need to ultimately make clean, renewable energy the profitable kind of energy. . . . So tonight, I challenge you to join me in setting a new goal: By 2035, 80 percent of America's electricity will come from clean energy sources."

Just a few years ago, this plan would have seemed ludicrous, but increasingly it appears readily achievable (Key Concepts, p. 324). A 2017 report found that U.S. utility-scale solar costs as little as 4–5 cents per kWh, as compared to gas at 6.1 cents and coal at 6.6 cents. Wind was best of all at 1.4 cents per kWh. Analysts in 2010 didn't expect to see these prices for wind and solar until 2020. And renewables don't have the external costs of pollution, health impacts, and price volatility of fossil fuels. Using only those sites where energy facilities are socially, economically, and politically acceptable, there's more than enough renewable energy from the sun, wind, geothermal, biomass, and other sources to meet all our present energy needs (fig. 13.22).

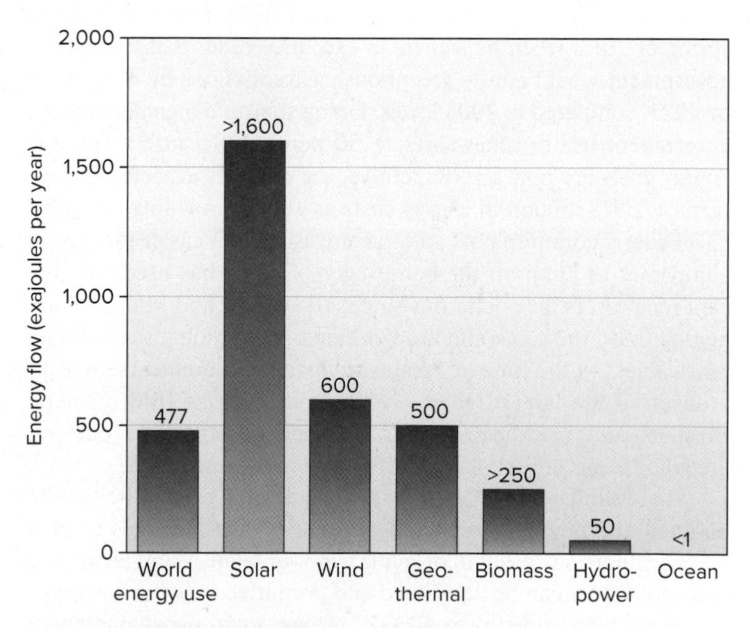

▲ **FIGURE 13.22** Potential energy available from renewable resources using currently available technology in presently accessible sites. Together, these sources could supply more than six times current world energy use. Source: UNDP and International Energy Agency.

▲ **FIGURE 13.23** China is now the world's largest producer of wind turbines and has about one-quarter of all installed wind power capacity. Together with solar, this clean technology provides more than 1 million jobs in China. ©Paul Springett 10/Alamy Stock Photo

President Obama made another big step for renewables in the spring of 2015, when he signed an executive order that the federal government would cut its greenhouse gas emissions by 40 percent by 2025, compared to 2008 levels. Doing so would mean increasing government use of renewables to 30 percent. Because renewable energy costs are now so competitive, the rule was expected to save taxpayers $18 million in energy costs as well as providing new jobs.

Federal commitments and incentives make it easier for private companies to jump on the bandwagon. Google has used 100 percent renewables for operations since 2017. Apple and Microsoft are aiming to do the same and are working to help global suppliers go renewables. At the time of President Obama's announcement, over a dozen of the largest federal contractors, such as IBM, General Electric, and Lockheed Martin, announced voluntary cuts on greenhouse gas emissions and efficiency improvements.

Renewable energy also has the advantage that it can be planned and deployed relatively quickly. It takes decades to plan and permit a coal, gas, oil, or nuclear power plant. Wind farms and solar collectors can be developed and permitted in just a few years. In 2017, China added over 50 GW of new solar, producing about 1.5 times as much energy as the Three Gorges Dam (about 80–100 GWh per year), but without a decade of construction and displacing millions of people.

Wind could meet all our energy needs

Wind power is the world's fastest-growing energy source. With 540 GW of globally installed capacity in 2017, wind power is producing over 1,000 TWh of electricity annually. The Wind Energy Association predicts that 1.5 million MW of capacity could be possible by 2020. Wind power could replace all the commercial energy we now use, if we chose to develop it.

China is the world's largest producer of wind turbines and has the second largest installed wind capacity, after Europe. Clean technology provides more than 1 million jobs in China, manufacturing equipment for domestic use and for export. China now has over 63 GW of wind power, or about one-quarter of the world total (fig. 13.23).

Most wind turbines produce 1–2 MW, but the largest ones now generate 8 MW of electricity, enough for 4,000 American homes. With less downtime for maintenance than many power plants, turbines can often produce power 90 percent of the time.

Prices for wind power have fallen sharply in the past few years, and this is currently the cheapest source of new electrical generation, costing 3 cents/kWh or less, compared to 6 cents/kWh for coal and five times that much for nuclear fuel.

The United States has a tremendous potential for wind power. Large areas of the Great Plains and mountain states, and all coastal states, have persistent winds suitable for commercial development (fig. 13.24). Texas, with 23.5 GW, leads the nation, in wind power, followed by Oklahoma, Iowa, California, and Illinois.

Wind power takes about one-third as much area and creates about five times as many jobs per MW as coal, when the land consumed by mining is taken into account (fig. 13.25). When wind power is installed on farmland, each tower uses only about 0.1 ha (0.25 acre) of cropland. Farmers can cultivate 90 percent of their land while getting $2,000 or more in annual rent for each wind machine. An even better return results if the landowner builds and operates the wind generator, selling the electricity to the local utility. Annual profits can be as much as $100,000 per turbine, a good addition to farm income.

Wind power provides local control of energy

Cooperatives are springing up to help landowners and communities finance, build, and operate their own wind generators. One gigawatt of wind power (equivalent to one large nuclear or fossil fuel plant) can create more than 3,000 permanent jobs, while paying about $4 million in rent to landowners and $3.6 million in tax payments to local governments. About 20 Native American tribes, for example, have formed a coalition to study wind power. Together their reservations (which are in the windiest, least productive parts of the Great Plains) could generate at least 350 GW of electrical power, equivalent to about half of the current total U.S. installed electrical capacity.

In Germany 47 percent of all wind and solar installations are owned by community cooperatives, in which members receive a share of the profits and the energy. This financing approach has greatly accelerated Germany's efforts to reduce dependence on imported fossil fuels. Energy independence has been a strong motivator for many European adopters of renewable energy.

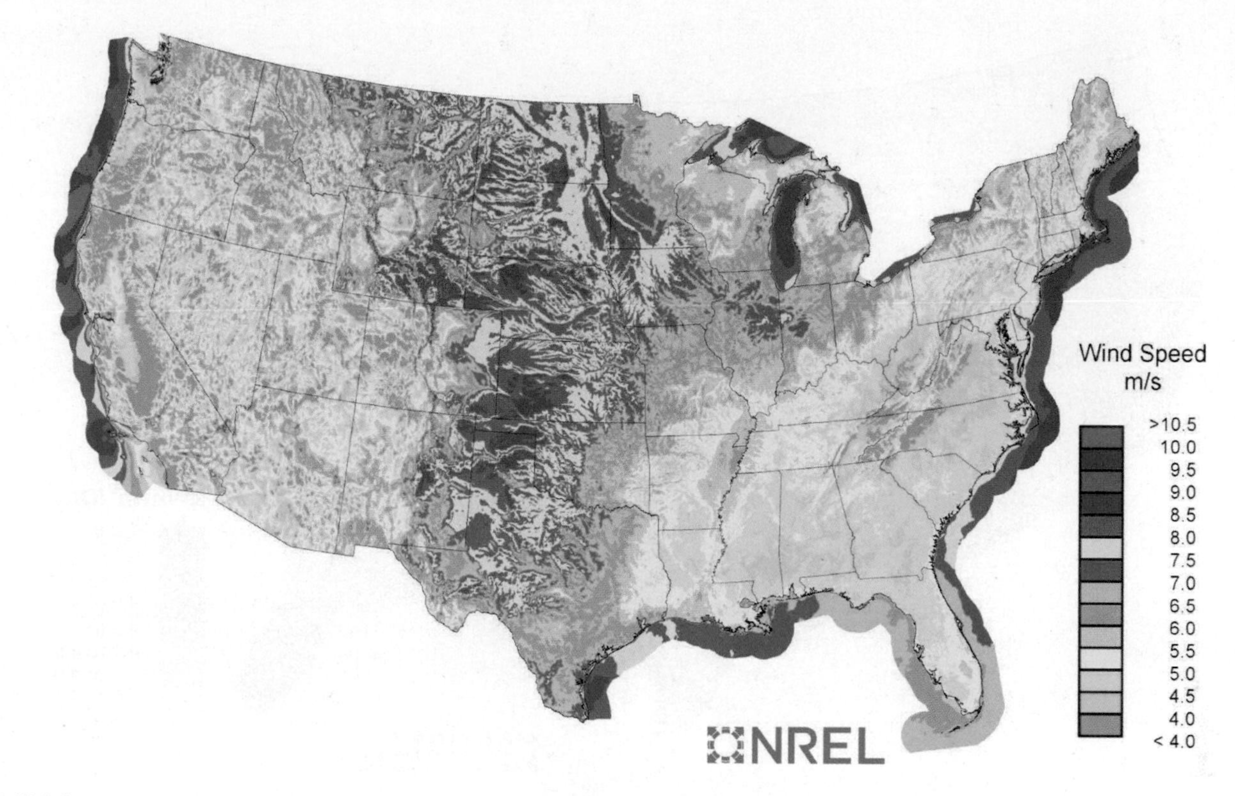

▲ **FIGURE 13.24** U.S. wind resources, including offshore wind, at 80 m altitude. Source: National Renewable Energy Laboratory (NREL).

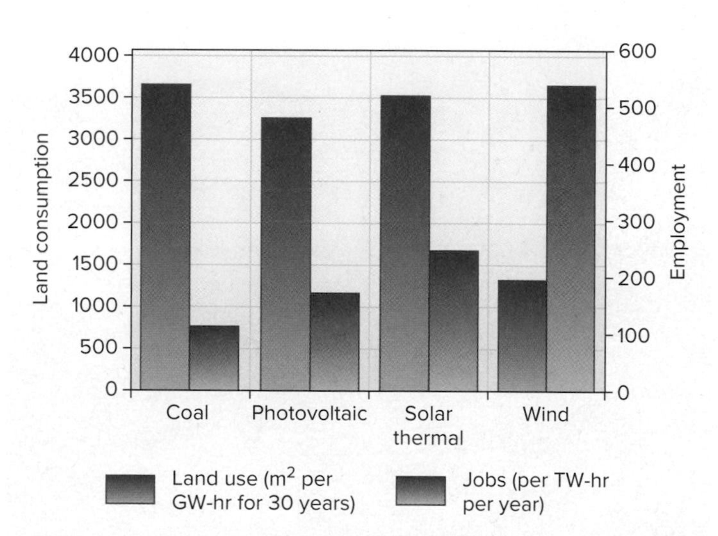

▲ **FIGURE 13.25** If you include the land required for mining, wind power takes about one-third as much area and creates about five times as many jobs to create the same amount of electrical energy as coal.

Bird and bat mortality around wind turbines has been a widespread concern. These are important questions, especially for early wind installations on California ridge tops, where migrating raptors have collided too frequently with turbine blades. On the other hand, coal and oil production is among the main causes of forest destruction and risks to endangered species worldwide. Natural gas developers in the western United States have led the charge to prevent protection of endangered sage grouse and other vulnerable species. Coal mining has eliminated mountaintop habitat for forest songbirds. Reliable numbers are hard to come by, but one estimate is that coal, oil, and gas are responsible for five to ten times as many bird deaths in the United States as wind turbines cause.

There have also been concerns about wind turbine noise, which can depend partly on context and on the design of the turbines. Many are less than 50 decibels at short range, about as noisy as light traffic. Resistance can also decline when local residents own a share in the energy production. Sometimes this is a "not in my backyard" problem. We are accustomed to getting coal, oil, and gas from remote places where we cannot see the environmental and social impacts of energy extraction, or watch conflicts over oil. What do you think, would greater awareness of those issues make you more tolerant of nearby wind power? or not?

Solar thermal systems collect usable heat

Solar power has always been our ultimate energy source. Solar energy drives winds and the hydrologic cycle, and it has produced all biomass, including fossil fuels and our food. The average amount of solar energy that reaches the earth's surface is about 10,000 times all the commercial energy used each year. This amount varies geographically (fig. 13.26), but solar energy is sufficient for economical production even at high latitudes, as in northern Europe.

Our simplest and oldest use of solar energy is passive heat absorption, using natural materials or absorptive structures with no moving parts to simply gather and hold heat. For thousands of years people have built thick-walled stone and adobe dwellings that collect heat during the day and gradually release that heat at night. After cooling at night, these massive building materials maintain a comfortable daytime temperature within the house, even as they absorb external warmth.

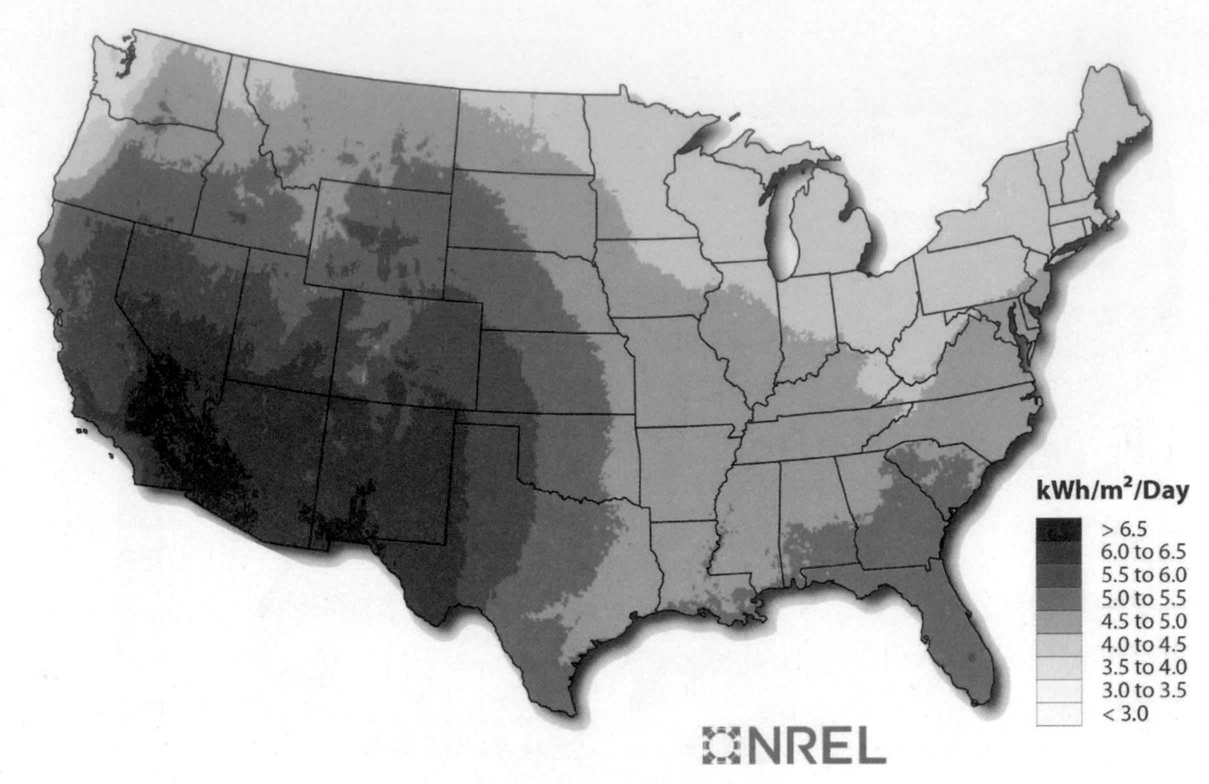

▲ **FIGURE 13.26** U.S. solar resources. Source: National Renewable Energy Laboratory (NREL).

Most modern solar thermal systems make hot water by running a liquid (such as water mixed with antifreeze) through thin pipes in a glass vacuum tube or in a glass-covered flat panel. The fluid reaches high temperatures, often 95°C (200°F). It then passes through a heat exchanger, which heats water for direct use or for space heating. A panel of about 5 m^2 can provide enough hot water for a family.

At least 330 GW of thermal energy have been installed globally, more than twice the installed solar-photovoltaic capacity. China both produces and uses about 80 percent of the world's solar thermal systems. The low cost, less than $200 each, makes them common there for household hot water. Over 30 million Chinese homes get hot water and/or space heat from solar energy.

Solar thermal systems are very efficient at capturing solar energy but they are much less common in the United States and Europe than solar-photovoltaic panels. In Europe, municipal solar systems do provide district heating (fig. 13.27).

▲ **FIGURE 13.27** Solar water heaters can be scaled up to provide hot water and space heating for large buildings or for whole towns, like this facility on Aero Island, Denmark. ©Mary Ann Cunningham

CSP makes electricity from heat

Solar thermal systems can produce extremely hot water, hot enough to make steam and run a turbine in **concentrating solar power** (CSP) systems. CSP often uses long, trough-shaped, parabolic mirrors to reflect and concentrate sunlight on a central tube containing a heat-absorbing fluid. Reaching extremely high temperatures, the fluid passes through a heat exchanger, where it heats water and generates steam. The steam turns a turbine to produce electricity. Heat from the transfer fluid also can be stored in a medium such as molten salt for later use. This allows the system to continue to generate electricity on cloudy days or at night.

Another approach is to arrange thousands of mirrors in concentric rings around a tall central tower (fig. 13.28). Driven by electric motors, the mirrors track the sun and focus light on a heat absorber at the top of the "power tower" where a transfer medium is heated to temperatures as high as 500°C (1,000°F). This heat is used to turn water to steam, which drives an electric generator. Under optimum conditions, a 50-ha (130-acre) mirror array should be able to generate 100 MW of clean, renewable power.

▲ **FIGURE 13.28** A "power tower" is a form of concentrated solar thermal electrical generation. Thousands of movable mirrors focus intense energy on the tower, where fluid is heated to drive a steam turbine. ©Mlenny/Getty Images

CSP also is attractive to energy utilities. Because installations are large, often hundreds of megawatts, CSP is a way to produce solar energy on a scale that makes economic sense to investors. Acquiring land and permits is slow and expensive, but it's not necessarily harder than putting solar panels on millions of individual roofs, with separate contracts to negotiate for each one.

CSP plants can be controversial. Although they help free us from fossil fuels, they occupy desert land that can be easily scarred and slow to recover. CSP proposals sometimes threaten rare or endangered species, such as the desert tortoise, whose habitat is already reduced by development in the desert. In 2010, California regulators approved 13 large solar thermal facilities for the Mojave Desert. Most were eventually canceled or delayed by disputes over land use. Native American groups protested over threats to sacred cultural sites, and others objected to industrial activities in the pristine landscape. In response to these debates, millions of hectares of desert have been added to protected areas to forestall energy development.

Despite these challenges, CSP plants offer an important contribution to our energy future. According to the Renewable Energy Policy Network, there is now over 3.4 GW of installed capacity. The United States and Spain are the leading users. The largest plant in the world, which opened in 2014, is California's Ivanpah Solar Power Facility, at 377 MW capacity.

Photovoltaic cells generate electricity directly

Photovoltaic (PV) cells capture solar energy and convert it directly to electrical current by separating electrons from their parent atoms and accelerating them across a one-way electrostatic barrier, which is formed by the junction between two different types of semiconductor material (fig. 13.29). Over the past 25 years, the efficiency of energy captured by photovoltaic cells has increased from less than 1 percent of incident light to more than 16 percent in commercial installations and over 40 percent in the lab.

Global installed solar PV capacity is now over 300 GW of installed capacity. Installation is increasing worldwide, although China and India have been expanding most rapidly. Utility scale solar PV is also becoming widespread. Among the largest installations in the United States are two 550-MW solar PV stations in southern California, Desert Sunlight Solar Farm and Topaz Solar farm (fig. 13.30a). China is now leading on large solar, with the massive 1,500-MW Tengger Desert solar park and the 1,000-MW Datong solar park—which is expected to expand to 3,000 MW.

New types of photovoltaic cells are in development, including flexible kinds (fig. 13.30b) that could go on curved surfaces or attach to car rooftops. Thin, transparent films that could go on windows are even being explored. By some estimates, the United States could generate about three-quarters of present electrical consumption with rooftop PV, especially if roofs of warehouses and big-box retail stores are used (fig. 13.30c).

A variety of incentive structures can encourage rooftop systems. One of the most successful in Germany, which now has over

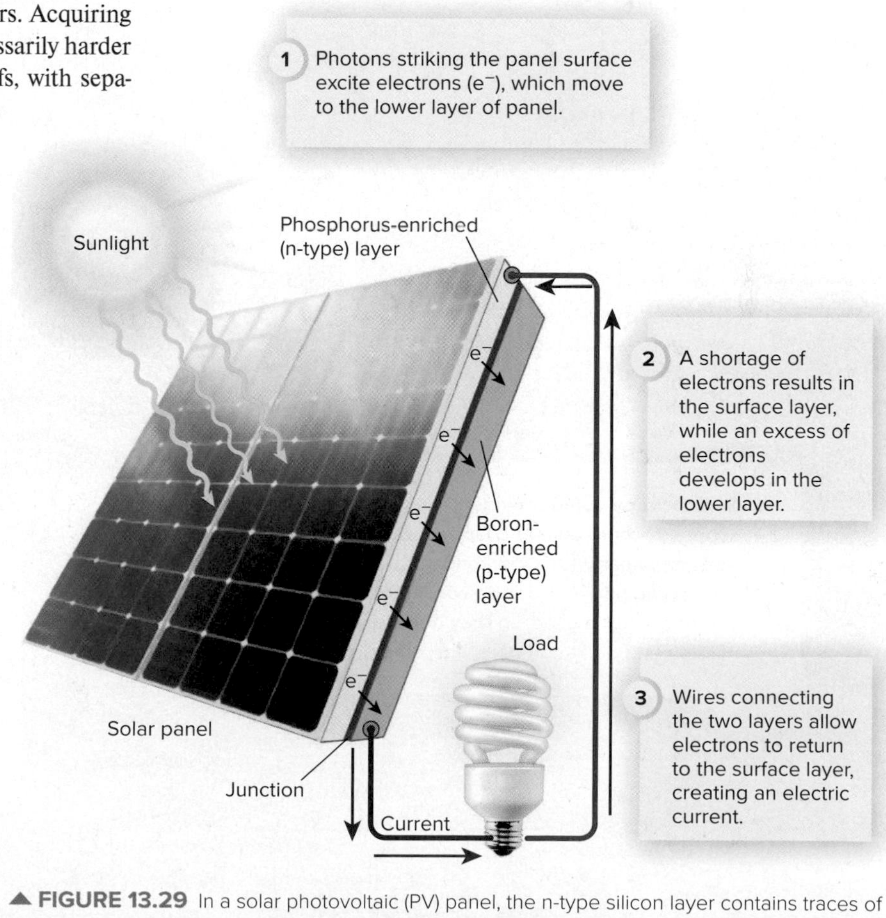

1 Photons striking the panel surface excite electrons (e^-), which move to the lower layer of panel.

Sunlight

Phosphorus-enriched (n-type) layer

2 A shortage of electrons results in the surface layer, while an excess of electrons develops in the lower layer.

Boron-enriched (p-type) layer

Load

3 Wires connecting the two layers allow electrons to return to the surface layer, creating an electric current.

Solar panel

Junction

Current

▲ **FIGURE 13.29** In a solar photovoltaic (PV) panel, the n-type silicon layer contains traces of an element such as phosphorus, with more outer-shell electrons than silicon, while the p-type silicon layer contains an element such as boron, with fewer outer-shell electrons. When sunlight strikes the panel, photons dislodge electrons, which travel through a circuit, creating an electrical current.

KEY CONCEPTS

How can we transition to alternative energy?

Current technologies are sufficient to produce all the energy we need, if we can decide to invest in them, according to studies from Stanford University and the University of California at Davis.* With existing technology, renewable sources could provide all the energy we need, including the fossil fuels we use now. And we could save money at the same time. **Land-based wind, water power, and solar potential exceed all global energy consumption.** Renewable energy supplies over the oceans are even larger, since oceans cover two-thirds of the earth's surface. Many studies suggest that renewables could meet future demand more economically and more safely than fossil-based energy plans. How would this energy future look?

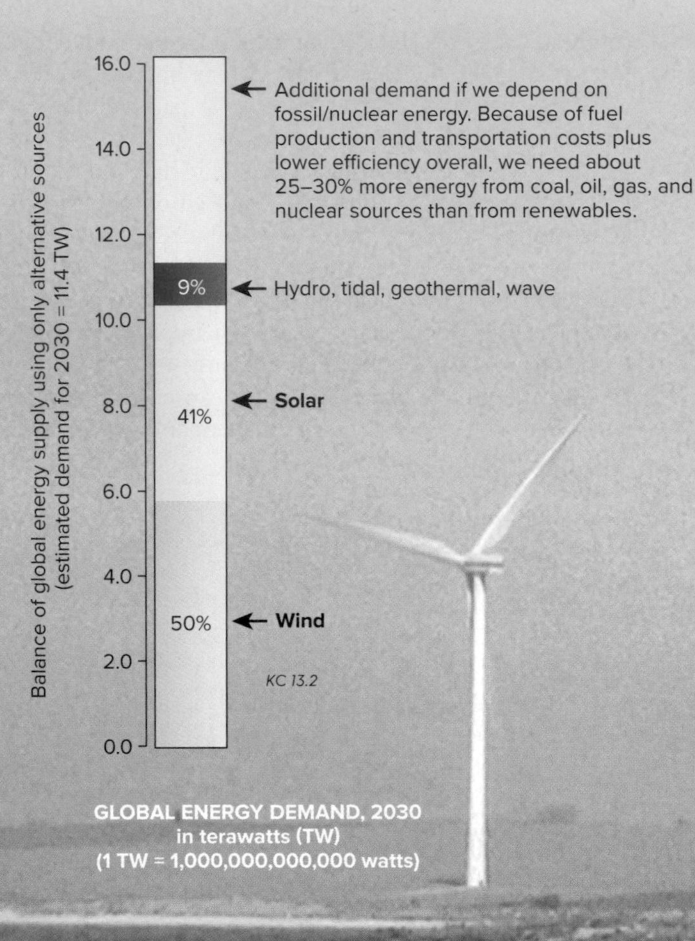

GLOBAL ENERGY DEMAND, 2030
in terawatts (TW)
(1 TW = 1,000,000,000,000 watts)

KC 13.1

1. **Wind** could supply 50 percent of our energy, according to this plan. It would take 3.8 million large wind turbines to supply electricity to the whole world. Isn't that an impossible task? Not necessarily: We build that many cars and trucks every year worldwide.

©Mary Ann Cunningham

2. **Solar energy** could provide 41 percent of our total energy supply. It would take 1.7 billion rooftop photovoltaic systems and nearly 100,000 concentrated solar power plants to provide 4.6 TW. Rooftop collectors can be located where energy is used, so they don't lose energy in transmission and don't compete with other land uses.

KC 13.3

©Doug Sherman/Geofile

◀ Solar thermal collectors already are price competitive with fossil fuels, but they generally can't be located close to consumers, and they may require scarce cooling water in arid lands where sunshine is plentiful.

3. **Hydropower (dams, tidal, geothermal, wave energy)** could supply about 9 percent of our energy. Most major rivers are already dammed, but underwater turbines in rivers and tidal areas could be effective. Deep wells could tap geothermal energy, but there are worries about triggering earthquakes and contaminating aquifers.

KC 13.4

*For more information: see Jacobson, M. Z., and M. A. Delucchi. 2009. A path to sustainable energy. *Scientific American* 301(5) 58–65.

Source: USGS.

Geothermal plant ▶

What about when the wind doesn't blow, or the sun doesn't shine?

Fortunately, the wind blows more at night to complement sunshine during the day. By balancing renewable sources, we can have just as reliable supplies as we now have with fossil fuels. Renewable sources also have a much better service record. Coal-burning power plants are out of production 46 days per year for maintenance. Solar panels and wind turbines average only 7 days down for repairs per year.

Solar, wind, and water power also solve two of our most pressing global problems: (1) the problem of climate change, perhaps the most serious and costly problem we face currently, as water shortages, crop failures, and refugee migrations destabilize developing regions; and (2) political conflict over fuel supplies, as in the oil fields of Iraq, Nigeria, and Ecuador, or nuclear fuel processing in Iran.

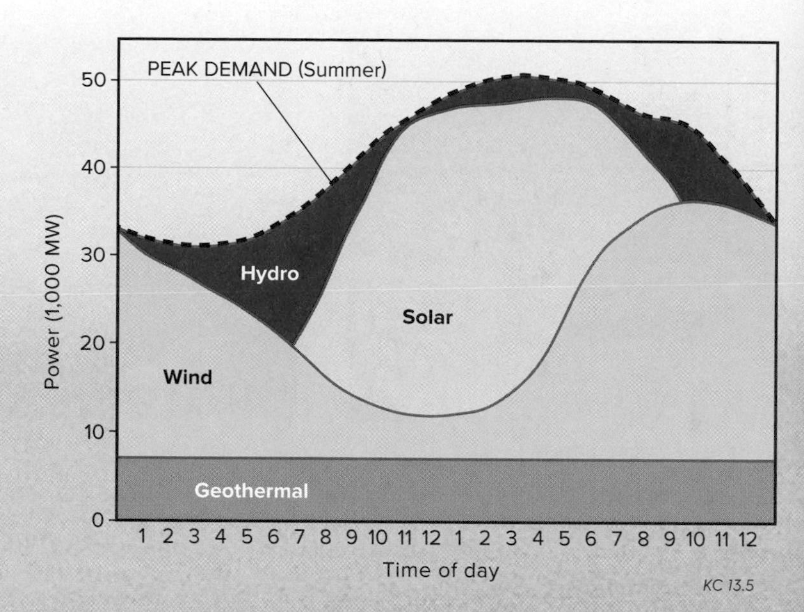

KC 13.5

What would renewable energy cost?

By 2020, wind and hydroelectricity should cost about half as much as fossil fuels or nuclear power, and because renewable energy sources are inherently more efficient than fossil fuels, it should take about one-third less energy to supply the same services with sun, wind, and water.

In addition to the energy we can obtain from renewable sources, conservation measures could save up to half the energy we now use. Mass transit, weather-proofing, urban in-fill, and efficient appliances are among the available strategies that can save money in the near term and in the long term.

Light Rail

KC 13.6

CAN YOU EXPLAIN?

1. **What would be the greatest benefits of switching to renewable energy?**
2. **Which of these sources is forecast to produce the most energy?**
3. **How many windmills would we need in this plan?**
4. **Who would benefit most and least from an alternative energy future? from a fossil and nuclear future? Why?**

(a) Base-load power facility

(b) Flexible, thin-film solar tiles

(c) Roof-top solar array

▲ **FIGURE 13.30** Solar-photovoltaic energy is highly versatile and can be used in a variety of dispersed settings. (a) Utility-scale PV arrays can provide base-load power. (b) Thin-film PV collectors can be printed on flexible backing and used like ordinary roof tiles. (c) Millions of square meters of roof tops on schools and commercial buildings could be fitted with solar panels. (a) Source: NREL/Harin Ullal, NREL staff; (b) Source: NREL/Stellar Sun Shop; (c) Source: NREL/U.S. Department of Energy, Craig Miller Productions

40 GW of installed solar, has been feed-in tariffs. These are set prices, subsidized by the government, that a utility must pay for excess power from homeowners.

Another approach is to have utility-owned panels. Building owners sign an agreement with a utility that owns and maintains the panels, and owners agree to purchase electricity at a given rate for a certain number of years from the utility. But all the energy from the panels goes directly to the grid. Because solar is about the same cost as other electricity these days, and because it seems a safe bet that fossil fuels will get more expensive over time, that agreement can make good economic sense. Homeowners also have the satisfaction of knowing they are contributing to carbon-free energy production.

13.6 BIOMASS AND GEOTHERMAL ENERGY

- *Ethanol is a major investment but has high costs.*
- *Methane from crop waste is an efficient energy source.*
- *Heat pumps and geothermal are key for heating and cooling.*

Plants capture immense amounts of solar energy by storing it in the chemical bonds of plant cells. As recently as 1850, wood supplied 90 percent of the heat used in the United States. For more than a billion people in developing countries, burning **biomass** (plant materials) remains the principal energy source for heating and cooking. An estimated 1,500 m^3 of fuelwood is gathered each year globally. This amounts to half of all wood harvested. Wood gathering and charcoal burning are important causes of deforestation in many rural areas. Providing efficient wood stoves can improve people's lives while saving forests.

Ethanol has been the main U.S. focus

Biofuels, such as ethanol and biodiesel, are by far the biggest recent news in biomass energy. Globally, production of these two fuels is booming–from Brazil, which gets about 40 percent of its transportation energy from ethanol generated from sugarcane, to Southeast Asia, where millions of hectares of tropical forest have been cleared for palm oil plantations, to the United States, where about 40 percent of the corn (maize) crop currently is used to make ethanol. Since 2007, it has been U.S. energy policy that ethanol production will increase from 9 billion to 36 billion gallons per year (30 billion to 136 billion liters) by 2022. Over 200 ethanol refineries have been built across the United States, with production capacity of 16 billion gallons (22 billion liters, fig. 13.31).

▲ **FIGURE 13.31** Liquid biofuels, mixed here with gasoline, make use of nearly half the U.S. corn crop and much of the soybean crop. Source: U.S. Department of Energy

Whether this is a good use of dollars and farm land is a much-debated question. Many analysts exclude ethanol from discussions of renewable energy because it relies on fossil fuels to run farm machinery and processing equipment. Ethanol provides just slightly more energy than is needed to produce it, according to most studies, so it does not dramatically improve energy resources. Planting, cultivating, irrigating, fertilizing, harvesting, drying, and processing crops take a great deal of fossil fuel energy. Environmental costs are also substantial, including soil erosion, water contamination, and biodiversity losses.

Crops with a high oil content, such as soybeans, sunflower seed, rape seed (usually called canola in the United States), and palm oil are relatively easy to make into biodiesel. In some cases the oil needs only minimal processing to be used in a standard diesel engine. However, it would take a very large land area to meet our transportation needs with soy or sunflowers. And diversion of these oils for vehicles deprives humans of important sources of edible oils.

Oil palms are considerably more productive per unit area than soy or sunflower (although palm fruit is more expensive to harvest and transport). Currently, millions of hectares of species-rich forests in Southeast Asia are being destroyed to create palm oil plantations (see chapter 6). Indonesia already has 6 million ha (15 million acres) of palm oil plantations, and Malaysia has nearly as much. Together these two countries produce nearly 90 percent of the world's palm oil.

Cellulosic ethanol remains mostly uneconomical

Producing 36 billion gallons of ethanol from corn would use the entire U.S. corn crop and a good portion of our water resources. So there has been great interest in developing ethanol from crop waste and other nonfood biomass.

A number of techniques have been proposed for extracting sugars from cellulosic materials. Most involve mechanical chopping or shredding followed by treatment with bacteria or fungi to break down cellulose into soluble sugars (fig. 13.32).

So far, there are no commercial-scale cellulosic ethanol factories operating in North America, but the Department of Energy has provided $385 million in grants for six cellulosic biorefinery plants. These pilot projects will test a wide variety of feedstocks, including rice and wheat straw, milo stubble, switchgrass hay, almond hulls, corn stover (stalks, leaves, and cobs), and wood chips.

Methane from biomass is efficient and clean

Just about any organic waste, but especially sewage and manure, can be used to produce methane. Methane gas, the main component of natural gas, is produced when anaerobic bacteria (bacteria living in an oxygen-free space) digest organic matter. The main byproduct of this digestion, CH_4, has no oxygen atoms because no oxygen was available in digestion. This molecule oxidizes, or burns, easily, producing CO_2 and H_2O (water vapor). Consequently, methane is a clean, efficient fuel. Today, as more cities struggle to manage urban sewage and feedlot manure, methane could be a rich source of energy. In China, in addition to solar and wind power, more than 6 million households use methane, also known as **biogas,** for cooking and lighting. Two large municipal facilities in Nanyang, China, for example, provide fuel for more than 20,000 families.

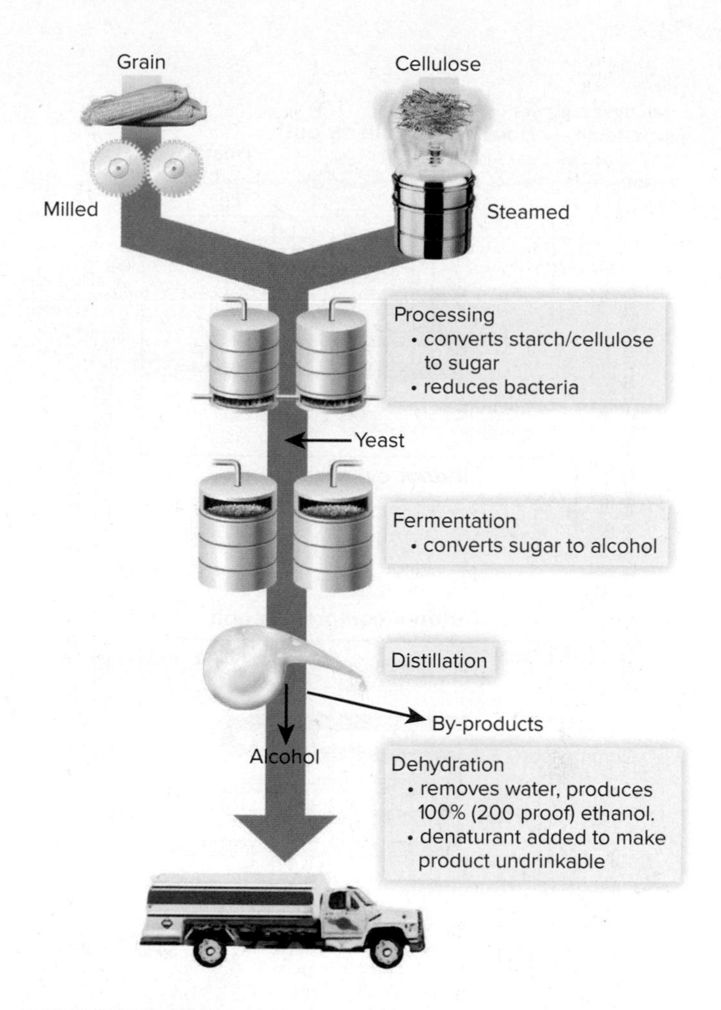

▲ **FIGURE 13.32** Ethanol (or ethyl alcohol) can be produced from a wide variety of sources. Maize (corn) and other starchy grains are milled (ground) and then processed to convert starch to sugar, which can be fermented by yeast into alcohol. Distillation removes contaminants and yields pure alcohol. Cellulosic crops, such as wood or grasses, can also be converted into sugars, but the process is more difficult. Steam blasting, alkaline hydrolysis, enzymatic conditioning, and acid pretreatment are a few of the methods for breaking up woody material. Once sugars are released, the processes are similar.

Methane is a promising resource where there is a reliable and pure source of organic waste. Dairy and other livestock farms are increasingly producing biogas from their manure. European countries have over 17,000 biogas facilities, about two-thirds of them in Germany. In addition to producing energy, this avoids a great deal of air and water pollution. Cities that have municipal household compost separation, as is common in some European countries, also use biogas digesters to produce usable biogas from urban waste. City sewage treatment plants and landfills also offer rich, and mostly untapped, potential for methane generation (see chapter 14).

In theory, photosynthetic algae rich in lipids (oils) could be another source of fuel from waste materials. Algae capture CO_2 to produce oils; in some single-celled species, half the mass is oil. Theoretically, an algae-producing facility could produce 30 times as much oil per hectare as the ethanol yield from a corn field.

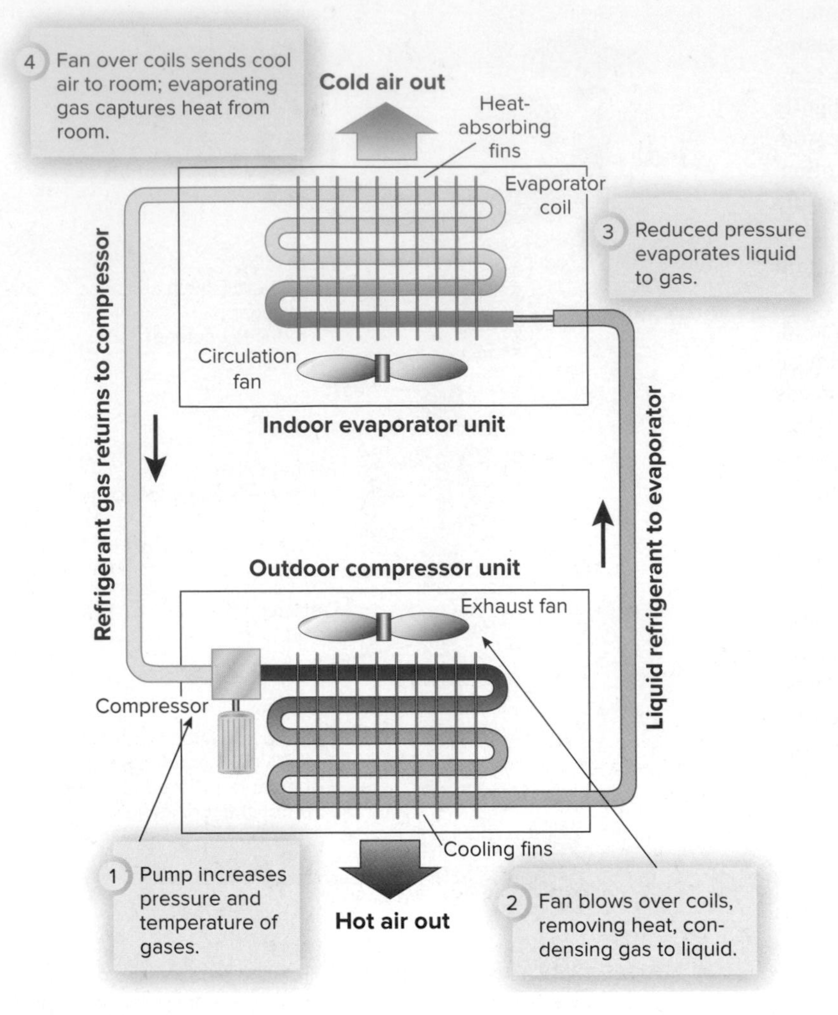

▲ **FIGURE 13.33** A heat pump compresses and expands gases to cool a building. Run in reverse, the heat pump provides heating. Running on electric pumps, this system can heat and cool buildings without fossil fuels.

Algae-growing facilities could use waste nutrients, as from a sewage treatment plant, and CO_2 emitted from conventional power plants. Unlike ethanol, they don't require prime farmland. An algae photobioreactor could occupy former industrial sites.

A number of algae bioenergy strategies have been explored—for example, developing algae strains that produce higher concentrations of lipids. Thus far, developers have found it hard to produce oil economically and at a marketable scale, and this approach remains a tantalizing idea.

Heat pumps provide efficient cooling and heating

The *Drawdown* study of 2017 (see section 9.4) argued that heat pumps could address most of the world's heating and cooling needs efficiently, and they could do it on renewable electricity. Heat pumps concentrate heat from the surrounding air (air-source heat pumps, fig. 13.33) or ground (ground-source heat pumps, fig. 13.34), a pond, or groundwater and deliver it to indoor space. Or they capture heat from a room and deliver it outside.

How do heat pumps work? In your refrigerator, cooling occurs as a refrigerant gas, such as CO_2, ammonia, or a chlorofluorocarbon (CFC) is compressed to liquid form; and is then evaporated. As a substance changes phase from liquid to vapor, it absorbs heat (think of water, which takes 540 calories of heat to evaporate a gram of H_2O). Conversely, as a substance changes from vapor to liquid, it releases heat.

In your refrigerator, or air conditioner, a pump compresses the refrigerant between two heat exchanger coils (that is, coiled tubes containing the refrigerant). In one coil, the electric pump compresses the gas, using pressure to heat it (see fig. 13.33). The warm gas travels to the condenser coils, where a fan blowing on the coils cools the heated gas, enough to condense it to a liquid state. The liquid is then pumped into expansion coils inside the refrigerator. As it expands (evaporates) to gas, the refrigerant absorbs heat from its surroundings. It is then pumped back to the compressor.

Heat exchangers are reversible: In summer, the refrigerant evaporates inside your home (absorbing heat) and is condensed outside (releasing heat). In winter, the refrigerant is condensed inside the home, releasing heat, and it is evaporated outside. Ground-source heat pumps use the ground's near-constant temperatures to provide or store heat (fig. 13.34). New air-source heat exchangers can capture heat even in cold weather—although the compressor pumps work harder and consume more electricity when winter air is cold.

This system is increasingly used for large buildings and campuses. Ball State University, in Indiana, was among the first to transition to geothermal. Skidmore College, in New York, has been 40 percent geothermal since 2014.

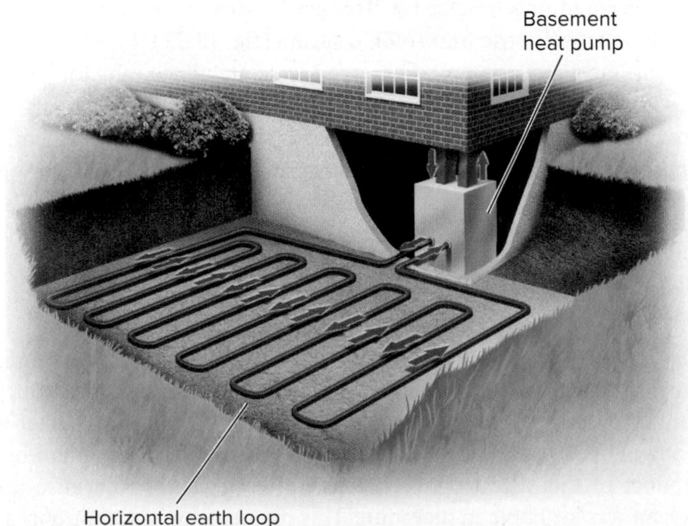

▲ **FIGURE 13.34** Geothermal energy can cut heating and cooling costs by half in many areas. In summer (shown here), warm water is pumped through buried tubing (earth loops), where it is cooled by constant underground temperatures. In winter, the system reverses and the relativity warm soil heats the house.

TABLE 13.3 How Can Colleges and Universities Transition to Renewables?

Sightlines and the University of New Hampshire Sustainability Institute have published assessments of the progress toward carbon reductions across U.S. colleges and universities. These are some recommendations in their 2017 report.

1. Identify efficiency as a policy priority and a funding priority.
2. Plan a transition to electricity-based heating systems (such as air-source or ground-source heat pumps).
3. Think of energy efficiency as an investment, which can yield three to five times the returns of traditional endowments.
4. Collect and assess data: Efficiency only improves if energy use is monitored.
5. Buildings are growing more than student populations. Focus on ways to right-size buildings by regularly evaluating usage and space needs.
6. Use power purchase agreements to buy renewable energy.
7. Work to develop on-site renewable energy, or buy renewables.
8. Where high-tech building equipment is required, focus on design and sustainability at the start.
9. Include life cycle cost assessments of building technology. Be clear about "nice-to-have" vs. "need-to-have" technology and spaces.

Source: UNH Sustainability Institute and Sightlines, 2017 State of Sustainability in Higher Education 2017.

Carleton College, in Minnesota, has recently installed geothermal heating and cooling for the entire campus. Using climate-safe refrigerants is critical, as more of the world demands air conditioning, and as heat pumps become more common for heating (see chapter 9).

Where high-temperature geothermal resources exist, earth-heated steam provides abundant electricity, as well as heat. Iceland is known for this, using geothermal steam to produce much of its electricity and nearly all its building heating. Costa Rica is also a leader in geothermal electricity, producing nearly all its electricity from steam around its many volcanoes.

13.7 WHAT DOES AN ENERGY TRANSITION LOOK LIKE?

- *A transition from fossil to renewable energy involves electrifying more energy uses.*
- *Energy storage, a key challenge, is growing and diversifying.*
- *Wind, water, and solar could meet all our needs with existing technology.*

Transitioning to new energy systems involves rethinking policy, energy use practices, and technology. It also has co-benefits in cost savings and cleaner air—and, of course, in working to avoid extreme climate change. This transition is expected to create millions of jobs and spur economic development. Reduced reliance on oil and gas, much of it sourced from politically unstable regions, means that repowering with carbon-free energy would also reduce security concerns. Colleges and universities can help lead the way in this transition (table 13.3).

Back in 2012, the U.S. National Renewable Energy Laboratory concluded that with available, affordable technology, renewable energy could supply 80 percent of total U.S. electricity generation by 2050 while meeting hourly electricity demand in every region of the country. By the end of the twenty-first century, it said, renewable sources could provide all our energy needs if we took the necessary steps soon.

The grid will need improvement

Electricity is delivered through vast networks of power lines—in the U.S. grid, more than 1 million km (600,000 mi) of transmission lines. The grid will need to expand as more energy uses, especially heating and transportation, become electrified. Germany, for example, anticipates expanding its long-distance, high-voltage grids by about 6 percent. The amount of expansion in different countries and regions depends on whether efficiency improvements reduce energy demand.

The grid also provides opportunities for balancing demand and production: The sun shines and the wind blows at different times in different regions, so exchanging between high- and low-producing regions can distribute supply and demand. U.S. policy divides the grid in two sections (fig. 13.35). Many renewable energy proponents argue increasing connectivity would greatly increase flexibility for renewables.

Upgrading to smart grid technologies with digital information flow will make it easier to monitor and maintain the grid; it will also make it possible for users to monitor their usage more easily, to be charged less for off-peak times, or to be paid for contributing electricity production or storage. Grid upgrades will be an expensive but much needed part of grid maintenance and modernization.

Storage options are changing rapidly

For decades, a principal means of storing electricity has been pumped hydropower. This approach uses excess electricity—for example from a solar plant at mid-day—to pump water up to an elevated reservoir. At night, that water can be released to run a hydropower generator. Thermal storage, such as heated water or salt, can provide heat to run steam turbines. Storage is an area of rapid change in renewable energy. The energy research group Irena reported in 2017 that electricity storage capacity appeared set to triple by 2030, to 12–15 TWh.

Battery storage is growing and diversifying rapidly. Commercial-scale battery banks, unheard of a few years ago, are becoming increasingly common: In 2018 the Hawaiian island of Kauai installed 52 MWh of batteries at a 13-MW solar park, at a cost of 11 cents/kWh, considerably less than the 15 cents/kWh that diesel cost on the island. A 65-MW solar farm in Arizona, coupled with a 50-MW battery storage system, is now able to provide electricity cheaper than conventional gas production. The U.S. Energy Information Agency reports that by 2016 U.S. utility-scale battery systems added

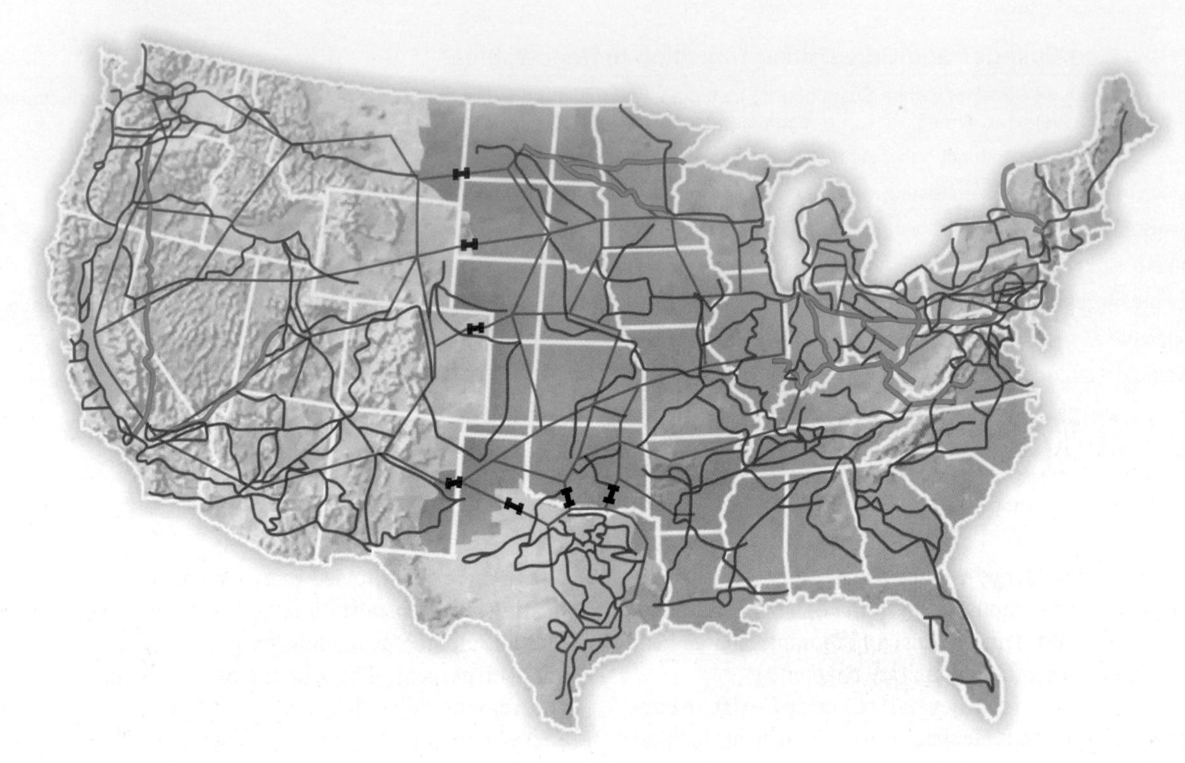

▲ **FIGURE 13.35** The green area is served by the western electrical grid, while the eastern electrical grid serves the pink area. Texas is largely independent of either grid. New high-voltage power lines with interlinks (black dots) are needed to tie these areas together for maximum efficiency.

up to over 600 mWh. These systems are far safer than storing and transporting gas, as well as being better for the climate. Battery systems are quick to install—just a few months for utility-scale systems—and now they are becoming cheaper than fossil fuels as well.

Lithium ion batteries have been standard for years because energy input and output are rapid. Battery research is active, however, including hydrogen, nanowires, solid-state lithium ion, graphene, and other technologies. The story is likely to keep changing. Batteries are also transforming the electric vehicle market, with new varieties promising to increase range and shorten recharge times dramatically.

Fuel cells release electricity from chemical bonding

Fuel cells are similar to batteries except that they are recharged by adding a fuel or a source of hydrogen atoms, while a battery is recharged by adding electrical current. For hydrogen fuel cells, the source of hydrogen could be hydrocarbon fuels such as natural gas (mainly CH_4) or ammonia (NH_3). Methane effluents from landfills and wastewater treatment plants can be used as a fuel source. The source can also be water (H_2O), whose molecules can be split (electrolyzed) with an electrical charge. If the source of electricity is wind or solar power, then a fuel cell can provide completely renewable power.

Fuel cells consist of a positive electrode (the cathode) and a negative electrode (the anode) separated by an electrolyte, a material that allows the passage of charged atoms, or ions, but is impermeable to electrons (fig. 13.36). In the most common systems, hydrogen or a hydrogen-containing fuel is passed over the anode while oxygen is passed over the cathode. At the anode, a reactive catalyst, such as platinum, strips an electron from each hydrogen atom, creating a positively charged hydrogen ion (a proton). The hydrogen ion can migrate through the electrolyte to the cathode, but the electron is excluded. Electrons flow through an external circuit, creating an electrical current. At the cathode, the electrons and protons are reunited and combined with oxygen to make water.

A fuel cell that runs on pure oxygen and hydrogen produces no waste products except drinkable water and heat. Other fuels create some pollutants (most commonly carbon dioxide), but the levels are typically far less than conventional fossil fuel combustion in a power plant or an automobile engine. Although the theoretical efficiency of electrical generation of a fuel cell can be as high as 70 percent, the actual yield is closer to 40 or 45 percent.

Wind, water, and solar are good answers

It is increasingly acknowledged that reducing climate change means ending reliance on fossil fuels. This transition may be hard to imagine when more than 80 percent of energy use is oil, gas, and coal. But the Paris Accord and the IPCC have both cited safe levels of atmospheric CO_2 and temperature change that are only possible if we shift entirely to renewables (see chapter 9).

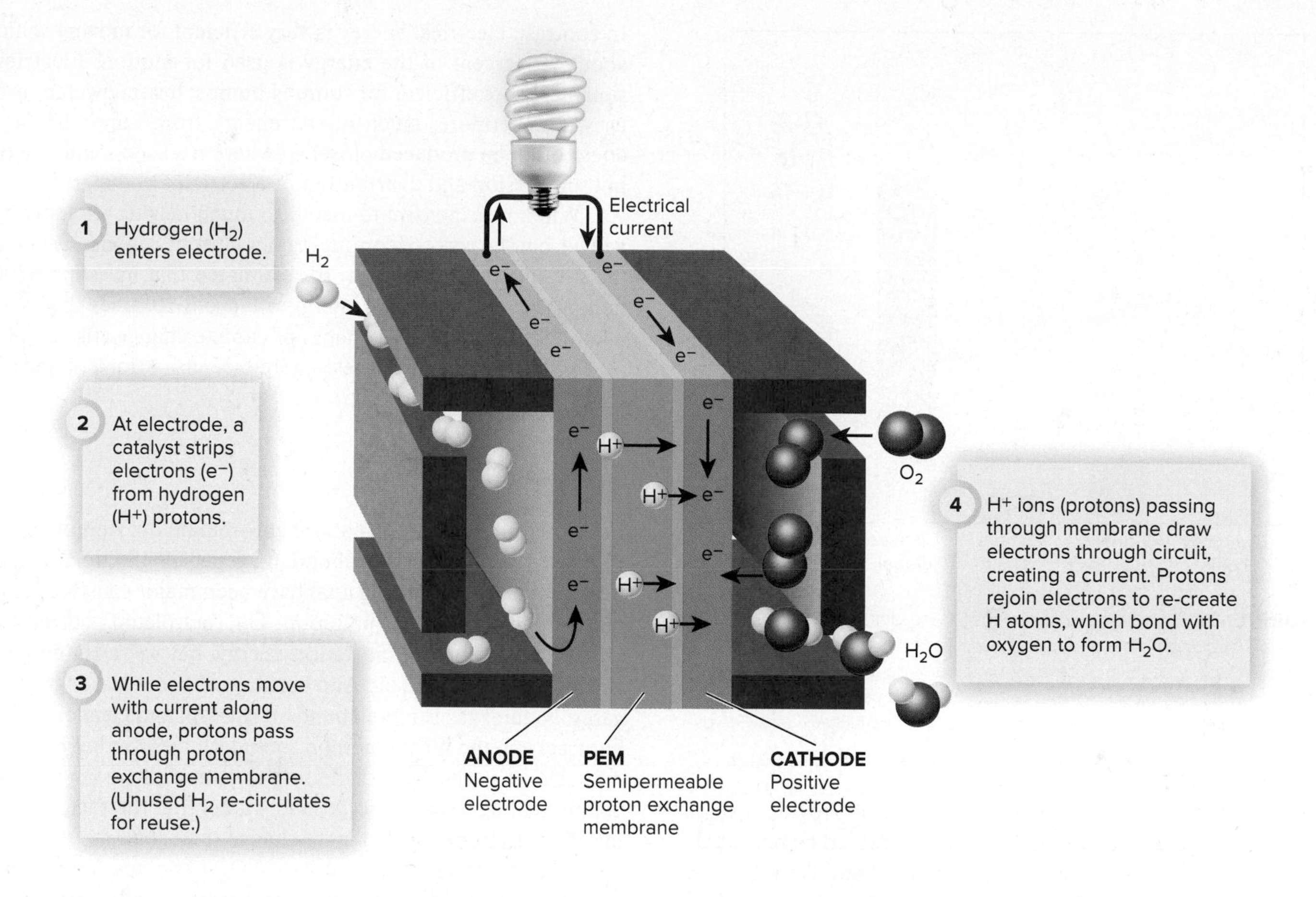

▲ **FIGURE 13.36** Fuel cell operation. Electrons are removed from hydrogen atoms at the anode to produce hydrogen ions (protons) that migrate through a semipermeable electrolyte medium to the cathode, where they reunite with electrons from an external circuit and oxygen atoms to make water. Electrons flowing through the circuit connecting the electrodes create useful electrical current.

How is this possible? Multiple strategies have been offered. Some of the most comprehensive and detailed have been done by Mark Jacobson, of Stanford University, and Mark Delucchi, of Santa Cruz, and their colleagues. They have detailed how a shift to just wind, water, and solar (WWS) would look, using renewables to meet all our heating, cooling, transportation, and electrical energy demands. For the United States, they find that currently available wind, water, and solar technologies could supply 80 percent of the world's energy by 2030 and 100 percent by 2050. They calculate that it would take 400,000 5-MW wind turbines, 75 million rooftop photovoltaic systems, 2.8 million commercial-scale PV facilities, and 2,000 concentrated solar power plants and industrial-sized photovoltaic arrays (fig. 13.37).

Wouldn't it be an overwhelming job to build and install all that technology? It would be a huge effort, but it's not without precedent. Jacobson points out that society has achieved massive transformations before. In 1956 the United States began building the Interstate Highway System, which now extends 47,000 mi (75,600 km) and has changed commerce, landscapes, and society. And every year roughly 60 million new cars and trucks are added to the world's highways. So there are precedents of dramatic shifts in technology and production.

The WWS models assume that most energy uses become electrified, including transportation, building heating and cooling, and manufacturing. Battery-electric vehicles, such as the Chevy Bolt or

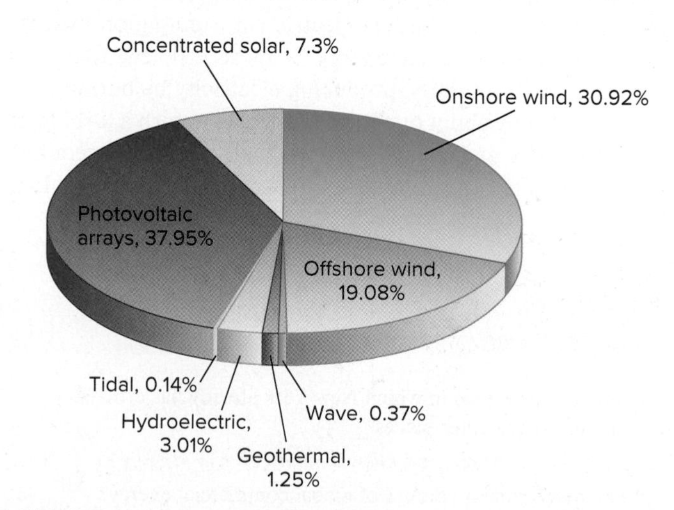

▲ **FIGURE 13.37** A 100 percent wind, water, and solar scenario for 2050.
Source: Jacobson, M. et al. 2015, 100% clean and renewable wind, water, and sunlight (WWS) all-sector energy roadmaps for the 50 United States. *Energy Environ. Sci.*, 2015, 8, 2093–2117.

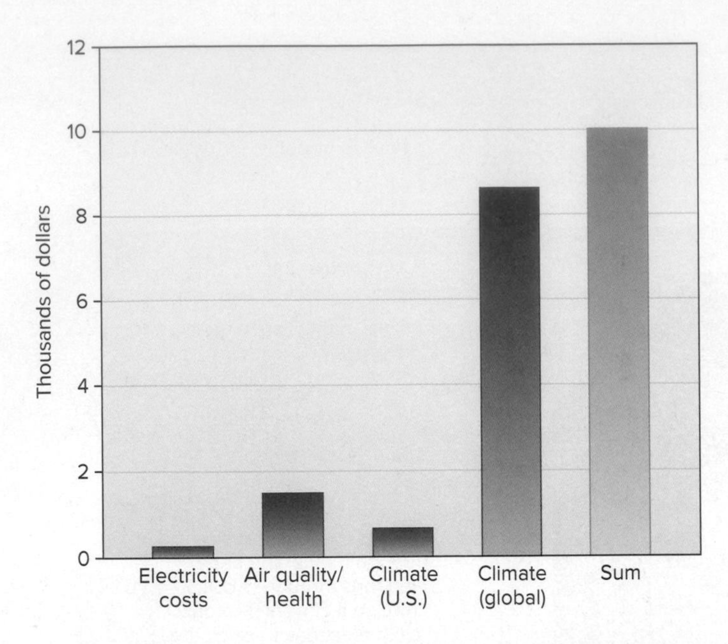

▲ **FIGURE 13.38** Calculated cost savings for the United States, per person per year, resulting from a transition from fossil fuels to wind, water, and solar energy. Climate costs include economic damage at both U.S. and global scales. Source: Jacobson, M. et al. 2015, 100% clean and renewable wind, water, and sunlight (WWS) all-sector energy roadmaps for the 50 United States. *Energy Environ. Sci.*, 2015, 8, 2093–2117.

Tesla's semitrailer truck, and electric trains can provide a large share of transportation. Hydrogen fuel cells, powered by renewable electricity, can support transportation, energy storage, or other electric uses. Building heating and cooling can be done with air-source or ground-source heat pumps.

Fortunately, electricity is efficient for most uses, so it would take a great deal less total energy to meet our needs with sun, wind, and water than to continue using fossil fuels—about 39 percent less. This is mainly because conventional energy production through fuel burning loses a great deal of energy as heat. Operating a car on gasoline uses only about 17 percent of the fuel energy for moving the vehicle. The rest is lost as heat. Jacobson and colleagues point out that a battery-electric car, running on hydrogen fuel cells powered by solar energy, is more efficient than a car burning gasoline. Similarly, producing electricity by burning coal, oil, or gas uses only about one-third of the fuel to turn a turbine; the rest is lost as "rejected energy"—that is, heat, which then requires a great deal of water and infrastructure for cooling (see fig. 13.2). In contrast, electrical energy is very efficient for moving vehicles—about 90 percent of the energy is used for motion. Electricity is similarly more efficient for running pumps, heating water, or other uses. Furthermore, much of the energy from renewable sources could often be produced closer to where it's used, reducing losses in transmission and distribution of electricity in the grid.

Will it be expensive to install so much new technology? Yes, it will be, but the costs of continuing our current dependence on fossil fuels would be much higher. It's estimated that investing $700 billion per year now in clean energy will avoid 20 times that much in a few decades from the damages of climate change (fig. 13.38).

To learn more about these energy scenarios, look online for the "Solutions Project."

CONCLUSION

Fossil fuels—oil, coal, and natural gas—remain our dominant energy sources. Coal is extremely abundant, especially in North America, but extracting and burning coal have been major causes of environmental damage and air pollution. Oil (petroleum) currently provides most of our transportation energy, but we're running out of cheap, easily extracted oil. And burning oil also releases greenhouse gases. Natural gas is more abundant than oil and cleaner than coal, but fracking, the most common method for extracting natural gas, may release so much methane into the atmosphere (in addition to contaminating surface and ground water) that the fuel it produces may be worse for global climate change than coal.

Nuclear power doesn't create CO_2 while operating, but mining, processing, and shipping fuel, together with perpetual storage of wastes, results in far more greenhouse gases than does wind energy. Risks of accidents, while remote, remain a worry, as do economic challenges. Hydroelectricity is also an important source of electricity, but large dams have large environmental and social impacts. Conservation and investment in renewables are key factors in a sustainable energy future. New designs in housing, office buildings, industrial production, and transportation can all save vast amounts of energy. Investments in wind and solar are growing rapidly, especially in China and in the developing world. Rapid innovations in solar, wind, and other renewable energy sources now make it possible to get all our energy from clean technologies. The choices we make about our energy sources and uses will have profound effects on our environment and society.

PRACTICE QUIZ

1. What are some ways in which New York already has efficiency advantages over other places?
2. Define *energy*, *power*, and *kilowatt-hour* (*kWh*).
3. What are the major sources of global commercial energy?
4. How does energy consumption in the United States compare to that in other countries?
5. Why don't we want to use all the coal in the ground?
6. Where is most liquid oil located? How long are supplies likely to last?
7. What are *tar sands* and *oil shales*? What are the environmental costs of their extraction?
8. How are nuclear wastes now being stored?
9. Explain how concentrating solar power (CSP) works.
10. How do photovoltaic cells work?

CRITICAL THINKING AND DISCUSSION

Apply the principles you have learned in this chapter to discuss these questions with other students.

1. If you were the energy czar of your state or country, where would you invest your budget? Why?
2. We have discussed a number of different energy sources and energy technologies in this chapter. Each has advantages and disadvantages. If you were an energy policy analyst, how would you compare such different problems as the risk of a nuclear accident versus air pollution effects from burning coal?
3. If your local utility company were going to build a new power plant in your community, what kind would you prefer? Why?
4. The nuclear industry is placing ads in popular magazines and newspapers, claiming that nuclear power is environmentally friendly because it doesn't contribute to the greenhouse effect. How do you respond to that claim?
5. How would you evaluate the debate about net energy loss or gain in biofuels? What questions would you ask the experts on each side of this question? What worldviews or hidden agendas do you think might be implicit in this argument?
6. It clearly will cost a lot of money to switch from fossil fuels to renewables. How would you respond to someone who says that future costs from climate change are no concern of theirs?

DATA ANALYSIS Personal Energy Use

For many college students, a car and a computer are essentials of life. Cars are also one of our most important single uses of energy, so differences in efficiency can greatly affect your energy consumption (and energy costs). This exercise asks you to modify an Excel spreadsheet in order to evaluate the impact of vehicle efficiency on energy use.

Suppose you were to buy a very efficient car, such as the Honda Insight, rather than a sport utility vehicle, such as a Ford Excursion. How much energy would that save, and how long could you run your computer with that energy? Go to Connect to find a spreadsheet that explores these questions, and to answer questions about personal energy use.

Design Elements: Active Learning (Toad): ©Gaertner/Alamy Stock Photo; Case Study (Globe): ©McGraw-Hill Education; Google Earth: ©McGraw-Hill Education; Abstract Background: ©Martin Kubat/Shutterstock; What do you think (Students using tablets): ©McGraw-Hill Education/Richard Hutchings, photographer; What can you do (Hand holding Globe): ©Christoph Weihs/Shutterstock

CHAPTER

14 Solid and Hazardous Waste

LEARNING OUTCOMES

After studying this chapter, you should be able to answer the following questions:

- What are the major components of the waste stream?
- How does a sanitary landfill operate? Why are we searching for alternatives to landfills?
- Why is ocean dumping a problem?
- What are the "three Rs" of waste reduction, and which is most important?
- How can biomass waste be converted to natural gas?
- What are toxic and hazardous waste? How do we dispose of them?
- What is bioremediation?
- What is the Superfund, and has it shown progress?

An endangered Hawaiian monk seal is disentangled from abandoned fishing nets in the Papahãnaumokuãkea Marine National Monument.

Source: NOAA Photo Library/NOAA's Fisheries Collection/National Oceanic and Atmospheric Administration (NOAA)

CASE STUDY

Plastic Seas

The Papahānaumokuākea Marine National Monument, one of the largest marine reserves in the world, is larger than all U.S. national parks combined. Established by President George W. Bush in 2006, the sanctuary was expanded by President Obama in 2016 to encompass a chain of islands, atolls, and reefs extending across 1.5 million km^2 (583,000 mi^2) northwest of Hawaii. The monument protects some of the most pristine and diverse deep coral reefs and over 7,000 marine species, including rare and endangered species, such as the Laysan albatross and the Hawaiian monk seal. The string of isolated islets and coral atolls makes up the world's largest tropical seabird rookery, supporting 14 million nesting seabirds. The preserve is also home to a wealth of cultural and historic heritage sites, including shipwrecks and World Heritage cultural sites for native Hawaiians.

Despite its remote location, Papahānaumokuākea* remains vulnerable to the flotsam and jetsam of modern life. The islands and reefs lie within the vast circulating currents known as the Pacific gyre. These swirling currents, driven by the rotation of the earth, concentrate nutrients, organic debris, and in recent decades an ocean of plastic trash. Often called the Great Pacific Garbage Patch, this region of floating plastic debris is a drifting cloud of plastic particles, soda bottles and caps, disposable shopping bags, packaging, discarded fishing nets, and other debris.

Much of the debris consists of tiny fragments floating just below the surface, but some pieces are large and recognizable, and some float 20–30 m deep. The greatest concentrations of plastic trash occur in the eastern Pacific, between California and Hawaii, and in the western Pacific near Japan. Similar garbage patches have been identified in the Atlantic and elsewhere in the world's oceans, but the Pacific cases are the best studied.

The Pacific garbage gyre is thought to contain more than 100 million tons of plastic and to cover an area larger than California. In some areas, this debris outweighs the living biomass. Researchers have found that many decomposing plastics emit dimethyl sulfide, which attracts fish and birds. Fish have been caught with stomachs full of plastic fragments. Seabirds gulp down plastic fragments, then regurgitate them for their chicks. With stomachs blocked by indigestible bottle caps, disposable lighters, and other refuse, chicks starve to death.

In one study of Laysan albatrosses, 90 percent of the carcasses of dead chicks contained plastic fragments (fig. 14.1). Seals, turtles, porpoises, and seabirds also become ensnared in ghost fishing nets and drown, or they die from ingesting indigestible materials. Even whales sometimes wash up dead on beaches, with stomachs full of plastic debris. Oceanographers worry that this debris is slowly starving ocean ecosystems.

Surveys at sea and on beaches indicate that 50 to 80 percent of the floating material originates onshore. The rest is discarded or lost at sea. Stray shopping bags, drink containers, fast-food boxes, and other refuse fall from dumpsters, escape from landfills, or are discarded on the street, then wash into storm sewers and streams. Eventually, these items travel to the sea, where they gradually break into smaller pieces as they join the great global masses of ocean plastic.

*Pronounced Pa-pa-ha-nao-Mo-kua-kea. To hear the pronunciation, visit the monument's website, www.papahanaumokuakea.gov.

The problem has been extraordinarily difficult to address because it is widespread, diffuse, abundant, and constantly replenished by careless or incomplete disposal of waste onshore and at sea. But growing awareness is starting to make a difference. Cleanup cruises in Papahānaumokuākea have collected more than 700 metric tons of discarded fishing gear that had clogged reefs.

In Papahānaumokuākea and elsewhere, marine debris has also caught the public's attention, and widespread beach cleanups are having a visible effect. And people are becoming active worldwide. According to the Ocean Conservancy, an organization that conducts regular coastal cleanups around the world and then measures and weighs the trash, nearly 9 million volunteers collected almost 66,000 tons of debris between 1990 and 2015.

Increasing awareness is also encouraging many fishing boats to reduce disposal of plastic garbage at sea. Because all this material fouls fishing gear, costing time and money, it is in their best interest to bring in the garbage they produce or collect in their nets. This is important because discarded fishing gear is the top accidental killer of marine life, followed by plastic shopping bags, balloons, and plastic bottle caps. You can help protect the oceans, no matter how far from them you are. Next time you see plastic debris that might wash into a storm sewer, remember that everything ends up eventually in the ocean. Pick it up if you can, and try to prevent your

▲ **FIGURE 14.1** A Laysan albatross chick, which died after being fed plastic debris rather than fish. Starvation after plastic ingestion is a leading cause of death for these chicks. Source: National Marine Sanctuary, photographer Claire Fackler

(continued)

CASE STUDY continued

own plastic trash from escaping into waterways. (And never release helium balloons into the sky.) You can also try to reduce the amount of disposable containers, bottles, and packaging you buy. Containing and minimizing loose garbage is one of the best ways to reduce marine debris.

The remote atolls of northwestern Hawaii show us that no place is too far away to be affected by our waste production and disposal. The materials we buy and the ways we manage our garbage can have dramatic impacts on living systems at home and far away. At the same time, responses to the problem have shown that people everywhere have an interest in taking care of the land and oceans and in keeping them beautiful. Often, the obstacles and volumes of waste seem insurmountable, but cleanup efforts in Hawaii have shown that progress can be real if we keep at it. In this chapter we examine the waste we produce, our methods to dispose of it, and strategies to reduce, reuse, and recycle it. ■

We are living in a false economy where the price of goods and services does not include the cost of waste and pollution.

—LYNN LANDES, FOUNDER AND DIRECTOR OF ZERO WASTE AMERICA

14.1 WHAT WASTE DO WE PRODUCE?

- *The waste stream is the steady flow of all forms of trash.*
- *Municipal solid waste is a special challenge because it is mixed.*
- *Americans produce over 2 kg per person per day, twice as much as Europe or Japan, and five to ten times as much as most developing countries.*

Waste is everyone's business, even though we don't think about it every day. We all produce unwanted by-products in nearly everything we do. According to the Environmental Protection Agency (EPA), the United States produces 11 billion tons of solid waste each year. That's roughly 3.6 tons per person. Nearly half of that amount consists of agricultural waste, such as crop residues and animal manure, which are generally recycled into the soil on the farms where they are produced. Agricultural wastes provide groundcover to reduce erosion, and they nourish new crops, but they are also the single largest source of nonpoint air and water pollution in the country. Another one-third of all solid wastes are mine tailings, overburden from strip mines, smelter slag, and other industrial waste from mining and metal processing. Much of this material is stored in or near its source of production. Improper disposal practices, however, can result in serious and widespread pollution.

Industrial waste—other than mining and mineral production—amounts to some 400 million metric tons per year in the United States. Most of this material is recycled, converted to other forms, sent to landfills, or disposed of in injection wells, which should be deep enough not to interact with usable groundwater. About 60 million metric tons of industrial waste fall into a special category of hazardous and toxic waste, which we will discuss later in this chapter.

Municipal solid waste, the garbage we produce in our houses, offices, and cities, accounts for a small percentage of total waste by weight, but it is one of our most important challenges in waste management. Municipal solid waste is hard to reuse and recycle because it contains many different kinds of materials, yet it amounts to about 250 million metric tons per year in the United States (fig. 14.2). That's just over 2 kg (4.6 lb) per person per day—twice as much per capita as Europe or Japan, and five to ten times as much as most developing countries.

Benchmark Data

Among the ideas and values in this chapter, these are a few worth remembering.

100 million tons	Plastic waste estimated in Pacific Garbage Patch
50–80%	Plastic in ocean that originated on shore
11 billion tons	Waste produced annually in U.S.
34%	Waste recycled in U.S.
60%	Waste recycled in Germany
71%	Portion of U.S. hazardous waste produced by chemical and petroleum industries
40 million tons	Contaminants released annually into air or water in U.S.
36,000	Superfund sites in U.S. identified by the EPA
400,000	Contaminated sites in U.S.

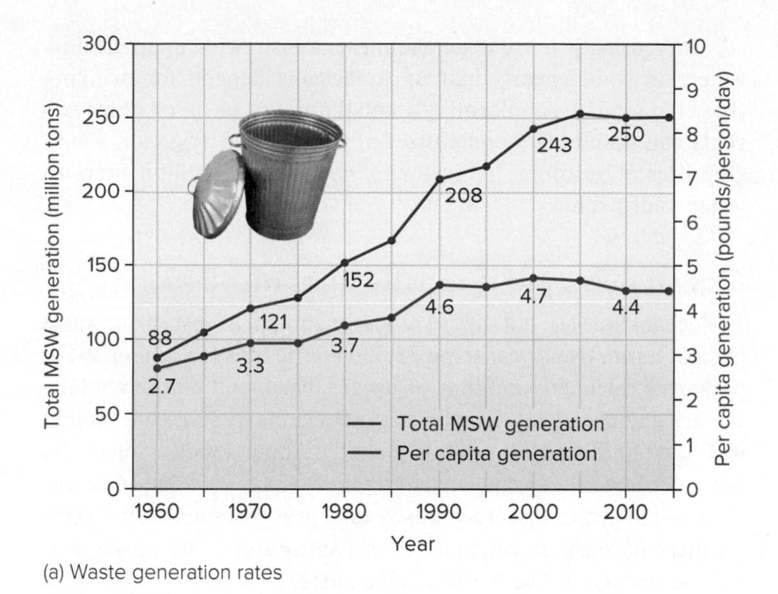

(a) Waste generation rates

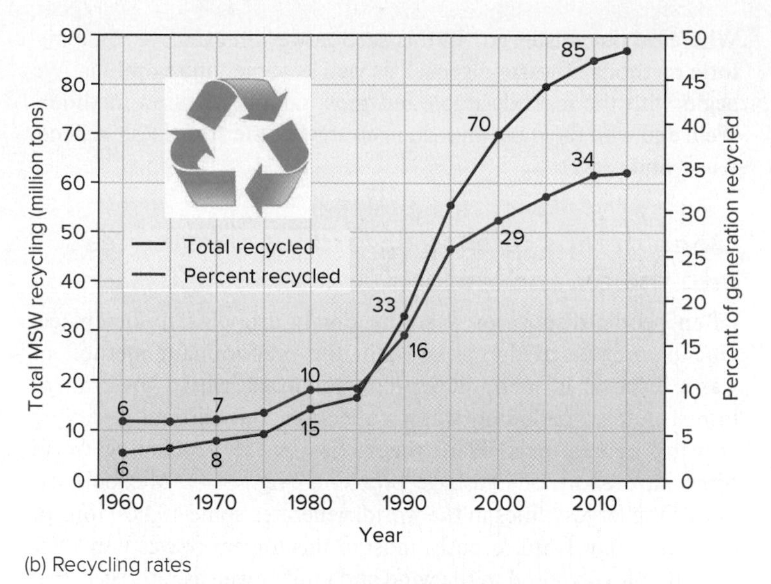

(b) Recycling rates

▲ **FIGURE 14.2** Bad news and good news in solid waste production. Per-capita waste has risen steadily to more than 2 kg/person/day (a). Recycling rates are also rising, however (b). Source: Data from U.S. EPA 2014.

Despite considerable progress in the past 20 years, we still recycle only about 30 percent of our glass bottles and jars, less than 50 percent of aluminum drink cans, and less than 7 percent of our plastic food and beverage containers. We could save money, energy, land, and many other resources if we could improve on these rates.

The waste stream is everything we throw away

What kinds of materials are in all that waste? There are organic materials, such as yard and garden wastes, food wastes, and sewage sludge from sewage treatment plants; junked cars; worn-out furniture; and consumer products of all types. Newspapers, magazines, packaging, and office refuse make paper one of our major wastes (fig. 14.3).

The **waste stream** is the steady flow of varied wastes that we all produce, from domestic garbage and yard wastes to industrial, commercial, and construction refuse. Many of the materials in our waste stream would be valuable resources if they were not mixed with other garbage. Unfortunately, our collecting and dumping processes mix and crush everything together, making separation an expensive and sometimes impossible task.

Despite improvements in recycling technology and composting programs we still send more than half of U.S. waste to landfills (fig. 14.3b). Recycling and combustion both increased from about 1980 to 2000, but neither has changed dramatically since then. Improving recycling rates remains an important challenge.

When hazardous materials get mixed into the waste stream, they are dispersed through thousands of tons of miscellaneous garbage. This mixing makes the disposal or burning of what might have been rather innocuous stuff a difficult, expensive, and risky business. When incinerated, batteries release cadmium, mercury, lead, or zinc into the air; plastics produce carcinogenic dioxins and PCBs (polychlorinated biphenyls). The best thing to do with household toxic and hazardous materials is to separate them for safe disposal or recycling, as we will see later in this chapter.

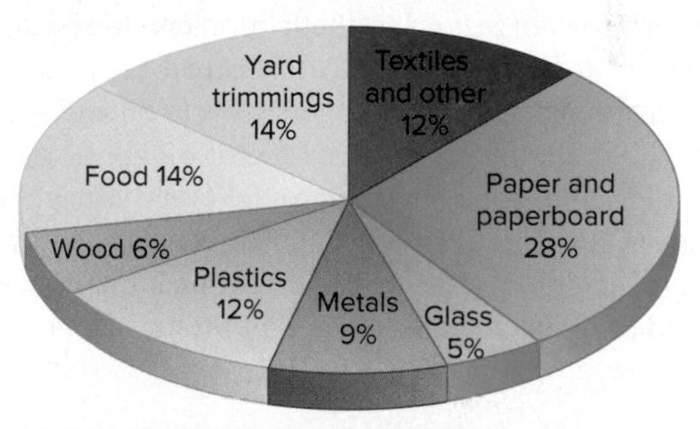

(a) Amount generated, by weight

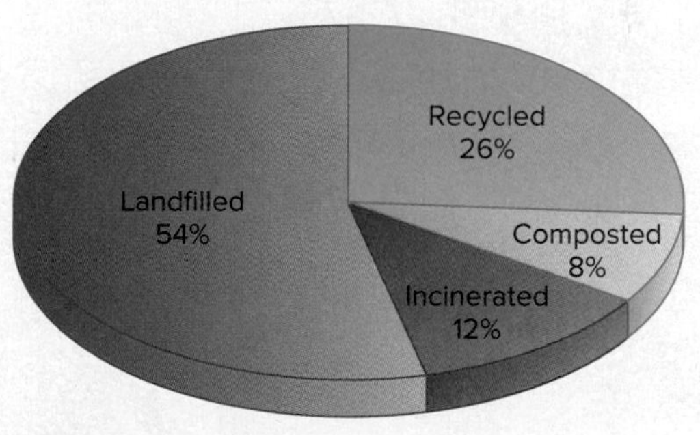

(b) Disposal methods

▲ **FIGURE 14.3** Composition of municipal solid waste in the United States, by weight and by disposal methods. Source: U.S. EPA Office of Solid Waste Management, 2015.

14.2 WASTE DISPOSAL METHODS

- *Open dumps remain common in developing countries.*
- *Sanitary landfills dominate U.S. waste management, but they are costly.*
- *We often export waste to countries ill-equipped to handle it.*

Where do our wastes go? In this section we will examine some historic methods of waste disposal, as well as some future options. We begin with the least desirable but most commonly used methods. We'll end with the most important strategies, the "three Rs": reduce, reuse, and recycle.

Open dumps release hazardous substances into the air and water

Often people dispose of waste by simply dropping it someplace. Open, unregulated dumps are still the predominant method of waste disposal in many developing countries, where government infrastructure, including waste collection, has difficulty serving growing populations. Giant megacities in the developing world often have enormous garbage problems (fig. 14.4). Mexico City, one of the largest cities in the world, generates some 10,000 tons of trash each day. Until recently, most of this torrent of waste was left in giant piles, exposed to the wind and rain, as well as rats, flies, and other scavengers.

Most developed countries forbid open dumping, at least in urban areas, but illegal dumping is still a problem. You have undoubtedly seen trash accumulating along roadsides and in vacant, weedy lots. This is not just an aesthetic problem. Much of this trash washes into sewers and then into the ocean (see next section). Often illegally dumped garbage includes waste oil and solvents. An estimated 200 million liters of waste motor oil are poured into the sewers or allowed to soak into the ground every year in the United States. This is about five times as much as was spilled by the landmark *Exxon Valdez* in Alaska in 1989! No one knows the volume of solvents and other chemicals disposed of by similar methods.

▲ **FIGURE 14.4** Trash disposal has become a crisis in the developing worlds, where people have adopted cheap plastic goods and packaging but lack good recycling and disposal options. ©William P. Cunningham

Increasingly, these toxic chemicals are showing up in groundwater, on which nearly half of Americans depend for drinking (see chapter 11). An alarmingly small amount of oil or other solvents can pollute large quantities of drinking or irrigation water. One liter of gasoline, for instance, can make a million liters of water undrinkable.

Ocean dumping is mostly uncontrolled

The oceans are vast, but they're not large enough to absorb our waste without harm. Every year some 25,000 metric tons (55 million lb) of packaging, including millions of bottles, cans, and plastic containers, are dumped at sea. Even in remote regions, beaches are littered with the nondegradable trash (fig. 14.5). About 150,000 tons (330 million lb) of fishing gear—including more than 1,000 km (660 mi) of nets—are lost or discarded at sea each year. An estimated 50,000 northern fur seals are entangled in this refuse and drown or starve to death every year in the North Pacific alone.

Until recently, many cities in the United States dumped municipal refuse, industrial waste, sewage, and sewage sludge into the ocean. Federal legislation now prohibits this dumping. New York City, the last to stop offshore sewage sludge disposal, finally ended this practice in 1992.

Plastic debris is a growing problem in all the world's oceans. Millions of tons of plastic drink bottles, bottle caps, plastic shopping bags, and other debris end up at sea. Most of these are probably carelessly discarded litter and uncontained garbage, but there is also deliberate disposal at sea, especially from cruise ships and container ships. All this debris, floating just below the surface, accumulates in vast regions of slowly swirling ocean currents.

Oceanographers are trying to find ways of collecting or controlling this debris that is overwhelming ocean ecosystems. Most material is too fine-grained to capture easily in nets, and it is distributed widely around the world's oceans. Growing awareness, however, is a first step toward resolving the problem. Increasingly, images and information about the Pacific Garbage Patch, and other garbage gyres, can be found online.

▲ **FIGURE 14.5** Plastic trash dumped on land and at sea ends up on remote beaches. ©sOulsurfing - Jason Swain/Getty Images

Landfills receive most of our waste

Currently, 54 percent of all municipal solid waste in the United States goes to landfills, 33 percent is recycled, and 13 percent is incinerated. While we have a long way to go in controlling waste, this is a dramatic change from 1960, when 94 percent was landfilled and only 6 percent was recycled.

A modern **sanitary landfill** is designed to contain waste. Operators are required to compact the refuse and cover it every day with a layer of dirt, to decrease smells and litter and to discourage insects and rats. This method helps control pollution, but the dirt fill also takes up as much as 20 percent of landfill space. Since 1994, all operating landfills in the United States have been required to control such hazardous substances as oil, chemical compounds, and toxic metals, which seep through piles of waste along with rain water. To prevent leakage to groundwater and streams, landfills require an impermeable clay and/or plastic lining (fig. 14.6). Drainage systems are installed in and around the liner to catch drainage and to help monitor chemicals that leak out. Modern municipal solid waste landfills now have many of the safeguards of hazardous waste repositories described later in this chapter.

Sanitary landfills also must manage methane, a greenhouse gas produced when organic material decomposes in the anaerobic conditions deep inside a landfill. Landfills are one of the largest anthropogenic sources of methane in the United States. Globally, landfills are estimated to produce more than 700 million metric tons of methane annually. Because methane is 20 times as potent at absorbing heat as CO_2, this represents about 12 percent of all greenhouse gas emissions. Until recently almost all this landfill methane was simply vented into the air. Now about half of all landfill gas in the United States is either flared (burned) on site or collected and used as fuel to generate electricity. Methane recovery in the United States produces 440 trillion Btu per year and is equivalent to removing 25 million vehicles from the highway. Some landfill operators are deliberately pumping water into their waste as a way of speeding up production of this valuable fuel.

Historically, landfills were convenient and cheap. This was because we had much less waste to deal with—we produced only a third as much in 1960 as today—and because there were few regulations about disposal sites and methods. Since 1984, when stricter financial and environmental protection requirements for landfills took effect, roughly 90 percent of landfills in the United States have closed. Nearly all of those landfills lacked environmental controls to keep toxic materials from leaking into groundwater and surface waters. Many areas now suffer groundwater and stream contamination from decades-old unregulated dump sites.

▲ **FIGURE 14.6** A plastic liner being installed in a sanitary landfill. This liner and a bentonite clay layer below it prevent leakage to groundwater. Trash is also compacted and covered with earth fill every day. ©Doug Sherman/Geofile

Active LEARNING

Life-Cycle Analysis

One step toward understanding your place in the waste stream is to look at the life cycle of the materials you buy. Here is a rough approximation of the process. With another student, choose one item that you use regularly. On paper, list your best guess for the following: (1) a list of the major materials in it; (2) the original sources (geographic locations and source materials) of those materials; (3) the energy needed to extract/convert the materials; (4) the distances the materials traveled; (5) the number of businesses involved in getting the item to you; (6) where the item will go when you dispose of it; (7) what kinds of reused/recycled products could be made from the materials in it.

With new rules for public health protection, landfills are becoming fewer, larger, and more expensive. Cities often truck their garbage hundreds of miles for disposal, a growing portion of the $10 billion per year that we spend to dispose of trash. A decade from now, it may cost us $100 billion per year to dispose of our trash and garbage. On the other hand, rising landfill costs make it more economical to pursue alternatives strategies, including waste reduction and recycling.

We often export waste to countries ill-equipped to handle it

Most industrialized nations agreed to stop shipping hazardous and toxic waste to less-developed countries in 1989, but the practice still continues. In 2006, for example, 400 tons of toxic waste were illegally dumped at 14 open dumps in Abidjan, the capital of the Ivory Coast. The black sludge—petroleum wastes containing hydrogen sulfide and volatile hydrocarbons—killed ten people and injured many others. At least 100,000 city residents sought medical treatment for vomiting, stomach pains, nausea, breathing difficulties, nosebleeds, and headaches. The sludge—which had been refused entry at European ports—was transported by an Amsterdam-based multinational company on a Panamanian-registered ship and handed over to an Ivoirian firm (thought to be connected to corrupt government officials) to be dumped in the Ivory Coast. The Dutch company agreed to clean up the waste and pay the equivalent of (U.S.) $198 million to settle claims.

Most of the world's obsolete ships are now dismantled and recycled in poor countries. The work is dangerous, and old ships often are full of toxic and hazardous materials, such as oil, diesel fuel, asbestos, and heavy metals. On India's Alang Beach, for

(a)

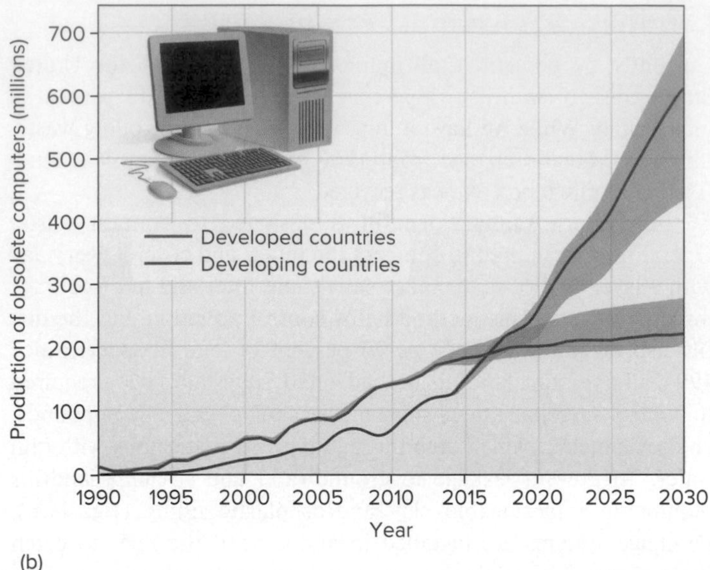

(b)

▲ **FIGURE 14.7** Demanufacturing of e-waste often occurs in places with weak health and environmental protections, as here in Accra, Ghana (a). Demanufacturing exposes workers and communities to hazardous materials, but production of e-waste is growing, with an increasing share from developing countries (b). (a) ©Pacific Press/Getty Images; (b) Source: Yu et al., *Environmental Science and Technology*, 2010.

example, more than 40,000 workers tear apart outdated vessels using crowbars, cutting torches, and even their bare hands. Metal is dragged away and sold for recycling. Organic waste is often simply burned on the beach, where ashes and oily residue wash back into the water.

Discarded electronics, or **e-waste,** is one of the greatest sources of toxic material currently going to developing countries. There are at least 2 billion television sets and personal computers in use globally. Televisions often are discarded after only about 5 years, while computers, PlayStations, cell phones, and other electronics become obsolete even faster. It's estimated that 50 million tons of e-waste are discarded every year worldwide. Only about 20 percent of the components are currently recycled. The rest generally goes to open dumps or landfills. This waste stream contains at least 2.5 billion kg of lead, as well as mercury, gallium, germanium, nickel, palladium, beryllium, selenium, arsenic, and valuable metals, such as gold, silver, copper, and steel.

Until recently, most of this e-waste went to China, where villagers, including young children, broke it apart to retrieve valuable metals. Often this scrap recovery was done under primitive conditions where workers had little or no protective gear. Health risks in this work are severe, especially for growing children. Soil, groundwater, and surface-water contamination at these sites is extremely high. Food grown in contaminated soils often contains toxic levels of lead and other metals.

Shipping e-waste to China is now officially banned, but illegal smuggling continues. In 2018, China extended its ban on trash imports to 24 categories, including many materials commonly collected for recycling in the United States (What Do You Think?, p. 341). With tighter regulation in China, informal e-waste recycling has shifted to India, Ghana, and other areas with weak environmental regulation (fig. 14.7a). Adding to the difficulty of this problem, these developing areas will soon be producing more e-waste than wealthier countries with better regulation (fig. 14.7b). Will those developing areas be able to defend public health with this increase in waste production?

One group monitoring global e-waste shipments and working conditions is the Basel Action Network. Their studies, including GPS tracking of waste shipments, continues to find increasing amounts of illicit e-waste exports around the world. This includes increasing shipments from China to Southeast Asia and elsewhere.

Incineration produces energy from trash

Faced with growing piles of garbage and a lack of available landfills at any price, many cities have built waste incinerators to burn municipal waste. Incineration reduces the volume of waste by 80 to 90 percent. Residual ash and unburnable residues representing 10 to 20 percent of the original volume are usually taken to a landfill for disposal, but landfilling costs are greatly reduced.

Most incinerators do some degree of **energy recovery,** using the heat derived from incinerated refuse to heat nearby buildings or to produce steam and generate electricity. Internationally, well over 1,000 waste-to-energy plants in Brazil, Japan, and Western Europe generate energy. In some areas, this is a major source of municipal heat and electricity production.

Municipal incinerators are specially designed plants capable of burning thousands of tons of waste per day. In some plants, refuse is sorted as it comes in to remove unburnable or recyclable materials before combustion. This sorted waste is called **refuse-derived fuel** because the enriched burnable fraction has a higher energy content than the raw trash. Another approach, called a **mass burn,** is to dump everything into a giant furnace, unsorted, and burn as much as possible (fig. 14.8). This technique avoids the expensive and unpleasant job of sorting, but it produces more unburned ash, and often it produces more air pollution as plastics, batteries, and other mixed substances are burned.

What Do YOU THINK?

Who Will Take Our Waste?

Have you ever thrown something into recycling that you shouldn't have—perhaps a greasy pizza box or a partially filled soda can? If you have, you aren't alone. According to industry professionals, wishful recyclers are their biggest problem. A high percentage—too high, unfortunately—of our recycling stream is contaminated with items that shouldn't be there.

This problem is having international repercussions. In 2018, China announced that it would no longer accept 24 kinds of low-quality, or dangerous, solid waste, including mining slag, unsorted paper, plastic scrap, discarded textiles, and a variety of commonly accepted recycling products. China is trying to clean up its environment, and it wants to show its citizens that progress is being made. Furthermore, China, which has become a world power, no longer wants to be the world's garbage dump.

▲ Thousands of Chinese have found informal jobs collecting and recycling waste but the effects on environmental quality and public health have often been severe. ©William P. Cunningham

To enforce these new restrictions, China is requiring much stricter examination of waste coming into the country. Less than 1 percent contamination will be required. In the United States, 15 percent contamination is considered standard. Removing unacceptable material is difficult and expensive once it's comingled in the waste stream.

Since the 1990s, China has been both the world's manufacturer and its leading waste importer. Trash recycling provided valuable resources for China's manufacturing sector along with thousands of low-skill (and low-paying) jobs. Recyclable waste also fills container ships that would otherwise return empty to China, after delivering consumer goods to North America or Europe. But recycling often releases toxic or dangerous materials. Air and water quality along with workers' health have suffered.

Until this ban, China accepted more than half of all the world's used plastic and paper. The United States, which is the world's leading waste source, shipped about 19 million tons of waste paper and 2 million tons of plastic to China in 2017. China's restrictions on foreign garbage imports have thrown American waste collectors and shippers into panic. If we can't send our waste to China, what will we do with it? In many places, landfills are filling up rapidly, and new sites are in short supply.

Could the American public be persuaded to use fewer disposable or single-use products? Could we start sorting recycling more carefully? China's new standards call for less than 1 percent contamination in waste. In general, the acceptable level in the United States is about 15 percent. Could we do better? Or will we simply start shipping our waste to some other poor country in Asia or Africa where standards are lower and people are more desperate? What are the ethics of waste imperialism?

▶ **FIGURE 14.8** A municipal garbage incinerator burns waste material, ideally sorted to remove polluting substances. Steam is used to generate electricity or to heat nearby buildings.

Stack
Cleaning system
Combustion chamber
Electrostatic precipitator
Filters
Feeding hopper
Boiler
Refuse bunker
Tipping area
Ash sent to landfill

The cost-effectiveness of garbage incinerators is the subject of heated debates. Initial construction costs are high–usually between \$100 million and \$300 million for a typical municipal facility. Ironically, one worry about incinerators is whether enough garbage will be available to feed them. Incinerators also compete with recyclers for paper and plastics, which are high-energy fuel. Cities usually have contracts guaranteeing certain amounts of waste daily. Some communities in which recycling has been really successful have had to buy garbage from neighbors to meet contractual obligations to waste-to-energy facilities.

Incinerators work best when regulatory agencies, such as the EPA, can ensure that they burn well and that emissions are closely monitored. Clean emissions require that municipalities sort their waste. Wet organic material (such as food) burns poorly; batteries from toys and electronics produce airborne metals; incompletely burned plastics produce carcinogenic dioxins and furans; gypsum board, which contains sulfur, produces sulfur dioxide.

European countries rely heavily on incineration, in part because they lack landfill space (fig. 14.9). Studies of hazardous air

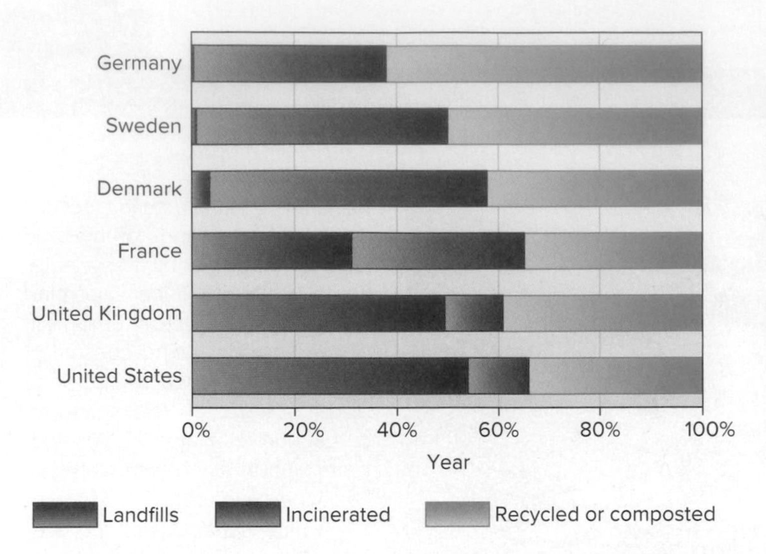

▲ **FIGURE 14.9** Reliance on landfills, recycling/composting, and incineration varies considerably among countries. Source: Data from Eurostat, U.S. EPA, 2014.

▲ **FIGURE 14.10** Separate bins for food waste (*left*) and other recyclables (*right*) make it easy to recycle organic waste in Kristianstad, Sweden. This system keeps both waste streams pure and easy to reuse. ©Mary Ann Cunningham

pollutants in Europe have shown that although municipal incinerators were major sources of hazardous emissions in 1990, modern incinerators are among the smallest sources. They now produce 1/80 as much emissions as metal processors do, and 1/20 as much as household fireplaces emit, because of tighter permitting processes, better inspections, and high burn temperatures.

In addition, municipal waste systems in many European countries sort waste to remove both recyclables and toxic materials. Increasingly, European cities are banning plastics from incinerator waste and requiring households to separate plastics from other garbage. This is expected to eliminate nearly all dioxins and other combustion by-products. Separation also helps avoid the expense of installing costly pollution-control equipment.

In places with weaker oversight, including many U.S. cities, it may be less clear that pollution-control equipment is correctly used or maintained. Opponents of incineration argue that we should be investing in more recycling rather than building expensive incinerators to deal with waste.

14.3 SHRINKING THE WASTE STREAM

- *Reducing waste is the most important of the three Rs.*
- *Recycling and reuse save materials and energy.*
- *Composting is an important component of recycling.*

Compared to landfilling and incineration, recycling saves money, energy, raw materials, and land space, and it reduces pollution. **Recycling,** as the term is used in solid waste management, is the reprocessing of discarded materials into new products. Usually recycling is easiest when materials are separated. Even food waste can be recycled if it's separated. Cities in Sweden, for example, routinely collect organic waste, which which can then be digested produce biogas (methane), a valuable fuel (fig. 14.10). This separation also makes it easier to recycle paper, plastic, and metals.

Sometimes recycled materials can be used again for their original purpose: Old aluminum cans and glass bottles are usually melted and recast into new cans and bottles, and the lead from car batteries can be made into new batteries. Sometimes entirely new products are made. Old tires are shredded and turned into rubberized walkways or road surfacing. Newspapers become cellulose insulation, and steel cans become new automobiles and construction materials.

There have been some dramatic successes in recycling in recent years. Nationally, the United States recycles or composts one-third of municipal solid waste. Minneapolis and Seattle claim a 60 percent recycling rate, something thought unattainable a decade ago. New Jersey, renowned for its waste sites, is a national leader at more than 60 percent recycling. San Francisco is aiming for 100 percent recycling. All this makes good environmental sense, but it also saves San Francisco the cost of waste disposal.

Making recycling pay for itself is often the critical challenge. Some materials are heavy and low-value, so they can be difficult to ship economically to recycling facilities. Aluminum is normally the easiest and most valuable material to recycle. Lightweight, high-value, and expensive to produce from raw materials, aluminum can be reused for thousands of purposes. Even so, only half of aluminum cans are recycled in the United States. This rate is up from only 15 percent 20 years ago, but Americans still throw away nearly 350,000 metric tons of aluminum beverage containers each year (fig. 14.11). That is enough to make 3,800 Boeing 747 airplanes. This is especially unfortunate because producing new aluminum is extraordinarily energy-intensive, while recycling is relatively easy.

Low prices for new materials is a primary obstacle. New plastic, made from oil, is usually cheaper than the cost of collecting and transporting used plastics (when the cost of disposal and other expenses are not considered). Consequently, less than 7 percent of the United States' 30 million tons of plastic waste is recycled each year. Contamination is a major obstacle in plastics recycling. Most plastic soft drink bottles are made of PET (polyethylene terephthalate), which can be remanufactured into carpet, fleece clothing, plastic strapping, and nonfood

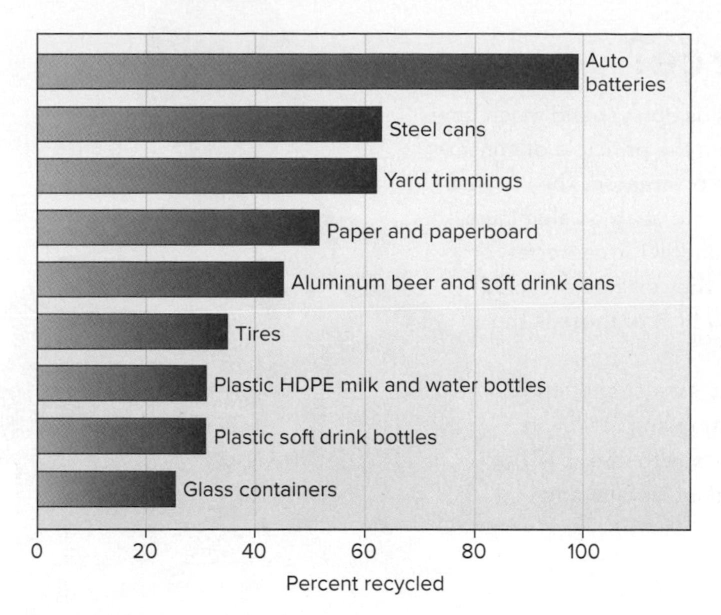

▲ **FIGURE 14.11** Recycling rates for selected materials in the United States. Car battery recycling is required by law, to keep lead out of the waste stream. Source: U.S. EPA.

packaging. However, even a trace of vinyl–a single PVC (polyvinyl chloride) bottle in a truckload, for example–can make PET useless. Because single-use beverage containers are so costly to recycle, they have been outlawed in Denmark and Finland, so that taxpayers don't have to bear the burden of cleanup and disposal.

The growing popularity of bottled water has created a serious waste disposal problem. Of the 300 billion bottles of water consumed each year globally, less than 20 percent are recycled. It takes around 75 billion liters (500 million barrels) of oil to manufacture and ship these bottles. In most American cities, tap water is safe and is subjected to more rigorous testing than bottled water. The best way to control this problem is through bottle deposits. States with deposit laws recover about 78 percent of all beverage containers, while those without generally have recycling rates of 20 percent or less.

Bottle deposits also help prevent the costs–and nuisance–of roadside litter. Americans pay an estimated 32¢ for each piece of litter picked up by crews along state highways, which adds up to $500 million every year. Bottle deposits have reduced littering in many states.

Recycling saves money, energy, and space

Some recycling programs cover their own expenses with materials sales; others have difficulty paying for themselves. Yet recycling is usually cheaper than other disposal methods, which aren't expected to pay for themselves. Curbside pickup of recyclables costs around $35 per ton, as opposed to the $80 paid to dispose of them at an average metropolitan landfill.

Philadelphia is investing in neighborhood collection centers that recycle 600 tons a day, enough to eliminate the need for a previously planned, high-priced incinerator. New York City's waste management became much more expensive when the city closed its last landfill in 2001, forcing it to export waste as far as Ohio and Pennsylvania. Since then, the city has worked aggressively to increase recycling and composting. The goal is to cut waste production in half by 2025 and to a quarter by 2030. They plan to eliminate landfilling entirely by 2030. This target supports environmental and climate plans, as well as budget goals.

Japan probably has the most successful recycling program in the world. Half of all household and commercial wastes in Japan are recycled, while the rest are about equally incinerated or landfilled. Some communities have raised recycling rates to 80 percent, and others aim to reduce waste altogether by 2020. This level of recycling takes a high level of participation and commitment. In Yokohama, a city of 3.5 million, there are now 10 categories of recyclables, including used clothing and sorted plastics. Some communities have 30 or 40 categories for sorting recyclables.

Recycling lowers demand for raw resources. The United States cuts down 2 million trees every day to produce newsprint and paper products, a heavy drain on its forests. Recycling the print run of a single Sunday issue of the *New York Times* would spare 75,000 trees (fig. 14.12). Every piece of plastic made in the United States reduces the reserve supply of petroleum and makes the country more dependent on foreign oil. Recycling 1 ton of aluminum saves 4 tons of bauxite (aluminum ore) and 700 kg of petroleum coke and pitch, as well as keeping 35 kg of aluminum fluoride out of the air.

Recycling also reduces energy consumption and air pollution. Plastic bottle recycling could save 50 to 60 percent of the energy needed to make new bottles. Making new steel from old scrap offers up to 75 percent energy savings. Producing aluminum from scrap instead of bauxite ore cuts energy use by 95 percent. If aluminum recovery were doubled worldwide, more than a million tons of air pollutants would be eliminated every year (Key Concepts, p. 344).

▼ **FIGURE 14.12** Keeping material out of landfills has multiple benefits. Consumers can help by buying recycled products. ©david trevor/Alamy Stock Photo

KEY CONCEPTS

Garbage: Liability or resource?

Municipal solid waste includes all our mixed refuse. Most of us don't spend much time thinking about where our waste ends up, but as you know from the principle of conservation of matter (chapter 2), materials are never destroyed or created; they're just transformed from one shape to another. Elements in our waste such as aluminum, lead, carbon, or nitrogen don't disappear. They may sit in a landfill for centuries, or they may be incinerated and emitted into the atmosphere, or they may be recycled and transformed into another useful object. The question is, which of these is the most efficient use of our resources and environment?

Materials in our municipal waste stream have been hard to extract and reuse because they're usually all mixed together, although new sorting and recycling systems can separate mixed waste. Waste can be a liability or a resource. It all depends on how much we produce, how much we landfill and incinerate, and how much we recycle.

KC 14.1

©Radius Images/Getty Images

What are the major types of waste we produce, and how much do we recycle?

We produce far more waste than our grandparents did. The EPA has tracked overall production of municipal solid waste in the United States since 1960. Disposable paper and plastic products have grown most dramatically in the past 50 years. Recovery rates are worst for plastics (because mixed waste contaminates plastics and new plastic is inexpensive) and food products (because these are relatively hard to store and transport to central recycling facilities). Metals, glass, and yard compost have relatively high recycling rates.

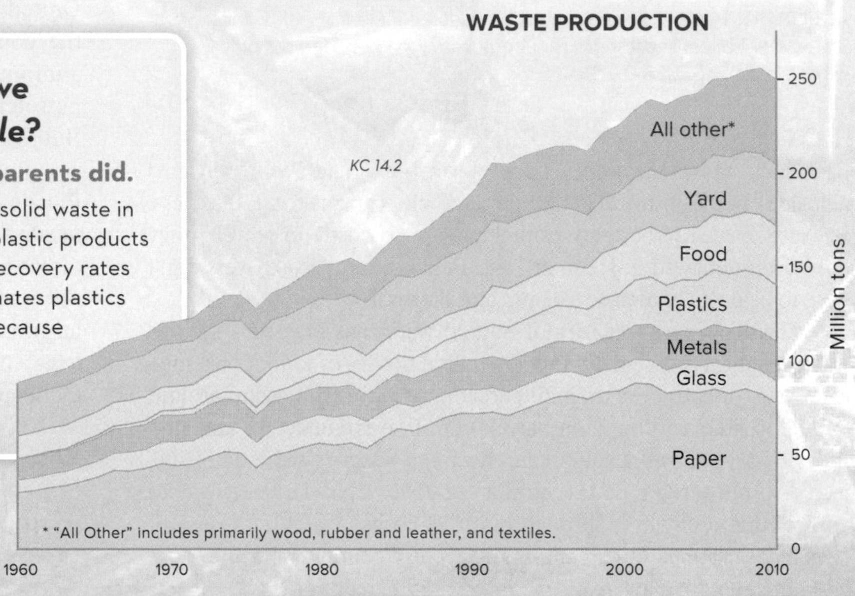

KC 14.2

Where does it all go?

As the available landfills decline, we are recycling and incinerating more municipal waste.

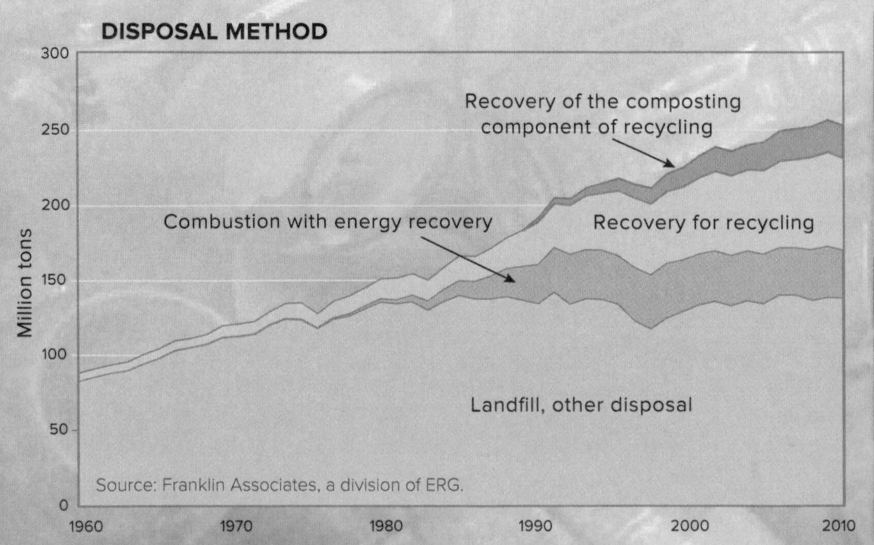

KC 14.3

©ShutterPNPhotography/Shutterstock.com

KC 14.4

©Stockbyte/Getty Images

KC 14.6

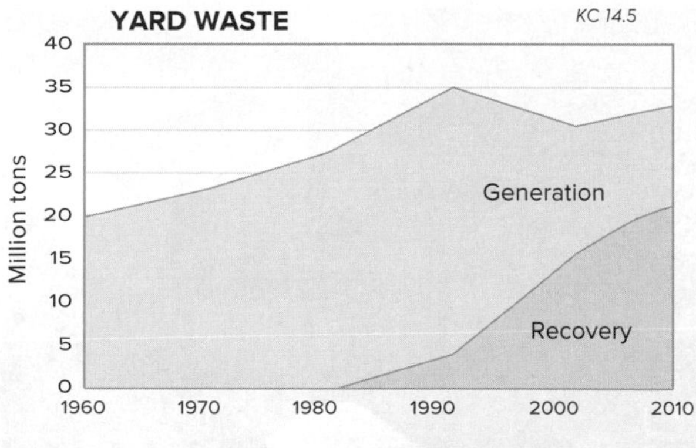

▲ Recycling rate = Yard waste: 65%
Food scraps: negligible

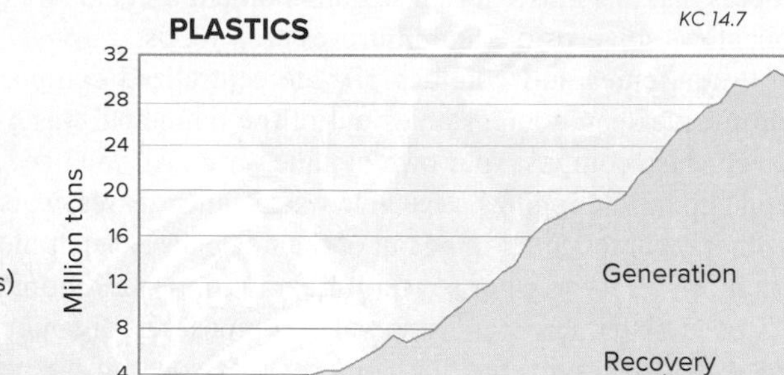

Recycling rate = negligible (all plastics)
PET bottles: 27%
HDPE bottles: 29% ▶

©teen00000/123RF

KC 14.8

METALS

KC 14.9

Million tons

22 20 18 16 14 12 10 8 6 4 2 0

1960 1970 1980 1990 2000 2010

Generation

Recovery

◀ Recycling rate = 35%
(Aluminum: 21%
Iron/steel: 34%
Other metals: 69%)

©George Doyle/Getty Images

KC 14.10

PAPER AND PAPERBOARD

KC 14.11

Million tons

100 90 80 70 60 50 40 30 20 10 0

1960 1970 1980 1990 2000 2010

Generation

Recovery

▲ Recycling rate = 56%

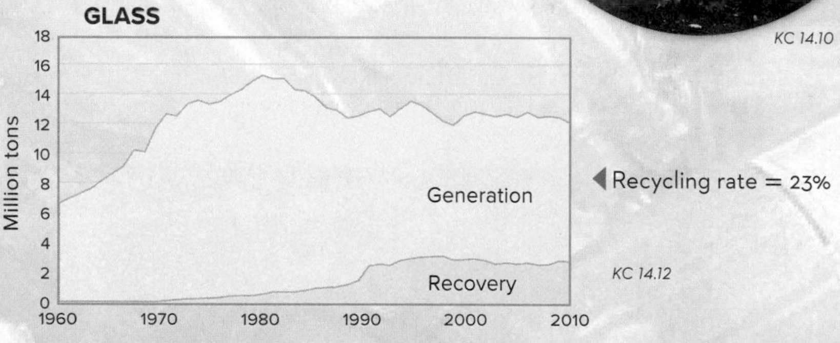

◀ Recycling rate = 23%

KC 14.12

©david trevor/Alamy Stock Photo

KC 14.13

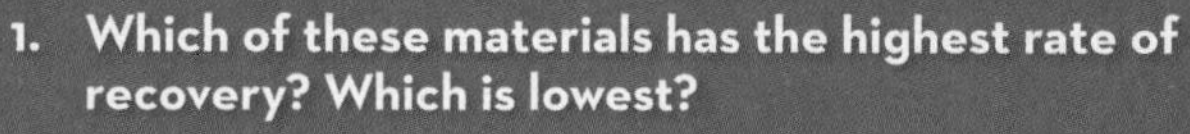

CAN YOU EXPLAIN?

1. **Which of these materials has the highest rate of recovery? Which is lowest?**
2. **Is there an approximate year in which recovery and recycling began to increase?**
3. **Why is recycling less common for plastics than for metals?**
4. **In the plot of production, which factors increased most from 1960 to 2008? Roughly what is the ratio of total production in 2008 to total production in 1960?**

©Stockbyte/Getty Images

Composting recycles organic waste

Pressed for landfill space, many cities have banned yard waste from municipal garbage. Rather than bury this valuable organic material, they are turning it into a useful product through **composting:** biological degradation or breakdown of organic matter under aerobic (oxygen-rich) conditions. The organic compost resulting from this process makes a nutrient-rich soil amendment that aids water retention, slows soil erosion, and improves crop yields.

Many cities and counties provide centralized composting, to help people keep compostables out of the municipal waste stream. You can also compost your own organic waste. All you need to do is to pile up lawn clippings, vegetable waste, fallen leaves, wood chips, or other organic matter in an out-of-the way place, keep it moist, and turn it over every week or so (fig. 14.13). Within a few months, naturally occurring microorganisms will decompose the organic material into a rich, pleasant-smelling compost that you can use to enrich your yard or garden.

There is tremendous potential in these systems, since about 30 percent of municipal waste is biodegradable organic material. Methane biogas plants are increasingly common ways for European cities to dispose of organic waste. Over 14,500 biogas digesters convert organic waste to gas that can be burned for heat and electricity. In addition, over 280 plants purify biogas to "biomethane," which is 96 percent CH_4 and is pure enough to send to natural gas networks or to power vehicles.

Some U.S. farms now produce biogas for their own consumption. Most are dairies, which have to dispose of manure somehow, and which can use the fuel on site. Most U.S. cities don't yet separate household organic waste, but they are making progress at sewage plants. Around 9 percent of U.S. sewage treatment facilities produce some biogas from sewage.

▼ **FIGURE 14.13** Composting keeps tremendous volumes of material out of the waste stream, as well as providing rich nutrients for gardens. ©Gabor Havasi/123RF

▲ **FIGURE 14.14** Reusing discarded products is a creative and efficient way to reduce waste. This recycling center in Berkeley, California, provides building supplies and saves money for the community. Courtesy of Urban Ore, Inc. Berkeley, CA

Reuse is even better than recycling

Even better than recycling or composting is cleaning and reusing materials in their present form, thus saving the cost and energy of remaking them into something else. We do this already with some specialized items. Auto parts are regularly sold from junk yards, especially for older car models. In some areas stained-glass windows, brass fittings, fine woodwork, and bricks salvaged from old houses bring high prices. Some communities sort and reuse a variety of materials received in their dumps (fig. 14.14).

Reusing glass and plastic beverage containers is far more efficient than producing new bottles. A reusable glass container makes an average of 15 round-trips between factory and customer before it becomes so scratched and chipped that it has to be recycled. Reusable containers also favor local bottling companies and help preserve regional businesses. Since the advent of cheap, lightweight, disposable food and beverage containers, many small, local breweries, canneries, and bottling companies have been forced out of business by national or multinational conglomerates. Big companies can afford to ship food and beverages thousands of miles, as long as it is a one-way trip. National companies favor recycling, rather than collection and reuse, because they don't want the responsibility for collecting and reusing containers.

In many less-affluent nations, reusing manufactured goods is an established tradition. If manufactured products are expensive and labor is cheap, it pays to salvage, clean, and repair products. Cairo, Manila, Mexico City, and many other cities have large populations of poor people who make a living by scavenging. Entire ethnic populations may survive on scavenging, sorting, and reprocessing scraps from city dumps. These people provide essential but unpaid services to society by helping to reduce and reuse the mountains of waste products that governments can't afford to manage.

What Can YOU DO?

Reducing Waste

1. Buy foods that come with less packaging; shop at farmers' markets or co-ops, using your own containers. Bring your own shopping bag.
2. Take your own washable, refillable beverage container to meetings.
3. When you have a choice at the grocery store among plastic, glass, or metal containers for the same food, buy the reusable or easier-to-recycle glass or metal.
4. Separate your cans, bottles, papers, and plastics for recycling.
5. Avoid single-use packaging. Wash and reuse bottles, aluminum foil, plastic bags, and so on for your personal use.
6. Compost yard and garden wastes, leaves, and grass clippings.
7. Help your school develop responsible systems for disposing of electronics and other waste.
8. Write to your senators and representatives, and urge them to vote for container deposits, recycling, and safe incinerators or landfills.

Source: Minnesota Pollution Control Agency.

Reducing waste is the cheapest option

Most of our attention in waste management focuses on recycling, but slowing the production of throw-away products is by far the most effective way to save energy, materials, and money. Among the "three Rs"—reduce, reuse, recycle—the most important strategy is the first. Industries are increasingly finding that reducing saves money. Soft-drink makers use less aluminum per can than they did 20 years ago, and plastic bottles use less plastic. 3M has saved over $500 million in the past 30 years by reducing its use of raw materials, reusing waste products, and increasing efficiency. Rethinking consumption habits can be done at any scale, from the nation, city, or corporation to the individual (What Can You Do?, above).

In recent decades, we have greatly increased our waste production rather than reducing it. As consumer goods have multiplied and as global economies have grown, all of us have done our part for the economy by consuming, and discarding, more things. Moreover, as developing countries become wealthier they are catching up with the high levels of consumption and waste production in wealthier countries.

Excessive packaging of food and consumer products is one of our greatest sources of unnecessary waste. Paper, plastic, glass, and metal packaging materials make up 50 percent of our domestic trash by volume. Much of that packaging is primarily for marketing and has little to do with product protection. Manufacturers and retailers can reduce these practices if consumers ask for products with less packaging. Canada's National Packaging Protocol (NPP) recommends that packaging minimize depletion of virgin resources and production of toxins in manufacturing. The preferred hierarchy is (1) no packaging, (2) minimal packaging, (3) reusable packaging, and (4) recyclable packaging. This plan sets an ambitious target of 50 percent reduction in excess packaging.

Many countries, including China, Bangladesh, India, Eritrea, Rwanda, Somalia, Tanzania, Kenya, and Uganda, have banned ultrathin (less than 0.025 mm thick) plastic bags and called for a return to reusable cloth bags for shopping. This could eliminate up to 3 billion plastic bags used every day in these countries. Japan, Ireland, South Africa, and Taiwan also have discouraged single-use plastic bags through taxes or prohibitions. Since 2008, over 200 U.S. cities and counties, and the State of California, have limited or banned single-use bags or instituted charges for them.

Where disposable packaging is necessary, we still can reduce the volume of waste in our landfills by using materials that are compostable or degradable. **Photodegradable plastics** break down when exposed to ultraviolet radiation. **Biodegradable plastics** incorporate materials, such as cornstarch, that microorganisms can decompose. Several states have introduced legislation requiring biodegradable or photodegradable six-pack beverage yokes, fast-food packaging, and disposable diapers. These degradable plastics often don't decompose completely, however; many kinds only break down to small particles that remain in the environment.

14.4 HAZARDOUS AND TOXIC WASTES

- *Hazardous waste is dangerous in small doses, easily combustible, or corrosive.*
- *Handling and cleaning up hazardous waste are expensive, so federal legislation enforces cleanup and safe handling.*
- *Hazardous waste can be used, recycled, decontaminated, or stored permanently.*

The most dangerous aspect of the waste stream is that it often contains highly toxic and hazardous materials that are injurious to both human health and environmental quality (fig. 14.15). We now produce and use a vast array of flammable, explosive, caustic, acidic,

▲ **FIGURE 14.15** Hazardous waste is dangerous even with small exposure. Here a worker tests for radioactive soil. ©Arthur S Aubry /Photodisc/ Getty Images

and highly toxic chemical substances for industrial, agricultural, and domestic purposes. According to the EPA, U.S. industries generate about 900 million metric tons of officially classified hazardous wastes each year, about 3 metric tons for each person in the country. In addition, considerably more toxic and hazardous waste material is generated by industries or processes not regulated by the EPA. Shockingly, at least 40 million metric tons (22 billion lb) of toxic and hazardous wastes are released into the air, water, and land in the United States each year. The biggest sources of these toxins are the chemical and petroleum industries (fig. 14.16).

Hazardous waste includes many dangerous substances

Legally, a **hazardous waste** is any discarded material, liquid or solid, that contains substances known to be (1) fatal to humans or laboratory animals in low doses; (2) toxic, carcinogenic, mutagenic, or teratogenic to humans or other life-forms; (3) ignitable with a flash point less than 60°C; (4) corrosive; or (5) explosive or highly reactive (undergoes violent chemical reactions either by itself or when mixed with other materials). Notice that this definition includes both toxic and hazardous materials, as defined in chapter 8. Certain compounds are exempt from regulation as hazardous waste if they are accumulated in less than 1 kg (2.2 lb) of commercial chemicals or 100 kg of contaminated soil, water, or debris. Even larger amounts (up to 1,000 kg) are exempt when stored at an approved waste treatment facility for the purpose of being beneficially used, recycled, reclaimed, detoxified, or destroyed.

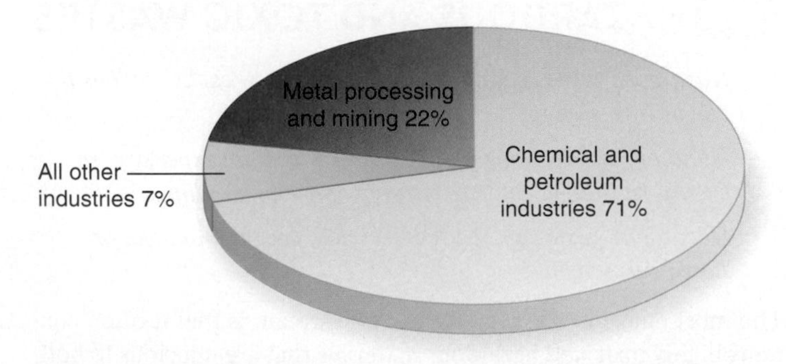

▲ **FIGURE 14.16** Producers of hazardous wastes in the United States. Source: U.S. EPA.

Active LEARNING

A Personal Hazardous Waste Inventory

Make a survey of your house or apartment to see how many toxic materials you can identify. If you live in a dorm, you may need to survey your parents' house. If you are in an art class, consider paints and solvents. Read the cautionary labels. Which of the products are considered hazardous? When you use them, do you usually follow all the safety procedures recommended? Where is the nearest hazardous waste collection site for disposal of unwanted household products? If you don't know, call your city administrator or mayor's office to find out.

Most hazardous waste is recycled, converted to nonhazardous forms, stored, or otherwise disposed of on-site by the generators—chemical companies, petroleum refiners, and other large industrial facilities—so that it doesn't become a public problem. Still, the hazardous waste that does enter the waste stream or the environment represents a serious environmental problem. And "orphan" wastes left behind by abandoned industries remain a serious threat to both environmental quality and human health. For years little attention was paid to this material. Wastes stored on private property, buried, or allowed to soak into the ground were considered of little concern to the public. An estimated 5 billion metric tons of highly poisonous chemicals were improperly disposed of in the United States between 1950 and 1975 before regulatory controls became more stringent.

Federal legislation regulates hazardous waste

Two important federal laws regulate hazardous waste management and disposal in the United States. The Resource Conservation and Recovery Act (RCRA, pronounced "rickra") of 1976 is a comprehensive program that requires rigorous testing and management of toxic and hazardous substances. At every step, generators, shippers, users, and disposers of these materials must keep an account of everything they handle and what happens to it from generation (cradle) to ultimate disposal (grave) (fig. 14.17).

The Comprehensive Environmental Response, Compensation, and Liability Act (CERCLA or Superfund Act), passed in 1980, is aimed at rapid containment, cleanup, or remediation of abandoned toxic waste sites. The act establishes a National Priority

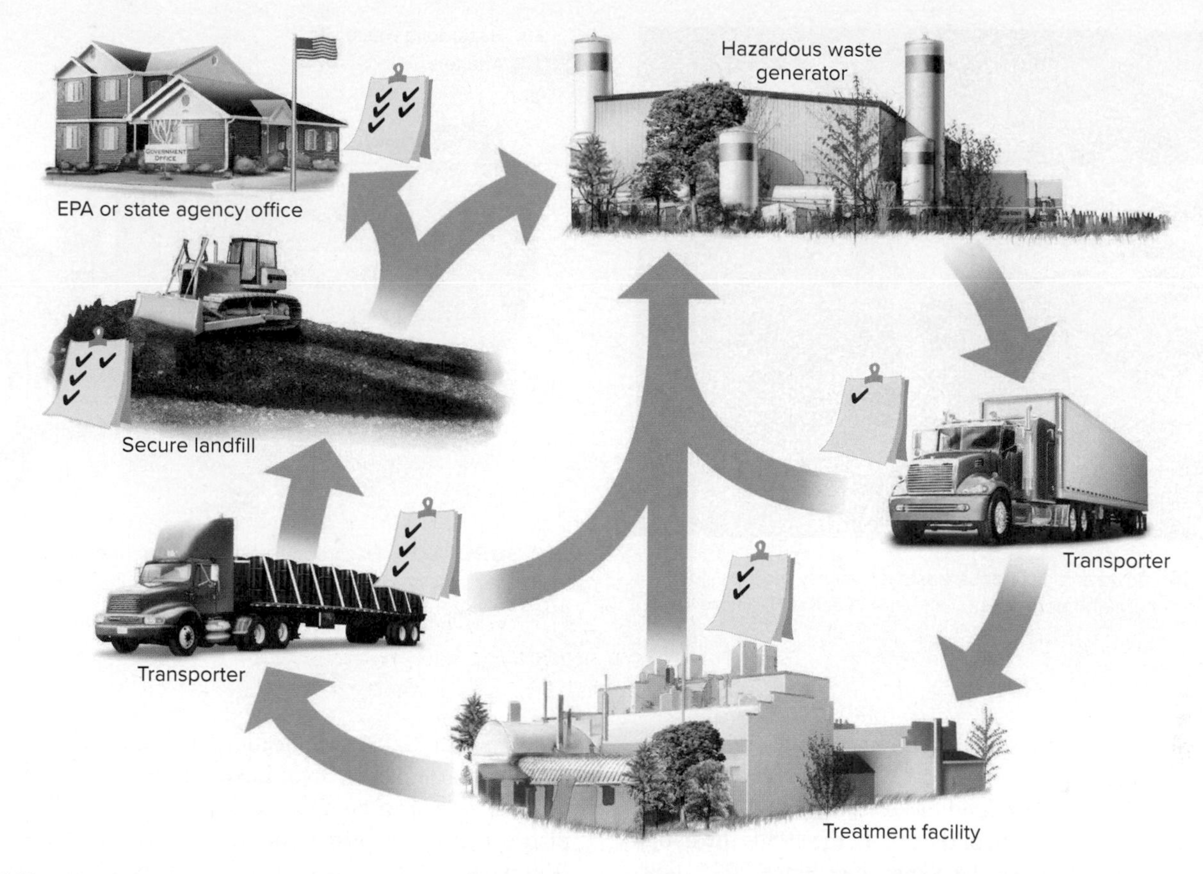

▲ **FIGURE 14.17** Toxic and hazardous wastes must be tracked from "cradle to grave" by detailed shipping manifests (records).

List (NPL) of sites most in need of remediation. In 2010 there were approximately 1,280 sites on the list, with 60 waiting for a decision and over 1,000 sites designated as "completed." CERCLA authorizes the EPA to take emergency actions when there is a threat that toxic material could leak into the environment. The EPA also is empowered to sue responsible parties for the recovery of treatment costs.

About 30 percent of sites on the NPL are orphan sites, whose owners have disappeared or gone out of business. For these, CERCLA established a "Superfund," a pool of money to cover remediation costs until responsible parties can be located.

The government does not have to prove that anyone violated a law or what role he or she played in a Superfund site. Rather, liability under CERCLA is "strict, joint, and several," meaning that anyone associated with a site can be held responsible for the entire cost of cleaning it up, no matter how much of the mess he or she made. In some cases property owners have been assessed millions of dollars for removal of wastes left there years earlier by previous owners. This strict liability has been a headache for the real estate and insurance businesses.

CERCLA was amended in 1995 to make some of its provisions less onerous. In cases where treatment is unavailable or too costly and it is likely that a less costly remedy will become available within a reasonable time, interim containment is now allowed. The EPA also now has the discretion to set site-specific cleanup levels, rather than adhere to rigid national standards.

CERCLA was modified in 1984 by the Superfund Amendments and Reauthorization Act (SARA). SARA also established a community "right to know," the notion that the public had a right to know about production, use, or transportation of toxic materials where they live. A key part of public information and emergency planning is the **Toxic Release Inventory,** a listing of addresses where regulated materials are handled. This inventory requires 20,000 manufacturing facilities to report annually on releases of more than 300 toxic materials. The EPA publishes this list, and from it you can find specific information in the inventory about what is in your neighborhood.

Superfund sites are listed for federally funded cleanup

The EPA estimates that there are at least 36,000 seriously contaminated sites in the United States. The General Accounting Office (GAO) places the number much higher, perhaps more than 400,000 when all are identified. Originally, about 1,671 sites were placed on the National Priority List for cleanup with financing from the federal Superfund program. The **Superfund** is a revolving pool designed to (1) provide an immediate response to emergency situations that pose imminent hazards and (2) to clean up or remediate abandoned or inactive sites. Without this fund, sites would languish for years or decades while the courts decided who was responsible for paying for the cleanup (fig. 14.18). Originally a $1.6 billion pool, the fund peaked at $3.6 billion. From its inception, the fund was financed by taxes on producers of toxic and hazardous wastes. Industries opposed this "polluter pays" tax,

▲ **FIGURE 14.18** Toxic and hazardous waste is a messy business, but disposal must be secure and permanent. ©Michael Greenlar/The Image Works

▲ **FIGURE 14.19** Hazardous waste sites are often located on aquifer recharge zones, making groundwater contamination a common risk. Source: U.S. EPA.

because current manufacturers are often not the ones responsible for the original contamination. In 1995 Congress agreed to let the tax expire. Since then the Superfund has dwindled, and the public has picked up an increasing share of the bill. In the 1980s the public covered less than 20 percent of the Superfund. Since 2004, however, general revenues (public tax dollars) have paid the entire cost of a greatly reduced program, and the industry share has been zero.

Total costs for hazardous waste cleanup in the United States are estimated to be between $370 billion and $1.7 trillion, depending on how clean sites must be and what methods are used. For years, Superfund money was spent mostly on lawyers and consultants, and cleanup efforts were often bogged down in disputes over liability and best cleanup methods. During the 1990s, however, progress improved substantially, with a combination of rule adjustments and administrative commitment to cleanup. By 2004, more than half (over 1,000) of the original NPL sites were listed as completed in cleanup or containment.

What qualifies a site for the NPL? These sites are considered to be especially hazardous to human health and environmental quality because they are known to be leaking or have a potential for leaking supertoxic, carcinogenic, teratogenic, or mutagenic materials (see chapter 8). The ten substances of greatest concern or most commonly detected at Superfund sites are lead, trichloroethylene, toluene, benzene, PCBs, chloroform, phenol, arsenic, cadmium, and chromium. These and other hazardous or toxic materials are known to have contaminated groundwater at 75 percent of the sites now on the NPL. In addition, 56 percent of these sites have contaminated surface waters, and airborne materials are found at 20 percent of the sites. Seventy million Americans, including 10 million children, live within 6 km of a Superfund site.

Where are these thousands of hazardous waste sites, and how did they get contaminated? Old industrial facilities, such as smelters, mills, petroleum refineries, and chemical manufacturing plants, are highly likely to have been sources of toxic wastes. Regions of the country with high concentrations of aging factories, such as the "rust belt" around the Great Lakes or the Gulf Coast petrochemical centers, have large numbers of Superfund sites (fig. 14.19). Mining districts also are prime sources of toxic and hazardous waste. Within cities, factories and places such as railroad yards, bus repair barns, and filling stations, where solvents, gasoline, oil, and other petrochemicals were spilled or dumped on the ground, often are highly contaminated.

Some of the most infamous toxic waste sites were old dumps where many different materials were mixed together indiscriminately. For instance, Love Canal in Niagara Falls, New York, was an open dump that both the city and nearby chemical factories used as a disposal site. More than 20,000 tons of toxic chemical waste were buried under what later became a housing development. Another infamous example occurred in Hardeman County, Tennessee, where about a quarter of a million barrels of chemical waste were buried in shallow pits, which subsequently leaked toxins into the groundwater.

Brownfields present both liability and opportunity

Among the biggest problems in cleaning up hazardous waste sites are questions of liability and the degree of purity required. In many cities, these problems have created large areas of contaminated properties, known as **brownfields,** that have been abandoned or are not being used to their potential because of real or suspected pollution. Up to one-third of all commercial and industrial sites in the urban core of many big cities fall in this category. In heavy industrial corridors the percentage typically is higher.

For years no one was interested in redeveloping brownfields because of liability risks. Who would buy a property, knowing that they might be forced to spend years in litigation and negotiations and be forced to pay millions of dollars for pollution they didn't create? Even if a site has been cleaned to current standards, there is a worry that additional pollution might be found in the future or that more stringent standards might be applied.

In many cases, property owners complain that unreasonably high levels of purity are demanded in remediation programs. Consider the case of Columbia, Mississippi. For many years a 35-ha (81-acre) site in Columbia was used for turpentine and pine tar manufacturing. Soil tests showed concentrations of phenols and other toxic organic compounds exceeding federal safety standards. The site was added to the Superfund NPL, and remediation was ordered. Some experts recommended that the best solution was to simply cover the surface with clean soil and enclose the property with a fence to keep people out. The total costs would have been about $1 million. Instead, the EPA ordered Reichhold Chemical, the last known property owner, to excavate more than 12,500 tons of soil and haul it to a commercial hazardous waste dump in Louisiana at a cost of some $4 million. The intention is to make the site safe enough to be used for any purpose, including housing—even though no one has proposed building anything there. According to the EPA, the dirt must be clean enough for children to play in—even eat—without risk.

Similarly, in places where contaminants have seeped into groundwater, the EPA generally demands that cleanup be carried to drinking-water standards. Many critics believe that these pristine standards are unreasonable. Former congressman Jim Florio, a principal author of the original Superfund Act, says, "It doesn't make any sense to clean up a rail yard in downtown Newark so it can be used as a drinking water reservoir." Depending on where the site is, what else is around it, and what its intended uses are, much less stringent standards may be perfectly acceptable.

Brownfield redevelopment is increasingly seen as an opportunity for rebuilding cities, creating jobs, increasing the tax base, and preventing needless destruction of open space at urban margins. In 2002 the EPA established a new brownfields revitalization fund designed to encourage restoration of more sites, as well as more kinds of sites. In some communities former brownfields are being turned into "eco-industrial parks" that feature environmentally friendly businesses and bring in much-needed jobs to inner-city neighborhoods (see chapter 15).

Hazardous waste must be processed or stored permanently

What shall we do with toxic and hazardous wastes? In our homes, we can reduce waste generation and choose less toxic materials. Buy only what you need for the job at hand. Use up the last little bit, or share leftovers with a friend or neighbor. Many common materials that you probably already have make excellent alternatives to commercial products.

Produce Less Waste As with other wastes, the safest and least expensive way to avoid hazardous waste problems is to avoid creating the wastes in the first place. Manufacturing processes can be modified to reduce or eliminate waste production. In Minnesota, the 3M Company reformulated products and redesigned manufacturing processes to eliminate more than 140,000 metric tons of solid and hazardous wastes, 4 billion liters (1 billion gal) of wastewater, and 80,000 metric tons of air pollution each year. It frequently found that these new processes not only spared the environment but also saved money by using less energy and fewer raw materials.

Recycling and reusing materials also eliminates hazardous wastes and pollution. Many waste products of one process or industry are valuable commodities in another. Already, about 10 percent of the wastes that would otherwise enter the waste stream in the United States are sent to surplus material exchanges, where they are sold as raw materials for use by other industries. This figure could probably be raised substantially with better waste management. In Europe at least one-third of all industrial wastes are exchanged through clearinghouses, where beneficial uses are found. This represents a double savings: The generator doesn't have to pay for disposal, and the recipient pays little, if anything, for raw materials.

Convert to Less Hazardous Substances Several processes are available to make hazardous materials less toxic. *Physical treatments* tie up or isolate substances. Charcoal or resin filters absorb toxins. Distillation separates hazardous components from aqueous solutions. Precipitation and immobilization in ceramics, glass, or cement isolate toxins from the environment, so that they become essentially nonhazardous. One of the few ways to dispose of metals and radioactive substances is to fuse them in silica at high temperatures to make a stable, impermeable glass that is suitable for long-term storage. Plants, bacteria, and fungi can also concentrate or detoxify contaminants (Exploring Science, p. 352).

Incineration is applicable to mixtures of wastes. A permanent solution to many problems, it is quick and relatively easy, but not necessarily cheap—or always clean—unless done correctly. Wastes must be heated to over 1,000°C (2,000°F) for a sufficient period of time to complete destruction. The ash resulting from thorough incineration is reduced in volume up to 90 percent and often is safer to store in a landfill or another disposal site than the original wastes.

Chemical processing can transform materials to make them nontoxic. Included in this category are neutralization, removal of metals or halogens (chlorine, bromine, etc.), and oxidation. The Sunohio Corporation of Canton, Ohio, for instance, has developed a process called PCBx, in which chlorine in such molecules as PCBs is replaced with other ions that render the compounds less toxic. A portable unit can be moved to the location of the hazardous wastes, eliminating the need for shipping them.

Store Permanently Inevitably, there are some materials we can't destroy, make into something else, or otherwise eliminate. We will have to store them out of harm's way (fig. 14.20).

Permanent retrievable storage involves placing waste storage containers in a secure place such as a salt mine or bedrock cavern, where they can be inspected periodically and retrieved if necessary. This approach is expensive because it requires monitoring, but it has the advantage that we don't completely lose control of highly toxic substances that could eventually leak into groundwater if they were buried in a landfill. If we learn someday that our disposal

EXPLORING Science — Bioremediation

Cleaning up the thousands of hazardous waste sites at factories, farms, and gas stations is an expensive project. In the United States alone, waste cleanup is projected to cost at least $700 billion. Usually hazardous waste remediation (cleanup) involves digging up soil and incinerating it, potentially releasing toxins into the air, or trucking it to a secure landfill. Contaminated groundwater is frequently pumped out of the ground; hopefully, contaminants are retrieved at the same time.

How do plants, bacteria, and fungi do all this? Many of the biophysical details are poorly understood, but in general, plant roots are designed to efficiently extract nutrients, water, and trace minerals from soil and groundwater. The mechanisms involved may aid extraction of metallic and organic contaminants. Some plants also use toxic elements as a defense against herbivores: Locoweed, for example, selectively absorbs elements such as selenium, concentrating toxic levels in its leaves. Absorption can be extremely effective. Bracken fern growing in Florida has been found to contain arsenic at concentrations more than 200 times higher than in the soil in which it was growing.

Genetically modified plants are also being developed to process toxins. Poplars have been developed to process toxins, using a gene borrowed from bacteria that transforms a toxic compound of mercury into a safer form. In another experiment, a gene for producing mammalian liver enzymes, which specialize in breaking down toxic organic compounds, was inserted into tobacco plants. The plants succeeded in producing the liver enzymes and breaking down toxins absorbed through their roots.

These remediation methods are not without risks. Insects could consume leaves containing concentrated substances, allowing contaminants to enter the food chain. Some absorbed contaminants are volatilized, or emitted in gaseous form, through pores in plant leaves. Once contaminants are absorbed into plants, the plants themselves are usually toxic and must be landfilled. But the cost of phytoremediation (cleanup using plants) can be less than half the cost of landfilling or treating toxic soil, and the volume of plant material requiring secure storage is a fraction of the volume of the contaminated dirt.

Cleaning up hazardous and toxic waste sites will be a big business for the foreseeable future, in North America and around the world. Innovations such as **bioremediation,** or biological waste treatment, offer promising prospects for business development, as well as for environmental health and saving taxpayer money.

A promising alternative to these methods involves bioremediation. Microscopic bacteria and fungi can absorb, accumulate, and detoxify a remarkable variety of toxic compounds. They can also accumulate heavy metals, and some have been developed that can metabolize (break down) PCBs. Aquatic plants such as water hyacinths and cattails can also be used to purify contaminated effluent.

An increasing variety of plants have been used in phytoremediation. Some types of mustard can extract lead, arsenic, zinc, and other metals from contaminated soil. Radioactive strontium and cesium were extracted from soil near the Chernobyl nuclear power plant using common sunflowers. Poplar trees can absorb and break down toxic organic chemicals. Natural bacteria in groundwater, when provided with plenty of oxygen, can neutralize contaminants in aquifers. Experiments have shown that pumping air *into* groundwater can be a more effective cleanup method than pumping water *out*.

▲ **FIGURE 14.20** Hazardous substances we can't decontaminate must be catalogued, contained, and stored permanently. Here workers retrieve improperly buried waste from the Hanford nuclear site in Washington State.
Source: DOE Photo/U.S. Department of Energy

methods were bad, we can retrieve waste and treat it more effectively. Retrieving waste from storage in a mine is much cheaper and more effective than digging up and remediating buried pollutants from a landfill.

Secure landfills are the most common solutions for hazardous waste disposal, however. Although many landfills have been environmental disasters, newer techniques make it possible to create safe, secure, modern landfills that can contain many hazardous wastes. As with a modern solid waste landfill, the first line of defense in a secure landfill is a thick bottom cushion of compacted clay that surrounds the pit like a bathtub (fig. 14.21). Moist clay is flexible and resists cracking if the ground shifts. It is impermeable to groundwater and will safely contain wastes. A layer of gravel is spread over the clay liner, and perforated drainpipes are laid in a grid to collect any seepage that escapes from the stored material. A thick polyethylene liner, protected from punctures by soft padding materials, covers the gravel bed. A layer of soil or absorbent sand cushions the inner liner, and the wastes are packed in drums, which then are placed into the pit, separated into small units by thick berms of soil or packing material.

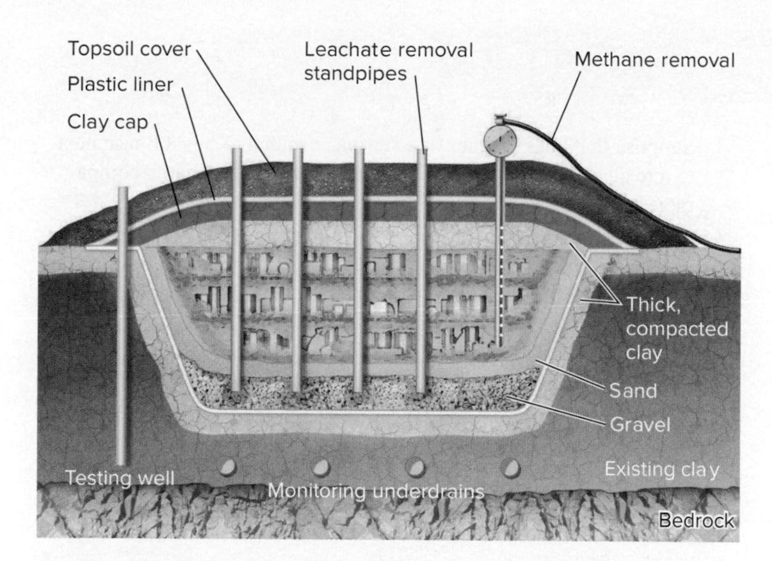

▲ **FIGURE 14.21** A secure landfill has a thick plastic liner, and two or more layers of compacted clay and a gravel bed, from which drains collect material leaching from the landfill. Testing wells allow monitoring for escaping contaminants or combustible methane.

When the landfill has reached its maximum capacity, a cover much like the bottom sandwich of clay, plastic, and soil–in that order–caps the site. Vegetation stabilizes the surface and improves its appearance. Sump pumps collect any liquids that filter through the landfill, from either rainwater or leaking drums. This leachate is treated and purified before being released. Monitoring wells check groundwater around the site to ensure that no toxins have escaped.

Most landfills are buried below ground level to be less conspicuous; however, in areas where the groundwater table is close to the surface, it is safer to build aboveground storage. The same protective construction techniques are used as in a buried pit. An advantage to such a facility is that leakage is easier to monitor because the bottom is at ground level.

Transportation of hazardous wastes to disposal sites is of concern because of the risk of accidents. Emergency-preparedness officials conclude that the greatest risk in most urban areas is not nuclear war or natural disaster but crashes involving trucks or trains carrying hazardous chemicals through densely packed urban corridors. Another worry is who will bear financial responsibility for abandoned waste sites. Hazardous wastes remain toxic long after the businesses that created them are gone. As is the case with nuclear wastes (see chapter 13), we may need new institutions for perpetual care of these wastes.

CONCLUSION

Modern society produces a prodigious amount of waste. Each of us produces far more waste, and more toxic substances, than our grandparents did. Growing amounts of packaging, including plastics, are among our largest-volume problems; e-waste, with its mixed materials, and hazardous waste are growing and expensive challenges in waste management. Government policies and economies of scale make it cheaper and more convenient to extract virgin raw materials to make new consumer products than to reuse or recycle used materials. But we all bear the cost of disposing of this constantly increasing stream of waste products.

The rising cost and declining availability of landfills have led to new and creative strategies to reduce, reuse, and recycle waste. Reuse and recycling programs, and incentives to reduce packaging materials are increasingly widespread are among the approaches that have proven both successful and economically beneficial. Incinerators, which have a mixed economic and public health record in the United States, have been improved in other countries and are contributing to waste-management solutions.

Controlling the production and disposal of hazardous waste requires careful oversight by government agencies and waste-management regulations. These rules and agencies are put in place to defend public health. Without them, disastrous cases like Love Canal, with cleanup paid for by the public rather than by polluters, develop all too easily.

A first step toward reducing our waste production is to understand how much we produce and what we produce. Another key step is to make our waste disposal more visible. Paying attention to recycling, reusing, and reducing household and hazardous waste can greatly improve our awareness of our environmental liabilities.

PRACTICE QUIZ

1. What is the "Pacific Garbage Patch," and what is the main source of plastic in it?
2. What are *solid wastes* and *hazardous wastes*? What is the difference between them?
3. Describe the difference between an open dump, a sanitary landfill, and a modern, secure hazardous waste disposal site.
4. Describe some concerns about waste incineration.
5. What are the major types of materials recycled from municipal waste, and how are they used?
6. What is *e-waste*? How is most of it disposed of, and what are some strategies for improving recycling rates?
7. What is *composting*, and how does it fit into solid waste disposal?
8. What materials are most recycled in the United States?
9. What are *brownfields,* and why do cities want to redevelop them?
10. What are *bioremediation* and *phytoremediation*? What are some advantages to these methods?

CRITICAL THINKING AND DISCUSSION

Apply the principles you have learned in this chapter to discuss these questions with other students.

1. A toxic waste disposal site has been proposed for the Pine Ridge Indian Reservation in South Dakota. Many tribal members oppose this plan, but some favor it because of the jobs and income it will bring to an area with 70 percent unemployment. If local people choose immediate survival over long-term health, should we object or intervene?
2. Should industry officials be held responsible for dumping chemicals that were legal when they did it but are now known to be extremely dangerous? At what point can we argue that they should have known about the hazards involved?
3. Suppose that your brother or sister has decided to buy a house next to a toxic waste dump because it costs $20,000 less than a comparable house elsewhere. What do you say to him or her?
4. Is there a fundamental difference between incinerating municipal, medical, or toxic industrial waste? Would you oppose an incinerator in your neighborhood for one type of waste but not others? Why or why not?
5. Some scientists argue that permanent, retrievable storage of toxic and hazardous wastes is preferable to burial. How can we be sure that material that will be dangerous for thousands of years will remain secure? If you were designing such a repository, how would you address this question?

DATA ANALYSIS How Much Waste Do You Produce, and How Much Do You Know How to Manage?

As people become aware of waste disposal problems in their communities, more people are recycling more materials. Some things are easy to recycle, such as newsprint, office paper, or aluminum drink cans. Other things are harder to classify. Most of us give up pretty quickly and throw things in the trash if we have to think too hard about how to recycle them. This exercise asks you to collect your own data by surveying your class about recycling know-how. Go to Connect for details and to assess your data.

CHAPTER

15 Economics and Urbanization

LEARNING OUTCOMES

After studying this chapter, you should be able to answer the following questions:

- How have the size and location of the world's largest cities changed over the past century?
- Define *slum* and *shantytown*, and describe the conditions you might find in them.
- What is urban sprawl? How have automobiles contributed to sprawl?
- What are some principles of smart growth and new urbanism?
- What value do we get from free ecological services?
- What's the difference between GNP and GPI?
- What do we mean by internalizing external costs?

British Columbians are justifiably proud of their environment and concerned about the effects of climate change. The province is using tax policy to move toward a low-carbon future.

©Pierre Leclerc Photography/Getty Images

CASE STUDY

Using Economics to Fight Climate Change

British Columbians are justifiably proud of their spectacular environment, and they want to protect it. One of the greatest challenges they—and all of us—face is global climate change. How can we move our economy to a low-carbon future?

The province has taken a bold and important step in that direction by adopting a tax on carbon fuels. In 2008, British Columbia (BC) passed the first broad-based carbon tax in North America. The idea of a carbon tax is to provide a price incentive to find alternatives, such as investing in efficiency or renewable energy. The tax is small for most fuel consumers, less than 8 cents per liter of gasoline, for example, but it adds up for institutional consumers of fuel.

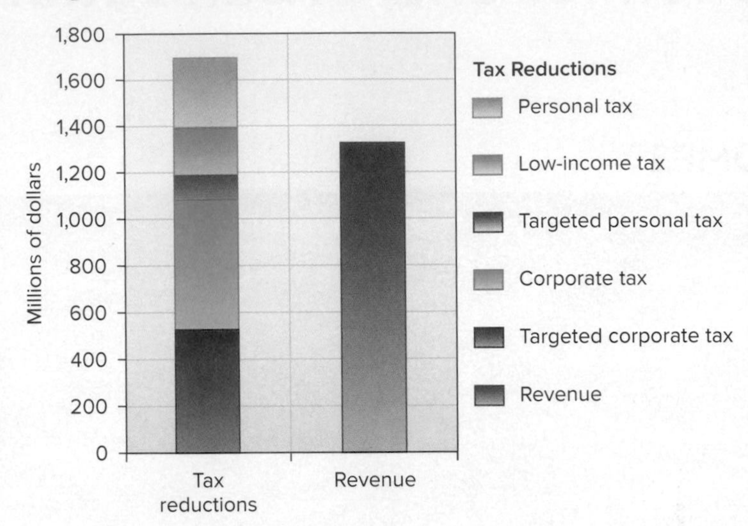

▲ **FIGURE 15.1** Revenues from British Columbia's carbon tax and personal and corporate tax reductions made possible by this revenue. Source: Data from Duke/Nicholas Institute

Opponents have claimed the tax would "destroy jobs and growth," but the opposite has been true. The latest numbers from Statistics Canada show that the carbon tax has been a tremendous environmental and economic success. Since the tax came into effect, fossil fuel use has dropped by 16 percent in BC compared to a 3.5 percent increase in the rest of Canada. At the same time, the BC economy has grown faster and unemployment has been lower than any other province.

The BC carbon tax is designed to be revenue neutral. That is, every penny taken in by the tax is used to reduce other taxes. Roughly 65 percent of the income is used to decrease corporate taxes. About 35 percent goes to lower personal taxes. Half that amount is targeted to low-income families and individuals for whom the carbon tax creates a particular disadvantage. And about half the corporate tax reductions are targeted for industries, such as clean energy, digital media and film, international businesses, and investment capital that the government wants to encourage. The result is that BC has the lowest personal income tax rate in Canada, and one of the lowest corporate rates in North America. Tax reductions have actually been greater than the amount taken in by the carbon tax. The difference is made up by increased taxes in sectors stimulated by tax shifts (fig. 15.1).

Passing this carbon tax was easier in BC than in other places because approximately 95 percent of the province's electricity currently comes from hydroelectric dams, which don't need carbon fuels. Still, most people heat their homes with coal or natural gas, and fuel their vehicles with petroleum-based fuels. The carbon tax started out low (C$10 per metric ton of CO_2) to ease people into the new system, and then rose gradually, reaching C$35 per ton in 2018. It will continue to rise until it reaches C$50 per ton in 2021.

Observing the economic and climate benefits of the BC carbon tax, the rest of Canada has followed suit. A similar federal tax, again starting at C$10 per ton, is slated to increase to C$50 per ton by 2022.

Nearly 60 percent of BC's residents approved of the carbon tax when it was first proposed. Support has varied over time and among constituencies, but most people are aware of the urgency of climate change. They have seen their forests devastated by drought, insect infestations, and increasing forest fires. They know their hydropower depends on winter snowpack and that droughts, heat waves, and storms undermine regional and national economies.

The idea of using tax policy to discourage harmful actions (pollution, climate change) and to reward beneficial sectors (such as renewable energy) is a fundamental idea in economic theory. Economists generally hold that a tax on carbon is the most transparent and efficient approach to reducing emissions. It provides a direct price incentive to find ways to reduce emissions, rather than imposing regulatory rules or indirect incentives. Economists also approve of revenue-neutral taxes, which don't just increase the size of government. This fiscal reform is especially appealing when it may bring a double dividend: reducing pollution and stimulating economic growth in renewable energy and conservation.

There still are many people who object to new taxes of any sort. Even when climate change threatens our economy and life as we know it, the concept of a new tax is difficult for many to swallow. Despite these concerns, most British Columbians have found that their total tax bill has gone down significantly with this measure. And poorer people, as well as those in remote areas, have benefited from programs supported by the tax.

In this chapter we explore two related issues of how we occupy and make decisions about our environment. We examine cities, where most of us live, and the environmental challenges and opportunities associated with the ways we choose to build our cities. We then explore the economic concepts and incentives that shape the decisions we make. Because while concern for environmental quality and biodiversity frequently drive priorities in resource use, final decisions often involve economic questions of costs, benefits, and trade-offs. ■

What kind of world do you want to live in? Demand that your teachers teach you what you need to know to build it.

—PETER KROPOTKIN

15.1 CITIES ARE PLACES OF CRISIS AND OPPORTUNITY

- *The world is becoming more urban, and cities are growing larger.*
- *The largest cities are now in developing countries.*
- *Economic policies can push people toward cities.*

More than half of us now live in cities, and in the next quarter century that proportion will approach two-thirds of all humans. This is a dramatic change from our previous history, in which most people lived by hunting and gathering, farming, or fishing. Since the beginning of the Industrial Revolution about 300 years ago, cities have grown rapidly in both size and power (fig. 15.2). In 1950, 38 percent of the world's population lived in cities; by 2030 that proportion will reach 60 percent (table 15.1).

▼ **FIGURE 15.2** In less than 20 years, Shanghai, China, built Pudong, a new city of 1.5 million residents and 500 skyscrapers on former marshy farmland across the Huang Pu River from the historic city center. This kind of rapid urban growth is occurring in many developing countries. ©Steve Allen/Getty Images

TABLE 15.1 Urban Share of Total Population (Percentage)

	1950	2000	2030*
Africa	18.4	35.0	48.4
Asia	19.3	37.5	56.7
Europe	56.0	71.1	77.5
Latin America	40.0	75.5	83.6
North America	63.9	79.1	84.7
Oceania	32.0	68.3	68.9
World	38.3	46.7	60.4

*Projected.

Source: Data from United Nations Population Division, 2018.

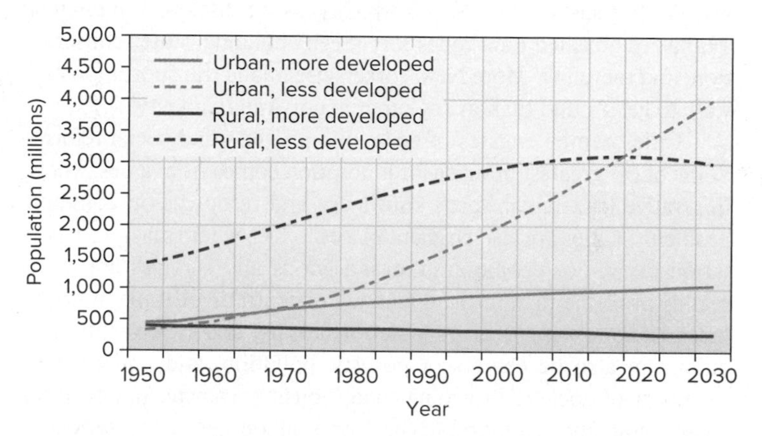

▲ **FIGURE 15.3** Growth of urban and rural populations in more-developed regions and in less-developed regions. Source: United Nations Population Division, World Urbanization Prospects, 2004.

The vast majority of urban growth will occur in less-developed countries (fig. 15.3). Populations in these cities are expanding far

Benchmark Data

Among the ideas and values in this chapter, these are a few worth remembering.

60	Percentage of world population expected to live in cities by 2030
10	Number of world's 10 largest cities in Europe and North America, 1900
0	Number in 2100
$173 trillion	2012 estimate of total biosphere ecological services
$78 trillion	Annual market value of global GDP in 2012
$350,000/ha/yr	Ecosystem services from an "average" hectare of coral reefs
50,000	Coal mining jobs in the U.S. in 2018
1 million	Jobs associated with renewable energy in the U.S. in 2018

faster than infrastructure, such as roads, transportation, housing, water supplies, sewage treatment, and schools, can adapt. Building new infrastructure is especially hard in poor countries, where incomes are low and tax collection is insufficient to support public services. Despite these challenges, cities are also places where innovation occurs. Ideas mix and experimentation happens in urban areas. In cities there arise diverse employment opportunities and new economies, as well as concentrations of poverty. Huge **urban agglomerations** (merging of multiple municipalities) are forming throughout the world. Some have become **megacities** (with populations of over 10 million people). Though these cities pave over vast landscapes and consume large amounts of resources, their per-capita resource use is relatively efficient. Environmental degradation would probably be much worse if that many people were spread across the countryside. Cities, for all their ills, are one of the places where we can learn new ways to live sustainably. New York City, as we discussed in chapter 13, has established new codes for "green" building, water conservation, and recycling. More New Yorkers use public transportation and walk to get around than in any other major American city.

Cities can be engines of economic progress and social reform. Some of the greatest promise for innovation comes from cities, where innovative leaders can focus knowledge and resources on common problems. Cities are also efficient places to live, where mass transportation can move people around and goods and services are more readily available than in the country. Concentrating people in urban areas leaves open space available for farming and biodiversity. But cities can also be centers of poverty, pollution, and undervalued members of society. Providing food, housing, transportation, jobs, clean water, and sanitation to the 2 or 3 billion new urban residents expected to crowd into cities—especially those in the developing world—may be one of the preeminent challenges of this century.

Large cities are expanding rapidly

You can already see the dramatic shift in size and location of big cities. In 1900 only 13 cities in the world had populations over 1 million (table 15.2). All of those cities except Tokyo and Beijing were in Europe or North America.

By 2018 there were more than 525 cities in the world—160 of them in China alone—with more than 1 million residents. And there were 37 megacities with populations above 10 million. Of the 13 largest of these metropolitan areas today, none are in Europe, and only New York City is in a developed country. It's expected that by 2025 at least 1,000 cities will have populations over 1 million, while the number of megacities is expected to grow to 50. Three-fourths of those cities will be in developing countries (fig. 15.4). Determining the exact boundaries of a city, and thus its population, can be difficult. Most cities have a relatively compact historic core. Beyond this are suburbs, exurbs, or satellite cities that are socioeconomically tied to the urban core. The economic reach of megacities can be very large, sometimes extending for hundreds of square kilometers.

China has seen the largest demographic shift in human history. Since the end of collectivized farming and factory work in 1986, at least 500 million people have moved from rural areas into China's 662 main cities. A once largely rural country has become nearly 60 percent urban. It's expected that, by 2025, more than 1 billion Chinese—two-thirds of the population—will be urban and China will have more than 220 cities with at least 1 million residents.

Many fast-growing countries urbanize before they have the infrastructure or economies to support large cities. It took centuries for industrialization and urbanization to occur in Europe and the Americas. Today, we're seeing a doubling of population in poorer countries in 25 years. Most cities in the developing world don't have the infrastructure, employment base, or productivity to manage this growth. These cities develop as a maze of informal settlements with overloaded infrastructure and extreme levels of income inequality. Many of these megacities are projected to reach astonishing sizes by the end of this century (table 15.2). Cities in developing countries could grow to more than 50 million residents. What will the living conditions be in those enormous urban conglomerations?

Consider the case of Niamey, Niger, for example. It may be the largest city of which you've never heard. For centuries Niamey was a sleepy fishing village on the Niger River about 500 mi (800 km) upstream from Timbuktu. When Niger became independent in 1926 Niamey became its capital, even though its population was only a

TABLE 15.2 The World's Largest Urban Areas (Populations in Millions)

1900		2000		2100 (est)	
London, England	6.6	Mexico City, Mexico	26.3	Lagos, Nigeria	88.3
New York, USA	4.2	Sao Paulo, Brazil	24.0	Kinshasa, Congo	83.5
Paris, France	3.3	Tokyo, Japan	17.1	Dar Es Salaam, Tanzania	73.7
Berlin, Germany	2.4	Calcutta, India	16.6	Mumbai, India	67.2
Chicago, USA	1.7	Bombay, India	16.0	Delhi, India	57.3
Vienna, Austria	1.6	New York, USA	15.5	Khartoum, Sudan	56.6
Tokyo, Japan	1.5	Seoul, Korea	13.5	Niamey, Niger	56.1
St. Petersburg, Russia	1.4	Shanghai, China	13.5	Dhaka, Bangladesh	54.2
Philidelphia, USA	1.4	Rio de Janeiro, Brazil	13.3	Kolkata, India	50.3
Manchester, England	1.3	Delhi, India	13.3	Kabul, Afghanistan	50.2

Source: Data from T. Chandler, *Three Thousand Years of Urban Growth*, 1974; UN Population Division, *The World's Cities* in 2016.

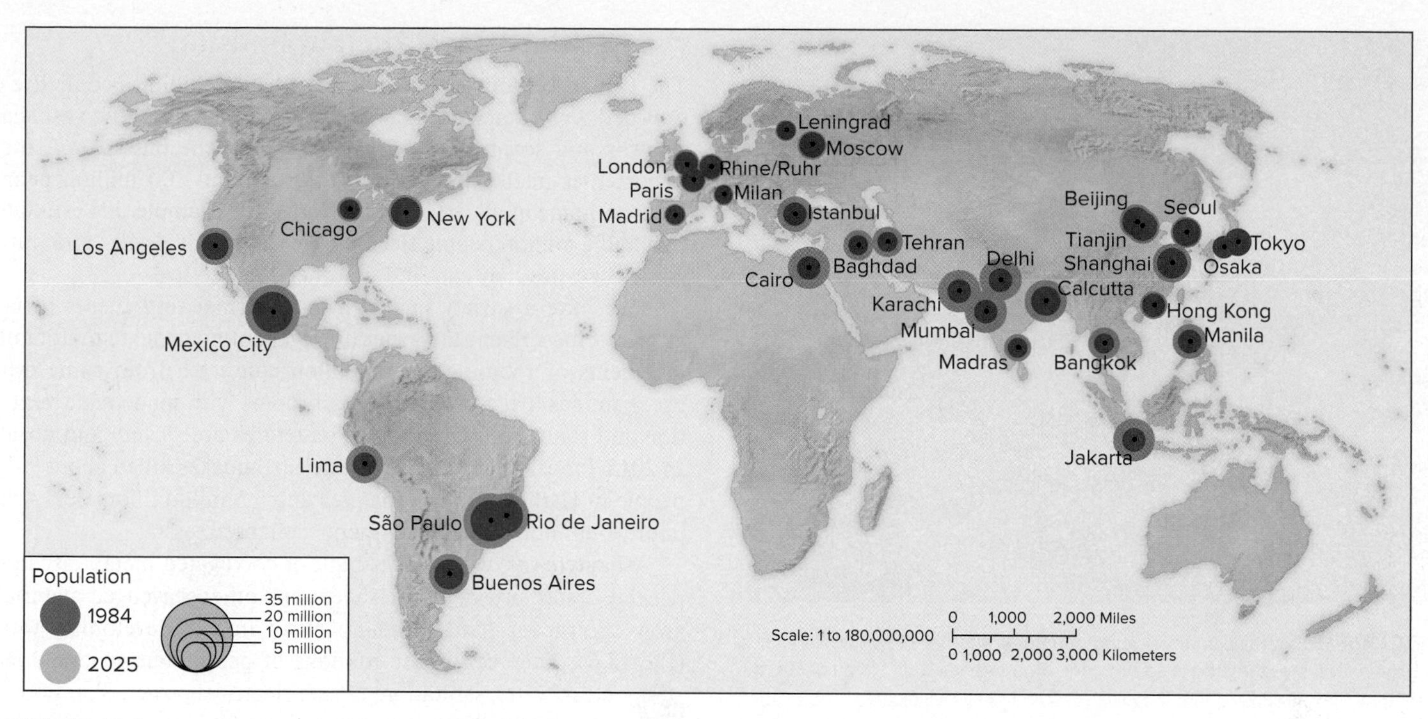

▲ **FIGURE 15.4** By 2025, it's expected that more than 1,000 cities will have populations over 1 million, and at least 50 megacities will have more than 10 million residents. Three-fourths of the world's largest cities will be developing countries that already have trouble housing, feeding, and employing their people.

few thousand people. The population grew slowly to about 30,000 by 1960. Severe droughts in the 1970s and 1980s, together with civil wars and famine, triggered a mass migration of rural people to the city. By 2000 Niamey had grown to a vast tent city of nearly 1 million migrants. Although there are rich neighborhoods, most of the city sprawls across the desert in impoverished chaos. Many people live in crowded refugee camps and transient settlements around the outskirts of the city. The great majority of dwellings aren't connected to piped water or sanitation. The city's dusty streets are choked with traffic, and food shortages are perennial.

By 2100 Niamey is expected to have more than 50 million residents. The city has practically no rain from mid-October to April and it is one of the hottest major cities on the planet. Temperatures reach 38°C (100°F) at least some days in every month in the year. One of the poorest countries in the world, Niger is currently next to last on the UN human development list. How will a city of such huge size with so little resources provide for such a huge population?

Immigration is driven by push and pull factors

People migrate to cities for many reasons. In China over the past 20 years—or in America during the twentieth century—mechanization eliminated jobs and drove people off the land. Where cropland is owned by a minority of wealthy landlords, as is the case in many developing countries, subsistence farmers are often ejected when new cash crops or cattle grazing become economically viable. Many people also move to the city for the opportunities and independence offered there. Cities offer jobs, better housing, entertainment, and freedom from the constraints of village traditions. Possibilities exist in the city for upward social mobility, prestige, and power not ordinarily available in the country. Cities support specialization in arts, crafts, and professions for which markets don't exist elsewhere.

Government policies often favor urban over rural areas in ways that both push and pull people into cities. Developing countries commonly spend most of their budgets on improving urban areas (especially around the capital city, where leaders live). This gives the major cities a virtual monopoly on new jobs, housing, education, and finance, all of which bring in rural people searching for a better life. Lima, for example, has only 20 percent of Peru's population but has 50 percent of the national wealth, 60 percent of the manufacturing, 65 percent of the retail trade, 73 percent of the industrial wages, and 90 percent of all banking in the country. Similar statistics pertain to many national capitals.

Congestion, pollution, and water shortages plague many cities

First-time visitors to a megacity—particularly in a developing country—often are overwhelmed by the immense crush of pedestrians and vehicles of all sorts jostling for space in the streets. The noise, congestion, and confusion of traffic make it seem suicidal to venture onto the street. Delhi, India, for instance, is one of the most densely populated cities in the world (fig. 15.5). Traffic is chaotic and often grid-locked. People often spend 3 or 4 hours each way commuting to work from outlying areas.

In 2018, the World Health Organization (WHO) declared that Delhi had the worst air quality of any of the 1,600 major cities it surveyed around the world. Beijing, China, is infamous for bad air quality, but, on average, Delhi's air is twice as bad as Beijing's.

▲ **FIGURE 15.5** Traffic in developing countries, such as New Delhi, India, has grown far faster than the road network. Monumental traffic jams occur at almost any time of day. ©Igor Ovsyannykov/EyeEm/Getty Images

India has the world's highest death rate from chronic respiratory diseases and asthma, according to the WHO. It's estimated that 2.5 million people die every year in India from diseases related to air pollution (see chapter 10).

Few cities in developing countries can afford to build modern waste treatment systems for their rapidly growing populations. The World Bank estimates that only one-third of urban residents in developing countries have satisfactory sanitation services. In Egypt, Cairo's sewer system was built about 50 years ago to serve a population of 2 million people. It's now being overwhelmed by more than five times that many residents. Less than 1 percent of India's 500,000 towns and villages have even partial sewage systems or water treatment facilities.

It's often difficult to find clean drinking water for urban areas. Cape Town, South Africa, for example, is running out of water. Drought, extravagant use patterns, rapid growth, and poor planning threaten to leave the city almost entirely dry. What that means for a major world city is hard to imagine. Part of the problem is climate change: Between 2016 and 2018, the city experienced a severe drought that should occur only once in a millennium. But severe drought is expected to be increasingly common. At the same time, the city has not done enough to invest in water storage and conservation efforts.

In other places, water is available but is too polluted for most uses. In China, for example, experts report that 70 percent of the country's surface water is so polluted by industrial toxins, human waste, and agricultural chemicals that it is unsuited for human consumption. Worldwide, according to the United Nations, at least 1 billion people don't have safe drinking water, and twice that many lack adequate sanitation. Scarcity of clean water is one of our greatest environmental health crises.

Many cities lack sufficient housing

The United Nations estimates that at least 1 billion people live in crowded, unsanitary slums of the central cities or in the vast shantytowns and squatter settlements that ring the outskirts of most major cities in the developing world. Around 100 million people have no home at all. In Mumbai, India, for example, it's estimated that half a million people sleep on the streets, sidewalks, and traffic circles because they can find no other place to live.

We have a variety of terms for informal settlements in fast-growing cities. **Slums** are generally legal but inadequate multifamily tenements or rooming houses, often converted from some other use. Families live crowded in small rooms with inadequate ventilation and sanitation. Often these structures are rickety and unsafe. In 2015, for example, 7.8-magnitude earthquakes killed about 9,000 people in Kathmandhu, Nepal, leaving 3.5 million homeless when poorly built homes and apartments collapsed.

Shantytowns, with shacks built of corrugated metal, discarded packing crates, brush, plastic sheets, and other scavenged materials, grow on vacant land in many cities in the developing world (fig. 15.6). They can house millions of people but generally lack clean water, sanitation, or safe electrical power. Shantytowns are usually illegal, but they quickly fill in the empty space in towns where squatters can build shelters close to jobs. With little or no public services, such as waste collection and sanitation, shantytowns often fill with trash and debris. Many governments try to clean out illegal settlements by torching or bulldozing the huts and sending riot police to drive out residents, but people either move back in or relocate to another shantytown. In 2005 the government of Zimbabwe destroyed the homes of some 700,000 people in

▲ **FIGURE 15.6** Homeless people have built shacks along this busy railroad track in Jakarta. It's a dangerous place to live, with many trains per day using the tracks, but for the urban poor there are few other choices. ©Helga Leitner

shantytowns around the capital of Harare. Families were evicted in the middle of the night during the coldest weather of the year, often with only minutes to gather their belongings. President Robert Mugabe justified this blitzkrieg as necessary to control crime, but critics claimed it was mainly to remove political opponents.

Two-thirds of the population of Kolkata are thought to live in unplanned squatter settlements, and nearly half the 25 million residents of Mexico City occupy the unauthorized *colonias* around the city. Many shantytowns occupy the most polluted, dangerous parts of cities where no one else wants to live. In Bhopal, India, and Mexico City, for example, squatter settlements were built next to deadly industrial sites. In Brazil, shantytowns called *favelas* perch on steep hillsides unwanted for other building. As desperate and inhumane as conditions are, though, many people do more than merely survive there. They work hard, raise families, educate their children, and often improve their living standard little by little.

Many countries are recognizing that the only way they can house all their citizens is to cooperate with shantytown dwellers. Recognizing land rights, providing financing for home improvements, and supporting community efforts to provide water, sewers, and power can greatly improve living conditions for many poor people.

15.2 URBAN PLANNING

- *Better environmental planning is good for health as well as our environment.*
- *Aging cities have many opportunities to rebuild and redesign.*
- *Sustainable urbanism emphasizes renewable energy, public transportation, and compact growth.*

How can we live together in cities in ways that are environmentally sound, socially just, and economically sustainable? Often infrastructure and planning are central to this question. The ways we organize transportation, housing, and public space shape the ways we experience a city. The ways we develop this infrastructure also largely controls our environmental impact, including our carbon footprint, consumption of water and materials, destruction of biodiversity, and other factors.

Transportation is crucial in city development

Getting people around within a large urban area has become one of the most difficult problems that many city officials face. Freeway construction, which began in America in the 1950s, allowed people to move out into the countryside. Cities that were once compact began to spread over the landscape, consuming space and wasting resources. This pattern of development is known as **sprawl.** While there is no universally accepted definition of the term, sprawl generally includes the characteristics outlined in table 15.3.

In most American metropolitan areas, the bulk of new housing is in large, tract developments that leapfrog out beyond the city edge in a search for inexpensive rural land with few restrictions on land use or building practices (fig. 15.7).

The U.S. Department of Housing and Urban Development calculates that urban sprawl consumes some 200,000 ha (roughly 500,000 acres) of farmland and open space every year. Although the price of raw farmland generally is less than comparable urban property, there are external costs in the form of new roads, sewers, water mains, power lines, schools, shopping centers, and other infrastructure required by this low-density development.

Because many Americans live far from where they work, shop, or recreate, they consider it essential to own a private automobile. The average U.S. driver spends 443 hours per year behind a steering wheel, or the equivalent of one full 8-hour day per week in an automobile. The freeway system was designed to allow drivers to travel

TABLE 15.3 | Characteristics of Urban Sprawl

1. Unlimited outward extension
2. Low-density residential and commercial development
3. Leapfrog development that consumes farmland and natural areas
4. Fragmentation of power among many small units of government
5. Dominance of freeways and private automobiles
6. No centralized planning or control of land uses
7. Widespread strip-malls and "big-box" shopping centers
8. Great fiscal disparities among localities
9. Reliance on deteriorating older neighborhoods for low-income housing
10. Decaying city centers as new development occurs in previously rural areas

Source: Excerpted from "Some Realities about Sprawl and Urban Decline" by Anthony Downs, The Brookings Institution, *Housing Policy Debate*, Volume 10, Issue 4, p. 956.

▲ **FIGURE 15.7** Huge houses on sprawling lots consume land, alienate us from our neighbors, and make us ever more dependent on automobiles. They also require a lot of watering and lawn mowing!

KEY CONCEPTS

What makes a city green?

©Mary Ann Cunningham

KC 15.1

Efficiency. Over half of humans now live in cities. Environmental scientists have often criticized cities for expanding into farmland and for their tremendous consumption of energy, water, food, concrete, and land. But the environmental cost per person is usually lower for urban living than for suburban or rural living, especially in wealthy countries. Because they are compact, cities require fewer miles of roads, water and sewer lines, less heating, and fewer private cars per household. Because distances are shorter, roads and utility infrastructure are shared, apartments or row houses share heat, and public transportation reduces the need for driving to work.

Polluted cities can be unhealthy, but well-organized cities can provide cultural resources and preserve environmental resources in many beneficial ways.

Here are ten features that make cities healthy for people and the environment.

KC 15.2

©William P. Cunningham

1. Public transit
High-density areas can afford to support reliable, efficient transit systems, where many riders share the cost of getting around, such as this bus-rapid transit system in Curitiba, Brazil. Public transit uses far less space, energy, and materials than does private travel.

©Mary Ann Cunningham

KC 15.3a

KC 15.3b

©Mary Ann Cunningham

2. Safe walking and bike routes
Freedom from dependence on cars increases mobility for young people, old people, and others without cars. Cities with separated walkways and bikeways are friendly for children and families; they also provide exercise and save money.

KC 15.4

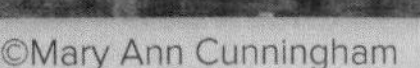

©Mary Ann Cunningham

3. Compact building
A compact urban design greatly increases efficiency of land use, reduces transit distances, and increases heating or cooling efficiency, as buildings share walls. Reduced dependence on cars, and car sharing, can help control the problem of parking shortages.

◀ Amsterdam's row houses give the city its historic identity as well as efficiency.

4. Mixed-use planning
Integrating housing with shopping, entertainment, and office space provides jobs and services where people live. These neighborhoods can encourage walking and build community, as people spend less time in travel to shopping and work.

A used bookstore and cafe share space with housing in the historic city center of Trondheim, Norway. ▶

KC 15.5

©William P. Cunningham

©Mary Ann Cunningham

KC 15.6

10. Farmland conservation
Sprawling suburbs gobble up farmland, woodlands, and wetlands. This is the fastest type of land use change in most developing countries. Compact cities minimize destruction of farmland, habitat, recreational space, and watersheds.

KC 15.8

©David Frazier/Photolibrary/Getty Images

©William P. Cunningham

KC 15.7

9. Local food
Local farm economies are more viable if farmers can sell directly to consumers—something that is much easier where there are lots of buyers in one place. Cities have become an essential income source for many produce farmers.

▲ The St. Johnsbury, Vermont, farmers' market provides fresh, locally grown food for urban residents.

©Mary Ann Cunningham

KC 15.9

8. Energy efficiency
Alternative energy is easier to use right at the source. Rooftop solar energy, district heating, and other strategies aid efficiency.

▲ This biomass-burning plant in Copenhagen, and others like it, provides nearly all heating for Denmark's major cities.

©Mary Ann Cunningham

KC 15.10

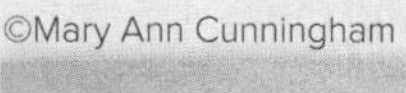

7. Green infrastructure
New techniques moderate the impact of impervious surfaces, including permeable pavement, green roofs, and better building design.

▲ A "green" parking lot in Chengdu, China, supports both traffic and vegetation, allowing rainfall to percolate into the ground.

6. Recycling programs
Recycling collection is easiest where transportation is minimal and where recyclable materials are abundant.

KC 15.11

©Josie Elias/Getty Images

◀ These bins in Kuala Lumpur, Malaysia, accept all kinds of recyclables.

5. Green space
Recreational space has physical and emotional benefits for urban residents. Living vegetation and soils cool the local microclimate, store nutrients and moisture, and provide habitat for birds and other wildlife.

◀ Here a visitor watches skaters in New York's Central Park.

CAN YOU EXPLAIN?

1. **What factors can make per-capita energy use low in urban areas?**
2. **Which of the green factors listed would be easiest to enhance where you live? Why?**
3. **Which do you find most and least appealing? Why?**

▲ **FIGURE 15.8** Freeways give us the illusion of speed and privacy, but they consume land, encourage sprawl, and create congestion as people move farther from the city to get away from traffic and then have to drive to get anywhere. ©Comstock Images/Alamy Stock Photo

at high speeds from source to destination without ever having to stop (fig. 15.8). As more and more vehicles clog highways, however, the reality is far different.

Altogether, the average driver spends about 100 hours per year (2.5 weeks of work) in bumper-to-bumper traffic, and congestion costs the United States $78 billion per year in wasted time and fuel. Some people argue that the existence of traffic jams in cities shows that more highways are needed. Often, however, building more traffic lanes simply encourages more people to drive farther and put more cars on the road. Meanwhile, about one-third of Americans are too young, too old, or too poor to drive. For these people, car-oriented development causes isolation and makes daily tasks like grocery shopping difficult. Parents spend long hours transporting young children. Teenagers and aging grandparents are forced to drive, often presenting a hazard on public roads.

It is possible to build cities for people, rather than for cars. Most European urban areas have good mass transit systems that have allowed them to preserve historic city centers and remain relatively compact while avoiding the sprawl engendered by an American-style freeway system. Many American cities are now rebuilding the public transportation systems that were abandoned in the 1950s (fig. 15.9).

A famous example of successful mass transit is found in Curitiba, Brazil. High-speed, bi-articulated buses, each of which can carry 270 passengers, travel on dedicated roadways closed to all other vehicles (Key Concepts, p. 362). This bus-rapid transit system is linked to 340 feeder routes extending throughout the city. Everyone in the city is within walking distance of a bus stop that has frequent, convenient, affordable service. Curitiba's buses carry some 1.9 million passengers per day, or about three-quarters of all personal trips within the city. Working with existing roadways for the most part, the city was able to construct this system for one-tenth the cost of a light-rail system or freeway system, and one-hundredth the cost of a subway.

▲ **FIGURE 15.9** Many American cities are now restoring light-rail systems that were abandoned in the 1950s when freeways were built. Light rail is energy efficient and popular, but it can cost up to $100 million per mile ($60 million per kilometer). ©William P. Cunningham

Rebuilding cities

There's an opportunity—some would say a necessity—to rebuild and redesign many aging cities in America's industrial "rust belt." A declining manufacturing base coupled with middle-class flight to the suburbs has hollowed-out many cities. Detroit's population, for example, has declined from 1.85 million in the 1950s to about 700,000 today. At least 100,000 abandoned homes and businesses have been demolished (fig. 15.10). Whole blocks are now uninhabited. It costs the city at least $3 million per year just to mow the weeds on empty lots. At least 40 mi^2 (104 km^2) of land inside the city limits are vacant. Urban pioneers are starting farms and

▲ **FIGURE 15.10** Many old, industrial cities in America's "rust belt" are tearing down tens of thousands of abandoned, derelict houses. This opens up an opportunity for redevelopment with gardens, green belts, diverse flexible housing, walkable neighborhoods, and other features of smart growth. ©Aaron Roeth/Aaron Roeth Photography

establishing a subsistence, barter-based economy. And the city plans to tear down at least 10,000 more homes and resettle residents into compact, more easily served neighborhoods.

The question is, how will those new urban spaces be designed? Are there alternative mixed uses that could help revitalize cities? One widely noted project in Detroit was initiated by John Hantz, a wealthy money manager who has invested millions of dollars in an urban tree farm on abandoned residential blocks. Thousands of young trees have replaced empty houses and yards overrun with weeds. There has been controversy over land sales to a wealthy white financier, but supporters point out that a lot of capital is needed to clean up large areas of collapsing and dangerous structures. The hope is that clearing some neighborhoods will make others more valuable, and trees can eventually be sold for local revenue.

Empty business and industrial districts are centers of revitalization in many cities. Denver, for example has turned two square blocks in Lower Downtown into a human-scale neighborhood with walking streets, bike paths, rooftop gardens, solar panels, and energy retrofits that have cut energy consumption in half. The area has become a nightlife epicenter drawing crowds of shoppers, diners, and partygoers. Another successful revitalization project is on Manhattan's West Side in New York City. An abandoned elevated rail line running along 10th Avenue from 14th Street to 34th Street has been converted into a linear park called the High Line. Landscaping, public gathering spaces, food stands, performance areas, and a 2-mile (3.2 km) promenade turn this into welcome open space in the crowded city.

We can make our cities more livable

Are there alternatives to unplanned sprawl and wasteful resource use? One option proposed by many urban planners is **smart growth,** which makes effective use of land resources and existing infrastructure by encouraging in-fill development that avoids costly duplication of services and inefficient land use (table 15.4). Smart growth aims to provide a mix of land uses to create a variety of affordable housing choices and opportunities. It also attempts to provide a variety of transportation choices, including pedestrian-friendly neighborhoods. This approach to planning also seeks to maintain a unique sense of place by respecting local cultural and natural features.

By making land use planning open and democratic, smart growth makes urban expansion fair, predictable, and cost-effective.

TABLE 15.4 Goals for Smart Growth
1. Create a positive self-image for the community.
2. Make the downtown vital and livable.
3. Alleviate substandard housing.
4. Solve problems with air, water, toxic waste, and noise pollution.
5. Improve communication between groups.
6. Improve community member access to the arts.

Source: Data from Vision 2000, Chattanooga, Tennessee.

All stakeholders are encouraged to participate in creating a vision for the city and to collaborate with rather than confront each other. Goals are established for staged and managed growth in urban transition areas with compact development patterns. This approach is not opposed to growth. It recognizes that the goal is not to block growth but to channel it to areas where it can be sustained over the long term. Smart growth strives to enhance access to equitable public and private resources for everyone and to promote the safety, livability, and revitalization of existing urban and rural communities.

Smart growth protects environmental quality. It tries to reduce traffic and to conserve farmlands, wetlands, and open space. As cities grow and transportation and communications enable more community interaction, the need for regional planning becomes greater and more pressing. Community and business leaders must make decisions based on a clear understanding of regional growth needs and how infrastructure can be built most efficiently and for the greatest good.

Sustainable urbanism incorporates smart growth

Rather than abandon the cultural history and infrastructure investment in existing cities, many architects and urban planners are attempting to redesign metropolitan areas to make them more appealing, efficient, and livable. The aim is to focus urban design on improving the urban experience for people, encouraging them to live in the city and contribute to civic life, rather than commute to far-flung suburbs.

Some urban designers focus on a neotraditionalist approach, these designers attempt to recapture some of the best features of small towns and livable cities of the past. They prioritize modest-sized urban communities and neighborhoods that integrate houses, offices, shops, and civic buildings. Ideally, no house should be more than a 5-minute walk from a neighborhood center with a convenience store, a coffee shop, a bus stop, and other amenities. A mix of apartments, townhouses, and detached houses in a variety of price ranges ensures that neighborhoods will include a diversity of ages and income levels. Other "sustainable urbanists" are concerned less with city size and more with ensuring sustainable transportation, energy efficiency, and urban green space. These are some of the principles of many of these planners:

- Maintain greenbelts in and around cities. These provide recreational space and promote efficient land use, as well as help improve air quality, moderate temperature, and reduce water pollution.
- Determine in advance where development will take place. This protects property values and prevents chaotic development. Planning can also protect historical sites, agricultural resources, and ecological services of wetlands, clean rivers, and groundwater replenishment.
- Locate everyday shopping and services so people can meet daily needs with greater convenience, less stress, less automobile dependency, and less use of time and energy (fig. 15.11). This might be accomplished by encouraging small-scale commercial development in or close to residential areas.

▶ **FIGURE 15.11** This walking street in Queenstown, New Zealand, provides opportunities for shopping, dining, and socializing in a pleasant outdoor setting. ©William P. Cunningham

- Encourage walking and bicycling for many local trips now performed in private automobiles. Creating special traffic lanes, reducing the number or size of parking spaces, and closing shopping streets to cars might encourage such alternatives.
- Promote more diverse, flexible housing as an alternative to conventional detached, single-family houses. In-fill building between existing houses saves energy, reduces land costs, and might help provide a variety of living arrangements, including cooperatives and co-housing, where households share yards, public space, and community.
- Make cities more self-sustainable by growing food locally, recycling wastes and water, using renewable energy sources, reducing noise and pollution, and creating a cleaner, safer environment. Encourage community gardening (fig. 15.12). Reclaimed inner-city space or a greenbelt of agricultural and forestland around the city provides food and open space, and it contributes valuable ecological services, such as purifying air, supplying clean water, and protecting wildlife habitat and recreation land.
- Equip buildings with "green roofs" or rooftop gardens that improve air quality, conserve energy, reduce stormwater runoff, reduce noise, and help reduce urban heat island effects. Intensive gardens can include large trees, shrubs, and flowers, and may require regular maintenance (fig. 15.13). Extensive gardens require less soil, add less weight to the building, and usually have simple plantings of prairie plants or drought-resistant species, such as sedum, that require minimum care. They can last twice as long as conventional roofs. In Europe more than 1 million m^2 of green roofs are installed every year. Urban roofs are also a good place for solar collectors or wind turbines.
- Plan cluster housing, or open-space zoning, which preserves at least half of a subdivision as natural areas, farmland, or other forms of open space. Studies have shown that people who move to the country don't necessarily want to live miles from the nearest neighbor; what they most desire is long views across an interesting landscape and an opportunity to see wildlife. By carefully clustering houses on smaller lots, a conservation subdivision can provide the same number of buildable lots as a conventional subdivision and still preserve 50 to 70 percent of the land as open space (fig. 15.14). This not only reduces development costs (less distance to build roads, lay telephone lines, sewers, power cables, and so on) but also helps foster a greater sense of community among new residents.
- Preserve urban habitat. It can make a significant contribution toward saving biodiversity as well as improving mental health and giving us access to nature.

▲ **FIGURE 15.12** Many cities have large amounts of unused open space that could be used to grow food. Residents often need help decontaminating soil and gaining access to the land. ©William P. Cunningham

▲ **FIGURE 15.13** This award-winning green roof on the Chicago City Hall is functional as well as beautiful. It reduces rain runoff by about 50 percent and keeps the surface as much as 30°F cooler than a conventional roof on hot summer days. ©Roofscapes, Inc. Used by permission; all rights reserved.

▲ **FIGURE 15.14** Jackson Meadows, an award-winning cluster development near Stillwater, Minnesota, groups houses at sociable distances and preserves surrounding open space for walking, gardening, and scenic views from all houses. ©William P. Cunningham

These planning principles aren't just a matter of aesthetics. Dr. Richard Jackson, former director of the National Center for Environmental Health in Atlanta, points out a strong association between urban design and our mental and physical health. As our cities have become ever more spread out and impersonal, we have fewer opportunities for healthful exercise and socializing. Chronic diseases, such as cardiovascular diseases, asthma, diabetes, obesity, and depression, are becoming the predominant health concerns in the United States.

"Despite common knowledge that exercise is healthful," Dr. Jackson says, "fewer than 40% of adults are regularly active, and 25% do no physical activity at all. The way we design our communities makes us increasingly dependent on automobiles for the shortest trip, and recreation has become not physical but observational." Long commutes and a lack of reliable mass transit and walkable neighborhoods mean that we spend more and more time in stressful road congestion. "Road rage" isn't imaginary. Every commuter can describe unpleasant encounters with rude drivers. Urban design that offers the benefits of more walking, more social contact, and surroundings that include water and vegetation can provide healthful physical exercise and psychic respite.

15.3 ECONOMICS AND RESOURCE MANAGEMENT

- *Ecological economics incorporates ecological principles to resource management.*
- *Ecosystem services regulate, supply, provide, or produce resources we need.*
- *Resources are often successfully managed by cooperative agreements among users.*

Like many of our environmental issues, improving urban conditions will ultimately be decided by economics and policy decisions. We'll discuss policy in chapter 16. In the next part of this chapter, we'll review some of the principles of environmental economics.

Our definitions of resources influence how we use them

To understand the problems and promise of sustainability, you need to understand the different kinds of resources we use. The way we treat resources depends largely on how we view and define them. **Classical economics,** developed in the 1700s by philosophers such as Adam Smith (1723–1790) and Thomas Malthus (1766–1834), assumes that natural resources are finite—that resources such as oil, gold, water, and land exist in fixed amounts. According to this view, as populations grow, scarcity of these resources reduces quality of life, increases competition, and ultimately causes populations to fall again. In a free market, where buyers and sellers make free, independent decisions to buy and sell, the price of a commodity depends on the available supply (it's cheap when plenty is available) and the demand for it (buyers pay more when they must compete to get the resource). Perhaps the purest expression of this system is a farmers' market, where the price of goods is determined primarily by supply and demand (fig. 15.15).

The nineteenth-century economist John Stuart Mill agreed that most resources are finite, but he developed the idea of a **steady-state economy.** Rather than boom-and-bust cycles of population and resource use, as envisioned by Malthus, Mill proposed that economies can achieve an equilibrium of resource use and production. Intellectual and moral development continues, he argued, once this stable, secure state is achieved.

Neoclassical economics, developed in the nineteenth century, expanded the idea of resources to include labor, knowledge, and capital. Labor and knowledge are resources because they are necessary to create goods and services; they are not finite because every new person can add more labor and energy to an economy. **Capital** is any form of wealth that contributes to the production of more wealth. Money can be invested to produce more money. Mineral

▲ **FIGURE 15.15** A farmers' market is a good example of classical economics. When supplies are abundant, prices fall, and farmers don't plant those crops. When there's a shortage of a particular commodity, prices rise, and farmers will work to bring more of that crop to market. ©William P. Cunningham

resources can be developed and manufactured into goods that return more money. Economists distinguish several kinds of capital:

1. Natural capital: goods and services provided by nature
2. Human capital: knowledge, experience, human enterprise
3. Manufactured (built) capital: tools, buildings, roads, technology

To this list some theorists would add social capital, the shared values, trust, cooperation, and organization that can develop in a group of people but cannot exist in one individual alone.

Because the point of capital is the production of more capital (that is, wealth), neoclassical economics emphasizes the idea of growth. Growth results from the flow of resources, goods, and services (fig. 15.16). Continued growth is always necessary for continued prosperity, according to this view. Natural resources contribute to production and growth, but they are not critical supplies that limit growth. They are not limiting because resources are considered to be interchangeable and substitutable. Neoclassical economics predicts that as one resource becomes scarce, a substitute will be found.

Because production of wealth is central to neoclassical economics, an important measure of growth and wealth is consumption. If a society consumes more oil, minerals, and food, it is presumably becoming wealthier. This idea has extended to the idea of *throughput,* the amount of resources a society uses and discards. More throughput is a measure of greater consumption and greater wealth, according to this view. Throughput is commonly measured in terms of **gross national product (GNP),** the sum of all products and services bought and sold in an economy. Because GNP includes activities of offshore companies, economists sometimes prefer **gross domestic product (GDP),** which more accurately reflects the local economy by accounting for only those goods and services bought and sold locally.

Natural resource economics extends the neoclassical viewpoint to treat natural resources as important waste sinks (absorbers), as well as sources of raw materials. Natural capital (resources) is considered more abundant, and therefore cheaper, than built or human-made capital.

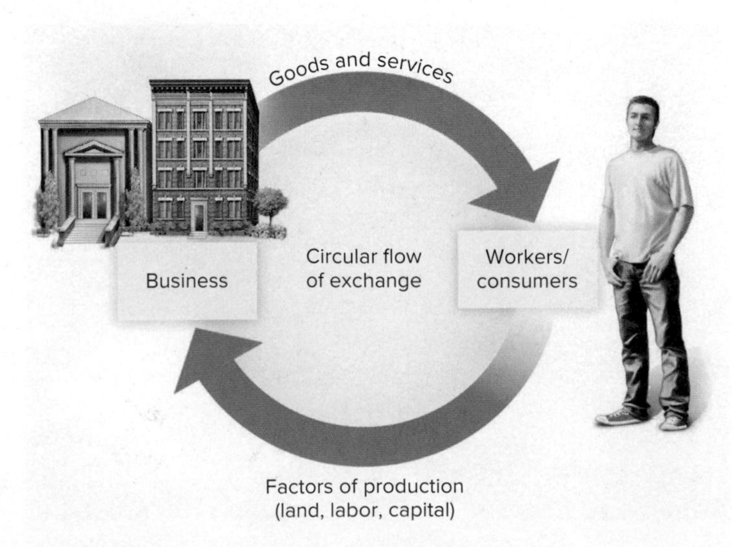

▲ **FIGURE 15.16** The neoclassical model of the economy focuses on the flow of goods, services, and factors of production (land, labor, capital) between businesses and individual workers and consumers. The social and environmental consequences of these relationships are irrelevant in this view.

TABLE 15.5 | **Important Ecological Services**

We depend on our environment to continually provide

1. A regulated global energy balance and climate; chemical composition of the atmosphere and oceans; water catchment and groundwater recharge; production and recycling of organic and inorganic materials; maintenance of biological diversity.
2. Space and suitable substrates for human habitation, crop cultivation, energy conversion, recreation, and nature protection.
3. Oxygen, fresh water, food, medicine, fuel, fodder, fertilizer, building materials, and industrial inputs.
4. Aesthetic, spiritual, historic, cultural, artistic, scientific, and educational opportunities and information.

Source: Data from R. S. de Groot, *Investing in Natural Capital*, 1994.

Ecological economics incorporates principles of ecology

Ecological economics applies ecological ideas of system functions and recycling to the definition of resources. This school of thought also recognizes efficiency in nature, and it acknowledges the importance of ecosystem functions for the continuation of human economies and cultures. In nature, one species' waste is another's food, so that nothing is wasted. We need an economy that recycles materials and uses energy efficiently, much as a biological community does. Ecological economics also treats the natural environment as part of our economy, so that natural capital becomes a key consideration in economic calculations. Ecological functions, such as absorbing and purifying wastewater, processing air pollution, providing clean water, carrying out photosynthesis, and creating soil, are known as **ecological services** (table 15.5). These services are free: We don't pay for them directly (although we often pay indirectly when we suffer from their absence). Therefore, they are often excluded from conventional economic accounting, a situation that ecological economists attempt to rectify (fig. 15.17).

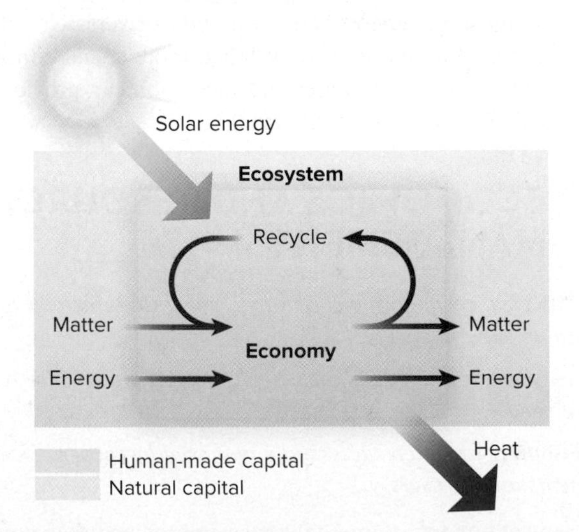

▲ **FIGURE 15.17** Ecological economics includes services such as recycling and resource provision and economic accounting. An effort is made to internalize, rather than externalize, natural resources and services. Source: Herman Daly in A. M. Jansson et al., *Investing in Natural Capital*. ISEE.

FIGURE 15.18 Living organisms are unique resources in that they reproduce themselves indefinitely, and yet, once lost through overexploitation or habitat destruction, they will probably never be replaced. ©Robert Brown/DesignPics

Many ecological economists also promote the idea of a steady-state economy. As with John Stuart Mill's original conception of steady states, these economists argue that economic health can be maintained without constantly growing consumption and throughput. Instead, efficiency and recycling of resources can allow steady prosperity where there is little or no population growth. Low birth rates and death rates (like *K*-adapted species; see chapter 3), political and social stability, and reliance on renewable energy would characterize such a steady-state economy. Like Mill, these economists argue that human and social capital—knowledge, happiness, art, life expectancies, and cooperation—can continue to grow even without constant expansion of resource use.

Both ecological economics and neoclassical economics distinguish between renewable and nonrenewable resources. **Nonrenewable resources** exist in finite amounts: Minerals, fossil fuels, and groundwater that recharges extremely slowly are all fixed, at least on a human time scale. **Renewable resources** are naturally replenished and recycled at a fairly steady rate. Fresh water, living organisms, air, and food resources are all renewable (fig. 15.18).

These categories are important, but they are not as deterministic as you might think. Nonrenewable resources, such as iron and gold, can be extended through more efficient use: Cars now use less steel than they once did, and gold is mixed in alloys to extend its use. Substitution also reduces demand for these resources: Car parts once made of iron are now made of plastic and ceramics; copper wire, once stockpiled to provide phone lines, is now being replaced with cheap, lightweight fiber-optic cables made from silica (sand). Recycling also extends supplies of nonrenewable resources. Aluminum, platinum, gold, silver, and many other valuable metals are routinely recycled now, further reducing the demand for extracting new sources.

The main limit to recycling is usually the low cost of extracting new resources compared with collecting used materials. Recoverable sources of nonrenewable resources are also expanded by technological improvements. New methods make it possible to mine very dilute metal ores, for example. Gold ore of extremely low concentrations is now economically recoverable—that is, you can make money on it—even though this greater efficiency, more discoveries, and resource substitution can make the price of such metals fall. Scarcity of resources, seen by classical economists as the trigger for conflict and suffering, can actually provide the impetus for much of the innovation that leads to substitution, recycling, and efficiency.

On the other hand, renewable resources can become exhausted if they are managed badly. This is especially apparent in biological resources, such as the passenger pigeon, American bison, and Atlantic cod. All these species once existed in extraordinary numbers, but within a few years, each was brought to the brink of extinction (or eliminated entirely) by overharvesting.

Scarcity can lead to innovation

Are we about to run out of essential natural resources? In the 1970s a team of scientists from MIT, headed by Donnela Meadows, created a complex computer model of the world economy. They examined different rates of resource depletion, growing population, pollution, and industrial output. Most of their models predicted that our population will boom and bust as we run out of natural resources and drown in pollution, more or less like the population dynamics models we discussed in chapter 3. A typical run of their computer program from *The Limits to Growth* (1972) is shown in figure 15.19.

The prospect that unchecked population growth and overconsumption would lead to a crisis has had a powerful impact on environmentalists for many years. However, an underlying assumption of this view is the idea that the natural resources and ecological services on which we depend are irreplaceable. Ultimately, of course, many of them are irreplaceable, but that ultimate limit may be far off.

Many economists have criticized this model because it underestimated technical progress and innovation that could mitigate the effects of scarcity. They point out that scarcity can stimulate research and development that result in new processes and materials that extend our resources significantly.

As we discussed in chapter 4, Thomas Malthus warned two centuries ago that food production couldn't continue to keep up with population growth, and that starvation, poverty, crime, and war were inevitable unless we slowed birth rates drastically. Malthus didn't

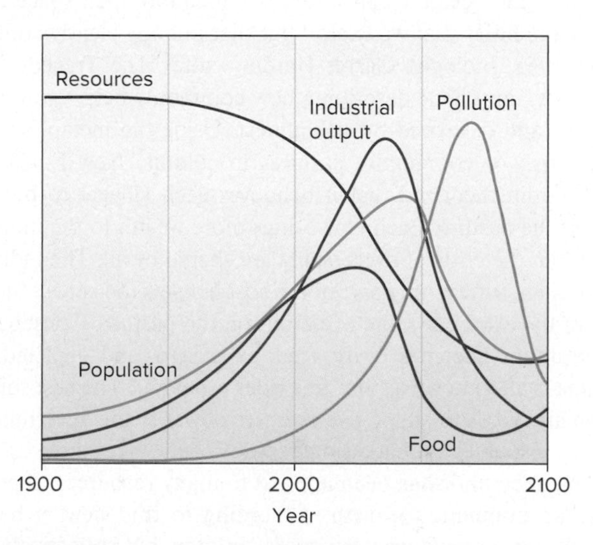

▲ **FIGURE 15.19** A "business as usual" model from *The Limits to Growth* predicts that as population and consumption grow, resources decline and pollution increases until a crash occurs. Notice that pollution continues to climb well after industrial output, food supplies, and population all have plummeted.

▲ **FIGURE 15.20** Scarcity stimulates innovation. As we deplete fossil fuel supplies, and our atmosphere's ability to absorb CO_2, renewable energy sources, such as those being demonstrated here at a solar fair, become more feasible. ©William P. Cunningham

foresee agricultural progress that would enable us to produce more food per person for today's world population of 7 billion than was available for the 1 billion in his day. Similarly, one of the greatest worries in recent years is that we are approaching (or may have already passed) peak oil. How will a society so dependent on oil manage if our supply dries up? Or is that the wrong question, and will we develop renewable energy available to meet all our current needs? The fact that we're using up the cheap, easily extracted fossil fuel supplies is now making solar, wind, and other renewable energy technologies competitive (fig. 15.20). As Sheik Yamani, the former Saudi oil minister said, "The stone age didn't end because we ran out of stones, and the oil age won't end because we have run out of oil."

Communal property resources are a classic problem in economics

One of the difficulties of economics and resource management is that there are many resources we all share but nobody owns. Clean air, fish in the ocean, clean water, wildlife, and open space are all natural amenities that we exploit but that nobody clearly controls.

In 1968, biologist Garret Hardin wrote "The Tragedy of the Commons," an article describing how commonly held resources are degraded and destroyed by self-interest. Using the metaphor of the "commons," or community pastures in colonial New England villages, Hardin theorized that it behooves each villager to put more cows on the pasture. Each cow brings more wealth to the individual farmer, but the costs of overgrazing are shared by all. The individual farmer, then, suffers only part of the cost but gets to keep all the profits from the extra cows she or he put on the pasture. Consequently, the commons becomes overgrazed, exhausted, and depleted. This dilemma is also known as the "free rider problem." The best solution, Hardin argued, is to either give coercive power to the government or privatize resources so that a single owner controls resource use.

This metaphor has been applied to many resources, especially to human population growth. According to this view, it benefits poor villagers to produce a few more children, but collectively these children consume all the resources available, making us all poorer in the end. The same argument has been applied to many resource overuse problems, such as depletion of ocean fisheries, pollution, African famines, and urban crime.

▲ **FIGURE 15.21** Communal irrigation systems on the island of Bali have been managed for centuries by village cooperatives called Subaks. A complex system coordinated by Hindu temple priests regulates water delivery so everyone shares. ©William P. Cunningham

Recent critics have pointed out that what Hardin was really describing was not a commons, or collectively owned and managed resource, but an **open access system,** in which there are no rules to manage resource use. The work of Nobel laureate Elinor Ostrom and others shows that many common resources have been managed successfully for centuries by cooperative agreements among users. Native American management of wild rice beds, Swiss village-owned mountain forests and pastures, Maine lobster fisheries, and communal irrigation systems in Spain, Bali, Laos, and many other countries have all remained viable for centuries under communal management (fig. 15.21).

Each of these "commons," or **communal resource management systems,** shares a number of features: (1) Community members have lived on the land or used the resource for a long time and anticipate that their children and grandchildren will as well, thus giving them a strong interest in sustaining the resource and maintaining bonds with their neighbors; (2) the resource has clearly defined boundaries; (3) the community group size is known and enforced; (4) the resource is relatively scarce and highly variable, so that the community is forced to be interdependent; (5) the management strategies appropriate for local conditions have evolved over time and are collectively enforced—that is, those affected by the rules have a say in them; (6) the resource and its use are actively monitored, discouraging anyone from cheating or taking too much; (7) conflict-resolution mechanisms reduce discord; and (8) incentives encourage compliance with rules, while sanctions for noncompliance keep community members in line.

Rather than being the only workable solution to problems in common-pool resources, privatization and increasing external controls often prove to be disastrous. Where villages have owned and operated local jointly held forests and fishing grounds for generations, nationalization and commodification of resources generally have led to rapid destruction of both society and ecosystems. Where communal systems once enforced restraint over harvesting, privatization encouraged narrow self-interest and allowed outsiders to take advantage of the weakest members of the community.

15.4 NATURAL RESOURCE ACCOUNTING

- *Cost-benefit analysis allows quantitative evaluation of a project.*
- *Accounting for externalized costs is an important challenge.*
- *The Genuine Progress Index and the Human Development Index measure real changes in human welfare.*

Decision making about sustainable resource use often entails **cost-benefit analysis (CBA),** the process of accounting and comparing the costs of a project and its benefits. Ideally this process assigns values to social and environmental effects of a given undertaking, as well as the value of the resources consumed or produced. However, the results of CBA often depend on how resources are accounted for and measured in the first place. CBA is one of the main conceptual frameworks of resource economics, and it is used by decision makers around the world as a way of justifying the building of dams, roads, and airports, as well as in considering what to do about biodiversity loss, air pollution, and global climate change. CBA is a useful way of rational decision making about these projects. It is also widely disputed because it tends to discount the value of natural resources, ecological services, and human communities, and it is used to justify projects that jeopardize all these resources.

In CBA the monetary value of all benefits of a project are counted up and compared with the monetary costs of the project. Usually the direct expenses of a project are easy to ascertain: How much will you have to pay for land, materials, and labor? The monetary worth of lost opportunities–to swim or fish in a river or to see birds in a forest–on the other hand, is much harder to appraise, as are inherent values of the existence of wild species or wild rivers. What is a bug or a bird worth, for instance, or the opportunity for solitude or inspiration? Eventually the decision maker compares all the costs and benefits to see whether the project is justified or whether an alternative action might bring greater benefit at less cost.

Critics of CBA point out its absence of standards, inadequate attention to alternatives, and the placing of monetary values on intangible and diffuse or distant costs and benefits. Who judges how costs and benefits will be estimated? How can we compare things as different as the economic gain from cheap power with loss of biodiversity or the beauty of a free-flowing river? Critics claim that placing monetary values on everything could lead to a belief that only money and profits matter and that any behavior is acceptable as long as you can pay for it. Sometimes speculative or even hypothetical results are given specific numerical values in CBA and then treated as if they were hard facts.

Figure 15.22 shows a hypothetical example of a cost-benefit analysis for reducing air pollution. Initial pollution controls have low cost (orange line) and high social benefits (blue line). As easy controls are completed, further controls have higher costs, and added benefits decline. Ideally there is a cross-over point, where costs equal benefits. Beyond this optimum point, the costs of pollution removal are greater than the benefits. As we've already noted, however, benefits can be intangible and widely dispersed, so they're often discounted by those who bear the costs.

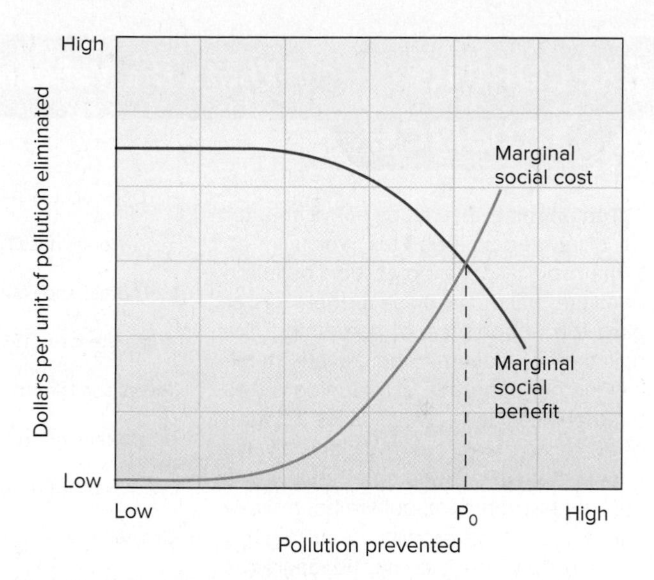

▲ **FIGURE 15.22** Early stages of pollution control have low costs and high benefits. Theoretically, optimum economic efficiency is at (P_0), where the cost of preventing pollution equals the benefits of doing so.

Values such as wildlife, nonhuman ecological systems, and ecological services can be incorporated with natural resource accounting. In theory this accounting contributes to sustainable resource use because it can put a value on long-term or intangible goods that are necessary but often disregarded in economic decision making. One important part of natural resource accounting is assigning a value to ecological services (Exploring Science, p. 372). The total value of nature's services was estimated to be $33 trillion, which was more than the entire global GDP at the time (table 15.6). A 2017 update revised the estimate to $173 trillion, or more than twice the current global GDP. As mentioned earlier, GDP is a

TABLE 15.6 Estimated Annual Value of Ecological Services

ECOSYSTEM SERVICES	VALUE (TRILLION $ U.S.)
Soil formation	17.1
Recreation	3.0
Nutrient cycling	2.3
Water regulation and supply	2.3
Climate regulation (temperature and precipitation)	1.8
Habitat	1.4
Flood and storm protection	1.1
Food and raw materials production	0.8
Genetic resources	0.8
Atmospheric gas balance	0.7
Pollination	0.4
All other services	1.6
Total value of ecosystem services	33.3

Source: Adapted from R. Costanza, et al., "The Value of the World's Ecosystem Services and Natural Capital," *Nature*, vol. 387, 1997.

EXPLORING Science What's the Value of Nature?

The opening case study of this chapter discussed a carbon tax, which aimed to help put a price on carbon pollution because without a price, people disregard the importance of something like pollution, or clean air and climate stability. Another approach to assigning prices to otherwise intangible values is the idea of ecosystem services. We have always known that we depend on a healthy and resilient environment, but we have rarely had a means of knowing how much it was worth, when it comes to decisions about preventing pollution, protecting forests, or defending biodiversity.

Sometimes calculating value is easy: loss of clean water or air has health impacts that we can quantify. Sometimes calculating value is hard: how can we quantify the value of biodiversity? Of butterflies? Of the cooling effect of standing forests?

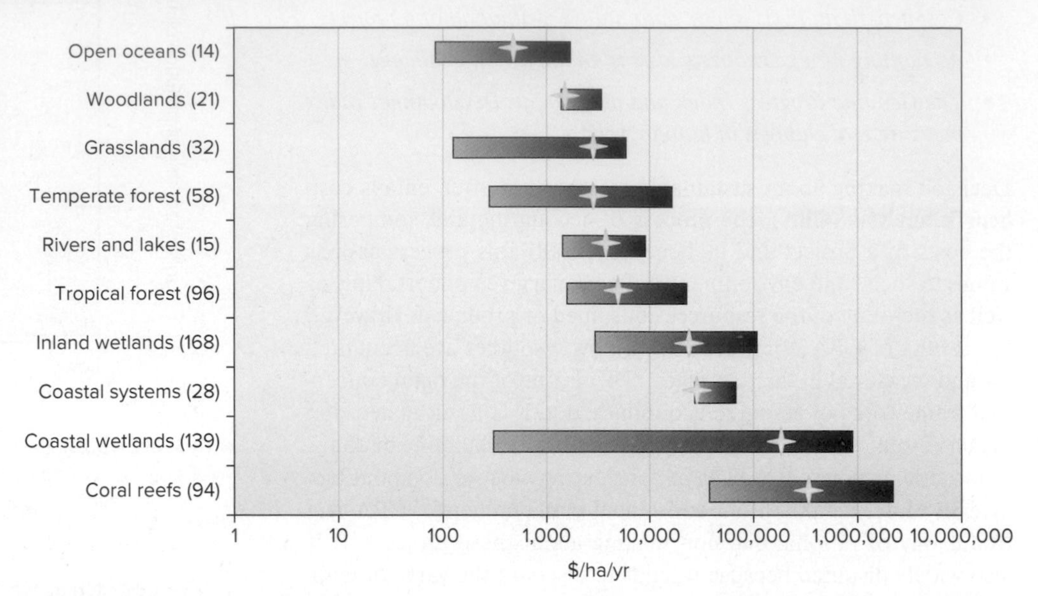

▲ **FIGURE 1** Estimated value of ecosystem services for major biomes in 2007 dollars/hectare. The number of estimates is shown in parentheses. Bars show the range of values, and stars show average estimates.
Source: de Groot, et al. 2012. *Ecosystem Services* 1: 50–61.

One of the first efforts to estimate the economic value of the entire biosphere was carried out by Robert Costanza and his colleagues in 1997. They divided nature's contributions into three categories: (1) Provisioning services include food, timber, fish, game, fresh water, and other resources we need to live. (2) Regulating services are factors that make our environment safer and more livable, for example wetlands that absorb flood waters, forests that stabilize climate, or soils that detoxify waste. (3) Cultural services involve religious, artistic, or aesthetic values, including recreation or education that contribute to well-being.

Human, built, and social capital are needed to make many of these ecosystem services valuable to us. For example, to make seafood available requires boats and fishing gear (built capital) as well as fishing knowledge (human capital) and fishing communities (social capital).

We often don't appreciate or understand many of the ecosystem-regulating services from which we benefit. For example, natural predators that control insect pests may be mostly invisible to us. Or flood protection provided by wetlands or coastal barrier islands may be underappreciated or misunderstood. Climatic forces that regulate temperature, rainfall, winds, and sunshine may be unknown or misinterpreted by the public that benefits from their operation.

How do we estimate the value of ecosystem services? For some natural resources we can simply gather data from current consumption. How much do we pay presently for seafood or wood products? For intangible benefits or services that are now free, however, the calculation is more difficult. What's the value of a walk in the woods or the ability to gaze at a beautiful sunset? One approach is to question people about their "willingness-to-pay" for these opportunities. How much compensation would you accept if those benefits were denied to you? However, individuals may not have an accurate assessment of how valuable a resource or an opportunity is until it's actually gone.

In some cases, we can measure the economic losses from storms or floods when coastal wetlands are destroyed or damaged, or we can calculate what it would cost to build substitute protections. But some natural resources, such as clean air or water, have no realistic substitute. We simply can't live without them. They're worth everything we have. Some people argue that it's offensive to try to put a price on existential resources. They are literally priceless. Nevertheless, it can help us make better decisions if we acknowledge how important these ecosystem services are.

The 1997 Constanza study estimated the value for the entire biosphere to be in the range of U.S.$16–54 trillion per year, with an average estimate of U.S.$33 trillion. This was considerably more than the entire global GDP at the time. A 2012 update written by many of the authors of the original 1997 article estimated the value of all ecosystem services to be equivalent to U.S.$173 trillion in current dollars. This study estimated values per unit area of ten major biomes based on a database drawn from 320 publications from a number of research groups, including the UN's program on The Economics of Ecosystems and Biodiversity (TEEB).

Drawing on more than 600 individual estimates, the study suggests that the value of the bundle of services that could be provided by these biomes can vary from $490/ha/yr for open oceans to almost $350,000/ha/yr for an "average" hectare of coral reefs (fig. 1). There is a large uncertainty in these estimates because most of the benefits are intangible and have no market price. Still, the "average" unit value for each of these biomes gives us some measure of the importance of protecting nature.

For more information, see Constanza, R., et al. 2017. "Twenty years of ecosystem services: How far have we come and how far do we still need to go?" *Ecosystem Services* 28 (2017) 1–16.

Active LEARNING

Costs and Benefits

Figure 15.22 shows a hypothetical example of the costs and benefits of reducing air pollution.

1. For the first 25 percent (or so) of reduction (on the X axis), the benefits are uniformly high. What are some of the economic benefits that might be represented by this part of the graph?
2. What might be some of the low-cost pollution reduction steps represented by the "costs" curve in the first part of the graph?
3. Why is the P_0 considered the optimum point?

ANSWERS: 1. Benefits include improved health; less damage to buildings, crops, and materials; and improved quality of life. 2. Conservation, improved planning and efficiency in transportation, buildings, and energy production; replacing old, inefficient industrial plants and vehicles. 3. Because it's the point at which the costs just equal the benefits.

widely used measure of wealth that is based on rates of consumption and throughput. It doesn't account, however, for natural resource depletion or ecosystem damage.

The World Resources Institute, for example, estimates that soil erosion in Indonesia reduces the value of crop production there about 40 percent per year. If natural capital were taken into account, Indonesian GDP would be reduced by at least 20 percent annually. Similarly, Costa Rica experienced impressive increases in timber, beef, and banana production between 1970 and 1990. But decreased natural capital during this period, represented by soil erosion, forest destruction, biodiversity losses, and accelerated water runoff, added up to at least $4 billion, or about 25 percent of annual GDP. A number of countries, including Canada and China, now use a "green GDP" that measures environmental costs as part of economic accounting.

Internalizing external costs

One of the factors that can make resource-exploiting enterprises look good in cost-benefit analysis is externalizing costs. **Externalizing costs** is the act of disregarding or discounting resources or goods that contribute to producing something but for which the producer does not actually pay. Usually external costs are diffuse and difficult to quantify. Generally they are costs to society at large, not to the individual user. When a farmer harvests a crop in the fall, for example, the values of seeds, fertilizer, and the sale of the crop are tabulated; the values of soil lost to erosion, water quality lost to nonpoint-source pollution, and depleted fish populations are not accounted for. These are most often costs shared by the whole society, rather than borne by the resource user. They are external to the accounting system, and they are generally ignored in cost-benefit analysis—or when the farmer evaluates whether the year was profitable. Larger enterprises, such as dam building, logging, and road building, generally externalize the cost of ecological services lost along the way.

One way to use the market system to optimize resource use is to make sure that those who reap the benefits of resource use also bear all the external costs. This is referred to as **internalizing costs.** Calculating the value of ecological services or diffuse pollution is not easy, but it is an important step in sustainable resource accounting.

New approaches measure real progress

A number of systems have been proposed as alternatives to GNP that reflect genuine progress and social welfare. The economist Herman Daly and philosopher John Cobb proposed a **genuine progress index (GPI),** which takes into account real per-capita income, quality of life, distributional equity, natural resource depletion, environmental damage, and the value of unpaid labor. They point out that, while per-capita GDP in the United States nearly doubled between 1970 and 2005, per-capita GPI increased only 4 percent (fig. 15.23). Some social service organizations would add to this index the costs of social breakdown and crime, which would decrease real progress even further over this time span. Bhutan measures Gross Domestic Happiness as their indicator of progress.

The United Nations Development Programme (UNDP) uses a benchmark called the **human development index (HDI)** to track social progress. HDI incorporates life expectancy, educational attainment, and standard of living as critical measures of development. Gender issues are accounted for in the gender development index (GDI), which is simply HDI adjusted or discounted for inequality or achievement between men and women.

In its annual Human Development Report, the UNDP compares country-by-country progress. As you might expect, the highest development levels are generally found in North America, Europe, and Japan. In 2018 Norway ranked first in the world in both HDI and GDI. The United States was tenth in HDI but it was 49th in a measure of environmental protection and governance. In 2018, 19 of the 20 countries with the lowest HDI were in Africa.

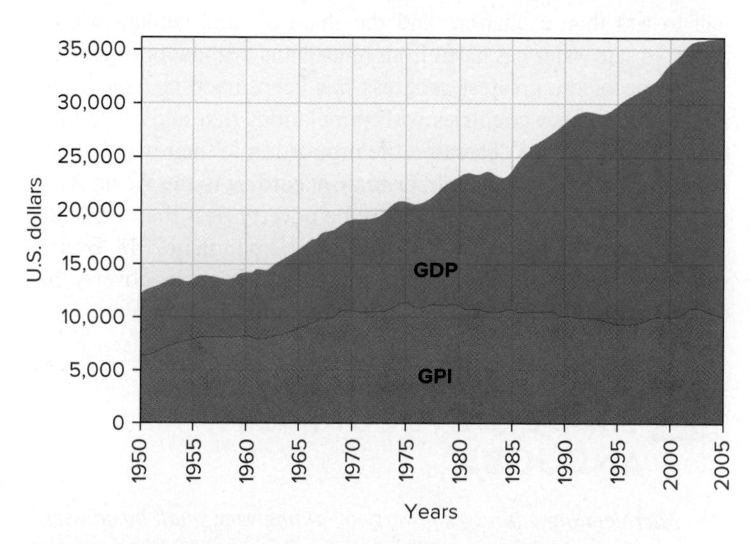

▲ **FIGURE 15.23** Although per-capita GDP in the United States nearly doubled between 1970 and 2000 in inflation-adjusted dollars, a genuine progress index that takes into account natural resource depletion, environmental damage, and options for future generations hardly increased at all.

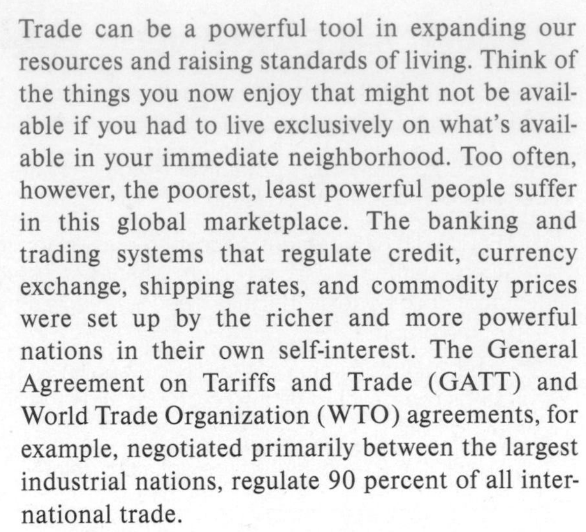

What Can YOU DO?

Personally Responsible Consumerism

Each of us can do many things to support "green" businesses through responsible consumerism and ecological economics.

- Practice living simply. Invest in experiences, rather than material goods.
- Rent, borrow, or barter when you can. Can you reduce the amount of stuff you consume by renting, instead of buying, machines and equipment you actually use only rarely?
- Recycle or reuse building materials: doors, windows, cabinets, appliances. Shop at thrift stores and yard sales, to reuse clothes, dishes, appliances, etc.
- Buy "green" products. Look for efficient, high-quality materials that will last and that are produced in the most environmentally friendly manner possible. Subscribe to clean-energy programs if they are available in your area.
- Buy products made under humane conditions by workers who receive a fair wage.
- Think about the total life-cycle costs of the things you buy, especially big purchases, such as cars. Try to account for the environmental impacts, energy use, and disposal costs, as well as initial purchase price.
- Invest in socially and environmentally responsible mutual funds or "green" businesses when you have money for investment.

As mentioned earlier in this chapter, some of these countries also have the largest and most rapidly growing megacities.

Although poverty remains widespread in many places, encouraging news also can be found in development statistics. Poverty has fallen more in the past 50 years, the UNDP reports, than in the previous 500 years. Child death rates in developing countries as a whole have been more than halved. Average life expectancy has increased by 30 percent, while malnutrition rates have declined by almost a third. The proportion of children who lack primary school has fallen from more than half to less than a quarter. And the share of rural families without access to safe water has fallen from nine-tenths to about one-quarter.

Some of the greatest progress has been made in Asia. China and a dozen other countries with populations that add up to more than 1.6 billion have decreased the proportion of their people living below the poverty line by half. Overall, according to the World Bank, the number of people living in extreme poverty (less than US$1.25/day) fell from 1.8 billion in 1990 to about 1.3 billion in 2018. Even in sub-Saharan Africa, the number of people living in poverty has dropped below 50 percent of the population for the first time ever.

15.5 TRADE, DEVELOPMENT, AND JOBS

- *Microlending addresses poverty by aiding very small businesses.*
- *Green business practices are showing ways to cut environmental impacts and costs at the same time.*
- *Renewable energy is an area of rapid job growth.*

Trade can be a powerful tool in expanding our resources and raising standards of living. Think of the things you now enjoy that might not be available if you had to live exclusively on what's available in your immediate neighborhood. Too often, however, the poorest, least powerful people suffer in this global marketplace. The banking and trading systems that regulate credit, currency exchange, shipping rates, and commodity prices were set up by the richer and more powerful nations in their own self-interest. The General Agreement on Tariffs and Trade (GATT) and World Trade Organization (WTO) agreements, for example, negotiated primarily between the largest industrial nations, regulate 90 percent of all international trade.

These systems often keep the less-developed countries in a perpetual role of resource suppliers to the more-developed countries. The producers of raw materials, such as mineral ores or agricultural products, get very little of the income generated from international trade. As a prerequisite for international development loans, the IMF frequently requires debtor nations to adopt harsh "structural adjustment" plans that slash welfare programs and impose cruel hardships on poor people. The WTO has issued numerous rulings that favor international trade over pollution prevention or protection of endangered species.

Microlending helps the poorest of the poor

No single institution has more influence on financing and policies of developing countries than the World Bank. Of some $25 billion loaned each year for development projects by international agencies, about two-thirds comes from the World Bank. Founded in 1945 to fund reconstruction of Europe and Japan, the World Bank shifted its emphasis to aid developing countries in the 1950s. Many of its projects have had adverse environmental and social effects, however. Its loans often go to corrupt governments and fund ventures such as nuclear power plants, huge dams, and giant water diversion schemes that displace local populations. Former U.S. treasury secretary Paul O'Neill said that these loans have driven poor countries "into a ditch" by loading them with unpayable debt. He said that funds should not be loans, but rather grants to fight poverty.

Global aid often supports banks and industries more than it helps the impoverished populations who most need assistance. Often costs are shifted to the poorest to pay back loans negotiated by their governments and industries. These concerns led Dr. Muhammad Yunus of Bangladesh to initiate the microlending plan of the Grameen Bank.

Microlending programs have assisted billions of people—most of them low-status women who have no other way to borrow money at reasonable interest rates. This model is now being used by hundreds of other development agencies around the world. Even in the United States, organizations assist micro-enterprises with loans, grants, and training. The Women's Self-Employment Project in Chicago, for instance, teaches job skills to single mothers in housing projects. Similarly, "tribal circle" banks on Native American reservations successfully finance microscale economic development ventures.

One of the most important innovations of the Grameen Bank is that borrowers take out loans in small groups. Everyone in the group is responsible for the others' performance. The group not only guarantees loan repayment, it helps businesses succeed by offering support, encouragement, and advice. Where banks depend on the threat of foreclosure and a low credit rating to ensure debt repayment, the Grameen Bank has something at least as powerful for poor villagers–the threat of letting down your neighbors and relatives. Becoming a member of a Grameen group also requires participation in a savings program that fosters self-reliance and fiscal management.

The process of running a successful business and repaying the loan transforms many individuals. Women who previously had little economic power, influence, or self-esteem are empowered with a sense of pride and accomplishment. Dr. Yunus also discovered that money going to families through women helped the families much more than the same amount of money in men's salaries. Women were more likely to spend money on children's food or education, producing generational benefits with the increased income.

The most recent venture for the Grameen Bank is providing mobile phone service to rural villages. Supplying mobile phones to poor women not only allows them to communicate, it provides another business opportunity. They rent out their phones to neighbors, giving the owner additional income and linking the whole village to the outside world. Suddenly people who had no access to communication can talk with their relatives, order supplies from the city, check on prices at the regional market, and decide when and where to sell their goods and services. This is a great example of "bottom-up development." Founded in 1996, Grameen Phone now has 55 million subscribers and is Bangladesh's largest mobile phone company and offers coverage to 99 percent of the country.

Active LEARNING

Try Your Hand at Microlending

The best way to observe microlending at work is to try it out. Collect donations of $1–$5 from people in your class, until the total is $25. Go to www.kiva.org, a microlending organization that pools small loans, and select a business to support. You can use PayPal to send the money, and for the next year, you will receive periodic reports on how the business is going. To evaluate whether you're getting a good rate of return, consider that for most stock market investments, about 5–10 percent annual return is reasonable.

1. What is 5–10 percent of $25?

Also poll the class to get these averages:

2. What percentage interest do you earn from your bank accounts?
3. What percentage do you pay to credit card companies?

ANSWERS: 1. $1.25–$2.50; 2., 3. Answers will vary, but probably <2% and >15%.

Market mechanisms can reduce pollution

As noted in the opening case study, climate change is an economic threat, as well as an environmental one. In 2006 the British economist Sir Nicolas Stern warned us that the damage caused by climate change 50 years from now could be equivalent to losing as much as 20 percent of the global GDP every year. Or to put it another way, every dollar we invest now to reduce greenhouse gas emissions will save $20 in the future.

But how can we bring about a transition to a low-carbon economy? Many economists believe market forces can reduce pollution more efficiently than rules and regulations. Assessing a tax on each ton of carbon emitted, as in British Columbia, encourages businesses to reduce greenhouse gases, but still allows them to search for the most cost-effective ways to achieve this goal. And the incentives promote a continuing search for even better ways to reduce emissions. The more you reduce your discharges, the more you save.

Another approach is to set up a **cap-and-trade** system. The first step is to mandate upper limits (the cap), on how much each country, sector, or specific industry is allowed to emit. Companies that can cut pollution by more than they're required to can sell the credit to other companies that have more difficulty meeting their mandated levels. This sets up the same kind of incentives as a tax but doesn't require as much government oversight.

We have a good model for cap-and-trade in sulfur emission. A market created in 1990 has been remarkably effective in reducing sulfur pollution. Most observers agree that this market has been much more cost-effective than rigid rules and directives would have been.

Green business and green design

Businesspeople and consumers are increasingly aware of the unsustainability of producing the goods we use every day. Recently, business innovators have tried to develop green enterprises that produce environmentally and socially sound products. "Green" companies, such as the Body Shop, Patagonia, Aveda, Malden Mills, and Johnson and Johnson, have shown that operating according to the principles of sustainable development and environmental protection can be good for public relations, employee morale, and sales (table 15.7).

Green business works because consumers are becoming aware of the ecological consequences of their purchases. Increasing interest in environmental and social sustainability has caused an explosive growth of green products. The *National Green Pages* published by Co-Op America currently lists more than 2,000 green companies. You

TABLE 15.7 Goals for an Eco-Efficient Economy

- Introduce no hazardous materials into the air, water, or soil.
- Measure prosperity by how much natural capital we can accrue in productive ways.
- Measure productivity by how many people are gainfully and meaningfully employed.
- Measure progress by how many buildings have no smokestacks or dangerous effluents.
- Make the thousands of complex governmental rules that now regulate toxic or hazardous materials unnecessary.
- Produce nothing that will require constant vigilance from future generations.
- Celebrate the abundance of biological and cultural diversity.
- Live on renewable solar income rather than fossil fuels.

TABLE 15.8 McDonough Design Principles

Inspired by the way living systems actually work, Bill McDonough offers three simple principles for redesigning processes and products:

- *Waste equals food.* This principle encourages elimination of the concept of waste in industrial design. Every process should be designed so that the products themselves, as well as leftover chemicals, materials, and effluents, can become "food" for other processes.
- *Rely on current solar income.* This principle has two benefits: First, it diminishes, and may eventually eliminate, our reliance on hydrocarbon fuels. Second, it means designing systems that sip energy rather than gulping it down.
- *Respect diversity.* Evaluate every design for its impact on plant, animal, and human life. What effects do products and processes have on identity, independence, and integrity of humans and natural systems? Every project should respect the regional, cultural, and material uniqueness of its particular place.

can find eco-travel agencies, telephone companies that donate profits to environmental groups, entrepreneurs selling organic foods, shade-grown coffee, straw-bale houses, paint thinner made from orange peels, sandals made from recycled auto tires, and a plethora of hemp products, including burgers, ale, clothing, shoes, rugs, and shampoo.

Although these eco-entrepreneurs represent a tiny sliver of the $7 trillion per year U.S. economy, they often are pioneers in developing new technologies and offering innovative services. Markets also grow over time: Organic food marketing has grown from a few funky local co-ops to a $49 billion market segment. Most supermarket chains now carry some organic food choices. Similarly, natural-care health and beauty products reached $10 billion in sales in 2014 out of a $57 billion industry. By supporting these products, you can ensure that they will continue to be available and, perhaps, even help expand their penetration into the market.

Corporations are increasingly aware of the value of environmental action, both for their image and for their bottom line. Many are hiring sustainability officers to reduce impacts and improve efficiency. Applying the famous three Rs–reduce, reuse, recycle–these firms have saved money and gotten welcome publicity. Savings can be substantial. Pollution-prevention programs at 3M, for example, have saved $857 million over the past 25 years. Microsoft used an internal carbon tax to cut its energy costs by $10 million a year. The company is also cutting water waste has committed to diverting 90 percent of its waste from landfills.

Architects also are starting to get on board the green bandwagon. Acknowledging that heating, cooling, lighting, and operating buildings is one of our biggest uses of energy and resources, architects such as William McDonough are designing "green office" projects. Among McDonough's projects are the Environmental Defense Fund headquarters in New York City; and the Gap corporate offices in San Bruno, California (fig. 15.24). Each uses a combination of energy-efficient designs and technologies, including natural lighting and efficient water systems (table 15.8).

The Gap office building, for example, is intended to promote employee well-being and productivity, as well as efficiency. It has high ceilings, abundant skylights, windows that open, a full-service fitness center (including pool), and a landscaped atrium for each office bay that brings the outside in. The roofs are covered with native grasses. Warm interior tones and natural wood surfaces (all wood used in the building was harvested by certified sustainable methods) give a friendly feeling. Paints, adhesives, and floor coverings are low-toxicity, and the building is one-third more energy-efficient than strict California laws require. The pleasant environment helps improve employee effectiveness and retention.

▼ FIGURE 15.24 The award-winning Gap, Inc., corporate offices in San Bruno, California, demonstrate some of the best features of environmental design. A roof covered with native grasses provides insulation and reduces runoff. Natural lighting, an open design, and careful relation to its surroundings all make this a pleasant place to work. ©Mark Luthringer

SCIENCE AND Citizenship — Green Energy Jobs Versus Fossil Fuels

©Elena Elisseeva/123RF

President Trump has said that he "loves fossil fuels" and promises to bring back "so many beautiful jobs in coal mining," but the evidence suggests that he's betting on the wrong horse in this race. In 2018, solar energy employed more than twice as many Americans as coal. Wind and bioenergy engaged half again the total number in coal. Mechanization and competition from cheap natural gas have eliminated thousands of jobs in coal mining. In 1923, the United States employed nearly 863,000 coal miners. But in 2016, although total coal production had nearly doubled, the number of miners had fallen 95 percent to only about 50,000.

More than 373,000 Americans work part or full time in solar energy. These jobs are distributed in communities across the country, making them available to a wide variety of people. The solar sector has more jobs across the country, but coal is still a more important employer in some places. Five states—Wyoming, West Virginia, Kentucky, Illinois, and Pennsylvania—produce almost three-fourths of all coal mined in the United States. Not surprisingly, those states represent some of the strongest political support for President Trump. According to a Columbia University study, Kentucky, a state that voted overwhelmingly for Trump, lost 64 percent of its coal jobs between 2011 and 2016. Most of Trump's attempts to restore those jobs have centered on reducing regulations (such as mine safety, pollution prevention, and medical care for workers), but critics claim that overturning rules helps mine owners while reducing security for employees.

Wind power is the fastest-growing energy sector in the United States, increasing by about 20 percent to more than 100,000 jobs in 2018. The U.S. Department of Energy projects that by 2050 more than 600,000 wind-related jobs, including manufacturing, installation, and maintenance, will be available. Although a majority of these jobs will be in the Midwest and Great Plains where winds are strongest, some workers will be needed in every state to support this rapidly growing sector. Wind turbine technician isn't just one of the fastest-growing jobs in clean energy—it's the single fastest-growing occupation in America, according to the Bureau of Labor Statistics. And it pays well, with a median income of $52,000 per year, which is set to double by 2024, the agency estimates.

Solar installation is another rapidly growing U.S. job class. More than 73,000 jobs in this industry were added in 2016, a 25 percent increase over the previous year. A benefit of solar power is that jobs exist wherever the sun shines. You aren't limited to a few coal-producing states for where you live. Growth in solar has sparked an array of opportunities for other supporting industries that make or distribute solar equipment or batteries. Solar power is expanding 17 times faster than the total U.S. economy, according to the International Renewable Power Agency. Although men have most of the jobs currently in solar installation, that is starting to change, especially in the sales business. Women now hold 28 percent of solar jobs, up from 19 percent in 2013.

It's always the case that technology destroys some jobs while creating others. Buggy whip factories suffered a catastrophic decline with the advent of automobiles. Ice delivery jobs disappeared when the electric refrigerator appeared. How much should we subsidize jobs in disappearing industries? And how do workers and communities transition to a new system? These are questions to evaluate as we think about how to prioritize public subsidies and which policies have the greatest public benefits.

Gap, Inc., estimates that the increased energy and operational efficiency will have a 4- to 8-year payback.

Green business creates jobs

For years, many business leaders and politicians have portrayed environmental protection and jobs as mutually exclusive. They claim that pollution control, protection of natural areas and endangered species, and limits on use of nonrenewable resources will strangle the economy and throw people out of work (Science and Citizenship, above). Ecological economists dispute this claim, however. Their studies show that only 0.1 percent of all large-scale layoffs in the United States in recent years were due to government regulations (fig. 15.25). Environmental protection, they argue, is not only necessary for a healthy economic system; it actually creates jobs and stimulates business.

Green businesses often create far more jobs and stimulate local economies far more than environmentally destructive ones. Wind energy, for example, provides about five times as many jobs per kilowatt-hour of electricity than does coal-fired power (see chapter 13).

As chapter 13 shows, China has emerged as the world leader in sustainable energy. Recognizing the multibillion-dollar economic potential of "green" business, China is investing at least $8 billion per year on research and development, and now it is selling about

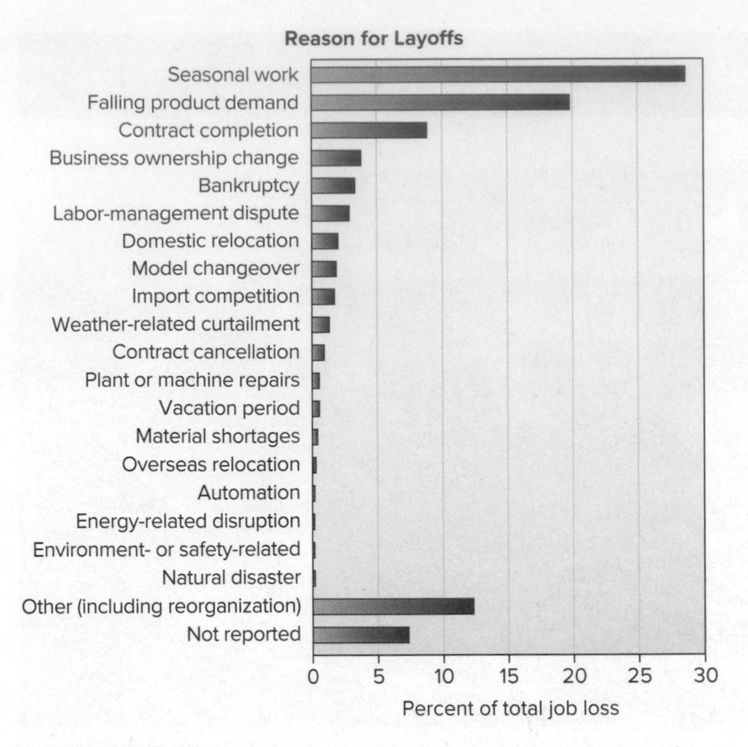

▲ **FIGURE 15.25** Although opponents of environmental regulation often claim that protecting the environment costs jobs, studies by economist E. S. Goodstein show that only 0.1 percent of all large-scale layoffs in the United States were the result of environmental laws. Source: Data from E. S. Goodstein, Economic Policy Institute, Washington, D.C.

$12 billion worth of equipment and services per year worldwide. Japan, also, is marketing advanced waste incinerators, pollution-control equipment, alternative energy sources, and water treatment systems. Superefficient "hybrid" gas-electric cars are helping Japanese automakers flourish, while U.S. corporations that once dominated the world slide into bankruptcy. Letting go of fossil-based industries, economies, and ideas from past generations is always a challenge, and resistance to change has strongly influenced U.S. politics in recent years. But it is increasingly clear that moving into a green economy is where the future is. The question is how we will negotiate this transition.

CONCLUSION

More than half of us now live in cities, and in a generation two-thirds of us probably will be urban dwellers. Most of this growth will occur in the large cities of developing countries, where resources are already strained. Cities draw immigrants from the countryside by offering jobs, social mobility, education, and other opportunities unavailable in rural areas.

Urban planning tries to minimize the strains of urbanization. Good planning saves money, because providing roads and services to sprawling suburbs is very expensive. Transportation, which is key to our economy, is a critical part of planning because it determines how far-flung urban development will be. Smart growth, cluster development, and improved standards for environmentally conscious building are also important in urban planning.

Economic policies are often at the root of the success or failure of cities. These policies build from some basic sets of assumptions about the nature of resources. Classical economics assumes that resources are finite, so that we compete to control them. Neoclassical economics assumes that resources are based on capital, which can include knowledge and social capital as well as resources, and that constant growth is both possible and essential. Natural resource economics extends neoclassical ideas to internalize the value of ecological services in economic accounting.

The "tragedy of the commons" is a classic description of our inability to take care of public resources. Subsequent explanations have pointed out that collective rules of ownership are necessary for the survival of shared resources. Our ability to agree on these rules appears to depend on a number of factors, including the scarcity of the resource and our ability to monitor its use.

Because classic productivity indices count many social and environmental ills as positive growth, alternatives have been proposed, including the genuine progress index (GPI). Measures like the GPI can be used to help ensure fair and responsible growth in developing areas. Microlending is another innovative strategy that promotes equity in economic growth. Green business and green design are fast-growing parts of many economies. These approaches save money by minimizing consumption and waste.

PRACTICE QUIZ

1. How many people now live in urban areas?
2. How many cities were over 1 million in 1900? How many are now?
3. Why do people move to urban areas?
4. What is the difference between a shantytown and a slum?
5. Define *sprawl*.
6. In what ways are cities ecosystems?
7. Define *smart growth*.
8. Describe a "green" roof.
9. Describe the main features of the British Columbia carbon tax.
10. Define *sustainable development*.
11. Briefly summarize the differences in how neoclassical and ecological economics view natural resources.
12. What is the estimated economic value of all the world's ecological services?
13. How is it that nonrenewable resources can be extended indefinitely, while renewable resources are exhaustible?
14. In your own words, describe what is shown in figure 15.22.
15. What's the difference between open access and communal resource management?
16. Describe the genuine progress index (GPI).
17. What is *microlending*?

CRITICAL THINKING AND DISCUSSION

Apply the principles you have learned in this chapter to discuss these questions with other students.

1. Some people—especially automakers—claim that Americans will never give up their cars. Do you agree? What might persuade you to change to a car-free lifestyle?
2. This chapter presents a number of proposals for suburban redesign. Which of them would be appropriate or useful for your community? Try drawing up a plan for the ideal design of your neighborhood.
3. A city could be considered an ecosystem. Using what you learned in chapters 2 and 3, describe the structure and function of a city in ecological terms.
4. If you were doing a cost-benefit study, how would you assign a value to the opportunity for good health or the existence of rare and endangered species in faraway places? Is there a danger or cost in simply saying some things are immeasurable and priceless and therefore off limits to discussion?
5. What would be the effect on the developing countries of the world if we were to change to a steady-state economic system? How could we achieve a just distribution of resource benefits while still protecting environmental quality and future resource use?
6. When an ecologist warns that we are using up irreplaceable natural resources and an economist rejoins that ingenuity and enterprise will find substitutes for most resources, what underlying premises and definitions shape their arguments?

DATA ANALYSIS Plotting Trends in Urbanization and Economic Indicators

Urbanization and economic growth are two closely related changes going on in societies today. How are these two processes related? How do they compare in different parts of the world? Is the United States more urbanized or less so than other regions? How do economic growth rates compare in different regions?

Gapminder.org is a rich source of data on global population, health, and development that we examined in the Data Analysis exercise for chapter 4. The site includes animated graphs showing change over time. Go to Connect to find a link to Gapminder graphs, and answer questions about what they tell you.

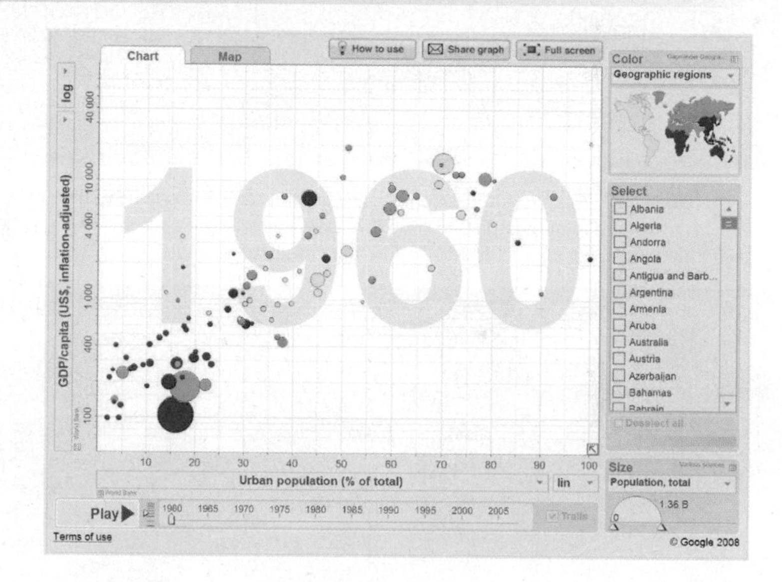

Design Elements: Active Learning (Toad): ©Gaertner/Alamy Stock Photo; Case Study (Globe): ©McGraw-Hill Education; Google Earth: ©McGraw-Hill Education; Abstract Background: ©Martin Kubat/Shutterstock; What do you think (Students using tablets): ©McGraw-Hill Education/Richard Hutchings, photographer; What can you do (Hand holding Globe): ©Christoph Weihs/Shutterstock

CHAPTER

16 Environmental Policy and Sustainability

LEARNING OUTCOMES

After studying this chapter, you should be able to answer the following questions:

- What is environmental policy, and how is it formed?
- What is NEPA, and what does it do?
- Describe some of the important U.S. environmental laws.
- Describe several international environmental laws and conventions.
- How do the different branches of government influence environmental policy?
- Explain some of the ways students can contribute to environmental protection.
- What are the Sustainable Development Goals?

Direct damage in New York City and the surrounding region from Hurricane Sandy amounted to at least $70 billion. This crisis inspired dramatic action on climate change, including challenging the major industries responsible for it.

CASE STUDY

Fossil Fuel Divestment

In December 2017 New York governor Andrew Cuomo announced that the state of New York would divest its $200 billion pension fund of all fossil fuel–related assets. Shortly after this announcement, New York City mayor Bill DeBlasio announced that the city, too, would "decarbonize" its pension fund, which is roughly the same size as the state's, by ceasing all new acquisitions in fossil fuels, selling current holdings in fossil fuel companies, and increasing investments in clean energy. Bill McKibben, an author and co-founder of the climate advocacy group 350.org, praised the city's actions. "I've been watching the climate fight for the last 30 years," McKibben said. "This is one of the handful of most important moments in that 30-year fight. It sends a signal that in the very center of world finance, sentiment is turning sharply against fossil-fuel investment."

These steps were part of a global movement to shed fossil fuel investments, since investment capital supports the operations of oil, gas, and coal companies. The World Bank has said it would no longer lend money for oil and gas exploration. Shortly before the New York announcements, Norway declared it would strip fossil fuel investments from its trillion-dollar sovereign-wealth fund, the largest in the world. Even more surprising, the Norwegian oil and gas giant Statoil declared in 2018 that it was changing its name to Equinor and moving its investment to offshore wind energy.

Divestment is spreading globally. Some 800 colleges, universities, cities, counties, states, and national governments representing $6 trillion in assets have pledged to eliminate their fossil fuel investments.

What's the point of selling these assets? Many economists argue that fossil fuel divestment is a futile exercise in political correctness, a pointless and possibly expensive moral crusade. Academic administrators often reject divestment as philosophically inappropriate—arguing that endowments are strictly financial and not policy-related, and that it is risky to restrict financial decision making. Assuming it's impossible to remove carbon-based fuels from our lives, they argue, it is hypocritical to remove them from investment portfolios. And it would be foolhardy to sell energy stocks if they're still profitable.

But some economists and pension fund managers are persuaded by economic arguments that align with ethical reasons for dumping fossil fuel investments. Climate change associated with fossil fuels is not just another social issue; it is an existential threat to society and the global economy, with trillions of dollars at stake in agriculture, forest products, water resources, and other factors. Studies of fuel stocks suggest that 80 percent of all coal, oil, and gas reserves must stay in the ground if we are to avoid catastrophic levels of climate change. Already companies like Shell and BP are "writing down" the value of their in-the-ground assets—lowering estimates of their value in the company's net worth. As these concerns become more widely recognized, coal, oil, and gas companies are likely to become bad investments. And renewable energy is rapidly replacing fossil fuels and now offers an affordable alternative energy source.

Even more notable than its divestment pledge, New York City mayor DeBlasio announced that the city would sue the world's five largest publicly traded oil companies, seeking to hold them responsible for past and future damage caused by climate change. Hurricane Sandy, in 2012, cost about $70 billion, making it the fourth-costliest weather disaster in American history at that time (fig. 16.1). The storm damaged or destroyed at least 650,000 homes and caused 72 deaths directly, with another 87 deaths from hypothermia during power outages in the blizzard that followed Sandy. Now New York is spending billions more to protect coastlines and power systems from future storms. The city claims the companies covered up the role their products played in global warming and made it more difficult to prepare for emergencies.

Environmental policy can result from actions at many scales, from individuals to cities and states, to national and international levels. Most environmental policies involve disputes about whether social costs of environmental harm should be borne by society or by polluters. The field of environmental policy is constantly evolving, but in this chapter we examine some of the factors that shape environmental laws and policies, as well as some of the important benchmark policies that affect environmental quality today. ■

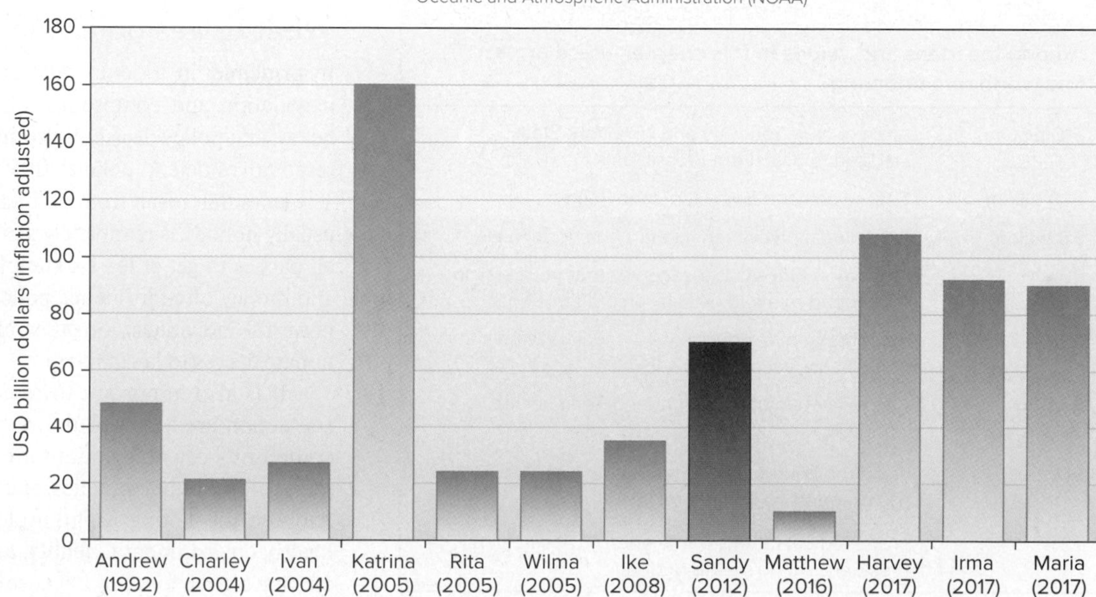

▼ **FIGURE 16.1** Direct damage from Hurricane Sandy amounted to at least $70 billion and infrastructure changes to prevent or reduce further climate change damage could cost an equal amount. Source: National Oceanic and Atmospheric Administration (NOAA)

Never doubt that a small, highly committed group of individuals can change the world; indeed, it is the only thing that ever has.

—MARGARET MEAD

16.1 ENVIRONMENTAL POLICY AND SCIENCE

- *Environmental policy includes rules, laws, and public opinion.*
- *The policy cycle describes how we create, implement, and evaluate policies.*
- *The precautionary principle says we should understand risks before starting an activity.*

In general, policy includes the rules and decisions that influence how we act, as individuals and as a society. As an individual, you probably have policies that guide your actions. Your school establishes policies: It may be one of the many that have, like New York City, established plans to divest from fossil fuels (fig. 16.1). On a national level, we have policies to protect property, individual rights, public health, and other priorities. International policies and agreements regulate international trade in hazardous chemicals or in endangered species. Recently, there have been efforts to enact policies aimed at slowing climate change–such as EPA rules on carbon dioxide emissions and incremental increases in funding for alternative energy development.

Environmental science underlies policies regarding environmental quality and environmental health. Passing a Clean Air Act (see chapter 10), for example, required years of data collection to understand pollutant levels in unhealthy air and in cleaner air. Accumulated measurements over many places and years made it possible to quantify target contaminant levels, and to compare air quality to accepted standards. If we say Delhi has contaminant levels twice World Health Organization standards, we are comparing measured contaminant concentrations to accepted standard quantities of particulates, sulfur dioxide, or other measures. Enforcing standards and environmental policies requires ongoing monitoring.

Benchmark Data	
Among the ideas and values in this chapter, these are a few worth remembering.	
$10 billion	Amount New York City and New York State pledged to divest from fossil fuels
$70 billion	Cost of damage from Hurricane Sandy
$6 trillion	Total global divestment (so far) from fossil fuels
80%	Proportion of fossil fuel reserves that must stay in the ground to avoid catastrophic climate change
77%	Percentage of residents of 40 countries who support climate action
24,000	Approximate number of endangered species worldwide
50	Years the court case of *Ecuador vs Chevron/Texaco* has dragged on without payment
72%	Proposed cut in renewable energy research in President Trump's 2018 budget

▲ **FIGURE 16.2** This Moving Planet march was one of 2,000 events organized by 350.org in 175 countries in 2011. This mobilization, initiated by a group of students at Middlebury College, helped shift global debates on climate policy. ©William P. Cunningham

Environmental policy includes rules, practices, and laws designed to protect human health and well-being, natural resources, and environmental quality. The safety of our air, drinking water, and food is protected by laws developed by past generations of voters and policymakers. Access to public lands and public waterways is also protected by laws. Most of the time we forget about these rules, and we take these protections for granted.

The best way to ensure that policies serve the general public interest is to ensure that the public is active, well informed, and involved in policymaking. By studying environmental science you are taking an important step. It is also important to follow the news, preferably from a variety of sources, to be in touch with elected representatives, and to vote. You can also practice being involved in policymaking by engaging in governance in your school or in your community (see section 16.5).

What drives policymaking?

In principle, in a democratic system, policy is established through negotiation and compromise. Open debate allows all voices to be heard, and policy decisions promote collective well-being. Elected representatives defend policies they think will benefit their constituents. The rules that result from political wrangling are often imperfect, and usually nobody is completely satisfied, but ideally compromises allow all parties to get at least some of what they want. In practice, power and money often influence policymaking. Industry associations, and powerful individuals, or other interest groups often have disproportionate access to lawmakers.

It is also important to remember that public interest and citizen action are a powerful force in policy formation. Altruistic and community-oriented actions are widespread, and they are evident in many of our public policies, such as the Clean Water Act, the Clean Air Act, the Voting Rights Act, and other laws that defend public health, environmental quality, and social equity. Usually, these policies have gone forward because civic or environmental groups have

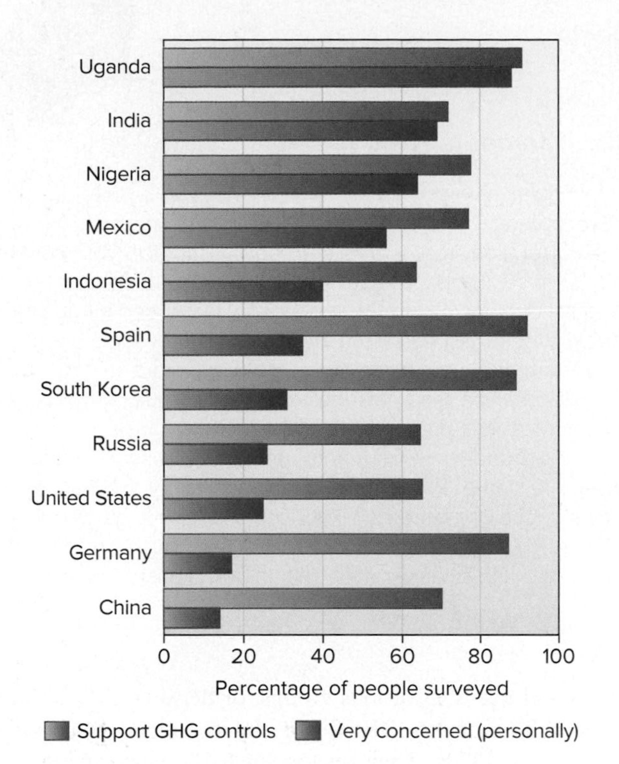

▲ **FIGURE 16.3** Support for international agreements to curb greenhouse gas (GHG) emissions is strong, even though relatively few people in most countries are very concerned about personal impacts. Source: Data from Pew Research Center, 2015.

organized voters, staged letter-writing campaigns, or taken actions to draw attention to an issue (fig. 16.2). The divestment campaign (opening case study) is another example of public-spirited policy action. Most divestment advocates are educated and economically secure; they are not among those who will suffer most from climate change, but still they advocate for change.

This support for environmental protections is strong globally as well. A 2015 survey by the Pew Research Center found that among the 40 countries surveyed, 77 percent of respondents favored actions to control greenhouse gas emissions, and 84 percent considered climate change a serious or very serious threat. Even though only 43 percent of people polled were very concerned about impacts to themselves personally, most were in favor of action to curb climate change (fig. 16.3).

Economic arguments also motivate environmental policy. Companies, cities, and states all recognize that environmental protection is often necessary for their own prosperity. Insurance companies, for example, were among the first corporations to call for climate action, because they stand to lose hundreds of billions of dollars in worsening storms and droughts. Insurance giants Swiss Re and Allianz have even joined the divestment movement and renounced fossil fuel assets. Energy companies are starting to follow suit, as they seek to ride out a transition to renewable energy. Danish Oil and Natural Gas has divested its fossil holdings, reinvested in wind energy, and renamed itself Ørsted, to represent the end of its relationship with fossil fuels. The Norwegian oil giant Statoil has also begun transitioning to offshore wind and other renewables.

A combination of long-term economic interests and a desire to take the moral high ground motivates many academic institutions, cities, and countries to enact environmental policies. These two motivators have led many institutions to divest from fossil fuels: Among the academic institutions joining in this drive are Oxford University in England, Trinity College in Ireland, and Stanford University in California. Similarly, cities including Berlin, Germany, and Washington, D.C., have divestment policies. And the same day that New York was announcing its divestment plans, the president of France, Emmanuel Macron, announced that his country was joining the "keep it in the ground" effort by no longer granting licenses for oil and gas exploration in its territories.

Policy creation is ongoing and cyclic

How do policy issues and options make their way onto the stage of public debate? We often describe policy development as a cycle (fig. 16.4): A problem is identified, usually by concerned groups of citizens or interest groups. Plans to resolve the problem are developed through discussion, and new rules are proposed. Popular support is built to gather votes for the new rule. If the rule is agreed to, then it is implemented. Evaluation then leads to identification of flaws in the rule, and the cycle starts again.

Building support is central to policy development. Proponents often do media campaigns, public education, and personal lobbying of decision makers. Often groups hire a lobbyist who can dedicate weeks, months, or even years to develop the support of legislators. Special interests with money to spend work to sway public opinion through media. They may spend millions of dollars on campaigns to convince the public to support a particular view on a public policy issue.

When watching the media, it's always a good idea to "follow the money" if you want to learn more about the views expressed. The tremendous amount of money spent on news media and advertising indicates how important it is to enlist public support in order to pass (or block) new policies.

The next step in a policy cycle is implementation, or carrying out the new rules. Ideally, government agencies faithfully carry out policy directives as they provide services and enforce rules and

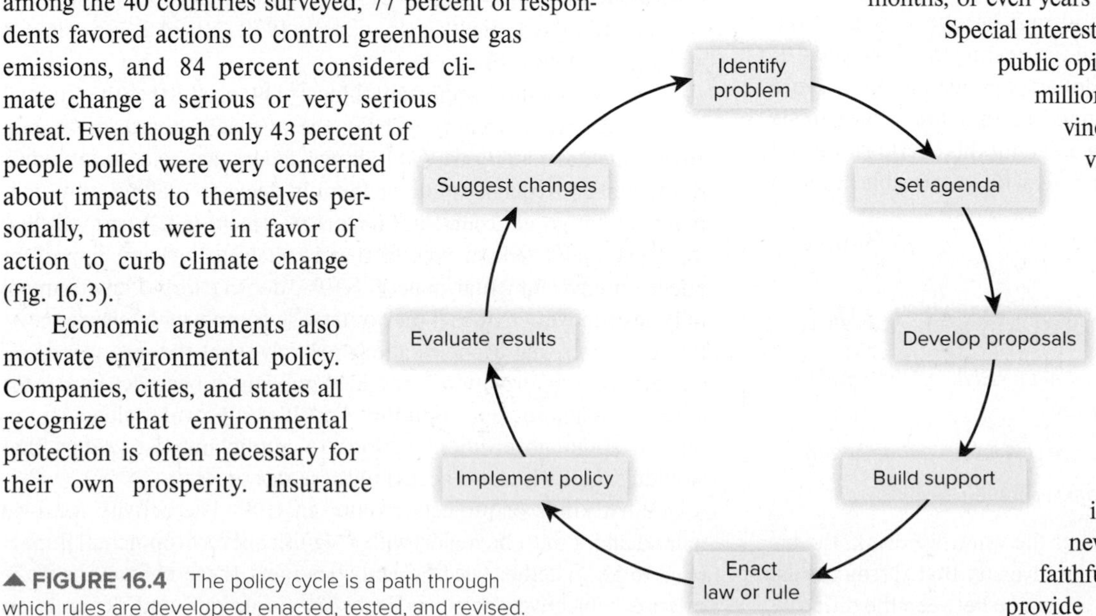

▲ **FIGURE 16.4** The policy cycle is a path through which rules are developed, enacted, tested, and revised.

regulations. Continued public attention is needed to make sure the government enforces its own rules. Many of our worst problems in air pollution, and related respiratory ailments, for example, could be controlled if we enforced existing air quality rules more effectively.

Once a rule is enacted, it almost invariably requires reevaluation and improvement after time. Often laws expire after a designated number of years, so that it is necessary to "reauthorize" a law, or vote again to continue it.

Are we better safe than sorry?

A fundamental concept in policy is the **precautionary principle,** the idea that if an activity threatens to harm health or the environment, we should fully understand risks before initiating that activity. According to this principle, for example, we shouldn't mass-market new chemicals, new cars, or new children's toys until we're sure they are safe. These are four widely accepted tenets of this principle:

- People have a duty to take steps to prevent harm. If you suspect something bad might happen, you have an obligation to try to stop it.
- The burden of proof for a new technology, process, activity, or chemical lies with its developers, not with the general public.
- Before using a new technology, process, or chemical, or starting a new activity, there is an obligation to examine a full range of alternatives, including the alternative of not using it.
- Decisions using the precautionary principle must be open and democratic and must include the affected parties.

The European Union has adopted this precautionary principle as the basis of its environmental policy. In the United States, opponents of this approach claim that it threatens productivity and innovation. However, many American firms that do business in Europe–including virtually all of the largest corporations–are having to change their manufacturing processes to adapt to more careful EU standards. For example, lead, mercury, and other hazardous materials must be eliminated from electronics, toys, cosmetics, clothing, and a variety of other consumer products. A proposal being debated by the EU would require testing of thousands of chemicals, cost industry billions of dollars, and lead to many more products and compounds being banned as they are shown to be unsafe to the public. What would you do about this? Is this proposal just common sense, or does it interfere with reasonable trade if people want to buy those products?

Active LEARNING

Environment, Science, and Policy in Your Community

Take several minutes with your classmates and make a list of five current, environment-related policy issues in your community or state. (Hint: Consider energy, economic developments, population changes, traffic congestion, water and air quality, or public health issues in the news.) Discuss which branch of government—legislative, executive, or judicial—you think should take the lead in addressing each issue. Why did you designate these authorities as you did?

Select one issue from your list, and discuss the following: What are some of the sources of uncertainty involved in the decision-making process? How could scientific research help identify or solve this policy issue? Where in the rule-making process could scientific data help inform decision makers?

16.2 MAJOR ENVIRONMENTAL LAWS

- *The National Environmental Policy Act (NEPA) of 1969 underlies many key environmental policies.*
- *The Clean Air Act and other laws are critical for public health.*
- *Environmental laws also have major economic benefits.*

We depend on countless laws to protect the water we drink, the air we breathe, the food we eat, and the biodiversity that surrounds us. Most of these laws work to negotiate a balance between the differing interests and needs of various groups, or between private interests and the public interest. The foundational policies discussed here date from the 1970s. They responded to decades of unregulated post-war environmental degradation, including rapid industrialization and the growth of polluting petrochemical and mining industries. They also occurred at a time of strong bipartisan action in Washington: The lobbying industry was still small, and corporate money in politics was legally restricted. Environmental policy since then has been expanded, refined, modified, and in some cases reduced, but the policies noted here remain the backbone of much environmental law in the United States. Many other countries have similar legal frameworks, some of them influenced by U.S. laws.

NEPA (1969) establishes public oversight

The cornerstone of U.S. environmental policy is the **National Environmental Policy Act (NEPA).** Signed into law by President Nixon, and taking effect on January 1, 1970, NEPA is a model for many other countries.

NEPA does three important things: (1) It establishes the Council on Environmental Quality (CEQ), the oversight board for general environmental conditions, (2) it directs federal agencies to take environmental consequences into account in decision making, and (3) it requires that an **environmental impact statement (EIS)** be published for every major federal project that is likely to have an important effect on environmental quality. NEPA doesn't forbid environmentally destructive activities if they otherwise comply with relevant laws, but it requires that agencies admit publicly what they plan to do. If embarrassing information is revealed publicly, it becomes harder for agencies to ignore public opinion. An EIS can provide valuable information about government actions to public interest groups that wouldn't otherwise have access to these resources.

What kinds of projects require an EIS? The activity must be federal and it must be major, with a significant environmental impact (fig. 16.5). Whether specific activities meet these characteristics is often a subjective decision. Each case is unique and depends on

▲ FIGURE 16.5 Every major federal project in the United States must be preceded by an environmental impact statement. ©Russell Illig/Getty Images

context, geography, the balance of beneficial versus harmful effects, and whether any areas of special cultural, scientific, or historical importance might be affected. A complete EIS for a project is usually time-consuming and costly. The final document is often hundreds of pages long and generally takes 6 to 9 months to prepare. Sometimes just requesting an EIS is enough to sideline a questionable project. In other cases, the EIS process gives adversaries time to rally public opposition and information with which to criticize what's being proposed. If agencies don't agree to prepare an EIS voluntarily, citizens can petition the courts to force them to do so.

Every EIS must contain the following elements: (1) purpose and need for the project, (2) alternatives to the proposed action (including taking no action), and (3) a statement of positive and negative environmental impacts of the proposed activities. In addition, an EIS should make clear the relationship between short-term resource effects and long-term productivity, as well as any irreversible impacts on resources resulting from the project.

The Clean Air Act (1970) regulates air emissions

The first major environmental legislation to follow NEPA was the Clean Air Act (CAA) of 1970 Air quality has been a public concern since the beginning of the Industrial Revolution, when coal smoke, airborne sulfuric acid, and airborne metals such as mercury became common in urban and industrial areas around the world (fig. 16.6). Sometimes these conditions produced public health crises: In India, for example, the World Health Organization estimates that 2.5 million people die every year from diseases related to dirty air.

Although crises of this magnitude have been rare, chronic exposure to bad air has long been a leading cause of illness in many areas. The Clean Air Act provided the first nationally standardized rules in the United States to identify, monitor, and reduce air contaminants. The core of the act is an identification and regulation of seven major "criteria pollutants," also known as "conventional pollutants": sulfur oxides, lead, carbon monoxide, nitrogen oxides (NO_x), particulates (dust), volatile organic compounds, and metals and halogens (such as mercury and bromine compounds).

Most of these pollutants have declined dramatically since 1970. An exception is NO_x, which derives from internal combustion engines such as those in our cars. Further details on these pollutants are covered in chapter 10.

▲ FIGURE 16.6 The Clean Air Act has greatly reduced the health and economic losses associated with air pollution. ©William P. Cunningham

The Clean Water Act (1972) protects surface water

Water protection has been a goal with wide public support, in part because clean water is both healthy and an aesthetic amenity. The act aimed to make the nation's waters "fishable and swimmable," that is, healthy enough to support propagation of fish and shellfish that could be consumed by humans and low in contaminantsso that they were safe for swimming and recreation.

The first goal of the Clean Water Act (CWA) was to identify and control point-source pollutants—end-of-the-pipe discharges from factories, municipal sewage treatment plants, and other sources. Discharges are not eliminated, but effluent at pipe outfalls must be tested, and permits are issued that allow moderate discharges of low-risk contaminants such as nutrients or salts. Metals, solvents, oil, high counts of fecal bacteria, and other more serious contaminants must be captured before water is discharged from a plant.

By the late 1980s, point sources were increasingly under control, and the CWA was used to address nonpoint sources, such as runoff from urban storm sewers. The act has also been used to promote watershed-based planning, in which communities and agencies collaborate to reduce contaminants in their surface waters. As with the CAA, the CWA provides funding to aid pollution-control projects. Those funds have declined in recent years, however, leaving many municipalities struggling to pay for aging and deteriorating sewage treatment facilities. For more detail on the CWA and water pollution control, see chapter 11 and Key Concepts (p. 388).

The Endangered Species Act (1973) protects wildlife

The Endangered Species Act (ESA) provides a structure for identifying and listing species that are vulnerable, threatened, or endangered. Once a species is listed as endangered, this act provides rules for protecting it and its habitat, ideally in order to help make

▶ **FIGURE 16.7** The Endangered Species Act is charged with protecting species and their habitat. The black-footed ferret was declared extinct in 1979, but a remnant population was discovered in 1981, and captive breeding programs have restored the species to more than 1,000 mature, wild-born animals in eight states. ©Jeff Vanuga/Digital Stock/Corbis

recovery possible (fig. 16.7). Listing of a species has become a controversial process, because habitat conservation tends to get in the way of land development. For example, many ESA controversies arise when developers want to put new housing developments in scenic areas where the last remnants of a species occur. To reduce disputes, the ESA provides habitat and land-use planning assistance and grants, as well as guaranteeing landowner rights when an effective habitat conservation plan has been developed. Landowners can also get a tax reduction in exchange for habitat conservation. These strategies increasingly allow for both development and species protection.

The ESA maintains a worldwide list of endangered species. In 2018 the list included 24,431 species, 1,950 of them in the United States. The responsibility for studying and attempting to restore threatened and endangered species lies mainly with the Fish and Wildlife Service and the National Oceanic and Atmospheric Administration. You can read more about endangered species, biodiversity, and the ESA in chapter 5.

The Superfund Act (1980) addresses hazardous sites

Most people know this law as the Superfund Act because it created a giant fund to help remediate abandoned toxic sites. The proper name of this law is informative, though: the Comprehensive Environmental Response, Compensation, and Liability Act (CERCLA). The act aims to be comprehensive, addressing abandoned sites, emergency spills, or uncontrolled contamination, and it allows the EPA to try to establish liability, so that polluters help to pay for cleanup. Because it's much cheaper to make toxic waste than to clean it up, we have thousands of chemical plants, gas stations, and other sites that have been abandoned because they were too expensive to clean properly. The EPA is responsible for finding a private party to do cleanup, and the Superfund was established to cover the costs, which can be in the billions of dollars.

Originally the fund was supplied mainly by contributions from industrial producers of hazardous wastes. In 1995, however, Congress voted to end that source, and the Superfund was allowed to dwindle to negligible levels. Since then, taxpayers have picked up the bill and corporations have had little responsibility for cleanups. According to the EPA, one in four Americans lives within 3 miles of a hazardous waste site. The Superfund program has identified more than 47,000 sites that may require cleanup. The most serious of these (or the most serious for which proponents have been sufficiently vigorous) have been put on a National Priorities List. About 1,600 sites have been put on this list, and over 1,000 cleanups have been completed. The total cost of remediation is expected to be somewhere between $370 billion and $1.7 trillion.

16.3 HOW ARE POLICIES IMPLEMENTED?

- *Legislative bodies vote on statutory laws; judges decide case law.*
- *Executive agencies make rules and enforce laws.*
- *Regulatory capture is installing an agency leader antagonistic to the agency's mission.*

The laws discussed above are among our most important environmental protections (see also table 16.1). Each of these laws resulted from action at many local and national levels. Individual citizens lobbied for change. State and federal representatives negotiated policies. Courts tested the legitimacy of the law. Local, state, and federal agencies worked to implement the law.

We examine here the ways that laws are enacted in the United States at the federal level. Environmental law can be established or modified in each of the three branches of government—legislative, judicial, and executive. Many of the 50 states have bodies with similar structure and function.

The legislative branch establishes statutes (laws)

Federal laws, also called **statutes,** are enacted by Congress and signed by the president. Thousands of proposed laws, or bills, are introduced every year in Congress. Some are very narrow, providing funds to build a specific section of road or to help a particular person, for instance. Others are extremely broad, perhaps overhauling the Social Security system or changing the tax code. Sometimes a bill will have 100 or more coauthors and more than 1,000 pages of legal text if it is a prominent issue.

All bills and all public laws (bills passed by Congress) are available for you to examine, as part of the public record. You can find the details of any national law by looking at www.thomas.gov. The EPA also provides access to the text of environmental legislation at www.epa.gov/lawsregs/.

Citizens can be involved in the legislative process by writing or calling their elected representatives and by appearing at public hearings. A personal letter or statement is always more persuasive

TABLE 16.1 Major U.S. Environmental Laws

LEGISLATION	PROVISIONS
Wilderness Act of 1964	Established the national wilderness preservation system.
National Environmental Policy Act of 1969	Declared national environmental policy, required environmental impact statements, created Council on Environmental Quality.
Clean Air Act of 1970	Established national primary and secondary air quality standards. Required states to develop implementation plans. Major amendments in 1977 and 1990.
Clean Water Act of 1972	Set national water quality goals and created pollutant discharge permits. Major amendments in 1977 and 1996.
Federal Pesticides Control Act of 1972	Required registration of all pesticides in U.S. commerce. Major modifications in 1996.
Marine Protection Act of 1972	Regulated dumping of waste into oceans and coastal waters.
Coastal Zone Management Act of 1972	Provided funds for state planning and management of coastal areas.
Endangered Species Act of 1973	Protected threatened and endangered species. Directed FWS to prepare recovery plans.
Safe Drinking-Water Act of 1974	Set standards for safety of public drinking-water supplies and to safeguard groundwater. Major changes made in 1986 and 1996.
Toxic Substances Control Act of 1976	Authorized EPA to ban or regulate chemicals deemed a risk to health or the environment.
Federal Land Policy and Management Act of 1976	Charged the BLM with long-term management of public lands. Ended homesteading and most sales of public lands.
Resource Conservation and Recovery Act of 1976	Regulated hazardous-waste storage, treatment, transportation, and disposal. Major amendments in 1984.
National Forest Management Act of 1976	Gave statutory permanence to national forests. Directed USFS to manage forests for "multiple use."
Surface Mining Control and Reclamation Act of 1977	Limited strip-mining on farmland and steep slopes. Required restoration of land to original contours.
Alaska National Interest Lands Act of 1980	Protected 40 million ha (100 million acres) of parks, wilderness, and wildlife refuges.
Comprehensive Environmental Response, Compensation, and Liability Act of 1980	Created $1.6 billion "Superfund" for emergency response, spill prevention, and site remediation for toxic wastes. Established liability for cleanup costs.
Superfund Amendments and Reauthorization Act of 1994	Increased Superfund to $8.5 billion. Shared responsibility for cleanup among potentially responsible parties. Emphasized remediation and public "right to know."

Source: Data from N. Vig and M. Kraft, *Environmental Policy in the 1990s*, 3rd Congressional Quarterly Press.

than simply signing a petition. A petition with a million signatures will probably catch the attention of a legislator—especially if they are all potential voters.

Being involved in local election campaigns can greatly increase your access to legislators. Media attention can also sway the opinions of decision makers. Drawing public attention to an issue or campaign can also help influence public opinion and enthusiasm for an issue (fig. 16.8). It's hard for a single individual to have much impact, but if you join a group, you can have a collective impact on public policy.

The judicial branch resolves legal disputes

The judicial branch of government decides (1) what the precise meaning of a law is, (2) whether or not laws have been broken, and (3) whether a law violates the Constitution. The cumulative body of legal opinions from court cases is known as case law. Often this law involves interpreting what a law really means. Interpretation is needed because legislation is frequently written in vague and general terms so as to make it widely enough accepted to gain passage. When trying to interpret a law, the courts depend on the legislative record from hearings, such as what was said by whom, to determine congressional intent.

When a law may have been broken, it becomes a matter of **criminal law.** Serious crimes such as murder or theft, as well as

▲ **FIGURE 16.8** Nonviolent protests at the Oceti Sakowin camp at Standing Rock in North Dakota. Protesters won attention and sympathy from many in the general public when officials responded with heavy-handed tactics. ©Robyn Beck/Getty Images

KEY CONCEPTS

How does the Clean Water Act benefit you?

Environmental policies are rules we establish to protect public health and resources. The 1972 Clean Water Act is one of the most important and effective environmental laws in America. Dramatic improvements in water quality have resulted from the billions of dollars in public and private investment in pollution control. While we still haven't met all the goals of this momentous act, the water in your neighborhood is almost certainly cleaner than it was 40 years ago.

To learn the details for yourself, find further explanation, and the text of the law, at **http://water.epa.gov/lawsregs/lawsguidance**.

What does the Clean Water Act do?

- Establishes rules for regulating discharges of pollutants into water
- Charges the EPA with establishing and regulating standards for water quality
- Makes it illegal to pollute navigable waters from a point source (such as a discharge pipe) without a permit

©Jennifer Kivioja

KC 16.1

◀ One of the first steps in cleaning up surface waters was to stop the dumping of untreated sewage and industrial effluents. Over the past 40 years, the United States has spent at least $2 trillion in public and private funds on point-source pollution control.

KC 16.2

◀ Nonpoint-source pollution, such as runoff from city streets and random dumping, is harder to control, but there has been dramatic improvement in this area as well.

©Mary Ann Cunningham

©Steve Allen/Brand X Picture/Alamy Stock Photo

KC 16.3

◀ A major goal of the Clean Water Act was to make all U.S. surface waters "fishable and swimmable." This aim has been partly a success and partly a failure. The EPA has reported that over 90 percent of all monitored river miles and 87 percent of assessed lake acres meet this goal. However, many waters remain impaired. You can't necessarily eat all the fish you catch, and it may be unsafe to drink the water in which you swim.

©William P. Cunningham

Regular monitoring is an important part of protecting and improving water quality ▶. The EPA monitors 4,000 watersheds for bacteria, nutrients, metals, clarity, and other standard measures. States are required to develop total maximum daily loads for each expected pollutant and each listed water body. Biological organisms often make good indicators of water quality. You can learn about water quality in your area from **www.epa.gov/owow/tmdl/**.

KC 16.4

©Mary Ann Cunningham

Agricultural pollution remains a serious problem. A majority of remaining water pollution in the United States is thought to come from soil erosion, nutrient runoff from farm fields, feedlots, and other farm operations. Hundreds of millions of tons of fertilizers, pesticides, and manure wash into rivers and lakes. Excess fertilizer causes vast dead zones or explosive growth of harmful algae in the estuaries of major rivers. ▶

Novel pollutants—chemicals the EPA doesn't yet test for—are also a growing concern.

KC 16.5

©Image Source/Getty Images

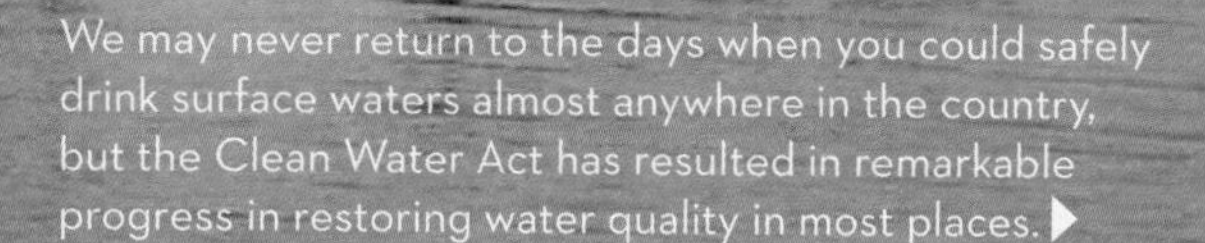

We may never return to the days when you could safely drink surface waters almost anywhere in the country, but the Clean Water Act has resulted in remarkable progress in restoring water quality in most places. ▶

Like other environmental laws, this one is imperfect and incomplete, but it provides safeguards we depend on, usually without even knowing they're there.

KC 16.6

©M. L. Heinselman

CAN YOU EXPLAIN?

1. Ask your parents or grandparents what water quality was like in your area 40 years ago. What progress has been made?
2. What water quality problems do you think remain most troublesome?
3. What do you regard as the most valuable provisions of the Clean Water Act?

criminal violations of environmental laws, are matters of criminal law. Charges are usually brought by the state, rather than by individuals, and usually proof of intent or willful neglect of responsibilities is needed for a criminal conviction. **Civil law,** on the other hand, aims to resolve disputes between individuals and corporations. Civil disputes can involve issues such as property rights, injury, or personal freedom, and proof of criminal intent is not necessarily required. Fines can result from either civil disputes or criminal suits, but jail time results only from criminal cases. Civil suits can also be used to stop activities, such as endangering water resources, endangered species, or public health.

Damages can be difficult to collect after a successful suit. For example, when Texaco stopped oil drilling in Ecuador in 1992, after nearly 30 years of operations, it left behind what critics describe as a toxic dump of 6.8 million liters (1.8 million gallons) of spilled crude oil and drilling wastes–almost twice as much as Alaska's *Exxon Valdez* spill. In 2003, lawyers representing 30,000 indigenous Ecuadorians filed a lawsuit against Texaco claiming environmental damage and adverse health effects from more than 350 open toxic waste pits. The case dragged on until 2011, when an Ecuadoran court ordered Chevron, which had bought Texaco (and its liabilities) in 2001, to pay $18 billion in damages. A district judge in Manhattan immediately issued an injunction to block payment, but a U.S. Appeals Court threw out that injunction. Still, in 2018, more than 50 years after the drilling started, no payments had been made to those who suffered damage (or their heirs).

If a lawsuit questions the legitimacy of a law itself, the suit may be decided by the U.S. Supreme Court (fig. 16.9). The nine justices of the federal Supreme Court decide whether a law is consistent with the Constitution of the United States. States also have supreme courts, which determine whether a law is consistent with the state's constitution.

Because the justices of the court interpret the Constitution and its meaning, they have far-reaching influence over our laws, policies, and practices. An important example is carbon dioxide regulation. How or whether greenhouse gases could be regulated has been debated for decades, but finally in 2007, the Supreme Court issued a landmark decision that greenhouse gases endanger public health and welfare. Some observers worry, however, that with a new conservative majority, that finding may not stand.

▲ **FIGURE 16.9** The Supreme Court decides pivotal cases, many of them bearing on resources or environmental health. ©Bloomberg/Getty Images

Perhaps the most widely debated Supreme Court case in recent years was the 2010 decision *Citizens United v. Federal Election Commission*. This decision determined that the government may not limit political spending by corporations. Reversing 70 years of precedent and multiple previous court rulings, the *Citizens United* case was decided on a 5-to-4 vote, and it unleashed a torrent of political spending in the 2012 elections.

The 2012 election quickly became the most expensive in U.S. history, and subsequent elections have been more costly still. Political action committees now spend billions of dollars to attack opponents. Corporations, millionaires, and political organizations fund countless TV ads and media campaigns. Victory isn't necessarily guaranteed to these financers, but historically the better-funded candidate usually wins. Many regard this Supreme Court decision as a serious threat to democracy, because it makes politicians responsive primarily to campaign donors, rather than to voters.

The executive branch oversees administrative rules

Within the executive branch various agencies, boards, and commissions oversee environmental policies. These bodies set rules, decide disputes, and investigate misconduct. At the federal level, executive rules are made that can have far-reaching environmental consequences. Executive rules can be made quickly and with little interference from Congress. They are often a strategy to change public policy with little public oversight or discussion. For example, President Obama's administration established the Clean Power Plan in 2015 that limited greenhouse gas emissions from power plants. But shortly after taking office in 2017, the Trump administration withdrew that rule. So, while rules can be made relatively rapidly, they can be overturned quickly, too.

The executive branch includes administrative agencies, such as the Environmental Protection Agency (EPA), which oversee and enforce public laws. The EPA is the primary agency with responsibility for protecting environmental quality in the United States. The EPA was created in 1970 at the same time as NEPA.

Other agencies that have profound environmental impacts are the Department of the Interior, which oversees public lands and national parks, and the Department of Agriculture, which oversees the nation's forests and grasslands, as well as agricultural issues. The Department of the Interior is home to the U.S. Fish and Wildlife Service, which operates more than 500 national wildlife refuges and administers endangered species protection. The Department of Agriculture is home to the U.S. Forest Service, which manages about 175 national forests and grasslands, totaling some 78 million ha (193 million acres). With 39,000 employees, the Forest Service is nearly twice as large as the EPA (fig. 16.10).

How much government do we want?

In his 1981 inaugural address, President Ronald Reagan famously said, "Government is not the solution to our problem; government is the problem." In this, he invoked a perennial debate in American politics: Is government a power that undermines personal liberties? Or is government a defender of liberties and rights, in a world where bullies too easily have their way? Do governments defend their own interests? Or are they a mechanism for defending the public interest and the common good?

The answer sometimes depends on when you ask. During political campaigns, many people decry the size and cost of government agencies. During times of crisis, most of us assume the government will step in to help (fig. 16.11). Notably, in some surveys, more than half the people who receive food stamps, Social Security payments, public schooling, and police and fire protection, or who drive on public highways, claim they have never benefited from a government program. Similarly, corporations eager to avoid taxes nonetheless rely on the state to provide educated workers, roads, communication infrastructure, and professional police forces to maintain order.

Debates about the proper size and role of government are always present. We value self-reliance and individualism. Yet we also want someone to protect us from contaminated food and drugs, to educate our children, and to provide roads, bridges, and safe drinking water. Many conservatives favor "free market" capitalism, with businesses unfettered by rules such as the Clean Air Act or the Clean Water Act. Liberals, on the other hand, argue that anti-government rhetoric focuses on freedom for individuals and small business owners, but the elimination of public health and safety regulations undermines the interests of those it claims to protect. Advantages are often enjoyed by the biggest players in a rule-less game.

◀ **FIGURE 16.10** Smokey Bear is the public face of an executive agency, the U.S. Forest Service. ©William P. Cunningham

▲ **FIGURE 16.11** When tornadoes, floods, hurricanes, or other natural disasters afflict us, we expect government agencies to help us. Source: Andrea Booher/FEMA

With the inauguration of President Trump in 2017 the small-government philosophy gained influence in the U.S. government. This resulted in a dramatic shift in U.S. environmental policy. One of the strategies used to disable agencies was to designate an agency head who is antagonistic to the agency's mission, a process often descried as **regulatory capture.** The new head can eliminate staff, choose not to pursue policies, or actively change rules. The Trump administration also nominated agency staff members who had been lobbyists or employees of the industries they were to regulate. For example, Scott Pruitt, who became head of the EPA, had been attorney general of Oklahoma, where he sued the EPA more than a dozen times to overturn clean air and water rules. In his first 4 months in office, Pruitt reversed or delayed dozens of air, water, and climate regulations. He dismissed members of his scientific review board and replaced them with industry representatives whose pollution the agency is supposed to regulate.

Similarly, Rick Perry, head of the Department of Energy, said as a candidate that he wanted to eliminate the whole agency, and Ryan Zinke, head of the Department of the Interior, promised to open coastal waters to oil drilling, shrink national monuments, lift fossil fuel regulations, and reduce wildlife protection, all steps he implemented rapidly upon taking office. In his first budget proposal, Trump cut renewable energy research by 72 percent and reduced the EPA budget by one-third. And his $200 billion infrastructure plan released in 2018 was much less about rebuilding roads and bridges than about reducing environmental reviews for fossil fuel developments and privatizing public lands.

Regulatory capture can lead directly to public crises. The 2007–08 financial collapse on Wall Street, which caused widespread business failures and unemployment, resulted from the dismantling of rules that were intended to control risky gambling by banks. The dismantling was overseen by President George W. Bush's chair of the Securities and Exchange Commission. President Obama's administration implemented new restrictions on risky trading, to avoid further bank failures, but President Trump installed another anti-regulatory Wall Street lawyer, Walter Clayton, to once again weaken bank regulation.

Debates about the importance of rule-making probably will always be with us. Which view is most correct depends on many factors—your interest group, life experience, philosophical perspective, and economic position; the time frame you analyze; and other priorities.

16.4 INTERNATIONAL POLICIES

- *The Montreal Protocol shows the power of global cooperation.*
- *The 1994 UNFCCC seeks to slow climate change.*
- *Self-reporting on progress is often the only means of enforcement among states.*

With growing recognition of the interconnections in our global environment, nations have become increasingly interested in signing on to international agreements (often called "treaties" or "conventions") for environmental protection (table 16.2). A principal motivation in these treaties is the recognition that countries can no longer act alone to protect their own resources and interests. Water resources, the atmosphere, trade in endangered species, and many other concerns cross international boundaries. Over time, the number of parties taking part in negotiations has grown, and the speed at which agreements take force has increased. The Convention on International Trade in Endangered Species (CITES), for example, was not enforced until 14 years after its ratification in 1973, but the Convention on Biological Diversity (1991) was enforceable after just 1 year and had 160 signatories only 4 years after introduction.

Over the past 25 years, more than 170 treaties and conventions have been negotiated to protect our global environment. Designed to regulate activities ranging from intercontinental shipping of hazardous waste to deforestation, overfishing, trade in endangered species, global warming, and wetland protection, these agreements theoretically cover almost every aspect of human impacts on the environment.

Many of these policies have emerged from major international meetings. One of the first of these was the 1972 UN Conference on the Human Environment in Stockholm, which set an agenda for subsequent meetings. This conference gathered representatives of 113 countries and several dozen nongovernmental organizations. A much larger gathering was the "Earth Summit" 20 years later. Officially called the UN Conference on Environment and Development (UNCED), this meeting was held in Rio de Janeiro in 1992. This time over 110 nations participated, as well as 2,400 nongovernmental organizations. The Rio meeting was such a watershed event that it was repeated in 2012, on its 20th anniversary. The Agenda 21 document produced from the first Rio meeting has laid out principles of sustainability and equity that have guided much policymaking since 1992.

TABLE 16.2 Some Important International Treaties

CBD: Convention on Biological Diversity 1992 (1993)*
CITES: Convention on International Trade on Endangered Species of Wild Fauna and Flora 1973 (1987)
CMS: Convention on the Conservation of Migratory Species of Wild Animals 1979 (1983)
Basel: Basel Convention on the Transboundary Movements of Hazardous Wastes and Their Disposal 1989 (1992)
Ozone: Vienna Convention for the Protection of the Ozone Layer and Montreal Protocol on Substances That Deplete the Ozone Layer 1985 (1988)
UNFCCC: United Nations Framework Convention on Climate Change 1992 (1994)
CCD: United Nations Convention to Combat Desertification in Those Countries Experiencing Serious Drought and/or Desertification, Particularly in Africa 1994 (1996)
Ramsar: Convention on Wetlands of International Importance Especially as Waterfowl Habitat 1971 (1975)
Heritage: Convention Concerning the Protection of the World Cultural and Natural Heritage 1972 (1975)

*The first date listed is when the treaty was enacted. The second date is when it went into force.

Major international agreements

International accords and conventions have emerged slowly but fairly steadily from meetings such as those in Stockholm and Rio (table 16.2). A few of the important benchmark agreements are discussed here.

The **Convention on International Trade in Endangered Species (CITES, 1973)** declared that wild flora and fauna are valuable, irreplaceable, and threatened by human activities. To protect disappearing species, CITES maintains a list of threatened and endangered species that may be affected by trade. As with most international agreements, this one takes no position on movement or loss of species within national boundaries, but it establishes rules to restrict unauthorized or illegal trade across boundaries. In particular, an export permit is required specifying that a state expert declares an export is legal, that it is not cruel, and that it will not threaten a wild population.

The **Montreal Protocol (1987)** protects stratospheric ozone. This treaty committed signatories to phase out the production and use of several chemicals that break down ozone in the atmosphere. The ozone "hole," a declining concentration of ozone (O_3) molecules over the South Pole, threatened living things: Ozone high in the atmosphere blocks cancer-causing ultraviolet radiation, keeping it from reaching the earth's surface. The stable chlorine- and fluorine-based chemicals at fault for reducing ozone are used mainly as refrigerants. Alternative refrigerants have since been developed, and the use of chlorofluorocarbons (CFCs) and related molecules has plummeted. Although the ozone "hole" has not disappeared, it has declined as predicted by atmospheric scientists since the phase-out of CFCs. The Montreal Protocol is often held up as an example of a highly successful and effective international environmental agreement.

The Montreal Protocol was effective because it bound signatory nations not to purchase CFCs or products made using them from countries that refused to ratify the treaty. This trade restriction put substantial pressure on producing countries. Initially the protocol called for only a 50 percent reduction in CFC production, but subsequent research showed that ozone was being depleted faster than previously thought (see chapter 10). The protocol was strengthened to an outright ban on CFC production, in spite of the objection of a few countries.

The **Basel Convention (1992)** restricts shipment of hazardous waste across boundaries. The aim of this convention, which has 172 signatories, is to protect health and the environment, especially in developing areas, by stating that hazardous substances should be disposed of in the states that generated them. Signatories are required to prohibit the export of hazardous wastes unless the receiving state gives prior informed consent, in writing, that a shipment is allowable. Parties are also required to minimize production of hazardous materials and to ensure that there are safe disposal facilities within their own boundaries. This convention establishes that it is the responsibility of states to make sure that their own corporations comply with international laws. The Basel Convention was enhanced by the Rotterdam Convention (1997), which places similar restrictions on unauthorized transboundary shipment of industrial chemicals and pesticides.

The 1994 **UN Framework Convention on Climate Change (UNFCCC)** directs governments to share data on climate change, to develop national plans for reducing greenhouse gas (GHG) emissions, and to cooperate in planning for adaptation to climate change. Under the UNFCCC, the Kyoto Protocol (1997) set binding targets for signatories to reduce greenhouse gas emissions to less than 1990 levels by 2012. The idea of binding targets was strong, but few countries achieved their goals, and some, such as the United States, declined to be subjected to international policies. A key dispute was that restrictions were stronger for industrialized countries, which are responsible for roughly 90 percent of GHG emissions up to the present, than for developing countries. In 2012, Canada also withdrew from the agreement, in part because energy-intensive development of oil shale made it impossible for Canada to meet its targets.

In 2015, the UNFCCC led to the Paris Agreement, which set a climate change target of "well below" 2°C. Within a year, 195 countries agreed to establish their own national determined contributions to this goal (fig. 16.12). As with Kyoto, plans and reporting are to be transparent, with regular updates as technology improves (see chapter 9).

▲ **FIGURE 16.12** President Barack Obama met with Chinese premier Xi Jinping at the United Nations Climate Change Conference in Paris in 2015. Their two countries are the largest emitters of GHGs. ©JIM WATSON/Getty Images

Some observers consider Paris to be a historic turning point in the effort to combat climate change, while others criticize the fact that pledges remain insufficient to meet the target. A particularly important question is progress in the United States, where Congress has blocked clean power legislation, and in China, which continues to build coal power plants to meet rising demand for electricity. Early in his presidency, Donald Trump announced that the U.S. would reject the Paris accord, but that process takes at least 3 years. Meanwhile, many U.S. cities and states are working to set their own goals (see section 9.4).

Enforcement often relies on national pride

Enforcement of international agreements frequently depends on how much countries care about their international reputation. Except in extreme cases such as genocide, the global community is unwilling to send an external police force into a country, because states are wary of interfering with the internal sovereignty of other states. However, most countries also are reluctant to appear irresponsible or immoral in the eyes of the international community, so moral persuasion and public embarrassment can be effective enforcement strategies. Shining a spotlight on transgressions will often push a country to comply with international agreements.

But national pride can also stand in the way of species protection. In 2010, Japan almost single-handedly derailed CITES protection for bluefin tuna, despite the fact that these beautiful, highly evolved, long-lived top predators are now at less than 5 percent of historic levels. Part of Japan's objections to fishing limits is that they don't like to be told what they can or can't eat. But economics plays a key role. This iconic fish is extremely lucrative, and its high value, often thousands of dollars for a single fish, signifies great wealth (fig. 16.13). Dwindling populations give it still more prestige, and companies are now stockpiling frozen bluefins, which will be worth far more on the sushi market once they are extinct.

▲ **FIGURE 16.13** Bluefin tuna, shown here at Tokyo's Tsukiji market, are endangered or threatened worldwide. Japan has almost single-handedly blocked CITES listing, although 96 percent may be gone. ©liorpt/123RF

Often international negotiators aim for unanimous agreement to ensure strong acceptance of international policies. While this approach makes for a strong agreement, a single recalcitrant nation can have veto power over the wishes of the vast majority. For instance, more than 100 countries at the UN Conference on Environment and Development (UNCED), held in Rio de Janeiro in 1992, agreed to restrictions on the release of greenhouse gases. At the insistence of U.S. negotiators, however, the climate convention was reworded so that it only urged—but did not require—nations to stabilize their emissions.

When a consensus cannot be reached, negotiators may seek an agreement acceptable to a majority of countries. This approach was used in negotiating the Kyoto Protocol on climate change, which sought, and eventually achieved, agreement from a majority of countries. Only signing countries are bound by such a treaty, but nonsigning countries may comply anyway, to avoid international embarrassment.

When strong accords with meaningful sanctions cannot be passed, sometimes the pressure of world opinion generated by revealing the sources of pollution can be effective. Activists can use this information to expose violators. For example, the environmental group Greenpeace discovered monitoring data in 1990 showing that Britain was disposing of coal ash in the North Sea. Although not explicitly forbidden by the Oslo Convention on ocean dumping, this evidence proved to be an embarrassment, and the practice was halted.

Trade sanctions can be an effective tool to compel compliance with international treaties. The Montreal Protocol used the threat of trade sanctions very effectively to cut CFC production dramatically (see chapter 10). On the other hand, trade agreements also can work against environmental protection. The World Trade Organization (WTO) was established in 1995 to promote free international trade and to encourage economic development. The WTO's emphasis on unfettered trade, however, has sometimes led to weakening of local environmental rules.

16.5 WHAT CAN STUDENTS DO?

- *Environmental groups are a voice for resource conservation.*
- *Environmental literacy includes knowledge about ecological systems, local environments, and current environmental issues.*
- *Campus activities give you a voice and experience.*

The most important step you can take is what you are doing right now: learning to understand environmental science and the ways it informs environmental practices. But students and citizens can also join groups to work on policy issues, develop skills that make them more effective advocates, work locally on school policy issues, and more. Working with student organizations is a good place to start: These give you the opportunity to learn leadership skills, which are good for employment as well as for activism. If your school has a campus sustainability office, that is also a place to gain experience and learn about influencing practices and standards locally. You should also be in regular contact with elected representatives. Despite the importance of money in politics, policymakers do record and respond to voters' opinions (What Can You Do?, p. 395).

If you are interested in a career in environmental policy, law, science, journalism, or other related fields, getting involved in student groups is a good way to build knowledge and experience. Many of today's environmental and political leaders gained experience in policy, leadership, and organizing by working with student groups (fig. 16.14). These activities can be as informative and life-changing as the courses you take, because they give you the experience of being in charge and making decisions.

▲ **FIGURE 16.14** We can learn about environmental systems almost anywhere. The first step is getting outside to have fun. ©William P. Cunningham

Working together gives you influence, and it's fun

Working together is a good way to get encouragement, energy, and useful information about issues you're interested in. It's easy to get discouraged by the slow pace of change, so having a group for support is important.

Public organizations often arise in response to environmental and social challenges. Many of the groups we consider establishment institutions today started as associations of young radicals fighting powerful political and economic forces. Although the largest organizations have sometimes become less radical, they also have important influence in policy circles. In the United States, these include the National Wildlife Federation, the World Wildlife Fund, the Audubon Society, the Sierra Club, Greenpeace, the Natural Resources Defense Council, and The Wilderness Society. The Audubon Society organized to protect egrets and other birds, which were being slaughtered for their plumes to decorate ladies' hats. The Sierra Club was formed to protect the giant redwood trees of California, which were rapidly being logged for timber. The Sierra Club then went on to fight for National Parks throughout the American West. The Wilderness Society was created to protect wild lands from development.

As these groups have grown and become more established, they have often become less radical, but many have also become more effective on policy questions. Often, professional staff work with lawmakers and research issues. Most members are only occasionally involved in organizational activities, but they depend on organization staff to promote their interests and keep them informed.

Volunteers rarely have the time or expertise to give effective leadership in policymaking, but they can step up when urgent action is needed. In addition, many of these groups have local chapters that sponsor social gatherings, outings, volunteer work, or political activities. You could join one of these groups to learn about issues, have fun, and meet people.

Established groups are influential because their mass membership, professional staffs, and long history give them a degree of respectability and influence not found in newer, smaller groups. The Sierra Club, for instance, with more than 2.5 million members in 64 chapters, has a national staff of about 200, an annual budget of about $100 million, and up to 20 full-time professional lobbyists in Washington, D.C. These national groups are a potent force in Congress, especially when they band together to push important legislation, such as the Endangered Species Act or the Clean Air Act.

What Can YOU DO?

Actions to influence environmental policy

In addition to building environmental literacy, the following are important steps individuals can take.

1. Contact your elected representatives regularly and tell them what you think about policy matters.
2. Vote.
3. Encourage others to vote.
4. Join civic organizations to multiply the power of your voice.
5. Practice leadership locally—join student governance groups or community groups, attend city council meetings, organize activities in your neighborhood.
6. Apply your education: Become an environmental scientist, journalist, artist, engineer, or other practitioner.
7. Find creative ways to build sustainability into any job you do: Encourage recycling, smart energy conservation, and other practices.
8. Practice conservation at home: Drive less, walk more, avoid disposable goods, air-dry your clothes, eat a little less meat, avoid watering your lawn, buy efficient appliances.
9. Stay informed: Read the paper, listen to the news, and not just the weather and murder reports.
10. Go outside and learn about your local environment.

New groups and approaches are emerging

International organizations are also important. International **nongovernmental organizations (NGOs),** that is, organizations not affiliated with governments, can be vital in the struggle to protect areas of outstanding biological value. Without this help, most local groups could never mobilize the public interest or financial support for major projects. Greenpeace, for instance, carries out well-publicized confrontations with whalers, seal hunters, toxic waste dumpers, and others who threaten very specific and visible resources. Greenpeace may well be the largest environmental organization in the world, claiming some 2.8 million contributing members.

New environmental organizations are increasingly focused on combined social and environmental causes. The 350.org network, for example, which played an important role in fossil fuel divestment, campaigns for social justice for the poorest people in the world, who will suffer the most from climate change. They also fight for intergenerational equity. Emerging leadership from historically disadvantaged movements often leads this effort. A prominent example of an environmental justice issue that changed many perceptions was the encampment at Standing Rock in North Dakota (see Science and Citizenship p. 397).

Find your own niche

Organizations, legislative offices, and other groups that influence policy are all composed of individuals who have committed their energy, education, or careers to causes they find important. Whatever your skills and interests, you can participate in policy formation and help in environmental protection.

If you enjoy science, there are many disciplines that contribute to environmental science. As you know by now, biology, chemistry, geology, ecology, climatology, geography, hydrology, and other sciences all provide essential ideas and data to environmental science.

Skills in art, writing, communication, working with children, history, politics, economics, and many other areas are also critical for developing ideas and engaging the public. Communicating the ideas of environmental science requires educators, policymakers, artists, and writers. Lawyers and other specialists are needed to develop and improve environmental laws and regulations. Engineers are needed to develop technologies and products to clean up pollution and to prevent its production in the first place. Economists and social scientists are needed to evaluate the costs of pollution and resource depletion and to develop equitable and appropriate solutions for different parts of the world. In addition, businesses will be looking for a new class of environmentally literate and responsible leaders who can help improve the green credentials of their products and services.

Environmental educators are also needed to help train an environmentally literate populace. We urgently need more teachers at every level who are trained in environmental education, and who are familiar with the outdoors. Environmental education is often strongest if it draws on your own experience. If you enjoy being outdoors and learning about your local environment, that is also a key part of environmental action (fig. 16.14). As author Edward Abbey wrote,

> It is not enough to fight for the land; it is even more important to enjoy it. While it is still there. So get out there and mess around with your friends, ramble out yonder and explore the forests, encounter the grizz, climb the mountains. Run the rivers, breathe deep of that yet sweet and lucid air, sit quietly for a while and contemplate the precious stillness, that lovely mysterious and awesome space.

TABLE 16.3 | Themes in Environmental literacy

Systems: An understanding of the earth as a physical system, including humans, societies, and their living environment

Science: A familiarity with basic modes of inquiry, critical thinking, and problem-solving skills; an ability to interpret and synthesize information

Citizenship: An understanding of the ideals, principles, and practices of citizenship in order to participate in resolving issues

Action: Empowerment and awareness of how to take action, as individuals and with groups

Source: North American Association of Environmental Education.

TABLE 16.4 | Most Influential Books on the Environment

What are some of the most influential readings that have informed environmental scientists? Surveys[1] of environmental leaders frequently put the same books at the top of the list. Here are votes for the top ten.

A Sand County Almanac by Aldo Leopold (100)[2]

Silent Spring by Rachel Carson (81)

State of the World by Lester Brown and the Worldwatch Institute (31)

The Population Bomb by Paul Ehrlich (28)

Walden by Henry David Thoreau (28)

Wilderness and the American Mind by Roderick Nash (21)

Small Is Beautiful: Economics as If People Mattered by E. F. Schumacher (21)

Desert Solitaire: A Season in the Wilderness by Edward Abbey (20)

The Closing Circle: Nature, Man, and Technology by Barry Commoner (18)

The Limits to Growth: A Report for the Club of Rome's Project on the Predicament of Mankind by Donella H. Meadows, et al. (17)

[1]Robert Merideth, 1992, G. K. Hall/Macmillan, Inc.

[2]Indicates number of votes for each book. Because the preponderance of respondents were from the United States (82 percent), American books are probably overrepresented.

Environmental literacy integrates science and policy

The National Environmental Education Act, passed by Congress in 1990, identified environmental education as a national priority. The act established two broad goals: (1) to improve public understanding of our environment and (2) to encourage postsecondary students to pursue careers related to the environment. Learning objectives (table 16.3) include awareness and appreciation of our environment, knowledge of basic ecological concepts, acquaintance with a broad range of current environmental issues, and experience in using investigative, critical thinking, and problem-solving skills in solving environmental problems (fig. 16.15). **Environmental literacy** is a working knowledge of our environment and its systems. Happily, pursuing environmental literacy is enjoyable as well as important. There are thousands of excellent books you can read. Some of them have been persistently important for many readers over time (table 16.4).

Practicing science is also a way to contribute while building environmental literacy. Internships in agencies or environmental organizations are one way of doing this. Another is to get involved in organized **citizen science** projects in which ordinary people join with established scientists to answer real scientific questions. The Audubon Christmas Bird Count, for example, gets community groups out birding together. The Cornell Lab of Ornithology organizes a wide variety of bird counts, from backyard feeders to urban parks and more. Many of these records contribute to the "ebird" database, which has provided key insights into migration patterns and timing. Countless community organizations also organize coastal restoration, river and lake monitoring, biodiversity studies, and more. All of these are good ways to become more literate about your local environment, and to have fun doing it (fig. 16.16).

▲ **FIGURE 16.15** Environmental education helps develop awareness and appreciation of ecological systems and how they work. ©William P. Cunningham

Colleges and universities are powerful catalysts for change

Educational institutions are centers of innovation in sustainability. Participating in campus environmental planning and research is a valuable way for students to engage and gain experience. A good

▲ **FIGURE 16.16** Working together with others can give you energy, inspiration, and a sense of accomplishment. ©William P. Cunningham

SCIENCE AND Citizenship

Water Protectors at Standing Rock

How do we learn to engage in the political process? Often we are too busy or too discouraged to take action; often we don't know how to engage. But sometimes a particular challenge galvanizes people, and new strategies for citizen action are learned.

The 2016 protest at the Standing Rock Sioux Reservation was one of these watershed moments. Thousands of protesters camped on the route of the Dakota Access Pipeline, being built from the Bakken oil fields in western North Dakota to southern Illinois. At issue: The pipeline threatened ancient burial grounds and sacred sites, and it crossed under the Missouri River just upstream from the reservation's water supply. A pipeline rupture would endanger drinking water and could poison vast reaches of the river.

The pipeline's original route crossed the Missouri just upstream of Bismark, North Dakota, but residents there objected to its proximity to residential areas and the risk of spills. The Army Corps of Engineers ordered the Texas-based pipeline company, Energy Transfer Partners, to move the route downstream, to just above the Standing Rock water intake. The tribe sued, arguing that they hadn't been consulted, as required by federal law. But construction continued apace.

In April 2016, a group of young people from multiple tribes, the International Indigenous Youth Council, decided to take action. They set up a prayer camp, which they called the Seven Council Fires camp in honor of seven major Dakota and Lakota groups, in the path of the pipeline construction. Their rallying cry was "*Mni Wiconi!*" (water is life!).

Through the summer and fall, members of more than 300 Native American tribes and indigenous people from as far away as Peru, Norway, and New Zealand joined the Water Protectors' camp. Thousands of supporters, drawn through social media networks, came from all over North America. This was the largest, most diverse tribal gathering since the battle of the Little Bighorn, which took place 141 years earlier and 1,000 km (600 mi) west of the site. The camps were remarkably harmonious. Many of the tribes were traditional enemies, but participants emphasized peace, spirituality, respect, and nonviolent civil disobedience. (fig. 1)

▲ **FIGURE 1** Although the Standing Rock Camp contained thousands of people from hundreds of tribes and indigenous groups from across North America and around the world, the encampment was remarkably peaceful and orderly. Protestors were committed to nonviolent resistance and cooperative decisionmaking. ©Helen H. Richardson/Denver Post/Getty Images

Reactions from Energy Transfer Partners was neither peaceful nor respectful. Tensions exploded on September 3, when protestors blocked a road to protect sacred cultural sites. Private security guards attacked the unarmed protesters with police dogs, pepper spray, and clubs. Videos, seen around the world on YouTube, showed a grim similarity to earlier civil rights, labor, and anti-war protests. Police departments from across the Dakotas and neighboring states sent more armed soldiers and more police with riot gear and military equipment. In early November, in freezing weather, police attacked protestors with water cannons, tear gas, rubber bullets, and concussion grenades, injuring hundreds.

The power of peaceful protest was demonstrated in December, when President Obama finally declared a stop to construction. But a month later, two days after his inauguration in January 2017, President Donald Trump ordered immediate completion of the pipeline. Within six months of starting operation, the pipeline had eight spills. There are no plans for what to do if, or when, crude oil leaks into the Missouri River.

Oil interests won this battle, but the protests focused global attention on the need for indigenous rights and environmental protection. Most important, a movement like this teaches a new generation that they can—and must—influence the political process. It teaches organization, social networking, media, and policy. Will some run for office in a few years? It's too early to say, but movements like this often produce new candidates, policymakers, writers, and leaders, all key elements to the democratic process.

---

starting point is to explore what kind of sustainability plan your university has. You can contribute to pursuing its aims, or updating it, or if your institution has no sustainability plan, you can help start writing one. Many institutions also have greenhouse gas reduction plans or green building guidelines. Working on these can teach you a great deal about both technology and policymaking.

Green building is one of the areas in which colleges and universities often show leadership. It is increasingly common for new campus buildings to meet LEED (Leadership and Energy and Environmental Design) standards for sustainability. This is partly because high-performing buildings can be more efficient to operate, thus saving money, and it's partly because students press their administrations to do better. Some schools are striving for even higher standards–The College of the Atlantic has student housing built to extremely low-energy Passive House standards, and Hampshire College has a Living Building, which seek to provide all water, energy, and waste management needs on site.

Often student initiatives help to kick-start green innovation. Oberlin's Center for Environmental Studies, for example, was designed through planning and design seminars. The center features 370 m^2 of photovoltaic panels on its roof; a geothermal well to help heat and cool the building; large, south-facing windows for

▲ **FIGURE 16.17** The Adam Joseph Lewis Center for Environmental Studies at Oberlin College is designed to be self-sustaining even in northern Ohio's cool, cloudy climate. Large, south-facing windows let in sunlight, while 370 m^2 of solar panels on the roof generate electricity. Natural wastewater treatment, including a constructed wetland, purifies wastewater. Source: National Renewable Energy Laboratory/NREL/PIX

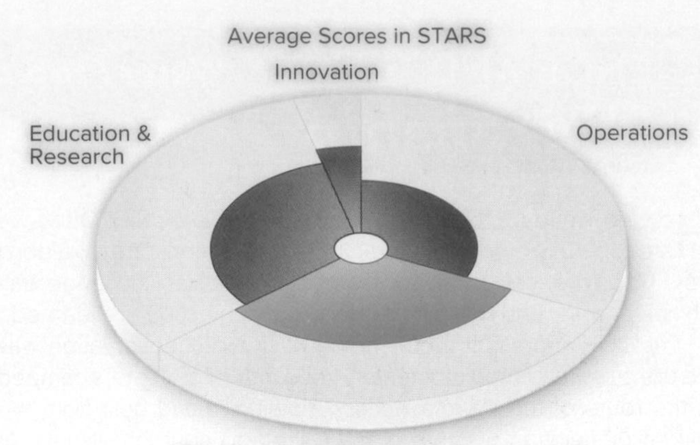

▲ **FIGURE 16.18** This pie chart shows the proportion of a STARS score contributed by different categories (slice width) and overall average score (length of slice) for all reporting institutions. Operations tend to score low, while innovation and engagement tend to score higher, on average. Source: Association for the Advancement of Sustainability in Higher Education.

passive solar gain; and a "living machine" for water treatment, including plant-filled tanks in an indoor solarium and a constructed wetland outside (fig. 16.17).

Students also do important work in public interest research or community-engaged research. A network of public interest research groups (PIRGs), for example, has been contributing to policy research for decades on campuses across the United States. These groups often make a key difference in local and national policy. Law schools often have public interest law programs, in which students gain experience while assisting public needs. Business schools, accounting, even public interest programming projects all use students' abilities to respond to community needs.

Audits help reduce resource consumption

Campus facilities provide important educational opportunities, particularly in terms of collecting and analyzing data that describe how well a campus is doing. Collecting building energy use data, for example, is the first step toward saving energy and reducing carbon emissions. Monitoring waste is the first step toward reducing it. These steps can reduce an institution's operating costs, as well as its environmental impacts.

Students, faculty, and staff can work together in a wide variety of campus audits. An audit involves collecting data to assess consumption rates. Usually, an audit provides baseline information to guide new strategies, or it measures progress on those plans. Water and energy use, waste production and disposal, paper consumption, recycling, local food purchasing, and other factors can all yield important consumption information. Students can also audit courses for sustainability content, sustainable practices in labs and classrooms, lighting practices, and other factors.

Audits show the importance of data collection in environmental science: We always need meaningful data to evaluate how a system is working. Audits show where a campus is performing well and where to focus efforts. They also allow comparison to other institutions, through a school ranking and assessment system.

One of the leading campus audit programs is run by the Association for Advancement of Sustainability in Higher Education (AASHE). Institutions that do an AASHE-based audit can compare their performance to that of nearly 900 other educational institutions, and they can plot progress over time. After more than a decade of data collection, AASHE records also show where most schools are doing well (fig. 16.18). Curriculum and administration generally do well—they require decision making but no costly infrastructure. The area where most schools struggle most is in operations, especially improving energy use and building operation. This remains a challenge, but without auditing data, we wouldn't know this was where the main challenges lie.

Campus rankings motivate progress

Sustainability ranking systems provide both visibility for well-performing schools and incentives for schools to do better. AASHE's rating system, known as the Sustainability Tracking and Reporting System (STARS), assigns categories of bronze, silver, gold, and platinum. Three institutions have achieved platinum: Stanford, Colorado State, and the University of New Hampshire (UNH). These high rankings represent a commitment of institutional efforts to innovative research in sustainability, ground-breaking energy use and building strategies, and other efforts. Stanford, for example, now meets 88 percent of its heating load by capturing waste heat from air conditioning—which occurs simultaneously with heating in many institutions. The same system has reduced water waste, and it incorporates solar panels for electricity production. Colorado State and UNH have made exemplary progress in energy efficiency, waste management, and research programs. Of the 445 institutions with published STARS scores, nearly half are silver, and about one-quarter have achieved gold ranking. Having a rating system gives a structure for many of these to aim for the next step up in rankings.

Small institutions can also be leaders. The Sierra Club's "Cool Schools" ranking lists several small innovators as top environmental performers. The College of the Atlantic (COA), in Bar Harbor,

▲ **FIGURE 16.19** Students work on landscaping at Furman University's Shi Center for Sustainability. Gaining experience in sustainable practices is a good way to develop ecological literacy. ©Furman University's of the Shi Center

Maine, was ranked first in 2017 and 2018. COA has divested from fossil fuel companies and is working toward 100 percent renewable energy, including high-efficiency student housing. Much of the food students eat is grown on one of two organic farms. Second in the Sierra Club rankings was Green Mountain College (GMC), in Poultney, Vermont. All GMC students take environmental classes, and everyone contributes to a campus greening fund, which has produced a biomass facility, a bike-sharing program, and a bat habitat. Classes are available on resilient and sustainable communities and local food systems. Green Mountain College has also divested from fossil fuels, and the school produces most of its own energy with photovoltaic cells.

Middlebury College is also a pioneer, having officially become carbon neutral at the end of 2016. Among Middlebury's strategies are heating the campus with wood pellets, sourced from waste wood products in the nearby region, and installing a methane (gas) digester at a local dairy farm, which now provides gas to the campus. Rankings also recognize schools where students grow their own food or have persuaded the administration to buy locally produced food and to provide organic, vegetarian, and fair-trade options in campus cafeterias, or where campus landscaping includes native or low-maintenance plantings (fig. 16.19).

As a final gesture with potentially powerful effects, graduating students at more than 100 U.S. colleges and universities have taken this pledge at graduation time:

> I pledge to explore and take into account the social and environmental consequences of any job I consider and will try to improve these aspects of any organization for which I work.

How much is enough?

An important thing any individual can do is to think carefully about consumption. Technology has made consumer goods and services so cheap and readily available that it's hard to grasp the impacts of our consumption patterns. But we do know that we in the industrialized world use vastly disproportionate amounts of resources, and we know that everyone in the world cannot consume at our level. Can we consume less and still be happy? This is partly a philosophical question. Some people say no. Others say we can become *happier* by consuming less—and by reducing the credit card debt we use to pay for the things we buy.

A century ago, economist and social critic Thorstein Veblen wrote *The Theory of the Leisure Class,* in which he coined the term **conspicuous consumption.** He used the term to describe things we buy just to impress others, things we don't really want or need. Veblen's ideas are more relevant today than ever. The average American consumes twice as many goods and services as in 1950. The average house is now more than twice as big as it was 50 years ago, even though the typical family is half as large. Shopping shapes our identity and consumes our time. Social observers frequently point out that the things we buy don't really make us young, beautiful, smart, and interesting. By giving so much attention to earning and spending money, we lose the time to have real friends, to cook real food, to have creative hobbies, or to do work that makes us feel we have accomplished something with our lives.

Some social critics call this accelerated consumerism **affluenza.** A growing number of people find themselves stuck in a vicious circle: They work frantically at an unfulfilling job, to buy things they don't need, so they can save time to work even longer hours (fig. 16.20). Seeking a measure of balance in their lives, some opt

▲ **FIGURE 16.20** Is this our highest purpose? ©1990 Bruce von Alten.

out of the rat race and adopt simpler, less-consumptive lifestyles. As Thoreau wrote in *Walden*, "Our life is frittered away by detail . . . simplify, simplify."

Choosing to consume less can be an easy way to reduce your global environmental footprint and save money. Cook simple foods with friends instead of eating prepared foods. Grow a garden. Spend less time shopping and more time talking and having fun with family and friends. Although individual choices may make a small impact, collectively they have global effects.

16.6 THE CHALLENGES OF SUSTAINABLE DEVELOPMENT

- *Sustainable development means meeting the needs of the present without compromising the ability of future generations to meet their own needs.*
- *Sustainable Development Goals identify and quantify 17 global targets.*
- *Fundamentally, sustainability is about justice and equity.*

The idea of sustainability is that if we intend to be here for the long term, we can't deplete the natural systems we depend on for food, water, energy, fiber, waste disposal, and other life-support services. The idea of **sustainable development** is to share opportunity by improving people's lives in impoverished regions, and to extend human well-being over many generations.

Often we might question whether these ideals are even possible. Absolute sustainability may be an unachievable goal. Historically, improving livelihoods has involved increased consumption. Yet sustainability may also be a goal worth aiming for, like world peace or universal access to potable water and safe food. The idea has been a persistently useful organizing idea for efforts to conserve life-support systems for our grandchildren and great-grandchildren. Generally, *development* implies economic growth (increasing wealth) accompanied by improved living conditions. *Sustainable development* implies that growth can be based on nonconsumptive activities, such as education or arts, as well as on carefully managed renewable resources, such as soils, forests, and fisheries (fig. 16.21).

▲ **FIGURE 16.21** A model for integrating ecosystem health, human needs, and sustainable economic growth.

Sustainable Development Goals aim to improve conditions for all

In 1987 the World Commission on Environment and Development provided a definition of "sustainable development" in the report *Our Common Future.* The core of the definition was "meeting the needs of the present without compromising the ability of future generations to meet their own needs." In 2016, the United Nations initiated a program to promote 17 **Sustainable Development Goals** (SDGs). The goals are ambitious and global, and they include eliminating the most severe poverty and hunger; promoting health, education, and gender equality; providing safe water and clean energy; and preserving biodiversity (fig. 16.22). This global effort, developed by representatives of the member states of the UN, seeks to coordinate data gathering and reporting, so that countries can monitor their progress, share resources, and promote sustainable investment in developing areas.

For each of the 17 goals, organizers identified targets—some quantifiable, some more general. For example, Goal 1, "End poverty," includes eradicating extreme poverty, ensuring rights to basic services, and protecting ownership and inheritance of property. Goal 7, "Ensure access to affordable, sustainable energy," includes targets of doubling energy efficiency and enhancing international investment in clean energy. Goal 12, "Ensure sustainable consumption and production," calls for cutting per-capita food waste in half, as well as phasing out fossil fuel subsidies that encourage wasteful consumption. The UN aims to meet these targets by 2030, in a span of 15 years (fig. 16.23).

The SDGs also include a number of targets for economic and social equity, and for better governance. To most economists and policymakers, it seems obvious that economic growth is the only way to bring about a long-range transformation to more advanced and productive societies and to provide resources to improve the lot of all people. As former U.S. president John F. Kennedy said, "A rising tide lifts all boats." But equity is also essential. Political stability, democracy, and fair access to resources and opportunity are needed to ensure that the poor will get a fair share of the benefits of greater wealth in a society. According to a study released in 2006 by researchers at Yale and Columbia Universities, environmental sustainability tends to occur where there are open political systems and good government.

These ambitious goals might appear unrealistic, but they build on the remarkable (though not complete) successes of the **Millennium Development Goals.** These eight goals were a 15-year

▲ **FIGURE 16.22** The United Nations Sustainable Development Goals are intended to improve well-being of the world's poorest people while also protecting biodiversity, natural resources, and climate. These goals follow the largely successful Millennium Development Goals. Source: UN Development Programme

FIGURE 16.23 Sustainable Development Goals aim to improve conditions for future generations, with greater equity, reduced poverty, and protections for a healthy environment. ©Jerry Alexander/Getty Images

effort, from 2000 to 2015, to improve literacy, health, access to safe water, child survival, and other goals. Targets included ending poverty and hunger, universal education, gender equity, child health, maternal health, combating HIV/AIDS, environmental sustainability, and global cooperation in development efforts. While only modest progress was achieved on some goals, UN Secretary General Ban Ki-Moon called that effort the most successful anti-poverty movement in history. Extreme poverty dropped from nearly half the population of developing countries to just 14 percent in only 15 years. The proportion of undernourished people dropped by almost half, from 23 percent to 13 percent. Primary school enrollment rates climbed from 83 percent to 91 percent in developing countries, and women gained access to education, employment, and more political representation in national parliaments.

CONCLUSION

The policy choices we make now, at scales from national policy to individual consumption patterns, will influence environmental quality and natural resources for future generations. In this chapter we have examined some basic environmental policies and the ways policies are formed. There are many players in the formation of environmental policy: Legislative, administrative, and judicial branches of governments, at local, state, and federal scales, all have power to shape policy. Individuals also have opportunities to influence policy, especially when they act together.

Student activism has long been essential both for shaping policy and for building leadership skills and experience in the leaders of tomorrow. Working with local groups on local issues is an essential opportunity to gain skills that can be useful in your career and life. You can apply these skills in a variety of fields, at national, local, and international scales.

Sustainable development is one of our greatest challenges. People everywhere wish to have the same comfort and opportunities enjoyed by those in wealthier regions. Finding ways to share opportunity without destroying resources requires creativity and commitment. The UN Sustainable Development Goals note these key principles:

- All of us depend on nature and ecosystem services to provide the conditions for a decent, healthy, and secure life.
- Growing demands for food, fresh water, and energy have made unprecedented demands on ecosystems and resources. Dramatic acceleration of species extinctions, pressure on ecosystems, and consumption of resources threatens our own well-being.
- Currently available technology and knowledge can greatly reduce our impacts on ecosystems. Better accounting for the value of services and the cost of lost resources is important for helping us rethink our use of these systems. A central remaining question is whether we can find ways to collaborate across sectors of governments, businesses, international institutions, and communities in order to find creative solutions for the future.

PRACTICE QUIZ

1. What is a *policy*? How are policies formed?
2. Describe three important provisions of NEPA.
3. List four important U.S. environmental laws (besides NEPA), and briefly describe what each does.
4. Why are international environmental conventions and treaties often ineffective? What can make them more successful?
5. Why is the U.S. Supreme Court ruling that greenhouse gases endanger public health and welfare controversial?
6. List two broad goals of environmental education identified by the National Environmental Education Act.
7. What is the UNFCCC, and what does it seek to accomplish?
8. List five things each of us could do to help preserve our common environment.
9. Describe some things schools and students have done to promote sustainable living.
10. Define *sustainable development*, and describe some of its principal tenets.

CRITICAL THINKING AND DISCUSSION

Apply the principles you have learned in this chapter to discuss these questions with other students.

1. In your opinion, how much environmental protection is too much? Think of a practical example in which some stakeholders may feel oppressed by governmental regulations. How would you justify or criticize these regulations?
2. Which is a better choice at the grocery store, paper or plastic? How would you evaluate the trade-offs between packaging choices? What evidence would you look for to make this decision in your life?
3. Do you agree with Margaret Mead that a small group of committed individuals is the only thing that can change the world? What do you think she meant? Think of some examples of groups of individuals who have changed the world. How did they do it?
4. Suppose that you were going to make a presentation on divestment from fossil fuels to your school administrators. What suggestions would you make for changes on your campus? What information would you need to support these proposals?
5. Debate this question: Is sustainability an achievable goal? What do you think are the main barriers to attaining this objective? What ideas in this book give you reason to believe sustainability is possible? What are the alternatives?

DATA ANALYSIS Campus Environmental Audit

If you want to understand how sustainable your campus is, the first step is to gather data. A campus environmental audit can be a huge, professionally executed task, but you can make a reasonable approximation if you work with your class to gather some basic information. Among the many places to find established listings of factors to consider, one widely used audit is that of the Association for the Advancement of Sustainability in Higher Education. You can find details of their full audit system online, but you can do a simplified version yourself or in collaboration with others in your class. Factors involved in an audit include building efficiency, transportation alternatives, waste management, dining facilities, administrative actions, and other factors. How sustainable is your school? Go to Connect to find a basic campus environmental audit. Then you can collect and evaluate your own data to find out.

Design Elements: Active Learning (Toad): ©Gaertner/Alamy Stock Photo; Case Study (Globe): ©McGraw-Hill Education; Google Earth: ©McGraw-Hill Education; Abstract Background: ©Martin Kubat/Shutterstock; What do you think (Students using tablets): ©McGraw-Hill Education/Richard Hutchings, photographer; What can you do (Hand holding Globe): ©Christoph Weihs/Shutterstock

Appendixes

©McGraw-Hill Education

Vegetation

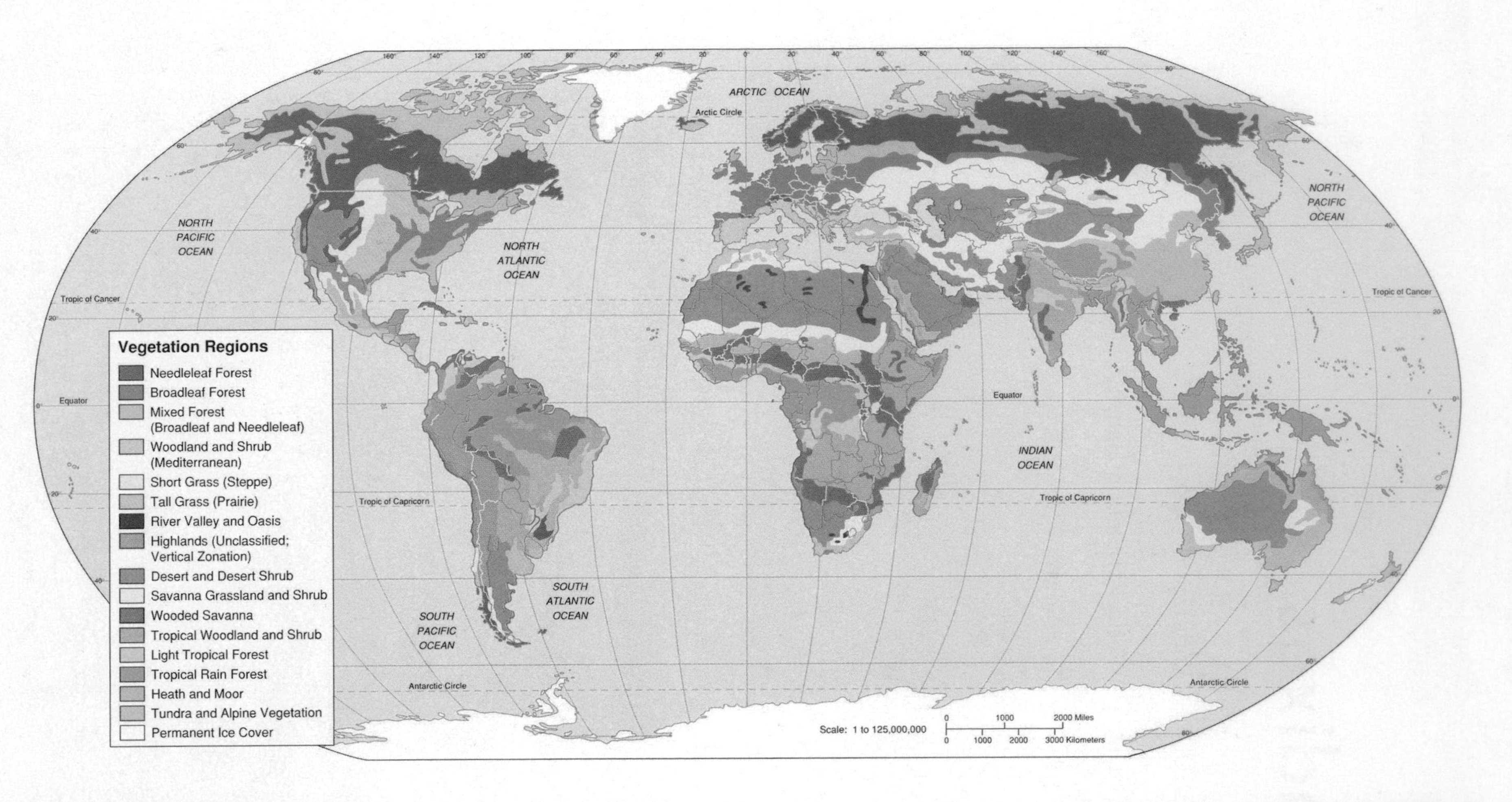

Vegetation is the most visible consequence of the distribution of temperature and precipitation. The global distribution of vegetation types and the global distribution of climate are closely related, but not all vegetation types are the consequence of temperature and precipitation or other climatic variables. Many types of vegetation, in many areas of the world, are the consequence of human activities, particularly the grazing of domesticated livestock, burning, and forest clearance.

World Population Density

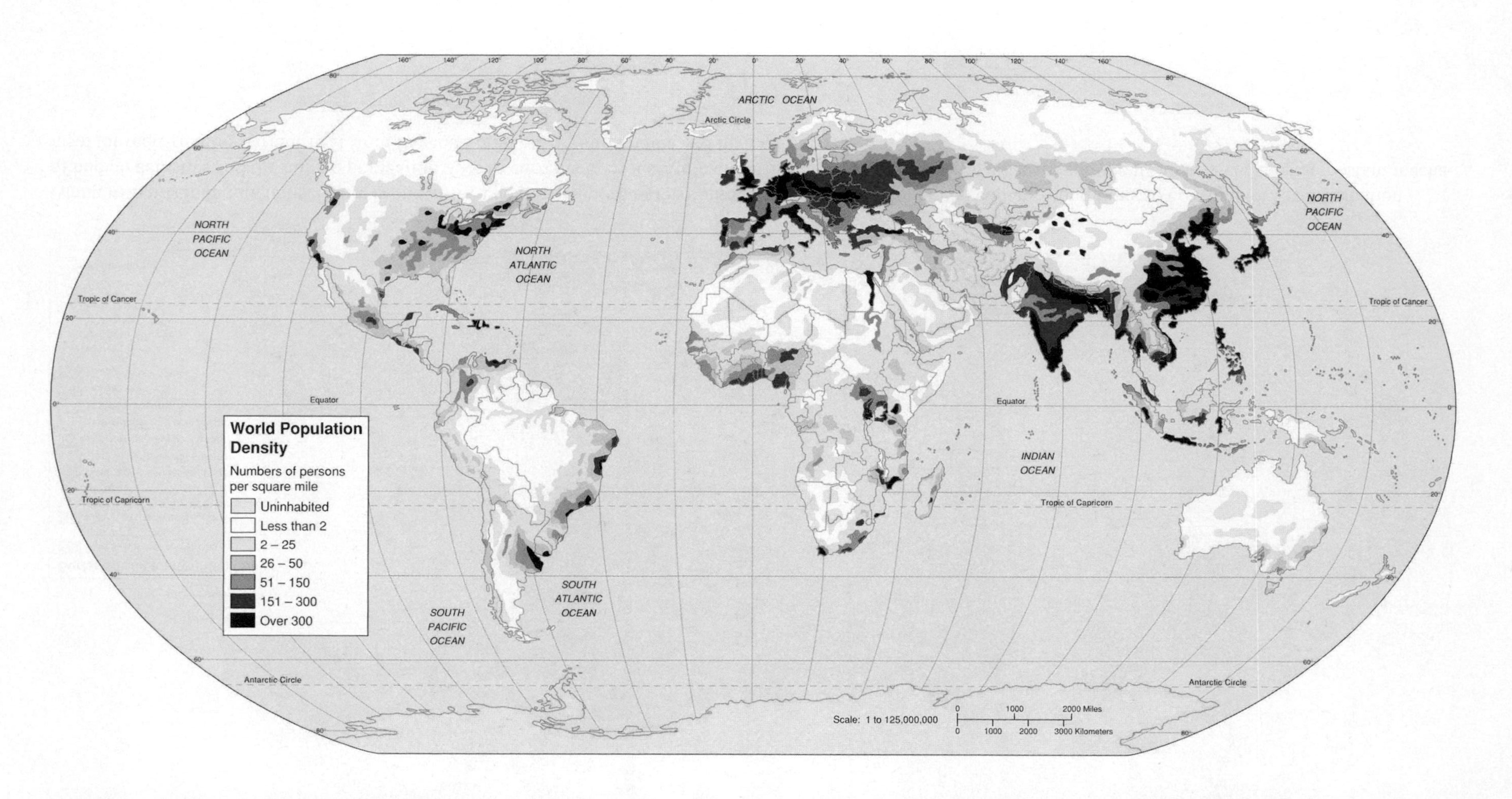

No feature of human activity is more reflective of environmental conditions than where people live. In the areas of densest population, a mixture of natural and human factors have combined to allow maximum food production, maximum urbanization, and especially concentrated economic activity. Three such great concentrations appear on the map–East Asia, South Asia, and Europe–with a fourth lesser concentration in eastern North America (the "Megalopolis" region of the United States and Canada). The areas of future high density (in addition to those already existing) are likely to be in Middle and South America and Africa, where population growth rates are well above the world average.

Temperature Regions and Ocean Currents

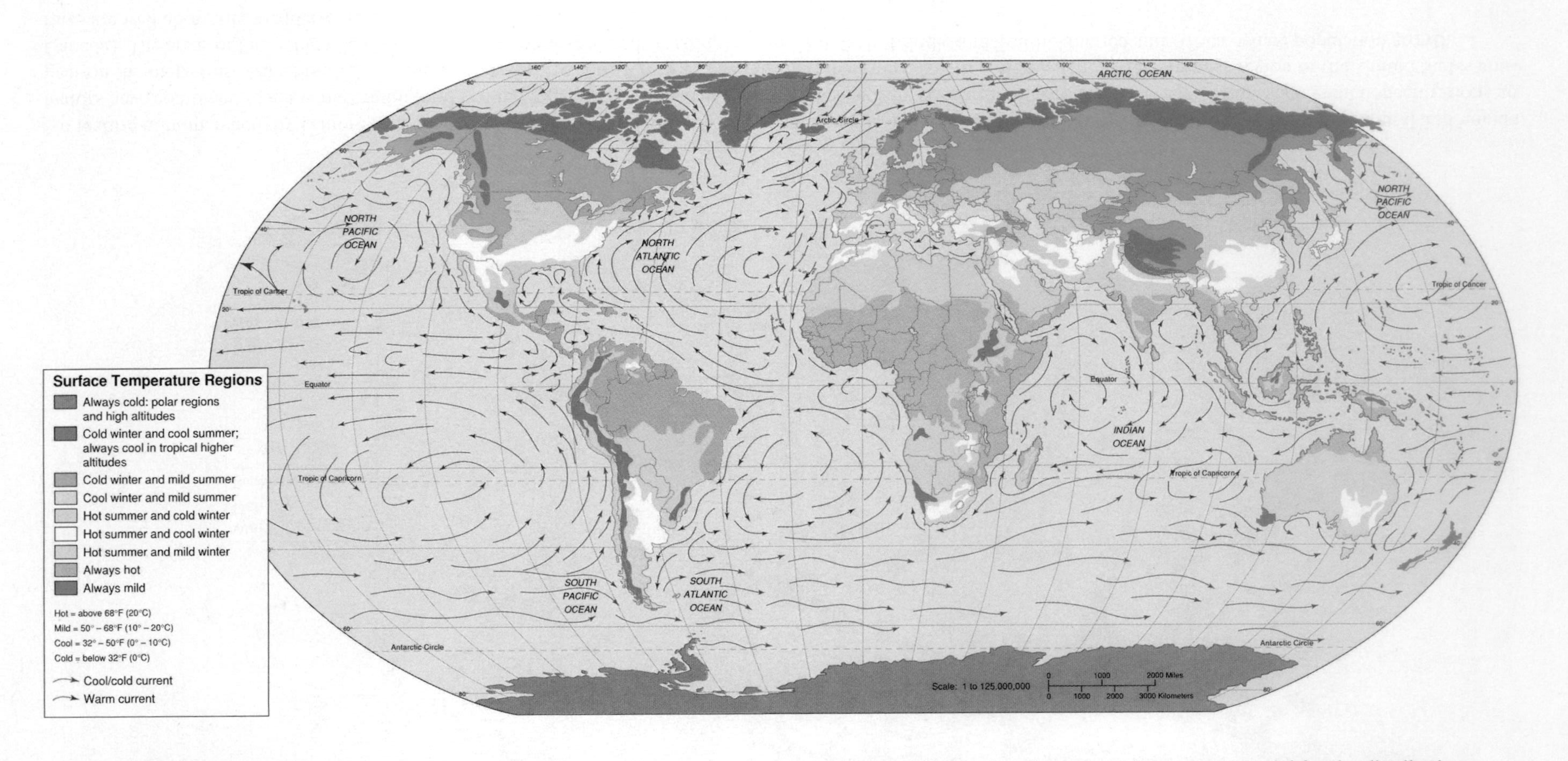

Along with precipitation, temperature is one of the two most important environmental variables, defining the climatic conditions so essential for the distribution of human activities and the human population. Ocean currents exert a significant influence over the climate of adjacent continents and are the most important mechanism for redistributing surplus heat from the equatorial region into middle and high latitudes.

Glossary

A

abundance The number of individuals of a species in an area.

acid precipitation Acidic rain, snow, or dry particles deposited from the air due to increased acids released by anthropogenic or natural resources.

acids Substances that release hydrogen atoms in water.

active solar systems Mechanical systems that use moving substances to collect and transfer solar energy.

acute effects A sudden onset of symptoms or effects of exposure to some factor.

acute poverty Insufficient income or access to resources needed to provide the basic necessities for life, such as food, shelter, sanitation, clean water, medical care, and education.

adaptation Physical changes that allow organisms to survive in a given environment.

adaptive management A management plan designed from the outset to "learn by doing" and to actively test hypotheses and adjust treatments as new information becomes available.

administrative law Executive orders, administrative rules and regulations, and enforcement decisions by administrative agencies and special administrative courts.

aerosols Minute particles or liquid droplets suspended in the air.

affluenza An addiction to spending and consuming beyond one's needs.

albedo A description of a surface's reflective properties.

allergens Substances that activate the immune system and cause an allergic response; may not be directly antigenic themselves but may make other materials antigenic.

allopatric speciation Species that arise from a common ancestor through geographic isolation or some other barrier to reproduction.

ambient air The air immediately around us.

amorphous silicon collectors Photovoltaic cells made from randomly assembled silicon molecules rather than silicon crystals. Amorphous collectors are less efficient but far cheaper than crystalline collectors.

analytical thinking A way of systematic analysis that asks, "How can I break this problem down into its constituent parts?"

anemia Low levels of hemoglobin due to iron deficiency or lack of red blood cells.

anthropocentric Believing that humans hold a special place in nature; being centered primarily on humans and human affairs.

antigens Substances that stimulate the production of, and react with, specific antibodies.

aquifers Porous, water-bearing layers of sand, gravel, and rock below the earth's surface; reservoirs for groundwater.

arithmetic scale A pattern of growth that increases at a constant amount per unit time, such as 1, 2, 3, 4 or 1, 3, 5, 7.

atmospheric deposition Sedimentation of solids, liquids, or gaseous materials from the air.

atom The smallest particle that exhibits the characteristics of an element.

atomic number The characteristic number of protons per atom of an element.

autotroph An organism that synthesizes food molecules from inorganic molecules by using an external energy source, such as light energy.

B

barrier islands Low, narrow, sandy islands that form offshore from a coastline.

Basel Convention Restricts shipment of hazardous waste across international boundaries.

bases Substances that readily bond with hydrogen ions in an aqueous solution.

Batesian mimicry Evolution by one species to resemble another species that is protected from predators by a venomous stinger, bad taste, or some other defensive adaptation.

benthic The bottom of a sea or lake.

binomials Scientific or Latin names that combine the genus and species, e.g., *Zea mays.*

bioaccumulation The selective absorption and concentration of molecules by cells.

biocentrism The belief that all creatures have rights and values; being centered on nature rather than humans.

biochemical oxygen demand (BOD) A standard test for measuring the amount of dissolved oxygen utilized by aquatic microorganisms.

biodegradable plastics Plastics that can be decomposed by microorganisms.

biodiversity The genetic, species, and ecological diversity of the organisms in a given area.

biofuel Fuel made from biomass.

biogeochemical cycles Movement of matter within or between ecosystems; caused by living organisms, geologic forces, or chemical reactions. The cycling of nitrogen, carbon, sulfur, oxygen, phosphorus, and water are examples.

biological community The populations of plants, animals, and microorganisms living and interacting in a certain area at a given time.

biological controls Use of natural predators, pathogens, or competitors to regulate pest populations.

biomagnification Increase in concentration of certain stable chemicals (for example, heavy metals or fat-soluble pesticides) in successively higher trophic levels of a food chain or web.

biomass The accumulated biological material produced by living organisms.

biomass fuel Organic material produced by plants, animals, or microorganisms that can be burned directly as a heat source or converted into a gaseous or liquid fuel.

biomass pyramid A metaphor or diagram that explains the relationship between the amounts of biomass at different trophic levels.

biomes Broad, regional types of ecosystems characterized by distinctive climate and soil conditions and distinctive kinds of biological community adapted to those conditions.

bioremediation Use of biological organisms to remove pollution or restore environmental quality.

biosphere The zone of air, land, and water at the surface of the earth that is occupied by organisms.

biosphere reserves World heritage sites identified by the IUCN as worthy for national park or wildlife refuge status because of high biological diversity or unique ecological features.

biotic potential The maximum reproductive rate of an organism, given unlimited resources and ideal environmental conditions.

birth control Any method used to reduce births, including celibacy, delayed marriage, contraception; devices or medications that prevent implantation of fertilized zygotes and induced abortions.

blind experiments A design in which researchers don't know which subjects were given experimental treatment until after data have been gathered and analyzed.

bogs Areas of waterlogged soil that tend to be peaty; fed mainly by precipitation; low productivity; some bogs are acidic.

boreal forest A broad band of mixed coniferous and deciduous trees that stretches across northern North America (and Europe and Asia); its northernmost edge, the taiga, intergrades with the arctic tundra.

brownfields Abandoned or underused urban areas in which redevelopment is blocked by liability or financing issues related to toxic contamination.

C

cancer Invasive, out-of-control cell growth that results in malignant tumors.

cap-and-trade agreement A policy to set pollution limits, then allow companies to buy and sell their allotted rights to emit pollutants.

capital Any form of wealth, resources, or knowledge available for use in the production of more wealth.

carbohydrate An organic compound consisting of a ring or chain of carbon atoms with hydrogen and oxygen attached; examples are sugars, starches, cellulose, and glycogen.

carbon cycle The circulation and reutilization of carbon atoms, especially via the processes of photosynthesis and respiration.

carbon management Projects to reduce carbon dioxide emissions from fossil fuel or to ameliorate their effects.

carbon monoxide Colorless, odorless, nonirritating but highly toxic gas produced by incomplete combustion of fuel, incineration of biomass or solid waste, or partially anaerobic decomposition of organic material.

carbon neutral Producing no net carbon dioxide emissions.

carbon sink Places of carbon accumulation, such as in large forests (organic compounds) or ocean sediments (calcium carbonate).

carcinogens Substances that cause cancer.

carnivores Organisms that mainly prey upon animals.

carrying capacity The maximum number of individuals of any species that can be supported by a particular ecosystem on a long-term basis.

case law Precedents from both civil and criminal court cases.

catalytic converter The device on an automobile that uses platinum-palladium and rhodium catalysts to remove NO_x, hydrocarbons, and carbon monoxide from the exhaust.

cell Minute compartments surrounded by semipermeable membranes within which the processes of life are carried out by all living organisms.

cellular respiration The process in which a cell breaks down sugar or other organic compounds to release energy used for cellular work; may be anaerobic or aerobic, depending on the availability of oxygen.

chain reaction A self-sustaining reaction in which the fission of nuclei produces subatomic particles that cause the fission of other nuclei.

chaparral A biological community characterized by thick growth of thorny, evergreen shrubs typical of a Mediterranean climate.

chemical bond The force that holds molecules together.

chemical energy Potential energy stored in chemical bonds of molecules.

chemosynthesis Extracting energy for life from inorganic chemicals, such as hydrogen sulfide, rather than from sunlight.

chlorinated hydrocarbons Hydrocarbon molecules to which chlorine atoms are attached. Often used as pesticides and are both highly toxic and long-lasting in the environment.

chlorofluorocarbons Chemical compounds with a carbon skeleton and one or more attached chlorine and fluorine atoms. Commonly used as refrigerants, solvents, fire retardants, and blowing agents.

chloroplasts Chlorophyll-containing organelles in eukaryotic organisms; sites of photosynthesis.

chronic effects Long-lasting results of exposure to a toxin; can be a permanent change caused by a single, acute exposure or a continuous, low-level exposure.

citizen science Projects in which trained volunteers work with scientific researchers to answer real-world questions.

civil law A body of laws regulating relations between individuals or between individuals and corporations concerning property rights, personal dignity and freedom, and personal injury.

classical economics Modern, Western economic theories of the effects of resource scarcity, monetary policy, and competition on supply and demand of goods and services in the marketplace. This is the basis for the capitalist market system.

clear-cutting Cutting every tree in a given area, regardless of species or size; an appropriate harvest method for some species; can be destructive if not carefully controlled.

climate A description of the long-term pattern of weather in a particular area.

climax community A long-lasting, self-sustaining community resulting from ecological succession that is resistant to disturbance.

closed-canopy A forest where tree crowns spread over 20 percent of the ground; has the potential for commercial timber harvests.

closed system A system in which there is no exchange of energy or matter with its surroundings.

cloud forests High mountain forests where temperatures are uniformly cool and fog or mist keeps vegetation wet all the time.

coevolution The process in which species exert selective pressure on each other and gradually evolve new features or behaviors as a result of those pressures.

cogeneration The simultaneous production of electricity and steam or hot water in the same plant.

coliform bacteria Bacteria that live in the intestines (including the colon) of humans and other animals; used as a measure of the presence of feces in water or soil.

commensalism A symbiotic relationship in which one member is benefited and the second is neither harmed nor benefited.

communal resource management systems Resources managed by a community for long-term sustainability.

community (ecological) structure The patterns of spatial distribution of individuals, species, and communities.

competitive exclusion A theory that no two populations of different species will occupy the same niche and compete for exactly the same resources in the same habitat for very long.

complexity The number of species at each trophic level and the number of trophic levels in a community.

composting The biological degradation of organic material under aerobic (oxygen-rich) conditions to produce compost, a nutrient-rich soil amendment and conditioner.

compound Substance composed of different kinds of atoms.

confidence limits Upper and lower values in which the true value (such as a mean) is likely to fall.

confined animal-feeding operation Feeding large numbers of livestock at a high density in pens or barns.

conifer A needle-bearing tree that produces seeds in cones.

coniferous See *conifer*.

conservation medicine Attempts to understand how changes we make in our environment threaten our health as well as that of natural communities on which we depend.

conservation of matter In any chemical reaction, matter changes form; it is neither created nor destroyed.

conspicuous consumption A term coined by economist and social critic Thorstein Veblen to describe buying things we don't want or need to impress others.

constructed wetlands Artificially constructed wetlands.

consumers Organisms that obtain energy and nutrients by feeding on other organisms or their remains. See also *heterotroph*.

consumption The fraction of withdrawn water that is lost in transmission or that is evaporated, absorbed, chemically transformed, or otherwise made unavailable for other purposes as a result of human use.

contour plowing Plowing along hill contours; reduces erosion.

control rods Neutron-absorbing material inserted into spaces between fuel assemblies in nuclear reactors to regulate fission reaction.

controlled studies Comparisons made between two populations that are identical (as far as possible) in every factor except the one being studied.

convection currents Rising or sinking air currents that stir the atmosphere and transport heat from one area to another. Convection currents also occur in water.

Convention on International Trade in Endangered Species (CITES) An international convention to protect endangered species.

conventional (criteria) pollutants The seven substances (sulfur dioxide, carbon monoxide, particulates, hydrocarbons, nitrogen oxides, photochemical oxidants, and lead) identified by the Clean Air Act as the most serious threat of all pollutants to human health and welfare.

convergent evolution Species evolve from different origins but under similar environmental conditions to have similar traits.

coral bleaching Whitening of corals when stressors, such as high temperatures, induce corals to expel their colorful single-celled protozoa, known as zooxanthellae, or when zooxanthellae die. Death of the coral reef may result.

coral reefs Prominent oceanic features composed of hard, limy skeletons produced by coral animals; usually formed along edges of shallow, submerged ocean banks or along shelves in warm, shallow, tropical seas.

core The dense, intensely hot mass of molten metal, mostly iron and nickel, thousands of kilometers in diameter at the earth's center.

core habitat A habitat patch large enough and with ecological characteristics suitable to support a critical mass of the species that make up a particular community.

Coriolis effect The tendency for air above the earth to appear to be deflected to the right (in the Northern Hemisphere) or the left (in the South) because of the earth's rotation.

corridors Strips of natural habitat that connect two adjacent nature preserves to allow migration of organisms from one place to another.

cost-benefit analysis (CBA) An evaluation of large-scale public projects by comparing the costs and benefits that accrue from them.

cover crops Plants, such as rye, alfalfa, or clover, that can be planted immediately after harvest to hold and protect the soil.

creative thinking Original, independent thinking that asks, "How might I approach this problem in new and inventive ways?"

criminal law A body of court decisions based on federal and state statutes concerning wrongs against persons or society.

criteria pollutants See *conventional pollutants*.

critical factor The single environmental factor closest to a tolerance limit for a given species at a given time.

critical thinking An ability to evaluate information and opinions in a systematic, purposeful, efficient manner.

crude birth rate The number of births per thousand persons in a given year (using the midyear population).

crude death rate The number of deaths per thousand persons in a given year; also called crude mortality rate.

crust The cool, lightweight, outermost layer of the earth's surface that floats on the soft, pliable underlying layers; similar to the "skin" on a bowl of warm pudding.

cultural eutrophication An increase in biological productivity and ecosystem succession caused by human activities.

D

debt-for-nature swaps Forgiveness of international debt in exchange for nature protection in developing countries.

deciduous Trees and shrubs that shed their leaves at the end of the growing season.

decomposer Fungus or bacterium that breaks complex organic material into smaller molecules.

deductive reasoning "Top down" reasoning in which we start with a general principle and derive a testable prediction about a specific case.

deforestation Removing trees from a forest.

demographic transition A pattern of falling death rates and birth rates in response to improved living conditions; typically leads to rapid then stabilizing population growth.

demography The statistical study of human populations relating to growth rate, age structure, geographic distribution, etc., and their effects on social, economic, and environmental conditions.

density-dependent factors Either internal or external factors that affect growth rates of a population depending on the density of the organisms in the population.

density-independent A population is affected no matter what its size.

deoxyribonucleic acid (DNA) The double helix of genetic material that determines heredity.

dependency ratio The number of nonworking members compared with working members for a given population.

dependent variable Also known as the response variable; is one affected by other variables.

desalinization (or desalination) Removal of salt from water by distillation, freezing, or ultrafiltration.

desertification Denuding and degrading a once fertile land, initiating a desert-producing cycle that feeds on itself and causes long-term changes in soil, climate, and biota of an area.

deserts Biomes characterized by low moisture levels and infrequent and unpredictable precipitation. Daily and seasonal temperatures fluctuate widely.

detritivore Organism that consumes organic litter, debris, and dung.

dieback A sudden population decline; also called a population crash.

disability-adjusted life years (DALYs) A health measure that assesses the total burden of disease by combining premature deaths and loss of a healthy life that result from illness or disability.

discharge The amount of water that passes a fixed point in a given amount of time; usually expressed as liters or cubic feet of water per second.

discount rate The amount we discount or reduce the value of a future payment. When you borrow money from the bank at 10 percent annual interest, you are in effect saying that having the money now is worth 10 percent more to you than having the same amount 1 year from now.

disease A deleterious change in the body's condition in response to destabilizing factors, such as nutrition, chemicals, or biological agents.

dissolved oxygen (DO) content Amount of oxygen dissolved in a given volume of water at a given temperature and atmospheric pressure; usually expressed in parts per million (ppm).

disturbance Any force that disrupts the established patterns and processes, such as species diversity and abundance, community structure, community properties, or species relationships.

disturbance-adapted species Species that depend on repeated disturbance for their survival and propagation.

diversity The number of species present in a community (species richness), as well as the relative abundance of each species.

DNA Deoxyribonucleic acid; the long, double-helix molecule in the nucleus of cells that contains the genetic code and directs the development and functioning of all cells.

double-blind experiment Neither the subject (participant) nor the experimenter knows which participants are receiving the experimental or the control treatments until after data have been gathered and analyzed.

dust domes High concentrations of dust and aerosols in the air over cities.

E

earthquakes Sudden, violent movements of the earth's crust.

ecological diseases Sudden, widespread epidemics among livestock and wild species.

ecological economics Application of ecological insights to economic analysis; incorporating ecological principles and priorities into economic accounting systems.

ecological footprint An estimate of our individual and collective environmental impacts. It is usually calculated and expressed as the area of bioproductive land required to support a particular lifestyle.

ecological niche The functional role and position of a species in its ecosystem, including what resources it uses, how and when it uses the resources, and how it interacts with other species.

ecological services Processes or materials, such as clean water, energy, climate regulation, and nutrient cycling, provided by ecosystems.

ecological succession The process by which organisms gradually occupy a site, alter its ecological conditions, and are eventually replaced by other organisms.

ecosystem A specific biological community and its physical environment interacting in an exchange of matter and energy.

ecosystem management An integration of ecological, economic, and social goals in a unified systems approach to resource management.

ecosystem restoration To reinstate an entire community of organisms to as near its natural condition as possible.

ecosystem services Resources or services provided by environmental systems.

ecotones Boundaries between two types of ecological communities.

ecotourism A combination of adventure travel, cultural exploration, and nature appreciation in wild settings.

edge effects A change in species composition, physical conditions, or other ecological factors at the boundary between two ecosystems.

electron A negatively charged subatomic particle that orbits around the nucleus of an atom.

element A substance that cannot be broken into simpler units by chemical means.

El Niño A climatic change marked by shifting of a large warm-water pool from the western Pacific Ocean toward the east. Wind direction and precipitation patterns are changed over much of the Pacific and perhaps around the world.

emergent disease A new disease or one that has been absent for at least 20 years.

emergent properties Properties that make a system more than the sum of its parts.

emigration The movement of members from a population.

emission standards Regulations for restricting the amounts of air pollutants that can be released from specific point sources.

endangered species A species considered to be in imminent danger of extinction.

endemic species A species that is restricted to a single region, country, or other area.

endocrine hormone disrupters Chemicals that interfere with the function of endocrine hormones such as estrogen, testosterone, thyroxine, adrenaline, or cortisone.

energy The capacity to do work, such as moving matter over a distance.

energy intensity The amount of energy needed to provide the goods and services consumed in an economy.

energy recovery The incineration of solid waste to produce useful energy.

entropy A measure of disorder and usefulness of energy in a system.

environment The circumstances or conditions that surround an organism or a group of organisms as well as the complex of social or cultural conditions that affect an individual or a community.

environmental health The science of external factors that cause disease, including elements of the natural, social, cultural, and technological worlds in which we live.

environmental impact statement (EIS) An analysis of the effects of any major program or project planned by a federal agency; required by provisions in the National Environmental Policy Act of 1970.

environmental law Legal rules, decisions, and actions concerning environmental quality, natural resources, and ecological sustainability.

environmental literacy A basic understanding of ecological principles and the ways society affects, or responds to, environmental conditions.

environmental policy The official rules or regulations concerning the environment adopted, implemented, and enforced by some government agency.

environmental science The systematic, scientific study of our environment as well as our role in it.

epigenetics Effects (both positive and negative) expressed in future generations that are not caused by nuclear mutations and are not inherited by normal Mendelian genetics.

epigenome DNA and its associated proteins and other small molecules that regulate gene function in ways that can affect multiple generations.

epiphyte A plant that grows on a substrate other than the soil, such as the surface of another organism.

equilibrium A system in a stable balance.

estuaries Bays or drowned valleys where a river empties into the sea.

eutrophic Rivers and lakes rich in organic material (*eu* = well; *trophic* = nourished).

evolution A theory that explains how random changes in genetic material and competition for scarce resources cause species to change gradually.

evolutionary species concept A definition of species that depends on evolutionary relationships.

e-waste Discarded electronic equipment, including TVs, cell phones, computers, etc.

exotic organisms Alien species introduced by human agency into biological communities where they would not naturally occur.

explanatory variables Independent variables that help explain differences in the dependent variable.

exponential growth Growth at a constant rate of increase per unit of time; can be expressed as a constant fraction or exponent. See also *geometric growth*.

externalizing costs Shifting expenses, monetary or otherwise, to someone other than the individuals or groups who use a resource.

extinction The irrevocable elimination of species; can be a normal process of the natural world as species outcompete or kill off others or as environmental conditions change.

F

family planning Controlling reproduction; planning the timing of birth and having only as many babies as are wanted and can be supported.

famines Acute food shortages characterized by large-scale loss of life, social disruption, and economic chaos.

fauna All of the animals present in a given region.

fecundity The physical ability to reproduce.

federal laws (statutes) Laws passed by the federal legislature and signed by the chief executive.

fens Wetlands fed mainly by groundwater.

feral A domestic animal that has taken up a wild existence.

fetal alcohol syndrome A tragic set of permanent physical, mental, and behavioral birth defects that result when mothers drink alcohol during pregnancy.

first law of thermodynamics States that energy is conserved; that is, it is neither created nor destroyed under normal conditions.

flood An overflow of water onto land that normally is dry.

floodplains Low lands along riverbanks, lakes, and coastlines subjected to periodic inundation.

food security The ability of individuals to obtain sufficient food on a day-to-day basis.

food web A complex, interlocking series of individual food chains in an ecosystem.

fossil fuels Petroleum, natural gas, and coal created by geologic forces from organic wastes and dead bodies of formerly living biological organisms.

fragmentation Disruption of habitat into small, isolated fragments.

fuel assembly A bundle of hollow metal rods containing uranium oxide pellets; used to fuel a nuclear reactor.

fuel cells Mechanical devices that use hydrogen or hydrogen-containing fuel, such as methane, to produce an electric current. Fuel cells are clean, quiet, and highly efficient sources of electricity.

fugitive emissions Substances that enter the air without going through a smokestack, such as dust from soil erosion, strip-mining, rock crushing, construction, and building demolition.

fungi Nonphotosynthetic, eukaryotic organisms with cell walls, filamentous bodies, and absorptive nutrition.

fungicide A chemical that kills fungi.

G

gap analysis A biogeographical technique of mapping biological diversity and endemic species to find gaps between protected areas that leave endangered habitats vulnerable to disruption.

gene A unit of heredity; a segment of DNA nucleus of the cell that contains information for the synthesis of a specific protein, such as an enzyme.

generalists Species that tolerate a wide range of conditions or exploit a wide range of resources.

genetic engineering Laboratory manipulation of genetic material using molecular biology.

genetically modified organisms (GMOs) Organisms created by combining natural or synthetic genes using the techniques of molecular biology.

genuine progress index (GPI) An alternative to GNP or GDP for economic accounting that measures real progress in quality of life and sustainability.

geographic isolation Geographical changes that isolate populations of a species and prevent reproduction or gene exchange for a long enough time so that genetic drift changes the populations into distinct species.

geometric growth Growth that follows a geometric pattern of increase, such as 2, 4, 8, 16, etc. See also *exponential growth.*

geothermal energy Energy drawn from the internal heat of the earth, either through geysers, fumaroles, hot springs, or other natural geothermal features or through deep wells that pump heated groundwater.

GIS Geographical information systems that use computers to combine and analyze geographical data.

global environmentalism The extension of modern environmental concerns to global issues.

grasslands Biomes dominated by grasses and associated herbaceous plants.

Great Pacific Garbage Patch A vast area of the Pacific Ocean containing plastic debris concentrated by global ocean circulation currents. One of several oceanic garbage gyres.

green pricing Plans in which consumers can voluntarily pay premium prices for renewable energy.

green revolution Dramatically increased agricultural production brought about by "miracle" strains of grain; usually requires high inputs of water, plant nutrients, and pesticides.

greenhouse effect Trapping of heat by the earth's atmosphere, which is transparent to incoming visible light waves but absorbs outgoing longwave infrared radiation.

greenhouse gas A gas that traps heat in the atmosphere.

gross domestic product (GDP) The total economic activity within national boundaries.

gross national product (GNP) The sum total of all goods and services produced in a national economy. Gross domestic product (GDP) is used to distinguish economic activity within a country from that of offshore corporations.

gully erosion Removal of layers of soil, creating channels or ravines too large to be removed by normal tillage operations.

H

habitat The place or set of environmental conditions in which a particular organism lives.

half-life The time required for one-half of a sample to decay or change into some other form.

hazardous air pollutants (HAPs) A special category of toxins that cause cancer and nerve damage and disrupt hormone function and fetal development. These persistent substances remain in ecosystems for long periods of time and accumulate in animal and human tissues.

hazardous waste Any discarded material containing substances known to be toxic, mutagenic, carcinogenic, or teratogenic to humans or other life-forms; ignitable, corrosive, explosive, or highly reactive alone or with other materials.

health A state of physical and emotional well-being; the absence of disease or ailment.

heap-leach extraction A technique for separating gold from extremely low-grade ores. Crushed ore is piled in huge heaps and sprayed with a dilute alkaline-cyanide solution, which percolates through the pile to extract the gold.

heat Total kinetic energy of atoms or molecules in a substance not associated with the bulk motion of the substance.

heat islands Areas of higher temperatures around cities.

herbicide A chemical that kills plants.

herbivores Organisms that eat only plants.

heterotroph An organism that is incapable of synthesizing its own food and, therefore, must feed upon organic compounds produced by other organisms.

high-level waste repository A place where intensely radioactive wastes can be buried and remain unexposed to groundwater and earthquakes for tens of thousands of years.

high-quality energy Intense, concentrated, and high-temperature energy that is considered high-quality because of its usefulness in carrying out work.

HIPPO Habitat destruction, Invasive species, Pollution, Population (human), and Overharvesting, the leading causes of extinction.

histogram A bar graph, generally with upright bars.

holistic science The study of entire, integrated systems rather than isolated parts. Often takes a descriptive or an interpretive approach.

homeostasis A dynamic, steady state in a living system maintained through opposing, compensating adjustments.

hormesis Nonlinear effects of toxic materials.

human development index (HDI) A measure of quality of life using data for life expectancy, child survival, adult literacy, education, gender equity, access to clean water and sanitation, and income.

hydrologic cycle The natural process by which water is purified and made fresh through evaporation and precipitation. This cycle provides all the fresh water available for biological life.

hypothesis A conditional explanation that can be verified or falsified by observation or experimentation.

I

I = PAT Our environmental impacts (I) are the product of our population size (P) times affluence (A) and the technology (T) used to produce the goods and services we consume.

igneous rocks Crystalline minerals solidified from molten magma from deep in the earth's interior; basalt, rhyolite, andesite, lava, and granite are examples.

independent variable One that does not respond to other variables in a particular test.

indicator species Species that tell us something about the health or condition of a biological community.

indicators Species that have very specific environmental requirements and tolerance levels that make them good indicators of pollution or other environmental conditions.

indigenous people Natives or original inhabitants of an area, those who have lived in a particular place for a very long time.

inductive reasoning "Bottom-up" reasoning in which we study specific examples and try to discover patterns and derive general explanations from collected observations.

insecticide A chemical that kills insects.

insolation Incoming solar radiation.

integrated gasification combined cycle (IGCC) A process in which a fuel (coal or biomass) is heated in the presence of high oxygen levels to produce a variety of gases, mostly hydrogen and carbon dioxide. Impurities, including CO_2, can easily be removed and the synthetic hydrogen gas, or syngas, is burned in a turbine to produce electricity. Superheated gas from the turbine is used to generate steam that produces more electricity, raising the efficiency of the system.

integrated pest management (IPM) An ecologically based pest-control strategy that relies on natural mortality factors, such as natural enemies, weather, cultural control methods, and carefully applied doses of pesticides.

Intergovernmental Panel on Climate Change (IPCC) A large group of scientists from many nations and a wide variety of fields assembled by the United Nations Environment Programme and World Meteorological Organization to assess the current state of knowledge about climate change.

internalizing costs Planning so that those who reap the benefits of resource use also bear all the external costs.

international treaties and conventions Agreements between nations on important issues.

interspecific competition In a community, competition for resources between members of different species.

intraspecific competition In a community, competition for resources among members of the same species.

invasive species Organisms that thrive in new territory where they are free of predators, diseases, or resource limitations that may have controlled their population in their native habitat.

ionosphere The lower part of the thermosphere.

ions Electrically charged atoms that have gained or lost electrons.

island biogeography The study of rates of colonization and extinction of species on islands or other isolated areas based on size, shape, and distance from other inhabited regions.

isotopes Forms of a single element that differ in atomic mass due to a different number of neutrons in the nucleus.

J

J curve A growth curve that depicts exponential growth; called a J curve because of its shape.

joule A unit of energy. One joule is the energy expended in 1 second by a current of 1 amp flowing through a resistance of 1 ohm.

K

keystone species A species whose impacts on its community or ecosystem are much larger and more influential than would be expected from mere abundance. This could be a top predator, a plant that shelters or feeds other organisms, or an organism that plays a critical ecological role.

kinetic energy Energy contained in moving objects, such as a rock rolling down a hill, the wind blowing through the trees, or water flowing over a dam.

***K*-selected species** Organisms whose population growth is regulated by internal (or intrinsic) as well as external factors. Large animals, such as whales and elephants, as well as top predators, generally fall in this category. They have relatively few offspring and often stabilize their population size near the carrying capacity of their environment.

Kyoto Protocol An international treaty adopted in Kyoto, Japan, in 1997, in which 160 nations agreed to roll back CO_2, methane, and nitrous oxide emissions to reduce the threat of global climate change.

L

La Niña The opposite of El Niño.

landscape ecology The study of the reciprocal effects of spatial pattern on ecological processes.

landslides Mass wasting or mass movement of rock or soil downhill. Often triggered by seismic events or heavy rainfall.

latent heat Stored energy in a form that is not sensible (detectable by ordinary senses).

LD50 A chemical dose lethal to 50 percent of a test population.

life expectancy The average age that a newborn infant can expect to attain in a particular time and place.

limiting factors Chemical or physical factors that limit the existence, growth, abundance, or distribution of an organism.

limits to growth A belief that the world has a fixed carrying capacity for humans.

Living Machine® A wastewater treatment system composed of tanks or beds or constructed wetlands in which living organisms remove contaminants, nutrients, and pathogens from water.

logarithmic scale One that uses logarithms as units in a sequence that progresses by a factor of 10 in each step.

logical thinking A rational way of thought that asks, "How can orderly, deductive reasoning help me think clearly?"

logistic growth Growth rates regulated by internal and external factors that establish an equilibrium with environmental resources. See also *S curve*.

LULUs Locally Unwanted Land Uses, such as toxic waste dumps, incinerators, smelters, airports, freeways, and other sources of environmental, economic, or social degradation.

M

magma Molten rock from deep in the earth's interior; called lava when it spews from volcanic vents.

malnourishment A nutritional imbalance caused by lack of specific dietary components or inability to absorb or utilize essential nutrients.

Malthusian growth A population explosion followed by a population crash; also called irruptive growth.

Man and Biosphere (MAB) program A design for nature preserves that divides protected areas into zones with different purposes. A highly protected core is surrounded by a buffer zone and peripheral regions in which multiple-use resource harvesting is permitted.

mangrove forests Diverse groups of salt-tolerant trees and other plants that grow in intertidal zones of tropical coastlines.

manipulative experiment Altering a particular factor for a test or experiment while holding all others (as much as possible) constant.

mantle A hot, pliable layer of rock that surrounds the earth's core and underlies the cool outer crust.

marasmus A widespread human protein deficiency disease caused by a diet low in calories and protein or imbalanced in essential amino acids.

marginal costs The cost to produce one additional unit of a good or service.

marshes Wetlands without trees; in North America, this type of land is characterized by cattails and rushes.

mass burn The incineration of unsorted solid waste.

matter Anything that takes up space and has mass.

mean Average.

megacities See *megalopolis.*

megalopolis Also known as a megacity or supercity; megalopolis indicates an urban area with more than 10 million inhabitants.

mesosphere The atmospheric layer above the stratosphere and below the thermosphere; the middle layer; temperatures are usually very low.

metamorphic rocks Igneous and sedimentary rocks modified by heat, pressure, and chemical reactions.

methane hydrate Small bubbles or individual molecules of methane (natural gas) trapped in a crystalline matrix of frozen water.

microlending Small loans made to poor people who otherwise don't have access to capital.

midocean ridges Mountain ranges on the ocean floor where magma wells up through cracks and creates new crust.

Milankovitch cycles Periodic variations in tilt, eccentricity, and wobble in the earth's orbit; Milutin Milankovitch suggested these are responsible for cyclic weather changes.

millennium assessment A set of ambitious environmental and human development goals established by the United Nations in 2000.

mineral A naturally occurring, inorganic, crystalline solid with definite chemical composition, a specific internal crystal structure, and characteristic physical properties.

minimills Mills that use scrap metal as their starting material.

minimum viable population The number of individuals needed for long-term survival of rare and endangered species.

modern environmentalism A fusion of conservation of natural resources and preservation of nature with concerns about pollution, environmental health, and social justice.

molecules Combinations of two or more atoms.

monoculture forestry Intensive planting of a single species; an efficient wood production approach, but one that encourages pests and disease infestations and conflicts with wildlife habitat or recreation uses.

Montreal Protocol An international treaty to eliminate chloroflurocarbons that destroy stratospheric ozone.

morbidity Illness or disease.

mortality Death rate in a population, such as number of deaths per thousand people per year.

Müllerian (or Muellerian) mimicry Evolution of two species, both of which are unpalatable and have poisonous stingers or some other defense mechanism, to resemble each other.

municipal solid waste The mixed refuse produced by households and businesses.

mutagens Agents, such as chemicals or radiation, that damage or alter genetic material (DNA) in cells.

mutation A change, either spontaneous or by external factors, in the genetic material of a cell; mutations in the gametes (sex cells) can be inherited by future generations of organisms.

mutualism A symbiotic relationship between individuals of two different species in which both species benefit from the association.

N

National Environmental Policy Act (NEPA) The law that established the Council on Environmental Quality and that requires environmental impact statements for all federal projects with significant environmental impacts.

natural experiment Observation of natural events to deduce causal relationships.

natural increase Crude death rate subtracted from crude birth rate.

natural resource economics Economics that takes natural resources into account as valuable assets.

natural resources Goods and services supplied by the environment.

natural selection The mechanism for evolutionary change in which environmental pressures cause certain genetic combinations in a population to become more abundant; genetic combinations best adapted for present environmental conditions tend to become predominant.

negative feedback loop A signal or factor that tends to decrease a process or component.

negative feedbacks Factors that result from a process and, in turn, reduce that same process.

neoclassical economics The branch of economics that attempts to apply the principles of modern science to economic analysis in a mathematically rigorous, noncontextual, abstract, predictive manner.

net primary productivity The amount of biomass produced by photosynthesis and stored in a community after respiration, emigration, and other factors that reduce biomass.

neurotoxins Toxic substances, such as lead or mercury, that specifically poison nerve cells.

neutron A subatomic particle, found in the nucleus of the atom, that has no electromagnetic charge.

new source review A permitting process required by 1977 amendments to the Clean Air Act, required when industries expand or modify facilities. The rule is contentious because vague language in the law allows industries to avoid oversight.

NIMBY Not-In-My-Back-Yard: the position of those opposed to LULUs.

nitrogen cycle The circulation and reutilization of nitrogen in both inorganic and organic phases.

nitrogen-fixing bacteria Bacteria that convert nitrogen from the atmosphere or soil solution into ammonia that can then be converted to plant nutrients by nitrite- and nitrate-forming bacteria.

nitrogen oxides (NO_x) Highly reactive gases formed when nitrogen in fuel or combustion air is heated to over 650°C (1,200°F) in the presence of oxygen or when bacteria in soil or water oxidize nitrogen-containing compounds.

noncriteria pollutants See *unconventional pollutants.*

nongovernmental organizations (NGOs) Pressure and research groups, advisory agencies, political parties, professional societies, and other groups concerned about environmental quality, resource use, and many other issues.

nonpoint sources Scattered, diffuse sources of pollutants, such as runoff from farm fields, golf courses, and construction sites.

nonrenewable resources Minerals, fossil fuels, and other materials present in essentially fixed amounts (within human time scales) in our environment.

normal distribution A bell-shaped curve or a Gaussian distribution.

nuclear fission The radioactive decay process in which isotopes split apart to create two smaller atoms.

nuclear fusion A process in which two smaller atomic nuclei fuse into one larger nucleus and release energy; the source of power in a hydrogen bomb.

nucleic acids Large, organic molecules made of nucleotides that function in the transmission of hereditary traits, in protein synthesis, and in control of cellular activities.

nucleus The center of the atom; occupied by protons and neutrons. In cells, the organelle that contains the chromosomes (DNA).

O

obese Pathologically overweight, having a body mass greater than 30 kg/m^2, or roughly 30 pounds above normal for an average person.

oil shales Fine-grained sedimentary rock rich in solid organic material called kerogen. When heated, the kerogen liquefies to produce a fluid petroleum fuel.

old-growth forests Forests free from disturbance for long enough (generally 150 to 200 years) to have mature trees, physical conditions, species diversity, and other characteristics of equilibrium ecosystems.

oligotrophic Condition of rivers and lakes that have clear water and low biological productivity (*oligo* = little; *trophic* = nourished); are usually clear, cold, infertile headwater lakes and streams.

omnivores Organisms that eat both plants and animals.

open access system A commonly held resource for which there are no management rules.

open canopy A forest where tree crowns cover less than 20 percent of the ground; also called woodland.

open system A system that exchanges energy and matter with its environment.

organic compounds Complex molecules organized around skeletons of carbon atoms arranged in rings or chains; include biomolecules, molecules synthesized by living organisms.

organophosphates Organic molecules to which a phosphate group is attached. A group of highly toxic pesticides that are primarily neurotoxins.

overgrazing Allowing domestic livestock to eat so much plant material that it degrades the biological community.

overharvesting Harvesting so much of a resource that it threatens its existence.

overnutrition Receiving too many calories.

oxygen sag Oxygen decline downstream from a pollution source that introduces materials with high biological oxygen demands.

ozone A highly reactive molecule containing three oxygen atoms; a dangerous pollutant in ambient air. In the stratosphere, however, ozone forms an ultraviolet absorbing shield that protects us from mutagenic radiation.

P

paradigms Overarching models of the world that shape our worldviews and guide our interpretation of how things are.

parasite An organism that lives in or on another organism, deriving nourishment at the expense of its host, usually without killing it.

parasitism A relationship in which one organism feeds on another without immediately killing it.

particulate material Atmospheric aerosols, such as dust, ash, soot, lint, smoke, pollen, spores, algal cells, and other suspended materials; originally applied only to solid particles but now extended to droplets of liquid.

passive solar absorption The use of natural materials or absorptive structures without moving parts to gather and hold heat; the simplest and oldest use of solar energy.

pastoralists People who live by herding domestic animals.

pathogens Organisms that produce disease in host organisms, disease being an alteration of one or more metabolic functions in response to the presence of the organisms.

peat Deposits of moist, acidic, semidecayed organic matter.

pelagic Zones in the vertical water column of a water body.

permafrost A permanently frozen layer of soil that underlies the arctic tundra.

permanent retrievable storage Placing waste storage containers in a secure location where they can be inspected periodically and retrieved, if necessary, for repacking or for transfer if a better means of disposal or reuse is developed.

persistent organic pollutants (POPs) Chemical compounds that persist in the environment and retain biological activity for a long time.

pest Any organism that reduces the availability, quality, or value of a useful resource.

pesticide Any chemical that kills, controls, drives away, or modifies the behavior of a pest.

pesticide treadmill A situation in which farmers must use increasingly complex and expensive cocktails of pesticides to combat pests: similar to addiction.

pH A value that indicates the acidity or alkalinity of a solution on a scale of 0 to 14, based on the proportion of H^+ ions present.

phosphorus cycle The movement of phosphorus atoms from rocks through the biosphere and hydrosphere and back to rocks.

photochemical oxidants Products of secondary atmospheric reactions. See also *smog*.

photodegradable plastics Plastics that break down when exposed to sunlight or to a specific wavelength of light.

photosynthesis The biochemical process by which green plants and some bacteria capture light energy and use it to produce chemical bonds. Carbon dioxide and water are consumed while oxygen and simple sugars are produced.

photovoltaic cell An energy-conversion device that captures solar energy and directly converts it to electrical current.

phylogenetic species concept A definition of species that depends on genetic similarities (or differences).

phytoplankton Microscopic, free-floating, autotrophic organisms that function as producers in aquatic ecosystems.

pioneer species In primary succession on a terrestrial site, the plants, lichens, and microbes that first colonize the site.

plankton Primarily microscopic organisms that occupy the upper water layers in both freshwater and marine ecosystems.

point sources Specific locations of highly concentrated pollution discharge, such as factories, power plants, sewage treatment plants, underground coal mines, and oil wells.

policy A societal plan or statement of intentions intended to accomplish some social or economic goal.

policy cycle The process by which problems are identified and acted upon in the public arena.

pollution To make foul, unclean, dirty; any physical, chemical, or biological change that adversely affects the health, survival, or activities of living organisms or that alters the environment in undesirable ways.

pollution charges Fees assessed per unit of pollution based on the "polluter pays" principle.

population All members of a species that live in the same area at the same time.

population crash A sudden population decline caused by predation, waste accumulation, or resource depletion; also called a dieback.

population explosion Growth of a population at exponential rates to a size that exceeds environmental carrying capacity; usually followed by a population crash.

population momentum A potential for increased population growth as young members reach reproductive age.

positive feedback loop A signal or factor that tends to increase a process or component.

positive feedbacks Factors that result from a process and, in turn, increase that same process.

potential energy Stored energy that is latent but available for use. A rock poised at the top of a hill and water stored behind a dam are examples of potential energy.

power The rate of energy delivery; measured in horsepower or watts.

precautionary principle The rule that we should leave a margin of safety for unexpected developments. This principle implies that we should strive to prevent harm to human health and the environment even if risks are not fully understood.

predator An organism that feeds directly on other organisms in order to survive; live-feeders, such as herbivores and carnivores.

predator-mediated competition A situation in which the effects of a predator dominate population dynamics.

preservation A philosophy that emphasizes the fundamental right of living organisms to exist and to pursue their own ends.

primary pollutants Chemicals released directly into the air in a harmful form.

primary producers Photosynthesizing organisms.

primary productivity Synthesis of organic materials (biomass) by green plants using the energy captured in photosynthesis.

primary standards Regulations of the 1970 Clean Air Act; intended to protect human health.

primary succession Ecological succession that begins in an area where no biotic community previously existed.

primary treatment A process that removes solids from sewage before it is discharged or treated further.

principle of competitive exclusion A result of natural selection whereby two similar species in a community occupy different ecological niches, thereby reducing competition for food.

probability The likelihood that a situation, a condition, or an event will occur.

producer An organism that synthesizes food molecules from inorganic compounds by using an external energy source; most producers are photosynthetic.

productivity The amount of biomass (biological material) produced in a given area during a given period of time.

prokaryotic Cells that do not have a membrane-bounded nucleus or membrane-bounded organelles.

pronatalist pressures Influences that encourage people to have children.

prospective study A study in which experimental and control groups are identified before exposure to some factor. The groups are then monitored and compared for a specific time after the exposure to determine any effects the factor may have.

proteins Chains of amino acids linked by peptide bonds.

proton A positively charged subatomic particle found in the nucleus of an atom.

R

radioactive decay A change in the nuclei of radioactive isotopes that spontaneously emit high-energy electromagnetic radiation and/or subatomic particles while gradually changing into another isotope or different element.

random sample A subset of a collection of items or observations chosen at random.

rational choice Public decision making based on reason, logic, and science-based management.

recharge zones Areas where water infiltrates into an aquifer.

reclamation Chemical, biological, or physical cleanup and reconstruction of severely contaminated or degraded sites to return them to something like their original topography and vegetation.

recycling Reprocessing of discarded materials into new, useful products; not the same as reuse of materials for their original purpose, but the terms are often used interchangeably.

reduced tillage systems Farming methods that preserve soil and save energy and water through reduced cultivation; include minimum till, conserve-till, and no-till systems.

reflective thinking A thoughtful, contemplative analysis that asks, "What does this all mean?"

reformer A device that strips hydrogen from fuels such as natural gas, methanol, ammonia, gasoline, or vegetable oil so they can be used in a fuel cell.

refuse-derived fuel Processing of solid waste to remove metal, glass, and other unburnable materials; organic residue is shredded, formed into pellets, and dried to make fuel for power plants.

regenerative farming Farming techniques and land stewardship that restore the health and productivity of the soil by rotating crops, planting ground cover, protecting the surface with crop residue, and reducing synthetic chemical inputs and mechanical compaction.

regulatory capture A regulatory agency comes to be dominated by, and to benefit, the industry it is intended to regulate; or an appointed agency head is antagonistic to the agency's mission.

relative humidity At any given temperature, a comparison of the actual water content of the air with the amount of water that could be held at saturation.

remediation Cleaning up chemical contaminants from a polluted area.

renewable resources Resources normally replaced or replenished by natural processes; resources not depleted by moderate use; examples include solar energy, biological resources such as forests and fisheries, biological organisms, and some biogeochemical cycles.

renewable water supplies Annual freshwater surface runoff plus annual infiltration into underground freshwater aquifers that are accessible for human use.

replacement rate The number of children per couple needed to maintain a stable population. Because of early deaths, infertility, and nonreproducing individuals, this is usually about 2.1 children per couple.

replication Repeating studies or tests.

reproducibility Making an observation or obtaining a particular result consistently.

residence time The length of time a component, such as an individual water molecule, spends in a particular compartment or location before it moves on through a particular process or cycle.

resilience The ability of a community or an ecosystem to recover from disturbances.

resource partitioning In a biological community, various populations sharing environmental resources through specialization, thereby reducing direct competition. See also *ecological niche.*

resources In economic terms, anything with potential use in creating wealth or giving satisfaction.

restoration ecology Seeks to repair or reconstruct ecosystems damaged by human actions.

retrospective study A study that looks back in history at a group of people (or other organisms) who suffer from some condition to try to identify something in their past life that the whole group shares but that is not found in the control group as near as possible to those being studied but who do not suffer from the same condition.

riders Amendments attached to bills in conference committee, often completely unrelated to the bill to which they are added.

rill erosion The removing of thin layers of soil as little rivulets of running water gather and cut small channels in the soil.

risk The probability that something undesirable will happen as a consequence of exposure to a hazard.

risk assessment Evaluation of the short-term and long-term risks associated with a particular activity or hazard; usually compared with benefits in a cost-benefit analysis.

rock A solid, cohesive aggregate of one or more crystalline minerals.

rock cycle The process whereby rocks are broken down by chemical and physical forces; sediments are moved by wind, water, and gravity; sedimented and reformed into rock; and then crushed, folded, melted, and recrystallized into new forms.

rotational grazing Confining grazing animals in a small area for a short time to force them to eat weedy species as well as the more desirable grasses and forbes.

***r*-selected species** Organisms whose population growth is regulated mainly by external factors. They tend to have rapid reproduction and high mortality of offspring. Given optimum environmental conditions, they can grow exponentially. Many "weedy" or pioneer species fit in this category.

runoff The excess of precipitation over evaporation; the main source of surface water and, in broad terms, the water available for human use.

S

S curve A curve that depicts logistic growth; called an S curve because of its shape.

salinity The amount of dissolved salts (especially sodium chloride) in a given volume of water.

salinization A process in which mineral salts accumulate in the soil, killing plants; occurs when soils in dry climates are irrigated profusely.

salt marsh A wetland with saltwater and salt tolerant plants, usually coastal.

saltwater intrusion The movement of saltwater into freshwater aquifers in coastal areas where groundwater is withdrawn faster than it is replenished.

sample To analyze a small but representative portion of a population to estimate the characteristics of the entire class.

sanitary landfills Landfills in which garbage and municipal waste are buried every day under enough soil or fill to eliminate odors, vermin, and litter.

savanna An open prairie or grassland with scattered groves of trees.

scavengers In biology, organisms that consume carrion, or organisms not killed by the scavenger.

science The orderly pursuit of knowledge, relying on observations that test hypotheses in order to answer questions.

scientific consensus A general agreement among informed scholars.

scientific method A systematic, precise, objective study of a problem. Generally this requires observation, hypothesis development and testing, data gathering, and interpretation.

scientific theory An explanation or idea accepted by a substantial number of scientists.

sea-grass beds Large expanses of rooted, submerged, or emergent aquatic vegetation, such as eel grass or salt grass.

second law of thermodynamics States that, with each successive energy transfer or transformation in a system, less energy is available to do work.

secondary pollutants Chemicals modified to a hazardous form after entering the air or that are formed by chemical reactions as components of the air mix and interact.

secondary succession Succession on a site where an existing community has been disrupted.

secondary treatment Bacterial decomposition of suspended particulates and dissolved organic compounds that remain after primary sewage treatment.

secure landfills Solid waste disposal sites lined and capped with an impermeable barrier to prevent leakage or leaching.

sedimentary rocks Rocks composed of accumulated, compacted mineral fragments, such as sand or clay; examples include shale, sandstone, breccia, and conglomerates.

sedimentation The deposition of organic materials or minerals by chemical, physical, or biological processes.

selection pressure Limited resources or adverse environmental conditions that tend to favor certain adaptations in a population. Over many generations, this can lead to genetic change, or evolution.

selective cutting Harvesting only mature trees of certain species and size; usually more expensive than clear-cutting but less disruptive for wildlife and often better for forest regeneration.

shade-grown coffee and cocoa Plants grown under a canopy of taller trees, which provides habitat for birds and other wildlife.

shantytowns Groups of shacks built of inexpensive materials on empty land.

sheet erosion Peeling off thin layers of soil from the land surface; accomplished primarily by wind and water.

shelterwood harvesting Mature trees are removed from the forest in a series of two or more cuts, leaving young trees and some mature trees as a seed source for future regeneration.

sick building syndrome A cluster of allergies and other illnesses caused by sensitivity to molds, synthetic chemicals, or other harmful compounds trapped in insufficiently ventilated buildings.

sinkhole A large surface crater caused by the collapse of an underground channel or cavern; often triggered by groundwater withdrawal.

sludge A semisolid mixture of organic and inorganic materials that settles out of wastewater at a sewage treatment plant.

slums Legal but inadequate multifamily tenements or rooming houses.

smart growth The efficient use of land resources and existing urban infrastructure that encourages in-fill development, provides a variety of affordable housing and transportation choices, and seeks to maintain a unique sense of place by respecting local cultural and natural features.

smart metering A system of meters that give information about the source and price of electricity used by individual appliances and that can time usage to take advantage of the lowest cost power.

smelting Roasting ore to release metals from mineral compounds.

smog The combination of smoke and fog in the stagnant air of London; now often applied to photochemical pollution.

social justice Equitable access to resources and the benefits derived from them; a system that recognizes inalienable rights and adheres to what is fair, honest, and moral.

soil creep The slow, downhill movement of soil due to erosion.

sound science Although definitions differ, this generally means valid science according to basic scientific principles.

Southern Oscillation The combination of El Niño and La Niña cycles.

specialists Species that require a narrow range of conditions or exploit a very specific set of resources.

speciation Evolution of new species.

species All the organisms genetically similar enough to breed and produce live, fertile offspring in nature.

species diversity The number and relative abundance of species present in a community.

specific heat The amount of heat energy needed to change the temperature of a body. Water has a specific heat of 1, which is higher than most substances.

sprawl Unlimited, unplanned growth of urban areas that consumes open space and wastes resources.

stability In ecological terms, a dynamic equilibrium among the physical and biological factors in an ecosystem or a community; relative homeostasis.

state shift An abrupt response to a disturbance that causes a persistent change in a system to a new set of conditions and relationships.

statutory law Rules passed by a state or national legislature.

steady-state economy Characterized by low birth and death rates, use of renewable energy sources, recycling of materials, and emphasis on durability, efficiency, and stability.

stratosphere The zone in the atmosphere extending from the tropopause to about 50 km (30 mi) above the earth's surface; temperatures are stable or rise slightly with altitude; has very little water vapor but is rich in ozone.

streambank erosion Erosion along the edges of a stream.

stress Physical, chemical, or emotional factors that place a strain on an animal. Plants also experience physiological stress under adverse environmental conditions.

strip-cutting Harvesting trees in strips narrow enough to minimize edge effects and to allow natural regeneration of the forest.

strip-farming Planting different kinds of crops in alternating strips along land contours; when one crop is harvested, the other crop remains to protect the soil and prevent water from running straight down a hill.

strip-mining Extracting shallow mineral deposits (especially coal) by scraping off surface layers with giant earth-moving equipment; creates a huge open pit; an alternative to underground or deep open-pit mines.

Student Environment Action Coalition (SEAC) A grassroots coalition of student and youth environmental groups, working together to protect our planet and our future.

subduction (subducted) Where the edge of one tectonic plate dives beneath the edge of another.

subsidence Settling of the ground surface caused by the collapse of porous formations that result from withdrawal of large amounts of groundwater, oil, or other underground materials.

subsoil A layer of soil beneath the topsoil that has lower organic content and higher concentrations of fine mineral particles; often contains soluble compounds and clay particles carried down by percolating water.

sulfur cycle The chemical and physical reactions by which sulfur moves into or out of storage and through the environment.

sulfur dioxide (SO_2) A colorless, corrosive gas directly damaging to both plants and animals.

Superfund A fund established by Congress to pay for containment, cleanup, or remediation of abandoned toxic waste sites. The fund is financed by fees paid by toxic waste generators and by cost recovery from cleanup projects.

surface mining Mineral extraction from open pits or strip mines, as opposed to underground mining. See also *strip-mining.*

surface soil The A horizon in a soil profile; the soil just below the litter layer.

surface tension The tendency for a surface of water molecules to hold together, producing a surface that resists breaking.

sustainability Ecological, social, and economic systems that can last over the long term.

sustainable agriculture (regenerative farming) Ecologically sound, economically viable, socially just agricultural system. Stewardship, soil conservation, and integrated pest management are essential for sustainability.

sustainable development A real increase in well-being and standard of life for the average person that can be maintained over the long term without degrading the environment or compromising the ability of future generations to meet their own needs.

sustained yield Utilization of a renewable resource at a rate that does not impair or damage its ability to be fully renewed on a long-term basis.

swamps Wetlands with trees, such as the extensive swamp forests of the southern United States.

symbiosis The intimate living together of members of two species; includes mutualism, commensalism, and, in some classifications, parasitism.

sympatric speciation A gradual change (generally through genetic drift) so that offspring are genetically distinct from their ancestors even though they live in the same place.

synergism When an injury caused by exposure to two environmental factors together is greater than the sum of exposure to each factor individually.

synergistic effects The combination of several processes or factors is greater than the sum of their individual effects.

systems Networks of interdependent components and processes.

T

taiga The northernmost edge of the boreal forest, including species-poor woodland and peat deposits; intergrading with the arctic tundra.

tailings Mining waste left after mechanical or chemical separation of minerals from crushed ore.

taking The unconstitutional confiscation of private property.

tar sands Geologic deposits composed of sand and shale particles coated with bitumen, a viscous mixture of long-chain hydrocarbons.

tectonic plates Huge blocks of the earth's crust that slide around slowly, pulling apart to open new ocean basins or crashing ponderously into each other to create new, larger landmasses.

telemetry Locating or studying organisms at a distance using radio signals or other electronic media.

temperate rainforest The cool, dense, rainy forest of the northern Pacific coast; enshrouded in fog much of the time; dominated by large conifers.

temperature A measure of the speed of motion of a typical atom or molecule in a substance.

temperature inversions Atmospheric conditions in which a layer of warm air lies on top of cooler air and blocks normal convection currents. This can trap pollutants and degrade air quality.

teratogens Chemicals or other factors that specifically cause abnormalities during embryonic growth and development.

terracing Shaping the land to create level shelves of earth to hold water and soil; requires extensive hand labor or expensive machinery, but it enables farmers to farm very steep hillsides.

tertiary treatment The removal of inorganic minerals and plant nutrients after primary and secondary treatment of sewage.

thermal pollution Artificially raising or lowering of the temperature of a water body in a way that adversely affects the biota or water quality.

thermocline In water, a distinctive temperature transition zone that separates an upper layer that is mixed by the wind (the epilimnion) and a colder deep layer that is not mixed (the hypolimnion).

thermodynamics The branch of physics that deals with transfers and conversions of energy.

thermohaline circulation A large-scale oceanic circulation system in which warm water flows from equatorial zones to higher latitudes, where it cools, evaporates, and becomes saltier and more dense, which causes it to sink and flow back toward the equator in deep ocean currents.

threatened species A species that is still abundant in some areas but has declined significantly in total numbers and may be on the verge of extinction in certain regions or localities.

thresholds Conditions where sudden change can occur in a system.

throughput The flow of energy and/or matter into and out of a system.

tide pools Small pools of water left behind by falling tides.

tolerance limits See *limiting factors.*

topsoil A layer of mixed organic and mineral soil material, also called the A horizon.

total fertility rate The number of children born to an average woman in a population during her entire reproductive life.

total growth rate The net rate of population growth resulting from births, deaths, immigration, and emigration.

total maximum daily loads (TMDL) The amount of particular pollutant that a water body can receive from both point and nonpoint sources and still meet water quality standards.

Toxic Release Inventory A program created by the Superfund Amendments and Reauthorization Act of 1984 that requires manufacturing facilities and waste handling and disposal sites to report annually on releases of more than 300 toxic materials. You can find out from the EPA whether any of these sites are in your neighborhood and what toxins they release.

toxins Poisonous chemicals that react with specific cellular components to kill cells or to alter growth or development in undesirable ways; often harmful, even in dilute concentrations.

tradable permits Pollution quotas or variances that can be bought or sold.

"Tragedy of the Commons" An inexorable process of degradation of communal resources due to selfish self-interest of "free riders" who use or destroy more than their fair share of common property. See *open access system.*

transpiration The evaporation of water from plant surfaces, especially through stomates.

trophic level Step in the movement of energy through an ecosystem; an organism's feeding status in an ecosystem.

tropical rainforests Forests near the equator in which rainfall is abundant—more than 200 cm (80 in.) per year—and temperatures are warm to hot year-round.

tropical seasonal forests Semi-evergreen or partly deciduous forests tending toward open woodlands and grassy savannas dotted with scattered, drought-resistant trees.

tropopause The boundary between the troposphere and the stratosphere.

troposphere The layer of air nearest to the earth's surface; both temperature and pressure usually decrease with increasing altitude.

tsunami Far-reaching waves caused by earthquakes or undersea landslides.

tundra Treeless arctic or alpine biome characterized by cold, dark winters; a short growing season; and potential for frost any month of the year; vegetation includes low-growing perennial plants, mosses, and lichens.

U

UN Framework Convention on Climate Change Directs governments to share data on climate change, to develop national plans for controlling greenhouse gases, and to cooperate in planning for adaptation to climate change.

unconventional pollutants Toxic or hazardous substances, such as asbestos, benzene, beryllium, mercury, polychlorinated biphenyls, and vinyl chloride, not listed in the original Clean Air Act because they were not released in large quantities; also called noncriteria pollutants.

urban agglomerations Urban areas where several cities or towns have coalesced.

utilitarian conservation The philosophy that resources should be used for the greatest good for the greatest number for the longest time.

V

vertical stratification The vertical distribution of specific subcommunities within a community.

vertical zonation Vegetation zones determined by climate changes brought about by altitude changes.

volatile organic compounds Organic chemicals that evaporate readily and exist as gases in the air.

volcanoes Vents in the earth's surface through which molten lava (magma), gases, and ash escape to create mountains.

vulnerable species Naturally rare organisms or species whose numbers have been so reduced by human activities that they are susceptible to actions that could push them into threatened or endangered status.

W

warm front A long, wedge-shaped boundary caused when a warmer advancing air mass slides over neighboring cooler air parcels.

waste stream The steady flow of varied wastes, from domestic garbage and yard wastes to industrial, commercial, and construction refuse.

waterlogging Water saturation of soil that fills all air spaces and causes plant roots to die from lack of oxygen; a result of overirrigation.

water scarcity Having less than 1,000 m^3 (264,000 gal) of clean, fresh water available per person per year.

watershed The land surface and groundwater aquifers drained by a particular river system.

water stress Countries that consume more than 10 percent of renewable water supplies.

water table The top layer of the zone of saturation; undulates according to the surface topography and subsurface structure.

watt One joule per second.

weather The physical conditions of the atmosphere (moisture, temperature, pressure, and wind).

weathering Changes in rocks brought about by exposure to air, water, changing temperatures, and reactive chemical agents.

wetlands Ecosystems of several types in which rooted vegetation is surrounded by standing water during part of the year. See also *swamps, marshes, bogs, fens.*

withdrawal A description of the total amount of water taken from a lake, a river, or an aquifer.

work The application of force through a distance; requires energy input.

world conservation strategy A proposal for maintaining essential ecological processes, preserving genetic diversity, and ensuring that utilization of species and ecosystems is sustainable.

Z

zero population growth (ZPG) A condition in which births and immigration in a population just balance deaths and emigration.

zone of aeration Upper soil layers that hold both air and water.

zone of saturation Lower soil layers where all spaces are filled with water.

©Martin Kubat/Shutterstock

Index

A

B

D

E

F

G

H

N

O

P

Q

R

S

T

U

V

W

Y

Z